Heat and Mass Transfer

Springer

Berlin
Heidelberg
New York
Barcelona
Budapest
Hong Kong
London
Milan
Paris
Santa Clara
Singapore
Tokyo

Hans Dieter Baehr · Karl Stephan

Heat and Mass Transfer

Translated by Nicola Jane Park

With 327 Figures

Springer

Professor Dr. Hans Dieter Baehr
Institut für Thermodynamik
Universität Hannover
Callinstraße 36
D-30167 Hannover, Germany

Professor Dr. Karl Stephan
Institut für Thermodynamik
und Thermische Verfahrenstechnik
Universität Stuttgart
Pfaffenwaldring 9
D-70569 Stuttgart, Germany

Translated from the second German Edition "Wärme- und Stoff-übertragung" (Springer-Verlag, 1996) by Nicola Jane Park, MEng., University of London, Imperial College of Science, Technology and Medicine.

ISBN 3-540-63695-1 Springer-Verlag Berlin Heidelberg New York

Library of Congress Cataloging-in-Publication Data

Die Deutsche Bibliothek – Cip-Einheitsaufnahme
Baehr, Hans Dieter:
Heat and mass transfer / H. D. Baehr ; K. Stephan. Transl. by Nicola Jane Park.
Berlin ; Heidelberg ; New York ; Barcelona ; Budapest ; Hong Kong ; London ;
Milan ; Paris ; Santa Clara ; Singapore ; Tokyo : Springer, 1998
Dt. Ausg. u. d. T.: Baehr, Hans Dieter: Wärme- und Stoffübertragung
ISBN 3-540-63695-1

© Springer-Verlag Berlin Heidelberg 1998
Printed in Germany

Production: ProduServ GmbH Verlagsservice, Berlin
Typesetting: Camera-ready by authors
SPIN: 10548563   60/3020-5 4 3 2 1 0 - Printed on acid -free paper

Preface

This book is the English translation of our German publication, which appeared in 1994 with the title "Wärme und Stoffübertragung" (2nd edition Berlin: Springer Verlag 1996). The German version originated from lecture courses in heat and mass transfer which we have held for many years at the Universities of Hannover and Stuttgart, respectively. Our book is intended for students of mechanical and chemical engineering at universities and engineering schools, but will also be of use to students of other subjects such as electrical engineering, physics and chemistry. Firstly our book should be used as a textbook alongside the lecture course. Its intention is to make the student familiar with the fundamentals of heat and mass transfer, and enable him to solve practical problems. On the other hand we placed special emphasis on a systematic development of the theory of heat and mass transfer and gave extensive discussions of the essential solution methods for heat and mass transfer problems. Therefore the book will also serve in the advanced training of practising engineers and scientists and as a reference work for the solution of their tasks. The material is explained with the assistance of a large number of calculated examples, and at the end of each chapter a series of exercises is given. This should also make self study easier.

Many heat and mass transfer problems can be solved using the balance equations and the heat and mass transfer coefficients, without requiring too deep a knowledge of the theory of heat and mass transfer. Such problems are dealt with in the first chapter, which contains the basic concepts and fundamental laws of heat and mass transfer. The student obtains an overview of the different modes of heat and mass transfer, and learns at an early stage how to solve practical problems and to design heat and mass transfer apparatus. This increases the motivation to study the theory more closely, which is the object of the subsequent chapters.

In the second chapter we consider steady-state and transient heat conduction and mass diffusion in quiescent media. The fundamental differential equations for the calculation of temperature fields are derived here. We show how analytical and numerical methods are used in the solution of practical cases. Alongside the Laplace transformation and the classical method of separating the variables, we have also presented an extensive discussion of finite difference methods which are very important in practice. Many of the results found for heat conduction can be transferred to the analogous process of mass diffusion. The mathematical solution formulations are the same for both fields.

The third chapter covers convective heat and mass transfer. The derivation of the mass, momentum and energy balance equations for pure fluids and multi-component mixtures are treated first, before the material laws are introduced and the partial differential equations for the velocity, temperature and concentration fields are derived. As typical applications we consider heat and mass transfer in flow over bodies and through channels, in packed and fluidised beds as well as free convection and the superposition of free and forced convection. Finally an introduction to heat transfer in compressible fluids is presented.

In the fourth chapter the heat and mass transfer in condensation and boiling with free and forced flows is dealt with. The presentation follows the book, "Heat Transfer in Condensation and Boiling" (Berlin: Springer-Verlag 1988) by K. Stephan. Here, we consider not only pure substances; condensation and boiling in mixtures of substances are also explained to an adequate extent.

Thermal radiation is the subject of the fifth chapter. It differs from many other presentations in so far as the physical quantities needed for the quantitative description of the directional and wavelength dependency of radiation are extensively presented first. Only after a strict formulation of Kirchhoff's law, the ideal radiator, the black body, is introduced. After this follows a discussion of the material laws of real radiators. Solar radiation and heat transfer by radiation are considered as the main applications. An introduction to gas radiation, important technically for combustion chambers and furnaces, is the final part of this chapter.

As heat and mass transfer is a subject taught at a level where students have already had courses in calculus, we have presumed a knowledge of this field. Those readers who only wish to understand the basic concepts and become familiar with simple technical applications of heat and mass transfer need only study the first chapter. More extensive knowledge of the subject is expected of graduate mechanical and chemical engineers. The mechanical engineer should be familiar with the fundamentals of heat conduction, convective heat transfer and radiative transfer, as well as having a basic knowledge of mass transfer. Chemical engineers also require, in addition to a sound knowledge of these areas, a good understanding of heat and mass transfer in multiphase flows. The time set aside for lectures is generally insufficient for the treatment of all the material in this book. However, it is important that the student acquires a broad understanding of the fundamentals and methods. Then it is sufficient to deepen this knowledge with selected examples and thereby improve problem solving skills.

In the preparation of the manuscript we were assisted by a number of our colleagues, above all by Nicola Jane Park, MEng., University of London, Imperial College of Science, Technology and Medicine. We owe her sincere thanks for the translation of our German publication into English, and for the excellent cooperation.

Hannover and Stuttgart,   **H.D. Baehr**
        Spring 1998   **K. Stephan**

Contents

Nomenclature

Symbol	Meaning	SI units
A	area	m^2
A_m	average area	m^2
A_q	cross sectional area	m^2
A_f	fin surface area	m^2
a	thermal diffusivity	m^2/s
a	hemispherical total absorptivity	—
a_λ	spectral absorptivity	—
a'_λ	directional spectral absorptivity	—
a_t	turbulent thermal diffusivity	m^2/s
a^*	specific surface area	m^2/m^3
b	thermal penetration coefficient, $b = \sqrt{\lambda c \varrho}$	$W\,s^{1/2}/(m^2\,K)$
b	Laplace constant, $b = \sqrt{2\sigma/g\,(\varrho_L - \varrho_G)}$	m
C	circumference, perimeter	m
C	heat capacity flow ratio	—
c	specific heat capacity	$J/(kg\,K)$
c	concentration	mol/m^3
c	propagation velocity of electromagnetic waves	m/s
c_0	velocity of light in a vacuum	m/s
c_f	friction factor	—
c_p	specific heat capacity at constant pressure	$J/(kg\,K)$
c_R	resistance factor	—
D	binary diffusion coefficient	m^2/s
D_t	turbulent diffusion coefficient	m^2/s
d	diameter	m
d_A	departure diameter of vapour bubbles	m
d_h	hydraulic diameter	m
E	irradiance	W/m^2
E_0	solar constant	W/m^2
E_λ	spectral irradiance	$W/(m^2\,\mu m)$
e	unit vector	—
F	force	N

F_B	buoyancy force	N
F_f	friction force	N
F_R	resistance force	N
F_{ij}	view factor between surfaces i and j	—
$F(0, \lambda T)$	fraction function of black radiation	—
f	frequency of vapour bubbles	1/s
f_j	force per unit volume	N/m^3
g	acceleration due to gravity	m/s^2
H	height	m
H	radiosity	W/m^2
H	enthalpy	J
$\dot{H}$	enthalpy flow	J/s
h	Planck constant	J s
h	specific enthalpy	J/kg
h_{tot}	specific total enthalpy, $h_{tot} = h + w^2/2$	J/kg
h_i	partial specific enthalpy	J/kg
Δh_v	specific enthalpy of vaporisation	J/kg
$\Delta \tilde{h}_v$	molar enthalpy of vaporisation	J/mol
I	momentum	kg m/s
I	directional emissive power	W/(m^2 sr)
j	diffusional flux	mol/(m^2 s)
j^*	diffusional flux in a centre of gravity system	kg/(m^2 s)
$_u j$	diffusional flux in a particle based system	mol/(m^2 s)
K	incident intensity	W/(m^2 sr)
K_λ	incident spectral intensity	W/(m^2 μm)
k	overall heat transfer coefficient	W/(m^2 K)
k	extinction coefficient	—
k	Boltzmann constant	J/K
k_G	spectral absorption coefficient	1/m
k_H	Henry coefficient	N/m^2
k_j	force per unit mass	N/kg
k_1	rate constant for a homogeneous first order reaction	1/s
k_1', k_1''	rate constant for a homogeneous (heterogeneous) first order reaction	m/s
k_n''	rate constant for a heterogeneous n-th order reaction	$\dfrac{\text{mol/(m}^2\text{s)}}{(\text{mol/m}^3)^n}$
L	length	m
L	total intensity	W/(m^2 sr)
L_λ	spectral intensity	W/(m^2 μm sr)
L_0	reference length	m
L_S	solubility	mol/(m^3 Pa)

l	length, mixing length	m
M	mass	kg
M	modulus, $M = a\Delta t/\Delta x^2$	—
M	(hemispherical total) emissive power	W/m^2
M_λ	spectral emissive power	W/(m^2 μm)
$\dot{M}$	mass flow rate	kg/s
$\tilde{M}$	molecular mass, molar mass	kg/mol
m	optical mass	kg/m^2
m_r	relative optical mass	—
$\dot{m}$	mass flux	kg/(m^2 s)
N	amount of substance	mol
N_i	dimensionless transfer capability (number of transfer units) of the material stream i	—
$\dot{N}$	molar flow rate	mol/s
n	refractive index	—
$\boldsymbol{n}$	normal vector	—
$\dot{n}$	molar flux	mol/(m^2 s)
P	power	W
P_diss	dissipated power	W
p	pressure	Pa
p^+	dimensionless pressure	—
Q	heat	J
$\dot{Q}$	heat flow	W
$\dot{q}$	heat flux	W/m^2
R	radius	m
R_cond	resistance to thermal conduction	K/W
R_m	molar (universal) gas constant	J/(mol K)
r	radial coordinate	m
r	hemispherical total reflectivity	—
r_λ	spectral reflectivity	—
r'_λ	directional spectral reflectivity	—
r_e	electrical resistivity	Ω m
r^+	dimensionless radial coordinate	—
$\dot{r}$	reaction rate	mol/(m^3 s)
S	suppression factor in convective boiling	—
S	entropy	J/K
s	specific entropy	J/(kg K)
s	Laplace transformation parameter	1/s
s	beam length	m
s	slip factor, $s = w_\mathrm{G}/w_\mathrm{L}$	—
s_l	longitudinal pitch	m

s_q	transverse pitch	m
T	thermodynamic temperature	K
T_e	eigentemperature	K
T_{St}	stagnation point temperature	K
t	time	s
t^+	dimensionless time	—
t_k	cooling time	s
t_j	stress vector	N/m^2
t_R	relaxation time, $t_R = 1/k_1$	s
t_D	relaxation time of diffusion, $t_D = L^2/D$	s
U	internal energy	J
u	average molar velocity	m/s
u	specific internal energy	J/kg
u	Laplace transformed temperature	K
V	volume	m^3
V_A	departure volume of a vapour bubble	m^3
v	specific volume	m^3/kg
W	work	J
$\dot{W}$	power density	W/m^3
$\dot{W}_i$	heat capacity flow rate of a fluid i	W/K
w	velocity	m/s
w_0	reference velocity	m/s
w_S	velocity of sound	m/s
w_τ	shear stress velocity, $w_\tau = \sqrt{\tau_0/\varrho}$	m/s
w'	fluctuation velocity	m/s
w^+	dimensionless velocity	—
X	moisture content; Lockhart-Martinelli parameter	—
$\tilde{X}$	molar content in the liquid phase	—
x	coordinate	m
$\tilde{x}$	mole fraction in the liquid	—
x^+	dimensionless x-coordinate	—
x^*	quality, $x^* = \dot{M}_G/\dot{M}_L$	—
x^*_{th}	thermodynamic quality	—
$\tilde{Y}$	molar content in the gas phase	—
y	coordinate	m
$\tilde{y}$	mole fraction in the gas phase	—
y^+	dimensionless y-coordinate	—
z	number	—
z	axial coordinate	m
z^+	dimensionless z-coordinate	—
z_R	number of tube rows	—

Greek letters

Symbol	Meaning	SI units
α	heat transfer coefficient	$W/(m^2\ K)$
α_m	mean heat transfer coefficient	$W/(m^2\ K)$
β	mass transfer coefficient	m/s
β_m	mean mass transfer coefficient	m/s
β	thermal expansion coefficient	1/K
β	polar angle, zenith angle	rad
β_0	base angle	rad
$\dot{\Gamma}$	mass production rate	$kg/(m^3\ s)$
$\dot{\gamma}$	molar production rate	$mol/(m^3\ s)$
Δ	difference	—
δ	thickness; boundary layer thickness	m
δ_{ij}	Kronecker symbol	—
ε	volumetric vapour content	—
ε^*	volumetric quality	—
ε	hemispherical total emissivity	—
ε_λ	hemispherical spectral emissivity	—
ε'_λ	directional spectral emissivity	—
ε_D	turbulent diffusion coefficient	m^2/s
ε_i	dimensionless temperature change of the material stream i	—
$\dot{\varepsilon}_{ii}$	dilatation	1/s
$\dot{\varepsilon}_{ji}$	strain tensor	1/s
ε_p	void fraction	—
ε_t	turbulent viscosity	m^2/s
ζ	resistance factor	—
ζ	bulk viscosity	$kg/(m\ s)$
η	dynamic viscosity	$kg/(m\ s)$
η_f	fin efficiency	—
Θ	overtemperature	K
ϑ	temperature	K
ϑ^+	dimensionless temperature	—
κ	isentropic exponent	—
κ_G	optical thickness of a gas beam	—
Λ	wave length of an oscillation	m
λ	wave length	m
λ	thermal conductivity	$W/(K\ m)$
λ_t	turbulent thermal conductivity	$W/(K\ m)$
μ	diffusion resistance factor	—
ν	kinematic viscosity	m^2/s

ν	frequency	1/s
ϱ	density	kg/m^3
σ	Stefan-Boltzmann constant	W/(m^2 K^4)
σ	interfacial tension	N/m
ξ	mass fraction	—
τ	transmissivity	—
τ_λ	spectral transmissivity	—
τ	shear stress	N/m^2
τ_{ji}	shear stress tensor	N/m^2
Φ	radiative power, radiation flow	W
Φ	viscous dissipation	W/m^3
φ	angle, circumferential angle	rad
Ψ	stream function	m^2/s
ω	solid angle	sr
ω	reference velocity	m/s
$\dot\omega$	power density	W/m^2

Subscripts

Symbol	Meaning
A	air, substance A
abs	absorbed
B	substance B
C	condensate, cooling medium
diss	dissipated
E	excess, product, solidification
e	exit, outlet
eff	effective
eq	equilibrium
F	fluid, feed
f	fin, friction
G	gas
g	geodetic, base material
I	at the phase interface
i	inner, inlet
id	ideal
in	incident radiation, irradiation
K	substance K
L	liquid
lam	laminar
m	mean, molar (based on the amount of substance)
max	maximum
min	minimum

n	normal direction
o	outer, outside
P	particle
ref	reflected, reference state
S	solid, bottom product, sun, surroundings
s	black body, saturation
tot	total
trans	transmitted
turb	turbulent
u	in particle reference system
V	boiler
W	wall, water
α	start
δ	at the point $y = \delta$
λ	spectral
ω	end
0	reference state; at the point $y = 0$
∞	at a great distance; in infinity

Dimensionless numbers

$Ar = [(\varrho_S - \varrho_F)/\varrho_F]\,(d_P^3 g/\nu^2)$	Archimedes number
$Bi = \alpha L/\lambda$	Biot number
$Bi_D = \beta L/D$	Biot number for mass transfer
$Bo = \dot{q}/(\dot{m}\Delta h_v)$	boiling number
$Da = k_1'' L/D$	Damköhler number (for 1st order heterogeneous reaction)
$Ec = w^2/(c_p \Delta\vartheta)$	Eckert number
$Fo = at/L^2$	Fourier number
$Fr = w^2/(gx)$	Froude number
$Ga = gL^3/\nu^2$	Galilei number
$Gr = g\beta\Delta\vartheta L^3/\nu^2$	Grashof number
$Ha = \left(k_1 L^2/D\right)^2$	Hatta number
$Le = a/D$	Lewis number
$Ma = w/w_S$	Mach number
$Nu = \alpha L/\lambda$	Nusselt number
$Pe = wL/a$	Péclet number
$Ph = h_E/[c\,(\vartheta_E - \vartheta_0)]$	phase change number
$Pr = \nu/a$	Prandtl number
$Ra = GrPr$	Rayleigh number
$Re = wL/\nu$	Reynolds number
$Sc = \nu/D$	Schmidt number
$Sh = \beta L/D$	Sherwood number
$St = \alpha/(w\varrho c_p)$	Stanton number
$St = 1/Ph$	Stefan number

1 Introduction. Technical Applications

In this chapter the basic definitions and physical quantities needed to describe heat and mass transfer will be introduced, along with the fundamental laws of these processes. They can already be used to solve technical problems, such as the transfer of heat between two fluids separated by a wall, or the sizing of apparatus used in heat and mass transfer. The calculation methods presented in this introductory chapter will be relatively simple, whilst a more detailed presentation of complex problems will appear in the following chapters.

1.1 The different types of heat transfer

In thermodynamics, heat is defined as the energy that crosses the boundary of a system when this energy transport occurs due to a temperature difference between the system and its surroundings, cf. [1.1], [1.2]. The second law of thermodynamics states that heat always flows over the boundary of the system in the direction of falling temperature.

However, thermodynamics does not state how the heat transferred depends on this temperature driving force, or how fast or intensive this irreversible process is. It is the task of the science of heat transfer to clarify the laws of this process. Three modes of heat transfer can be distinguished: conduction, convection, and radiation. The following sections deal with their basic laws, more in depth information is given in chapter 2 for conduction, 3 and 4 for convection and 5 for radiation. We limit ourselves to a phenomenological description of heat transfer processes, using the thermodynamic concepts of temperature, heat, heat flow and heat flux. In contrast to thermodynamics, which mainly deals with homogeneous systems, the so-called *phases*, heat transfer is a continuum theory which deals with fields extended in space and also dependent on time.

This has consequences for the concept of heat, which in thermodynamics is defined as energy which crosses the system boundary. In heat transfer one speaks of a heat flow also within the body. This contradiction with thermodynamic terminology can be resolved by considering that in a continuum theory the mass and volume elements of the body are taken to be small systems, between which energy can be transferred as heat. Therefore, when one speaks of heat flow within a solid body or fluid, or of the heat flux vector field in conjunction with the temperature field, the thermodynamic theory is not violated.

As in thermodynamics, the thermodynamic temperature T is used in heat transfer. However with the exception of radiative heat transfer the zero point of the thermodynamic temperature scale is not needed, usually only temperature *differences* are important. For this reason a thermodynamic temperature with an adjusted zero point, an example being the Celsius temperature, is used. These thermodynamic temperature differences are indicated by the symbol ϑ, defined as

$$\vartheta := T - T_0 \qquad (1.1)$$

where T_0 can be chosen arbitrarily and is usually set at a temperature that best fits the problem that requires solving. When $T_0 = 273.15$ K then ϑ will be the Celsius temperature. The value for T_0 does not normally need to be specified as temperature differences are independent of T_0.

1.1.1   Heat conduction

Heat conduction is the transfer of energy between neighbouring molecules in a substance due to a temperature gradient. In metals also the free electrons transfer energy. In solids which do not transmit radiation, heat conduction is the only process for energy transfer. In gases and liquids heat conduction is superimposed by an energy transport due to convection and radiation.

The mechanism of heat conduction in solids and fluids is difficult to understand theoretically. We do not need to look closely at this theory; it is principally used in the calculation of thermal conductivity, a material property. We will limit ourselves to the phenomenological discussion of heat conduction, using the thermodynamic quantities of temperature, heat flow and heat flux, which are sufficient to deal with most technically interesting conduction problems.

The transport of energy in a conductive material is described by the vector field of heat flux

$$\dot{q} = \dot{q}(x, t) \ . \qquad (1.2)$$

In terms of a continuum theory the heat flux vector represents the direction and magnitude of the energy flow at a position indicated by the vector x. It can also be dependent on time t. The heat flux $\dot{q}$ is defined in such a way that the heat flow $\mathrm{d}\dot{Q}$ through a surface element $\mathrm{d}A$ is

$$\mathrm{d}\dot{Q} = \dot{q}(x, t)n\,\mathrm{d}A = |\dot{q}| \cos \beta \,\mathrm{d}A \ . \qquad (1.3)$$

Here n is the unit vector normal (outwards) to the surface, which with $\dot{q}$ forms the angle β, Fig. 1.1. The heat flow $\mathrm{d}\dot{Q}$ is greatest when $\dot{q}$ is perpendicular to $\mathrm{d}A$ making $\beta = 0$. The dimension of heat flow is energy/time (thermal *power*), with

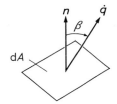

Fig. 1.1: Surface element with normal vector n and heat flux vector $\dot{q}$.

SI unit $J/s = W$. Heat flux is the heat flow per surface area with units $J/s\ m^2 = W/m^2$.

The transport of energy by heat conduction is due to a temperature gradient in the substance. The temperature ϑ changes with both position and time. All temperatures form a temperature field

$$\vartheta = \vartheta(\boldsymbol{x}, t) \ .$$

Steady temperature fields are not dependent on the time t. One speaks of unsteady or transient temperature fields when the changes with time are important. All points of a body that are at the same temperature ϑ, at the same moment in time, can be thought of as joined by a surface. This isothermal surface or isotherm separates the parts of the body which have a higher temperature than ϑ, from those with a lower temperature than ϑ. The greatest temperature change occurs normal to the isotherm, and is given by the temperature gradient

$$\operatorname{grad} \vartheta = \frac{\partial \vartheta}{\partial x} \boldsymbol{e}_x + \frac{\partial \vartheta}{\partial y} \boldsymbol{e}_y + \frac{\partial \vartheta}{\partial z} \boldsymbol{e}_z \qquad (1.4)$$

where $\boldsymbol{e}_x$, $\boldsymbol{e}_y$ and $\boldsymbol{e}_z$ represent the unit vectors in the three coordinate directions. The gradient vector is perpendicular to the the isotherm which goes through the point being considered and points to the direction of the greatest temperature increase.

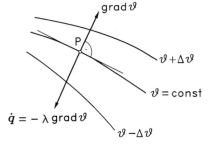

Fig. 1.2: Point P on the isotherm $\vartheta = \text{const}$ with the temperature gradient $\operatorname{grad} \vartheta$ from (1.4) and the heat flux vector $\dot{q}$ from (1.5).

Considering the temperature gradients as the cause of heat flow in a conductive material, it suggests that a simple proportionality between cause and effect may be asumed, allowing the heat flux to be written as

$$\dot{q} = -\lambda \operatorname{grad} \vartheta \ . \qquad (1.5)$$

This is J. B. Fourier's[1] basic law for the conduction of heat, from 1822. The minus sign in this equation is accounting for the 2nd law of thermodynamics: heat flows in the direction of *falling* temperature, Fig. 1.2. The constant of proportion in (1.5) is a property of the material, the thermal conductivity

$$\lambda = \lambda(\vartheta, p) \ .$$

It is dependent on both the temperature ϑ and pressure p, and in mixtures on the composition. The thermal conductivity λ is a scalar as long as the material is *isotropic*, which means that the ability of the material to conduct heat depends on position within the material, but for a given position not on the direction. All materials will be assumed to be isotropic, apart from a few special examples in Chapter 3, even though several materials do have thermal conductivities that depend on direction. This can be seen in wood, which conducts heat across its fibres significantly better than along them. In such non-isotropic medium λ is a tensor of second order, and the vectors $\dot{q}$ and grad ϑ form an angle in contrast to Fig. 1.2. In isotropic substances the heat flux vector is always perpendicular to the isothermal surface. From (1.3) and (1.5) the heat flow $d\dot{Q}$ through a surface element dA oriented in any direction is

$$d\dot{Q} = -\lambda \left(\operatorname{grad} \vartheta \right) \boldsymbol{n} \, dA = -\lambda \frac{\partial \vartheta}{\partial n} \, dA \ . \tag{1.6}$$

Here $\partial\vartheta/\partial n$ is the derivative of ϑ with respect to the normal (outwards) direction to the surface element.

Table 1.1: Thermal conductivity of selected substances at 20 °C und 100 kPa

Substance	λ in W/K m	Substance	λ in W/K m
Silver	427	Water	0.598
Copper	399	Hydrocarbons	0.10...0.15
Aluminium 99.2 %	209	CO_2	0.0162
Iron	81	Air	0.0257
Steel Alloys	13...48	Hydrogen	0.179
Brickwork	0.5...1.3	Krypton	0.0093
Foam Sheets	0.02...0.09	R 123	0.0090

The thermal conductivity, with SI units of W/K m, is one of the most important properties in heat transfer. Its pressure dependence must only be considered for gases and liquids. Its temperature dependence is often not very significant and can then be neglected. More extensive tables of λ are available in Appendix B,

[1]Jean Baptiste Fourier (1768–1830) was Professor for Analysis at the Ecole Polytechnique in Paris and from 1807 a member of the French Academy of Science. His most important work "Théorie analytique de la chaleur" appeared in 1822. It is the first comprehensive mathematical theory of conduction and cointains the "Fourier Series" for solving boundary value problems in transient heat conduction.

Tables B1 to B8, B10 and B11. As shown in the short Table 1.1, metals have very
high thermal conductivities, solids which do not conduct electricity have much
lower values. One can also see that liquids and gases have especially small values
for λ. The low value for foamed insulating material is because of its structure. It
contains numerous small, gas-filled spaces surrounded by a solid that also has low
thermal conductivity.

1.1.2   Steady, one-dimensional conduction of heat

As a simple, but practically important application, the conduction of heat independent of time, so called steady conduction, in a flat plate, in a hollow cylinder
and in a hollow sphere will be considered in this section. The assumption is made
that heat flows in only one direction, perpendicular to the plate surface, and radially in the cylinder and sphere, Fig. 1.3. The temperature field is then only
dependent on one geometrical coordinate. This is known as one-dimensional heat
conduction.

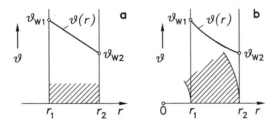

Fig. 1.3: Steady, one dimensional conduction. **a** Temperature profile in a flat plate of
thickness $\delta = r_2 - r_1$, **b** Temperature profile in a hollow cylinder (tube wall) or hollow
sphere of inner radius r_1 and outer radius r_2.

   The position coordinate in all three cases is designated by r. The surfaces
$r = $ const are isothermal surfaces; and therefore $\vartheta = \vartheta(r)$. We assume that ϑ has
the constant values $\vartheta = \vartheta_{W1}$, when $r = r_1$, and $\vartheta = \vartheta_{W2}$, when $r = r_2$. These
two surface temperatures shall be given. A relationship between the heat flow
$\dot{Q}$ through the flat or curved walls, and the temperature difference $\vartheta_{W1} - \vartheta_{W2}$,
must be found. For illustration we assume $\vartheta_{W1} > \vartheta_{W2}$, without loss of generality.
Therefore heat flows in the direction of increasing r. The heat flow $\dot{Q}$ has a certain
value, which on the inner and outer surfaces, and on each isotherm $r = $ const is
the same, as in steady conditions no energy can be stored in the wall.
   Fourier's law gives the following for the heat flow

$$\dot{Q} = \dot{q}(r)A(r) = -\lambda(\vartheta)\frac{\mathrm{d}\vartheta}{\mathrm{d}r}A(r) \ . \tag{1.7}$$

In the *flat wall* A is not dependent on r: $A = A_1 = A_2$. If the thermal conductivity
is constant, then the temperature gradient $\mathrm{d}\vartheta/\mathrm{d}r$ will also be constant. The steady

temperature profile in a plane wall with constant λ is linear. This is not true in the case of both the cylinder and the sphere, and also if λ changes with temperature. In these more general cases (1.7) becomes

$$-\lambda(\vartheta)\,\mathrm{d}\vartheta = \dot{Q}\frac{\mathrm{d}r}{A(r)}$$

and after integrating over the wall thickness $\delta = r_2 - r_1$

$$-\int_{\vartheta_{W1}}^{\vartheta_{W2}} \lambda(\vartheta)\,\mathrm{d}\vartheta = \dot{Q}\int_{r_1}^{r_2}\frac{\mathrm{d}r}{A(r)} \ .$$

From the mean value theorem for integration comes

$$-\lambda_{\mathrm{m}}\left(\vartheta_{W2} - \vartheta_{W1}\right) = \dot{Q}\frac{\delta}{A_{\mathrm{m}}}$$

or

$$\dot{Q} = \frac{\lambda_{\mathrm{m}}}{\delta}A_{\mathrm{m}}\left(\vartheta_{W1} - \vartheta_{W2}\right) \ . \tag{1.8}$$

The heat flow is directly proportional to the difference in temperature between the two surfaces. The driving force of temperature difference is analogous to the potential difference (voltage) in an electric circuit and so $\lambda_{\mathrm{m}}A_{\mathrm{m}}/\delta$ is the thermal conductance and its inverse

$$R_{\mathrm{cond}} := \frac{\delta}{\lambda_{\mathrm{m}}A_{\mathrm{m}}} \tag{1.9}$$

the thermal resistance. In analogy to electric circuits we get

$$\dot{Q} = \left(\vartheta_{W1} - \vartheta_{W2}\right)/R_{\mathrm{cond}} \ . \tag{1.10}$$

The average thermal conductivity can easily be calculated using

$$\lambda_{\mathrm{m}} := \frac{1}{\left(\vartheta_{W2} - \vartheta_{W1}\right)}\int_{\vartheta_{W1}}^{\vartheta_{W2}} \lambda(\vartheta)\,\mathrm{d}\vartheta \ . \tag{1.11}$$

In many cases the temperature dependence of λ can be neglected, giving $\lambda_{\mathrm{m}} = \lambda$. If λ changes *linearly* with ϑ then

$$\lambda_{\mathrm{m}} = \frac{1}{2}\left[\lambda\left(\vartheta_{W1}\right) + \lambda\left(\vartheta_{W2}\right)\right] \ . \tag{1.12}$$

This assumption is generally sufficient for the region $\vartheta_{W1} \leq \vartheta \leq \vartheta_{W2}$ as λ can rarely be measured with a relative error smaller than 1 to 2%.

The average area A_{m} in (1.8) is defined by

$$\frac{1}{A_{\mathrm{m}}} := \frac{1}{r_2 - r_1}\int_{r_1}^{r_2}\frac{\mathrm{d}r}{A(r)} \tag{1.13}$$

We have

$$A(r) = \begin{cases} A_1 = A_2 & \text{for a flat plate} \\ 2\pi L r & \text{for a cylinder of length } L \\ 4\pi r^2 & \text{for a sphere.} \end{cases} \quad (1.14)$$

From (1.13) we get

$$A_{\mathrm{m}} = \begin{cases} A_1 = A_2 = \frac{1}{2}\left(A_1 + A_2\right) & \text{flat plate} \\ \left(A_2 - A_1\right)/\ln\left(A_2/A_1\right) & \text{cylinder} \\ \sqrt{A_1 A_2} & \text{sphere .} \end{cases} \quad (1.15)$$

The average area A_{m} is given by the average of both surface areas $A_1 = A(r_1)$ and $A_2 = A(r_2)$. One gets the arithmetic mean for the flat plate, the logarithmic mean for the cylinder and the geometric mean for the sphere. It is known that

$$\sqrt{A_1 A_2} \le \frac{A_2 - A_1}{\ln\left(A_2/A_1\right)} \le \frac{1}{2}\left(A_1 + A_2\right) \ .$$

For the thermal resistance to conduction it follows

$$R_{\mathrm{cond}} = \begin{cases} \dfrac{\delta}{\lambda_{\mathrm{m}} A} & \text{flat plate} \\[2mm] \dfrac{\ln\left(d_2/d_1\right)}{2\pi L \lambda_{\mathrm{m}}} & \text{cylinder} \\[2mm] \dfrac{\left(d_2/d_1\right) - 1}{2\pi d_2 \lambda_{\mathrm{m}}} & \text{sphere} \end{cases} \quad (1.16)$$

The wall thickness for the cylinder (tube wall) and sphere is

$$\delta = r_2 - r_1 = \frac{1}{2}\left(d_2 - d_1\right)$$

so that R_{cond} can be expressed in terms of both diameters d_1 and d_2.

The temperature profile in each case shall also be determined. We limit ourselves to the case $\lambda = \text{const}$. With $A(r)$ from (1.14) and integrating

$$-\,\mathrm{d}\vartheta = \frac{\dot{Q}}{\lambda}\frac{\mathrm{d}r}{A(r)}$$

the dimensionless temperature ratio is

$$\frac{\vartheta(r) - \vartheta_{\mathrm{W2}}}{\vartheta_{\mathrm{W1}} - \vartheta_{\mathrm{W2}}} = \begin{cases} \dfrac{r_2 - r}{r_2 - r_1} & \text{flat plate} \\[2mm] \dfrac{\ln\left(r_2/r\right)}{\ln\left(r_2/r_1\right)} & \text{cylinder} \\[2mm] \dfrac{1/r - 1/r_2}{1/r_1 - 1/r_2} & \text{sphere} \end{cases} \quad (1.17)$$

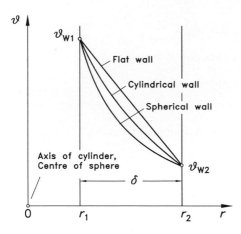

Fig. 1.4: Steady temperature profile from (1.17) in a flat, cylindrical and spherical wall of the same thickness δ and with $r_2/r_1 = 3$.

As already mentioned the temperature change is linear in the flat plate. The cylinder has a logarithmic, and the sphere a hyperbolic temperature dependence on the radial coordinates.

Fig. 1.4 shows the temperature profile according to (1.17) in walls of equal thickness. The largest deviation from the straight line by the logarithmic and hyperbolic temperature profiles appears at the point $r = r_{\mathrm{m}}$, where the cross sectional area $A(r)$ assumes the value $A(r_{\mathrm{m}}) = A_{\mathrm{m}}$ according to (1.15).

Example 1.1: A flat wall of thickness $\delta = 0.48$ m, is made out of fireproof stone whose thermal conductivity changes with temperature. With the Celsius temperature ϑ, between 0 °C and 800 °C it holds that

$$\lambda(\vartheta) = \frac{\lambda_0}{1 - b\vartheta} \tag{1.18}$$

where $\lambda_0 = 0.237$ W/K m and $b = 4.41 \cdot 10^{-4}$ K^{-1}. The surface temperatures are $\vartheta_{\mathrm{W}1} = 750$ °C and $\vartheta_{\mathrm{W}2} = 150$ °C. The heat flux $\dot{q} = \dot{Q}/A$ and the temperature profile in the wall need to be calculated.

From (1.8) the heat flux is

$$\dot{q} = \frac{\lambda_{\mathrm{m}}}{\delta}\left(\vartheta_{\mathrm{W}1} - \vartheta_{\mathrm{W}2}\right) \tag{1.19}$$

with the average thermal conductivity

$$\lambda_{\mathrm{m}} = \frac{1}{\vartheta_{\mathrm{W}2} - \vartheta_{\mathrm{W}1}} \int\limits_{\vartheta_{\mathrm{W}1}}^{\vartheta_{\mathrm{W}2}} \lambda(\vartheta)\,\mathrm{d}\vartheta = \frac{\lambda_0}{b\left(\vartheta_{\mathrm{W}1} - \vartheta_{\mathrm{W}2}\right)} \ln\frac{1 - b\vartheta_{\mathrm{W}2}}{1 - b\vartheta_{\mathrm{W}1}} \quad .$$

Putting as an abbreviation $\lambda(\vartheta_{\mathrm{W}i}) = \lambda_i\ (i = 1, 2)$ we get

$$\lambda_m = \frac{\ln(\lambda_1/\lambda_2)}{\dfrac{1}{\lambda_2} - \dfrac{1}{\lambda_1}} = \frac{\ln(\lambda_1/\lambda_2)}{\lambda_1 - \lambda_2}\lambda_1\lambda_2 \quad . \tag{1.20}$$

The average thermal conductivity λ_{m} can be calculated using the λ values for both surfaces. It is the square of the geometric mean divided by the logarithmic mean of the two values λ_1 and λ_2. This yields from (1.18)

$$\lambda_1 = \lambda(\vartheta_{\mathrm{W}1}) = 0.354\,\mathrm{W/K\,m} \quad , \qquad \lambda_2 = \lambda(\vartheta_{\mathrm{W}2}) = 0.254\,\mathrm{W/K\,m} \quad ,$$

and from that $\lambda_m = 0.298$ W/K m. The heat flux follows from (1.19) as $\dot{q} = 373$ W/m^2.

Under the rather inaccurate assumption that λ varies linearly with the temperature, it would follow that

$$\lambda_m = \frac{1}{2}(\lambda_1 + \lambda_2) = 0.304 \text{ W/K m}$$

Although this value is 1.9% too large it is still a useful approximation, as its deviation from the exact value is within the bounds of uncertainty associated with the measurement of thermal conductivity.

To calculate the temperature profile in the wall we will use (1.7) as the starting point,

$$-\lambda(\vartheta)\, d\vartheta = \dot{q}\, dr \quad,$$

and with $x = r - r_1$ this gives

$$-\lambda_0 \int_{\vartheta_{W1}}^{\vartheta} \frac{d\vartheta}{1 - b\vartheta} = \frac{\lambda_0}{b} \ln \frac{1 - b\vartheta}{1 - b\vartheta_{W1}} = \dot{q}x \quad.$$

With $\dot{q}$ from (1.19) and λ_m from (1.20) it follows that

$$\ln \frac{1 - b\vartheta}{1 - b\vartheta_{W1}} = \frac{x}{\delta} \ln \frac{1 - b\vartheta_{W2}}{1 - b\vartheta_{W1}}$$

or

$$\frac{1 - b\vartheta}{1 - b\vartheta_{W1}} = \left(\frac{1 - b\vartheta_{W2}}{1 - b\vartheta_{W1}}\right)^{x/\delta} \quad.$$

Finally using (1.18) we get

$$\vartheta(x) = \frac{1}{b}\left[1 - \frac{\lambda_0}{\lambda_1}\left(\frac{\lambda_1}{\lambda_2}\right)^{x/\delta}\right] \tag{1.21}$$

for the equation to calculate the temperature profile in the wall.

Fig. 1.5 shows $\vartheta(x)$ and the deviation $\Delta\vartheta(x)$ from the linear temperature profile between ϑ_{W1} and ϑ_{W2}. At high temperatures, where the thermal conductivity is large, the temperature

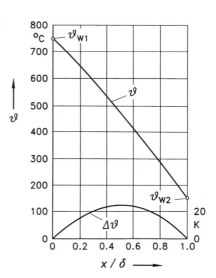

Fig. 1.5: Steady temperature profile $\vartheta = \vartheta(x/\delta)$ from (1.21) in a flat wall-with temperature dependent thermal conductivity according to (1.18). $\Delta\vartheta$ is the deviation of the temperature profile from the straight line which is valid for a constant value of λ, right hand scale.

gradient is smaller than at lower temperatures, where $\lambda(\vartheta)$ is smaller. At each point in the wall the product

$$\dot{q} = -\lambda(\vartheta)\frac{\mathrm{d}\vartheta}{\mathrm{d}x}$$

has to be the same. Smaller values of the thermal conductivity are 'compensated' by larger temperature gradients.

1.1.3 Convective heat transfer. Heat transfer coefficient

In a flowing fluid, energy is transferred not only through heat conduction but also by the macroscopic movement of the fluid. When we imagine an area located at a given position within the fluid, heat flows through this area by conduction due to the temperature gradient and in addition energy as enthalpy and kinetic energy of the fluid which crosses the area. This is known as convective heat transfer which can be described as the superposition of thermal conduction and energy transfer by the flowing fluid.

Heat transfer between a solid wall and a fluid, e.g. in a heated tube with a cold gas flowing inside it, is of special technical interest. The fluid layer close to the wall has the greatest effect on the amount of heat transferred. It is known as the *boundary layer* and boundary layer theory founded by L. Prandtl[2] in 1904 is the area of fluid dynamics that is most important for heat and mass transfer. In the boundary layer the velocity component parallel to the wall changes, over a small distance, from zero at the wall to almost the maximum value occurring in the core fluid, Fig. 1.6. The temperature in the boundary layer also changes from that at the wall ϑ_W to ϑ_F at some distance from the wall.

Heat will flow from the wall into the fluid as a result of the temperature difference $\vartheta_W - \vartheta_F$, but if the fluid is hotter than the wall, $\vartheta_F > \vartheta_W$, the fluid will be cooled as heat flows into the wall. The heat flux at the wall $\dot{q}_W$ depends on the temperature and velocity fields in the fluid; their evaluation is quite complex and can lead to considerable problems in calculation. One puts therefore

$$\dot{q}_W = \alpha\left(\vartheta_W - \vartheta_F\right) \tag{1.22}$$

with a new quantity, the local *heat transfer coefficient*, defined by

$$\alpha := \frac{\dot{q}_W}{\vartheta_W - \vartheta_F} \ . \tag{1.23}$$

This definition replaces the unknown heat flux $\dot{q}_W$, with the heat transfer coefficient, which is also unknown. This is the reason why many researchers see the

---

[2]Ludwig Prandtl (1875–1953) was Professor for Applied Mechanics at the University of Göttingen from 1904 until his death. He was also Director of the Kaiser-Wilhelm-Institut for Fluid Mechanics from 1925. His boundary layer theory, and work on turbulent flow, wing theory and supersonic flow are fundamental contributions to modern fluid mechanics.

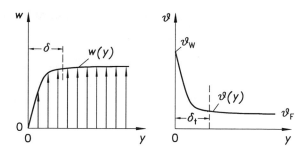

Fig. 1.6: Velocity w (left) and temperature ϑ (right) profiles in a fluid as a function of distance from the wall y. δ and δ_t represent the velocity and temperature boundary layer thicknesses.

introduction of α as unnecessary and superfluous. Nevertheless the use of heat transfer coefficients seems to be reasonable, because when α is known both the basic questions in convective heat transfer can be easily answered: What is the heat flux $\dot{q}_W$ for a given temperature difference $\vartheta_W - \vartheta_F$, and what difference in temperature $\vartheta_W - \vartheta_F$ causes a given heat flux $\dot{q}_W$ between the wall and the fluid?

In order to see how the heat transfer coefficient and the temperature field in the fluid are related, the immediate neighbourhood of the wall ($y \to 0$) is considered. Here the fluid adheres to the wall, except in the case of very dilute gases. Its velocity is zero, and energy can only be transported by heat conduction. So instead of (1.22) the physically based relationship (Fourier's Law) is valid:

$$\dot{q}_W = -\lambda \left(\frac{\partial \vartheta}{\partial y} \right)_W , \qquad (1.24)$$

where λ, or to be more exact $\lambda(\vartheta_W)$, is the thermal conductivity of the fluid at the wall temperature. The heat flux $\dot{q}_W$ is found from the gradient of the temperature profile of the fluid at the wall, Fig. 1.7. From the definition (1.23), it follows for the heat transfer coefficient

$$\alpha = -\lambda \frac{\left(\frac{\partial \vartheta}{\partial y} \right)_W}{\vartheta_W - \vartheta_F} . \qquad (1.25)$$

Fig. 1.7: Fluid temperature $\vartheta = \vartheta(y)$ as a function of distance from the wall y and illustration of the ratio λ/α as a subtangent.

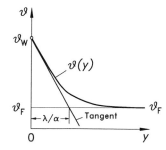

From this it is clear that α is determined by the gradient of the temperature profile at the wall and the difference between the wall and fluid temperatures. Therefore, to calculate the heat transfer coefficient, knowledge of the temperature field in the fluid is required. This is, in turn, influenced by the velocity field within the fluid. So, in addition to the energy balances from thermodynamics, the equations of fluid motion from fluid mechanics furnish the fundamental relationships in the theory of convective heat transfer.

A simple graphical illustration of α follows from (1.25). As shown in Fig. 1.7 the ratio λ/α is the distance from the wall at which the tangent to the temperature profile crosses the $\vartheta = \vartheta_F$ line. The length of λ/α is of the magnitude of the (thermal) boundary layer thickness which will be calculated in sections 3.5 and 3.7.1 and which is normally a bit larger than λ/α. A thin boundary layer indicates good heat transfer whilst a thick layer leads to small values of α.

The temperature of the fluid ϑ_F far away from the wall, appears in (1.23), the definition of the local heat transfer coefficient. If a fluid flows around a body, so called external flow, the temperature ϑ_F is taken to be that of the fluid so far away from the surface of the body that it is hardly influenced by heat transfer. ϑ_F is called the *free flow temperature*, and is often written as ϑ_∞. However, when a fluid flows in a channel, (internal flow), e.g. in a heated tube, the fluid temperature at each point in a cross-section of the channel will be influenced by the heat transfer from the wall. The temperature profile for this case is shown in Figure 1.8. ϑ_F is defined here as a *cross sectional average* temperature in such a way that ϑ_F is also a characteristic temperature for energy transport in the fluid along the channel axis. This definition of ϑ_F links the heat flow from the wall characterised by α and the energy transported by the flowing fluid.

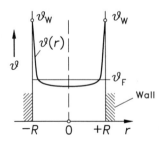

Fig. 1.8: Temperature profile in a channel cross section. Wall temperature ϑ_W and average fluid temperature ϑ_F.

To define ϑ_F we will take a small section of the channel, Fig. 1.9. The heat flow from the wall area dA to the fluid is

$$d\dot{Q} = \alpha \left(\vartheta_W - \vartheta_F \right) dA \ . \tag{1.26}$$

From the first law of thermodynamics, neglecting the change in kinetic energy, we have

$$d\dot{Q} = \left(\dot{H} + d\dot{H} \right) - \dot{H} = d\dot{H} \ . \tag{1.27}$$

The flow of heat causes a change in the enthalpy flow $\dot{H}$ of the fluid. The cross sectional average

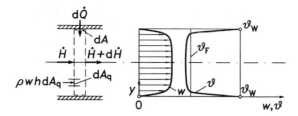

Fig. 1.9: Energy balance for a channel section (left); fluid velocity w and temperature ϑ profiles in channel cross section (right).

fluid temperature ϑ_F is now defined such that the enthalpy flow can be written as

$$\dot{H} = \int_{(A_q)} \varrho\, w\, h(\vartheta)\, \mathrm{d}A_q = \dot{M} h(\vartheta_F) \qquad (1.28)$$

as the product of the mass flow rate

$$\dot{M} = \int_{(A_q)} \varrho\, w\, \mathrm{d}A_q$$

and the specific enthalpy $h(\vartheta_F)$ at the average temperature ϑ_F.

ϑ_F is also called the *adiabatic mixing temperature*. This is the average temperature of the fluid when all elements in a cross section are mixed adiabatically in a container leaving it with the constant temperature ϑ_F. According to the first law, the enthalpy flow $\dot{H}$ with which the unmixwd fluid enters the adiabatic container must be equal to the enthalpy flow $\dot{M} h(\vartheta_F)$ of the fluid as it leaves the container. This is implied by (1.28) where ϑ_F has been implicitly defined.

To calculate the adiabatic mixing temperature ϑ_F the pressure dependence of the specific enthalpy is neglected. Then setting

$$h(\vartheta) = h_0 + [c_p]_{\vartheta_0}^{\vartheta} (\vartheta - \vartheta_0)$$

and

$$h(\vartheta_F) = h_0 + [c_p]_{\vartheta_0}^{\vartheta_F} (\vartheta_F - \vartheta_0)$$

with $[c_p]_{\vartheta_0}^{\vartheta}$ as the average specific heat capacity of the fluid between ϑ and the reference temperature ϑ_0 at which $h(\vartheta_0) = h_0$, we get from (1.28)

$$\vartheta_F = \vartheta_0 + \frac{1}{\dot{M}[c_p]_{\vartheta_0}^{\vartheta_F}} \int_{(A_q)} \varrho\, w\, [c_p]_{\vartheta_0}^{\vartheta} (\vartheta - \vartheta_0)\, \mathrm{d}A_q \ . \qquad (1.29)$$

For practical calculations a constant specific heat capacity c_p is assumed, giving

$$\vartheta_F = \frac{1}{\dot{M}} \int_{(A_q)} \varrho\, w\, \vartheta\, \mathrm{d}A_q \qquad (1.30)$$

as well as

$$\mathrm{d}\dot{H} = \dot{M} c_p\, \mathrm{d}\vartheta_F \ . \qquad (1.31)$$

The adiabatic mixing temperature from (1.30) is the link between the local heat transfer coefficient α from (1.23) and the enthalpy flow for every cross section, because from (1.26), (1.27) and (1.31) follows

$$\mathrm{d}\dot{Q} = \alpha\, (\vartheta_W - \vartheta_F)\, \mathrm{d}A = \dot{M} c_p\, \mathrm{d}\vartheta_F \ . \qquad (1.32)$$

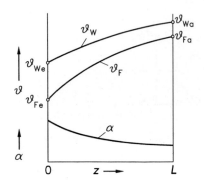

Fig. 1.10: Average fluid temperature ϑ_F, wall temperature ϑ_W and local heat transfer coefficient as functions of the axial distance z, when heating a fluid in a tube of length L.

The adiabatic mixing temperature ϑ_F is different from the integrated average of the cross sectional temperature

$$\vartheta_m = \frac{1}{A_q} \int\limits_{(A_q)} \vartheta \, dA_q \ .$$

Both temperatures are only equal if the velocity at each point in the cross section is the same, i.e. in plug flow with $w = \mathrm{const}$.

So far we have considered the local heat transfer coefficient, which can be different at every point of the wall. In practice generally only an *average heat transfer coefficient* α_m is required in order to evaluate the flow of heat $\dot{Q}$ from an area A into the fluid:

$$\dot{Q} = \alpha_m A \Delta\vartheta$$

or

$$\alpha_m := \frac{\dot{Q}}{A\Delta\vartheta} \ . \tag{1.33}$$

In this definition for α_m the temperature difference $\Delta\vartheta$ can still be chosen at will; a reasonable choice will be discussed later on.

If the local heat transfer coefficient α is known, α_m can be found by integration. This gives for the flow of heat transferred

$$\dot{Q} = \int\limits_{(A)} \dot{q}(A) \, dA = \int\limits_{(A)} \alpha \, (\vartheta_W - \vartheta_F) \, dA \ . \tag{1.34}$$

The three quantities — α, ϑ_W and ϑ_F — all change over the area. Fig. 1.10 shows this behaviour for a fluid being heated in a tube. It is assumed that all three quantities change with the axial coordinate z, but not radially. From equations (1.33) and (1.34), the average heat transfer coefficient is then

$$\alpha_m = \frac{1}{A\Delta\vartheta} \int\limits_{(A)} \alpha \, (\vartheta_W - \vartheta_F) \, dA \ .$$

In *external flow* the free flow temperature $\vartheta_F = \vartheta_\infty$ is normally constant. Therefore $\Delta\vartheta$ will be defined using ϑ_∞ and the characteristic wall temperature

$\vartheta_{\mathrm{W}}^{\star}$: $\Delta\vartheta = \vartheta_{\mathrm{W}}^{\star} - \vartheta_{\infty}$. If the wall temperature of the body around which the fluid is flowing is constant, then $\Delta\vartheta = \vartheta_{\mathrm{W}} - \vartheta_{\infty}$ gives the following for the average heat transfer cofficient:

$$\alpha_{\mathrm{m}} = \frac{1}{A} \int\limits_{(A)} \alpha \, \mathrm{d}A \ .$$

In *channel flow*, $\dot{Q}$ can either be calculated by integration over the heat transfer area A, or more simply, by using (1.32):

$$\dot{Q} = \dot{M} c_p \left(\vartheta_{\mathrm{Fa}} - \vartheta_{\mathrm{Fe}} \right) \ .$$

From this and (1.33)

$$\alpha_{\mathrm{m}} = \frac{\dot{M} c_p}{A \Delta\vartheta} \left(\vartheta_{\mathrm{Fa}} - \vartheta_{\mathrm{Fe}} \right) \ ,$$

where ϑ_{Fa} and ϑ_{Fe} are the mean fluid temperatures at the channel entrance and exit. Certain means of the temperature difference $\vartheta_{\mathrm{W}} - \vartheta_{\mathrm{F}}$ at entrance and exit are used for $\Delta\vartheta$; the main one is the logarithmic mean

$$\Delta\vartheta_{\mathrm{log}} = \frac{(\vartheta_{\mathrm{W}} - \vartheta_{\mathrm{F}})_e - (\vartheta_{\mathrm{W}} - \vartheta_{\mathrm{F}})_a}{\ln \dfrac{(\vartheta_{\mathrm{W}} - \vartheta_{\mathrm{F}})_e}{(\vartheta_{\mathrm{W}} - \vartheta_{\mathrm{F}})_a}} = \frac{\vartheta_{\mathrm{Fa}} - \vartheta_{\mathrm{Fe}}}{\ln \dfrac{(\vartheta_{\mathrm{W}} - \vartheta_{\mathrm{F}})_e}{(\vartheta_{\mathrm{W}} - \vartheta_{\mathrm{F}})_a}} \qquad (1.35)$$

With $\Delta\vartheta = \Delta\vartheta_{\mathrm{log}}$ we get

$$\alpha_{\mathrm{m}} = \frac{\dot{M}}{A} c_p \ln \frac{(\vartheta_{\mathrm{W}} - \vartheta_{\mathrm{F}})_e}{(\vartheta_{\mathrm{W}} - \vartheta_{\mathrm{F}})_a} \ . \qquad (1.36)$$

To determine α_{m} experimentally only $\dot{M}$, the wall temperature and average fluid temperature at entry and exit need to be measured The use of the logarithmic mean value is also suggested by the following: when both the wall temperature ϑ_{W} and the local heat transfer coefficient α are constant, then $\alpha_{\mathrm{m}} = \alpha$ is only true when $\Delta\vartheta_{\mathrm{log}}$ is used in the definition for α_{m}. Then from (1.32) it follows that

$$\frac{\mathrm{d}\vartheta_{\mathrm{F}}}{\vartheta_{\mathrm{W}} - \vartheta_{\mathrm{F}}} = \frac{\alpha \, \mathrm{d}A}{\dot{M} c_p} \ .$$

Integration at constant α and wall temperature ϑ_{W} gives

$$\ln \frac{(\vartheta_{\mathrm{W}} - \vartheta_{\mathrm{F}})_e}{(\vartheta_{\mathrm{W}} - \vartheta_{\mathrm{F}})_a} = \frac{\alpha A}{\dot{M} c_p} \ .$$

Putting this into (1.36), it follows that $\alpha_{\mathrm{m}} = \alpha$.

1.1.4 Determining heat transfer coefficients. Dimensionless numbers

Knowledge of the temperature field in the fluid is a prerequisite for the calculation of the heat transfer coefficient using (1.25). This, in turn, can only be determined when the velocity field is known. Only in relatively simple cases, exact values for the heat transfer coefficient can be found by solving the fundamental partial differential equations for the temperature and velocity. Examples of this include heat transfer in fully developed, laminar flow in tubes and parallel flow over a flat plate with a laminar boundary layer. Simplified models are required for turbulent flow, and the more complex problems such as nucleate boiling cannot be handled theoretically at all.

An important method for finding heat transfer coefficients was and still is the experiment. By measuring the heat flow or flux, as well as the wall and fluid temperatures the local or mean heat transfer coefficient can be found using (1.25) and (1.33). To completely solve the heat transfer problem all the quantities which influence the heat transfer must be varied when these measurements are taken. These quantities include the geometric dimensions (e.g. tube length and diameter), the characteristic flow velocity and the properties of the fluid, namely density, viscosity, thermal conductivity and specific heat capacity.

The number of these variables is generally between five and ten. To quantify the effect of one particular property, experiments must be done with at least n (e.g. $n = 5$) different values whilst keeping all other variables constant. With m different variables to consider in all, the number of individual experiments required will be n^m. With six variables and $n = 5$ then $5^6 = 15625$ experimental runs will have to be done. Obviously this demands a great deal of time and expense.

The use of similarity or model theory, cf. [1.19],[1.20], can reduce the number of experiments significantly. Similarity theory utilises the principle that the temperature and velocity fields (like other physical correlations) can be described by dimensionless quantities, namely dimensionless variables and dimensionless groups of physical quantities. This fact is a consequence of the general principle that the solution of a physical problem has to be independent of any system of units and can therefore be represented by *dimensionless* variables. This is done by dividing the position coordinate by a characteristic length, the velocity component by a constant reference velocity and the temperature by a characteristic temperature difference. Temperature and velocity fields corresponding in their dimensionless coordinates are known as similar fields. They can be transformed into each other by a change in scale, namely a change in the reference quantities.

Velocity and temperature fields are therefore only similar when also the dimensionless groups or numbers concur. These numbers contain geometric quantities, the decisive temperature differences and velocities and also the properties of the heat transfer fluid. The number of dimensionless quantities is notably smaller than the total number of all the relevant physical quantities. The number of experiments is significantly reduced because only the functional relationship between the dimensionless numbers needs to be investigated. Primarily, the values of the dimensionless numbers are varied rather than the individual quantities which make up the dimensionless numbers.

The theoretical solution of a heat transfer problem is structured more clearly when dimensionless variables are used. It is therefore recommended that dimensionless variables should be introduced at the beginning of the problem solving process. The evaluation and representation of the solution will also be simplified by keeping the number of independent variables as small as possible through the use of dimensionless variables and groups.

The partial differential equations for the velocity and temperature are the starting point for finding the characteristic quantities of heat transfer. The variables which appear, space coordinates, velocity components and temperature, are made dimensionless by dividing respectively by a characteristic length, velocity and temperature. The original partial differential equations are then transformed into partial differential equations with dimensionless variables and groups which consist of dimensionless power products of the charateristic quantities (length, velocity and temperature), and the fluid properties such as density, viscosity and thermal conductivity.

This procedure can be explained by using (1.25), which links the local heat transfer coefficient α to the temperature field, as an example. With L_0 as the characteristic length e.g. the tube diameter, the dimensionless distance from the wall is

$$y^+ := y/L_0 \ .$$

The temperature ϑ is made dimensionless by dividing by a characteristic temperature difference $\Delta\vartheta_0$. As only temperature differences or derivatives are present, by subtracting a reference temperature ϑ_0 from ϑ one gets

$$\vartheta^+ := \frac{\vartheta - \vartheta_0}{\Delta\vartheta_0} \ . \tag{1.37}$$

The choice of ϑ_0 is adapted to the problem and fixes the zero point of the dimensionless temperature ϑ^+. Then (1.25) gives

$$\alpha = -\frac{\lambda}{L_0}\frac{(\partial\vartheta^+/\partial y^+)_\mathrm{W}}{\vartheta_\mathrm{W}^+ - \vartheta_\mathrm{F}^+}$$

or

$$\frac{\alpha L_0}{\lambda} = -\frac{(\partial\vartheta^+/\partial y^+)_\mathrm{W}}{\vartheta_\mathrm{W}^+ - \vartheta_\mathrm{F}^+} \tag{1.38}$$

The right hand side of (1.38) is a dimensionless expression, this also holds for the left hand side. The power product of the heat transfer coefficient α, the characteristic length of the particular problem L_0 and the thermal conductivity λ of the fluid is known as the Nusselt number

$$Nu := \alpha L_0/\lambda \ . \tag{1.39}$$

This number and those that follow were named after eminent researchers — in this case W. Nusselt[3] — and are abbreviated to the first two letters of their surnames

---

[3]Wilhelm Nusselt (1882–1957) was nominated Professor of Theoretical Mechanical Engineering at the Technische Hochschule, Karlsruhe in 1920. Between 1925 and 1952 he taught at the Technische Hochschule, Munich. In 1915 he published his fundamental work "The Fundamental Laws of Heat Transfer", in which he introduced dimensionless groups for the first time. Further important investigations included heat transfer in film condensation, cross current heat transfer and the analogy between heat and mass transfer in evaporation.

when given as symbols in formulae.

The calculation of α leads back to the determination of the Nusselt number. According to (1.38) Nu is dependent on the dimensionless temperature field, and therefore it must be clarified which dimensionless numbers determine the dimensionless temperature ϑ^+. To this end, instead of using the fundamental differential equations — this will be done in chapter 3 — we establish a list of the physical quantities and then use this to derive the dimensionless numbers.

The dimensionless temperature ϑ^+ from (1.37) depends on the dimensionless space coordinates

$$x^+ := x/L_0, \qquad y^+ := y/L_0, \qquad z^+ := z/L_0$$

and a series of other dimensionless numbers K_i :

$$\vartheta^+ = \vartheta^+ \left(x^+, y^+, z^+, K_1, K_2, \ldots \right) \ . \tag{1.40}$$

Some of these dimensionless numbers are purely geometric parameters. In heat transfer between a flowing fluid and the inner wall of a tube with diameter d and length L one of these numbers is the ratio L/d (or its inverse d/L). Only tubes with the same value of L/d can be said to be geometrically similar. These geometry based dimensionless numbers will not be explicitly considered here, but those which independently of the geometry, determine the velocity and temperature fields, will be derived.

The velocity field is determined by the characteristic length L_0, and velocity w_0 e.g. the entry velocity in a tube or the undisturbed velocity of a fluid flowing around a body, along with the density ϱ and viscosity η of the fluid. While density already plays a role in frictionless flow, the viscosity is the fluid property which is characteristic in friction flow and in the development of the boundary layer. The two material properties, thermal conductivity λ and specific heat capacity c_p, of the fluid are important for the determination of the temperature field in conjunction with the characteristic temperature difference $\Delta\vartheta_0$. The specific heat capacity links the enthalpy of the fluid to its temperature.

With this we have seven quantities, namely

$$L_0, w_0, \varrho, \eta, \Delta\vartheta_0, \lambda \quad \text{and} \quad c_p \ ,$$

on which the temperature field and therefore, according to (1.38), the heat transfer coefficient and its dimensionless counterpart the Nusselt number depend. The dimensionless groups K_i will be made up of power products of these seven quantities. With suitable chosen exponents from a to g it follows that

$$K_i = L_0^a \cdot w_0^b \cdot \varrho^c \cdot \eta^d \cdot \Delta\vartheta_0^e \cdot \lambda^f \cdot c_p^g \ , \quad i = 1, 2, \ldots \tag{1.41}$$

The dimension of any of these seven quantities can be written as a power product of the four fundamental dimensions length **L**, time **Z**, mass **M** and temperature **T**, which are sufficient for describing thermodynamics and heat transfer by physical quantities. For example, density is defined as the quotient of mass and volume, with dimensions mass divided by length cubed:

$$\dim \varrho = \mathsf{M}/\mathsf{L}^3 \; .$$

Expressing the other six quantities in the same manner gives the following for the dimension of K_i from (1.41)

$$\dim K_i = \mathsf{L}^a \left(\mathsf{L}^b \cdot \mathsf{Z}^{-b} \right) \left(\mathsf{M}^c \mathsf{L}^{-3c} \right) \left(\mathsf{M}^d \mathsf{L}^{-d} \mathsf{Z}^{-d} \right) \mathsf{T}^e \left(\mathsf{M}^f \mathsf{L}^f \mathsf{Z}^{-3f} \mathsf{T}^{-f} \right) \left(\mathsf{L}^{2g} \mathsf{Z}^{-2g} \mathsf{T}^{-g} \right) \; .$$

For K_i to be a *dimensionless* number

$$\dim K_i = 1$$

must hold. In order to fulfill this condition the exponents of the fundamental dimensions L, Z, M and T must be zero. This produces four homogeneous equations for the seven exponents as shown below

$$
\begin{aligned}
\dim \mathsf{L} = 1: &\quad a + b - 3c - d && + f + 2g = 0 \\
\dim \mathsf{Z} = 1: &\quad - b - d && - 3f - 2g = 0 \\
\dim \mathsf{M} = 1: &\quad c + d && + f = 0 \\
\dim \mathsf{T} = 1: &\quad e - f - g = 0 \; .
\end{aligned}
\tag{1.42}
$$

Four homogeneous equations for the exponents of the seven quantities give $7 - 4 = 3$ *independent* dimensionless numbers, which are found by choosing arbitrary values for three exponents. This can be done in an infinite number of ways leading to an infinite amount of dimensionless numbers, but only three are independent of each other, the rest are all products of these three and will not give any new description of the temperature field. The three most used characteristic numbers are found by taking the values for the exponents a, e and f from Table 1.2 and then calculating the values for b, c, d and g from (1.42). From (1.41) one gets the the values for K_1, K_2, and K_3 given in Table 1.2.

Table 1.2: Values for the exponents a, e and f in (1.41) and (1.42) and the resulting characteristic numbers K_i.

i	a	e	f	K_i
1	1	0	0	$w_0 \varrho L_0 \eta^{-1}$
2	0	0	-1	$\eta c_p \lambda^{-1}$
3	0	-1	0	$w_0^2 c_p^{-1} \Delta \vartheta_0^{-1}$

The number K_1 is well known in fluid dynamics and is called the Reynolds[4] number

$$Re := \frac{w_0 \varrho L_0}{\eta} = \frac{w_0 L_0}{\nu} \; . \tag{1.43}$$

---

[4]Osborne Reynolds (1842–1912) was Professor of Engineering in Manchester, England, from 1868 until 1905. He was well known for his fundamental work in fluid mechanics especially his investigation of the transition between laminar and turbulent flow. He also developed the mathematical basis for the description of turbulent flow.

In place of η the kinematic viscosity

$$\nu := \eta/\varrho$$

with SI units m^2/s, is introduced. The Reynolds number characterises the influence of the frictional and inertial forces on the flow field. The second number K_2 contains only properties of the fluid. It is called the Prandtl number

$$Pr := \frac{\eta c_p}{\lambda} = \frac{\nu}{a} \; , \tag{1.44}$$

where

$$a := \lambda/c_p \varrho$$

is the thermal diffusivity of the fluid. The Prandtl number links the temperature field to the velocity field.

The third characteristic number K_3 is known as the Eckert[5] number:

$$Ec := \frac{w_0^2}{c_p \Delta \vartheta_0} \; . \tag{1.45}$$

It only affects the temperature field and only has to be taken into account when friction gives rise to a noticeable warming of the fluid. This only occurs at high velocities, of the order of the speed of sound, and with large velocity gradients, such as those that appear in flow through narrow slots.

The dimensionless temperature field depends on the dimensionless coordinates and these three numbers Re, Pr and Ec, as well as the numbers which are necessary for describing the geometry of the heat transfer problem. The dimensionless numbers for the geometry are represented by the abbreviation K_{geom} giving

$$\vartheta^+ = \vartheta^+ \left(x^+, y^+, z^+, Re, Pr, Ec, K_{\text{geom}} \right) \; .$$

The local Nusselt number is yielded from ϑ^+ with (1.38). It is not dependent on y^+, when y^+ is interpreted as the dimensionless distance from the wall, so that the temperature gradient at the wall is calculated at $y^+ = 0$. Therefore

$$Nu = f \left(x^+, z^+, Re, Pr, Ec, K_{\text{geom}} \right) \; . \tag{1.46}$$

The mean heat transfer coefficient α_m from (1.33) is also independent of x^+ and z^+. The mean Nusselt number Nu_m, which contains α_m, is only a function of characteristic numbers:

$$Nu_m = \frac{\alpha_m L_0}{\lambda} = F \left(Re, Pr, Ec, K_{\text{geom}} \right) \; . \tag{1.47}$$

---

[5] Ernst Rudolph Georg Eckert (born 1904 in Prague) investigated the radiation properties of solid bodies and the gases CO_2 and H_2O between 1935 and 1938. In 1938 he became a lecturer at the Technische Hochschule in Braunschweig, Germany, and he also worked at the Aeronautics Research Institute there on heat transfer at high velocities. He left for the USA in 1945 and became a Professor at the University of Minnesota in 1951. He and his students explored numerous problems of heat transfer.

As already mentioned the influence of the Eckert number only has to be considered in exceptional cases. Therefore Ec can normally be left out of (1.46) and (1.47). This will be done throughout what follows.

The type of relationship between the Nusselt number and the other characteristic numbers, or the form of the functions in (1.46) and (1.47), has to be determined either through theory, the development of a suitable model or on the basis of experiments. It must also be noted that it varies from problem to problem. In the case of flow in a tube, with $L_0 = d$, the tube diameter we get

$$\frac{\alpha_{\mathrm{m}} d}{\lambda} = F_{\mathrm{tube}} \left(\frac{w_0 d}{\nu}, \frac{\nu}{a}, \frac{L}{d} \right)$$

or

$$Nu_{\mathrm{m}} = F_{\mathrm{tube}} \left(Re, Pr, L/d \right) \quad .$$

For heat transfer between a sphere and the fluid flowing around it it follows that

$$\frac{\alpha_{\mathrm{m}} d}{\lambda} = F_{\mathrm{sphere}} \left(\frac{w_0 d}{\nu}, \frac{\nu}{a} \right)$$

or

$$Nu_{\mathrm{m}} = F_{\mathrm{sphere}} \left(Re, Pr \right) \quad .$$

A geometric dimensionless number does not appear here, as a sphere is already geometrically characterised by its diameter d. The functions F_{tube} and F_{sphere} have different forms because flow fields and heat transfer conditions in flow through a tube differ from those in a fluid flowing around a sphere.

The geometric and flow conditions are not the only parameters which have a considerable influence on the relationship between the Nusselt number and the other dimensionless numbers. The thermal boundary conditions also affect heat transfer. An example of this is, with the same values of Re and Pr in parallel flow over a plate, we have different Nusselt numbers for a plate kept at constant wall temperature ϑ_{W}, and for a plate with a constant heat flux $\dot{q}_{\mathrm{W}}$ at the wall, where the surface temperature adjusts itself accordingly.

As already mentioned, Nu, Re, Pr and Ec can be replaced by other dimensionless numbers which are power products of these four numbers. The Péclet[6] number can be used in place of the Reynolds number,

$$Pe := \frac{w_0 L_0}{a} = \frac{w_0 \varrho c_p L_0}{\lambda} = Re\, Pr \quad , \tag{1.48}$$

and can be written as the product of the Reynolds and Prandtl numbers. In american literature the Nusselt number is often replaced by the Stanton[7] number

$$St := \frac{\alpha}{w_0 \varrho c_p} = \frac{Nu}{Re\, Pr} \quad . \tag{1.49}$$

---

[6]Jean Claude Eugene Péclet (1793–1857) became Professor of Physics in Marseille in 1816. He moved to Paris in 1827 to take up a professorship. His famous book "Traité de la chaleur et de ses applications aux arts et aux manufactures" (1829) dealt with heat transfer problems and was translated into many different languages.

[7]Thomas Edward Stanton (1865–1931) was a student of Reynolds in Manchester. In 1899 he became a Professor for Engineering at the University of Bristol. Stanton researched momentum and heat tranport in friction flow. He also worked in aerodynamics and aeroplane construction.

The Stanton number is useful for describing heat transfer in channels, and can be interpreted in a plausible way. This is demonstrated by using a Stanton number which contains the mean heat transfer coefficient

$$St_m := \alpha_m/w_0 \varrho c_p \ ,$$

in which

$$w_0 = \dot{M}/A_q \varrho$$

is the mean flow velocity in a channel with constant cross sectional area A_q. The heat transferred between the channel wall (area A) and the fluid is

$$\dot{Q} = \alpha_m A \Delta\vartheta = \dot{M} c_p \left(\vartheta_{Fa} - \vartheta_{Fe}\right) \ ,$$

from which the Stanton number

$$St_m = \frac{A_q}{A} \frac{\vartheta_{Fa} - \vartheta_{Fe}}{\Delta\vartheta}$$

is obtained. This gives (multiplied by the ratio of areas A_q/A) the change in temperature between entry and exit divided by the 'driving' temperature difference $\Delta\vartheta$ used in the defintion of α_m. If the logarithmic temperature difference from (1.35) is used it follows that

$$St_m = \frac{A_q}{A} \ln \frac{(\vartheta_W - \vartheta_F)_e}{(\vartheta_W - \vartheta_F)_a} \ .$$

The ratio of areas for a tube with circular cross section (diameter d) is $A_q/A = d/4L$. It is clear from this relationship that St_m is directly linked to the data required in the design of heat exchangers.

Up until now the material properties ϱ, η, λ and c_p have been taken to be constant. This is an approximation which can only be made with small changes in the fluid temperature. Although the pressure dependence of these quantities can be neglected, they vary noticeably with temperature. In order to account for this temperature dependence, further characteristic numbers must be introduced. These appear often in the form of ratios of the material properties, e.g. $\eta(\vartheta_F)/\eta(\vartheta_W)$. This can be avoided by the use of a suitable mean temperature for calculating the values of the material properties.

The temperature dependence of the fluid density is especially important in heat transfer caused by *natural* or *free convection*. In contrast to the previous case of forced convection, where the fluid is forced at speed by a blower or pump, free convective flow exists due to density changes in the earth's gravitational field, which originate from the variation in temperature. This is how, for example, a quiescent fluid next to a hot wall is heated. The density of the fluid adjacent to the wall is lowered, causing an upward flow in the gravitational field to develop.

In free convection, the characteristic (forced) velocity w_0 does not apply. The new physical quantity is the acceleration due to gravity g. The Reynolds number Re, which contains the reference velocity w_0, is replaced by a new group, the dimensionless acceleration due to gravity

$$Ga := \frac{gL_0^3}{\nu_0^2} = \frac{g\varrho_0^2 L_0^3}{\eta^2} \ , \tag{1.50}$$

known as the Galilei number. As the density changes with temperature $\nu = \eta/\varrho$ must be calculated with a constant density $\varrho_0 = \varrho(\vartheta_0)$ using a fixed reference temperature ϑ_0. All other fluid properties are assumed to be constant.

At least one new characteristic number K_ϱ is required to describe the density changes with temperature. This gives the following for the average Nusselt number

$$Nu_\mathrm{m} = f\left(Ga, K_\varrho, Pr, K_\mathrm{geom}\right) \ . \tag{1.51}$$

The density based characteristic number K_ϱ and other such numbers $K_{\varrho 1}, K_{\varrho 2}, \ldots$ characterise the dimensionless temperature dependence of the density:

$$\varrho(\vartheta^+)/\varrho_0 = f_\varrho\left(\vartheta^+, K_{\varrho 1}, K_{\varrho 2}, \ldots\right) \ .$$

Here, ϑ^+ is the dimensionless temperature introduced by (1.37). With relatively small temperature differences a Taylor series

$$\varrho(\vartheta) = \varrho_0\left[1 + \frac{1}{\varrho_0}\left(\frac{\partial \varrho}{\partial \vartheta}\right)_{\vartheta_0}(\vartheta - \vartheta_0) + \ldots\right] \tag{1.52}$$

requiring only the linear term with the expansion coefficient

$$\beta_0 = \beta(\vartheta_0) := -\frac{1}{\varrho_0}\left(\frac{\partial \varrho}{\partial \vartheta}\right)_{\vartheta_0} \tag{1.53}$$

can be used. Then with ϑ^+ from (1.37), it follows that

$$\varrho\left(\vartheta^+\right)/\varrho_0 = 1 - \beta_0 \Delta\vartheta_0 \vartheta^+ \ . \tag{1.54}$$

Provided that the temperature differences are not too large the temperature dependence of the density can be determined by the product of the expansion coefficient β_0 and a characteristic temperature difference $\Delta\vartheta_0$. One density based characteristic group is sufficient for this case, namely

$$K_\varrho = \beta_0 \Delta\vartheta_0 \ .$$

If the temperature difference is large further terms in (1.52) have to be considered. Instead of the one characteristic number K_ϱ, two or more are needed.

The density variation due to the temperature generates a buoyancy force in the gravitational field. However, this has little influence on the other forces, including inertia and friction, which affect the fluid particles. As a good approximation it is sufficient to consider the temperature dependence of the density in buoyancy alone. This assumption is known as the *Boussinesq-Approximation*[8]. A characteristic (volume related) lift force is

$$g\left[\varrho\left(\vartheta_\mathrm{W}\right) - \varrho\left(\vartheta_\mathrm{F}\right)\right] = g\left(\varrho_\mathrm{W} - \varrho_\mathrm{F}\right) \ .$$

---

[8]Joseph Valentin Boussinesq (1842–1929) gained his PhD in 1867 with a thesis on heat propagation, even though he was never a student and had taught himself science. In 1873 he became a Professor in Lille, later moving to Paris. He published over 100 scientific works, including two volumes entitled "Théorie analytique de la chaleur", which appeared in Paris in 1901 and 1903.

Dividing this expression by ϱ_F and putting it into the Gallilei number in place of g produces a new number

$$Gr := \frac{gL_0^3}{\nu_F^2} \frac{\varrho_W - \varrho_F}{\varrho_F} \quad , \tag{1.55}$$

called the Grashof[9] number. This combines the Galilei number and the characteristic quantity $(\varrho_W - \varrho_F)/\varrho_F$ for the temperature dependence of the density, in one number. When the Boussinesq-Approximation is valid the Nusselt number can be written as

$$Nu_m = f\left(Gr, Pr, K_{\text{geom}}\right) \quad . \tag{1.56}$$

The change of density with temperature can be found using (1.54), as long as the temperature difference between wall and fluid further away from the wall is not too great. With $\vartheta_0 = \vartheta_F$ and $\Delta\vartheta_0 = (\vartheta_W - \vartheta_F)$ one obtains

$$\frac{\varrho_W - \varrho_F}{\varrho_F} = \beta_F \left(\vartheta_W - \vartheta_F\right) \quad .$$

The Grashof Number then becomes

$$Gr = Ga\,\beta_F\left(\vartheta_W - \vartheta_F\right) = \frac{g\,\beta_F\left(\vartheta_W - \vartheta_F\right)L_0^3}{\nu^2} \quad . \tag{1.57}$$

The expansion coefficient β has to be calculated at the fluid temperature ϑ_F. In order to account for the temperature dependence of ν, it must be determined at an average temperature between ϑ_W and ϑ_F.

According to (1.56) and (1.57) the Nusselt number depends on the temperature difference $(\vartheta_W - \vartheta_F)$. Although the heat transfer coefficient α is found by dividing the heat flux $\dot{q}_W$ by this temperature difference, cf. (1.24), in free convection α is not independent of $(\vartheta_W - \vartheta_F)$. In other words the transferred heat flux, $\dot{q}_W$, does not increase proportionally to $\vartheta_W - \vartheta_F$. This is because $\vartheta_W - \vartheta_F$ is not only the 'driving force' for the heat flow but also for the buoyancy, and therefore the velocity field in free convection. In contrast, the heat transfer coefficient for forced convection is not expected to show any dependence on the temperature difference.

---

[9]Franz Grashof (1826–1893) taught from 1863 until 1891 as Professor of Theoretical Mechanical Engineering at the Technische Hochschule in Karlsruhe, Germany. His work written with mathematical rigor, especially dealt with mechanics and technical thermodynamics. His major publication "Theoretische Maschinenlehre", consisted of three volumes and appeared between 1875 and 1890. It is a comprehensive, expert portrayal of mechanical engineering. In 1856 Grashof founded the "Verein Deutscher Ingenieure" (VDI), the association of german engineers along with 22 other young engineers. He was its first director and for many years also the editor of the association's technical magazine in which he published 42 papers in total.

1.1.5 Thermal radiation

All forms of matter emit energy to their surroundings through electromagnetic waves. This already happens because the matter has a positive thermodynamic temperature. This type of energy release is known as thermal radiation or heat radiation. The emission of radiation is due to the conversion of the body's internal energy into energy which is transported by electromagnetic waves. When electromagnetic waves hit any matter, part of the energy is absorbed, the rest is then either reflected or transmitted. The radiation energy which is absorbed by the body is converted into internal energy. Thermal radiation causes a special type of heat transfer which is known as radiative exchange. The transport of radiation does not require any matter as electromagnetic waves can also travel through a vacuum. This allows heat to be transferred between two bodies over great distances. In this way the earth receives a large energy flow from the sun.

Gases and liquids are partly transparent to thermal radiation. Therefore emission and absorption of radiation takes place inside the gas or liquid space. In gases and liquids emission and absorption are volumetric effects. In contrast, on the surface of a solid object, radiation is completely absorbed within a very thin layer (a few micrometres). Radiation from the interior of a solid body does not penetrate the surface, emission is limited to a thin surface layer. It can therefore be said that emission and absorption of radiation by a solid body are surface effects. This means that it is allowed to speak of radiating and absorbing surfaces rather than correctly of radiating solid bodies.

There is an upper limit for the emission of heat radiation, which only depends on the thermodynamic temperature T of the radiating body. The maximum heat flux from the surface of a radiating body is given by

$$\dot{q}_{\mathrm{s}} = \sigma T^4 \ . \tag{1.58}$$

This law was found 1879 by J. Stefan[10] as a result of many experiments and was derived in 1884 by L. Boltzmann[11] from the electromagnetic theory of radiation using the second law of thermodynamics. It contains an universal constant, known as the Stefan-Boltzmann constant σ, which has a value of

$$\sigma = (5.67051 \pm 0.00019)10^{-8} \ \mathrm{W/m^2 \ K^4} \ .$$

---

[10] Josef Stefan (1835–1893) became Professor of Physics at the University of Vienna in 1863. He was an excellent researcher and published numerous papers on heat conduction and diffusion in fluids, ice formation, and the connection between surface tension and evaporation. He suggested the T^4-law after careful evaluation of lots of earlier experiments on the emission of heat from hot bodies.

[11] Ludwig Boltzmann (1844–1906) gained his PhD in 1867 as a scholar of J. Stefan in Vienna. He was a physics professor in Graz, Munich, Leipzig and Vienna. His main area of work was the kinetic theory of gases and its relationship with the second law of thermodynamics. In 1877 he found the fundamental relation between the entropy of a system and the logarithm of the number of possible molecular distributions which make up the macroscopic state of the system.

An emitter, whose emissive power, or heat flux emitted by radiation, reaches the maximum value $\dot{q}_s$ in (1.58), is called a *black body*. This is an ideal emitter whose emissive power cannot be surpassed by any other body at the same temperature. On the other hand, a black body absorbs *all* incident radiation, and is, therefore, an ideal absorber. The emissive power of real radiators can be described by using a correction factor in (1.58). By putting

$$\dot{q} = \varepsilon(T)\sigma T^4 \ , \tag{1.59}$$

the *emissivity* $\varepsilon(T) \leq 1$ of the radiator is defined. This material property, emissivity, does not only depend on the material concerned but also on the condition of the surface, for example its roughness. Some values for ε are compiled in Table 1.3.

Table 1.3: Emissivity $\varepsilon(T)$ of some materials

Material	T in K	ε	Material	T in K	ε
Concrete, rough	293	0.94	Nickel, polished	373	0.053
Wood, oak	293	0.90	Iron, shiny corroded	423	0.158
Brick, red	293	0.93	Copper, oxidised	403	0.725
Aluminium, rolled	443	0.049			

When radiation hits a body, some of it will be reflected, some absorbed and some transmitted. These portions are represented by the reflectivity r, the absorptivity a, and the transmissivity τ. These three dimensionless quantities are not purely properties of the irradiated material, they also depend on the kind of radiation which strikes the body. Of main influence is the distribution of radiation over the wave length spectrum of the electromagnetic waves incident on the material. However it can always be said that

$$r + a + \tau = 1 \ . \tag{1.60}$$

Most solid bodies are opaque, they do not allow any radiation to be transmitted, so with $\tau = 0$ the absorptivity from (1.60) is $a = 1 - r$.

The absorption of thermal radiation will be treated in more depth in chapter 5. The connection between emission and absorption will also be looked at; this is known as Kirchhoff's law, see section 5.1.6. It basically says that a good emitter of radiation is also a good absorber. For the ideal radiator, the black body, both absorptivity a and emissivity ε are equal to the maximum value of one. The black body, which absorbs all incident radiation ($a = 1$), also emits more than any other radiator, agreeing with (1.58), the law from Stefan and Boltzmann.

1.1.6   Radiative exchange

In heat transfer, the heat transmitted by radiation between two bodies at different temperatures is very important. It is not only the hotter body that radiates heat to warm the body at the lower temperature, the colder one also emits electromagnetic waves which transfer energy to the body at the higher temperature. It is therefore applicable here to talk of an exchange of radiation. Finally the net heat flow from the body at the higher temperature to that at the lower temperature is of interest. The evaluation of the heat flow is difficult for a number of reasons. In general, other bodies will also play a part in radiative exchange. A body may not absorb all the radiation that hits it, part of it may be reflected and could hit the original emitter. This complicated interplay between two radiators becomes even more complex when the medium between the two bodies absorbs part of the radiation passing through it and emits energy itself. This is the case in so called gas radiation and must be considered, for example, in heat transfer in a furnace.

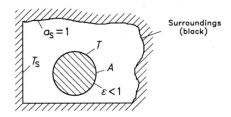

Fig. 1.11: Radiative exchange between a body at temperature T and black surroundings at temperature T_S.

As an introduction, a simple case of radiative exchange will be looked at. A radiator with area A, and at temperature T, is located in surroundings which are at temperature T_S, see Fig. 1.11. The medium between the two shall have no effect on the radiation transfer; it shall be completely transparent for radiation, which is a very good approximation for atmospheric air. The surroundings shall behave like a black body, absorbing all radiation, $a_S = 1$.

The heat flow emitted by the radiator

$$\dot{Q}_{em} = A \, \varepsilon \, \sigma \, T^4$$

reaches the black surroundings and is completely absorbed. The black radiation emitted by the surroundings will be partly absorbed by the radiator at temperature T, the rest will be reflected back to, and absorbed by, the surroundings. The heat absorbed by the radiator is

$$\dot{Q}_{ab} = A \, a \, \sigma \, T_S^4 \ ,$$

where a is the absorptivity of the radiator at temperature T, for the black body radiation of temperature T_S. The absorptivity a is not a property of the material alone, as it is not only dependent on the properties of the absorbing surface, but also on the source and type of incident radiation. In the example, this is the black body radiation coming from the surroundings which is completely characterised by the temperature T_S.

The net flow of heat $\dot{Q}$, from the radiator to the surroundings enclosing it is

$$\dot{Q} = \dot{Q}_{em} - \dot{Q}_{ab} = A\sigma\left(\varepsilon T^4 - aT_S^4\right) \ . \tag{1.61}$$

In many cases a simple assumption is made about the radiator: it is treated as a *grey* radiator. This is only an approximation but it simplifies matters greatly. The absorptivity of a grey radiator does not depend on the type of incident radiation, and it always agrees with the emissivity, such that $a = \varepsilon$.

For a grey radiator in black surroundings (1.61) simplifies to

$$\dot{Q} = A\sigma\varepsilon\left(T^4 - T_S^4\right) \ . \tag{1.62}$$

The difference between the fourth power of the temperature of the emitter and that of the body which receives the radiation, is characteristic of radiative exchange. This temperature dependence is found in numerous radiative heat transfer problems involving grey radiators.

In many applications heat transfer by convection must be considered in addition to radiative heat transfer. This is, for example, the case where a radiator releases heat to a room which is at a lower temperature. Radiative heat exchange takes place between the radiator and the walls of the room, whilst at the same time heat is transferred to the air by convection. These two kinds of heat transfer are parallel to each other and so the heat flow by convection and that by radiation are added together in order to find the total heat exchanged. The heat flux then becomes

$$\dot{q} = \dot{q}_{conv} + \dot{q}_{rad}$$

or

$$\dot{q} = \alpha(T - T_A) + \varepsilon\sigma\left(T^4 - T_S^4\right) \ . \tag{1.63}$$

Here, α is the heat transfer coefficient for the convective heat transfer to air at temperature T_A. Equation (1.62) was used to evaluate $\dot{q}_{rad}$.

Normally $T_A \approx T_S$, allowing the convective and radiative parts of heat transfer to be put together. This gives

$$\dot{q} = \left(\alpha + \alpha_{rad}\right)\left(T - T_S\right) \ . \tag{1.64}$$

The radiative heat transfer coefficient defined above, becomes from (1.63)

$$\alpha_{rad} = \varepsilon\sigma\frac{T^4 - T_S^4}{T - T_S} = \varepsilon\sigma(T^2 + T_S^2)(T + T_S) \ . \tag{1.65}$$

This quantity is dependent on the emissivity ε, and both temperatures T and T_S. The introduction of α_{rad} allows the influence of radiation to be compared to convection. As $\varepsilon \leq 1$ is always true, it is immediately obvious that an upper limit for the radiative part of heat transfer exists.

Example 1.2: A poorly insulated horizontal pipe (outer diameter $d = 0.100$ m), with a surface temperature $\vartheta_W = 44\,°C$, runs through a large room of quiescent air at $\vartheta_A = 18\,°C$. The heat loss per length L of the pipe, $\dot{Q}/L$ has to be determined. The pipe is taken to behave as a grey radiator with emissivity $\varepsilon = 0.87$. The walls of the room are treated as black surroundings which are at temperature $\vartheta_S = \vartheta_A = 18\,°C$.

The tube gives off heat to the air by free convection and to the surrounding walls by radiation. Then from (1.64) with $\vartheta_S = \vartheta_A$ comes

$$\dot{Q}/L = \pi d\,(\dot{q}_{conv} + \dot{q}_{rad}) = \pi d\,(\alpha_m + \alpha_{rad})\,(\vartheta_W - \vartheta_A) \quad , \tag{1.66}$$

where α_m is the mean heat transfer coefficient for free convection. The heat transfer coefficient for radiation, using the data given and (1.65) is

$$\alpha_{rad} = \varepsilon\,\sigma \frac{T_W^4 - T_A^4}{T_W - T_A} = 0.87 \cdot 5.67 \cdot 10^{-8}\,\frac{W}{m^2 K^4}\,\frac{317^4 - 291^4}{317 - 291}\,K^3 = 5.55\,\frac{W}{m^2 K} \quad .$$

For heat transfer by free convection from a horizontal pipe S.W. Churchill and H.H.S. Chu [1.3] give the dimensionless relationship

$$Nu_m = \frac{\alpha_m d}{\lambda} = \left\{ 0.60 + \frac{0.387(GrPr)^{1/6}}{\left[1 + (0.559/Pr)^{9/16}\right]^{8/27}} \right\}^2 \tag{1.67}$$

which has the same form as Eq. (1.56), that of $Nu_m = f(Gr, Pr)$. According to (1.57) the Grashof number is

$$Gr = \frac{g\beta\,(\vartheta_W - \vartheta_L)\,d^3}{\nu^2} \quad .$$

The expansion coefficient β has to be calculated for the air temperature ϑ_A. As air can be treated as an ideal gas, therefore $\beta = 1/T_A = 0.00344\,K^{-1}$. To take the temperature dependence of the material properties into account, ν, λ and Pr all have to be calculated at the mean temperature $\vartheta_m = \frac{1}{2}\,(\vartheta_W + \vartheta_A)$. For air at $\vartheta_m = 31\,°C$, $\nu = 16.40 \cdot 10^{-6}\,m^2/s$, $\lambda = 0.0265\,W/Km$ and $Pr = 0.713$. This gives

$$Gr = \frac{9.81\,(m/s^2)\,0.00344\,K^{-1}\,(44 - 18)\,K \cdot 0.100^3\,m^3}{16.40^2 \cdot 10^{-12}\,m^4/s^2} = 3.26 \cdot 10^6 \quad .$$

Using (1.67) the Nusselt number is $Nu_m = 18.48$, out of which comes the heat transfer coefficient

$$\alpha_m = Nu_m \frac{\lambda}{d} = 18.48\,\frac{0.0265\,W/Km}{0.100\,m} = 4.90\,\frac{W}{m^2 K} \quad .$$

Eq.(1.66) gives the heat loss as $\dot{Q}/L = 85.4\,W/m$.

The heat transfer coefficients α_m and α_{rad} are approximately equal. This infers that free convection to the air and radiative exchange transport almost the same amount of heat. This is not true for forced convection, where α_m, depending on the flow velocity, is one to two powers of ten larger than the value calculated here. However, α_{rad} remains unaffected and when compared to α_m can generally be neglected.

1.2 Overall heat transfer

In many applications of heat transfer two fluids at different temperatures are separated by a solid wall. Heat is transferred from the fluid at the higher temperature to the wall, conducted through the wall, and then finally transferred from the cold side of the wall into the fluid at the lower temperature. This series of convective and conductive heat transfer processes is known as *overall heat transfer.*

Overall heat transfer takes place, above all in heat exchangers, which will be dealt with in section 1.3. Here, for example, a hot fluid flowing in a tube gives heat up, via the wall, to the colder fluid flowing around the outside of the tube. House walls are also an example for overall heat transfer. They separate the warm air inside from the colder air outside. The resistance to heat transfer should be as large as possible, so that despite the temperature difference between inside and outside, only a small amount of heat will be lost through the walls. In contrast to this case, the heat transfer resistances present in a heat exchanger should be kept as small as possible; here a great amount of heat shall be transferred with a small temperature difference between the two fluids in order to keep thermodynamic (exergy) losses as small as possible.

As these examples show, the calculation of the overall heat transfer is of significant technical importance. This problem is dealt with in the next sections.

1.2.1 The overall heat transfer coefficient

The following analysis is based on the situation shown in Fig. 1.12: A flat or curved wall separates a fluid at temperature ϑ_1 from another with a temperature $\vartheta_2 < \vartheta_1$. At steady state heat $\dot{Q}$, flows from fluid 1 through the wall to fluid 2, as a result of the temperature difference $\vartheta_1 - \vartheta_2$. The heat flow $\dot{Q}$ is transferred from fluid 1 to the wall which has an area A_1 and is at temperature ϑ_{W1}. With α_1 as the heat transfer coefficient, it follows from section 1.1.3 that

$$\dot{Q} = \alpha_1 A_1 \left(\vartheta_1 - \vartheta_{W1} \right) \; . \tag{1.68}$$

For the conduction through the wall, according to section 1.1.2

$$\dot{Q} = \frac{\lambda_m}{\delta} A_m \left(\vartheta_{W1} - \vartheta_{W2} \right) \; . \tag{1.69}$$

Here λ_m is the mean thermal conductivity of the wall according to (1.11), δ its thickness and A_m the average area calculated from (1.15). Finally an analogous relationship to (1.68) exists for the heat transfer from the wall to fluid 2

$$\dot{Q} = \alpha_2 A_2 \left(\vartheta_{W2} - \vartheta_2 \right) \; . \tag{1.70}$$

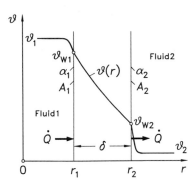

Fig. 1.12: Temperature profile for heat transfer through a tube wall bounded by two fluids with temperatures ϑ_1 and $\vartheta_2 < \vartheta_1$.

The unknown wall temperatures ϑ_{W1} and ϑ_{W2}, can be eliminated from the three equations for $\dot{Q}$. This means that $\dot{Q}$ can be calculated by knowing only the fluid temperatures ϑ_1 and ϑ_2. This results in

$$\dot{Q} = kA\left(\vartheta_1 - \vartheta_2\right) \tag{1.71}$$

where

$$\frac{1}{kA} = \frac{1}{\alpha_1 A_1} + \frac{\delta}{\lambda_m A_m} + \frac{1}{\alpha_2 A_2} \tag{1.72}$$

is valid. The *overall heat transfer coefficient* k, for the area A is defined in (1.71), where A is the size of any reference area. Equation (1.72) shows that kA can be calculated using the quantities already introduced for convective heat transfer and conduction.

As shown by equn. (1.71) and (1.72), only the product kA is used. Giving k alone, without stating anything about the size of the area A is meaningless. As A can be chosen to be any area, k takes on the value associated with the choice of A. In practice values for k are often given and used. This can be seen for example in german building regulations (Norm DIN 4108) where a minimum value for k is set for house walls. This is to guarantee a sufficient degree of insulation in each house that is built. This sort of statement of k is tacitly related to a certain area. For flat walls this is the area of the wall $A_1 = A_2 = A_m$; for tubes mostly the outer surface A_2, which does not normally differ greatly from A_1 or A_m. In what follows, only the product kA will be used so that A will not need to be specified. In exceptional cases a value for k will be given along with a value for the related area.

In (1.72), $(1/kA)$ represents the *resistance* to overall heat transfer. It is made up of the single resistances of each transfer process in the series; the resistance to convective transfer between fluid 1 and the wall, $(1/\alpha_1 A_1)$, the conduction resistance in the wall, $(\delta/\lambda_m A_m)$ and the resistance to convective transfer between the wall and fluid 2, $(1/\alpha_2 A_2)$. This series approach for overall heat transfer resistance is analogous to that in electrical circuits, where the total resistance to the current is found by the addition of all the single resistances in series. Therefore, the three resistances which the heat flow $\dot{Q}$ must pass through, are added together. These three are the resistance due to the boundary layer in fluid 1, the conduction resistance in the wall and the resistance to transfer associated with the boundary layer in fluid 2.

The temperature drop due to these thermal resistances behaves in exactly the same manner as the voltage drop in an electrical resistor, it increases as the resistance goes up and as the current becomes stronger. From (1.68) to (1.72) it follows that

$$\dot{Q} = \frac{\vartheta_1 - \vartheta_{W1}}{\dfrac{1}{\alpha_1 A_1}} = \frac{\vartheta_{W1} - \vartheta_{W2}}{\dfrac{\delta}{\lambda_m A_m}} = \frac{\vartheta_{W2} - \vartheta_2}{\dfrac{1}{\alpha_2 A_2}} = \frac{\vartheta_1 - \vartheta_2}{\dfrac{1}{kA}} \ . \tag{1.73}$$

From this the temperature drop in the wall and the boundary layers on both sides can be calculated. To find the wall temperatures the equations

$$\vartheta_{W1} = \vartheta_1 - \frac{kA}{\alpha_1 A_1}(\vartheta_1 - \vartheta_2) = \vartheta_1 - \frac{\dot{Q}}{\alpha_1 A_1} \tag{1.74}$$

and

$$\vartheta_{W2} = \vartheta_2 + \frac{kA}{\alpha_2 A_2}(\vartheta_1 - \vartheta_2) = \vartheta_2 + \frac{\dot{Q}}{\alpha_2 A_2} \tag{1.75}$$

are used.

For the *overall heat transfer through a pipe*, (1.72) can be applied, when it is taken into account that a pipe of diameter d and length L has a surface area of $A = \pi d L$. Then from (1.72) it follows with A_m from (1.15) that

$$\frac{1}{kA} = \frac{1}{\pi L}\left(\frac{1}{\alpha_1 d_1} + \frac{\ln d_2/d_1}{2\lambda_m} + \frac{1}{\alpha_2 d_2}\right) \tag{1.76}$$

where d_1 is the inner and d_2 the outer diameter of the pipe.

1.2.2   Multi-layer walls

The analogy to electrical circuits is also used to extend the relationships derived in section 1.2.1 for overall heat transfer, to walls with several layers. Walls with two or more layers are often used in technical practise. A good example of these multi-layer walls is the addition of an insulating layer made from a material with low thermal conductivity λ_{is}. Fig. 1.13 shows a temperature profile for a wall that consists of a number of layers. The resistance to heat transfer for each layer in series is added together and this gives the overall heat transfer resistance for the wall as

$$\frac{1}{kA} = \frac{1}{\alpha_1 A_1} + \sum_i \frac{\delta_i}{\lambda_{mi} A_{mi}} + \frac{1}{\alpha_2 A_2} \ . \tag{1.77}$$

In curved walls the average area of a layer A_{mi} is calculated using the inner and outer areas of the section using (1.15). Applying (1.77), for the overall resistance to heat transfer, it is assumed that each layer touches its neighbour so closely that there is no noticeable temperature difference between the layers. If this was not

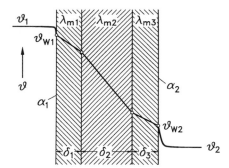

Fig. 1.13: Temperature profile for overall heat
transfer through a flat wall of three layers of different materials.

true a thermal contact resistance, similar to the contact resistance that appears
in electric circuits, would have to be considered.

The temperature drop $\vartheta_i - \vartheta_{i+1}$ in the i-th layer is proportional to the heat
flow and the resistance to conduction. This is analogous to a voltage drop across
a resistor in an electric circuit. It follows that

$$\vartheta_i - \vartheta_{i+1} = \frac{\delta_i}{\lambda_{mi} A_{mi}} \dot{Q} \; . \tag{1.78}$$

Using $\dot{Q}$ from (1.71) and (1.77), $\vartheta_i - \vartheta_{i+1}$ is fairly simple to calculate. The surface
temperatures ϑ_{W1} and ϑ_{W2} are calculated in exactly the same way as before using
(1.74) and (1.75)

In tubes which consist of several layers e.g. the actual tube plus its insulation,
(1.77) is extended to

$$\frac{1}{kA} = \frac{1}{\pi L} \left(\frac{1}{\alpha_1 d_1} + \frac{1}{2} \sum_{i=1}^{n} \frac{1}{\lambda_{mi}} \ln \frac{d_{i+1}}{d_i} + \frac{1}{\alpha_2 d_{n+1}} \right) \; . \tag{1.79}$$

The i-th layer is bounded by the diameters d_i and d_{i+1}. The first and last layers,
which are in contact with the fluid, can also be layers of dirt or scale which develop
during lengthy operation and represent an additional conductive resistance to the
transfer of heat.

1.2.3   Overall heat transfer through walls with extended surfaces

The overall resistance to heat transfer $(1/kA)$ is found by adding all the resistances
to convective heat transfer and conduction, as shown in (1.72). Therefore the
largest single resistance determines the value for $(1/kA)$. This is especially distinct
when the other resistances are very small. It is therefore possible to significantly
improve the insulation of a wall by adding a layer which has high resistance to
conduction $\delta/\lambda_m A_m$, or in other words a thick layer of a material with low thermal
conductivity.

If, however, the heat transfer must be as good as possible, a good example being a heat exchanger, a high convective resistance $(1/\alpha A)$ can be an impediment. The main cause of a large resistance is a small heat transfer coefficient α. Unfortunately even with significant increases in flow velocity, α barely increases. This leaves only one option for reducing $(1/\alpha A)$, which is to increase the area A available for heat transfer by the addition of fins, rods, or pins on the side where heat transfer is 'bad'. Examples of these extended surfaces are shown in Fig. 1.14. Surface extension will lead to a remarkable increase of the original area, in some cases by 10-100 times.

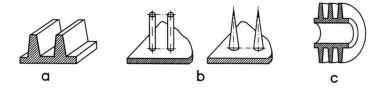

a   b   c

Fig. 1.14: Examples of extended surface, **a** straight fins, **b** pins, **c** circular fins.

The overall heat transfer resistance is not reduced by the same degree. The increased heat transfer area brings with it an additional conduction resistance. The heat which is absorbed by the fluid from an area near the fin tip, has to be transported to the vicinity of the tip by conduction. This requires a temperature difference between the fin base and its tip as a driving force. As a consequence of this the fins (or any other forms of extended surface) have an average temperature below that of the finless base material. Therefore the fins are not completely effective, because the heat transfer between fin and fluid takes place with a smaller temperature difference than that between the base material and the fluid.

To calculate the effectiveness of extended surfaces we consider Fig. 1.15. The

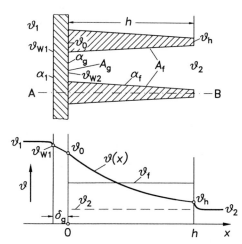

Fig. 1.15: Temperature in a finned wall along the line $\overline{AB}$. ϑ_f is the average temperature of the fin.

heat flow $\dot{Q}$ into Fluid 2 is made up of two parts:

$$\dot{Q} = \dot{Q}_g + \dot{Q}_f \ .$$

The heat flow $\dot{Q}_g$ removed from the surface A_g of the finless base material (ground material) is

$$\dot{Q}_g = \alpha_g A_g \left(\vartheta_{W2} - \vartheta_2\right) \ , \tag{1.80}$$

where α_g is the associated heat transfer coefficient. The temperature in the fin falls from ϑ_0 at the base of the fin $(x = 0)$ to ϑ_h at its tip $(x = h)$. With ϑ_f as the average fin temperature the heat flow $\dot{Q}_f$ transferred to the fluid from the fin surface area A_f will be

$$\dot{Q}_f = \alpha_f A_f \left(\vartheta_f - \vartheta_2\right) \ ,$$

where α_f is the (average) heat transfer coefficient between fin and fluid.

If the fin was at the same temperature all over as at its base ϑ_0, then the heat flow would be given as

$$\dot{Q}_{f0} = \alpha_f A_f \left(\vartheta_0 - \vartheta_2\right) \ .$$

The effectiveness of the fin is described by the *fin efficiency*

$$\eta_f := \frac{\dot{Q}_f}{\dot{Q}_{f0}} = \frac{\vartheta_f - \vartheta_2}{\vartheta_0 - \vartheta_2} \tag{1.81}$$

and it then follows that

$$\dot{Q}_f = \alpha_f \eta_f A_f \left(\vartheta_0 - \vartheta_2\right) \ . \tag{1.82}$$

The fin efficiency is always less than one. It depends on both the conduction processes in the fin and the convective heat transfer, which influence each other. Therefore, the geometry of the fin, the thermal conductivity λ_f, and the heat transfer coefficient α_f play a role in the calculation of the fin efficiency. This will be discussed in detail in 2.2.4.

The temperature of the fin base ϑ_0 is different from that of the surface where no fins are present, ϑ_{W2}. The heat flux through the fin base into the fin is significantly higher than the flux from the base material, i.e. the wall, into the fluid. The temperature drop which appears underneath the fin causes a periodic temperature profile to develop in the base material. This is shown schematically in Fig. 1.16. As a simplification this complicated temperature change is neglected, such that

$$\vartheta_0 = \vartheta_{W2} \ , \tag{1.83}$$

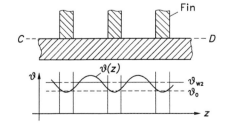

Fig. 1.16: Periodic temperature profile along the line $\overline{CD}$. ϑ_0 average temperature of the fin base, ϑ_{W2} average temperature of the surface of the base material between the fins.

thereby assuming an isothermal temperature distribution under the fins and on the surface between the fins. This simplification leads to an overestimation of the heat flow which is transferred. The heat flow calculated can be as much as 25% bigger than it actually is, as shown first by O. Krischer and W. Kast [1.4], and later by E.M. Sparrow and D.K. Hennecke [1.5] as well as E.M. Sparrow und L. Lee [1.6]. In many cases this error will be less than 5%, particularly if the fins are thick and placed very close together. We assume (1.83) to be valid and from (1.80) and (1.82), using this simplification, we obtain

$$\dot{Q} = \dot{Q}_g + \dot{Q}_f = \left(\alpha_g A_g + \alpha_f \eta_f A_f \right) \left(\vartheta_0 - \vartheta_2 \right) \ . \tag{1.84}$$

The fin is not effective over its total surface area A_f, only over the smaller surface area $\eta_f A_f$, reduced with the fin efficiency. Normally we have $A_f \gg A_g$, so in the first bracket of (1.84) the second term is the most important, despite $\eta_f < 1$. It is therefore possible to set $\alpha_g \approx \alpha_f$ without incurring large errors. This then gives

$$\dot{Q} = \alpha_f \left(A_g + \eta_f A_f \right) \left(\vartheta_0 - \vartheta_2 \right) \ . \tag{1.85}$$

The overall resistance of a finned wall, taking into account (1.68), (1.69) and (1.71), is given by

$$\frac{1}{kA} = \frac{1}{\alpha_1 A_1} + \frac{\delta}{\lambda_m A_m} + \frac{1}{\alpha_f \left(A_g + \eta_f A_f \right)} \ . \tag{1.86}$$

Here δ is the thickness, λ_m the average thermal conductivity, and A_m the average area of the wall without fins. Overall heat transfer for finned walls can be calculated using the same relationships as for an unfinned wall. The only change being that the fin area multiplied by the fin efficiency replaces the surface area of the fins in the equations.

Example 1.3: A pipe made from an aluminium alloy ($\lambda_m = 205 \, \text{W/K m}$) has an inner diameter of $d_1 = 22 \, \text{mm}$ and an outer diameter $d_2 = 25 \, \text{mm}$. Water at $\vartheta_1 = 60 \, ^\circ\text{C}$ flows inside the pipe, whilst air at $\vartheta_2 = 25 \, ^\circ\text{C}$ flows around the outside of the pipe perpendicular to its axis. Typical heat transfer coefficients are $\alpha_1 = 6150 \, \text{W/m}^2\text{K}$ and $\alpha_2 = 95 \, \text{W/m}^2\text{K}$. The heat flow per length of the pipe $\dot{Q}/L$ is to be calculated.
From (1.76)

$$
\begin{aligned}
\frac{1}{kA} &= \frac{1}{\pi L} \left(\frac{1}{\alpha_1 d_1} + \frac{1}{2\lambda_m} \ln \frac{d_2}{d_1} + \frac{1}{\alpha_2 d_2} \right) \\
&= \frac{1}{\pi L} \left(0.0074 + 0.0003 + 0.4211 \right) \frac{\text{K m}}{\text{W}} = \frac{0.1365}{L} \frac{\text{K m}}{\text{W}}
\end{aligned}
$$

and from (1.71)

$$\dot{Q}/L = (kA/L) \left(\vartheta_1 - \vartheta_2 \right) = 256 \, \text{W/m} \ .$$

The overall heat transfer resistance $(1/kA)$ is determined by the convection resistance on the outside of the pipe as this is the largest value. This resistance can be reduced by attaching fins.

Fins in the form of annular discs with outer diameter $d_f = 60 \, \text{mm}$, thickness $\delta_f = 1 \, \text{mm}$ and a separation of $t_f = 6 \, \text{mm}$ have been chosen for this case. The number of fins is $n = L/t_f$. The finless outer surface of the tube is

$$A_g = \pi d_2 \left(L - n\delta_f \right) = \pi d_2 L \left(1 - \delta_f / t_f \right)$$

and the surface area of the fins is

$$A_f = 2n\frac{\pi}{4}\left(d_f^2 - d_2^2\right) = \frac{\pi}{2}\frac{L}{t_f}\left(d_f^2 - d_2^2\right) \quad .$$

The narrow surface of width δ_f at the top of the fin has been neglected because it is only slightly hotter than the air and therefore its contribution to the heat transfer is insignificant. The surface area of the finned tube in comparison to that without fins $A_0 = \pi d_2 L$, is increased by a factor of $(A_g + A_f)/A_0 = 10.75$.

The fin surfaces are not competely effective for heat transfer. Taking $\alpha_f = \alpha_2$ and a fin efficiency of $\eta_f = 0.55$ gives according to (1.86)

$$
\begin{aligned}
\frac{1}{kA} &= \frac{1}{\pi L}\left(\frac{1}{\alpha_1 d_1} + \frac{1}{2\lambda_m}\ln\frac{d_2}{d_1} + \frac{\pi L}{\alpha_f\left(A_g + \eta_f A_f\right)}\right) \\
&= \frac{1}{\pi L}\left(0.0074 + 0.0003 + 0.0670\right)\frac{\mathrm{K\,m}}{\mathrm{W}} = \frac{0.0238}{L}\frac{\mathrm{K\,m}}{\mathrm{W}} \quad .
\end{aligned}
$$

The resistance to heat transfer on the outside of the tube is still the greatest resistance, but with the attachment of the fins it has been significantly reduced. The heat flow is $\dot{Q}/L = 1472\,\mathrm{W/m}$. The surface area enlargement by a factor of 10.75 has increased the heat flow by a factor of 5.75.

1.2.4 Heating and cooling of thin walled vessels

The relationships for steady heat flow can also be applied to the solution of a transient heat transfer problem, namely to the calculation of the temperature change with time during the heating and cooling of a thin walled vessel filled with a liquid. Two simplifications have to be made:

1. The temperature of the liquid inside the vessel is the same throughout, it only changes with time: $\vartheta_F = \vartheta_F(t)$.
2. The heat stored in the vessel wall, or more precisely the change in its internal energy, can be neglected.

The first assumption is true for most cases, as free or forced convection due to an agitator in the vessel, lead to almost the same temperature throughout the liquid. The second is only correct when the heat capacity of the contents is much larger than the heat capacity of the vessel wall. This happens in the heating and cooling of liquids in thin walled vessels, but may not be applied to vessels containing gases, which have either thick or well insulated walls.

When both these assumptions are valid, at every point in time the temperature of the fluid is spatially uniform, and the wall temperature will be predicted by the equations valid for steady state. In a flat vessel wall the temperature changes linearly, however the straight line moves according to time.

The process of cooling, as in Fig. 1.17, will now be considered. The heat flow $\dot{Q}(t)$ transferred from the liquid at temperature $\vartheta_F(t)$, through the vessel wall, to the surroundings which are at a constant temperature ϑ_S, is given by

$$\dot{Q}(t) = kA\left[\vartheta_F(t) - \vartheta_S\right] \tag{1.87}$$

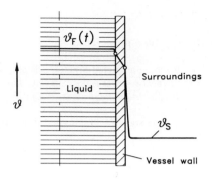

Fig. 1.17: Temperature profile for the cooling of a thin walled vessel.

The overall heat transfer coefficient k can be calculated using (1.72). According to the first law, the heat $\dot{Q}$ flowing out of the liquid causes a reduction of the internal energy U_F of the fluid in the vessel:

$$\dot{Q}(t) = -\frac{dU_F}{dt} = -M_F c_F \frac{d\vartheta_F}{dt} \ . \tag{1.88}$$

Here, M_F is the mass and c_F is the specific heat capacity of the liquid, which is assumed to be constant.

The ordinary differential equation for the liquid temperature follows on from (1.87) and (1.88):

$$\frac{d\vartheta_F}{dt} + \frac{kA}{M_F c_F}(\vartheta_F - \vartheta_S) = 0 \ .$$

The solution with the initial conditions

$$\vartheta_F = \vartheta_{F0} \qquad \text{at time} \quad t = 0$$

becomes, in dimensionless form

$$\vartheta_F^+ := \frac{\vartheta_F - \vartheta_S}{\vartheta_{F0} - \vartheta_S} = \exp\left(-\frac{kA}{M_F c_F}t\right) \ . \tag{1.89}$$

The liquid temperature falls exponentially from its initial value ϑ_{F0} to the temperature of the surroundings ϑ_S. Fig. 1.18 shows temperature plots for different values of the decay time

$$t_0 := M_F c_F / kA \ . \tag{1.90}$$

This decay time appears in Fig. 1.18 as the subtangent to the curve at any time, in particular at the time $t = 0$.

The *heating* of the vessel contents shall begin at time $t = 0$, when the whole container is at the same temperature as the surroundings:

$$\vartheta_F = \vartheta_S \qquad \text{for} \quad t = 0 \ . \tag{1.91}$$

A heat flow $\dot{Q}_H = \dot{Q}_H(t)$, that can be an arbitrary function of time, shall be added to the liquid for all $t \geq 0$. The heat flow $\dot{Q}(t)$ lost through the the thin vessel walls is found using (1.87). These two quantities enter the balance equation

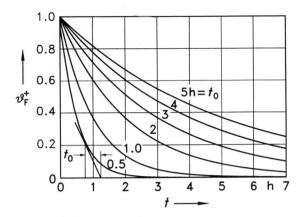

Fig. 1.18: Liquid temperature ϑ_F^+ variation over time according to (1.89) with t_0 from (1.90) during the cooling of a vessel.

$$\frac{dU_F}{dt} = -\dot{Q}(t) + \dot{Q}_H(t) \ ,$$

from which the differential equation follows as

$$\frac{d\vartheta_F}{dt} + \frac{kA}{M_F c_F}(\vartheta_F - \vartheta_S) = \frac{\dot{Q}_H(t)}{M_F c_F} \ .$$

Its general solution, with the initial condition from (1.91), is

$$\vartheta_F = \vartheta_S + \exp\left(-t/t_0\right) \int_0^t \frac{\dot{Q}_H(t)}{M_F c_F} \exp\left(t/t_0\right) dt \qquad (1.92)$$

with t_0 from (1.90).

If the heat load $\dot{Q}_H$ is assumed to be *constant*, it follows from (1.92) that

$$\vartheta_F = \vartheta_S + \frac{\dot{Q}_H}{kA}\left[1 - \exp\left(-t/t_0\right)\right] \ .$$

After a long period of time has elapsed ($t \to \infty$), the temperature of the liquid reaches the value

$$\vartheta_{F\infty} = \vartheta_S + \frac{\dot{Q}_H}{kA} \ .$$

Then the heat flow added just counterbalances the heat loss through the wall $\dot{Q}$ from (1.87): a steady state is reached.

1.3 Heat exchangers

When energy, as heat, has to be transferred from one stream of fluid to another both fluids are directed through an apparatus known as a heat exchanger. The two streams are separated by a barrier, normally the wall of a tube or pipe, through which heat is transferred from the fluid at the higher temperature to the colder one. Calculations involving heat exchangers use the equations derived in section 1.2 for overall heat transfer. In addition to these relationships, the energy balances of the first law of thermodynamics link the heat transferred with the enthalpy changes and therefore the temperature changes in both the fluids.

Heat exchangers exist in many different forms, and can normally be differentiated by the flow regimes of the two fluids. These different types will be discussed in the first part of this section. This will be followed by a section on the equations used in heat exchanger design. These equations can be formulated in a favourable manner using dimensionless groups. The calculation of countercurrent, cocurrent and cross current exchangers will then be explained. The final section contains information on combinations of these three basic flow regimes which are used in practice.

The calculation, design and application of heat exchangers is covered comprehensively in other books, in particular the publications from H. Hausen [1.7], H. Martin [1.8] as well as W. Roetzel [1.9] should be noted. The following sections serve only as an introduction to this extensive area of study, and particular emphasis has been placed on the thermal engineering calculation methods.

1.3.1 Types of heat exchanger and flow configurations

One of the simplest designs for a heat exchanger is the *double pipe heat exchanger* which is schematically illustrated in Fig. 1.19. It consists of two concentric tubes, where fluid 1 flows through the inner pipe and fluid 2 flows in the annular space between the two tubes. Two different flow regimes are possible, either countercurrent where the two fluids flow in opposite directions, Fig. 1.19a, or cocurrent as in Fig. 1.19b.

Fig. 1.19 also shows the cross-sectional mean values of the fluid temperatures ϑ_1 and ϑ_2 over the whole length of the heat exchanger. The entry temperatures are indicated by one dash, and the exit temperature by two dashes. At every cross-section $\vartheta_1 > \vartheta_2$, when fluid 1 is the hotter of the two. In countercurrent flow the two fluids leave the tube at opposite ends, and so the exit temperature of the hot fluid can be lower than the exit temperature of the colder fluid ($\vartheta_1'' < \vartheta_2''$), because only the conditions $\vartheta_1'' > \vartheta_2'$ and $\vartheta_1' > \vartheta_2''$ must be met. A marked cooling of fluid 1 or a considerable temperature rise in fluid 2 is not possible with cocurrent flow. In this case the exit temperatures of both fluids occur at the same end of the exchanger and so $\vartheta_1'' > \vartheta_2''$ is always the case, no matter how long the exchanger is. This is the first indication that countercurrent flow is superior to cocurrent flow: not all heat transfer tasks carried out in countercurrent flow can be realised in cocurrent flow. In addition to this fact, it will be shown in section 1.3.3, that for the transfer of the same heat flow, a countercurrent heat exchanger always has

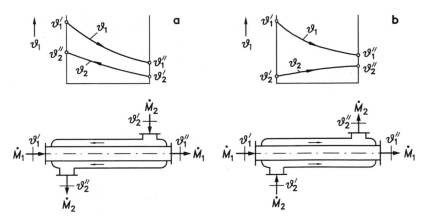

Fig. 1.19: Fluid temperatures ϑ_1 and ϑ_2 in a double-pipe heat exchanger. **a** countercurrent flow, **b** cocurrent flow.

a smaller area than a cocurrent exchanger, assuming of course, that the both flow regimes are suitable to fulfill the task. Therefore, cocurrent flow is seldom used in practise.

In practical applications the *shell-and-tube heat exchanger*, as shown in Fig. 1.20 is the most commonly used design. One of the fluids flows in the many parallel tubes which make up a tube bundle. The tube bundle is surrounded by a shell. The second fluid flows around the outside of the tubes within this shell. Countercurrent flow can be realised here except at the ends at of the heat exchanger where the shell side fluid enters or leaves the exchanger. The addition of baffles, as in Fig. 1.21, forces the shell side fluid to flow perpendicular to the tube bundle, which leads to higher heat transfer coefficients than those found in flow along the tubes. In the sections between the baffles the fluid is neither in

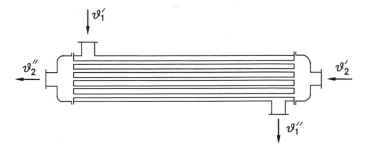

Fig. 1.20: Shell-and-tube heat exchanger (schematic)

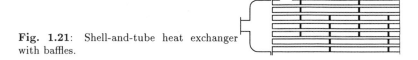

Fig. 1.21: Shell-and-tube heat exchanger with baffles.

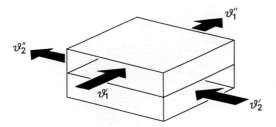

Fig. 1.22: Scheme of a plate exchanger with crossflow.

counter or cocurrent flow but in *crossflow*.

Pure crossflow is found in *flat plate heat exchangers*, as indicated by Fig. 1.22. The temperatures of both fluids also change perpendicular to the flow direction. This is schematically shown in Fig. 1.23. Each fluid element that flows in a crossflow heat exchanger experiences its own temperature change, from the entry temperature ϑ_i' which is the same for all particles to its individual exit temperature. Crossflow is often applied in a shell-and-tube heat exchanger when one of the fluids is gaseous. The gas flows around the rows of tubes crosswise to the tube axis. The other fluid, normally a liquid, flows inside the tubes. The addition of fins to the outer tube walls, cf. 1.2.3 and 2.2.3, increases the area available for heat transfer on the gas side, thereby compensating for the lower heat transfer coefficient.

Fig. 1.24 shows a particularly simple heat exchanger design, a *coiled tube* inside a vessel, for example a boiler. One fluid flows through the tube, the other one is in the vessel and can either flow through the vessel or stay there while it is being

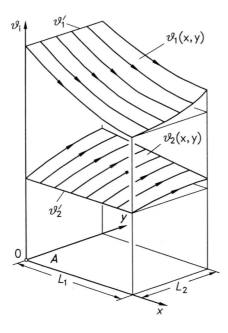

Fig. 1.23: Fluid temperatures $\vartheta_1 = \vartheta_1(x,y)$ and $\vartheta_2 = \vartheta_2(x,y)$ in crossflow.

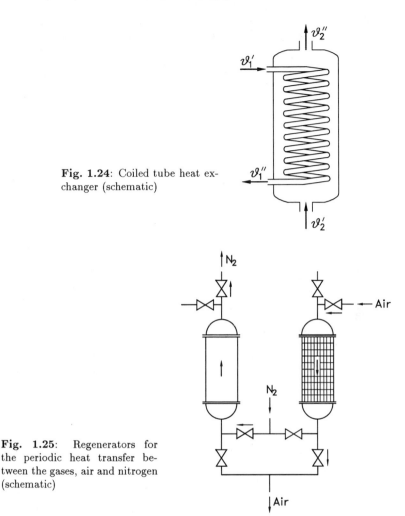

Fig. 1.24: Coiled tube exchanger (schematic)

Fig. 1.25: Regenerators for the periodic heat transfer between the gases, air and nitrogen (schematic)

heated up or cooled down. The vessel is usually equipped with a stirrer that mixes the fluid, improving the heat transfer to the coiled tube.

There are also numerous other special designs for heat exchangers which will not be discussed here. It is possible to combine the three basic flow regimes of countercurrent, cocurrent and cross flow in a number of different ways, which leads to complex calculation procedures.

The heat exchangers dealt with so far have had two fluids flowing steadily through the apparatus at the same time. They are always separated by a wall through which heat flows from the hotter to the colder fluid. These types of heat exchangers are also known as recuperators, which are different from *regenerators*. They contain a packing material, for example a lattice of bricks with channels for the gas or a packed bed of stone or metal strips, that will allow gases to pass through it. The gases flow alternately through the regenerator. The *hot* gas transfers heat to the packing material, where it is stored as internal energy. Then the *cold* gas flows through the regenerator, removes heat from the packing and leaves at a higher temperature. Continuous operation requires at least two regenerators, so that one gas can be heated whilst the other one is being cooled, Fig. 1.25. Each of the regenerators will be periodically heated and cooled by switching the gas flows around. This produces a periodic change in the exit temperatures of the gases.

Regenerators are used as air preheaters in blast furnaces and as heat exchangers in low temperature gas liquefaction plants. A special design, the Ljungström preheater, equipped with a rotating packing material serves as a preheater for air in firing equipment and gas turbine plants. The warm gas in this case is the exhaust gas from combustion which should be cooled as much as possible for energy recovery.

The regenerator theory was mainly developed by H. Hausen [1.10]. As it includes a number of complicated calculations of processes that are time dependent no further study of the theory will be made here. The summary by H. Hausen [1.7] and the VDI-Heat Atlas [1.11] are suggested for further study on this topic.

1.3.2　General design equations. Dimensionless groups

Fig. 1.26 is a scheme for a heat exchanger. The temperatures of the two fluids are dentoted by ϑ_1 and ϑ_2, as in section 1.3.1, and it will be assumed that $\vartheta_1 > \vartheta_2$. Heat will therefore be transferred from fluid 1 to fluid 2. Entry temperatures are indicated by one dash, exit temperatures by two dashes.

The first law of thermodynamics is applied to for both fluids. The heat transferred causes an enthalpy increase in the cold fluid 2 and a decrease in the warm fluid 1. This gives

$$\dot{Q} = \dot{M}_1(h_1' - h_1'') = \dot{M}_2(h_2'' - h_2') \ , \qquad (1.93)$$

where $\dot{M}_i$ is the mass flow rate of fluid i. The specific enthalpies are calculated at the entry and exit temperatures ϑ_i' and ϑ_i'' respectively. These temperatures are averaged over the relevant tube cross section, and can be determined using the explanation in section 1.1.3 for calculating adiabatic mixing temperatures. Equation (1.93) is only valid for heat exchangers which are *adiabatic* with respect to their environment, and this will always be assumed to be the case.

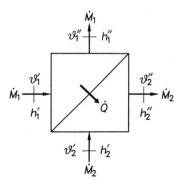

Fig. 1.26: Heat exchanger scheme, with the mass flow rate $\dot{M}_i$, entry temperatures ϑ_i', exit temperatures ϑ_i'', entry enthalpy h_i' and exit enthalpy h_i'' of both fluids ($i = 1, 2$).

The two fluids flow through the heat exchanger without undergoing a phase change, i.e. they do not boil or condense. The small change in specific enthalpy with pressure is neglected. Therefore only the temperature dependence is important, and with

$$\bar{c}_{pi} := \frac{h_i' - h_i''}{\vartheta_i' - \vartheta_i''} \quad , \qquad i = 1, 2 \tag{1.94}$$

the mean specific heat capacity between ϑ_i' and ϑ_i'' it follows from (1.93) that

$$\dot{Q} = \dot{M}_1 \bar{c}_{p1}(\vartheta_1' - \vartheta_1'') = \dot{M}_2 \bar{c}_{p2}(\vartheta_2'' - \vartheta_2') \ .$$

As an abbreviation the *heat capacity flow rate* is introduced by

$$\dot{W}_i := \dot{M}_i \bar{c}_{pi} \quad , \qquad i = 1, 2 \tag{1.95}$$

which then gives

$$\dot{Q} = \dot{W}_1(\vartheta_1' - \vartheta_1'') = \dot{W}_2(\vartheta_2'' - \vartheta_2') \ . \tag{1.96}$$

The temperature changes in both fluids are linked to each other due to the first law of thermodynamics. They are related inversely to the ratio of the heat capacity flow rates.

The heat flow $\dot{Q}$ is transferred from fluid 1 to fluid 2 because of the temperature difference $\vartheta_1 - \vartheta_2$ inside the heat exchanger. This means that the heat flow $\dot{Q}$ has to overcome the overall resistance to heat transfer $1/kA$ according to section 1.2.1. The quantity kA will from now on be called the *transfer capability* of the heat exchanger, and is a caracteristic quantity of the apparatus. It is calculated using (1.72) from the transfer resistances in the fluids and the resistance to conduction in the wall between them. The value for kA is usually taken to be an apparatus constant, where the overall heat transfer coefficient k is assumed to have the same value throughout the heat exchanger. However this may not always happen, the fluid heat transfer coefficient can change due to the temperature dependence of some of the fluid properties or by a variation in the flow conditions. In cases such as these, k and kA must be calculated for various points in the heat exchanger and a suitable mean value can be found, cf. W. Roetzel and B. Spang [1.12], to represent the characteristic transfer capability kA of the heat exchanger.

Before beginning calculations for heat exchanger design, it is useful to get an overview of the quantities which have an effect on them. Then the number of these quantities will be reduced by the introduction of dimensionless groups. Finally the relevant relationships for the design will be determined. Fig. 1.27 contains the seven quantities that influence the design of a heat exchanger. The effectiveness of the heat exchanger is characterized by its transfer capability kA, the two fluid flows by their heat capacity flow rates $\dot{W}_i$, entry temperatures ϑ_i' and exit temperatures ϑ_i''. As the temperature level is not important only the three temperature differences $(\vartheta_1' - \vartheta_1'')$, $(\vartheta_2'' - \vartheta_2')$ and $(\vartheta_1' - \vartheta_2')$, as shown in Fig. 1.28, are of influence. This reduces the number of quantities that have any effect by one so that six quantities remain:

$$kA, \ (\vartheta_1' - \vartheta_1''), \ \dot{W}_1, \ (\vartheta_2'' - \vartheta_2'), \ \dot{W}_2 \quad \text{and} \quad (\vartheta_1' - \vartheta_2') \ .$$

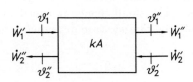

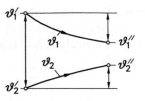

Fig. 1.27: Heat exchanger with the seven quantites which affect its design

Fig. 1.28: The three decisive temperature differences (arrows) in a heat exchanger

These belong to only two types of quantity either temperature (unit K) or heat capacity flow rate (units W/K). According to section 1.1.4, that leaves four ($= 6 - 2$) characterisic quantites to be defined. These are the dimensionless temperature changes in both fluids

$$\varepsilon_1 := \frac{\vartheta_1' - \vartheta_1''}{\vartheta_1' - \vartheta_2'} \quad \text{and} \quad \varepsilon_2 := \frac{\vartheta_2'' - \vartheta_2'}{\vartheta_1' - \vartheta_2'} \;, \tag{1.97}$$

see Fig. 1.29, and the ratios

$$N_1 := \frac{kA}{\dot{W}_1} \quad \text{and} \quad N_2 := \frac{kA}{\dot{W}_2} \;. \tag{1.98}$$

These are also known as the Number of Transfer Units or *NTU* for short. We suggest N_i be characterised as the dimensionless transfer capability of the heat exchanger. Instead of N_2 the ratio of the two heat capacity flow rates

$$C_1 := \frac{\dot{W}_1}{\dot{W}_2} = \frac{N_2}{N_1} \tag{1.99}$$

or its inverse

$$C_2 := \frac{\dot{W}_2}{\dot{W}_1} = \frac{1}{C_1} \tag{1.100}$$

is often used.

The four groups in (1.97) and (1.98), are not independent of each other, because applying the first law of thermodynamics gives

$$\frac{\varepsilon_1}{N_1} = \frac{\varepsilon_2}{N_2} \quad \text{or} \quad \varepsilon_2 = C_1 \varepsilon_1 \;. \tag{1.101}$$

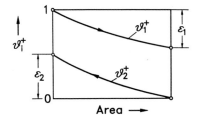

Fig. 1.29: Plot of the dimensionsless fluid temperatures $\vartheta_i^+ = (\vartheta_i - \vartheta_2') / (\vartheta_1' - \vartheta_2')$ over the area and illustration of ε_1 and ε_2 according to (1.97)

The relationship which exists between the three remaining characteristic quantities

$$F(\varepsilon_1, N_1, N_2) = 0 \quad \text{oder} \quad F(\varepsilon_1, N_1, C_1) = 0 \qquad (1.102)$$

is the *operating characteristic* of the heat exchanger. It depends on the flow configuration and is found from the temperature pattern of both fluids, that will be discussed in detail in the following sections.

Heat exchanger design mainly consists of two tasks:

1. Calculating the heat flow transferred in a given heat exchanger.
2. Design of a heat exchanger for a prescribed performance.

In the first case $(\vartheta_1' - \vartheta_2')$, $\dot{W}_1$, $\dot{W}_2$ and kA will all be given. The temperature changes in both fluids have to be found so that $\dot{Q}$, the heat flow transferred, can be determined from (1.96). As the charateristic numbers, N_1 and N_2 or N_1 and C_1 are given this problem can be solved immediately, if the operating characteristic in (1.102) can be explicitly resolved for ε_1:

$$\varepsilon_1 = \varepsilon_1(N_1, C_1) \ .$$

The dimensionless temperature change ε_2 of the other fluid follows from (1.101).

In the calculations for the design of a heat exchanger kA has to be found. Either the temperature changes in both fluids or the two values for the heat capacity flow and the temperature change in one of the fluids must be known, in order to determine kA. An operating characteristic which can be explicitly solved for N_1 or N_2 is desired:

$$N_1 = N_1(\varepsilon_1, C_1) \ .$$

This gives for the transfer capability

$$kA = N_1 \dot{W}_1 = N_2 \dot{W}_2 \ .$$

In Fig. 1.30 an operating characteristic for a heat exchanger with given flow configuration is shown. The solutions to both tasks, heat transfer calculation and design calculation are indicated. In many cases the explicit solution of the operating characteristic for ε_1 and N_1 is not possible, even if an analytical expression is available. If this arises a diagram similar to Fig. 1.30 should be used. Further details are given in section 1.3.5.

When the heat capacity flow rate $\dot{W}_i$ was introduced in (1.95) boiling and condensing fluids were not considered. At constant pressure a pure substance which is boiling or condensing does not undergo a change in temperature, but $c_{pi} \to \infty$. This leads to $\varepsilon_i = 0$, whilst $\dot{W}_i \to \infty$ resulting in $N_i = 0$ and $C_i \to \infty$. This simplifies the calculations for the heat exchanger, as the operating characteristic is now a relationship between only two rather than three quantities, namely ε and N of the other fluid, which is neither boiling nor condensing.

In heat exchanger calculations another quantity alongside those already introduced is often used, namely the *mean temperature difference* $\Delta\vartheta_\mathrm{m}$. This is found by integrating the local

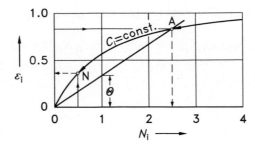

Fig. 1.30: Schematic representation of the operating characteristic for a heat exchanger with C_i = const. N is the assumed operating point for the heat transfer calculations: $\varepsilon_i = \varepsilon_i(N_i, C_i)$, A is the assumed operating point for the design: $N_i = N_i(\varepsilon_i, C_i)$. The determination of the mean temperature difference Θ for point A is also shown.

temperature difference $(\vartheta_1 - \vartheta_2)$ between the two fluids over the whole transfer area.

$$\Delta\vartheta_m = \frac{1}{A}\int\limits_{(A)}(\vartheta_1 - \vartheta_2)\,\mathrm{d}A \ . \tag{1.103}$$

In analogy to (1.71) the heat flow transferred is

$$\dot{Q} = kA\Delta\vartheta_m \ . \tag{1.104}$$

This equation can only strictly be used if the heat transfer coefficient k is the same at each point on A. If this is not true then (1.104) can be considered to be a definition for a mean value of k.

The introduction of $\Delta\vartheta_m$ in conjunction with (1.104), gives a relationship between the heat flow transferred and the transfer capability kA, and therefore with the area A of the heat exchanger. This produces the following equations

$$\dot{Q} = kA\Delta\vartheta_m = \dot{W}_1(\vartheta_1' - \vartheta_1'') = \dot{W}_2(\vartheta_2'' - \vartheta_2') \ .$$

With the dimensionless mean temperature difference

$$\Theta = \frac{\Delta\vartheta_m}{\vartheta_1' - \vartheta_2'} \tag{1.105}$$

the following relationship between the dimensionless groups is found:

$$\Theta = \frac{\varepsilon_1}{N_1} = \frac{\varepsilon_2}{N_2} \ . \tag{1.106}$$

The mean temperature $\Delta\vartheta_m$ and its associated dimensionless quantity Θ can be calculated using the dimensionless numbers that have already been discussed. The introduction of the mean temperature difference does not provide any information that cannot be found from the operating characteristic. This is also illustrated in Fig. 1.30, where Θ is the gradient of the straight line that joins the operating point and the origin of the graph.

1.3.3 Countercurrent and cocurrent heat exchangers.

The operating characteristic $F(\varepsilon_i, N_i, C_i) = 0$, for a countercurrent heat exchanger is found by analysing the temperature distribution in both fluids. The results can easily be transferred for use with the practically less important case of a cocurrent exchanger.

We will consider the temperature changes, shown in Fig. 1.31, in a countercurrent heat exchanger. The temperatures ϑ_1 and ϑ_2 depend on the z coordinate in the direction of flow of fluid 1. By applying the first law to a section of length $\mathrm{d}z$ the rate of heat transfer, $\mathrm{d}\dot{Q}$, from fluid 1 to fluid 2 through the surface element $\mathrm{d}A$ is found to be

$$\mathrm{d}\dot{Q} = -\dot{M}_1 c_{p1}\, \mathrm{d}\vartheta_1 = -\dot{W}_1\, \mathrm{d}\vartheta_1 \tag{1.107}$$

and

$$\mathrm{d}\dot{Q} = -\dot{M}_2 c_{p2}\, \mathrm{d}\vartheta_2 = -\dot{W}_2\, \mathrm{d}\vartheta_2 \ . \tag{1.108}$$

Now $\mathrm{d}\dot{Q}$ is eliminated by using the equation for overall heat transfer

$$\mathrm{d}\dot{Q} = k(\vartheta_1 - \vartheta_2)\, \mathrm{d}A = kA(\vartheta_1 - \vartheta_2)\frac{\mathrm{d}z}{L} \tag{1.109}$$

from (1.107) and (1.108) giving

$$\mathrm{d}\vartheta_1 = -(\vartheta_1 - \vartheta_2)\frac{kA}{\dot{W}_1}\frac{\mathrm{d}z}{L} = -(\vartheta_1 - \vartheta_2)N_1\frac{\mathrm{d}z}{L} \tag{1.110}$$

and

$$\mathrm{d}\vartheta_2 = -(\vartheta_1 - \vartheta_2)\frac{kA}{\dot{W}_2}\frac{\mathrm{d}z}{L} = -(\vartheta_1 - \vartheta_2)N_2\frac{\mathrm{d}z}{L} \tag{1.111}$$

for the temperature changes in both fluids.

The temperatures $\vartheta_1 = \vartheta_1(z)$ and $\vartheta_2 = \vartheta_2(z)$ will not be calculated from the two differential equations, instead the variation in the difference between the temperature of the two fluids $\vartheta_1 - \vartheta_2$ will be determined. By subtracting (1.111) from (1.110) and dividing by $(\vartheta_1 - \vartheta_2)$ it follows that

$$\frac{\mathrm{d}(\vartheta_1 - \vartheta_2)}{\vartheta_1 - \vartheta_2} = (N_2 - N_1)\frac{\mathrm{d}z}{L} \ . \tag{1.112}$$

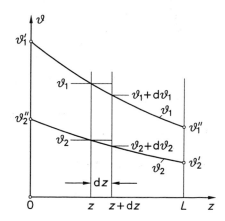

Fig. 1.31: Temperature pattern in a countercurrent heat exchanger.

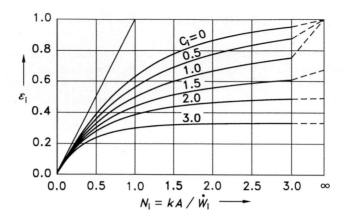

Fig. 1.32: Operating characteristic $\varepsilon_i = \varepsilon_i\,(N_i, C_i)$ for countercurrent flow from Tab. 1.4.

Integrating this differential equation between $z = 0$ and $z = L$ leads to

$$\ln \frac{(\vartheta_1 - \vartheta_2)_L}{(\vartheta_1 - \vartheta_2)_0} = \ln \frac{\vartheta_1'' - \vartheta_2'}{\vartheta_1' - \vartheta_2''} = N_2 - N_1 \ . \tag{1.113}$$

Now, we have

$$\frac{\vartheta_1'' - \vartheta_2'}{\vartheta_1' - \vartheta_2''} = \frac{\vartheta_1' - \vartheta_2' - (\vartheta_1' - \vartheta_1'')}{\vartheta_1' - \vartheta_2' - (\vartheta_2'' - \vartheta_2')} = \frac{1 - \varepsilon_1}{1 - \varepsilon_2} \ ,$$

which gives

$$\ln \frac{1 - \varepsilon_1}{1 - \varepsilon_2} = N_2 - N_1 \tag{1.114}$$

as the implicit form of the operating characteristic of a countercurrent heat exchanger. It is invariant with respect to an exchange of the indices 1 and 2. Using the ratios of C_1 and $C_2 = 1/C_1$ from (1.99) and (1.100), explicit equations are obtained,

$$\varepsilon_i = f(N_i, C_i) \quad \text{und} \quad N_i = f(\varepsilon_i, C_i) \quad , \qquad i = 1, 2$$

which have the same form for both fluids. These explicit fomulae for the operating characteristics are shown in Table 1.4. If the heat capacity flow rates are equal, $\dot{W}_1 = \dot{W}_2$, and because $C_1 = C_2 = 1$, it follows that

$$\varepsilon_1 = \varepsilon_2 = \varepsilon \quad \text{und} \quad N_1 = N_2 = N \ ,$$

and with a series development of the equations valid for $C_i \neq 1$ towards the limit of $C_i \to 1$, the simple relationships given in Table 1.4 are obtained.

Fig. 1.32 shows the operating characteristic $\varepsilon_i = f(N_i, C_i)$ as a function of N_i with C_i as a parameter. As expected the normalised temperature change ε_i grows monotonically with increasing N_i, and therefore increasing transfer capability kA. For $N_i \to \infty$ the limiting value is

$$\lim_{N_i \to \infty} \varepsilon_i = \begin{cases} 1 & \text{for} \quad C_i \leq 1 \\ 1/C_i & \text{for} \quad C_i > 1 \end{cases} \ .$$

Table 1.4: Equations for the calculation of the normalised temperature variation ε_i , the dimensionless transfer capability N_i and the mean temperature difference Θ in counter and cocurrent heat exchangers

Flow regime	$\varepsilon_i = \varepsilon_i\,(N_i, C_i)$	$N_i = N_i\,(\varepsilon_i, C_i)$	$\Theta = \Theta\,(\varepsilon_1, \varepsilon_2)$
counter current $C_i \neq 1$ $i = 1,2$	$\varepsilon_i = \dfrac{1 - \exp\left[(C_i - 1)\,N_i\right]}{1 - C_i \exp\left[(C_i - 1)\,N_i\right]}$	$N_i = \dfrac{1}{1 - C_i}\ln\dfrac{1 - C_i \varepsilon_i}{1 - \varepsilon_i}$	$\Theta = \dfrac{\varepsilon_1 - \varepsilon_2}{\ln\dfrac{1 - \varepsilon_2}{1 - \varepsilon_1}}$
$C = 1$	$\varepsilon = \dfrac{N}{1 + N}$	$N = \dfrac{\varepsilon}{1 - \varepsilon}$	$\Theta = 1 - \varepsilon$
co-current $i = 1,2$	$\varepsilon_i = \dfrac{1 - \exp\left[-\left(1 + C_i\right)N_i\right]}{1 + C_i}$	$N_i = -\dfrac{\ln\left[1 - \varepsilon_i\left(1 + C_i\right)\right]}{1 + C_i}$	$\Theta = \dfrac{-\left(\varepsilon_1 + \varepsilon_2\right)}{\ln\left[1 - \left(\varepsilon_1 + \varepsilon_2\right)\right]}$

Meaning of the characteristic numbers: $\quad \varepsilon_1 = \dfrac{\vartheta_1' - \vartheta_1''}{\vartheta_1' - \vartheta_2'} \quad , \quad \varepsilon_2 = \dfrac{\vartheta_2'' - \vartheta_2'}{\vartheta_1' - \vartheta_2'}$

$N_i = kA/\dot{W}_i \quad , \quad \Theta = \dfrac{\Delta\vartheta_m}{\vartheta_1' - \vartheta_2'} = \dfrac{\varepsilon_i}{N_i} \quad , \quad C_1 = \dfrac{\dot{W}_1}{\dot{W}_2} = \dfrac{\varepsilon_2}{\varepsilon_1} = \dfrac{N_2}{N_1} \quad , \quad C_2 = \dfrac{1}{C_1}$

If $C_i \leq 1$, then ε_i takes on the character of an efficiency. The normalised temperature change of the fluid with the smaller heat capacity flow is known as the *efficiency* or *effectiveness of the heat exchanger*. With an enlargement of the heat transfer area A the temperature difference between the two fluids can be made as small as desired, but only at one end of the countercurrent exchanger. Only for $\dot{W}_1 = \dot{W}_2$, which means $C_1 = C_2 = 1$, can an infinitely small temperature difference at at both ends, and therefore throughout the heat exchanger, be achieved by an enlargement of the surface area. The ideal case of reversible heat transfer between two fluids, often considered in thermodynamics, is thus only attainable when $\dot{W}_1 = \dot{W}_2$ in a heat exchanger with very high transfer capability.

As already mentioned in section 1.3.2, the function $\varepsilon_i = f(N_i, C_i)$ is used to calculate the outlet temperature and the transfer capability of a given heat exchanger. For the sizing of a heat exchanger for a required temperature change in the fluid, the other form of the operating characteristic, $N_i = N_i(\varepsilon_i, C_i)$, is used. This is also given in Table 1.4.

In a *cocurrent heat exchanger* the direction of flow is opposite to that in Fig. 1.31, cf. also Fig. 1.20b. In place of (1.108) the energy balance is

$$d\dot{Q} = \dot{M}_2 c_{p2}\,d\vartheta_2 = \dot{W}_2\,d\vartheta_2 \; ,$$

which gives the relationship

$$\frac{\mathrm{d}(\vartheta_1 - \vartheta_2)}{\vartheta_1 - \vartheta_2} = -(N_1 + N_2)\frac{\mathrm{d}z}{L} \qquad (1.115)$$

instead of (1.112). According to (1.114) the temperature difference between the two fluids in the direction of flow is always decreasing. Integration of (1.115) between $z = 0$ and $z = L$ yields

$$\ln \frac{\vartheta_1'' - \vartheta_2''}{\vartheta_1' - \vartheta_2'} = -(N_1 + N_2) \ ,$$

from which follows

$$\ln\left[1 - (\varepsilon_1 + \varepsilon_2)\right] = -(N_1 + N_2) = -\frac{\varepsilon_1 + \varepsilon_2}{\Theta} \qquad (1.116)$$

as the implicit form of the operating characteristic. This can be solved for ε_i and N_i giving the functions noted in Table 1.4. For $N_i \to \infty$ the normalised temperature variation reaches the limiting value of

$$\lim_{N_i \to \infty} \varepsilon_i = \frac{1}{1 + C_i} \ , \qquad i = 1, 2 \ .$$

With cocurrent flow the limiting value of $\varepsilon_i = 1$ is never reached except when $C_i = 0$, as will soon be explained.

The calculations for performance and sizing of a heat exchanger can also be carried out using a mean temperature difference Θ from (1.106) in section 1.3.2. In countercurrent flow, the difference $N_2 - N_1$ in (1.114) is replaced by Θ, ε_1 and ε_2 giving the expression $\Theta = \Theta(\varepsilon_1, \varepsilon_2)$ which appears in Table 1.4. Introducing

$$N_2 - N_1 = \frac{\varepsilon_2 - \varepsilon_1}{\Theta} = \frac{\vartheta_2'' - \vartheta_2' - (\vartheta_1' - \vartheta_1'')}{\Delta\vartheta_\mathrm{m}} = \frac{\vartheta_1'' - \vartheta_2' - (\vartheta_1' - \vartheta_2'')}{\Delta\vartheta_\mathrm{m}}$$

into (1.113) , gives

$$\Delta\vartheta_\mathrm{m} = \frac{\vartheta_1'' - \vartheta_2' - (\vartheta_1' - \vartheta_2'')}{\ln \dfrac{\vartheta_1'' - \vartheta_2'}{\vartheta_1' - \vartheta_2''}} \qquad (1.117)$$

for the mean temperature difference in a countercurrent heat exchanger. It is the *logarithmic mean* of the temperature difference between the two fluids at both ends of the apparatus.

The expression, from (1.116), for the normalised mean temperature difference Θ, in cocurrent flow is given in Table 1.4. Putting in (1.117) the defining equations for ε_1 and ε_2 yields

$$\Delta\vartheta_\mathrm{m} = \frac{\vartheta_1' - \vartheta_2' - (\vartheta_1'' - \vartheta_2'')}{\ln \dfrac{\vartheta_1' - \vartheta_2'}{\vartheta_1'' - \vartheta_2''}} \ . \qquad (1.118)$$

So $\Delta\vartheta_\mathrm{m}$ is also the logarithmic mean temperature difference at both ends of the heat exchanger in cocurrent flow.

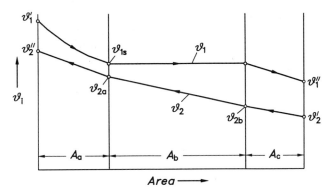

Fig. 1.33: Temperature in a condensor with cooling of superheated steam, condensation and subcooling of the condensate (fluid 1) by cooling water (fluid 2).

We will now compare the two flow configurations. For $C_i = 0$ the normalised temperature variation in Table 1.4 is

$$\varepsilon_i = 1 - \exp(-N_i)$$

and the dimensionsless transfer capability

$$N_i = -\ln(1 - \varepsilon_i)$$

both of which are independent of whether countercurrent or cocurrent flow is used. Therefore when one of the substances boils or condenses in the exchanger it is immaterial which flow configuration is chosen. However, if in a condensor, superheated steam is first cooled from ϑ_1' to the condensation temperature of ϑ_{1s}, then completely condensed, after which the condensate is cooled from ϑ_{1s} to ϑ_1'', more complex circumstances develop. In these cases it is not permissible to treat the equipment as simply one heat exchanger, using the equations that have already been defined, where only the inlet and outlet temperatures ϑ_i' and ϑ_i'' ($i = 1, 2$) are important, cf. Fig. 1.33. The values for the heat capacity flow rate $\dot{W}_1$ change significantly: During the cooling of the steam and the condensate $\dot{W}_1$ has a finite value, whereas in the process of condensation $\dot{W}_1$ is infinite. The exchanger has to be imaginarily split, and then be treated as three units in series. Energy balances provide the two unknown temperatures, ϑ_{2a} between the cooling and condensation section, and ϑ_{2b}, between the condensation and sub-cooling part. These in turn yield the dimensionless temperature differences ε_{ia}, ε_{ib} und ε_{ic} for the three sections cooler a, condensor b and sub-cooler c ($i = 1, 2$). The dimensionless transfer capabilities N_{ia}, N_{ib} and N_{ic} of the three equipment sections can then be calculated according to the relationships in Table 1.4. From N_{ij} the values for $(kA)_j$ can be found. Then using the relevant overall heat transfer coefficients k_j, we obtain the areas of the three sections A_j ($j = $ a, b, c), which together make up the total transfer area of the exchanger.

For $C_i > 0$ the countercurrent configuration is always superior to the cocurrent. A disadvantage of the cocurrent flow exists in that not all heat transfer tasks can be solved in such a system. A given temperature change ε_i is only realisable if the argument of the logarithmic term in

$$N_i^{co} = -\frac{1}{1 + C_i} \ln\left[1 - \varepsilon_i(1 + C_i)\right]$$

is positive. This is only the case for

$$\varepsilon_i < \frac{1}{1 + C_i} \ . \tag{1.119}$$

Larger normalised temperature changes cannot be achieved in cocurrent heat exchangers even in those with very large values for the transfer capability kA. In countercurrent exchangers this limitation does not exist. All values for ε_i are basically attainable and therefore all required heat loads can be transferred as long as the area available for heat transfer is made large enough.

A further disadvantage of cocurrent flow is that a higher transfer capability kA is required to fulfill the same task (same ε_i and C_i) when compared with a countercurrent system. This is shown in Fig. 1.34 in which the ratio

$$(kA)_{co}/(kA)_{cc} = N_i^{co}/N_i^{cc}$$

based on the equations in Table 1.4 is represented. This ratio grows sharply when ε_i approaches the limiting value according to (1.119). Even when a cocurrent exchanger is capable of fulfilling the requirements of the task, the countercurrent exchanger will be chosen as its dimensions are smaller. Only in a combination of small enough values of C_i and ε_i the necessary increase in the area of a cocurrent exchanger is kept within narrow limits.

Example 1.4: Ammonia, at a pressure of 1.40 MPa, is to be cooled in a countercurrent heat exchanger from $\vartheta_1' = 150.0\ °C$ to the saturation temperature $\vartheta_{1s}' = 36.3\ °C$, and then completely condensed. Its mass flow rate is $\dot{M}_1 = 0.200\ kg/s$. Specific enthalpies of $h(\vartheta_1') = 1797.1\ kJ/kg$, $h^g(\vartheta_{1s}) = 1488.8\ kJ/kg$, and $h^{fl}(\vartheta_{1s}) = 372.2\ kJ/kg$ are taken from the property tables for ammonia, [1.13]. Cooling water with a temperature of $\vartheta_2' = 12.0\ °C$ is available, and this can be heated to $\vartheta_2'' = 28.5\ °C$. Its mean specific heat capacity is $\bar{c}_{p2} = 4.184\ kJ/kgK$. The required transfer capabilities for the cooling $(kA)_{cooling}$ and $(kA)_{cond}$ for the condensation of the ammonia have to be determined.

At first the heat flow transferred $\dot{Q}$, and the required mass flow rate $\dot{M}_2$ of water have to be found. The heat flow removed from the ammonia is

$$\dot{Q} = \dot{M}_1 \left[h(\vartheta_1') - h^{fl}(\vartheta_{1s})\right] = 0.200\ \frac{kg}{s}\ (1797.1 - 372.2)\ \frac{kJ}{kg} = 285.0\ kW \ .$$

From that the mass flow rate of water is found to be

$$\dot{M}_2 = \frac{\dot{Q}}{\bar{c}_{p2}\left(\vartheta_2'' - \vartheta_1'\right)} = \frac{285.0\ kW}{4.184\ (kJ/kgK)\ (28.5 - 12.0)\ K} = 4.128\ \frac{kg}{s} \ .$$

The temperature ϑ_{2a} of the cooling water in the cross section between the cooling and condensation sections, cf. Fig. 1.35, is required to calculate the transfer capability. From the energy balance for the condensor section

$$\dot{M}_2 \bar{c}_{p2}\left(\vartheta_a - \vartheta_2'\right) = \dot{M}_1 \left[h^g(\vartheta_{1s}) - h^{fl}(\vartheta_{1s})\right] \ ,$$

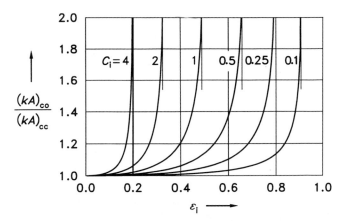

Fig. 1.34: Ratio $(kA)_{co} / (kA)_{cc} = N_i^{co}/N_i^{cc}$ of the transfer capabilities in cocurrent (index co) and countercurrent (index cc) flows as a function of ε_i and C_i.

it follows that

$$\vartheta_{2a} = \vartheta_2' + \frac{\dot{M}_1}{\dot{M}_2 \bar{c}_{p2}} \left[h^g(\vartheta_{1s}) - h^{fl}(\vartheta_{1s}) \right] = 24.9 \,^\circ C \ .$$

The transfer capability for the ammonia cooling section, using Table 1.4, is

$$\frac{(kA)_{cooling}}{\dot{W}_1} = N_1 = \frac{1}{1 - C_1} \ln \frac{1 - C_1 \varepsilon_1}{1 - \varepsilon_1} \ . \qquad (1.120)$$

The ratio of the heat capacity flow rates is found with

$$\dot{W}_1 = \dot{M}_1 \bar{c}_{p2} = \dot{M}_1 \frac{h(\vartheta_1') - h^g(\vartheta_s)}{\vartheta_1' - \vartheta_s} = 0.200 \, \frac{kg}{s} \frac{1797.1 - 1488.8}{150.0 - 36.3} \frac{kJ}{kgK} = 0.5423 \, \frac{kW}{K}$$

and with $\dot{W}_2 = \dot{M}_2 \bar{c}_{p2} = 17.272 \, kW/K$ giving $C_1 = 0.0314$. The dimensionless temperature variation of ammonia is

$$\varepsilon_1 = \frac{\vartheta_1' - \vartheta_s}{\vartheta_1' - \vartheta_{2a}} = \frac{150.0 - 36.3}{150.0 - 24.9} = 0.9089.$$

Then (1.120) yields $N_1 = 2.443$ and finally

$$(kA)_{cooling} = N_1 \dot{W}_1 = 1.325 \, kW/K \ .$$

For the *condensation section* of the heat exchanger $\varepsilon_1 = 0$, and because $\dot{W}_1 \to \infty$ this means $C_2 = \dot{W}_2/\dot{W}_1 = 0$. From Table 1.4 it follows that

$$(kA)_{cond}/\dot{W}_2 = N_2 = -\ln(1 - \varepsilon_2) \ .$$

With the normalised temperature change of the cooling water

$$\varepsilon_2 = \frac{\vartheta_{2a} - \vartheta_2'}{\vartheta_{1s} - \vartheta_2'} = \frac{24.9 - 12.0}{36.3 - 12.0} = 0.5309 \ ,$$

yielding $N_2 = 0.7569$, which then gives

$$(kA)_{cond} = N_2 \dot{W}_2 = 13.07 \, kW/K \ .$$

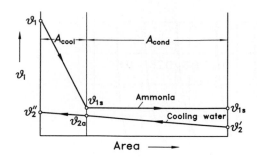

Fig. 1.35: Temperatures of ammonia and cooling water in a countercurrent heat exchanger (schematic).

In order to find the required area $A = A_{\text{cooling}} + A_{\text{cond}}$ for the countercurrent exchanger, from the values for $(kA)_{\text{cooling}}$ and $(kA)_{\text{cond}}$, the overall heat transfer coefficients for each part must be calculated. They will be different as the resistance to heat transfer in the cooling section is greatest on the gaseous ammonia side, whereas in the condensation section the greatest resistance to heat transfer is experienced on the cooling water side. The calculations for the overall heat transfer coefficients will not be done here as the design of the heat exchanger and the flow conditions have to be known for this purpose.

1.3.4   Crossflow heat exchangers

Before discussing pure crossflow as shown in Fig. 1.23, the operating characteristic for the simple case of cross flow where only the fluid on one side is laterally mixed will be calculated. In this flow configuration the temperature of one of the two fluids is only dependent on *one* position coordinate, e.g. x, while the temperature of the other fluid changes with both x and y. In Fig. 1.36 the laterally mixed fluid is indicated by the index 1. Its temperature ϑ_1 changes only in the direction of flow, $\vartheta_1 = \vartheta_1(x)$. Ideal mixing is assumed so that ϑ_1 does not vary with y. This assumption is closely met when fluid 1 flows through a single row of tubes and fluid 2 flows perpendicular to them, Fig. 1.37. This crossflow with a single row of tubes corresponds to one side laterally mixed crossflow. The mixed fluid 1 in the tubes does not have to be the fluid with the higher temperature, as was assumed before.

To determine the temperatures $\vartheta_1 = \vartheta_1(x)$ and $\vartheta_2 = \vartheta_2(x, y)$ of both fluids, the surface element, $dA = dx\,dy$ picked out in Fig. 1.36 will be considered. The heat flow transferred from fluid 1 to fluid 2 is given as

$$d\dot{Q} = [\vartheta_1(x) - \vartheta_2(x, y)]\, k\, dx\, dy \ .$$

The total heat transfer area is $A = L_1 L_2$, see Fig. 1.36. With the dimensionless coordinates

$$x^+ := x/L_1 \quad \text{and} \quad y^+ := y/L_2 \tag{1.121}$$

it follows that

$$d\dot{Q} = [\vartheta_1(x^+) - \vartheta_2(x^+, y^+)]\, kA\, dx^+\, dy^+ \ . \tag{1.122}$$

A second relationship for $\mathrm{d}\dot{Q}$ is yielded from the application of the first law on fluid 2, which flows over the surface element $\mathrm{d}A$. Its mass flow rate is

$$\mathrm{d}\dot{M}_2 = \dot{M}_2\,\mathrm{d}x/L_1 = \dot{M}_2\,\mathrm{d}x^+ \quad ,$$

which gives

$$\mathrm{d}\dot{Q} = \dot{M}_2\,\mathrm{d}x^+ c_{p2}\left(\vartheta_2 + \frac{\partial\vartheta_2}{\partial y^+}\,\mathrm{d}y^+ + \ldots - \vartheta_2\right) = \dot{M}_2 c_{p2}\left(\frac{\partial\vartheta_2}{\partial y^+}\,\mathrm{d}y^+ + \ldots\right)\mathrm{d}x^+$$

or

$$\mathrm{d}\dot{Q} = \dot{W}_2\frac{\partial\vartheta_2}{\partial y^+}\,\mathrm{d}x^+\,\mathrm{d}y^+ \quad . \tag{1.123}$$

The differential equation

$$\frac{\partial\vartheta_2}{\partial y^+} = N_2(\vartheta_1 - \vartheta_2) \tag{1.124}$$

is found from the relationships (1.122) and (1.123). The solution

$$\vartheta_2(x^+, y^+) = \vartheta_1(x^+) - \left[\vartheta_1(x^+) - \vartheta_2'\right]e^{-N_2 y^+} \tag{1.125}$$

still contains the unknown temperature $\vartheta_1(x^+)$ of the laterally mixed fluid.

   In order to find $\vartheta_1(x^+)$ the first law has to be applied to fluid 1. As it flows through the strip with width $\mathrm{d}x$ it gives up a heat flow $\mathrm{d}\dot{Q}^*$, to fluid 2, which does not agree with $\mathrm{d}\dot{Q}$, cf. Fig. 1.38. For $\mathrm{d}\dot{Q}^*$ the following is valid

$$-\mathrm{d}\dot{Q}^* = \dot{M}_1 c_{p1}\left[\vartheta_1 + \frac{\mathrm{d}\vartheta_1}{\mathrm{d}x}\,\mathrm{d}x + \ldots - \vartheta_1\right] \quad .$$

With x^+ from (1.121) it follows that

$$-\mathrm{d}\dot{Q}^* = \dot{W}_1\frac{\mathrm{d}\vartheta_1}{\mathrm{d}x^+}\,\mathrm{d}x^+ \quad . \tag{1.126}$$

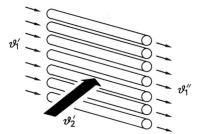

Fig. 1.37: Cross-flow with one tube row as a realisation of the one side laterally mixed crossflow.

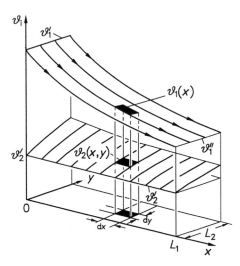

Fig. 1.36: Temperature variations in a one side laterally mixed crossflow. $\vartheta_1 = \vartheta_1(x)$ temperature of the laterally mixed fluid, $\vartheta_2 = \vartheta_2(x, y)$ temperature of the other fluid.

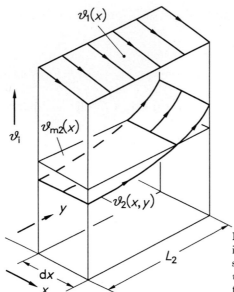

Fig. 1.38: Temperature changes in a strip of size $L_2 \, dx$, in one side laterally mixed crossflow. $\vartheta_{m2}(x)$ is the y direction average temperature of fluid 2.

A second relationship for $d\dot{Q}^*$ is the equation for the overall heat transfer:

$$d\dot{Q}^* = \left[\vartheta_1\left(x^+\right) - \vartheta_{m2}\left(x^+\right)\right] k L_2 \, dx = \left[\vartheta_1\left(x^+\right) - \vartheta_{m2}\left(x^+\right)\right] k A \, dx^+ \ . \tag{1.127}$$

In which

$$\vartheta_{m2}(x^+) = \int\limits_{y^+=0}^{1} \vartheta_2(x^+, y^+) \, dy^+ \tag{1.128}$$

is the temperature of fluid 2 averaged in the y direction, which is decisive for the overall heat transfer through the strip area $L_2 \, dx$. From (1.127) and (1.128) follows the ordinary differential equation

$$\frac{d\vartheta_1}{dx^+} = -N_1(\vartheta_1 - \vartheta_{m2}) \tag{1.129}$$

for the determination of $\vartheta_1(x^+)$.

Using (1.125) and (1.128) the average temperature of fluid 2 can be calculated, giving

$$\vartheta_{m2}(x^+) = \vartheta_1(x^+) - \frac{1}{N_2}\left[\vartheta_1(x^+) - \vartheta_2'\right]\left(1 - e^{-N_2}\right) \ . \tag{1.130}$$

Which then yields from (1.129) the differential equation

$$\frac{d\vartheta_1}{dx^+} = -\frac{N_1}{N_2}\left(1 - e^{-N_2}\right)\left(\vartheta_1 - \vartheta_2'\right) \ .$$

Integration between $x^+ = 0$ and $x^+ = 1$ delivers

$$\frac{\vartheta_1'' - \vartheta_2'}{\vartheta_1' - \vartheta_2'} = \exp\left[-\frac{N_1}{N_2}(1 - e^{-N_2})\right] = \exp\left[-\frac{1}{C_1}\left(1 - e^{-C_1 N_1}\right)\right] \ .$$

The temperature ratio on the left hand side agrees with $(1 - \varepsilon_1)$.

The operating characteristic for one side laterally mixed crossflow is

$$\varepsilon_1 = 1 - \exp\left[-\frac{1}{C_1}(1 - e^{-C_1 N_1})\right] \ . \tag{1.131}$$

It gives the temperature change in the fluid which is laterally mixed as a function of C_1 from (1.99) and the dimensionless transfer capability N_1 from (1.98). The operating characteristic (1.131) can be explicitly resolved for N_1. This gives

$$N_1 = -\frac{1}{C_1}\ln\left[1 + C_1\ln(1 - \varepsilon_1)\right] \ , \tag{1.132}$$

out of which the required value for kA can be immediately calculated. The temperature variation in the fluid which flows perpendicular to the tube row is found using (1.101) to be

$$\varepsilon_2 = \frac{\vartheta''_{m2} - \vartheta'_2}{\vartheta'_1 - \vartheta'_2} = C_1\varepsilon_1 \ . \tag{1.133}$$

Here ϑ''_{m2} is the mean outlet temperature, which could also be found by integrating (1.130) over x^+.

Crossflow with a single row of tubes was dealt with by D. M. Smith [1.14] in 1934. The extension of this work to n rows of tubes was first carried out by H. Schedwill [1.15] in 1968. This produced significantly more complex equations than those used in the case $n = 1$ handled here. The temperature change ε_1, with an outlet temperature averaged over all n rows of tubes, increases with the number of tube rows. The relevant equations can be found in [1.8] and [1.16].

An increase in the number n of tube rows in series approaches the case of *pure crossflow*, in which the temperatures of both fluids change in x and y or rather the dimensionless coordinates x^+ and y^+, from (1.121), cf. Fig. 1.23. The heat transferred, through a surface element of dimensions

$$dA = dx\,dy = A\,dx^+\,dy^+ \ ,$$

from fluid 1 to fluid 2, is found by the same reasoning which led to (1.123), giving the following equations:

$$d\dot{Q} = -\dot{W}_1\frac{\partial\vartheta_1}{\partial x^+}\,dx^+\,dy^+$$

(1st law applied to fluid 1),

$$d\dot{Q} = \dot{W}_2\frac{\partial\vartheta_2}{\partial y^+}\,dx^+\,dy^+$$

(1st law applied to fluid 2), and overall heat transfer

$$d\dot{Q} = kA(\vartheta_1 - \vartheta_2)\,dx^+\,dy^+ \ .$$

Elimination of $d\dot{Q}$ yields the two coupled differential equations

$$\frac{\partial\vartheta_1}{\partial x^+} = -N_1(\vartheta_1 - \vartheta_2) \tag{1.134a}$$

and

$$\frac{\partial \vartheta_2}{\partial y^+} = N_2(\vartheta_1 - \vartheta_2) \tag{1.134b}$$

for the tempratures $\vartheta_1 = \vartheta_1(x^+, y^+)$ and $\vartheta_2 = \vartheta_2(x^+, y^+)$. These have to fulfill the boundary conditions

$$\vartheta_1(0, y^+) = \vartheta_1' \quad \text{and} \quad \vartheta_2(x^+, 0) = \vartheta_2' . \tag{1.135}$$

W. Nusselt [1.17] used a power series to solve this problem. With

$$\xi := N_1 x^+ = (kA/\dot{W}_1)(x/L_1) \tag{1.136a}$$

and

$$\eta := N_2 y^+ = (kA/\dot{W}_2)(y/L_2) \tag{1.136b}$$

the solution has the form

$$\vartheta_1(\xi, \eta) = \left(\sum_{m=0}^{\infty} \frac{\eta^m}{m!} \sum_{j=0}^{m} \frac{\xi^j}{j!} \right) e^{-(\xi+\eta)} , \tag{1.137a}$$

$$\vartheta_2(\xi, \eta) = 1 - \left(\sum_{m=0}^{\infty} \frac{\xi^m}{m!} \sum_{j=0}^{m} \frac{\eta^j}{j!} \right) e^{-(\xi+\eta)} . \tag{1.137b}$$

With the mean values

$$\vartheta_{m1}'' = \frac{1}{N_2} \int_{\eta=0}^{N_2} \vartheta_1(N_1, \eta) \, d\eta$$

and

$$\vartheta_{m2}'' = \frac{1}{N_1} \int_{\xi=0}^{N_1} \vartheta_2(\xi, N_2) \, d\xi$$

the dimensionless temperature changes of both fluids are given by

$$\varepsilon_i = \frac{1}{C_i N_i} \sum_{m=0}^{\infty} \left\{ \left[1 - e^{-N_i} \sum_{j=0}^{m} \frac{N_i^j}{j!} \right] \left[1 - e^{-C_i N_i} \sum_{j=0}^{m} \frac{(C_i N_i)^j}{j!} \right] \right\} , \tag{1.138}$$

where due to the symmetry of the problem for $i = 1$ and $i = 2$, the same relationship is valid. It is not possible to solve this equation explicitly for N_i. H. Martin [1.8] gives a surprisingly short computer program which calculates the mean temperature difference $\Theta = \varepsilon_i/N_i$ and therefore ε_i.

For $N_i \to \infty$, pure crossflow delivers the limits

$$\lim_{N_i \to \infty} \varepsilon_i = \begin{cases} 1 & \text{for} \quad C_i \leq 1 \\ 1/C_i & \text{for} \quad C_i > 1 , \end{cases} \tag{1.139}$$

which are the same as those for countercurrent flow. The limit for $C_i = 0$ is

$$\varepsilon_i = 1 - e^{-N_i} , \qquad (C_i = 0) \tag{1.140}$$

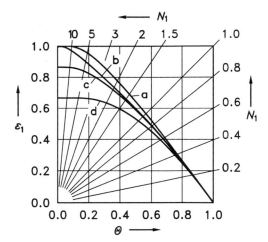

Fig. 1.39: Comparison of different flow configurations in a ε_1, Θ-diagram. a countercurrent, b pure crossflow, c one side laterally mixed crossflow, d cocurrent.

which also agrees with the result derived in section 1.3.3. This means that when one of the fluids is boiling or condensing the temperature change of the other one is not dependent on the flow configuration (cocurrent, countercurrent or crossflow).

For $C_i \neq 0$ and finite values of N_i the temperature changes attainable with cross flow are still significantly below those in countercurrent exchangers but are better than those in cocurrent flow. The comparison of the three cases of simple flow configurations is shown in Fig. 1.39. This diagram shows the dimensionless temperature change ε_1 against the dimensionless mean temperature difference $\Theta = \varepsilon_1/N_1$ for the constant ratio $C_1 = 0.5$. Lines of $N_1 = $ const appear as straight lines all of which go through the origin. A prescribed temperature difference ε_1, for example $\varepsilon_1 = 0.65$, requires the following values for the dimensionless transfer capability: countercurrent (curve a) $N_1 = 1.30$, crossflow (curve b) $N_1 = 1.50$, cocurrent (curve d) $N_1 = 2.44$. For a heat exchanger with $N_1 = 3.0$ the dimensionless temperature change ε_1 reached in countercurrent flow has the highest value of 0.874, with crossflow it is 0.816 and it goes down to 0.660 in cocurrent heat exchangers.

Example 1.5: The cooler in a motor vehicle is a cross-flow heat exchanger, with the cooling medium flowing through a row of parallel finned tubes. Air flows perpendicular to the tubes. At a certain state the volumetric flow rate of the cooling medium is $\dot{V}_1 = 1.25\,\mathrm{dm}^3/\mathrm{s}$; its density is $\rho_1 = 1.015\,\mathrm{kg/dm}^3$ and the mean specific heat capacity is $\bar{c}_{p1} = 3.80\,\mathrm{kJ/kgK}$. The air enters the cooler at $\vartheta_2' = 20.0\,^\circ\mathrm{C}$ with $\dot{V}_2 = 1.100\,\mathrm{m}^3/\mathrm{s}$, $\rho_2 = 1.188\,\mathrm{kg/m}^3$ and $\bar{c}_{p2} = 1.007\,\mathrm{kJ/kgK}$. The transfer capability of the cooler is $kA = 0.550\,\mathrm{kW/K}$. A heat flow of $\dot{Q} = 28.5\,\mathrm{kW}$ has to be transferred to the air. Determine the temperatures ϑ_1' and ϑ_1'' of the motor cooling medium, and what is the temperature ϑ_2'' of the air leaving the cooler?

From the energy balance equation (1.96), namely from

$$\dot{Q} = \dot{W}_1 \left(\vartheta_1' - \vartheta_1'' \right) = \dot{W}_2 \left(\vartheta_2'' - \vartheta_2' \right)$$

the exit temperature of the air

$$\vartheta_2'' = \vartheta_2' + \dot{Q}/\dot{W}_2 \tag{1.141}$$

and the temperature change in the cooling medium

$$\vartheta_1' - \vartheta_1'' = \dot{Q}/\dot{W}_1$$

can be obtained. Then from the defining equation (1.97) for the normalised temperature change ε_1 the entry temperature of the cooling medium is found to be

$$\vartheta_1' = \vartheta_2' + \frac{\vartheta_1' - \vartheta_1''}{\varepsilon_1} = \vartheta_2' + \frac{\dot{Q}}{\dot{W}_1 \varepsilon_1} \quad , \tag{1.142}$$

from which follows the cooling medium outlet temperature

$$\vartheta_1'' = \vartheta_1' - \dot{Q}/\dot{W}_1 \quad . \tag{1.143}$$

The operating characteristic for this case of cross-flow with a single row of tubes (one side laterally mixed cross-flow) according to (1.131), gives the value of ε_1 required in (1.142):

$$\varepsilon_1 = 1 - \exp\left[-\frac{1}{C_1}\left(1 - e^{-C_1 N_1}\right)\right] \quad . \tag{1.144}$$

To evaluate the equations, the heat capacity flow rates are calculated using the given data for the cooling medium,

$$\dot{W}_1 = \dot{V}_1 \rho_1 \bar{c}_{p1} = 1.25 \, \frac{\mathrm{dm}^3}{\mathrm{s}} \cdot 1.015 \, \frac{\mathrm{kg}}{\mathrm{dm}^3} \cdot 3.80 \, \frac{\mathrm{kJ}}{\mathrm{kgK}} = 4.821 \, \frac{\mathrm{kW}}{\mathrm{K}} \quad ,$$

and the air,

$$\dot{W}_2 = \dot{V}_2 \rho_2 \bar{c}_{p2} = 1.100 \, \frac{\mathrm{m}^3}{\mathrm{s}} \cdot 1.188 \, \frac{\mathrm{kg}}{\mathrm{m}^3} \cdot 1.007 \, \frac{\mathrm{kJ}}{\mathrm{kgK}} = 1.316 \, \frac{\mathrm{kW}}{\mathrm{K}} \quad .$$

From which the dimensionless numbers

$$C_1 = \dot{W}_1/\dot{W}_2 = 3.664$$

and

$$C_1 N_1 = N_2 = kA/\dot{W}_2 = 0.418$$

are found. From the operating characteristic according to (1.144) $\varepsilon_1 = 0.0890$ follows. This then yields the values for the temperatures of the cooling medium from (1.142) and (1.143) to be

$$\vartheta_1' = 86.4\,^\circ\mathrm{C} \quad \text{und} \quad \vartheta_1'' = 80.5\,^\circ\mathrm{C} \quad .$$

A relatively high temperature level of the cooling medium favourable for the working of the motor, is achieved. The exit temperature of the air, from (1.141) is $\vartheta_2'' = 41.7\,^\circ\mathrm{C}$.

1.3.5 Operating characteristics of further flow configurations. Diagrams

In addition to the cases of countercurrent, cocurrent and cross flow, further flow configurations are possible. These are also applied in industry and have been investigated by many different authors, cf. the compilation of W. Roetzel and

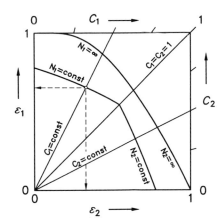

Fig. 1.40: Scheme for a $\varepsilon_1, \varepsilon_2$-diagram according to W. Roetzel and B. Spang [1.16] with lines of $N_1 = $ const and $N_2 = $ const.

B. Spang [1.16]. The operating characteristics $F(\varepsilon_i, N_i, C_i) = 0$ are often complex mathematical expressions, so it seems reasonable to represent the results graphically.

W. Roetzel and B. Spang [1.18] discussed the possibility of representing the operating characteristics in a graph, and came up with a clearly arranged diagram which can be found in the VDI-Heat Atlas [1.16]. This square shaped graph consists of two parts which are separated by the diagonal running from the bottom left to the top right hand corner, Fig. 1.40. The axes of the graph are the two dimensionless temperature changes ε_1 and ε_2 from (1.97). The area above the diagonal contains lines of constant dimensionless transfer capability $N_1 = kA/\dot{W}_1 = $ const as well as a host of straight lines that run through the origin, according to

$$C_1 = \frac{\dot{W}_1}{\dot{W}_2} = \frac{\varepsilon_2}{\varepsilon_1} = \frac{N_2}{N_1} = \text{const} . \tag{1.145}$$

These lines are not drawn on the graph, to prevent it from looking too crowded, only the end points are marked around the edges of the figure. Each point in the triangle above the diagonal corresponds to an operating state, for which the values for ε_1, N_1 and C_1 can be read off the graph. The accompanying value for ε_2 can be found from the point on the abcissa.

In the triangle below the diagonal, the operating characteristic is represented in the form of $F(\varepsilon_2, N_2, C_2) = 0$. The lines of equal transfer capability $N_1 = N_2$ meet at the diagonal $C_1 = C_2 = 1$ with a kink. This is due to (1.145). Only the line $N_1 = N_2 \rightarrow \infty$ does not have a kink at this point. In symmetrical flow configurations, e.g. counter or crosscurrent flow, the operating characteristics for $i = 1$ and $i = 2$ agree, so the two halves of the graph are symmetric about the diagonal. However in asymmetric flow, for example in one side laterally mixed cross flow, this is not the case. The indices for each fluid must be carefully checked so that the results from the graph are not mixed up. A comprehensive collection of these graphs for the design and construction of heat exchangers is available in the VDI-Heat Atlas [1.16].

1.4   The different types of mass transfer

Mass transfer is the transport of one or more components of a mixture, of fluid or solid material, within a phase[12] or over the phase boundary. Mass transport within a phase up to the phase boundary is called *mass transfer*. When this occurs over the phase boundary into another phase, it is then known as *overall mass transfer*. These terms correspond to those in heat transfer.

The driving forces for mass transfer are concentration, temperature or pressure gradients. We will explore the most common of these three, namely mass transfer due to a concentration gradient. As experience tells us, the components of a mixture move from regions of higher concentration to those with lower concentration. Equilibrium with respect to mass transfer is realised when the driving force, in this case the concentration difference, has disappeared.

Mass transfer processes can be found in various ways in both nature and technology. Highly developed plants and animals have circulatory systems which serve the supply of nutrition and energy, where processes of mass transfer are decisive. In plants the water taken up from the soil carries the products of photosynthesis, above all glucose, to the places where they are used or stored. Red blood corpuscles release carbon dioxide in the lungs, and take up oxygen which is needed by all cells in the body. Separation processes in chemical engineering such as the drying of solid materials, distillation, extraction and sorption are all affected by the processes of mass transfer. They also play a role in the production of materials in order to obtain the desired properties of a substance. Chemical reactions, including combustion processes, are often decisively determined by mass transfer.

As a simple example of mass transfer we will consider a glass filled with water in a room of dry air. Immediately above the liquid surface there is a large amount of water vapour, whilst further away there is far less. As a result of this concentration drop the air enrichs itself with water vapour. This flows in the direction of the concentration or partial pressure drop. In a volume element above the surface of the water, the velocity of the water molecules perpendicular to the liquid surface is greater than that of the air molecules. This leads to a perceptible, macroscopic relative movement between water vapour and air.

This sort of macroscopic relative movement of a single substance in a phase is known as *diffusion*. There are two different forms of diffusion. In quiescent fluids or solid bodies which are made up of different components, one substance can only be transferred if the average molecular velocities of some components in the

---

[12]The phase is understood here to be the area of the system, in which each volume element has values for the thermodynamic variables of pressure, temperature and concentration, among others. Only steady, rather than irregular changes in these quantities are permitted within a phase. In thermodynamics a phase is a *homogeneous* region in a system. In a phase in the sense of thermodynamics all the defined intensive variables of state are spatially constant.

mixture are different from each other. This is known as *molecular diffusion*. It also occurs in laminar flow, as the volume elements of the fluid move along defined flow paths and mass transfer between volume elements can only happen if the average molecular velocities of the components are different. In contrast, a turbulent flow is indicated by the irregular, random movements of the fluid elements, which can be different for each component. This overlaps the whole flow and therefore also the mass transport through molecular diffusion. The irregular fluid motion is often an order of magnitude greater than the molecular diffusion. Mass transfer due to irregular fluctuations in the fluid is known as *turbulent diffusion*. This can also be applied to the situation of the water in the glass, by inducing a significantly high flow velocity above the surface of the water by blowing or mixing. The amount of water vapour transferred to the air in a certain period of time will be far greater than that transferred in the same time interval in quiescent air.

In order to understand mass transfer due to moleular diffusion we will study the process in a vessel filled with a coloured solution, for example iodine solution. Water is carefully poured over the iodine solution, to avoid as far as possible convection currents. The coloured solution and the water are noticeably separate at the beginning of the experiment. After some time the upper layer becomes coloured, while the layer beneath it is clear enough to see through. Eventually, after long enough time has passed, the solution is the same colour throughout. So quite obviously despite there being no convection currents iodine molecules were transported from the lower to the upper part of the vessel. This can be explained by the diffusion of iodine molecules in water.

During the process the single iodine molecules penetrate partly into areas of higher and partly into areas of lower concentration, with out any preferred direction. Despite this transport of iodine molecules from regions of high to regions of low concentration take place. To help with understanding this concept, one should think of two thin, equally sized volume elements cut out from both sides of a horizontal cross section. Although the movements of a single iodine molecule in one of these volume elements cannot be predicted, we can say that after a certain time, on average a finite number of molecules from the lower element will pass through the cross section and penetrate the upper element. In the same manner a certain number of molecules will move in the opposite direction, from the upper to the lower element. As there were more iodine molecules in the element below the cross section, more molecules were going to move into the upper layer due to the random molecular motion. A balancing out of the concentration takes place, until enough time has passed and the concentration differences within the solution have been dismantled.

From a macroscopic standpoint molecular diffusion is mass transfer due to a concentration difference. Other types of diffusion, namely diffusion due to pressure differences (pressure diffusion) or temperature differences (thermal diffusion) will not be discussed here. The mechanism of molecular diffusion corresponds to that of heat conduction, whilst mass transfer in a flowing fluid, known for short as convective mass transfer correponds to convective mass transfer. Mass transfer by diffusion and convection are the only sorts of mass transfer. Radiative heat transfer has no corresponding mass transfer process.

1.4.1 Diffusion

The calculation of mass transfer by diffusion requires several definitions and relationships which will be outlined in the following.

1.4.1.1 Composition of mixtures

The composition of mixtures can be characterised in different ways. For a quantitative description the following quantities have to be introduced.

The *mass fraction* ξ_A is the mass M_A of component A over the total mass M in a volume element within a phase[13]:

$$\xi_A := \frac{M_A}{M} = \frac{M_A}{\sum_K M_K} \; . \tag{1.146}$$

The sum of all the mass fractions is

$$\sum_K \xi_K = 1 \; . \tag{1.146a}$$

The *mole fraction* $\tilde{x}_A$ is the number of moles N_A of component A over the total number of moles N in the mixture in a given phase:

$$\tilde{x}_A := \frac{N_A}{N} = \frac{N_A}{\sum_K N_K} \; . \tag{1.147}$$

The sum of all the mole fractions is

$$\sum_K \tilde{x}_K = 1 \; . \tag{1.147a}$$

The molar concentration of substance A is defined by

$$c_A := N_A/V \; . \tag{1.148}$$

The molar concentration of the mixture is

$$c := N/V = \sum_K c_K \; ,$$

---

[13]The letter K under the summation sign means that the sum is taken over all the components K.

which gives for the mole fraction of component A

$$\tilde{x}_A = c_A/c \; .$$

For ideal gases $c_A = p_A/R_m T$ and $c = p/R_m T$, is valid, where $p_A = \tilde{x}_A p$ is the partial pressure of component A and $R_m = 8{,}31451 \, \text{J}/(\text{mol K})$ is the molar gas constant.

These quantities in a composition are not independent from each other. To find a relationship between the mass and mole fractions, we multiply (1.147) by the molar mass $\tilde{M}_A = M_A/N_A$ of component A. This gives

$$\tilde{x}_A \tilde{M}_A = M_A/N \; .$$

Summation over all the components yields the average molar mass $\tilde{M} = M/N$:

$$\sum_K \tilde{x}_K \tilde{M}_K = \sum_K M_K/N = M/N = \tilde{M} \; . \tag{1.149}$$

This then gives the following relationship between the mole and mass fractions

$$\xi_A = \frac{M_A}{M} = \frac{M_A}{N_A} \frac{N_A}{N} \frac{N}{M} = \frac{\tilde{M}_A}{\tilde{M}} \tilde{x}_A \; . \tag{1.150}$$

In the reverse case, when the mass fractions are known, the mole fractions come from

$$\tilde{x}_A = \frac{\tilde{M}}{\tilde{M}_A} \xi_A \; , \tag{1.151}$$

in which the average molar mass is found from the mass fractions and the molar masses of the components as

$$\frac{1}{\tilde{M}} = \frac{N}{M} = \sum_K \frac{N_K}{M} = \sum_K \frac{M_K}{M} \frac{N_K}{M_K}$$

or

$$\frac{1}{\tilde{M}} = \sum_K \xi_K \frac{1}{\tilde{M}_K} \; . \tag{1.152}$$

1.4.1.2 Diffusive fluxes

In each volume element the average particle velocities of each substance can be different, so that the convection of the volume elements overlaps the relative movement of the particles of different substances. This macroscopic relative movement is known as diffusion. The average velocity of the particles of substance A is denoted by the vector $\boldsymbol{w}_A$. To describe diffusion we will introduce the relative velocity $\boldsymbol{w}_A - \boldsymbol{\omega}$, where $\boldsymbol{\omega}$ is a reference velocity which is yet to be defined. As the diffusional flux (SI unit mol/m²s) of substance A, the quantity

$$\boldsymbol{j}_A := c_A(\boldsymbol{w}_A - \boldsymbol{\omega}) \tag{1.153}$$

is defined.

The reference velocity $\boldsymbol{\omega}$ can be chosen to be the velocity $\boldsymbol{w}$ at the centre of gravity of the mass. This is defined as the mass average velocity of a volume element:

$$\varrho \boldsymbol{w} := \sum_K \varrho_K \boldsymbol{w}_K \quad \text{or} \quad \boldsymbol{w} = \sum_K \xi_K \boldsymbol{w}_K \ . \tag{1.154}$$

The diffusional flux is then

$$\boldsymbol{j}_A = c_A(\boldsymbol{w}_A - \boldsymbol{w}) \ .$$

Multiplication with the molar mass $\tilde{M}_A$ yields with $c_A \tilde{M}_A = \varrho_A$

$$\boldsymbol{j}_A \tilde{M}_A = \boldsymbol{j}_A^* = \varrho_A(\boldsymbol{w}_A - \boldsymbol{w}) \ , \tag{1.155}$$

where $\boldsymbol{j}_A^*$ is the mass based diffusional flux of component A (SI unit kg/m²s). From (1.154) and (1.155) it follows that

$$\sum_K \boldsymbol{j}_K^* = 0 \ . \tag{1.156}$$

The reference system using the mass average velocity $\boldsymbol{w}$ from (1.154) is called the centre of gravity system. The momentum and energy balances for this system are easily formulated.

The molar average velocity $\boldsymbol{u}$ can also be used as a further reference velocity. It is defined by

$$\boldsymbol{u} := \sum_K \tilde{x}_K \boldsymbol{w}_K \ . \tag{1.157}$$

The associated diffusional flux (SI units mol/m²s) is

$${}_u\boldsymbol{j}_A := c_A(\boldsymbol{w}_A - \boldsymbol{u}) \ . \tag{1.158}$$

Taking into account $c_A = \tilde{x}_A c$ and $\sum_K c_K = c$ it follows from (1.157) and (1.158) that

$$\sum_K {}_u\boldsymbol{j}_K = 0 \ .$$

A reference system with the molar average velocity is called the particle reference system. Other reference systems and velocities are available in the literature [1.21]. The diffusional flux in one system can be transferred to any other system, as is shown in the example which follows.

Example 1.6: The difffusional flux of component A in a '-reference system $\boldsymbol{j}_A' = c_A(\boldsymbol{w}_A - \boldsymbol{\omega}')$ is given for a reference velocity $\boldsymbol{\omega}' = \sum_K \zeta_K' \boldsymbol{w}_K$, where for the 'weighting factors' ζ_K', $\sum_K \zeta_K' = 1$ is valid.

The diffusional flux of component A $\boldsymbol{j}_A'' = c_A(\boldsymbol{w}_A - \boldsymbol{\omega}'')$ in "-reference system with a reference velocity

$$\boldsymbol{\omega}'' = \sum_K \zeta_K'' \boldsymbol{w}_K \quad \text{mit} \quad \sum_K \zeta_K'' = 1$$

has to be calculated. The general relationship between the diffusional fluxes $\boldsymbol{j}_A$ and $_u\boldsymbol{j}_A$ has to be derived.

It is $\boldsymbol{j}_A'' - \boldsymbol{j}_A' = c_A(\boldsymbol{\omega}' - \boldsymbol{\omega}'') = c_A \sum_K (\zeta_K' - \zeta_K'')\,\boldsymbol{w}_K$. It follows further from $\boldsymbol{j}_A' = c_A(\boldsymbol{w}_A - \boldsymbol{\omega}')$ that the velocity is $\boldsymbol{w}_A = \boldsymbol{j}_A'/c_A + \boldsymbol{\omega}'$. Therefore

$$
\begin{aligned}
\boldsymbol{j}_A'' - \boldsymbol{j}_A' &= c_A \sum_K (\zeta_K' - \zeta_K'') \left(\frac{\boldsymbol{j}_K'}{c_K} + \boldsymbol{\omega}' \right) \\
&= c_A \sum_K \zeta_K' \frac{\boldsymbol{j}_K'}{c_K} + c_A \boldsymbol{\omega}' \sum_K \zeta_K' - c_A \sum_K \zeta_K'' \frac{\boldsymbol{j}_K'}{c_K} - c_A \boldsymbol{\omega}' \sum_K \zeta_K'' \quad,
\end{aligned}
$$

from which, accounting for $\sum_K \zeta_K' = 1$ and $\sum_K \zeta_K'' = 1$ and because $\sum_K \zeta_K' \boldsymbol{j}_K'/c_K = 0$ the relationship

$$
\boldsymbol{j}_A'' - \boldsymbol{j}_A' = -c_A \sum_K \zeta_K'' \frac{\boldsymbol{j}_K'}{c_K}
$$

is yielded. To change from the particle to the gravitational system we put in $\boldsymbol{j}_A'' = \boldsymbol{j}_A$, $\zeta_A'' = \xi_A$ and $\boldsymbol{j}_A' = _u\boldsymbol{j}_A$ which give

$$
\boldsymbol{j}_A - _u\boldsymbol{j}_A = -c_A \sum_K \xi_K \frac{_u\boldsymbol{j}_K}{c_K} \quad.
$$

Correspondingly the conversion of the diffusional flux from the gravitational to the particle system gives

$$
_u\boldsymbol{j}_A - \boldsymbol{j}_A = -c_A \sum_K \tilde{x}_K \frac{\boldsymbol{j}_K}{c_K} \quad.
$$

In a mixture of *two substances*, components A and B, these relationships are

$$
\boldsymbol{j}_A - _u\boldsymbol{j}_A = -c_A \left(\frac{\xi_A}{c_A}\,_u\boldsymbol{j}_A + \frac{\xi_B}{c_B}\,_u\boldsymbol{j}_B \right) \quad.
$$

With $_u\boldsymbol{j}_A = -_u\boldsymbol{j}_B$ and $\xi_A/c_A - \xi_B/c_B = (V/M)(\tilde{M}_A - \tilde{M}_B)$ it follows that

$$
\boldsymbol{j}_A - _u\boldsymbol{j}_A = -c_A \frac{V}{M}\,_u\boldsymbol{j}_A (\tilde{M}_A - \tilde{M}_B)
$$

and therefore with $c_A V/M = \tilde{x}_A/\tilde{M}$

$$
\boldsymbol{j}_A = _u\boldsymbol{j}_A \left[1 - \frac{\tilde{x}_A}{\tilde{M}} (\tilde{M}_A - \tilde{M}_B) \right] = _u\boldsymbol{j}_A \frac{\tilde{M} - \tilde{x}_A \tilde{M}_A + \tilde{x}_A \tilde{M}_B}{\tilde{M}} \quad.
$$

Then because $\tilde{M} = \tilde{x}_A \tilde{M}_A + \tilde{x}_B \tilde{M}_B$ the equation given above simplifies to

$$
\boldsymbol{j}_A = _u\boldsymbol{j}_A \frac{\tilde{M}_B}{\tilde{M}} \quad.
$$

And correspondingly for component B

$$
\boldsymbol{j}_B = _u\boldsymbol{j}_B \frac{\tilde{M}_A}{\tilde{M}} \quad.
$$

1.4.1.3 Fick's Law

The diffusional flux of component A is proportional to the concentration gradient
grad c_A. For the time being we will limit ourselves to a mixture of two components
A and B. We will also assume that diffusion only takes place along one coordinate
axis, for example the y-axis. The diffusional flux can be described by an empirical
statement corresponding to Fourier's law

$$_u\dot{j}_A = -D_{AB}\frac{dc_A}{dy} \ , \tag{1.159}$$

which was first formulated by A. Fick[14] and is called Fick's first law after him.
The proportionality factor D_{AB} (SI units m^2/s) is the diffusion coefficient in a
mixture of two components A and B. Eq. (1.159) is valid when it is assumed that
the molar concentration c of the mixture is constant. This condition is fulfilled in
constant pressure, isothermal mixtures of ideal gases due to $c = N/V = p/R_m T$.
Disregarding this assumption of $c = $ const, the general equation for mixtures of
two substances is

$$_u\dot{j}_A = -cD_{AB}\frac{d\tilde{x}_A}{dy} \ , \tag{1.160}$$

as proved by de Groot [1.22]. In the gravitational system, the equivalent relationship to Eq. (1.160), for the diffusional flux in the y direction is

$$\dot{j}_A^* = -\varrho D_{AB}\frac{d\xi_A}{dy} \ . \tag{1.161}$$

In a multicomponent mixture consisting of N components the diffusional flux $\boldsymbol{j}_A^*$
of component A is given by [1.23]

$$\boldsymbol{j}_A^* = \varrho \sum_{\substack{K=1 \\ K \neq A}}^{N} \frac{\tilde{M}_A \tilde{M}_K}{\tilde{M}^2} D_{AK} \, \text{grad} \, \tilde{x}_K \ . \tag{1.162}$$

From this equation (1.161) is obtained for the special case of $N = 2$.

Example 1.7: It has to be shown that (1.161) is equivalent to (1.160).
The diffusional fluxes $_u\dot{j}_A$ and $\dot{j}_A^*$ in the y-direction are linked by

$$\dot{j}_A^* = \dot{j}_A \tilde{M}_A = _u\dot{j}_A \frac{\tilde{M}_B \tilde{M}_A}{\tilde{M}}$$

(see the solution to example 1.6). And with that

$$\dot{j}_A^* = -c\,D_{AB}\frac{\partial \tilde{x}_A}{\partial y}\frac{\tilde{M}_B \tilde{M}_A}{\tilde{M}} \ .$$

---

[14]Adolph Fick (1829–1901), Professor of Physiology in Zürich and Würzburg, discovered the
fundamental laws of diffusion.

Where $c = N/V = \varrho/\tilde{M}$, and by differentiation of $\tilde{x}_A = \tilde{M}\,\xi_A/\tilde{M}_A$ with $\tilde{M} = \tilde{x}_A\,\tilde{M}_A + (1 - \tilde{x}_A)\,\tilde{M}_B$ it follows that

$$d\tilde{x}_A = \frac{\tilde{M}^2}{\tilde{M}_A\tilde{M}_B}\,d\xi_A \ .$$

Then from the equation given above

$$j_A^* = -\varrho\,D_{AB}\,\frac{\partial\xi_A}{\partial y}$$

is obtained.

By exchanging the indices A and B in (1.161) the diffusional flux for component B in the binary mixture can be found. As the sum of the two diffusional fluxes disappears, according to (1.156), it follows that

$$j_A^* + j_B^* = -\varrho D_{AB}\frac{d\xi_A}{dy} - \varrho D_{BA}\frac{d\xi_B}{dy} = 0 \ .$$

Here $d\xi_A/dy = -d\xi_B/dy$, because $\xi_A + \xi_B = 1$, and therefore $D_{AB} = D_{BA}$. The coefficient of diffusion of component A through component B is the same as the diffusion coefficient of component B through component A. Therefore the indices will no longer be used, and we will simply write D instead of $D_{AB} = D_{BA}$.

Typical *values for the diffusion coefficient* are $5 \cdot 10^{-6}$ to 10^{-5} m^2/s in gases, 10^{-10} to 10^{-9} m^2/s in liquids and 10^{-14} to 10^{-10} m^2/s in solid bodies. In gases the molecules can move about more easily and therefore the diffusion coefficients are greater than those for liquids, which in turn are larger than those in solid bodies. Diffusion in solids is several magnitudes slower than in liquids, whilst in gases it is fastest.

The diffusional flux can be calculated with the help of Fick's law provided the concentrations are known. If however the flux is known the concentration field can be found by integration of Fick's law. As a simple example we will take a solid material, from which component A should be removed using a liquid solvent B, Fig. 1.41.

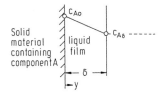

Fig. 1.41: Diffusion through a quiescent liquid film

The concentration of component A at the surface between the solid and the liquid film is given by c_{A0}, whilst the concentration of A in the bulk flow is represented by $c_{A\delta}$. We will presume $c = N/V = $ const . As the material only moves in the y-direction, there is no need to note the mass transfer direction in terms of vectors. The molar flux from the solid into the liquid, according to (1.158), is

$$\dot{n}_A = c_A w_A = {}_u\dot{j}_A + c_A u$$

with $_u j_A = -D \, dc_A/dy$ and $u = x_A w_A + x_B w_B$. The velocity w_B of the solvent in the y-direction is zero. With a small reference velocity u, and if the concentration c_A of substance A in the solvent is very low, the term $c_A u$ is neglible and so

$$\dot{n}_A = {}_u j_A = -D\frac{dc_A}{dy} \; . \tag{1.163}$$

At steady state, the same amount of A which flows into the liquid film also flows out of the film, which means $\dot{n}_A = \text{const}$. Under the assumption that the diffusion coefficient D is constant, by integrating the equation given above, it follows that

$$\dot{n}_A = -D\frac{c_{A0} - c_{A\delta}}{\delta} \; .$$

In a solid sphere of radius r_0, which is surrounded by a liquid film of thickness δ, the diffusional flux in the radial direction, according to (1.163), is

$$\dot{n}_A = {}_u j_A = -D\frac{dc_A}{dr} \; . \tag{1.164}$$

At steady state the molar flow which diffusives through the spherical shell is constant,

$$\dot{N}_A = \dot{n}_A 4\pi r^2 = -D\frac{dc_A}{dr}4\pi r^2 \; , \tag{1.165}$$

and with that

$$\frac{d\dot{N}_A}{dr} = 0 = \frac{d}{dr}\left(-D\frac{dc_A}{dr}r^2\right) \; .$$

Assuming $D = \text{const}$ and taking into account the boundary conditions $c_A(r = r_0) = c_{A0}$ and $c_A(r = r_0 + \delta) = c_{A\delta}$ the concentration profile

$$\frac{c_A - c_{A0}}{c_{A\delta} - c_{A0}} = \frac{1/r - 1/r_0}{1/(r_0 + \delta) - 1/r_0} \tag{1.166}$$

can be found by integration. It corresponds to the temperature profile for steady conduction in a hollow sphere according to (1.17). The diffusional flow, eq. (1.165), found by differentiation is

$$\dot{N}_A = D\frac{c_{A0} - c_{A\delta}}{1/r_0 - 1/(r_0 + \delta)}4\pi \; . \tag{1.167}$$

1.4.2 Diffusion through a semipermeable plane. Equimolar diffusion

In the previous example of the diffusion of component A from a solid substance into a solvent we presumed a low convection velocity u and a low concentration c_A of A in the solvent. As a result a negligible convection flow $c_A u$ was found. This supposition is not normally fulfilled. As a typical example we will consider a liquid A in a cylindrical vessel which is evaporating in a quiescent gas B, Fig. 1.42. The liquid level is kept at $y = y_1$, or it changes so slowly that we can take it to be quiescent. At the surface of the liquid $c_A(y = y_1) = c_{A1}$. At moderate total pressure, c_{A1} can be found from the thermal equation of state for ideal gases, as $c_{A1} = p_{A1}/R_mT$, where p_{A1} is the saturation partial pressure of A at temperature T of the liquid. The solubility of gas B in liquid A is negligible. This is a good approximation for water as the liquid A and air as gas B. A gaseous mixture of A and B flows over the top of the cylinder $y = y_2$, with a concentration $c_A(y = y_2) = c_{A2}$. The molar flux of component A in the direction of the y-axis is, as before

$$\dot{n}_A = c_A w_A = {}_u\dot{j}_A + c_A u$$

with $u = \tilde{x}_A w_A + \tilde{x}_B w_B$. As gas B is stagnant in the cylinder, $w_B = 0$, and therefore $u = \tilde{x}_A w_A$. This is called *diffusion through a semipermeable plane*. The plane in this case is the surface of the water which evaporates into the adjoining air. Only water passes through, the surface is, therefore, semipermeable. The last equation gives the relationship

$$\dot{n}_A = c_A w_A = \frac{1}{1 - \tilde{x}_A} \, {}_u\dot{j}_A \, , \tag{1.168}$$

which replaces eq. (1.163) for small mole fractions $\tilde{x}_A$ of the dissolved substance. In steady conditions $d\dot{n}_A/dy = 0$, and with Fick's law (1.160), it follows that

$$\frac{d}{dy}\left(\frac{cD}{1 - \tilde{x}_A}\frac{d\tilde{x}_A}{dy}\right) = 0 \, . \tag{1.169}$$

Here $D = D_{AB}$ is the binary diffusion coefficient. For mixtures of ideal gases at constant pressure and temperature $c = N/V = p/R_mT$ is constant. The diffusion coefficient only changes slightly with the composition of the mixture and can therefore be presumed to be constant. This gives the following differential equation for the concentration profile

$$\frac{d}{dy}\left(\frac{1}{1 - \tilde{x}_A}\frac{d\tilde{x}_A}{dy}\right) = 0 \, ,$$

which because $\tilde{x}_B = 1 - \tilde{x}_A$ can also be written as

$$\frac{d}{dy}\left(\frac{1}{\tilde{x}_B}\frac{d\tilde{x}_B}{dy}\right) = 0$$

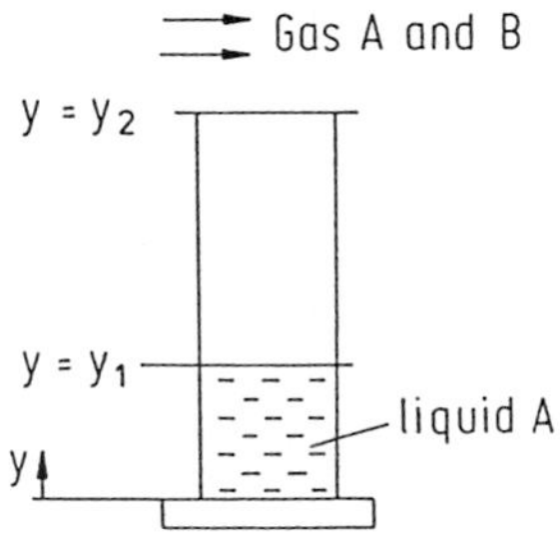

Fig. 1.42: Diffusion of component A in a gas mixture of components A and B.

or

$$\frac{d}{dy}\frac{d\ln\tilde{x}_B}{dy} = \frac{d^2\ln\tilde{x}_B}{dy^2} = 0 \qquad (1.170)$$

It has to be solved under the boundary conditions

$$\tilde{x}_B(y = y_1) = \tilde{x}_{B1} = 1 - \tilde{x}_{A1} \qquad \text{and}$$

$$\tilde{x}_B(y = y_2) = \tilde{x}_{B2} = 1 - \tilde{x}_{A2} \qquad .$$

The solution is

$$\frac{\tilde{x}_B}{\tilde{x}_{B1}} = \left(\frac{\tilde{x}_{B2}}{\tilde{x}_{B1}}\right)^{\frac{y - y_1}{y_2 - y_1}} \qquad . \qquad (1.171)$$

It is easy to check that the solution is correct: Taking logarithms of (1.171), gives the following expression

$$\ln\frac{\tilde{x}_B}{\tilde{x}_{B1}} = \frac{y - y_1}{y_2 - y_1}\ln\frac{\tilde{x}_{B2}}{\tilde{x}_{B1}} \qquad ,$$

which disappears by differentiating twice with repect to y. Differentiation of (1.171) gives the diffusional flux

$$_u\dot{j}_A = -cD\frac{d\tilde{x}_A}{dy} = cD\frac{d\tilde{x}_B}{dy} = cD\frac{\tilde{x}_B}{y_2 - y_1}\ln\frac{\tilde{x}_{B2}}{\tilde{x}_{B1}} \qquad .$$

The associated molar flux is found from (1.168),

$$\dot{n}_A = \frac{1}{1 - \tilde{x}_A}{_u\dot{j}_A} = \frac{1}{\tilde{x}_B}{_u\dot{j}_A}$$

to be

$$\dot{n}_A = \frac{cD}{y_2 - y_1}\ln\frac{\tilde{x}_{B2}}{\tilde{x}_{B1}} = \frac{cD}{y_2 - y_1}\ln\frac{1 - \tilde{x}_{A2}}{1 - \tilde{x}_{A1}} \qquad . \qquad (1.172)$$

The average mole fraction of component B between $y = y_1$ and $y = y_2$ is

$$\tilde{x}_{Bm} = \frac{1}{y_2 - y_1}\int_{y_1}^{y_2} \tilde{x}_B dy \qquad .$$

After insertion of $\tilde{x}_B$ from (1.171) and integration this yields

$$\tilde{x}_{Bm} = \frac{\tilde{x}_{B2} - \tilde{x}_{B1}}{\ln(\tilde{x}_{B2}/\tilde{x}_{B1})} \qquad . \qquad (1.173)$$

The average mole fraction is the logarithmic mean of the two values $\tilde{x}_{B1}$ and $\tilde{x}_{B2}$. It is therefore possible to write (1.172) as

$$\dot{n}_A = \frac{cD}{\tilde{x}_{Bm}}\frac{\tilde{x}_{B2} - \tilde{x}_{B1}}{y_2 - y_1} = \frac{cD}{(1 - \tilde{x}_{Am})}\frac{\tilde{x}_{A1} - \tilde{x}_{A2}}{y_2 - y_1} \qquad . \qquad (1.174)$$

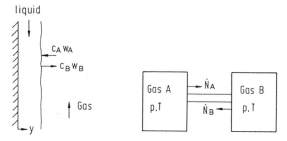

Fig. 1.43: Mass transfer in distillation   **Fig. 1.44**: Equimolar diffusion between two containers

For ideal gases (1.173) and (1.174) can be still be written with partial pressures p_B and the total pressure p,

$$\dot{n}_A = \frac{pD/R_mT}{y_2 - y_1}\ln\frac{p_{B2}}{p_{B1}} = \frac{pD/R_mT}{(y_2 - y_1)p_{Bm}}(p_{B2} - p_{B1}) \ , \qquad (1.175)$$

where p_{Bm} is the logarithmic mean partial pressure

$$p_{Bm} = \frac{p_{B2} - p_{B1}}{\ln(p_{B2}/p_{B1})} \quad .$$

Equimolar counter diffusion appears in the distillation of binary mixtures. In a distillation column the liquid falls downwards, and the vapour flows upwards, Fig. 1.43. As the liquid flowing down the column is colder than the vapour flowing upwards, chiefly the component with the higher boiling point, the so called least volatile component condenses, whilst the vapour from the boiling liquid mainly consists of the components with the lower boiling points, the more volatile components. The molar enthalpy of vaporisation is, according to Trouton's rule, approximately constant for all components. If a certain amount of the least volatile component condenses out from the vapour, then the same number of moles of the more volatile substance will be evaporated out of the liquid. At the phase boundary between liquid and vapour we have $c_A w_A = -c_B w_B$. The reference velocity u is zero because $cu = c_A w_A + c_B w_B$. The molar flux transported to the phase boundary from (1.158) and (1.160) is

$$\dot{n}_A = c_A w_A = {}_u j_A = -cD\frac{d\tilde{x}_A}{dy} \quad . \qquad (1.176)$$

Convective and diffusive flows are in agreement with each other.

Let us assume that two containers, each containing a different gas are linked by a thin pipe between them, Fig. 1.44. Equimolar counter diffusion will also take place in this case, if the pressure and temperature of both the gases are the same and obey the thermal equation of state for ideal gases.

We consider a volume of gas V in the pipe between the two containers. The thermal equation of state for this volume is

$$p = (N_A + N_B)R_mT/V \quad .$$

At steady state the total pressure is invariant with time. Therefore

$$\frac{\mathrm{d}p}{\mathrm{d}t} = (\dot{N}_A + \dot{N}_B)R_m T/V = 0$$

which means that $\dot{N}_A = -\dot{N}_B$.

1.4.3 Convective mass transfer

Mass transfer from a flowing fluid to the surface of another substance, or between two substances that are barely miscible, depends on the properties of the materials involved and the type of flow. As in the case of convective heat transfer the flow can be forced from outside through, for example, a compressor or a pump. This is known as mass transfer in forced convection. If however mass transfer is caused by a change in the density due to pressure or temperature variations, then we would speak of mass transfer in free convection.

Following on from the definition equation (1.23) for the heat transfer coefficient, the molar flow transferred to the surface (index 0) is described by

$$\dot{N}_{A0} := \beta_c A \Delta c_A \qquad (1.177)$$

and the molar flux is given by

$$\dot{n}_{A0} = \dot{N}_{A0}/A = \beta_c \Delta c_A \ . \qquad (1.178)$$

The mass transfer coefficient β_c with SI units of m/s or $\mathrm{m^3/(s\,m^2)}$ is defined using these equations. It is a measure of the volumetric flow transferred per area. The concentration difference Δc_A defines the mass transfer coefficient. A useful choice of the decisive concentration difference for mass transfer has to be made. A good example of this is for mass transfer in a liquid film, see Fig. 1.41 where the concentration difference $c_{A0} - c_{A\delta}$ between the wall and the surface of the film would be a a suitable choice. The mass transfer coefficient is generally dependent on the type of flow, whether it is laminar or turbulent, the physical properties of the material, the geometry of the system and also fairly often the concentration difference Δc_A. When a fluid flows over a quiescent surface, with which a substance will be exchanged, a thin layer develops close to the surface. In this layer the flow velocity is small and drops to zero at the surface. Therefore close to the surface the convective part of mass transfer is very low and the diffusive part, which is often decisive in mass transfer, dominates.

As the diffusive part is, according to Fick's law (1.159), proportional to the concentration gradient at the quiescent surface and therefore approximately proportional to the concentration difference Δc_A, the mass transfer coefficient can be usefully defined by

$$_u\dot{j}_{A0} := -\left(cD\frac{\mathrm{d}\tilde{x}_A}{\mathrm{d}y}\right)_0 = \beta \Delta c_A \ . \qquad (1.179)$$

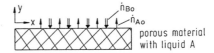

Fig. 1.45: Mass transfer from a porous
body into a gas flow

The diffusional flux $_u\dot{j}_{A0}$ is that at the surface (index 0). Therefore

$$\beta = \frac{-(cD\,\mathrm{d}\tilde{x}_A/\mathrm{d}y)_0}{\Delta c_A} \ . \tag{1.180}$$

The relationship between the molar flux according to (1.178) and the diffusional
flux from (1.179) is appropriately defined as

$$c_{A0}w_{A0} = \dot{n}_{A0} = {}_u\dot{j}_{A0} + c_{A0}u_0 = {}_u\dot{j}_{A0} + \tilde{x}_{A0}\dot{n}_0 \ . \tag{1.181}$$

For a vanishing convective flow $\dot{n}_0 \to 0$ we have

$$c_{A0}w_{A0} = \dot{n}_{A0} = {}_u\dot{j}_{A0}, \qquad \text{(valid for } \dot{n}_0 \to 0)$$

so that with help from the mass transfer coefficient from (1.179) the molar flux

$$\dot{n}_{A_0} = \beta \Delta c_A \qquad \text{(valid for } \dot{n}_0 \to 0) \tag{1.182}$$

can also be calculated.

In reality the concentration profile and the molar flux will differ from the values
calculated for $\dot{n}_0 \to 0$. As an example we will look at the scenario illustrated
in Fig. 1.45 where a porous body submerged in liquid A, for example water,
has a gas B e.g. an alcohol flowing over it. Liquid A evaporates in gas B, and
in reverse gas B mixes with liquid A in the porous body. The convective flux
$c_{A0}w_{A0} + c_{B0}w_{B0} = c_0u_0 = \dot{n}_0$ does not completely disappear at the surface of the
porous body, so that the molar flux of A at the surface is given by the familiar
equation

$$\dot{n}_{A0} = {}_u\dot{j}_{A0} + c_{A0}u_0 = {}_u\dot{j}_{A0} + \tilde{x}_{A0}\dot{n}_0 \ ,$$

where $c_{A0}u_0 = \tilde{x}_{A0}c_0u_0 = \tilde{x}_{A0}\dot{n}_0$.

The concentration profile is different as a result of the finite convective flux
when compared to the profile produced for vanishing convective flux $\dot{n}_0 \to 0$. The
greater the convective flux the larger the deviation in the concentration profile.
When the convection flux is in the direction of the wall, $\dot{n}_0 < 0$, such as when a
vapour is condensing out of a mixture, the concentration profile will be steeper
as a result of the convective flux, Fig. 1.47. In conjunction with this the material
transported out of the phase by diffusion will increase, whilst in the case of mass
transport into the phase, Fig. 1.46, the diffusional transport decreases. The mass
transfer coefficient for vanishing convection flux ($\dot{n}_0 \to 0$), defined in (1.180) is
therefore different from that with finite convection flux, and so it is valid that

$$\beta(\dot{n}_0 > 0) < \beta(\dot{n}_0 \to 0) \ ,$$

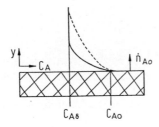
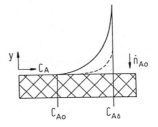

Fig. 1.46: Concentration profile $c_A(y)$ in mass transfer with a vanishing convective flow ($\dot{n}_0 \to 0$) —— and with finite convection flows ($\dot{n}_0 > 0$) - - -. Mass transfer into the (gaseous) phase

Fig. 1.47: Concentration profile $c_A(y)$ for vanishing convective flow ($\dot{n}_0 \to 0$) —— and for finite convective flows ($\dot{n}_0 < 0$) - - -. Mass transfer out of the (gaseous) phase

as can be seen recognised from Fig. 1.46, and

$$\beta(\dot{n}_0 < 0) > \beta(\dot{n}_0 \to 0) \ ,$$

as is shown in Fig. 1.47. As an abbreviation

$$\beta(\dot{n}_0 \neq 0) = \beta^{\bullet} \quad \text{and} \quad \beta(\dot{n}_0 = 0) = \beta$$

can be written. The superscript dot on mass transfer coefficients should indicate that the convection flow $\dot{n}_0$ is different from zero. Then the molar flux, according to (1.182) can be calculated to be

$$\dot{n}_{A0} = \beta^{\bullet}\Delta c_A + \tilde{x}_{A0}\dot{n}_0 \ . \tag{1.183}$$

At the limit $\dot{n}_0 \to 0$, of vanishing convective flow or by equimolar diffusion, we obtain $\beta^{\bullet} = \beta = \beta_c$, as shown by comparing (1.181) and (1.178).

The easiest of all the different mass transfer coefficients to investigate is the one for vanishing convective flows. It depends on the flow velocity, for example the speed at which gas B in Fig. 1.45 flows over the porous body, the physical properties of the gas and the geometric form of the porous body. These quantites can be combined,as will be shown later on, for certain geometries into dimensionless numbers,

$$\frac{\beta L}{D} = f\left(\frac{wL}{\nu}, \frac{\nu}{D}\right) \tag{1.184}$$

or

$$Sh = f(Re, Sc) \ . \tag{1.185}$$

The quantity $\beta L/D$, in which β is the mass transfer coefficient, L is a characteristic length, for example the the length of the plate that the gas is flowing over in Fig.

1.45, and D is the diffusion coefficient, is called the Sherwood[15] number Sh. ν/D is the Schmidt[16] Number Sc and wL/ν is the Reynolds number Re defined with the average velocity w and the kinematic viscosity ν of the gas. Relationships of the type shown in (1.185) are found either by experiments or in simple cases by solving the associated partial differential equations. This is the main subject of the theory of mass transfer yet to be discussed. The relationships like those in (1.185) for important practical geometries and flow regimes are available in the relevant literature [1.23] to [1.26]. As the convective flux is often small the mass transfer coefficient $\beta^{\bullet}$ can be calculated by multiplying β by a correction factor $\zeta = \beta^{\bullet}/\beta$, which in most practical applications is only a marginally different from 1.

1.5 Mass transfer theories

The calculation of the mass transfer coefficient can be carried out in different ways. Therefore a decision has to be made as to the type of problem, and which mass transfer theory is applicable to the solution of this problem, and therefore the determination of the mass transfer coefficient. The most important are the film, boundary layer and penetration theories. The essentials of these three theories will be introduced here.

1.5.1 Film theory

Film theory goes back to work by Lewis and Whitman [1.27] from 1924. In order to explain the principles we will assume that a substance A is transferred from

---

[15]Thomas Kilgore Sherwood (1903–1976) completed his PhD under the supervision of Warren K. Lewis, after whom the Lewis number was named, in 1929 at the Massachussetts Institute of Technology (MIT), Boston, USA. The subject of his thesis was 'The Mechanism of the Drying of Solids'. He was a professor at MIT from 1930 until 1969. His fundamental work on mass transfer in fluid flow and his book 'Absorption and Extraction' which appeared in 1937 made him famous worldwide.

[16]Ernst Schmidt (1892–1975) first studied civil engineering in Dresden and Munich, and then changed to electrical engineering. After working as an assistant to O. Knoblauch at the laboratory for applied physics at the Technische Hochschule in Munich he became a professor at the Technische Hochschule in Danzig (now Gdansk, Poland) in 1925. Following on from this, in 1937 he became the director of the Institute for Engine Research in the Aeronautics Research Establishment at Braunschweig, and later was made a professor at the Technische Hochschule in Braunschweig. In 1952 he took over from W. Nusselt the chair for Thermodynamics at the Technische Hochschule, Munich. His scientific works includ solutions of the unsteady heat conduction equation, the investigation of temperature fields in natural convection and methods to make the thermal boundary layer visible. He first used the number, which is now named after him, in a paper on the analogy between heat and mass transfer.

a quiescent solid or liquid surface, shown as a flat plate in Fig. 1.48, to flowing fluid B. The concentration of A drops from c_{A0} at the plate surface to $c_{A\delta}$ in the fluid. Film theory comes from the assumption that mass transfer takes place in a thin film of thickness δ near the wall, hence the name. Concentration and velocity should only change in the y direction, but not, as is further assumed, with time or in any other coordinate direction. In steady flow this results in a constant molar flux $\dot{n}_A = c_A w_A$ of A being transferred in the y direction. If this were not the case, more A would flow into a volume element of the fluid than out of it, and therefore the concentration of substance A would change with the time, or material A could also be flowing in the x direction which would therefore cause a concentration difference in another coordinate direction. However neither of these scenarios are admissable in terms of the prerequisites for the application of film theory. Therefore according to film theory

$$\frac{d\dot{n}_A}{dy} = 0 \ . \tag{1.186}$$

Then for vanishing convective flux $\dot{n} = \dot{n}_A + \dot{n}_B = 0$ in the y direction, because of

$$\dot{n}_A = \ _u\dot{j}_A + \tilde{x}_A\dot{n} = \ _u\dot{j}_A = -cD \ d\tilde{x}_A/dy$$

we also get

$$\frac{d^2\tilde{x}_A}{dy^2} = 0 \ ,$$

if constant values for cD are presupposed. The concentration profile in the film is a straight line

$$\frac{\tilde{x}_A - \tilde{x}_{A0}}{\tilde{x}_{A\delta} - \tilde{x}_{A0}} = \frac{y}{\delta} \ . \tag{1.187}$$

On the other hand for vanishing convection at the wall, $\dot{n}_0 \to 0$, according to (1.181)

$$\dot{n}_{A0} = \beta(c_{A0} - c_{A\delta}) = -cD \left(\frac{d\tilde{x}_A}{dy}\right)_0 \ . \tag{1.188}$$

Then from (1.187)

$$-cD \left(\frac{d\tilde{x}_A}{dy}\right)_0 = -c\frac{D}{\delta}(\tilde{x}_{A\delta} - \tilde{x}_{A0}) = \frac{D}{\delta}(c_{A0} - c_{A\delta})$$

and therefore

$$\beta = \frac{D}{\delta} \ . \tag{1.189}$$

As the film thickness δ is not normally known, the mass transfer coefficient β, cannot be calculated from this equation. However the values for the cases used most often in practice can be found from the relevant literature (i.e. [1.23] to [1.26]) which then allows the film thickness to be approximated using (1.189). In film theory the mass transfer coefficient β for vanishing convection flux $\dot{n}_0 \to 0$ is proportional to the diffusion coefficient D.

Fig. 1.48: Concentration profile at the surface over which a fluid is flowing.

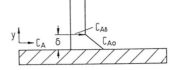

A different result is obtained when a finite convective flux $\dot{n}$ is permitted. As before

$$\frac{d\dot{n}_A}{dy} = 0 \ .$$

Then with the molar flux being

$$\dot{n}_A = {}_u \dot{j}_A + \tilde{x}_A \dot{n} = -cD d\tilde{x}_A/dy + \tilde{x}_A \dot{n}$$

we then, using the assumption that $cD = $ constant, get a second order differential equation

$$-cD \frac{d^2\tilde{x}_A}{dy^2} + \dot{n} \frac{d\tilde{x}_A}{dy} = 0 \ ,$$

which has to be solved under the boundary conditions $\tilde{x}_A(y = 0) = \tilde{x}_{A0}$ and $\tilde{x}_A(y = \delta) = \tilde{x}_{A\delta}$. We can then rearrange the equation into

$$\frac{d\ln\tilde{x}'_A}{dy} = \frac{\dot{n}}{cD}$$

with $\tilde{x}'_A = d\tilde{x}_A/dy$. Integration taking account of the boundary conditions gives the concentration profile $\tilde{x}_A(y)$ as

$$\frac{\tilde{x}_A - \tilde{x}_{A0}}{\tilde{x}_{A\delta} - \tilde{x}_{A0}} = \frac{\exp(\frac{\dot{n}}{cD}y) - 1}{\exp(\frac{\dot{n}}{cD}\delta) - 1} \ . \tag{1.190}$$

The concentration profile for vanishing convection $\dot{n} \to 0$ as given in (1.187) is obtained from (1.190). A Taylor series of the exponential function at $\dot{n} = 0$ can be developed to indicate this. The material flux transferred to the surface $y = 0$ (index 0 = wall) in the y-direction is

$$\dot{n}_A = -\left(cD \frac{d\tilde{x}_A}{dy}\right)_0 + \tilde{x}_{A0}\dot{n} \ . \tag{1.191}$$

According to film theory this is constant and equal to the value at the wall, where $\dot{n}_A = \dot{n}_{A0}$ and $\dot{n} = \dot{n}_0$. Differentiating (1.190) and introducing the result into (1.191) yields

$$\dot{n}_A = -(\tilde{x}_{A\delta} - \tilde{x}_{A0}) \frac{\dot{n}}{\exp(\frac{\dot{n}\delta}{cD}) - 1} + \tilde{x}_{A0}\dot{n} \ . \tag{1.192}$$

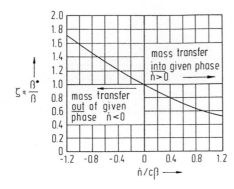

Fig. 1.49: Stefan correction factor $\zeta = \beta^\bullet/\beta$ from film theory

When this is compared to (1.183), we get

$$\beta^\bullet = \frac{\dot{n}/c}{\exp(\dfrac{\dot{n}\delta}{cD}) - 1} \; . \qquad (1.193)$$

Putting in the mass transfer coefficient $\beta = D/\delta$ for negligible convection from (1.189), and using the principles of film theory the following relationship between the mass transfer coefficients $\beta^\bullet$ and β, as shown in Fig. 1.49, can be found:

$$\frac{\beta^\bullet}{\beta} = \zeta = \frac{\dot{n}/c\beta}{\exp(\dfrac{\dot{n}}{c\beta}) - 1} = \frac{\dot{m}/\varrho\beta}{\exp(\dfrac{\dot{m}}{\varrho\beta}) - 1} \; . \qquad (1.194)$$

The factor ζ is known as the 'Stefan correction factor', [1.28]. In order to calculate the mass transferred using film theory, the mass transfer coefficient β has to be found. In cases where convection is negligible the mass transferred is calculated from equation (1.181), whilst where convection is significant the mass transferred is given by (1.183).

As a special case (1.192) is also used for the situation of *single side* mass transfer where $\dot{n}_A = \dot{n}$ and $\dot{n}_B = 0$. It then follows from (1.192) that

$$\frac{\tilde{x}_{A0} - \tilde{x}_{A\delta}}{1 - \tilde{x}_{A0}} = \exp\left(\frac{\dot{n}_A \delta}{cD}\right) - 1$$

or when solved for $\dot{n}_A$,

$$\dot{n}_A = \frac{cD}{\delta}\ln\frac{1 - \tilde{x}_{A\delta}}{1 - \tilde{x}_{A0}} = \frac{cD}{\delta}\ln\frac{\tilde{x}_{B\delta}}{\tilde{x}_{B0}} = c\beta\ln\frac{\tilde{x}_{B\delta}}{\tilde{x}_{B0}} \qquad (1.195)$$

it is in complete agreement with equation (1.172), which was found earlier.

In *drying a wet material* with air, the moisture content is calculated rather than the mole fraction. We will consider, as an example, a solid material, which contains water, that is to be dried by air. The equations and relationships already derived can be used: we give the water index A and the air index B. As the water content in moist air is very low, meaning $\ln(1 - \tilde{x}_A) \approx -\tilde{x}_A$, (1.195) can be also be written as

$$\dot{n}_A = c\beta(\tilde{x}_{A0} - \tilde{x}_{A\delta}) \ .$$

We then want to introduce the specific humidity or moisture content, which is defined by

$$X_A := M_A/M_B \ ,$$

where M_A is the mass of the water and M_B is the mass of the dry air, and therefore using the molar mass $\tilde{M} = M/N$ we can also write

$$X_A = \frac{\tilde{M}_A}{\tilde{M}_B}\frac{N_A}{N_B} = \frac{\tilde{M}_A}{\tilde{M}_B}\frac{\tilde{x}_A}{\tilde{x}_B} = \frac{\tilde{M}_A}{\tilde{M}_B}\frac{\tilde{x}_A}{1 - \tilde{x}_A} \ .$$

Solving for the mole fraction for water gives

$$\tilde{x}_A = \frac{X_A \tilde{M}_B/\tilde{M}_A}{1 + X_A \tilde{M}_B/\tilde{M}_A} \ .$$

The water content at ambient pressure is of the order $20 \cdot 10^{-3}$ kg/kg, which makes the term $X_A\,\tilde{M}_B/\tilde{M}_A \approx 0.03 \ll 1$. The approximation

$$\tilde{x}_A = X_A \tilde{M}_B/\tilde{M}_A$$

is then valid. The transferred molar flux of water will be

$$\dot{n}_A = c\beta\frac{\tilde{M}_B}{\tilde{M}_A}(X_{A0} - X_{A\delta})$$

with the mass flux as

$$\dot{m}_A = \dot{n}_A\tilde{M}_A = c\beta\tilde{M}_B(X_{A0} - X_{A\delta}) \ .$$

Using $c = p/R_m T$ and the gas constant for air $R_B = R_m/\tilde{M}_B$, we get

$$\dot{m}_A = \frac{p}{R_B T}\beta(X_{A0} - X_{A\delta}) \tag{1.195a}$$

for the mass flux of water.

1.5.2 Boundary layer theory

Boundary layer theory, just like film theory, is also based on the concept that mass transfer takes place in a thin film next to the wall as shown in Fig. 1.48. It differs from the film theory in that the concentration and velocity can vary not only in the y-direction but also along the other coordinate axes. However, as the change in the concentration profile in this thin film is larger in the y-direction than any of the other coordinates, it is sufficient to just consider diffusion in the direction of the y-axis. This simplifies the differential equations for the concentration significantly. The concentration profile is obtained as a result of this simplification, and from this the mass transfer coefficient β can be calculated according to the definition in (1.179). In practice it is normally enough to use the mean mass transfer coefficient

$$\beta_{\mathrm{m}} = \frac{1}{L} \int\limits_{x=0}^{L} \beta \, \mathrm{d}x \ .$$

This can be found, for forced flow from equations of the form

$$Sh_{\mathrm{m}} = f(Re, Sc)$$

with the mean Sherwood number $Sh_{\mathrm{m}} = \beta_{\mathrm{m}} L / D$. The function for the mean Sherwood number is practically identical to that for the average Nusselt number

$$Nu_{\mathrm{m}} = f(Re, Pr)$$

and is well known for surfaces over which the fluid flows. For example they can be taken from the [1.29], where the Nusselt number has to be replaced by the Sherwood number and the Prandtl number by the Schmidt number.

The determination of heat transfer coefficients with the assistance of dimensionless numbers has already been explained in section 1.1.4. This method can also be used for mass transfer, and as an example we will take the mean Nusselt number $Nu_{\mathrm{m}} = \alpha_{\mathrm{m}} L / \lambda$ in forced flow, which can be represented by an expression of the form

$$Nu_{\mathrm{m}} = c \, Re^n \, Pr^m \tag{1.196}$$

In (1.196) the quantities c, n, m still depend on the the type of flow, laminar or turbulent, and the shape of the surface or the channel over or through which the fluid flows. Correspondingly, the mean Sherwood number can be written as

$$Sh_{\mathrm{m}} = c \, Re^n \, Sc^m \ . \tag{1.197}$$

The average mass transfer coefficient can be calculated from (1.197). It can also be found by following the procedure outlined for average heat transfer coefficients, and dividing (1.197) by (1.196). This then gives a relationship between the heat and mass transfer coefficients,

$$\frac{Sh_{\mathrm{m}}}{Nu_{\mathrm{m}}} = \frac{\beta_{\mathrm{m}}}{D} \frac{\lambda}{\alpha_{\mathrm{m}}} \left(\frac{Sc}{Pr} \right)^m = \left(\frac{a}{D} \right)^m = Le^m$$

or

$$\beta_{\mathrm{m}} = \frac{\alpha_{\mathrm{m}}}{c_p \varrho} Le^{m-1} \ . \tag{1.198}$$

Here $m \approx 1/3$ and the dimensionless quantity $Le = a/D$ is called the *Lewis*[17] number, and for ideal gases is of the order of 1, so a general approximation

$$\beta_{\mathrm{m}} = \alpha_{\mathrm{m}} / \varrho c_p \tag{1.199}$$

Table 1.5: Lewis numbers of some gas mixtures at 0 °C and 1 bar

	in air	in hydrogen	in carbon dioxide
Water vapour	0.87	1.84	0.627
Carbon dioxide	1.33	2.34	—
Methanol	1.41	2.68	1.03
Ethanol	1.79	3.57	1.33
Benzene	2.40	4.24	1.74

is valid. Some values for the Lewis number are given in Table 1.5.

Equations (1.198) and (1.199) are also known as *Lewis' equations*. The mass transfer coefficents β_m calculated using this equation are only valid, according to the definition, for insignificant convective currents. In the event of convection being important they must be corrected. The correction factors $\zeta = \beta_m^\bullet/\beta_m$ for transverse flow over a plate, under the boundary layer theory assumptions are shown in Fig. 1.50. They are larger than those in film theory for a convective flow out of the phase, but smaller for a convective flow into the phase.

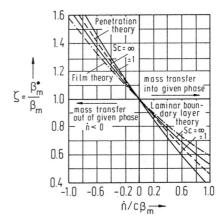

Fig. 1.50: Stefan correction $\zeta = \beta_m^\bullet/\beta_m$ for film, boundary layer and penetration theory, from [1.30]

1.5.3 Penetration and surface renewal theories

The film and boundary layer theories presuppose steady transport, and can therefore not be used in situations where material collects in a volume element, thus

---

[17]Warren Kendall Lewis (1882–1978) studied chemical engineering at the Massachussetts Institute of Technology (MIT) and gained his chemistry PhD in 1908 at the University of Breslau. Between 1910 and 1948 he was a professor at MIT. His research topics were filtration, distillation and absorption. In his paper "The evaporation of a liquid into a gas", Mech. Engineering 44 (1922) 445–448, he considered simultaneous heat and mass transfer during evaporation and showed how heat and mass transfer influence each other.

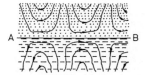

Fig. 1.51: Surface renewal theory. Possible flow patterns for contact between two liquids or between a liquid and a gas.

leading to a change in the concentration with time. In many mass transfer apparatus fluids come into contact with each other or with a solid material for such a short period of time that a steady state cannot be reached. When air bubbles, for example, rise in water, the water will only evaporate into the bubbles where it is contact with them. The contact time with water which surrounds the bubble is roughly the same as that required for the bubble to move one diameter further. Therefore at a certain position mass is transferred momentarily. The *penetration theory* was developed by Higbie in 1935 [1.31] for the scenario described here of momentary mass transfer. He showed that the mass transfer coefficient is inversely proportional to the square root of the contact (residence) time and is given by

$$\beta_\mathrm{m} = \frac{2}{\sqrt{\pi}} \sqrt{\frac{D}{t}} \; . \tag{1.200}$$

Here β_m is the mean mass transfer coefficient from time $t = 0$ to time t. Experience tells us that useful values for the mass transfer coefficient can be obtained when the contact time is calculated from $t = d/w$, where d is the diameter of the bubbles or droplets which are rising or falling, and w is the mean rise or fall velocity. It is more difficult to determine the contact time of a liquid falling through a packing material with a gas flowing through it.

Dankwerts' *surface renewal theory* from 1951 [1.32], represents an extension to penetration theory. Higbie always presupposed that the contact time between the phases was the same at all positions in the apparatus. Dankwerts went on to suggest that fluid elements which come into contact with each other, have different residence times which can be described by a residence time spectrum. One has to imagine that mass exchange between two different materials in the fluid phase takes place in individual fluid cells as indicated in Fig. 1.51. The contact time between the individual fluid elements obeys a distribution function, and after a certain amount of time an element can be dislodged from the the contact area and be replaced by another one. It is for this reason that we speak of a surface renewal theory. It has been successfully applied in the absorption of gases from agitated liquids. However the fraction of time for surface renewal is equally as unknown as the contact times in penetration theory, so while both theories are useful for the understanding of mass transfer processes, often neither is applicable for the calculation of the quantities involved in mass transfer.

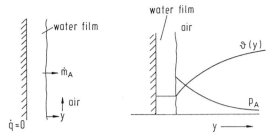

Fig. 1.52: Adiabatic evaporative cooling. Temperature ϑ and partial pressure p_A of the water in the air.

1.5.4   Application of film theory to evaporative cooling

We will look at evaporative cooling as an application of film theory. A solid, adiabatic, insulated wall is covered by a film of water, over which unsaturated humid air flows, as shown in Fig. 1.52.

The humid air takes water vapour from the film, by which the film of water and the air are cooled, until a time and position constant temperature is reached. It will be constant over the whole film because the adjoining wall is adiabatic and therefore no heat can be added to it. This adiabatic permanent temperature is called the *wet bulb temperature* The resistance to mass transfer is only on the gas side. Once the permanent temperature has been reached water still evaporates in the unsaturated air flowing over it. As the temperature of the water film is constant, the enthalpy of vaporisation required for the evaporation will be removed as heat from the air. Fig. 1.52 indicates how the temperature and partial pressure of the water vapour in the air changes at this permanent state. The wet bulb temperature is lower than the temperature of the humid air flowing over the water surface. Therefore a wet substance can be cooled down to its wet bulb temperature by evaporation.

We wish to find out the magnitude of the wet bulb temperature. It is determined by the amount of water transferred from the water surface into the humid air. As this is diffusion through a semipermeable plane the amount of water (substance A) being transferred to the air at the phase boundary I between the water and the air, according to (1.195) is given by

$$\dot{m}_A = \tilde{M}_A c\beta \ln\frac{1 - \tilde{x}_{A\delta}}{1 - \tilde{x}_{AI}} , \qquad (1.201)$$

when $\tilde{x}_{AI}$ is the mole fraction of water vapour in air at the surface of the water and $\tilde{x}_{A\delta}$ is the mole fraction at a large distance away from the water surface. As the amount of dry air (substance B) does not change it is useful to introduce the moisture content $X_A = M_A/M_B$. It is

$$\tilde{x}_A = \frac{N_A}{N} = \frac{N_A}{N_A + N_B} = \frac{X_A \tilde{M}_B/\tilde{M}_A}{1 + X_A \tilde{M}_B/\tilde{M}_A} .$$

Then (1.201) can also be written as

$$\dot{m}_A = \tilde{M}_A c\beta \ln\frac{1 + X_{AI}\tilde{M}_B/\tilde{M}_A}{1 + X_{A\delta}\tilde{M}_B/\tilde{M}_A} \ . \tag{1.202}$$

Here both the moisture content X_{AI} at the water surface and the mass flux $\dot{m}_A$ of the water being transferred are still unknowns. At the surface of the water saturation prevails and so the moisture content X_{AI}, as taught by thermodynamics [1.33], is given by

$$X_{AI} = 0{,}622\frac{p_s(\vartheta_I)}{p - p_s(\vartheta_I)} \tag{1.203}$$

where p_s is the saturation pressure of the water vapour at temperature ϑ_I at the surface.

The energy balance is available as a further equation. In order to set this up we will consider a balance region on the gas side, depicted by the dotted lines in Fig. 1.53, from the gas at the surface of the film to a point y. The energy balance

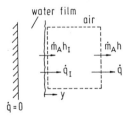

Fig. 1.53: Energy balance for evaporative cooling.

for a steady-state flow is

$$\dot{q}_I + \dot{m}_A h_I = \dot{q} + \dot{m}_A h = \text{const} \ .$$

In this equation the heat fluxes $\dot{q}_I$ and $\dot{q}$ are in the opposite direction to the y axis and therefore negative. From differentiation we get

$$\frac{d}{dy}(\dot{q} + \dot{m}_A h) = 0 \ .$$

This result and the introduction of $\dot{q} = -\lambda d\vartheta/dy$ and $dh = c_{pA}d\vartheta$ for the temperature pattern $\vartheta(y)$ produces the ordinary differential equation

$$-\lambda\frac{d^2\vartheta}{dy^2} + \dot{m}_A c_{pA}\frac{d\vartheta}{dy} = 0 \ , \tag{1.204}$$

which has to be solved for the boundary conditions

$$\vartheta(y = 0) = \vartheta_I \quad \text{and} \quad \dot{q}_I = -\lambda\left(\frac{d\vartheta}{dy}\right)_{y=0} \ ,$$

which gives

$$\vartheta - \vartheta_I = -\frac{\dot{q}_I}{\dot{m}_A c_{pA}} \left[\exp\left(\frac{\dot{m}_A c_{pA}}{\lambda} y\right) - 1 \right] \ . \tag{1.205}$$

The temperature ϑ_δ at a large distance $y = \delta$ away from the surface is therefore

$$\vartheta_\delta - \vartheta_I = -\frac{\dot{q}_I}{\dot{m}_A c_{pA}} \left[\exp\left(\frac{\dot{m}_A c_{pA}}{\lambda} \delta\right) - 1 \right] \ . \tag{1.206}$$

If the rate of evaporation is very small $\dot{m}_A \to 0$, heat will only be transferred by conduction perpendicular to the flow direction of the humid air. The heat flux at the water surface will then be

$$\dot{q}_I = \alpha(\vartheta_I - \vartheta_\delta) \qquad (\dot{m}_A \to 0) \ . \tag{1.207}$$

The heat transfer coefficient α is defined, just as in section 1.1.3, by this equation. On the other hand with the limit $\dot{m}_A \to 0$ we obtain, from (1.206):

$$\vartheta_\delta - \vartheta_I = -\frac{\dot{q}_I}{\dot{m}_A c_{pA}} \left[1 + \frac{\dot{m}_A c_{pA}}{\lambda} \delta + \ldots - 1 \right] \tag{1.208}$$

or

$$\dot{q}_I = \frac{\lambda}{\delta}(\vartheta_I - \vartheta_\delta) \qquad (\dot{m}_A \to 0) \ .$$

As the comparison with (1.207) shows, $\alpha = \lambda/\delta$. We put this into (1.206) and further note that the heat flow to the water surface is used for evaporation. It can then be said that

$$\dot{q}_I = -\dot{m}_A \Delta h_v \ ,$$

where Δh_v is the enthalpy of vaporisation of the water at temperature ϑ_I. Eq. (1.206) is then rearranged to form

$$\frac{\dot{m}_A c_{pA}}{\alpha} = \ln\left[1 + \frac{c_{pA}}{\Delta h_v}(\vartheta_\delta - \vartheta_I) \right] \ . \tag{1.209}$$

The quantites $\dot{m}_A$ and ϑ_I are unknowns, so by introducing the mass balance from eq. (1.202) we have a second equation which can be used to find these two unknowns. The mass flux, $\dot{m}_A$, can be eliminated from (1.209) with the the help of (1.202), leaving the following relationship for the wet bulb temperature ϑ_I

$$\vartheta_\delta - \vartheta_I = \frac{\Delta h_v}{c_{pA}} \left[\left(\frac{1 + X_{AI}\tilde{M}_B/\tilde{M}_A}{1 + X_{A\delta}\tilde{M}_B/\tilde{M}_A} \right)^{(\tilde{M}_A c\beta c_{pA}/\alpha)} - 1 \right] \ . \tag{1.210}$$

In cases of small evaporation rates $\dot{m}_A \to 0$ this can be simplified even more. We can then write (1.202) as

$$\frac{1 + X_{AI}\tilde{M}_B/\tilde{M}_A}{1 + X_{A\delta}\tilde{M}_B/\tilde{M}_A} = \exp\frac{\dot{m}_A}{\tilde{M}_A c\beta} = 1 + \frac{\dot{m}_A}{\tilde{M}_A c\beta} + \ldots \ .$$

Therefore when the evaporation rate is small, we get

$$\dot{m}_A = \tilde{M}_B c\beta \frac{X_{AI} - X_{A\delta}}{1 + X_{A\delta}\tilde{M}_B/\tilde{M}_A} \quad .$$

With this, equation (1.208) and $\dot{q}_I = -\dot{m}_A\Delta h_v$, we get the following relationship for small evaporation rates $\dot{m}_A \to 0$

$$\vartheta_\delta - \vartheta_I = \tilde{M}_B \frac{c\beta}{\alpha} \Delta h_v \frac{X_{AI} - X_{A\delta}}{1 + X_{A\delta}\tilde{M}_B/\tilde{M}_A} \quad . \qquad (1.211)$$

The wet bulb temperature ϑ_I, can be obtained from (1.210) or (1.211), and then the moisture content X_{AI}, which is determined by the temperature ϑ_I , can be found by using (1.203).

In reverse, by measuring the wet bulb temperature the moisture content $X_{A\delta}$ of the humid air can be calculated using the two equations. The wet bulb hygrometer, sometimes called *Assmann's Aspiration Psychrometer* works by this method. It consists of two thermometers, one of which is covered by a porous textile, the lower part of which is dipped into a container full of water, so that the bulb of the thermometer is always wet. Air is blown over the thermometer and after some time it reaches the wet bulb temperature ϑ_I. Meanwhile the second, so called dry thermometer, indicates the temperature ϑ_δ of the air flowing around it. With these measured values the moisture content $X_{A\delta}$ of the humid air can be found using equation (1.210) or (1.211).

Example 1.8: The dry thermometer of a wet bulb hygrometer shows a temperature ϑ_δ of 30 °C, whilst the wet thermometer indicates a temperature $\vartheta_I = 15$ °C. The total pressure is $p = 1000$ mbar. The moisture content $X_{A\delta}$ and the relative humidity of the air have to be calculated. For humid air, Lewis' equation (1.198) yields a value of $\tilde{M}_A c\beta_m c_{pA}/\alpha_m = 1.30$, the enthalpy of vaporisation of water at 15 °C is $\Delta h_v = 2466.1$ kJ/kg, the specific heat capacity $c_{pA} = 1.907$ kJ/kgK and the molar mass of water is $\tilde{M}_A = 18.015$ kg/kmol , that of the dry air is $\tilde{M}_B = 28.953$ kg/kmol.

According to (1.203) with $p_S(15$ °C$) = 17.039$ mbar, $X_{AI} = 0.622 \cdot 17.039/(1000 - 17.039) = 1.078 \cdot 10^{-2}$. It then follows from (1.210) that

$$(30 - 15)°C = \frac{2466.1 \cdot 10^3}{1907} \left[\left(\frac{1 + 1.078 \cdot 10^{-2} \cdot 28.953/18.015}{1 + X_{A\delta} \cdot 28.953/18.015} \right)^{1.3} - 1 \right] °C \quad .$$

This gives $X_{A\delta} = 5.189 \cdot 10^{-3}$. Almost exactly the same value is obtained using (1.211), where $X_{A\delta} = 5.182 \cdot 10^{-3}$. The relative humidity is

$$\varphi = \frac{p_A}{p_{As}} = \frac{X_{A\delta}}{0.622 + X_{A\delta}} \cdot \frac{p}{p_s(30 \text{ °C})} = \frac{5.189 \cdot 10^{-3}}{0.622 + 5.189 \cdot 10^{-3}} \frac{1000}{42.41} = 0.195 = 19.5\% \quad .$$

(This value can also be found from a Mollier's h_{1+X}, X-diagram).

1.6 Overall mass transfer

Whilst heat is often transferred from one fluid to another via a solid wall between them, in mass transfer one or more components in a phase are transferred to

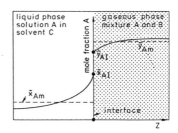

Fig. 1.54: Mole fractions during mass transfer from a gas to a liquid phase

another that is in direct contact with it, and not separated by a solid wall. In fluid flow, the phase boundary has an irregular shape due to the forces acting at the interface between the two phases. As an example we will study the transfer of component A, from a binary gaseous mixture of A and B, into a liquid C in which only A dissolves. The mole fractions of A in the gas phase $\tilde{y}_A$ and in the liquid phase $\tilde{x}_A$ are shown in Fig. 1.54. The integral mean values for the mole fractions in each phase are indicated by $\tilde{y}_{Am}$ and $\tilde{x}_{Am}$. Component A has to overcome three mass transfer resistances on its way from the gas to the liquid phase: the resistance in the gas phase itself, that at the interface between the two phases and finally the resistance within the liquid phase. However, mass transfer between the liquid and gas interface is very fast when compared to the much slower speed of the component within either the gas or liquid phases. The resistance at the interface can therefore be neglected, and it can also be assumed that an equilibrium state in terms of mass transfer exists at the interface. As mass transfer is determined by diffusion in the two phases, one speaks of a *two film theory of mass transfer*. Its fundamental assumption of equilibrium at the interface is no longer valid if a chemical reaction is taking place there or if traces of a surfactant have collected at the interface. In addition to this, equilibrium at the interface cannot be presumed if very large mass flow rates are to be transferred, as here the mass exchange in the phases is very quick.

In the case of equilibrium at the phase boundary, which we want to discuss here, with given values for pressure and temperature, the mole fractions $\tilde{y}_{AI}$ and $\tilde{x}_{AI}$ at the phase boundary are linked by the relationship

$$p_{AI} = \tilde{y}_{AI}p = f(\tilde{x}_{AI}) \qquad (1.212)$$

as taught in equilibrium thermodynamics. In the following it will be presumed that, close to the phase boundary, mass transfer is only due to diffusion and the convective flows are very small. This allows the use of eq. (1.181), for mass transfer at the phase boundary, for each of the two phases. Taking into account the relationship between the mole fraction and the molar concentration, $(c_A/c)_G = \tilde{y}_A$ and $(c_A/c)_L = \tilde{x}_A$ (G = gas, L = liquid), it follows that the mass flux in the gas phase is

$$\dot{n}_{AI} = (\beta c)_G(\tilde{y}_{Am} - \tilde{y}_{AI}) \qquad (1.213)$$

and in the liquid phase

$$\dot{n}_{AI} = (\beta c)_L(\tilde{x}_{AI} - \tilde{x}_{Am}) \ . \qquad (1.214)$$

Which then gives

$$\tilde{y}_{Am} - \tilde{y}_{AI} = \frac{\dot{n}_{AI}}{(\beta c)_G} \tag{1.215}$$

and

$$\tilde{x}_{AI} - \tilde{x}_{Am} = \frac{\dot{n}_{AI}}{(\beta c)_L} \;. \tag{1.216}$$

The mole fractions $\tilde{y}_{AI}$ and $\tilde{x}_{AI}$ at the interface are dependent on each other according to the thermodynamic relationship given in (1.212). As gases are only slightly soluble in liquids the mole fraction $\tilde{x}_{AI}$ is normally very small. Then because of *Henry's*[18] *law* eq. (1.212) can be approximated by the linear relationship

$$\tilde{y}_{AI} = k_H \tilde{x}_{AI}/p \tag{1.217}$$

in which the Henry coefficient for binary mixtures is only dependent on the temperature, $k_H(\vartheta)$. The mole fraction $\tilde{x}_{Am}$ in the liquid can, with the help of Henry's Law, be related to a mole fraction $\tilde{y}_{Aeq}$ in the gas phase, such that,

$$\tilde{y}_{Aeq} = k_H \tilde{x}_{Am}/p \;, \tag{1.218}$$

where $\tilde{y}_{Aeq}$ is the mole fraction of A in the gas which is in equilibrium with the liquid mole fraction $\tilde{x}_{Am}$. With (1.217) and (1.218), (1.216) is transformed into

$$\tilde{y}_{AI} - \tilde{y}_{Aeq} = \frac{\dot{n}_{AI} k_H}{(\beta c)_L p} \;. \tag{1.219}$$

Adding (1.215) and (1.219) gives

$$\tilde{y}_{Am} - \tilde{y}_{Aeq} = \dot{n}_{AI} \left[\frac{1}{(\beta c)_G} + \frac{k_H}{(\beta c)_L p} \right] \;.$$

This will then be written as

$$\dot{n}_{AI} = k_G(\tilde{y}_{Am} - \tilde{y}_{Aeq}) \;, \tag{1.220}$$

with the gas phase overall mass transfer coefficient

$$\frac{1}{k_G} = \frac{1}{(\beta c)_G} + \frac{k_H}{(\beta c)_L p} \;. \tag{1.221}$$

Eliminating the mole fraction $\tilde{y}_{Am}$ from (1.215) by $\tilde{y}_{Am} = k_H \tilde{x}_{Aeq}$ and the mole fraction $\tilde{y}_{AI}$ by (1.217), gives, with (1.216), a equivalent relationship equivalent to eq. (1.220)

$$\dot{n}_{AI} = k_L(\tilde{x}_{Aeq} - \tilde{x}_{Am}) \tag{1.222}$$

with the liquid phase overall mass transfer coefficient

$$\frac{1}{k_L} = \frac{p}{k_H(\beta c)_G} + \frac{1}{(\beta c)_L} \;. \tag{1.223}$$

---

[18]Named after William Henry (1775–1836), a factory owner from Manchester, who first put forward this law in 1803.

The resistance to mass transfer according to (1.221) and (1.223) is made up of the the individual resistances of the gas and liquid phases. Both equations show how the resistance is distributed among the phases. This can be used to decide whether one of the resistances in comparison to the others can be neglected, so that it is only necessary to investigate mass transfer in one of the phases. Overall mass transfer coefficients can only be developed from the mass transfer coefficients if the phase equilibrium can be described by a linear function of the type shown in eq. (1.217). This is normally only relevant to processes of absorption of gases by liquids, because the solubility of gases in liquids is generally low and can be described by Henry's Law (1.217). So called ideal liquid mixtures can also be described by the linear expression, known as Raoult's Law. However these seldom appear in practice. As a result of all this, the calculation of overall mass transfer coefficients in mass transfer play a far smaller role than their equivalent overall heat transfer coefficients in the study of heat transfer.

1.7 Mass transfer apparatus

In mass transfer apparatus one of two processes can take place. Multicomponent mixtures can either be separated into their individual substances or in reverse can be produced from these individual components. This happens in mass transfer apparatus by bringing the components into contact with each other and using the different solubilities of the individual components in the phases to separate or bind them together. An example, which we have already discussed, was the transfer of a component from a liquid mixture into a gas by evaporation. In the following section we will limit ourselves to mass transfer devices in which physical processes take place. Apparatus where a chemical reaction also influences the mass transfer will be discussed in section 2.5. Mass will be transferred between two phases which are in direct contact with each other and are not separated by a membrane which is only permeable for certain components. The individual phases will mostly flow countercurrent to each other, in order to get the best mass transfer. The separation processes most frequently implemented are absorption, extraction and rectification.

In *absorption* one or more components from a gaseous mixture are absorbed by either a liquid or a solid. In *extraction* individual components of a mixture of solids or liquids are dissolved in another liquid. Finally, *rectification* is used to separate a liquid mixture into individual components or mixtures, so-called fractions. In this case the liquid flows down the column where it is in contact with the vapour rising up the column and individual components are exchanged between the vapour and liquid.

Mass transfer apparatus are normally columns where one phase, generally the gas is introduced at the base of the column, whilst the other phase, usually a liquid, is fed in at the top. In order to intensify the mass transfer the internal surface area of the column should be as large as possible. This can be done by collecting the liquid on internals in the column and bubbling the gas through it. Distillation trays are such internals. A distillation tray is a plate inside the column that holds a pool of the downflowing liquid, usually a few centimetres deep, through which the upwards flowing vapour passes. An alternative is the addition of packing in the column which also hold a pool of the downflowing liquid which the rising vapour passes through, thereby encouraging good mixing between the two phases. The most frequently used packing include spheres, rings with flat or grooved surfaces, cross-partition rings, saddle packings, wire gauze or wire helices. Information on the different designs for mass transfer devices can be found in the literature [1.34]. The various types of mass transfer apparatus will not be discussed in this book as we are only interested in determining the size of the apparatus needed for a given separation process.

1.7.1 Material balances

A material balance is always required to determine the size of mass transfer apparatus, irrespective of the design. So as the energy balance links the temperatures of the fluid flows in a heat exchanger, the mass balance delivers the concentrations of the fluids.

As a typical example of an application we will look at an absorption column, Fig. 1.55, in which component A in a gas mixture flowing up the column is dissolved in a liquid flowing countercurrent to it.

We will presume that contact between two immiscible phases occurs, so that the following balance equations are valid, independent of whether a packed, falling film or plate column is being investigated. The raw gas consists of a carrier gas G and the component A which will be absorbed. The liquid is made up of the solvent L and the absorbed component A. The molar flow rate, in a cross section, of the rising gas is therefore $\dot{N}_G + \dot{N}_{GA}$, and that of the liquid flowing downwards

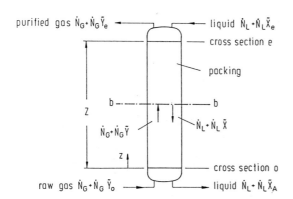

Fig. 1.55: Countercurrent absorber

is $\dot{N}_L + \dot{N}_{LA}$. As the molar flow rates $\dot{N}_G$ and $\dot{N}_L$ do not change through the column the composition of the streams can be described by the mole ratios of the component to the solvent or the carrier gas

$$\tilde{X} = \dot{N}_{LA}/\dot{N}_L \text{ and } \tilde{Y} = \dot{N}_{GA}/\dot{N}_G ,$$

where $\tilde{X}$ is the average mole ratio of component A in the liquid and $\tilde{Y}$ is the average mole ratio of A in the gas phase. The material balance over the total column between the inlet i and the outlet (exit) e of the liquid is

$$\dot{N}_L\tilde{X}_i - \dot{N}_G\tilde{Y}_i - \dot{N}_L\tilde{X}_e + \dot{N}_G\tilde{Y}_e = 0$$

or

$$\frac{\dot{N}_L}{\dot{N}_G} = \frac{\tilde{Y}_e - \tilde{Y}_i}{\tilde{X}_e - \tilde{X}_i} . \tag{1.224}$$

The changes in the gas and liquid compositions are determined by the ratio of the molar flow rates of the gas and liquid streams. The corresponding material balance for a control volume between stage e and any other stage b inside the column is,

$$\dot{N}_L\tilde{X}_i - \dot{N}_G\tilde{Y}_i - \dot{N}_L\tilde{X} + \dot{N}_G\tilde{Y} = 0$$

or

$$\tilde{Y} = \frac{\dot{N}_L}{\dot{N}_G}\tilde{X} + \left(\tilde{Y}_i - \frac{\dot{N}_L}{\dot{N}_G}\tilde{X}_i\right) . \tag{1.225}$$

Therefore at any stage in the column the average liquid mole ratio $\tilde{X}$ has an associated value for the average gas mole ratio $\tilde{Y}$. Equation (1.225) is represented by a straight line in a $\tilde{Y}, \tilde{X}$-diagram.

In a similar manner to that for the absorption column a linear relationship between the compositions of the two phases can be found for extraction and rectification. To illustrate this we will look at a rectification column. The basic process of rectification is when boiling a multicomponent mixture the vapour generated flows upwards countercurrent to the condensate which falls down the column. As the condensate is colder than the vapour, the components with higher boiling points, the least volatile, condense. They release their enthalpy of condensation to the components with the lower boiling points, the so called more volatile components, which are vaporised. This causes the vapour to become rich in the more volatile components while the less volatile components make up the liquid. The basic concept of a rectification column is shown in Fig. 1.56. The liquid mixture which is to be separated, called the *feed*, flows into a still or boiler V with a molar flow rate $\dot{N}_F$. In the boiler, heat $\dot{Q}_V$ is added to generate vapour which then flows towards the top of the column. After the vapour leaves the top of the column it is fed to a condensor C where heat $\dot{Q}_C$ is removed and hence the vapour is totally condensed. Part of the condensate is fed back into the top of the column as reflux and flows countercurrent back down the column, all the while undergoing heat

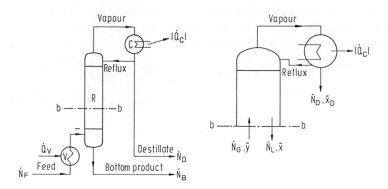

Fig. 1.56: Continuous rectification. Rectifying column R, condenser C and boiler V

and mass transfer with the rising vapour. The rest of the condensate, known as the *distillate*, is taken away as the product stream $\dot{N}_\mathrm{D}$. To ensure that the composition in the still is constant a further stream of flowrate $\dot{N}_\mathrm{B}$ must constantly be removed from the bottom of the column. This stream is called the bottom product. The material balance over the whole column is

$$\dot{N}_\mathrm{F} = \dot{N}_\mathrm{D} + \dot{N}_\mathrm{B}$$
$$\dot{N}_\mathrm{F}\tilde{x}_\mathrm{F} = \dot{N}_\mathrm{D}\tilde{x}_\mathrm{D} + \dot{N}_\mathrm{B}\tilde{x}_\mathrm{B} \ .$$

From which follows

$$\frac{\dot{N}_\mathrm{D}}{\dot{N}_\mathrm{F}} = \frac{\tilde{x}_\mathrm{F} - \tilde{x}_\mathrm{B}}{\tilde{x}_\mathrm{D} - \tilde{x}_\mathrm{B}} \ . \tag{1.226}$$

The mole fractions of the feed, bottom product and the distillate determine the ratio of feed to product flow rates. In a control volume from the top of the column to any cross section b, as shown in the right hand picture in Fig. 1.56, the material balance is as follows, when the mole fraction of the volatile components in the vapour is represented by $\tilde{y}$, and the fraction in the liquid is indicated by $\tilde{x}$,

$$\dot{N}_\mathrm{G} = \dot{N}_\mathrm{L} + \dot{N}_\mathrm{D}$$

and

$$\dot{N}_\mathrm{G}\tilde{y} = \dot{N}_\mathrm{L}\tilde{x} + \dot{N}_\mathrm{D}\tilde{x}_\mathrm{D} \ .$$

As shown previously both equations yield a linear relationship between $\tilde{y}$ and $\tilde{x}$:

$$\tilde{y} = \frac{\dot{N}_\mathrm{L}}{\dot{N}_\mathrm{L} + \dot{N}_\mathrm{D}}\tilde{x} + \frac{\dot{N}_\mathrm{D}}{\dot{N}_\mathrm{L} + \dot{N}_\mathrm{D}}\tilde{x}_\mathrm{D} \ . \tag{1.227}$$

1.7.2 Concentration profiles and heights of mass transfer columns

The global mass balance allows the average composition in a phase to be associated with that of the other phase. However, the local composition changes are still not known. In order to find this out a similar balance to that around a differential element in a heat exchanger, in this case a mass balance must be completed. Integration then gives the composition pattern along the whole exchanger. Once the concentration changes in the column are determined, the height of the column for a required outlet concentration of one of the components, can be calculated. The procedure for completing this task will be shown using, once again, the examples of a absorber and a packed rectification column.

This type of calculation does not have to be carried out for a plate column because the two phases are well mixed on each plate. This means that on each individual plate a state of equilibrium can be presumed. Therefore a volume element is identical to an equilibrium stage, and the height of the column can be obtained from the number of equilibrium stages required for a particular separation. This is a thermodynamic rather then mass transfer problem. This explains why a mass transfer device, such as a distillation column can be sized without any knowledge of the laws of mass transfer.

A packed column is filled with a packing material. The complex shape of the packing makes it difficult to ascertain the area of the phase interface where mass transfer takes place. Therefore only the product of the mass transfer coefficients and the interface area will be determined. The area of the interface A_I is related to the volume of the empty column V_K, and the interface area per volume a^* is defined by

$$a^* := A_I/V_K$$

with the volume $V_K = A_K Z$, where A_K is the cross sectional area of the empty column and Z is its height.

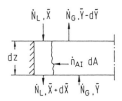

Fig. 1.57: Mass transfer in an absorber

We will first consider a control volume between two cross sections in an absorber, separated by a distance dz, Fig. 1.57. The amount of material transferred to the liquid over the interface area dA_I

$$\dot{n}_{AI} dA_I = \dot{n}_{AI} a^* A_K dz$$

is given up by the gas which effectively reduces the amount of component A in the gas phase by $\dot{N}_G d\tilde{Y}$. Simultaneously the amount of component A in the liquid increases by $-\dot{N}_G d\tilde{Y} = \dot{N}_L d\tilde{X}$. It is therefore valid that

$$\dot{n}_{AI} a^* A_K dz = -\dot{N}_G d\tilde{Y} \ . \tag{1.228}$$

The molar flux $\dot{n}_{AI}$ is given by (1.220), under the assumption that the amount of component A in the liquid and vapour phases is small, and so the values for the mole fractions $\tilde{y}_{Am} = \tilde{Y}$ and $\tilde{y}_{Aeq} = \tilde{Y}_{eq}$ can be replaced by mole ratios. It then follows from (1.228) that

$$k_G(\tilde{Y} - \tilde{Y}_{eq}) a^* A_K dz = -\dot{N}_G d\tilde{Y}$$

and

$$dz = \frac{-\dot{N}_G}{(k_G a^*) A_K} \frac{d\tilde{Y}}{\tilde{Y} - \tilde{Y}_{eq}} \tag{1.229}$$

which is comparable to the relationship for heat transfer in countercurrent flow, eq. (1.112). When integrating, the change in the mole ratio $\tilde{Y}_{eq}$ at the phase boundary along the length of the column, and therefore that of the average mole ratio $\tilde{Y}$ must be considered. The balance equation (1.225) gives

$$\tilde{X} = \frac{\dot{N}_G}{\dot{N}_L}(\tilde{Y} - \tilde{Y}_i) + \tilde{X}_i \ .$$

Then using Henry's Law, $\tilde{Y}_{eq} = k_H^+ \tilde{X}$ mit $k_H^+ = k_H/p$ it follows that

$$\tilde{Y}_{eq} = k_H^+ \frac{\dot{N}_G}{\dot{N}_L}(\tilde{Y} - \tilde{Y}_i) + k_H^+ \tilde{X}_i \ . \tag{1.230}$$

This allows $\tilde{Y} - \tilde{Y}_{eq}$ to be expressed as a linear function of the mole ratio $\tilde{Y}$ such that

$$\tilde{Y} - \tilde{Y}_{eq} = \left(1 - k_H^+ \frac{\dot{N}_G}{\dot{N}_L}\right) \tilde{Y} - k_H^+ \left(\tilde{X}_i - \frac{\dot{N}_G}{\dot{N}_L}\tilde{Y}_i\right) = a\tilde{Y} - b \ . \tag{1.231}$$

Equation (1.229) can then be transformed into

$$dz = \frac{-\dot{N}_G}{(k_G a^*) A_K} \frac{d\tilde{Y}}{a\tilde{Y} - b} \ . \tag{1.232}$$

The variation of the mole ratio $\tilde{Y}(z)$ over the height of the column can be found by integration. Using (1.225), the equation for the straight line from the material balance, the value for $\tilde{X}$, the mole ratio in the liquid, which corresponds to each $\tilde{Y}$, the mole ratio in the gas, can be calculated. It then follows from (1.232) that

$$z = \frac{-\dot{N}_G}{(k_G a^*) A_K} \int_{\tilde{Y}_e}^{\tilde{Y}} \frac{d\tilde{Y}}{a\tilde{Y} - b} = \frac{-\dot{N}_G}{(k_G a^*) A_K a} \ln \frac{a\tilde{Y} - b}{a\tilde{Y}_e - b}$$

or with (1.231) we get

$$z = \frac{\dot{N}_{\mathrm{G}}}{(k_{\mathrm{G}}a^*)A_{\mathrm{K}}\left(1 - k_{\mathrm{H}}^+\dot{N}_{\mathrm{G}}/\dot{N}_{\mathrm{L}}\right)}\ln\frac{(\tilde{Y} - \tilde{Y}_{\mathrm{eq}})_{\mathrm{o}}}{\tilde{Y} - \tilde{Y}_{\mathrm{eq}}} \ . \tag{1.233}$$

The necessary column height Z can be found from this equation, when proceeding to the liquid inlet cross section i at the top of the column

$$Z = \frac{\dot{N}_{\mathrm{G}}}{(k_{\mathrm{G}}a^*)A_{\mathrm{K}}(1 - k_{\mathrm{H}}^+\dot{N}_{\mathrm{G}}/\dot{N}_{\mathrm{L}})}\ln\frac{(\tilde{Y} - \tilde{Y}_{\mathrm{eq}})_{\mathrm{e}}}{(\tilde{Y} - \tilde{Y}_{\mathrm{eq}})_{\mathrm{i}}} \ . \tag{1.234}$$

The meaning of the mole ratio variations $(\tilde{Y} - \tilde{Y}_{\mathrm{eq}})_{\mathrm{e}}$ and $(\tilde{Y} - \tilde{Y}_{\mathrm{eq}})_{\mathrm{i}}$ are shown in Fig. 1.58.

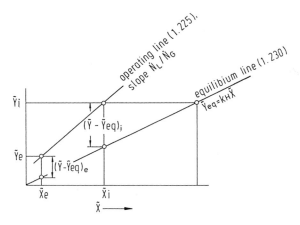

Fig. 1.58: Equilibrium and balance lines for an absorber.

The material balance from (1.228) is once again the starting point for investigating the concentration profile in a packed column being used to rectify a gaseous mixture. In this case the mole fractions will be used instead of the mole ratios. The amount tranferred from the gas to the liquid phase is $-\dot{N}_{\mathrm{G}}\mathrm{d}\tilde{y}$, wherein $\dot{N}_{\mathrm{G}}$ is now the molar flow rate of the gas mixture rather than that of the carrier gas as in (1.228). Equation (1.228) is therefore replaced by

$$\dot{n}_{\mathrm{AI}}a^*A_{\mathrm{K}}\mathrm{d}z = -\dot{N}_{\mathrm{G}}\mathrm{d}\tilde{y} \ . \tag{1.235}$$

Taking the molar flux from equation (1.220), in which we put the abbreviations $\tilde{y}_{\mathrm{Am}} = \tilde{y}$ and $\tilde{y}_{\mathrm{Aeq}} = \tilde{y}_{\mathrm{eq}}$, we get

$$\mathrm{d}z = \frac{-\dot{N}_{\mathrm{G}}}{(k_{\mathrm{G}}a^*)A_{\mathrm{K}}}\frac{\mathrm{d}\tilde{y}}{\tilde{y} - \tilde{y}_{\mathrm{eq}}} \ , \tag{1.236}$$

which corresponds to equation (1.229). Integration yields the concentration profile $\tilde{y}(z)$ over the height z to be

$$z = \frac{-\dot{N}_G}{(ka^*)A_K} \int_{\tilde{y}_e}^{\tilde{y}} \frac{d\tilde{y}}{\tilde{y} - \tilde{y}_{eq}} \, . \tag{1.237}$$

The required height Z for the column can be calculated by integrating to the point $\tilde{y} = \tilde{y}_e$. The evaluation of this integral is often only possible numerically as the mole fraction $\tilde{y}_{eq}$ is normally a complex function of the mole fraction $\tilde{x}$ of the liquid, and therefore according to the balance equation (1.227) is still dependent on the mole fractiom $\tilde{y}$. How the mole fractions at equilibrium are ascertained is dealt with in the thermodynamics of phase equilibria.

Example 1.9: A washing solution consisting of a mixture of high boiling hydrocarbons is fed into the top of a packed column. It is envisaged that it will remove benzene contained in the air that is rising up the column. The molar content of benzene at the base of the column (cross section a) is 3 %, and by the time the air reaches the top of the column (cross section e) 90 % of its benzene content must have been removed. The washing solution is fed into the column with a molar benzene ratio of 0,3 %.

Raoult's law $\tilde{y}_A = (p_{As}/p)\tilde{x}_A$ is valid for the solubility of benzene (substance A) in the washing solution, with $p_{As}(\vartheta)$ as the saturation pressure of pure benzene, which at the prevailing temperature of 30 °C, has a value of $p_{As} = 159.1\,\mathrm{mbar}$. The total pressure is $p = 1\,\mathrm{bar}$. The benzene free washing solution is fed to the column at a rate of $\dot{N}_L = 2.5\,\mathrm{mol/s}$, with a flow rate of $\dot{N}_G = 7.5\,\mathrm{mol/s}$ for the benzene free air. The inner diameter of the column is $d_K = 0.5\,\mathrm{m}$. The mass transfer coefficients are $(\beta c)_G a^* = 139.3\,\mathrm{mol/m^3s}$ and $(\beta c)_L a^* = 3.139\ \mathrm{mol/m^3s}$.

How high does the column have to be to meet the desired air purity conditions?

As the benzene content is very low, in Raoult's Law the mole fractions can be replaced by the mole ratios and can be written as $\tilde{Y}_A = k_H^+ \tilde{X}_A$, with $k_H^+ = p_{As}/p = 0.1591$. According to equation (1.221) we get

$$\frac{1}{k_G\,a^*} = \frac{1}{(\beta c)_G\,a^*} + \frac{k_H^+}{(\beta c)_L\,a^*} = \left(\frac{1}{139.3} + \frac{0.1591}{3.139}\right) \frac{\mathrm{m^3\,s}}{\mathrm{mol}}$$

$k_G\,a^* = 17.28\,\mathrm{mol/(m^3 s)}$. The benzene content of the washing solution leaving the column, from (1.224) is

$$\tilde{X}_e = \frac{\dot{N}_G}{\dot{N}_L}(\tilde{Y}_e - \tilde{Y}_i) + \tilde{X}_i = \frac{7.5}{2.5}(0.03 - 0.003) + 0.003 = 0.084.$$

Furthermore

$$\tilde{Y}_{eqi} = k_H^+ \tilde{X}_i = 0.1591 \cdot 0.003 = 4.773 \cdot 10^{-3}$$

$$\tilde{Y}_{eqe} = k_H^+ \tilde{X}_e = 0.1591 \cdot 0.084 = 1.336 \cdot 10^{-2} \, .$$

The required column height according to (1.234)is calculated to be

$$Z = \frac{7.5}{17.28 \cdot 0.196\,(1 - 0.1591 \cdot 3)} \ln \frac{0.03 - 0.001336}{0.03 - 0.000473}\,\mathrm{m} = 10.28\,\mathrm{m}.$$

1.8 Exercises

1.1: The outer wall of a room is made of brickwork with $\lambda = 0.75\,\text{W/K m}$. It has a thickness of $\delta = 0.36\,\text{m}$ and a surface area of $A = 15.0\,\text{m}^2$. Its surface temperatures are $\vartheta_1 = 18.0\,°\text{C}$ and $\vartheta_2 = 2.5\,°\text{C}$. The heat loss $\dot{Q}$ through the wall is to be calculated. How would the value for $\dot{Q}$ change if the wall was made of concrete blocks with $\lambda = 0.29\,\text{W/K m}$ and a wall thickness of $\delta = 0.25$?

1.2: The thermal conductivity is linearly dependent on the temperature, $\lambda = a + b\,\vartheta$. Prove equation (1.12) for the mean thermal conductivity λ_m.

1.3: The steady temperature profile $\vartheta = \vartheta(x)$ in a flat wall has a second derivative $\text{d}^2\vartheta/\text{d}x^2 > 0$. Does the thermal conductivity $\lambda = \lambda(\vartheta)$ of the wall material increase or decrease with rising temperature?

1.4: A copper wire with diameter $d = 1.4\,\text{mm}$ and specific electrical resistance of $r_\text{el} = 0.020 \cdot 10^{-6}\,\Omega\,\text{m}$ is surrounded by a plastic insulation of thickness $\delta = 1.0\,\text{mm}$, which has a thermal conductivity of $\lambda = 0.15\,\text{W/K m}$. The outer surface of the insulation will be maintained at a temprature of $\vartheta_\text{W2} = 20\,°\text{C}$. What is the largest current which can be applied to the wire if the inner temperature of the insulation is not allowed to rise above $30\,°\text{C}$?

1.5: A heat flow $\dot{Q} = 17.5\,\text{W}$ is generated inside a hollow sphere of dimensions $d_1 = 0.15\,\text{m}$, $d_2 = 0.25\,\text{m}$ and thermal conductivity $\lambda = 0.68\,\text{W/K m}$. The outer surface of the sphere has a temperature of $\vartheta_\text{W2} = 28\,°\text{C}$. What will the temperature of the inner surface be?

1.6: A flat body of thickness h and thermal conductivity λ has the shape of a right angled triangle whose short sides are of length l. The plane temperature profile of this body is given as $\vartheta(x, y) = \vartheta_0 + \vartheta_1 \left[(y/l)^2 - (x/l)^2\right]$; $0 \leq x \leq l$, $0 \leq y \leq x$.

a) Where do the highest and lowest temperatures ϑ_max and ϑ_min appear? How are they related to the given temperatures ϑ_0 und ϑ_1? $0 < \vartheta_1 < \vartheta_0$ can be assumed to be valid.

b) The values for $\text{grad}\,\vartheta$ and the vector $\dot{q}$ for the heat flux have to be calculated. At which point is $|\dot{q}|$ largest?

c) Calculate the heat flow through the three boundary surfaces indicated by $y = 0$, $x = l$ und $y = x$, and show that as much heat flows into the the triangle as flows out.

1.7: A saucepan contains water which boils at $\vartheta_\text{s} = 100.3\,°\text{C}$. The base of the saucepan (diameter $d = 18\,\text{cm}$) is electrically heated. It reaches a temperature of $\vartheta_\text{w} = 108.8\,°\text{C}$ with a thermal power input of $1.35\,\text{kW}$. How big is the heat transfer coefficient of the boiling water, based on the temperature difference $(\vartheta_\text{w} - \vartheta_\text{s})$?

1.8: The temperature profile $\vartheta = \vartheta(y)$ in the thermal boundary layer $(0 \leq y \leq \delta_\text{t})$ can be approximated to be a parabola

$$\vartheta(y) = a + by + cy^2$$

whose apex lies at the point $y = \delta_\text{t}$. What is the value of the local heat transfer coefficient, α, if $\delta_\text{t} = 11\,\text{mm}$ and $\lambda = 0.0275\,\text{W/K m}$ (air)?

1.9: In heat transfer in forced turbulent fluid flow through a tube, the approximation equation

$$Nu = C\,Re^{4/5}Pr^{1/3}\;.$$

can be applied. The characteristic length in the Nu and Re numbers is the inner diameter of the tube d. Calculate the ratio α_W/α_A of the heat transfer coefficients for water and air for the same flow velocity w and inner tube diameter d at the same mean temperature $\vartheta = 40\,°C$. The material properties required to solve this exercise are available in Table B1 for water and B2 for air in the appendix.

1.10: Heat is generated inside a very long hollow cylinder of length L due to radioactive decay. This heat flow per length of the cylinder, has a value of $\dot{Q}/L = 550\,W/m$. The hollow cylinder is made of a steel alloy ($\lambda = 15\,W/K\,m$), with inner diameter $d_i = 20\,mm$ and wall thickness $\delta = 10\,mm$. Heat is only given up from its outer surface into space ($T = 0\,K$) by radiation. The emissivity of the cylinder surface is $\varepsilon = 0.17$. Calculate the Celsius temperatures ϑ_i of the inner and ϑ_o of the outer surfaces. What is the value for the radiative heat transfer coefficient α_{rad}?

1.11: A house wall is made up of three layers with the following properties (inside to outside): inner plaster $\delta_1 = 1.5\,cm$, $\lambda_1 = 0.87\,W/K\,m$; wall of perforated bricks $\delta_2 = 17.5\,cm$, $\lambda_2 = 0.68\,W/K\,m$; outer plaster $\delta_3 = 2.0\,cm$, $\lambda_3 = 0.87\,W/K\,m$. The heat transfer coefficients are $\alpha_1 = 7,7\,W/m^2K$ inside and $\alpha_2 = 25\,W/m^2K$ outside. Calculate the heat flux, $\dot{q}$, through the wall, from inside at $\vartheta_1 = 22.0\,°C$ to the air outside at $\vartheta_2 = -12.0\,°C$. What are the temperatures ϑ_{W1} and ϑ_{W2} of the two wall surfaces?

1.12: In order to reduce the heat loss through the house wall in exercise 1.11, an insulating board with $\delta_3 = 6.5\,cm$ and $\lambda_3 = 0.040\,W/K\,m$ along with a facing of $\delta_4 = 11.5\,cm$ and $\lambda_4 = 0.79\,W/K\,m$ will replace the outer plaster wall. Calculate the heat flux $\dot{q}$ and the surface temperature ϑ_{W1} of the inner wall.

1.13: Steam at a temperature of $\vartheta_i = 600\,°C$ flows in a tube of inner diameter $d_1 = 0.25\,m$ and outer diameter $d_2 = 0.27\,m$ made of a steel alloy ($\lambda_1 = 16\,W/K\,m$). The heat transfer coefficient is $\alpha_i = 425\,W/m^2K$. The tube is insulated with a rock wool layer of thickness $\delta_2 = 0.05\,m$, on whose outer surface a hull of mineral fibres of thickness $\delta_3 = 0,02\,m$ is attached. The heat transfer coefficient between the hull and the air at temperature $\vartheta_o = 25\,°C$ is $\alpha_o = 30\,W/m^2K$. The thermal conductivity of the rock wool varies according to the temperature:

$$\frac{\lambda_2(\vartheta)}{W/K\,m} = 0.040 - 0.0005\frac{\vartheta}{100\,°C} + 0.0025\left(\frac{\vartheta}{100\,°C}\right)^2\;.$$

The mean thermal conductivity of the mineral fibre hull is $\lambda_{m3} = 0.055\,W/K\,m$. Calculate the heat lost per length L of the tube $\dot{Q}/L$ and check whether the temperature of the mineral fibre hull is below the maximum permissible value of $\vartheta_{max} = 250\,°C$.

1.14: A cylindrical drink can ($d = 64\,mm$, $h = 103\,mm$) is taken out of a fridge at $\vartheta_{F0} = 6\,°C$ and placed in a room where the air temperature is $\vartheta_A = 24\,°C$. Calculate the temperature ϑ_F of the can after it has been there for 2 hours. In addition to this find the time, t^*, it takes for the temperature of the can to reach $\vartheta_F = 20\,°C$. The only meaningful resistance to the overall heat transfer between the air and the contents of the can is that on the outside of the can which is given by $\alpha_o = 7.5\,W/m^2K$. The material properties of the drink are $\varrho_F = 1.0 \cdot 10^3\,kg/m^3$ and $c_F = 4.1 \cdot 10^3\,J/kgK$.

1.15: Derive the equations for the profile of the fluid temperatures $\vartheta_1 = \vartheta_1(z)$ and $\vartheta_2 = \vartheta_2(z)$ in a countercurrent heat exchanger, cf. section 1.3.3.

1.16: In a heat pump plant air from outside serves as the heat source, where it is cooled from $\vartheta_2' = 10.0\,°C$ to $\vartheta_2'' = 5.0\,°C$, thereby giving heat to a fluid which will be heated from $\vartheta_1' = -5.0\,°C$ to $\vartheta_1'' = 3.0\,°C$. Its mass flow rate is $M_1 = 0{,}125\,kg/s$, with a mean specific heat capacity $\overline{c}_{p1} = 3{,}56\,kJ/kgK$. For this heat transfer task three flow configurations will be compared: countercurrent, crossflow with one row of tubes and counter crossflow with two rows of tubes and two passageways (running in opposite directions), see Fig. 1.59. The operating characteristic of these configurations can be described by the following cf. [1.16]

$$\frac{1}{1 - \varepsilon_1} = \frac{f}{2} + \left(1 - \frac{f}{2}\right) \exp(2f/C_1) \quad \text{mit} \quad f = 1 - \exp(-C_1 N_1/2) \ .$$

Determine the required heat transfer capability kA for the heat exchangers with these three flow configurations.

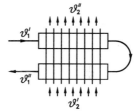

Fig. 1.59: Counter crossflow heat exchanger with two rows of tubes and two passageways.

1.17: A component A is to be dissolved out of a cylindrical rod of length L and radius r_0. It diffuses through a quiescent liquid film of thickness δ which surrounds the rod. Show that the diffusional flux is given by

$$\dot{N}_A = D\frac{c_{A0} - c_{A\delta}}{\ln(r_0 + \delta)/r_0}2\pi L \ ,$$

when $D = const$ and a low concentration of the dissolved component in the liquid film is assumed.

1.18: Ethanol (component A, molar mass $\tilde{M}_A = 46.07$ kg/kmol) is present in a cylinder. The liquid level is, according to Fig. 1.42 at $y_2 - y_1 = \delta_1 = 1\,cm$ below the top of the cylinder. Dry air (component B) at a pressure of $p = 1\,bar$ and temperature $T = 298\,K$ flows across the top of the cylinder.

a) Calculate the molar flux $\dot{n}_A$ at the beginning of the diffusion process.

b) After what period of time has the liquid level fallen by $\delta_2 = 1\,cm$?

As the diffusion is very slow this part of the exercise can be solved by presuming that each change in the liquid level caused by the diffusion can be found using the solution for steady flow. The diffusion coefficent for ethanol in air is $D = 1.19 \cdot 10^{-5}\,m^2/s$, the density of liquid ethanol is $\varrho_L = 875\,kg/m^3$ and the saturation pressure at $T = 298\,K$ is $p_{As} = 0.077\,bar$.

1.19: Ammonia, at 25 °C, flows through a pipe at a rate of $\dot{V} = 10\,m^3/h$. In order to maintain the pressure in the tube at 1 bar, a small tube of inner diameter 3 mm, has been welded onto the pipe, providing a link to the surrounding air, Fig. 1.60. This tube is spiral wound and 20 m long, so that the loss of ammonia by diffusion is very low. How much ammonia in m^3/h is lost and what is the mole fraction of air in the large pipe? The density of ammonia is $0.687\,kg/m^3$, the diffusion coefficient of ammonia in air at 25 °C is $D = 0.28 \cdot 10^{-4}\,m^2/s$

NH$_3$
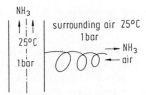

Fig. 1.60: Pressure maintenance in a pipe

1.20: Moist wood at 20 °C is to be dried by air at 0,1 MPa and the same temperature flowing over it. The mass transfer coefficient at the surface of the wood over which the air is flowing is $\beta = 2 \cdot 10^{-3}$ m/s, and the mole fraction of water vapour at the wood surface is $\tilde{x}_{A0} = 0,024$. How big is the mass flux $\dot{m}_A$ of the water into the air in kg/m²s?

1.21: Dry air at 15 °C flows over a water surface. Calculate the wet bulb temperature. Use the material properties given in example 1.8.

1.22: A hot, dry combustion gas at 600 °C and 1 bar flows over the wall of a combustion chamber. To prevent the wall temperature from rising too much, water is blown through the porous wall of the chamber, Fig. 1.61, and then evaporates, which reduces the wall temperature.

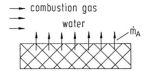

Fig. 1.61: Cooling of a hot, porous wall by blowing water through it

How low can the temperature of the wall become and how much water $\dot{m}_A$ in kg/m²s has to be fed into the chamber to achieve this temperature reduction?
Given: $\tilde{M}_A c \beta_m c_{pA}/\alpha_m = 1.3$, $\tilde{M}_B = 28.953$ kg/kmol, $\tilde{M}_A = 18.015$ kg/kmol and $\Delta h_v = 2346$ kJ/kg, $c_{pA} = 1.907$ kJ/kgK. The heat transfer coefficient between the combustion gas and the water surface is $\alpha_m = 120$ W/m²K.

1.23: Air with a flow rate of $\dot{N}_G = 250$ kmol/h which contains a molar ratio of benzene of $\tilde{Y}_e = 0.05$ is fed into an absorption column to remove the benzene from the air. A benzene free washing solution with a molar flow rate of $\dot{N}_L = 55$ kmol/h should remove 95 % of the benzene. What are the molar ratios of the washing solution and the air at the absorber outlet?

2 Heat conduction and mass diffusion

In this chapter we will deal with steady-state and transient (or non steady-state) heat conduction in quiescent media, which occurs mostly in solid bodies. In the first section the basic differential equations for the temperature field will be derived, by combining the law of energy conservation with Fourier's law. The subsequent sections deal with steady-state and transient temperature fields with many practical applications as well as the numerical methods for solving heat conduction problems, which through the use of computers have been made easier to apply and more widespread.

In conjunction with heat conduction we will also investigate mass diffusion. As a result of the analogy between these two molecular transport processes many results from heat conduction can be applied to mass diffusion. In particular the mathematical methods for the evaluation of concentration fields agree to a large extent with the solution methods for heat conduction problems.

2.1 The heat conduction equation

The basis of the solution of complex heat conduction problems, which go beyond the simple case of steady-state, one-dimensional conduction first mentioned in section 1.1.2, is the differential equation for the temperature field in a quiescent medium. It is known as the equation of conduction of heat or the heat conduction equation. In the following section we will explain how it is derived taking into account the temperature dependence of the material properties and the influence of heat sources. The assumption of constant material properties leads to linear partial differential equations, which will be obtained for different geometries. After an extensive discussion of the boundary conditions, which have to be set and fulfilled in order to solve the heat conduction equation, we will investigate the possibilities for solving the equation with material properties that change with temperature. In the last section we will turn our attention to dimensional analysis or similarity theory, which leads to the definition of the dimensionless numbers relevant for heat conduction.

2.1.1   Derivation of the differential equation for the temperature field

Solving a heat conduction problem requires that the spatial and time dependence of the temperature field

$$\vartheta = \vartheta(\boldsymbol{x}, t)$$

is determined. Once this is known the associated vector field of the heat flux $\dot{\boldsymbol{q}}$ can, as a result of Fourier's law, be calculated by

$$\dot{\boldsymbol{q}}(\boldsymbol{x}, t) = -\lambda \operatorname{grad} \vartheta(\boldsymbol{x}, t) \ , \tag{2.1}$$

which then means that the heat flux at any point in the body can be determined.

The temperature field can be obtained by solving a partial differential equation, the so called heat conduction equation, which will be derived now. This requires the application of the first law of thermodynamics to a closed system, namely a coherent region of any size, imaginarily taken from the conductive body, Fig. 2.1. The volume of this region is V and it has a surface area A. The first law produces the following power balance for the region:

$$\frac{\mathrm{d}U}{\mathrm{d}t} = \dot{Q}(t) + P(t) \ . \tag{2.2}$$

The change with time of the internal energy U is caused by two different influences: the heat flow $\dot{Q}$ and the (mechanical or electrical) power P, which cross the surface of the region.

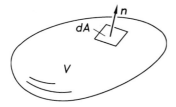

Fig. 2.1: Region of volume V in a thermal conductive body. Surface element $\mathrm{d}A$ of the region with the outward normal $\boldsymbol{n}$

As the heat conduction is being studied in a solid body, the small change in the density as a result of the temperature and pressure variations can be neglected. The model of an incompressible body $\varrho = \mathrm{const}$ is therefore used. Under this assumption

$$\frac{\mathrm{d}U}{\mathrm{d}t} = \frac{\mathrm{d}}{\mathrm{d}t} \int\limits_{(V)} \varrho u \, \mathrm{d}V = \varrho \int\limits_{(V)} \frac{\mathrm{d}u}{\mathrm{d}t} \, \mathrm{d}V \tag{2.3}$$

can be set, where the integral extends over the volume of the region. The specific internal energy u, introduced here, is dependent on the temperature of the incompressible body. So with $c(\vartheta)$ as the specific heat capacity

$$\mathrm{d}u = c(\vartheta) \, \mathrm{d}\vartheta$$

is valid. From (2.3), we obtain

$$\frac{\mathrm{d}U}{\mathrm{d}t} = \varrho \int\limits_{(V)} c(\vartheta)\frac{\partial\vartheta}{\partial t}\,\mathrm{d}V \tag{2.4}$$

for the change with time of the internal energy in the region under consideration.

In order to calculate the heat flow $\dot{Q}$ over the surface of the region, we will take a surface element $\mathrm{d}A$, which has a normal, $\boldsymbol{n}$, outward facing vector, Fig. 2.1. The heat flow through $\mathrm{d}A$, *into* the region is

$$\mathrm{d}\dot{Q} = -\dot{\boldsymbol{q}}\boldsymbol{n}\,\mathrm{d}A\ . \tag{2.5}$$

The heat flow according to (2.2) is positive when the flow is into the body. In the equation given above the heat flux vector $\dot{\boldsymbol{q}}$ is positive because its direction is into the region. In contrast the normal vector $\boldsymbol{n}$ is outward facing and so the scalar product is negative. Therefore a minus sign has to be introduced, thereby fulfilling the prequisite for positive heat flow into the region. By integrating all the heat flow rates $\mathrm{d}\dot{Q}$ from (2.5) the total heat flow $\dot{Q}$ is found to be

$$\dot{Q} = -\int\limits_{(A)} \dot{\boldsymbol{q}}\boldsymbol{n}\,\mathrm{d}A = -\int\limits_{(V)} \operatorname{div}\dot{\boldsymbol{q}}\,\mathrm{d}V\ . \tag{2.6}$$

The integral which extends over the area of the region has been converted to the volume integral of the divergence of $\dot{\boldsymbol{q}}$ according to Gauss' integral theorem.

The power P put into the region is made up of two distinct parts, firstly the power P_V which causes a change in the volume, and secondly the power P_{diss} dissipated inside the region. In an incompressible body $P_V \equiv 0$. P_{diss} is partly made up of the electrical power put into the body, which in a heat and electricity conducting material will be dissipated as a result of the body's electrical resistance. This is the so-called ohmic or resistance heating. Energy rich radiation, e.g. γ-rays, which penetrate a solid body from outside will also be absorbed, their energy will be dissipated, thereby contributing to an increase in the internal energy.

The power of this dissipative and therefore irreversible energy conversion inside the region is given by

$$P = P_{\mathrm{diss}} = \int\limits_{(V)} \dot{W}\,(\vartheta,\boldsymbol{x},t)\,\mathrm{d}V\ , \tag{2.7}$$

where $\dot{W}$ is the power per volume, the so-called *power density*. In a body, with a specific electrical resistance $r_e = r_e(\vartheta)$, which has a current flowing through it,

$$\dot{W}\,(\vartheta,\boldsymbol{x},t) = r_e(\vartheta)\,i_e^2$$

is obtained for the power density, where i_e is the electrical flux. Its SI units are $\mathrm{A/m^2}$, so that with $\Omega\,\mathrm{m}$ as the SI units for the specific electrical resistance, we obtain units of $\Omega\mathrm{mA^2/m^4}=\mathrm{VA/m^3}=\mathrm{W/m^3}$ for $\dot{W}$.

The dissipative and irreversible energy conversion inside the body releases thermal energy, in principle acting like an *internal heat source*. Similarly to the external heat Q, this internal heat source also contributes to an increase in the internal energy of the body according to (2.6). The same effect is created when chemical or nuclear reactions take place inside a body. These reactions are accompanied by the production of heat generated by the irreversible conversion of chemical or nuclear energy into thermal (internal) energy. With the exception of the less proliferous endothermic chemical reactions, they act as heat *sources* in solids. Although the chemical composition of the body is changed by a reaction, this will be neglected in the calculation of material properties, and the effect of the reaction is covered only as an internal heat source, which makes a contribution to the power density $\dot{W}$. Therefore we will consider chemical or nuclear reactions only in (2.7), and by neglecting the changes in the composition we can assume that the material properties λ and c only change with temperature ϑ and not with the composition of the solid.

Now taking the results for dU/dt from (2.4), $\dot{Q}$ from (2.6) and P from (2.7) and putting them into the power balance equation (2.2) of the first law, and combining all the volume integrals, gives

$$\int\limits_{(V)} \left[\varrho c(\vartheta) \frac{\partial \vartheta}{\partial t} + \mathrm{div}\dot{q} - \dot{W}(\vartheta, \boldsymbol{x}, t) \right] dV = 0 \ .$$

This volume integral only disappears for any chosen balance region if the integrand is equal to zero. This then produces

$$\varrho c(\vartheta)\frac{\partial \vartheta}{\partial t} = -\mathrm{div}\dot{q} + \dot{W}(\vartheta, \boldsymbol{x}, t) \ .$$

In the last step of the derivation we make use of Fourier's law and link the heat flux $\dot{q}$ according to (2.1) with the temperature gradient. This gives us

$$\varrho c(\vartheta) \frac{\partial \vartheta}{\partial t} = \mathrm{div}\left[\lambda(\vartheta)\,\mathrm{grad}\vartheta \right] + \dot{W}(\vartheta, \boldsymbol{x}, t) \ , \tag{2.8}$$

the differential equation for the temperature field in a quiescent, isotropic and incompressible material with temperature dependent material properties $c(\vartheta)$ and $\lambda(\vartheta)$. The heat sources within the thermally conductive body are accounted for by the power density $\dot{W}$.

In the application of the heat conduction equation in its general form (2.8) a series of simplifying assumptions are made, through which a number of special differential equations, tailor made for certain problems, are obtained. A significant simplification is the assumption of constant material properties λ and c. The linear partial differential equations which emerge in this case are discussed in the next section. Further simple cases are
 — no heat sources: $\dot{W} \equiv 0$,
 — steady-state temperature fields: $\partial \vartheta/\partial t \equiv 0$,

– geometric one-dimensional heat flow, e.g. only in the x-direction in cartesian coordinates or only in the radial direction with cylindrical and spherical geometries.

2.1.2 The heat conduction equation for bodies with constant material properties

In the derivation of the heat conduction equation in (2.8) we presumed an incompressible body, $\varrho = $ const. The temperature dependence of both the thermal
conductivity λ and the specific heat capacity c was also neglected. These assumptions have to be made if a mathematical solution to the heat conduction equation
is to be obtained. This type of closed solution is commonly known as the "exact" solution. The solution possibilities for a material which has temperature
dependent properties will be discussed in section 2.1.4.

With constant thermal conductivity the differential operator $\mathrm{div}\,[\lambda(\vartheta)\,\mathrm{grad}\vartheta]$
in (2.8) becomes the Laplace operator

$$\lambda\,\mathrm{div}\,\mathrm{grad}\vartheta = \lambda\nabla^2\vartheta \ ,$$

and the heat conduction equation assumes the form

$$\frac{\partial\vartheta}{\partial t} = a\nabla^2\vartheta + \frac{\dot{W}}{c\varrho} \ . \tag{2.9}$$

The constant which appears here is the thermal diffusivity

$$a := \lambda/c\varrho \tag{2.10}$$

of the material, with SI units m^2/s.

In the two most important coordinate systems, cartesian coordinates x, y, z,
and cylindrical coordinates r, φ, z the heat conduction equation takes the form

$$\frac{\partial\vartheta}{\partial t} = a\left(\frac{\partial^2\vartheta}{\partial x^2} + \frac{\partial^2\vartheta}{\partial y^2} + \frac{\partial^2\vartheta}{\partial z^2}\right) + \frac{\dot{W}}{c\varrho} \tag{2.11}$$

or

$$\frac{\partial\vartheta}{\partial t} = a\left(\frac{\partial^2\vartheta}{\partial r^2} + \frac{1}{r}\frac{\partial\vartheta}{\partial r} + \frac{1}{r^2}\frac{\partial^2\vartheta}{\partial\varphi^2} + \frac{\partial^2\vartheta}{\partial z^2}\right) + \frac{\dot{W}}{c\varrho} \tag{2.12}$$

respectively. With spherical coordinates we will limit ourselves to a discussion of
heat flow only in the radial direction. For this case we get from (2.9)

$$\frac{\partial\vartheta}{\partial t} = a\left(\frac{\partial^2\vartheta}{\partial r^2} + \frac{2}{r}\frac{\partial\vartheta}{\partial r}\right) + \frac{\dot{W}}{c\varrho} \ . \tag{2.13}$$

The simplest problem in transient thermal conduction is the calculation of a
temperature field $\vartheta = \vartheta(x, t)$, which changes with time and only in the x-direction.

A further requirement is that there are no sources of heat, i.e. $\dot{W} \equiv 0$. This is known as *linear* heat flow, governed by the partial differential equation

$$\frac{\partial \vartheta}{\partial t} = a \frac{\partial^2 \vartheta}{\partial x^2} \ . \tag{2.14}$$

This equation offers a clear interpretation of the thermal diffusivity a and the heat conduction equation itself. According to (2.14) the change in the temperature with time $\partial \vartheta / \partial t$ at each point in the conductive body is proportional to the thermal diffusivity. This material property, therefore, has an effect on how quickly the temperature changes. As Table 2.1 shows, metals do not only have high thermal conductivities, but also high values for the thermal diffusivity, which imply that temperatures change quickly in metals.

Table 2.1: Material properties of some solids at 20 °C

Material	λ W/K m	c kJ/kg K	ϱ kg/m^3	$10^6 a$ m^2/s
Silver	408	0.234	10497	166
Copper	372	0.419	8300	107
Aluminium	238	0.945	2700	93.4
Brass (MS 60)	113	0.376	8400	35.8
Cr-Ni-Steel	14.7	0.502	7800	3.75
Granite	2.9	0.890	2750	1.2
Concrete gravel	1.28	0.879	2200	0.662
Cork sheets	0.041	1.880	190	0.11
Fat	0.17	1.930	910	0.097

The differential equation (2.14) connects the temperature change with time at a certain point with the curvature of the temperature for the region surrounding this point. It is therefore possible to differentiate between the three cases shown in Fig 2.2. If $\partial^2 \vartheta / \partial x^2 > 0$, the temperature rises (heating); more heat flows in from the "right" than flows out to the "left". This means that energy has to be stored, and so the temperature rises with time. With an opposite sign for the curvature, $\partial^2 \vartheta / \partial x^2 < 0$ the temperature falls with time, whilst if $\partial^2 \vartheta / \partial x^2 = 0$ the temperature remains constant (steady-state limiting case).

If the thermal power $\dot{W}$ is linearly dependent or independent of the temperature ϑ, the heat conduction equation, (2.9), is a second order *linear*, partial differential equation of parabolic type. The mathematical theory of this class of equations was discussed and extensively researched in the 19th and 20th centuries. Therefore tried and tested solution methods are available for use, these will be discussed in 2.3.1. A large number of closed mathematical solutions are known. These can be found in the mathematically orientated standard work by H.S. Carslaw and J.C. Jaeger [2.1].

Steady-state temperature fields are independent of time, and are the end state of a transient cooling or heating process. It is then valid that $\partial \vartheta / \partial t = 0$, from

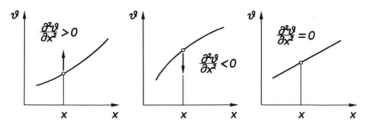

Fig. 2.2: Importance of the curvature for the temperature change with time according to (2.14)

which

$$\nabla^2 \vartheta + (\dot{W}/\lambda) = 0 \tag{2.15}$$

is obtained as the differential equation for a steady-state temperature field with heat sources. The thermal conductivity is the only material property which influences the temperature field. Eq. (2.15) is known as the Poisson differential equation. It passes for temperature fields without heat sources ($\dot{W} \equiv 0$) into the potential equation or Laplace differential equation

$$\nabla^2 \vartheta = 0 \ . \tag{2.16}$$

The Laplace operator ∇^2 takes the form given in equations (2.11), (2.12) and (2.13) for the different coordinate systems. The differential equations (2.15) and (2.16) for steady-state temperature fields, are linear and elliptical, as long as $\dot{W}$ is independent of, or changes linearly with ϑ. This leads to different methods of solution than those used in transient conduction where the differential equations are parabolic.

2.1.3 Boundary conditions

The heat conduction equation only determines the temperature inside the body. To completely establish the temperature field several boundary conditions must be introduced and fulfilled by the solution of the differential equation. These boundary conditions include an initial-value condition with respect to time and different local conditions, which are to be obeyed at the surfaces of the body. The temperature field is determined by the differential equation *and* the boundary conditions.

The *initial-value conditions* set, for a particular time, a temperature for each position in the body. Timekeeping generally starts at this moment, fulfilling the condition

$$\vartheta(x, y, z, t = 0) = \vartheta_0(x, y, z) \ . \tag{2.17}$$

The initial temperature profile $\vartheta_0(x, y, z)$, given in the problem, for example a constant initial temperature throughout a body that is to be cooled, changes during the transient conduction process.

The local *boundary conditions* can be devided into three different groups. At the surface of the body
1. the temperature can be given as a function of time and position on the surface, the so-called 1st type of boundary condition,
2. the heat flux normal to the surface can be given as a function of time and position, the 2nd type of boundary condition, or
3. contact with another medium can exist.

A given surface temperature is the most simple case to consider, especially when the surface temperature is constant. In the case of a prescribed heat flux $\dot{q}$, the condition

$$\dot{q} = -\lambda \frac{\partial \vartheta}{\partial n} \qquad (2.18)$$

must be fulfilled at every point on the surface, where the derivative for the outward normal direction is taken and λ is the value of the conductivity at the surface temperature. *Adiabatic* surfaces frequently appear, and because $\dot{q} = 0$, we get

$$\partial \vartheta / \partial n = 0 \ . \qquad (2.19)$$

This simple condition must also be fulfilled at planes of symmetry inside the body. Therefore, when formulating a conduction problem, it is often advantageous to only consider a section of the body that is bounded by one or more adiabatic planes of symmetry.

If the thermally conductive body is in contact with another medium, several different boundary conditions can apply, each depending on whether the other medium is a solid or fluid, and its respective material properties. When the other medium is another solid, the heat flux at the interface of body 1 to body 2 is the same for both bodies. According to (2.18), at the interface, index I, it is valid that

$$\lambda^{(1)} \left(\frac{\partial \vartheta^{(1)}}{\partial n} \right)_{\mathrm{I}} = \lambda^{(2)} \left(\frac{\partial \vartheta^{(2)}}{\partial n} \right)_{\mathrm{I}} \qquad (2.20)$$

and also

$$\vartheta_{\mathrm{I}}^{(1)} = \vartheta_{\mathrm{I}}^{(2)} \ . \qquad (2.21)$$

The temperature curve has a kink at the interface. The temperature gradient in the body with the lower thermal conductivity is larger, Fig. 2.3 a. Equation (2.21) is only valid if the two bodies are firmly joined. If this is not the case a contact resistance occurs, which results in a small temperature jump, Fig. 2.3 b. This resistance can be described by a contact heat transfer coefficient α_{ct}. In place of equation (2.21),

$$-\lambda^{(1)} \left(\frac{\partial \vartheta^{(1)}}{\partial n} \right)_{\mathrm{I}} = \alpha_{\mathrm{ct}} \left(\vartheta_{\mathrm{I}}^{(1)} - \vartheta_{\mathrm{I}}^{(2)} \right) \ , \qquad (2.22)$$

is valid. With constant α_{ct} the temperature drop at the interface is proportional to the heat flux.

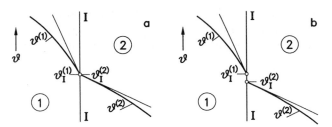

Fig. 2.3: Temperature at the interface between two bodies 1 and 2 in contact with each other. **a** no contact resistance, **b** contact resistance according to (2.22)

If the thermally conductive body is bounded by a fluid, a boundary layer develops in the fluid. The heat flux into the fluid, with α as the heat transfer coefficient is

$$\dot{q} = \alpha \left(\vartheta_W - \vartheta_F \right) \ ,$$

see Fig. 2.4 and section 1.1.3. As this heat flux must be transported to the surface of the body by conduction, the following boundary condition is obtained

$$- \lambda \left(\frac{\partial \vartheta}{\partial n} \right)_W = \alpha \left(\vartheta_W - \vartheta_F \right) \ . \tag{2.23}$$

Here λ is the thermal conductivity of the solid (not the fluid!) at the wall. Eq. (2.23) stipulates a linear relationship between the temperature ϑ_W and the slope of the temperature profile at the surface, this is also known as the 3rd type of boundary condition. As in (2.18), $\partial \vartheta / \partial n$ is the derivative in the normal direction outward from the surface. The fluid temperature ϑ_F, which can change with time, and the heat transfer coefficient α must be given for the solution of the heat conduction problem. If α is very large, the temperature difference $(\vartheta_W - \vartheta_F)$ will be very small and the boundary condition (2.23) can be replaced by the simpler boundary condition of a perscribed temperature $(\vartheta_W = \vartheta_F)$. The boundary condition in (2.23) is only linear as long as α is independent of ϑ_W or $(\vartheta_W - \vartheta_F)$, this factor is very important for the mathematical solution of heat conduction problems. In a series of heat transfer problems, for example in free convection, α changes with $(\vartheta_W - \vartheta_F)$, thereby destroying the linearity of the boundary condition. The same occurs when heat transfer by radiation is considered, as in this

Fig. 2.4: Temperature profile for the boundary condition (2.23). The tangent to the temperature curve at the solid surface meets the guidepoint R at the fluid temperature ϑ_F at a distance away from the surface $s = \lambda/\alpha = L_0/Bi$. The subtangent of the temperature profile in the fluid boundary layer is $s_u = \lambda_F/\alpha = L_0/Nu$.

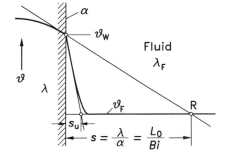

case $\dot{q}$ is dependent on T_W^4, see section 1.1.5. A closed solution for the thermal conduction problem is not possible in cases such as these, so numerical methods have to be applied and this will be discussed in 2.4.

In the cooling or heating of a vessel which contains a fluid, the temperature ϑ_F of this fluid can often be taken as spatially constant, as the temperatures in each individual part of the fluid will be equalised either by convection or mixing. The change in the temperature $\vartheta_F = \vartheta_F(t)$ over time depends on the manner in which heat is transferred from the vessel wall (the conductive body) to the fluid. This causes a change in the internal energy of the fluid and a corresponding variation in ϑ_F. The heat flux is then

$$\dot{q} = -\lambda \left(\frac{\partial \vartheta}{\partial n} \right)_W$$

(thermal conduction in the vessel wall) and also

$$\dot{q} = c_F \frac{M_F}{A} \frac{d\vartheta_F}{dt}$$

(heating the fluid) with c_F as the specific heat capacity of the fluid and M_F as the mass of the fluid, which is in contact with the vessel wall (area A). We then have the boundary condition

$$c_F \frac{M_F}{A} \frac{d\vartheta_F}{dt} + \lambda \left(\frac{\partial \vartheta}{\partial n} \right)_W = 0 \ . \tag{2.24}$$

This is to be supplemented by a heat transfer condition according to (2.23), or if a large heat transfer coefficient exists by the simplified boundary condition $\vartheta_W = \vartheta_F$.

2.1.4 Temperature dependent material properties

If the temperature dependence of the material properties $\lambda = \lambda(\vartheta)$ and $c = c(\vartheta)$ cannot be neglected then the heat conduction equation (2.8) must be the starting point for the solution of a conduction problem. We have a non-linear problem, that can only be solved mathematically in exceptional cases. With

$$\operatorname{div}\left[\lambda(\vartheta)\operatorname{grad}\vartheta\right] = \lambda(\vartheta)\operatorname{div}\operatorname{grad}\vartheta + \frac{d\lambda}{d\vartheta}\operatorname{grad}^2\vartheta$$

we obtain from (2.8)

$$c(\vartheta)\varrho\frac{\partial \vartheta}{\partial t} = \lambda(\vartheta)\nabla^2\vartheta + \frac{d\lambda}{d\vartheta}\operatorname{grad}^2\vartheta + \dot{W} \tag{2.25}$$

as the heat conduction equation. The non-linearity is clearly shown in the first two terms on the right hand side.

Equations (2.8) and (2.25) assume a simpler form when a new variable, the transformed temperature

$$\Theta = \Theta_0 + \frac{1}{\lambda_0}\int_{\vartheta_0}^{\vartheta} \lambda(\vartheta)\,d\vartheta \tag{2.26}$$

is introduced. Here λ_0 is the value for the thermal conductivity at the reference temperature ϑ_0 assigned to the transformed temperature Θ_0. From (2.26) we get

$$\frac{\partial \Theta}{\partial t} = \frac{\lambda}{\lambda_0}\frac{\partial \vartheta}{\partial t} \quad \text{and} \quad \text{grad}\Theta = \frac{\lambda}{\lambda_0}\text{grad}\vartheta \ .$$

Following on from (2.8), we obtain

$$c(\vartheta)\varrho\frac{\lambda_0}{\lambda(\vartheta)}\frac{\partial \Theta}{\partial t} = \lambda_0\nabla^2\Theta + \dot{W}$$

or

$$\frac{1}{a(\vartheta)}\frac{\partial \Theta}{\partial t} = \nabla^2\Theta + \frac{\dot{W}}{\lambda_0} \ , \qquad (2.27)$$

an equation which formally agrees with the heat conduction equation (2.9) for constant material properties. The thermal diffusivity a however depends on ϑ or Θ. Experience tells us that a varies less with temperature than λ, so $a(\vartheta)$ in (2.27) can be assumed to be approximately constant. A solution of the heat conduction equation for constant material properties can be applied to the case of temperature dependent properties by replacing ϑ by Θ. There is however, one important limitation: only boundary conditions with given temperature or heat flux (2.18) can be stipulated, as the heat transfer condition (2.23) is not retained by the transformation (2.26). The transformation (2.26) is used particularly in steady-state heat conduction problems, because the term which contains the temperature dependent a disappears due to $\partial\Theta/\partial t = 0$. Solving of the Poisson, or in the case of $\dot{W} = 0$ the Laplace equation can immediately be undertaken, provided that the boundary conditions of a given temperature or heat flux, and not the heat transfer condition according to (2.23), exist.

In most temperature dependent material property cases a closed solution can not be obtained. A numerical solution is the only answer in such cases and we will look more closely at this type of solution in section 2.4.

2.1.5 Similar temperature fields

The advantage of introducing dimensionless variables has already been shown in section 1.1.4. The dimensionless numbers obtained in that section provide a clear and concise representation of the physical relationships, due to the significant reduction in the influencing variables. The dimensionless variables for thermal conduction are easy to find because the differential equations and boundary conditions are given in an explicit form.

The starting point for the derivation of the dimensionless numbers in thermal conduction is the differential equation

$$\frac{\partial \vartheta}{\partial t} = a\frac{\partial^2 \vartheta}{\partial x^2} + \frac{\dot{W}}{c\varrho}$$

which contains all the significant terms, time dependence, local variations of the temperature field as well as the power density $\dot{W}$ of heat sources. Introducing a dimensionless position coordinate and a dimensionless time gives

$$x^+ := x/L_0 \quad , \quad t^+ := t/t_0 \ .$$

Here, L_0 is a characteristic length of the conductive body and t_0 is a characteristic time (interval) which still has to be determined. We choose

$$\vartheta^+ := (\vartheta - \vartheta_0)/\Delta\vartheta_0 \ , \tag{2.28}$$

as a dimensionless temperature, wherein ϑ_0 is a reference temperature (zero point of ϑ^+) and $\Delta\vartheta_0$ is a characteristic temperature difference of the problem.

The heat conduction equation takes the dimensionless form of

$$\frac{\partial\vartheta^+}{\partial t^+} = \frac{at_0}{L_0^2}\frac{\partial^2\vartheta^+}{\partial x^{+2}} + \frac{t_0}{\Delta\vartheta_0}\frac{\dot{W}}{c\varrho} \ . \tag{2.29}$$

The characteristic time is chosen to be

$$t_0 = L_0^2/a \ .$$

The dimensionless time is then

$$t^+ = at/L_0^2 \ . \tag{2.30a}$$

This dimensionless time t^+ is often called the Fourier number

$$Fo := at/L_0^2 = t^+ \ . \tag{2.30b}$$

However the Fourier number is not a dimensionless number in the usual sense, that it has a fixed value for a given problem, instead it is a dimensionless time variable which only has fixed values for fixed times.

Now introducing the characteristic power density

$$\dot{W}_0 = \Delta\vartheta_0 c\varrho/t_0 = \lambda\Delta\vartheta_0/L_0^2 \tag{2.31a}$$

as a reference quantity, and with

$$\dot{W}^+ = \dot{W}/\dot{W}_0 = \dot{W}L_0^2/\lambda\Delta\vartheta_0 \tag{2.31b}$$

as the dimensionless heat source function, we obtain from (2.28)

$$\frac{\partial\vartheta^+}{\partial t^+} = \frac{\partial^2\vartheta^+}{\partial x^{+2}} + \dot{W}^+ \ . \tag{2.32}$$

The heat conduction equation (2.32) does not contain any factors different from one when we use the dimensionless heat source function $\dot{W}^+$ from (2.31b), and so does not immediately force us to introduce dimensionless numbers. Only the dimensionless parameters, which specify the position, time and temperature dependence of $\dot{W}^+$, take over the role of dimensionless numbers. Further dimensionless numbers are expected when the boundary conditions are also made dimensionless.

The initial condition (2.17) and the boundary condition for a perscribed temperature are homogeneous in ϑ. They do not automatically lead to dimensionless numbers. However these boundary conditions should be observed for the choice of ϑ_0 and $\Delta\vartheta_0$ in (2.28), in order to obtain the most favourable dimensionless temperature ϑ^+, for example in the interval $0 \leq \vartheta^+ \leq 1$. If the heat flux $\dot{q}_W$ is stipulated as a boundary condition, we get from (2.18) the following dimensionless relationship

$$\dot{q}_W^+ = -\left(\frac{\partial\vartheta^+}{\partial x^+}\right)_W \tag{2.33a}$$

with

$$\dot{q}_W^+ := \dot{q}_W/\dot{q}_{W_0} = \dot{q}_W L_0/\lambda\Delta\vartheta_0 . \tag{2.33b}$$

The dimensionless function $\dot{q}_W^+$ contains one or more parameters which describe the position and time dependence of the heat flux $\dot{q}_W$, if the special case of $\dot{q}_W = 0$ is disregarded. These dimensionless parameters enter, in the same way as the parameters in the heat source function $\dot{W}^+$ from (2.31b), as dimensionless numbers into the solution of heat conduction problems.

Finally making the heat transfer condition (2.23) dimensionless, gives

$$\frac{\partial\vartheta^+}{\partial x^+} = \frac{\alpha L_0}{\lambda}\left(\vartheta_W^+ - \vartheta_F^+\right) . \tag{2.34}$$

A new dimensionless number appears here and it is known as the Biot[1] number

$$Bi := \alpha L_0/\lambda . \tag{2.35}$$

The temperature field is dependent on this number when heat transfer takes place into a fluid. The Biot number has the same form as the Nusselt number defined by (1.36). There is however one very significant difference, λ in the Biot number is the thermal conductivity of the solid whilst in the Nusselt number λ is the thermal conductivity of the fluid. The Nusselt number serves as a dimensionless representation of the heat transfer coefficient α useful for its evaluation, whereas the Biot number describes the boundary condition for thermal conduction in a solid body. It is the ratio of L_0 to the subtangent to the temperature curve within the solid body, cf. Fig. 2.4, whilst the Nusselt number is the ratio of a (possibly different choice of) characteristic length L_0 to the subtangent to the temperature profile in the boundary layer of the fluid.

---

[1] Jean Baptiste Biot (1774–1862) became Professor of Physics at the Collège de France, Paris in 1800. From 1804 onwards he investigated the cooling of heated rods and in 1816 published the differential equation for the temperature profile, without giving a clear derivation of the equation. In 1820 he discovered along with F. Savart, the Biot-Savart law for the strength of the magnetic field around a conductor with an electrical current passing through it.

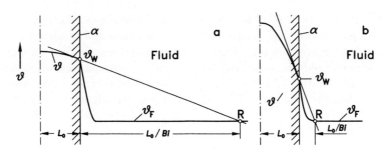

Fig. 2.5: Influence of the Biot number $Bi = \alpha L_0/\lambda$ on the temperature profile near to the surface. **a** small Biot number, **b** large Biot number

Interpreting L_0/λ as the specific thermal conduction resistance of the solid and $1/\alpha$ as the specific heat transfer resistance at its surface also allows the Biot number to be interpreted as the ratio of the two resistances

$$Bi = \frac{L_0/\lambda}{1/\alpha} \ .$$

A small Biot number means that the resistance to thermal conduction in the body, for example due to its high thermal conductivity, is significantly smaller than the heat transfer resistance at its boundary. With small Biot numbers the temperature difference in the body is small in comparison to the difference $(\vartheta_{\mathrm{W}} - \vartheta_{\mathrm{F}})$ between the wall and fluid temperatures. The reverse is valid for large Biot numbers. Examples of these two scenarios are shown for a cooling process in Fig. 2.5. Very large Biot numbers lead to very small values of $(\vartheta_{\mathrm{W}}^+ - \vartheta_{\mathrm{F}}^+)$, and for $Bi \to \infty$, according to (2.34) we get $(\vartheta_{\mathrm{W}}^+ - \vartheta_{\mathrm{F}}^+) \to 0$. The heat transfer condition (2.34) can be replaced by the simpler boundary condition $\vartheta_{\mathrm{W}}^+ = \vartheta_{\mathrm{F}}^+$.

Temperature fields in quiescent solid bodies can be represented in dimensionless form by

$$\vartheta^+ = \vartheta^+\left(x^+, y^+, z^+, t^+, \dot{W}^+, \dot{q}_{\mathrm{W}}^+, Bi, N_{\mathrm{Geom}}\right) \ . \qquad (2.36)$$

Here the dimensionless variables and numbers already discussed will be supplemented by dimensionless geometric numbers, which are indicated by the abbreviation N_{Geom}. Characteristic geometric numbers are for example the height to diameter H/D ratio of a thermally conductive cylinder or the ratios L_2/L_1 and L_3/L_1 of the edge lengths L_1 to L_3 of a rectangular body. The shape of the function in (2.36) depends on the geometry and the other conditions of the heat conduction problem. In general the dimensionless variables in (2.36) will not all appear at the same time. In steady-state heat conduction t^+ disappears; in bodies without any heat sources $\dot{W}^+ \equiv 0$; and when only perscribed temperaturs as boundary conditions are given $\dot{q}_{\mathrm{W}}^+$ and Bi in (2.36) are not present.

Finally, it should also be pointed out that in heat conduction problems the dimensionless representation and the combination of the influencing quantites into dimensionless numbers are not as significant as in the representation and determination of heat transfer coefficients in 1.1.4. In the following sections we will frequently refrain from making the heat conduction problem dimensionless and will only present the solution of a problem in a dimensionless form by a suitable combination of variables and influencing quantities.

2.2 Steady-state heat conduction

We speak of steady-state heat conduction when the temperature at every point in a thermally conductive body does not change with time. Some simple cases, which are of practical importance, have already been discussed in the introductory chapter, namely one dimensional heat flow in flat and curved walls, cf. section 1.1.2. In the following sections we will extend these considerations to geometric one-dimensional temperature distributions with internal heat sources. Thereafter we will discuss the temperature profiles and heat release of fins and we will also determine the fin efficiency first introduced in section 1.2.3. We will also investigate two- and three-dimensional temperature fields, which demand more complex mathematical methods in order to solve them, so that we are often compelled to make use of numerical methods, which will be introduced in section 2.4.6.

2.2.1 Geometric one-dimensional heat conduction with heat sources

In section 1.2.1 we dealt with geometric one-dimensional heat conduction without internal heat sources. The temperature in these cases was only dependent on one spatial coordinate. The equations for steady heat flow through a flat wall (plate), a hollow cylinder (tube wall) and a hollow sphere were obtained. In the following we will extend these considerations to thermally conductive materials with heat sources. Examples of this include an electrical conductor, through which a current is flowing, where the electrical energy is dissipated, and the cylindrical or spherical fuel elements in a nuclear reactor, in which the energy released during nuclear fission will be conducted as heat to the surface of the fuel elements.

As the heat conduction equation derived in 2.1.1 shows, the only material property which has an effect on the steady state temperature field, $\partial\vartheta/\partial t \equiv 0$, is the thermal conductivity $\lambda = \lambda(\vartheta)$. Assuming that λ is constant,

$$\nabla^2\vartheta + (\dot{W}/\lambda) = 0 \tag{2.37}$$

is the differential equation for the steady temperature field, in which $\dot{W}$ is the source term, namely the thermal power generated per volume. The function $\dot{W}$ can depend on either the temperature ϑ or the position coordinate. We will limit ourselves to one-dimensional heat flow. The temperature changes only with respect to the position coordinate which is indicated by r, even in cartesian coordinates. The Laplace operator is different for the three coordinate systems, cartesian, cylindrical and spherical. Combining the three cases gives from (2.37)

$$\frac{\mathrm{d}^2\vartheta}{\mathrm{d}r^2} + \frac{n}{r}\frac{\mathrm{d}\vartheta}{\mathrm{d}r} + \frac{\dot{W}(r,\vartheta)}{\lambda} = 0 \tag{2.38}$$

as the decisive differential equation with $n = 0$ for linear heat flow (plate), $n = 1$ for the cylinder and $n = 2$ for the sphere.

Solving the ordinary differential equation (2.38) for *constant* power density $\dot{W} = \dot{W}_0$, gives the solution

$$\vartheta(r) = \left\{ \begin{array}{ll} c_0 & + \quad c_1 (r - r_0) \\ c_0 & + \quad c_1 \ln (r/r_0) \\ c_0 & + \quad c_1 (1/r - 1/r_0) \end{array} \right\} - \frac{\dot{W}_0 r^2}{2(1+n)\lambda} \qquad (2.39)$$

for the three cases, plate ($n = 0$), cylinder ($n = 1$) and sphere ($n = 2$). Here c_0 and c_1 are constants, which have to be adjusted to the boundary conditions. As an example we will look at the simple case of heat transfer to a fluid with a temperature of $\vartheta = \vartheta_F$ taking place at the surface $r = \pm R$. We then have

$$- \lambda \left(\frac{\mathrm{d}\vartheta}{\mathrm{d}r} \right)_{r=R} = \alpha \left[\vartheta(R) - \vartheta_F \right] \qquad (2.40)$$

and due to symmetry

$$\left(\frac{\mathrm{d}\vartheta}{\mathrm{d}r} \right)_{r=0} = 0 \ . \qquad (2.41)$$

The condition (2.41) requires that $c_1 = 0$. Then from (2.40) the constant c_0 can be found, so that the solution to the boundary value problem is

$$\vartheta(r) = \vartheta_F + \frac{\dot{W}_0 R^2}{2\lambda(1+n)} \left(1 - \frac{r^2}{R^2} + \frac{2\lambda}{\alpha R} \right) \ .$$

With the dimensionless temperature

$$\vartheta^+ := \frac{\vartheta - \vartheta_F}{\dot{W}_0 R^2 / \lambda} \ , \qquad (2.42)$$

the dimensionless distance

$$r^+ = r/R \qquad (2.43)$$

from the central plane or point and with the Biot number

$$Bi := \alpha R / \lambda \qquad (2.44)$$

we obtain

$$\vartheta^+ = \frac{1}{2(1+n)} \left[1 - r^{+2} + \frac{2}{Bi} \right] \ . \qquad (2.45)$$

In the bodies a parabolic temperature profile exists, with the highest temperature at $r^+ = 0$, Fig. 2.6.

If the heat flux $\dot{q}_W$ at the surface ($r = R$) of the three bodies is given instead of the power density $\dot{W}_0$, then from the energy balance,

$$\dot{W}_0 V = \dot{q}_W A$$

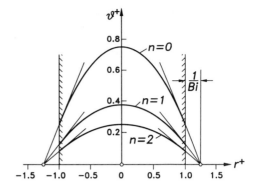

Fig. 2.6: Dimensionless temperature profile ϑ^+ according to (2.45) in a plate ($n = 0$), a cylinder ($n = 1$) and a sphere ($n = 2$) for $Bi = \alpha R/\lambda = 4.0$

with V as the volume and A as the surface area, we obtain the simple relationship

$$\dot{q}_{\mathrm{W}} = \dot{W}_0 \frac{V}{A} = \dot{W}_0 \frac{R}{1+n} \quad . \tag{2.46}$$

Equation (2.45) for the temperature profile then has a form independent of the shape of the body:

$$\vartheta(r) = \vartheta_{\mathrm{F}} + \frac{\dot{q}_{\mathrm{W}} R}{2\lambda} \left(1 - r^{+2} + \frac{2}{Bi} \right) \quad . \tag{2.47}$$

The considerations above can also be applied to the case of temperature dependent thermal conductivity $\lambda = \lambda(\vartheta)$. When $\dot{W}_0$ and the shape of the body are given, the heat flux $\dot{q}_{\mathrm{W}}$ at the wall ($r = R$) has to be calculated according to (2.46). The surface temperature ϑ_{W} is obtained from

$$\vartheta_{\mathrm{W}} = \vartheta_{\mathrm{F}} + \dot{q}_{\mathrm{W}}/\alpha \quad .$$

According to section 2.1.4 we introduce the transformed temperature

$$\Theta = \Theta_{\mathrm{W}} + \frac{1}{\lambda(\vartheta_{\mathrm{W}})} \int_{\vartheta_{\mathrm{W}}}^{\vartheta} \lambda(\vartheta)\, \mathrm{d}\vartheta \tag{2.48}$$

to account for the variations in λ. From (2.27) the following differential equation for Θ is obtained

$$\frac{\mathrm{d}^2\Theta}{\mathrm{d}r^2} + \frac{n}{r}\frac{\mathrm{d}\Theta}{\mathrm{d}r} + \frac{\dot{W}_0}{\lambda(\vartheta_{\mathrm{W}})} = 0 \quad ,$$

whose form agrees with (2.38). This is solved under the boundary conditions

$$\begin{aligned} r = 0 : &\quad \mathrm{d}\Theta/\mathrm{d}r = 0 \quad , \\ r = R : &\quad \Theta = \Theta_{\mathrm{W}} \quad (\text{corresponding to } \vartheta = \vartheta_{\mathrm{W}}) \quad . \end{aligned}$$

The solution is the parabola

$$\Theta - \Theta_{\mathrm{W}} = \frac{\dot{W}_0 R^2}{2(1+n)\lambda(\vartheta_{\mathrm{W}})} \left[1 - (r/R)^2 \right] = \frac{\dot{q}_{\mathrm{W}} R}{2\lambda(\vartheta_{\mathrm{W}})} \left[1 - (r/R)^2 \right] \quad . \tag{2.49}$$

For the calculation of the highest temperature $\vartheta_{\max}$ in the centre of the body, with $r = 0$ we obtain

$$\Theta_{\max} - \Theta_{\mathrm{W}} = \frac{\dot{W}_0 R^2}{2(1+n)\lambda(\vartheta_{\mathrm{W}})} = \frac{\dot{q}_{\mathrm{W}} R}{2\lambda(\vartheta_{\mathrm{W}})}$$

and then with (2.48) for $\vartheta_{\max}$ the equation

$$\int_{\vartheta_{\mathrm{W}}}^{\vartheta_{\max}} \lambda\left(\vartheta\right) \mathrm{d}\vartheta = \lambda\left(\vartheta_{\mathrm{W}}\right)\left(\Theta_{\max} - \Theta_{\mathrm{W}}\right) = \frac{\dot{W}_0 R^2}{2\left(1+n\right)} \quad . \tag{2.50}$$

M. Jakob [2.2] dropped the presumption that $\dot{W} = \dot{W}_0 = \mathrm{const}$ and considered heat development rising or falling linearly with the temperature. The first case occurs during the heating of a metallic electrical conductor whose electrical resistance increases with temperature.

Example 2.1: A cylindrical fuel rod of radius $r = 0.011\,\mathrm{m}$ is made of uranium dioxide ($\mathrm{UO_2}$). At a certain cross section in the element the power density is $\dot{W}_0 = 1.80 \cdot 10^5\,\mathrm{kW/m^3}$; the surface temperature has a value of $\vartheta_{\mathrm{W}} = 340\,^\circ\mathrm{C}$. The maximum temperature $\vartheta_{\max}$ in the centre of the element is to be calculated. The thermal conductivity of $\mathrm{UO_2}$, according to J. Höchel [2.3], is given by

$$\frac{\lambda\left(T\right)}{\mathrm{W/K\,m}} = \frac{3540\,\mathrm{K}}{T + 57\,\mathrm{K}} + 0{,}0747 \left(\frac{T}{1000\,\mathrm{K}}\right)^3 \quad ,$$

which is valid in the region $300\,\mathrm{K} < T < 3073\,\mathrm{K}$.

The maximum (thermodynamic) temperature $T_{\max}$ is obtained from (2.50) with $n = 1$ and $\vartheta = T$:

$$\int_{T_{\mathrm{W}}}^{T_{\max}} \lambda\left(T\right)\,\mathrm{d}T = \frac{\dot{W}_0 R^2}{4} = 5445\,\mathrm{W/m} \quad .$$

With the surface temperature $T_{\mathrm{W}} = \vartheta_{\mathrm{W}} + 273\,\mathrm{K} = 613\,\mathrm{K}$ this yields

$$\int_{T_{\mathrm{W}}}^{T_{\max}} \lambda\left(T\right)\,\mathrm{d}T = \left\{3540 \ln \frac{T_{\max} + 57\,\mathrm{K}}{670\,\mathrm{K}} + 18.675 \left[\left(\frac{T_{\max}}{1000\,\mathrm{K}}\right)^4 - 0.1412\right]\right\} \frac{\mathrm{W}}{\mathrm{m}} \quad .$$

From which the transcendental equation

$$\ln \frac{T_{\max} + 57\,\mathrm{K}}{670\,\mathrm{K}} = 1.5381 - 0.005275 \left[\left(\frac{T_{\max}}{1000\,\mathrm{K}}\right)^4 - 0.1412\right]$$

is obtained, with the solution $T_{\max} = 2491\,\mathrm{K}$, and correspondingly $\vartheta_{\max} = 2281\,^\circ\mathrm{C}$. This temperature lies well below the temperature at which $\mathrm{UO_2}$ melts, which is around $2800\,^\circ\mathrm{C}$.

2.2.2 Longitudinal heat conduction in a rod

When a rod shaped object, for example a bolt or a pillar, is heated at one end heat flows along the axial direction and is transferred to the environment through the outer surface of the object. The heat release from fins is a similar heat conduction problem, which we will look at in the next section. Finally there are a number of measuring procedures for the determination of the thermal conductivity which are based on the comparison of temperature drops in rods made of different materials, see [2.2].

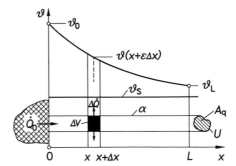

Fig. 2.7: Temperature profile in a rod of constant cross sectional area A_{q}

We will consider a rod of length L with constant cross sectional area A_{q} and constant circumference U. One end of the rod is kept at constant temperature ϑ_0 by the input of heat. Heat flows along the direction of the rod axis, and will be transferred to the surroundings via the outer surface of the rod, Fig. 2.7. The cross sectional area of the rod is so small that the temperature across the whole cross section can be assumed to be constant, it only changes in the x-coordinate direction. In a rod element with volume $\Delta V = A_{\mathrm{q}}\Delta x$, the heat flow through the outer surface of the rod released to the surroundings which have a constant temperature ϑ_{S} is

$$\Delta \dot{Q} = \alpha \left[\vartheta \left(x + \varepsilon \Delta x \right) - \vartheta_{\mathrm{S}} \right] U \Delta x \ , \quad 0 \leq \varepsilon \leq 1 \ .$$

This release of heat has the same effect as a heat sink in the rod material with the power density

$$\dot{W} = - \lim_{\Delta x \to 0} \frac{\Delta \dot{Q}}{\Delta V} = - \frac{\alpha U}{A_{\mathrm{q}}} \left[\vartheta(x) - \vartheta_{\mathrm{S}} \right] \ .$$

Putting this into the differential equation (2.38) for linear heat flow with a heat source $(n = 0)$ we obtain, with x instead of r

$$\frac{\mathrm{d}^2 \vartheta}{\mathrm{d} x^2} - \frac{\alpha U}{\lambda A_{\mathrm{q}}} \left(\vartheta - \vartheta_{\mathrm{S}} \right) = 0 \qquad (2.51)$$

as the differential equation which determines the temperature profile along the rod.

Assuming a constant heat transfer coefficient α and putting as an abbreviation

$$m^2 = \alpha U / \lambda A_{\mathrm{q}} \ , \qquad (2.52)$$

gives the general solution of (2.51) as

$$\begin{aligned} \vartheta(x) &= \vartheta_{\mathrm{S}} + c_1 \exp(-mx) + c_2 \exp(mx) \\ &= \vartheta_{\mathrm{S}} + C_1 \cosh(mx) + C_2 \sinh(mx) \ . \end{aligned}$$

The constants of integration c_1, c_2, C_1 and C_2 are determined from the boundary conditions.

At the left hand end ($x = 0$) of the rod the temperature is always assumed to be $\vartheta = \vartheta_0$, whilst at the other end ($x = L$) different boundary conditions are considered. The condition

$$\vartheta = \vartheta_0 \qquad \text{for} \quad x = 0$$

suggest that the dimensionless temperature

$$\vartheta^+ := \frac{\vartheta - \vartheta_{\mathrm{S}}}{\vartheta_0 - \vartheta_{\mathrm{S}}} \tag{2.53}$$

should be used. We then have $0 \le \vartheta^+ \le 1$.

We will now consider a rod which stretches into the surroundings Fig. 2.7, and releases heat through its end surface ($x = L$). Here

$$\dot{Q}_L = -\lambda A_{\mathrm{q}} \frac{\mathrm{d}\vartheta}{\mathrm{d}x} = \alpha_L A_{\mathrm{q}} (\vartheta - \vartheta_{\mathrm{S}}) \qquad \text{for} \quad x = L \tag{2.54}$$

is valid. The heat transfer coefficient α_L at the end surface does not have to be the same as α at the outer surface of the rod. After some calculations the temperature profile in the rod is found to be

$$\vartheta^+(x) = \frac{\cosh[m(L - x)] + (\alpha_L/m\lambda)\sinh[m(L - x)]}{\cosh[mL] + (\alpha_L/m\lambda)\sinh[mL]} \quad . \tag{2.55}$$

The heat flow $\dot{Q}$ released to the environment is equal to the heat flow through the rod cross section at $x = 0$:

$$\dot{Q} = \dot{Q}_0 = -\lambda A_{\mathrm{q}} \left(\frac{\mathrm{d}\vartheta}{\mathrm{d}x} \right)_{x=0} \quad .$$

By differentiating (2.55) we obtain

$$\dot{Q}_0 = \sqrt{\alpha \lambda A_{\mathrm{q}} U} \, (\vartheta_0 - \vartheta_{\mathrm{S}}) \frac{\tanh(mL) + (\alpha_L/m\lambda)}{1 + (\alpha_L/m\lambda)\tanh(mL)} \quad . \tag{2.56}$$

If $\alpha_L = \alpha$ then

$$\frac{\alpha_L}{m\lambda} = \frac{\alpha}{m\lambda} = \sqrt{\frac{\alpha A_{\mathrm{q}}}{\lambda U}} \quad .$$

A notable simplification of the equations is obtained when $\alpha_L = 0$. This is valid if the end of the rod is insulated or when the heat release through the small cross sectional area A_{q} can be neglected. Then for the temperature profile, we get

$$\vartheta^+(x) = \frac{\cosh[m(L - x)]}{\cosh(mL)} \quad . \tag{2.57}$$

The temperature at the free end falls to

$$\vartheta^+(L) = \frac{\vartheta_L - \vartheta_S}{\vartheta_0 - \vartheta_S} = \frac{1}{\cosh(mL)} \quad . \tag{2.58}$$

The heat released is yielded from (2.56) to be

$$\dot{Q}_0 = \sqrt{\alpha U \lambda A_q} \, (\vartheta_0 - \vartheta_S) \tanh(mL) \quad . \tag{2.59}$$

The functions $(\cosh mL)^{-1}$ and $\tanh mL$ are represented in Fig. 2.8.

If the simple equations (2.57) to (2.59) are also to be used for $\alpha_L \neq 0$, then the rod of length L, with heat release at $x = L$ has to be imaginarily replaced by a rod of length $L + \Delta L$ which is insulated at $x = L + \Delta L$. The extra length ΔL is determined such, that the heat flow $\dot{Q}_L$ released via the end surface is now released via the additional circumferential area $U\Delta L$. For small values of ΔL it is approximately valid that

$$\dot{Q}_L = \alpha_L A_q \, (\vartheta_L - \vartheta_S) = \alpha U \Delta L \, (\vartheta_L - \vartheta_S) \quad .$$

From which the corrected length L_C of the replacement rod is obtained as

$$L_C = L + \Delta L = L + \frac{\alpha_L}{\alpha} \frac{A_q}{U} \quad . \tag{2.60}$$

This value is to be used in place of L in equations (2.57) to (2.59).

We will now look at the case where the end of the rod $x = L$ has a given temperature ϑ_L, Fig. 2.9. After several calculations the temperature profile is found to be

$$\vartheta^+(x) = \frac{\sinh[m(L - x)]}{\sinh(mL)} + \vartheta^+(L)\frac{\sinh(mx)}{\sinh(mL)} \quad . \tag{2.61}$$

In order to calculate the heat released by the rod between $x = 0$ and $x = L$, the heat flow in the x-direction through the two cross sections $x = 0$ and $x = L$ has to be determined. For any cross section we obtain

$$\dot{Q}(x) = -\lambda A_q \frac{d\vartheta}{dx} = \frac{\lambda A_q m \, (\vartheta_0 - \vartheta_S)}{\sinh(mL)} \left[\cosh[m(L - x)] - \vartheta^+(L)\cosh(mx)\right] \quad .$$

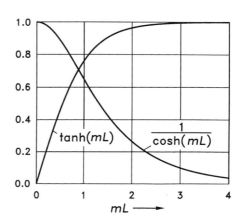

Fig. 2.8: Characteristic functions for calculation of the overtemperature at the free end of the rod according to (2.58) and of the heat flow released according to (2.59)

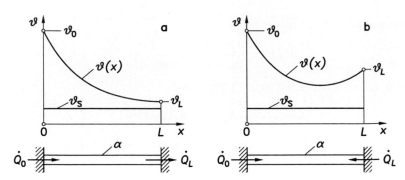

Fig. 2.9: Temperature profile for a rod, whose ends are maintained at the given temperatures ϑ_0 and ϑ_L. **a** at $x = L$ heat flows out ($\dot{Q}_L > 0$), **b** at $x = L$ heat flows in ($\dot{Q}_L < 0$)

From which we obtain, with m according to (2.52)

$$\dot{Q}_0 = \sqrt{\alpha \lambda A_q U} \frac{(\vartheta_0 - \vartheta_S)}{\sinh(mL)} \left[\cosh(mL) - \vartheta^+(L)\right]$$

and

$$\dot{Q}_L = \sqrt{\alpha \lambda A_q U} \frac{(\vartheta_0 - \vartheta_S)}{\sinh(mL)} \left[1 - \vartheta^+(L)\cosh(mL)\right] \quad .$$

The heat released between the two ends via the outer surface is then

$$\dot{Q} = \dot{Q}_0 - \dot{Q}_L = \sqrt{\alpha \lambda A_q U} \left(\vartheta_0 - \vartheta_S\right) \left[1 + \vartheta^+(L)\right] \frac{\cosh(mL) - 1}{\sinh(mL)} \quad . \tag{2.62}$$

Example 2.2: A cylindrical steel bolt ($\lambda = 52.5$ W/K m) with diameter $d = 0.060$ m and length $L = 0.200$ m protrudes from a plate, which has a temperature of $\vartheta_0 = 60.0$ °C. Heat is transferred to the air at temperature $\vartheta_S = 12.5$ °C, via the outer surface and the free end of the bolt. The heat transfer coefficient for the two surfaces is $\alpha = 8.0$ W/m^{2}K. The heat flow, $\dot{Q}_0$, transferred to the air, and the temperature ϑ_L of the free end shall be calculated.

For the heat flow from (2.56) with $U = \pi d$ and $A_q = \pi d^2/4$, we obtain

$$\dot{Q}_0 = \frac{\pi}{2} d \sqrt{\alpha \lambda d} \left(\vartheta_0 - \vartheta_S\right) \frac{\tanh(mL) + (\alpha/m\lambda)}{1 + (\alpha/m\lambda)\tanh(mL)} \quad .$$

Thereby according to (2.52)

$$mL = 2\sqrt{\frac{\alpha}{\lambda d}} L = 0.6375$$

and

$$\frac{\alpha}{m\lambda} = \frac{1}{2}\sqrt{\frac{\alpha d}{\lambda}} = 0.0478 \quad .$$

This gives $\dot{Q}_0 = 13.3$ W. The temperature of the free bolt end is yielded from (2.55) with $x = L$ to be

$$\vartheta^+ = \frac{1}{\cosh(mL) + (\alpha/mL)\sinh(mL)} = 0.8047$$

which produces

$$\vartheta_L = \vartheta_S + (\vartheta_0 - \vartheta_S)\vartheta_L^+ = 50.7 \text{ °C} \quad .$$

If the heat released via the free end of the bolt is neglected the value for the heat flow $\dot{Q}_0 = 12.7$ W obtained from (2.59) is too low. The temperature ϑ_L is found to have the larger value $\vartheta_L = 51.8\,°C$. Using the corrected bolt length

$$L_C = L + \frac{d}{4} = 0.215\,m$$

according to (2.60) we find that the values for $\dot{Q}_0$ from (2.59) and ϑ_L from (2.57) agree with the exact values to three significant figures.

2.2.3   The temperature distribution in fins and pins

As already shown in section 1.2.3, heat transfer between two fluids can be improved if the surface area available for heat transfer, on the side with the fluid which has the lower heat transfer coefficient, is increased by the addition of fins or pins. However this enlargement of the area is only partly effective, due to the existence of a temperature gradient in the fins without which heat could not be conducted from the fin base. Therefore the average overtemperature decisive for the heat transfer to the fluid is smaller than the overtemperature at the fin base. In order to describe this effect quantitatively, the *fin efficiency* was introduced in section 1.2.3. Its calculation is only possible if the temperature distribution in the fin is known, which we will cover in the following. Results for the fin efficiencies for different fin and pin shapes are given in the next section.

In order to calculate the temperature distribution some limiting assumptions have to be made:

1. The fin (or pin) is so thin that the temperature only changes in the direction from fin base to fin tip.
2. The fin material is homogeneous with constant thermal conductivity λ_f.
3. The heat transfer at the fin surface will be described by a constant heat transfer coefficient α_f.
4. The temperature ϑ_S of the fluid surrounding the fin is constant.
5. The heat flow at the tip of the fin can be neglected in comparison to that from its sides.

These assumptions are generally true with the exception of the constant α_f over the surface of the fin. The influence of a varying heat transfer coefficient has been investigated by S.-Y. Chen and G.L. Zyskowski [2.4], L.S. Han and S.G. Lefkowitz [2.5] and also H.C. Ünal [2.6].

Under the assumptions given above, the temperature distribution in fins can be described by a second order differential equation, which we will now derive. We cut a volume element, from any fin or pin as shown in Fig. 2.10, which has a thickness of Δx. The energy balance for this element is

$$\dot{Q}(x) = \dot{Q}(x + \Delta x) + \Delta \dot{Q}_S \ , \tag{2.63}$$

because the heat flowing, by conduction, into the element at point x, has to deliver the heat conducted further to $x + \Delta x$ and the heat flow $\Delta \dot{Q}_S$, which is transferred via the surface ΔA_f

of the element to the fluid surrounding it which has the temperature ϑ_S. Introducing the overtemperature as

$$\Theta(x) = \vartheta(x) - \vartheta_S \tag{2.64}$$

and taking ΔA_f as the surface area through which heat is released from the fin element, we obtain the following expression for $\Delta \dot{Q}_S$

$$\Delta \dot{Q}_S = \alpha_f \Delta A_f \Theta(x + \varepsilon \Delta x) \quad , \quad 0 \le \varepsilon \le 1 \ .$$

According to Fourier's law, the heat flow through the cross sectional area $A_q(x)$

$$\dot{Q}(x) = -\lambda_f A_q(x) \frac{d\Theta}{dx} \tag{2.65}$$

is transported by conduction. From this we obtain (Taylor series at point x)

$$
\begin{aligned}
\dot{Q}(x + \Delta x) &= \dot{Q}(x) + \frac{d\dot{Q}}{dx}\Delta x + O(\Delta x^2) \\
&= \dot{Q}(x) - \lambda_f \frac{d}{dx}\left[A_q(x)\frac{d\Theta}{dx}\right]\Delta x + O(\Delta x^2) \ ,
\end{aligned}
$$

where $O(\Delta x^2)$ indicates that the rest of the terms is of the order of Δx^2. Putting these relationships for the three heat flows into the balance equation (2.63) gives

$$0 = -\lambda_f \frac{d}{dx}\left[A_q(x)\frac{d\Theta}{dx}\right]\Delta x + O(\Delta x^2) + \alpha_f \Delta A_f \Theta(x + \varepsilon \Delta x) \ .$$

Division by Δx, yields for the limit $\Delta x \to 0$

$$\frac{d}{dx}\left[A_q(x)\frac{d\Theta}{dx}\right] - \frac{\alpha_f}{\lambda_f}\frac{dA_f}{dx}\Theta = 0 \tag{2.66}$$

as the desired differential equation for the overtemperature $\Theta(x)$.

This differential equation covers all forms of extended surfaces, as long as the aforementioned assumptions are met. The different fin or pin shapes are expressed by the terms $A_q = A_q(x)$ for the cross sectional area and $A_f = A_f(x)$ for the fin surface area over which the heat is released. So for a *straight fin* of width b perpendicular to the drawing plane in Fig. 2.11, with a profile function $y = y(x)$ we obtain the following for the two areas

$$A_q(x) = 2y(x)b \quad \text{and} \quad A_f(x) = 2bx \ .$$

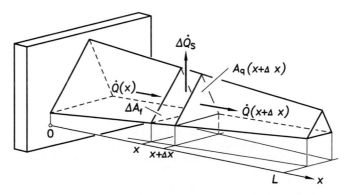

Fig. 2.10: Energy balance for a volume element of thickness Δx and surface area ΔA_f of any shape of fin or pin with a cross sectional area $A_q(x)$

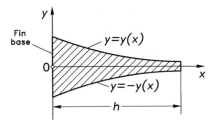

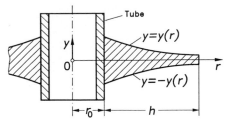

Fig. 2.11: Straight fin of width b (perpendicular to drawing plane) with any profile function $y = y(x)$

Fig. 2.12: Annular fin on a tube (outer radius r_0) with profile function $y = y(r)$

The narrow sides of the fin are not considered in A_f. In addition to this the difference between A_f and its projection on the plane formed by the fin width b and the x-axis will also be negelected. A thin fin is presumed. This leads to the differential equation

$$\frac{d^2\Theta}{dx^2} + \frac{1}{y}\frac{dy}{dx}\frac{d\Theta}{dx} - \frac{\alpha_f}{\lambda_f y}\Theta = 0 \ . \tag{2.67}$$

This has to be solved for a given fin profile $y = y(x)$ under consideration of the boundary conditions at the fin base,

$$\Theta = \Theta_0 = \vartheta_0 - \vartheta_S \quad \text{for} \quad x = 0 \ , \tag{2.68}$$

and the condition

$$\frac{d\Theta}{dx} = 0 \quad \text{for} \quad x = h \ . \tag{2.69}$$

This corresponds to the fifth assumption, the neglection of the heat released at the tip of the fin.

For thin *annular fins*, as in Fig. 2.12, with r as the radial coordinate we obtain

$$A_q(r) = 4\pi r y(r) \quad \text{and} \quad A_f(r) = 2\pi(r^2 - r_0^2) \ ,$$

where $y = y(r)$ describes the fin profile. The differential equation for annular fins is found from (2.66) to be

$$y\frac{d^2\Theta}{dr^2} + \left(\frac{y}{r} + \frac{dy}{dr}\right)\frac{d\Theta}{dr} - \frac{\alpha_f}{\lambda_f}\Theta = 0 \ . \tag{2.70}$$

The solution has to fulfill the boundary conditions

$$\Theta = \Theta_0 = \vartheta_0 - \vartheta_S \quad \text{for} \quad r = r_0$$

at the fin base and

$$d\Theta/dr = 0 \quad \text{for} \quad r = r_0 + h$$

at the tip of the fin.

Solutions to these differential equations, (2.67) and (2.70), for different profile functions $y = y(x)$ and $y = y(r)$ were first given by D.R. Harper and W.B. Brown [2.7] in 1922 and also by E. Schmidt [2.8] in 1926 respectively. In 1945 an extensive investigation of all profiles which lead to differential equations with solutions that are generalised Bessel functions was carried out by K.A. Gardner [2.9]. The differential equation derived from equation (2.66) for cone-shaped pins with various profiles was first given and then solved by R. Focke [2.10] in 1942. A summary of the temperature distributions in various elements of extended heat transfer surfaces can be found in the book by D.Q. Kern and A.D. Kraus [2.11].

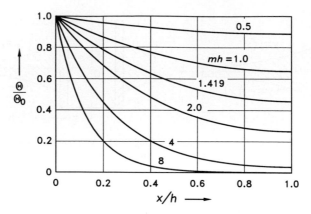

Fig. 2.13: Temperature distribution in a straight fin with a rectangular profile (fin thickness δ_f) as a function of x/h with mh as a parameter according to (2.72)

In the following we will deal with a straight fin with a rectangular profile. Its thickness will be indicated by δ_f, so that the profile function $y(x) = \delta_f/2$ in Fig. 2.11 is valid. From (2.67) follows the simple differential equation

$$\frac{d^2\Theta}{dx^2} - \frac{2\alpha_f}{\lambda_f\delta_f}\Theta = 0 \ .$$

It has the same form as equation (2.51) from section 2.2.2, which is valid for heat conduction in the axial direction of a rod. Using the abbreviation

$$m = \sqrt{\frac{2\alpha_f}{\lambda_f\delta_f}} \tag{2.71}$$

we obtain the solution already known from section 2.2.2

$$\Theta = \Theta_0\frac{\cosh\left[m(h-x)\right]}{\cosh(mh)} \ , \tag{2.72}$$

which fulfills the boundary conditions (2.68) and (2.69). If the heat release at the tip of the fin, neglected in (2.69), is to be considered, a corrected fin height, corresponding to (2.60), can be introduced, by which h will be replaced by $h_C = h + \Delta h = h + \delta_f/2$. Eq. (2.72) is also valid for a pin with constant cross sectional area A_q. This only requires that $\delta_f = A_q/2U$ with U as the circumference of the cross sectional area be put into (2.71).

Fig. 2.13 shows the temperature profile as a function of the dimensionless coordinate x/h for different values of the parameter mh, in which the dimensions of the fin, its thermal conductivity and the heat transfer coefficient are combined. Values in the region $0.7 < mh < 2$ should be chosen for practical applications. In long fins with $mh > 2$ the temperature quickly falls, and a significant portion of the fin transfers only a small amount of heat due to a too small overtemperature. Values of $mh < 0.7$ indicate that by lengthening the fin, a significantly larger heat flow could have been transferred.

An optimal value for mh can be found under the condition that the heat flow $\dot{Q}_\mathrm{f}$ released from the fin should be as large as possible, provided that the material usage, i.e. the volume of the fin is constant. We obtain $\dot{Q}_\mathrm{f}$ as the heat flow conducted through the fin base $(x = 0)$ to be

$$\dot{Q}_\mathrm{f} = -\lambda_\mathrm{f} b\, \delta_\mathrm{f} \left(\frac{\mathrm{d}\Theta}{\mathrm{d}x}\right)_{x=0} \quad . \tag{2.73}$$

Out of (2.72) comes

$$\dot{Q}_\mathrm{f} = \sqrt{2\alpha_\mathrm{f}\lambda_\mathrm{f}\delta_\mathrm{f}}\; b\, \Theta_0 \tanh\left(mh\right) \quad . \tag{2.74}$$

With the fin volume

$$V_\mathrm{f} = \delta_\mathrm{f} h b$$

we get

$$\dot{Q}_\mathrm{f} = \sqrt{2\alpha_\mathrm{f}\lambda_\mathrm{f}V_\mathrm{f} b}\; \frac{\Theta_0}{\sqrt{h}} \tanh\left(\sqrt{\frac{2\alpha_\mathrm{f} b}{\lambda_\mathrm{f}V_\mathrm{f}}}h^{3/2}\right) \quad .$$

In general, the fin width b is set by the dimensions of the apparatus which is to be cooled. $\dot{Q}_\mathrm{f}$ therefore depends on h and the condition $\mathrm{d}\dot{Q}_\mathrm{f}/\mathrm{d}h = 0$ leads to the equation

$$\tanh(mh) = \frac{3mh}{\cosh^2(mh)} = 3mh\left(1 - \tanh^2(mh)\right)$$

with the solution

$$mh = \sqrt{\frac{2\alpha_\mathrm{f}}{\lambda_\mathrm{f}\delta_\mathrm{f}}}h = 1.4192 \quad .$$

The chosen height h of the fin using this condition gives the maximum heat flow for a set fin volume V_f. A similar calculation for annular fins with different profile functions is available in A. Ullmann and H. Kalman [2.12].

E. Schmidt [2.8] determined the fin shape, which for a specific thermal power, required the least material. The profile of these fins is a parabola, with its vertex at the tip of the fin. These types of pointed fins are difficult to produce, which is why fins used in practice have either rectangular, trapezoidal or triangular cross sections.

2.2.4 Fin efficiency

The fin efficiency

$$\eta_\mathrm{f} := \frac{\dot{Q}_\mathrm{f}}{\dot{Q}_\mathrm{f0}} = \frac{\vartheta_\mathrm{f} - \vartheta_\mathrm{S}}{\vartheta_0 - \vartheta_\mathrm{S}} = \frac{\Theta_\mathrm{f}}{\Theta_0} \tag{2.75}$$

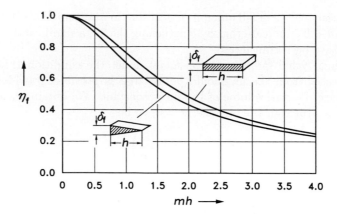

Fig. 2.14: Efficiency η_f of a straight fin with a rectangular profile according to (2.77) and with a triangular profile according to (2.79) as a function of mh from (2.78)

was introduced in section 1.2.3, as the ratio of the actual heat flow $\dot{Q}_f$ released by the fin to the heat flow

$$\dot{Q}_{f0} = \alpha_f A_f (\vartheta_0 - \vartheta_S) = \alpha_f A_f \Theta_0 \ ,$$

which would be released, if the temperature throughout the fin was the same as at its base ϑ_0, rather than the lower mean temperature ϑ_f. Here ϑ_S is the fluid temperature which was indicated by ϑ_2 in section 1.2.3. The heat flow released from the fin agrees with the heat flow $\dot{Q}(x = 0)$, which is conducted at the fin base into the fin itself:

$$\dot{Q}_f = \dot{Q}(x = 0) = -\lambda_f A_{q0} \left(d\Theta/dx \right)_{x=0} \ . \tag{2.76}$$

Here A_{q0} is the cross sectional area at the fin base.

The heat flow $\dot{Q}_f$ for a straight fin with a rectangular profile was calculated in the last section. Using (2.74) and $A_{q0} = b\,\delta_f$ it follows from (2.76) that

$$\eta_f = \frac{\tanh(mh)}{mh} \ . \tag{2.77}$$

The efficiency of a straight rectangular fin only depends on the dimensionless group

$$mh = \sqrt{\frac{2\alpha_f}{\lambda_f \delta_f}}\, h \tag{2.78}$$

Fig. 2.14 shows η_f according to (2.77). For the optimal value determined in 2.2.3 of $mh = 1.4192$ for the fin with the largest heat release at a given volume the efficiency is $\eta_f = 0.627$.

The efficiencies of other fin shapes can be found in the same manner from the temperature distribution in the fins. Fig. 2.14 shows η_f as a function of mh for a straight fin with a triangular profile. In this case δ_f is the thickness of the fin at its base. The plot of the efficiency η_f is similar to the plot of the fin efficiency for a straight rectangular fin. It is therefore possible to replace the expression for η_f containing Bessel functions with a similar function to that in equation (2.77):

$$\eta_f = \frac{\tanh\left(\varphi mh\right)}{\varphi mh} \ . \tag{2.79}$$

The correction factor φ is given by

$$\varphi = 0.99101 + 0.31484\frac{\tanh\left(0.74485mh\right)}{mh} \ , \tag{2.80}$$

from which we obtain a reproduction of the exact result, which for values of $mh < 5$ has an accuracy better than 0.05%.

For the frequently used annular fins of constant thickness δ_f, $y(r) = \delta_f/2$ has to be put into (2.70) for the profile function. The fin efficiency η_f is dependent on two dimensionless groups: mh according to (2.78) and the radius ratio $(r_0+h)/r_0 = 1+ h/r_0$, cf. Fig. 2.12. This yields a complicated expression containing modified Bessel functions. F. Brandt [2.13] found the rather accurate approximation equation

$$\eta_f = \frac{2r_0}{2r_0 + h}\frac{\tanh(mh)}{mh}\left[1 + \frac{\tanh(mh)}{2mr_0} - C\frac{[\tanh(mh)]^p}{(mr_0)^n}\right] \tag{2.81}$$

with $C = 0.071882$, $p = 3.7482$, $n = 1.4810$. It has a maximum error of 0.6% for all values of mh and for $mr_0 \geq 0.2$. A more simple approximation equation was found by Th.E. Schmidt [2.14]:

$$\eta_f = \frac{\tanh\left(mh\varphi\right)}{mh\varphi} \tag{2.82}$$

with

$$\varphi = 1 + 0.35\ln\left(1 + h/r_0\right) \ . \tag{2.82a}$$

For $\eta_f > 0.5$ it deviates by no more than $\pm 1\%$ from the exact values.

Frequently square, rectangular or hexagonal disk fins are attached to tubes, whereby several tubes can also be joined together by the use of sheets of fins through which the tubes pass, Fig. 2.15. In these fins the temperature does not only depend on one coordinate, but two-dimensional temperature fields must be reckoned with. As a first approach, the efficiency η_f of these disk fins can be calculated from (2.81) or (2.82) for an annular fin with the same surface area. Then for a rectangular fin as in Fig. 2.15 we get

$$h = \sqrt{s_1s_2/\pi} - r_0 = 0.564\sqrt{s_1s_2} - r_0 \tag{2.83}$$

and for the hexagonal fin with side length s

$$h = \sqrt{\frac{3\sqrt{3}}{2\pi}}s - r_0 = 1.211s - r_0 \tag{2.84}$$

to put into the equations. Somewhat more accurate values have been found by Th.E. Schmidt [2.14] with

$$h = 0.64\sqrt{s_2\left(s_1 - 0.2s_2\right)} - r_0 \tag{2.85}$$

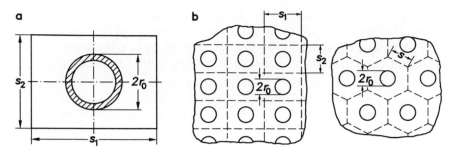

Fig. 2.15: Disk fins. **a** rectangular disks on a tube, **b** tube arrangement with disk fins (rectangular and hexagonal)

for the rectangular fin with s_1 taken as the larger side of the rectangle. E.M. Sparrow and S.H. Lin [2.15] provided analytical approximations for square ($s_1 = s_2$) fins and for the hexagonal shaped fins, which with more extensive calculations produce any exact value for η_f. After that the simple equations (2.83) with $s_1 = s_2 = s$ are sufficiently exact, as long as $h/r_0 \geq 0.5$. H.D. Baehr and F. Schubert [2.16] have experimentally determined the efficiency of a square disk fin with an electrical analogy method and have confirmed that the approximation equations (2.82) and (2.85) are in the main correct.

2.2.5   Geometric multi-dimensional heat flow

Plane and spatial steady temperature fields, which appear in geometric multi-dimensional heat flow, are significantly more difficult to calculate than the cases considered up until now, in which ϑ only changed in one coordinate direction. Solutions to the Laplace equation for plane temperature fields without heat sources

$$\nabla^2\vartheta = \frac{\partial^2\vartheta}{\partial x^2} + \frac{\partial^2\vartheta}{\partial y^2} = 0 \ , \tag{2.86}$$

can be obtained by applying different mathematical methods. The product or separation formula

$$\vartheta(x,y) = f_1(x) \cdot f_2(y)$$

yields two easy to solve ordinary differential equations from (2.86). The fulfillment of the boundary conditions is only possible in an easy way if the temperature field is sought in a rectangle with edges which run parallel with the x and y directions respectively. Examples of this method can be found in the book by S. Kakaç and Y. Yener [2.17].

A further solution method is *conformal mapping*. Its application has been described by H.S. Carslaw and J.C. Jaeger [2.1], as well as by U. Grigull and H. Sandner [2.18]. Regions with complex geometry can also be treated using this method. However, simple solutions are only obtained when as boundary conditions constant temperatures or adiabatic edges are prescribed. In general a heat transfer condition can only be considered approximately. As an example for this an investigation by K. Elgeti [2.19] should be mentioned, where he calculated the heat released from a pipe embedded in a wall. Finally the method of superposition of *heat sources and sinks* should be mentioned. This method also allows complex temperature fields between isothermally bounded bodies to be calculated. It corresponds to the singularity method which is used in fluid mechanics for calculating the potential flows around bodies with any given contour, cf. for example [2.20]. An application of this method will be shown in the next section.

2.2.5.1 Superposition of heat sources and heat sinks

In section 1.1.2 we calculated the temperature distribution $\vartheta = \vartheta(r)$ in a hollow cylinder of length L The heat flow in the radial direction is

$$\dot{Q} = 2\pi L\lambda \left(\vartheta - \vartheta_0\right) / \ln\left(r_0/r\right) \ ,$$

where ϑ is the temperature at distance r from the cylinder axis and ϑ_0 is the temperature at the radius r_0. The heat flow $\dot{Q}$ can be perceived as the strength of a linear heat source at $r = 0$, which runs parallel to the z-axis. This generates the temperature field which is only dependent on r,

$$\vartheta(r) = \vartheta_0 + \frac{\dot{Q}}{2\pi L\lambda} \ln\left(r_0/r\right) \ , \tag{2.87}$$

with temperatures which approach infinity for $r \to 0$.

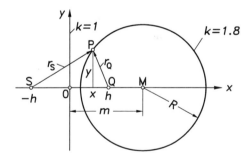

Fig. 2.16: Superposition of a linear heat source at point Q and a linear heat sink at point S

Such a linear or cylindrical heat source may lie on the x-axis at point Q at a distance h from the origin, Fig. 2.16. A heat sink of strength $(-\dot{Q})$ is located at point S at a distance $(-h)$ from the origin. We wish to determine the plane temperature field which is generated by the superposition of the source and the sink in the thermally conductive material.

At any particular point P in Fig. 2.16, the temperature is

$$\vartheta = \vartheta_{0Q} + \frac{\dot{Q}}{2\pi L\lambda} \ln \frac{r_0}{r_Q} + \vartheta_{0S} - \frac{\dot{Q}}{2\pi L\lambda} \ln \frac{r_0}{r_S} \ ,$$

from which, with $\vartheta_0 = \vartheta_{0Q} + \vartheta_{0S}$

$$\vartheta = \vartheta_0 + \frac{\dot{Q}}{2\pi L\lambda} \ln \frac{r_S}{r_Q} \tag{2.88}$$

is obtained. Here r_Q and r_S represent the distance of point P from the source or the sink respectively. An isotherm $\vartheta = \text{const}$ is, according to this equation, a line on which the distance ratio

$$k = r_S/r_Q \tag{2.89}$$

has a *constant* value. At all points on the y-axis, we have $r_S = r_Q$, and therefore the y-axis is the isotherm with $k = 1$, where $\vartheta = \vartheta_0$. All points to the right of the y-axis ($x > 0$) have a distance ratio $k > 1$ and temperatures $\vartheta > \vartheta_0$. For the points where $x < 0$, $0 \le k < 1$ and $\vartheta < \vartheta_0$ are valid.

We will now show that all isotherms $\vartheta \ne \vartheta_0$ form a set of circles with centres which all lie on the x-axis. According to Fig. 2.16

$$k^2 = \frac{r_S^2}{r_Q^2} = \frac{(x+h)^2 + y^2}{(x-h)^2 + y^2} \quad ,$$

is valid and from that follows

$$\left(x - \frac{k^2+1}{k^2-1}h\right)^2 + y^2 = \frac{4k^2}{(k^2-1)^2}h^2 \quad . \tag{2.90}$$

This is the equation for a circle with radius

$$R = \frac{2k}{k^2-1}h \quad , \tag{2.91}$$

whose centre M lies on the x-axis and at a distance

$$m = \frac{k^2+1}{k^2-1}h \tag{2.92}$$

from the origin. The isotherms $\vartheta > \vartheta_0$ lie to the right of the y-axis, because $k > 1$. As the temperature rises (increasing k) the circle radii get smaller, and the centre of the circle moves closer to the source point Q, Fig. 2.17. Isotherms with $\vartheta < \vartheta_0$ are circles which lie to the left of the y-axis. With falling temperatures the circles are drawn closer around sink S.

From (2.91) and (2.92) the following is obtained by squaring:

$$h^2 = m^2 - R^2 \quad . \tag{2.93}$$

By forming the ratio m/R, a quadratic equation for k with the solutions

$$k = \frac{m}{R} \pm \sqrt{\left(\frac{m}{R}\right)^2 - 1} \tag{2.94}$$

results. The positive root delivers distance ratios $k > 1$, and is therefore to be applied to circles to the right of the y-axis. The other root yields values of $k < 1$, which correspond to circles to the left of the y-axis.

The isotherms given by (2.88), (2.89) and (2.90) and represented in Fig. 2.17 allow the steady-state heat conduction in the medium between two isothermal circles, e.g. between two tubes with constant surface temperatures ϑ_1 and ϑ_2 to be calculated. This gives

$$\vartheta_1 - \vartheta_2 = \frac{\dot{Q}}{2\pi L\lambda}(\ln k_1 - \ln k_2)$$

or

$$\dot{Q} = \frac{2\pi L\lambda}{\ln(k_1/k_2)}(\vartheta_1 - \vartheta_2) = \frac{\vartheta_1 - \vartheta_2}{R_{\text{cond}}} \quad . \tag{2.95}$$

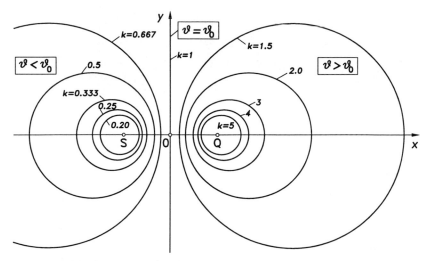

Fig. 2.17: Net of isotherms from (2.88) and (2.90) for different distance ratios k according to (2.89); Q heat source, S heat sink

This is the heat flow between the isotherms $\vartheta = \vartheta_1$ and $\vartheta = \vartheta_2$. It is inversely proportional to the resistance to heat conduction (cf. 1.2.2)

$$R_{\mathrm{cond}} = \frac{\ln (k_1/k_2)}{2\pi L \lambda} \qquad (2.96)$$

which depends on the position of the two isothermal circles (tubes). Thereby three cases must be distinguished:

1. $k_1 > k_2 > 1$. Two eccentric tubes, in Fig. 2.18 a, with axes a distance e away from each other. The larger tube 2 with surface temperature ϑ_2 surrrounds tube 1 with $\vartheta_1 > \vartheta_2$.

2. $k_1 > 1$; $k_2 = 1$. A tube with radius R, lying at a depth m under an isothermal plane e.g. under the earth's surface Fig. 2.18 b.

3. $k_1 > 1$; $k_2 < 1$. Two tubes with radii R_1 and R_2, whose axes have the separation $s > R_1 + R_2$. They both lie in an extensive medium, Fig. 2.18 c.

It is rather simple to calculate the resistance to heat conduction between a tube and an isothermal plane as shown in Fig. 2.18 b. With $k_2 = 1$ and k_1 according to (2.94) we obtain from (2.96)

$$R_{\mathrm{cond}} = \frac{1}{2\pi L \lambda} \ln \left(\frac{m}{R} + \sqrt{\left(\frac{m}{R}\right)^2 - 1} \right) = \frac{1}{2\pi L \lambda} \mathrm{arcosh}\,(m/R) \quad . \qquad (2.97)$$

If at the plane surface heat transfer to a fluid with temperature ϑ_2 is to be considered, this can be approximated by calculating R_{cond} with the enlarged distance

$$m^* = m + \lambda/\alpha \quad .$$

The actual surface at a depth λ/α is not isothermal, but shows a weak temperature maximum directly above the tube, as would be expected from a physical point of view. K. Elgeti [2.21] found an exact solution to this problem. The approximation is surprisingly accurate. Notable deviations from the exact solution only appear for $\alpha R/\lambda < 0.5$ and only with values of $m/R < 2$.

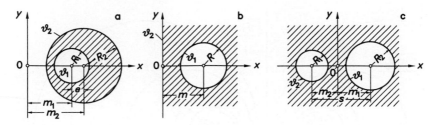

Fig. 2.18: Three arrangements of tubes with isothermal surfaces. **a** two eccentric tubes, one inside the other, **b** a tube and an isothermal plane, **c** two tubes with distance between their centres $s > R_1 + R_2$

The calculation of the conduction resistance between two eccentric tubes, as in Fig. 2.18 a, can only be carried out if both tube radii R_1 and R_2 and the distance between their two centres (eccentricity)

$$e = m_2 - m_1 \tag{2.98}$$

are given. The distances m_1 and m_2 are expressed using these three quantities. According to (2.93)

$$m_2^2 - m_1^2 = R_2^2 - R_1^2 \tag{2.99}$$

and from (2.98) it follows that

$$(m_2 - m_1)^2 = e^2 \ .$$

From these two equations we obtain

$$m_1 = \frac{R_2^2 - R_1^2 - e^2}{2e} \quad \text{and} \quad m_2 = \frac{R_2^2 - R_1^2 + e^2}{2e} \ .$$

This gives the thermal conduction resistance as

$$R_{\text{cond}} = \frac{1}{2\pi L\lambda} \left(\ln k_1 - \ln k_2 \right) = \frac{1}{2\pi L\lambda} \left[\text{arcosh} \, \frac{R_2^2 - R_1^2 - e^2}{2eR_1} - \text{arcosh} \, \frac{R_2^2 - R_1^2 + e^2}{2eR_2} \right]$$

or with the use of the addition theorem for inverse hyperbolic cosine functions

$$R_{\text{cond}} = \frac{1}{2\pi L\lambda} \text{arcosh} \, \frac{R_1^2 + R_2^2 - e^2}{2R_1 R_2} \ . \tag{2.100}$$

In the special case of concentric circles ($e = 0$) the result

$$R_{\text{cond}} = \frac{1}{2\pi L\lambda} \ln \left(R_2/R_1 \right)$$

already known from 1.1.2 will be obtained.

For the third case, two tubes in an extended medium as in Fig. 2.18 c, we express m_1 and m_2 through the tube radii R_1 and R_2 as well as the distance

$$s = m_1 + m_2 \ .$$

This gives, with (2.99)

$$m_1 = \frac{s^2 - \left(R_2^2 - R_1^2 \right)}{2s} \quad \text{and} \quad m_2 = \frac{s^2 + R_2^2 - R_1^2}{2s} \ .$$

The resistance to heat conduction, with

$$\ln k_2 = \ln \left(\frac{m_2}{R_2} - \sqrt{\left(\frac{m_2}{R_2} \right)^2 - 1} \right) = -\text{arcosh} \, \frac{m_2}{R_2}$$

is found to be

$$R_{\text{cond}} = \frac{1}{2\pi L\lambda} \left[\text{arcosh} \, \frac{s^2 - (R_2^2 - R_1^2)}{2sR_1} + \text{arcosh} \, \frac{s^2 + R_2^2 - R_1^2}{2sR_2} \right] \tag{2.101}$$

and finally

$$R_{\text{cond}} = \frac{1}{2\pi L\lambda} \text{arcosh} \, \frac{s^2 - R_1^2 - R_2^2}{2R_1 R_2} \, . \tag{2.102}$$

If the tubes have equal radii $R_1 = R_2 = R$, we obtain from (2.101)

$$R_{\text{cond}} = \frac{1}{\pi L\lambda} \text{arcosh} \, \frac{s}{2R} \, . \tag{2.103}$$

The superposition of several heat sources and sinks allows to calculate more complex temperature fields. W. Nusselt [2.22] replaced the tubes of a radiation heating system embedded in a ceiling by linear heat sources and so calculated the temperature distribution in the ceiling.

2.2.5.2 Shape factors

The heat flow $\dot{Q}$, from one isothermal surface at temperature ϑ_1 to another isothermal surface at temperature ϑ_2, can be calculated according to the simple relationship

$$\dot{Q} = (\vartheta_1 - \vartheta_2) / R_{\text{cond}}$$

if the resistance to conduction R_{cond} is known. Thereby R_{cond} is inversely proportional to the thermal conductivity λ of the material between the two isothermal surfaces. It is reasonable to define

$$R_{\text{cond}} = (\lambda S)^{-1} \tag{2.104}$$

or to put

$$\dot{Q} = \lambda S (\vartheta_1 - \vartheta_2) \tag{2.105}$$

and thereby introduce the *shape coefficient* S. The shape coefficient only depends on the geometrical arrangement of the two isothermal surfaces between which $\dot{Q}$ is transferred by conduction. It has the dimensions of length.

For bodies with two-dimensional (plane) temperature distributions, which have length L perpendicular to the plane of coordinates on which the temperature ϑ depends, a dimensionless shape coefficient $S_L := S/L$ can be defined. This dimensionless number is known as *shape factor*. Examples of these plane temperature fields were dealt with in the last section. The shape factor for a tube of radius R and length L, as in Fig. 2.18 b, lying at a depth of m under an isothermal surface is found, according to (2.97) and (2.104), to be

$$S_L = \frac{S}{L} = \frac{2\pi}{\text{arcosh} \, (m/R)} \, . \tag{2.106}$$

The shape factors for the tube arrangements shown in Figs. 2.18 a and 2.18 c can be found in the same manner from equations (2.100) and (2.102) respectively.

In order to calculate the shape coefficient S in general, the heat flow $\dot{Q}$ has to be determined by integration of the local heat flux

$$\dot{q} = -\lambda \frac{\partial \vartheta}{\partial n}$$

on the isothermal surfaces A_1 and A_2. The heat flow from A_1 to A_2 is given by

$$\dot{Q} = -\lambda \int\limits_{(A_1)} \frac{\partial \vartheta}{\partial n_1} \, dA_1 = \lambda \int\limits_{(A_2)} \frac{\partial \vartheta}{\partial n_2} \, dA_2 \ . \tag{2.107}$$

In which the surface normals n_1 and n_2 are directed into the conductive medium. Equation (2.105) gives the shape coefficient as

$$S = \frac{\dot{Q}}{\lambda \, (\vartheta_1 - \vartheta_2)} = -\frac{1}{\vartheta_1 - \vartheta_2} \int\limits_{(A_1)} \frac{\partial \vartheta}{\partial n_1} \, dA_1 = \frac{1}{\vartheta_1 - \vartheta_2} \int\limits_{(A_2)} \frac{\partial \vartheta}{\partial n_2} \, dA_2 \ . \tag{2.108}$$

This relationship enables S to be calculated from the known temperature field.

E. Hahne and U. Grigull [2.23] have compiled shape coefficients and shape factors for a large number of different geometrical arrangements and organised them systematically. A similar, extensive collection of shape factors for 45 different geometries is available in the VDI-Heat Atlas [2.24].

In the definition of the shape coefficient in (2.105) and its calculation according to (2.107) and (2.108), constant thermal conductivity λ was presumed. The temperature dependence of $\lambda = \lambda(\vartheta)$ is accounted for by the transformed temperature Θ from (2.26), which was introduced in section 2.1.4. It is found that a shape coefficient S calculated for constant λ can be used unaltered, for cases in which $\lambda = \lambda(\vartheta)$, thereby allowing the heat flow between two isothermal surfaces to be calculated. Equation (2.105) can be used for this, provided that λ is replaced by the integral mean value

$$\lambda_{\mathrm{m}} = \frac{1}{\vartheta_1 - \vartheta_2} \int\limits_{\vartheta_2}^{\vartheta_1} \lambda(\vartheta) \, d\vartheta \ . \tag{2.109}$$

This result is a generalisation of equations (1.8) to (1.11) exhibited in section 1.1.2 for geometric one-dimensional heat conduction between the two isothermal surfaces of flat and curved walls.

2.3 Transient heat conduction

Time dependent or transient temperature fields appear when the thermal conditions at the boundaries of the body change. If, for example, a body that initially has a constant temperature is placed in an environment at a different temperature then heat will flow over the surface of the body and its temperature will change over time. At the end of this time dependent process a new steady-state temperature distribution will develop.

In the following sections we will discuss simple solutions, which are also important for practical applications, of the transient heat conduction equation. The problems in the foreground of our considerations will be those where the temperature field depends on time and only one geometrical coordinate. We will discuss the most important mathematical methods for the solution of the equation. The solution of transient heat conduction problems using numerical methods will be dealt with in section 2.4.

2.3.1 Solution methods

The solution of a transient heat conduction problem can be found in three different ways:
1. by a closed solution of the heat conduction equation, fulfilling all the boundary conditions,
2. by a numerical solution of the differential equation (with boundary conditions) using either a finite difference or finite element method.
3. by an experimental method implementing an analogy process.

In order to find a closed solution by mathematical functions the material properties must be assumed to be temperature independent, which according to section 2.1.2 leads to the partial differential equation

$$\frac{\partial \vartheta}{\partial t} = a\nabla^2\vartheta + \frac{\dot{W}}{c\varrho} \ . \tag{2.110}$$

So that a *linear* differential equation is produced, the problem is limited to either conduction without internal heat sources ($\dot{W} \equiv 0$), or the power density $\dot{W}$ is presupposed to be independent of or only linearly dependent on ϑ. In addition the boundary conditions must also be linear, which for the heat transfer condition (2.23) requires a constant or time dependent, but non-temperature dependent heat transfer coefficient α.

The linear initial and boundary condition problems outlined here were solved for numerous cases in the 19th and 20th centuries. The classical solution methods of separating the variables, the superimposing of heat sources and sinks and Green's theorem were all used. In more recent times the Laplace transformation has become the more important solution method for transient heat conduction problems. The classical separation of the variables theory will be applied in sections 2.3.4 to 2.3.6. As the Laplace transformation is not very well known among engineers, a short introduction will be made in the following section. In section 2.3.3 we will show the applications of the Laplace transformation to problems which can easily be solved using this method. A more extensive exposition of the mathematical solution methods and a whole host of results can be found in the standard work by H.S. Carslaw and J.C. Jaeger [2.1].

The *numerical solution* of a transient heat conduction problem is of particular importance when temperature dependent material properties or bodies with irregular shapes or complex boundary conditions, for example a temperature dependent α, are present. In these cases a numerical solution is generally the only choice to solve the problem. The application possibilities for numerical solutions have considerably increased with the introduction of computers. These numerical methods will be discussed in section 2.4.

Experimental analogy procedures are based on the fact that different transient transport processes, in particular electrical conduction, lead to partial differential equations which have the same form as the heat conduction equation. We speak of analogous processes and use a model of such a process which is analogous to heat conduction, in order to transfer the results gained from the model to the thermal conduction process. A short description of this type of analogy model can be found in U. Grigull and H. Sandner [2.18]. As a result of the extensive progress in computer technology, this method has very little practical importance today and for this reason we will not look any further into analogy methods.

2.3.2 The Laplace transformation

The Laplace transformation has proved an effective tool for the solution of the linear heat conduction equation (2.110) with linear boundary conditions. It follows a prescribed solution path and makes it possible to obtain special solutions, for example for small times or at a certain position in the thermally conductive body, without having to determine the complete time and spatial dependence of its temperature field. An introductory illustration of the Laplace transformation and its application to heat conduction problems has been given by H.D. Baehr [2.25]. An extensive representation is offered in the book by H. Tautz [2.26]. The Laplace transformation has a special importance for one-dimensional heat flow, as in this case the solution of the partial differential equation leads back to the solution of a linear *ordinary* differential equation. In the following introduction we will limit ourselves to this case.

$\vartheta = \vartheta(x, t)$ is the temperature distribution that has to be calculated. Multiplying ϑ with the factor e^{-st}, in which s is a (complex) quantity with the same dimensions as frequency, and integrating the product from $t = 0$ to $t \to \infty$, produces a new function

$$u(x, s) = \mathcal{L}\{\vartheta(x, t)\} = \int_0^\infty \vartheta(x, t) e^{-st} \, dt \ . \tag{2.111}$$

This is known as the Laplace transform[2] of the temperature ϑ and depends on s and x. We use the symbol $\mathcal{L}\{\vartheta\}$ when we are stating theorems, while u is an abbreviation for $\mathcal{L}\{\vartheta\}$ in the solution of concrete problems. Often ϑ will be called

---

[2] In some books the Laplace parameter s is denoted by p, as in H.S. Carslaw and J.C. Jaeger [2.1] and in H. Tautz [2.26].

Table 2.2: Some general relationships for the Laplace transform $\mathcal{L}\{\vartheta\} = u$

1. $\mathcal{L}\{\vartheta_1 + \vartheta_2\} = \mathcal{L}\{\vartheta_1\} + \mathcal{L}\{\vartheta_2\}$

2. $\mathcal{L}\left\{\dfrac{\partial\vartheta}{\partial t}\right\} = s\mathcal{L}\{\vartheta\} - \vartheta_0 = s\,u - \vartheta_0(x)$

   with $\vartheta_0(x) = \lim\limits_{t \to +0} \vartheta\,(x,t)$, (Initial temperature profile)

3. $\mathcal{L}\left\{\dfrac{\partial^n\vartheta}{\partial x^n}\right\} = \dfrac{\partial^n u}{\partial x^n}$

4. $\mathcal{L}\left\{\displaystyle\int_0^t \vartheta\,(t')\,\mathrm{d}t'\right\} = \dfrac{1}{s}\mathcal{L}\{\vartheta\}$

5. If $\mathcal{L}\{\vartheta\,(t)\} = u\,(s)$ and k is a positive constant, then

   $$\mathcal{L}\{\vartheta\,(kt)\} = \frac{1}{k}\,u\left(\frac{s}{k}\right)$$

6. $\mathcal{L}\left\{\displaystyle\int_0^t f_1\,(\tau)\,f_2\,(t-\tau)\,\mathrm{d}\tau\right\} = \mathcal{L}\left\{\displaystyle\int_0^t f_2\,(\tau)\,f_1\,(t-\tau)\,\mathrm{d}\tau\right\}$

   $$= \mathcal{L}\{f_1\,(t)\} \cdot \mathcal{L}\{f_2\,(t)\}\ ,$$

   so-called convolution theorem, which is applied, in particular, to time dependent boundary conditions.

the object function with $\mathcal{L}\{\vartheta\}$ as the subfunction or transformed function. As s has the dimensions of frequency, it is said that through the Laplace transformation ϑ is transformed from the time region to the frequency region.

To apply the Laplace transformation several theorems are required, which have been put together, without proofs, in Table 2.2. In addition a table of functions of ϑ with their Laplace transforms u is also needed. This correspondence table is generated by the evaluation of the defining equation (2.111).

So for example, we obtain for

$$\vartheta\,(x,t) = f(x)\,\mathrm{e}^{-ct}$$

the Laplace transform

$$u\,(x,s) = \int_0^\infty f(x)\,\mathrm{e}^{-ct}\mathrm{e}^{-st}\,\mathrm{d}t = f\,(x)\int_0^\infty \mathrm{e}^{-(s+c)t}\,\mathrm{d}t$$

$$= -f\,(x)\left[\frac{\mathrm{e}^{-(s+c)t}}{s+c}\right]_0^\infty = \frac{f\,(x)}{s+c}\ .$$

We have therefore the general correspondence

$$u\,(x,s) \bullet\!\!-\!\!\circ \vartheta\,(x,t)$$

and in our example

$$\frac{f(x)}{s+c} \bullet\!\!-\!\!\circ f(x)\,e^{-ct} \ ,$$

where for the special case of $f(x) \equiv 1$ and $c = 0$

$$\frac{1}{s} \bullet\!\!-\!\!\circ 1$$

is valid.

Further correspondences of this type, which are important for the solution of the heat conduction equation are contained in Table 2.3. More extensive tables of correspondences can be found in the literature, e.g. [2.1], [2.26] to [2.28].

In order to explain the solution process we will limit ourselves to linear heat flow in the x-direction and write the heat conduction equation as

$$\frac{\partial^2 \vartheta}{\partial x^2} - \frac{1}{a}\frac{\partial \vartheta}{\partial t} = 0 \ . \tag{2.112}$$

Applying the Laplace transformation gives, according to Table 2.2, the ordinary differential equation

$$\frac{\mathrm{d}^2 u}{\mathrm{d}x^2} - \frac{s}{a}u = -\frac{1}{a}\vartheta_0(x) \ , \tag{2.113}$$

in which $\vartheta_0(x) = \vartheta(x, t = 0)$ represents the given initial temperature distribution at time $t = 0$. If $\vartheta_0(x) = 0$ a homogeneous linear differential equation is produced, whose solution can immediately be formulated:

$$
\begin{aligned}
u(x,s) &= c_1 \exp\left(-\sqrt{\tfrac{s}{a}}x\right) + c_2 \exp\left(\sqrt{\tfrac{s}{a}}x\right) \\
&= C_1 \cosh\left(\sqrt{\tfrac{s}{a}}x\right) + C_2 \sinh\left(\sqrt{\tfrac{s}{a}}x\right) \ .
\end{aligned}
\tag{2.114}
$$

The two integration constants c_1, c_2 or C_1, C_2 are found by fitting u to the boundary conditions, which are also subjected to a Laplace transform. If an initial condition of $\vartheta_0 \neq 0$ has to be accounted for, the solution given in (2.114) for the homogeneous differential equation has to be supplemented by a particular solution of the inhomogeneous equation.

Once the Laplace transform $u(x, s)$ of the temperature $\vartheta(x, t)$ which fits the initial and boundary conditions has been found, the back-transformation or so-called inverse transformation must be carried out. The easiest method for this is to use a table of correspondences, for example Table 2.3, from which the desired temperature distribution can be simply read off. However frequently $u(x, s)$ is not present in such a table. In these cases the Laplace transformation theory gives an inversion theorem which can be used to find the required solution. The temperature distribution appears as a complex integral which can be evaluated using Cauchy's Theorem. The required temperature distribution is yielded as an infinite series of exponential functions fading with time. We will not deal with the application of the inversion theorem, and so limit ourselves to cases where the

Table 2.3: Table of some correspondences. $u(s, x)$ is the Laplace transform of $\vartheta(t, x)$. The abbreviations $p = \sqrt{s/a}$ and $\xi = x/2\sqrt{at}$ should be recognised.

	$u(s, x)$	$\vartheta(t, x)$
1.	$\dfrac{c}{s}$	c
2.	$\dfrac{1}{s^{\nu+1}}, \nu > -1$	$\dfrac{t^{\nu}}{\Gamma(\nu+1)}$, Γ this is the Gamma function, see below
3.	$\dfrac{1}{s+c}$	e^{-ct}
4.	e^{-px}	$\dfrac{\xi}{\sqrt{\pi t}}e^{-\xi^2}$
5.	$\dfrac{e^{-px}}{p}$	$\sqrt{\dfrac{a}{\pi t}}e^{-\xi^2}$
6.	$\dfrac{e^{-px}}{s}$	$\operatorname{erfc}\xi$, complementary error function , see below
7.	$\dfrac{e^{-px}}{sp}$	$2\sqrt{at}\left(\dfrac{e^{-\xi^2}}{\sqrt{\pi}} - \xi\operatorname{erfc}\xi\right) = 2\sqrt{at}\operatorname{ierfc}\xi$
8.	$\dfrac{e^{-px}}{s^2}$	$t\left[(1+2\xi^2)\operatorname{erfc}\xi - \dfrac{2}{\sqrt{\pi}}\xi e^{-\xi^2}\right]$
9.	$\dfrac{e^{-px}}{p+h}$	$\sqrt{\dfrac{a}{\pi t}}e^{-\xi^2} - hae^{hx+ah^2t}\operatorname{erfc}\left(\xi + h\sqrt{at}\right)$
10.	$\dfrac{e^{-px}}{p(p+h)}$	$ae^{hx+ah^2t}\operatorname{erfc}\left(\xi + h\sqrt{at}\right)$
11.	$\dfrac{e^{-px}}{s(p+h)}$	$\dfrac{1}{h}\left[\operatorname{erfc}\xi - e^{hx+ah^2t}\operatorname{erfc}\left(\xi + h\sqrt{at}\right)\right]$

Special values of the Gamma function:

$$\Gamma(1) = 1 \quad ; \quad \Gamma(n) = (n-1)! = 1\cdot2\cdot3\cdot\ldots(n-1) \quad , \quad (n = 2, 3, \ldots)$$
$$\Gamma(\tfrac{1}{2}) = \sqrt{\pi} \quad ; \quad \Gamma(n+\tfrac{1}{2}) = \sqrt{\pi}\dfrac{1\cdot3\cdot5\cdot\ldots(2n-1)}{2^n} \quad , \quad (n = 1, 2, 3, \ldots)$$

Values for the complementary error functions $\operatorname{erfc}\xi$ and $\operatorname{ierfc}\xi$ can be found in Table 2.5 in section 2.3.3.1. A plot of these functions is shown in Fig. 2.22.

inverse transformation is possible using the correspondence tables. Applications of the inversion theorem are explained in the literature in [2.1], [2.25] and [2.26]. In addition to this method it is also possible to carry out the inverse transformation numerically using different algorithms, see [2.29].

The solution of transient heat conduction problems using the Laplace transformation consists of three steps:

1. Transformation of the differential equation with the initial and boundary conditions into the frequency region $(\vartheta \rightarrow u, t \rightarrow s)$.
2. Solution of the differential equation for the Laplace transform u considering the (transformed) boundary conditions.
3. Inverse transformation to the time region $(u \rightarrow \vartheta, s \rightarrow t)$ using a correspondence table $(u \bullet\!\!-\!\!\circ \vartheta)$ or the general inversion theorem.

Two advantageous properties of the Laplace transformation should be mentioned. If only the temperature change with time at a particular point in the thermally conductive body is required, rather than the temperature distribution for the whole body, then the total Laplace transform $u(s, x)$ does not have to be back-transformed. It is sufficient to just set the position variable in u as constant and then back-transform the simplified function $u(s)$ in $\vartheta(t)$. Besides this simplification for calculating only the required part of the temperature distribution, the possibility of obtaining solutions for small values of time also exists. This can be applied to the beginning of a heat conduction process, and the results assume an especially simple form. The Laplace transform u has to be developed into a series which converges for large values of s. A term by term back-transformation using a correspondence table gives a series for ϑ which converges for small values of t. These particular features and an illustration of the solution procedure are given in the following example.

Example 2.3: A flat wall of thickness δ has a constant temperature ϑ_0. At time $t = 0$ the temperature of the surface $x = \delta$ jumps to ϑ_S, whilst the other surface $x = 0$ is adiabatic, Fig. 2.19. Heat flows from the right hand surface into the wall. The temperature rises with time, whereby the temperature of the left hand surface of the wall rises at the slowest rate. The temperature increase at this point, i.e. the temperature $\vartheta(x = 0, t)$ is to be calculated.

This transient heat conduction problem can be used as a model for the following real process. A fire resistant wall is rapidly heated on its outer side $(x = \delta)$ as a result of a fire. We are interested in the temperature rise over time of the other side of the wall at $x = 0$. The assumption of an adiabatic surface at $x = 0$ results in a faster temperature rise than would be expected in reality. This assumption therefore leaves us on the safe side.

For the solution of this problem we will assume that heat only flows in the x-direction and that the thermal diffusivity a of the wall is constant. Introducing the overtemperature

$$\Theta(x, t) := \vartheta(x, t) - \vartheta_0$$

and leaving the use of dimensionless quantities until the results are presented gives us a clearer picture of the procedure. The heat conduction equation (2.112) takes the form

$$\frac{\partial^2 \Theta}{\partial x^2} - \frac{1}{a}\frac{\partial \Theta}{\partial t} = 0$$

with initial conditions

$$\Theta_0(x) = \Theta(x, 0) = 0 \ .$$

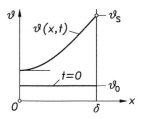

Fig. 2.19: Heating one side of a flat wall, by a sharp rise in the surface temperature at $x = \delta$ from ϑ_0 to ϑ_S

The boundary conditions are

$$\frac{\partial \Theta}{\partial x} = 0 \quad \text{at} \quad x = 0$$

(adiabatic surface) and

$$\Theta(\delta, t) = \Theta_S = \vartheta_S - \vartheta_0 \quad \text{at} \quad x = \delta .$$

Applying the Laplace transformation leads to the ordinary differential equation which corresponds to (2.113)

$$\frac{d^2 u}{dx^2} - \frac{s}{a} u = 0 \tag{2.115}$$

for the Laplace tranform $u = u(x, s)$ of the overtemperature $\Theta(x, t)$. The initial conditions have already been accounted for. Laplace transformation of the two boundary conditions yields

$$\frac{du}{dx} = 0 \qquad \text{for} \quad x = 0 \tag{2.116}$$

and

$$u(\delta, s) = \frac{\Theta_S}{s} \qquad \text{for} \quad x = \delta . \tag{2.117}$$

The general solution of the differential equation (2.115) is

$$u(x, s) = C_1 \cosh px + C_2 \sinh px ,$$

where the abbreviation $p = \sqrt{s/a}$ has been used. From the boundary condition (2.116) it follows that $C_2 = 0$ and (2.117) gives

$$C_1 \cosh p\delta = \frac{\Theta_S}{s} .$$

The required Laplace transform is therefore

$$u(x, s) = \frac{\Theta_S}{s} \frac{\cosh px}{\cosh p\delta} .$$

The inverse transformation can be achieved using the inversion theorem, which we do not need here, as we are only interested in the temperature at the point $x = 0$. For this we find

$$u(0, s) = \frac{\Theta_S}{s \cosh p\delta} = \frac{2\Theta_S}{s \left(e^{p\delta} + e^{-p\delta} \right)} = \frac{2\Theta_S e^{-p\delta}}{s \left(1 + e^{-2p\delta} \right)} .$$

In order to carry out the inverse transformation with the aid of correspondence tables, the numerator has to be expanded into a binomial series, giving

$$u(0, s) = \frac{2\Theta_S e^{-p\delta}}{s} \left[1 - e^{-2p\delta} + e^{-4p\delta} - e^{-6p\delta} + \cdots \right]$$

and finally

$$\frac{u(0, s)}{\Theta_S} = 2 \left[\frac{e^{-p\delta}}{s} - \frac{e^{-3p\delta}}{s} + \frac{e^{-5p\delta}}{s} - \cdots \right] .$$

This is a series development which for large values of s or $p = \sqrt{s/a}$ rapidly converges. The inverse transformation using correspondence tables (No. 6 from Table 2.3 with $x = \delta$, 3δ, 5δ, ...) yields a series that is particularly suitable for small times t. This gives

$$\frac{\Theta(0,t)}{\Theta_S} = 2\left[\text{erfc}\,\xi - \text{erfc}\,3\xi + \text{erfc}\,5\xi - \cdots\right] \ ,$$

in which the abbreviation

$$\xi = \frac{\delta}{2\sqrt{at}}$$

is utilised.

The function $\text{erfc}\,\xi$ that appears here is the error function complement, which will be discussed in section 2.3.3.1 and is also shown in Table 2.4. This transcendental function rapidly approaches zero for increasingly large arguments. This provides us with a series for large values of ξ, correspondingly small values of t, which converges very well. Introducing the dimensionles time

$$t^+ := \frac{at}{\delta^2} = (2\xi)^{-2}$$

produces, for the dimensionless temperature at $x = 0$, the series

$$\vartheta^+\left(t^+\right) := \frac{\vartheta(0,t^+) - \vartheta_0}{\vartheta_S - \vartheta_0} = \frac{\Theta(0,t)}{\Theta_S} = 2\left(\text{erfc}\,\frac{1}{2\sqrt{t^+}} - \text{erfc}\,\frac{3}{2\sqrt{t^+}} + \text{erfc}\,\frac{5}{2\sqrt{t^+}} - \cdots\right) \ .$$

(2.118)

Fig. 2.20 shows the pattern of $\vartheta^+(t^+)$. Only a few terms are required to give the temperature

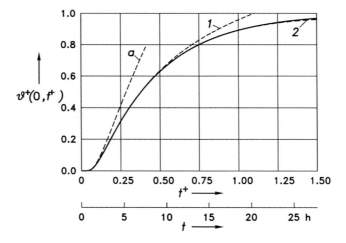

Fig. 2.20: Plot of the dimensionless temperature $\vartheta^+(0,t^+)$ over time at the adiabatic surface ($x = 0$) of the wall from Fig. 2.19. Curve 1: first term, Curve 2: Sum of the first two terms in (2.118), Curve a: asymptotic development according to (2.119) for very small times

even at rather large times t^+ accurately. The behaviour of $\vartheta^+(t^+)$ at very small times is described by the first term of the asymptotic development of $\text{erfc}(1/2\sqrt{t^+})$. According to (2.127) from section 2.3.3.1 we obtain for this

$$\vartheta^+\left(t^+\right) \approx 2\,\text{erfc}\,\frac{1}{2\sqrt{t^+}} \simeq \frac{4}{\sqrt{\pi}}\sqrt{t^+}\exp\left(-1/4t^+\right) \ .$$

(2.119)

This function is represented by curve a in Fig. 2.20.

As an example, a concrete wall with $a = 3.75 \cdot 10^{-6}\,\mathrm{m^2/s}$ and thickness $\delta = 0.50\,\mathrm{m}$ will be considered. For the "real" time t we get

$$t = (\delta^2/a)\, t^+ = 18.5\,\mathrm{h}\, t^+ \ .$$

The additional scale in Fig. 2.20 is drawn from this relationship. As an example, after 2.41 h we reach $\vartheta^+(t^+) = 0.1$, i.e. the temperature at $x = 0$ will rise by 10% of the temperature increase $(\vartheta_S - \vartheta_0)$ assumed at the other wall surface.

2.3.3 The semi-infinite solid

We will consider transient conduction in a very thick plate, whose free surface $x = 0$, cf. Fig. 2.21, is bounded by the surroundings and in all other directions is considered to be infinite. The ground at the earth's surface bordering the atmosphere is a good example of a semi-infinite solid. Bodies of finite thickness can also be treated as infinitely wide at the beginning of a heating or cooling process which occurs at the surface $x = 0$, thereby allowing the relatively simple results that follow to be applied to these cases.

2.3.3.1 Heating and cooling with different boundary conditions

The desired temperature distribution $\vartheta = \vartheta(x,t)$ has to fulfill the differential equation

$$\frac{\partial \vartheta}{\partial t} = a \frac{\partial^2 \vartheta}{\partial x^2}, \quad t \geq 0, \quad x \geq 0, \tag{2.120}$$

and the simple initial condition

$$\vartheta(x, t = 0) = \vartheta_0 = \mathrm{const} \ .$$

In addition ϑ for $x \to \infty$ has to be bounded. At the surface $x = 0$ different boundary conditions are possible. These are shown in Fig. 2.21:
  - a jump in the surface temperature to the value ϑ_S, which should remain constant for $t > 0$, Fig. 2.21a;
  - a constant heat flux $\dot{q}_0$, Fig. 2.21b;
  - a jump in the temperature of the surroundings to the constant value $\vartheta_S \neq \vartheta_0$ for $t > 0$, so that heat transfer with a heat transfer coefficient of α takes place, Fig. 2.21c.

For the calculation of the temperature distribution in all three cases we introduce the overtemperature

$$\Theta := \vartheta - \vartheta_0$$

as a new dependent variable. Then instead of (2.120) we have

$$\frac{\partial^2 \Theta}{\partial x^2} - \frac{1}{a} \frac{\partial \Theta}{\partial t} = 0 \tag{2.121}$$

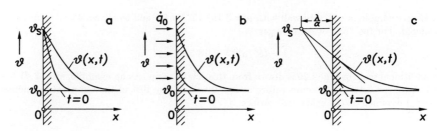

Fig. 2.21: Heating of a semi-infinite body with different boundary conditions at the surface ($x = 0$). **a** jump in the surface temperature to ϑ_S, **b** constant heat flux $\dot{q}_0$, **c** heat transfer from a fluid at $\vartheta = \vartheta_S$

with the initial condition

$$\Theta(x, t = 0) = 0 \ . \tag{2.122}$$

We apply the Laplace transformation to these equations and obtain the ordinary differential equation already discussed in section 2.3.2

$$\frac{d^2 u}{dx^2} - \frac{s}{a} u = 0$$

for the Laplace transform u of Θ with the solution

$$u(x, s) = C \exp\left(-\sqrt{\frac{s}{a}} x\right) = C \exp\left(-px\right) \ . \tag{2.123}$$

As Θ and therefore u must be bounded for $x \to \infty$ the second exponential term $\exp(+px)$ in (2.114) disappears.

Table 2.4 contains the boundary conditions at $x = 0$ for the three cases in Fig. 2.21, the Laplace transforms of the boundary conditions and the expressions yielded for the constant C. The inverse transformation of the Laplace transform u according to (2.123) with C from Table 2.4 succeeds without complication by finding the relevant correspondences in Table 2.3. We will now discuss successively the different temperature distributions.

With *a jump in the surface temperature* from ϑ_0 to ϑ_S (Case a) the temperature distribution looks like (No. 6 from Table 2.3)

$$\frac{\Theta}{\Theta_S} = \frac{\vartheta - \vartheta_0}{\vartheta_S - \vartheta_0} = \text{erfc}\,\frac{x}{2\sqrt{at}} = \text{erfc}\,\xi \ . \tag{2.124}$$

It is dependent on the dimensionless variable combination

$$\xi = \frac{x}{2\sqrt{at}} \ . \tag{2.125}$$

Here

$$\text{erfc}\,\xi = 1 - \text{erf}\,\xi = 1 - \frac{2}{\sqrt{\pi}} \int_0^\xi e^{-w^2} \, dw \tag{2.126}$$

Table 2.4: Boundary conditions at the free surface $x = 0$ of a semi-infinite solid as in Fig. 2.21 with their Laplace transforms and the constants C yielded from Eq. (2.123)

Case	Boundary condition at $x = 0$	Transformed boundary condition	Constant C in Eq. (2.123)
a	$\Theta = \vartheta_S - \vartheta_0 = \Theta_S$	$u = \dfrac{\Theta_S}{s}$	$C = \dfrac{\Theta_S}{s}$
b	$-\lambda \dfrac{\partial \Theta}{\partial x} = \dot{q}_0$	$-\lambda \dfrac{du}{dx} = \dfrac{\dot{q}_0}{s}$	$C = \dfrac{\dot{q}_0}{\lambda \, s \, p}$
c	$-\lambda \dfrac{\partial \Theta}{\partial x} = \alpha(\Theta_S - \Theta)$	$-\lambda \dfrac{du}{dx} = \alpha \left(\dfrac{\Theta_S}{s} - u \right)$	$C = \dfrac{\alpha \Theta_S}{\lambda s \, (p + \alpha/\lambda)}$

is the error function complement, whilst erf ξ is called the error function. Table 2.5 gives values of erfc ξ. More extensive tables are available in [2.30] to [2.32]. Series developments and equations for erf ξ and erfc ξ can be found in J. Spanier and K.B. Oldham [2.28]. For larger arguments ($\xi > 2{,}6$) the asymptotic development

$$\text{erfc}\,\xi = \frac{\exp(-\xi^2)}{\sqrt{\pi}\xi}\left(1 - \frac{1}{2\xi^2} + \frac{3}{4\xi^4} - \frac{15}{8\xi^6} + \cdots\right) \tag{2.127}$$

holds. The functions erf ξ and erfc ξ are represented in Fig. 2.22. The limit erfc$(\xi = 0) = 1$ corresponds to the temperature $\vartheta = \vartheta_S$. It will be reached for $x = 0$, according to the stipulated boundary condition, and also within the body for very large times ($t \to \infty$). The limit erfc$(\infty) = 0$ corresponds to $\vartheta = \vartheta_0$, and consequently to the initial condition for $t = 0$.

The heat flux at depth x at time t is obtained from (2.124) to be

$$\dot{q}(x,t) = -\lambda\frac{\partial\vartheta}{\partial x} = \frac{\lambda(\vartheta_S - \vartheta_0)}{\sqrt{\pi a t}}\exp\left(-x^2/4at\right) \ .$$

It is usual here to introduce the material dependent *thermal penetration coefficient*

$$b := \sqrt{\lambda c \varrho} = \lambda/\sqrt{a} \ , \tag{2.128}$$

which gives

$$\dot{q}(x,t) = \frac{b\,(\vartheta_S - \vartheta_0)}{\sqrt{\pi t}}\exp\left(-x^2/4at\right) \ . \tag{2.129}$$

At the surface $x = 0$

$$\dot{q}_0 = \dot{q}(0,t) = b\,(\vartheta_S - \vartheta_0)/\sqrt{\pi t} \ . \tag{2.130}$$

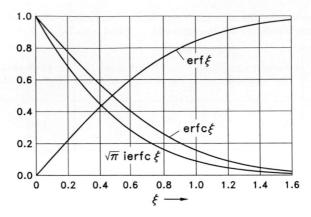

Fig. 2.22: Error functions $\operatorname{erf}\xi$ and $\operatorname{erfc}\xi$ from (2.126) along with integrated error function $\operatorname{ierfc}\xi$ according to (2.134)

The entering heat flux is proportional to the material property b and decays with time as $t^{-1/2}$. During the time interval $t = 0$ to $t = t^*$ heat

$$Q(t^*) = A \int_0^{t^*} \dot{q}(0,t)\,\mathrm{d}t = A \frac{2}{\sqrt{\pi}} b(\vartheta_{\mathrm{S}} - \vartheta_0)\sqrt{t^*} \qquad (2.131)$$

flows through the surface A.

The equations found here are equally valid, without any alteration, for cooling. In this case $\vartheta_{\mathrm{S}} < \vartheta_0$, and in (2.129) to (2.131) the sign is changed. The temperature distribution can, by rearrangement of (2.124), be written as

$$\frac{\vartheta - \vartheta_{\mathrm{S}}}{\vartheta_0 - \vartheta_{\mathrm{S}}} = \operatorname{erf}\xi = \frac{2}{\sqrt{\pi}} \int_0^\xi e^{-w^2}\,\mathrm{d}w \;. \qquad (2.132)$$

If the surface of the semi-infinite solid is heated with *constant heat flux* $\dot{q}_0$ (Case b), the temperature distribution, from No. 7 in Table 2.3 is found to be

$$\Theta(x,t) = \vartheta(x,t) - \vartheta_0 = \frac{\dot{q}_0}{\lambda} 2\sqrt{at}\;\operatorname{ierfc}\xi = 2\frac{\dot{q}_0}{b}\sqrt{t}\;\operatorname{ierfc}\xi \;. \qquad (2.133)$$

The function

$$\operatorname{ierfc}\xi = \frac{1}{\sqrt{\pi}} e^{-\xi^2} - \xi\operatorname{erfc}\xi \qquad (2.134)$$

is also called the integrated error function. It is shown in Fig. 2.22 and in Table 2.5. With $\operatorname{ierfc}(0) = \pi^{-1/2}$ the surface temperature is obtained as

$$\vartheta(0,t) = \vartheta_0 + \frac{\dot{q}_0}{b} \frac{2}{\sqrt{\pi}}\sqrt{t} \;. \qquad (2.135)$$

Table 2.5: Values of the error function complement erfc ξ from (2.126) and the integrated error function ierfc ξ in the form $\sqrt{\pi}$ ierfc $\xi = e^{-\xi^2} - \sqrt{\pi}\,\xi$ erfc ξ

ξ	erfc ξ	$\sqrt{\pi}$ ierfc ξ	ξ	erfc ξ	$\sqrt{\pi}$ ierfc ξ	ξ	erfc ξ	$\sqrt{\pi}$ ierfc ξ
0	1.00000	1.00000	0.75	0.28884	0.18581	1.5	0.03389	0.01528
0.05	0.94363	0.91388	0.80	0.25790	0.16160	1.6	0.02365	0.01023
0.10	0.88754	0.83274	0.85	0.22933	0.14003	1.7	0.01621	0.00673
0.15	0.83200	0.75655	0.90	0.20309	0.12088	1.8	0.01091	0.00436
0.20	0.77730	0.68525	0.95	0.17911	0.10396	1.9	0.00721	0.00277
0.25	0.72367	0.61874	1.00	0.15730	0.08907	2.0	0.00468	0.00173
0.30	0.67137	0.55694	1.05	0.13756	0.07602	2.1	0.00298	0.00107
0.35	0.62062	0.49970	1.10	0.11980	0.06463	2.2	0.00186	0.00064
0.40	0.57161	0.44688	1.15	0.10388	0.05474	2.3	0.00114	0.00038
0.45	0.52452	0.39833	1.20	0.08969	0.04617	2.4	0.00069	0.00022
0.50	0.47950	0.35386	1.25	0.07710	0.03879	2.5	0.00041	0.00013
0.55	0.43668	0.31327	1.30	0.06599	0.03246	2.6	0.00024	0.00007
0.60	0.39614	0.27639	1.35	0.05624	0.02705	2.7	0.00013	0.00004
0.65	0.35797	0.24299	1.40	0.04771	0.02246	2.8	0.00008	0.00002
0.70	0.32220	0.21287	1.45	0.04031	0.01856	2.9	0.00004	0.00001

It increases quickly at first and then slower as time t goes on. With large thermal penetration coefficients b the material can "swallow" large heat flows, so that the surface temperature increases more slowly than in a body with small b.

If *heat transfer* from a fluid at $\vartheta = \vartheta_S$ to the free surface, takes place at $x = 0$, Case c from Table 2.4, the temperature is also obtained from a correspondence table (No. 11 in Table 2.3). The solution is written in dimensionless form with the variables

$$\vartheta^+ := \frac{\vartheta - \vartheta_0}{\vartheta_S - \vartheta_0} \tag{2.136}$$

and

$$x^+ := \frac{x}{\lambda/\alpha} \quad , \quad t^+ := \frac{at}{(\lambda/\alpha)^2} \tag{2.137}$$

giving

$$\vartheta^+(x^+, t^+) = \operatorname{erfc}\frac{x^+}{2\sqrt{t^+}} - \exp\left(x^+ + t^+\right)\operatorname{erfc}\left(\frac{x^+}{2\sqrt{t^+}} + \sqrt{t^+}\right) \quad . \tag{2.138}$$

This temperature distribution is illustrated in Fig. 2.23. The heat fluxes at the surface and within the body are easily calculated by differentiating (2.138). The temperature distribution for cooling, $\vartheta_S < \vartheta_0$, is found to be

$$\frac{\vartheta - \vartheta_S}{\vartheta_0 - \vartheta_S} = 1 - \vartheta^+(x^+, t^+) \quad ,$$

where $\vartheta^+(x^+, t^+)$ is calculated according to (2.138).

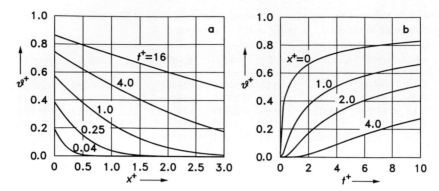

Fig. 2.23: Temperature field $\vartheta^+(x^+, t^+)$ from (2.138) in a semi-infinite solid with heat transfer from a fluid to the free surface $x^+ = 0$. **a** ϑ^+ as a function of x^+ with t^+ as a parameter, **b** ϑ^+ as a function of t^+ with x^+ as a parameter

2.3.3.2   Two semi-infinite bodies in contact with each other

We will consider the two semi-infinite bodies shown in Fig. 2.24, which have different, but constant, initial temperatures ϑ_{01} and ϑ_{02}. Their material properties λ_1, a_1 and λ_2, a_2 are also different. At time $t = 0$ both bodies are brought into (thermal) contact with each other along the plane indicated by $x = 0$. After a very short period of time an average temperature ϑ_m is reached along the plane. Heat flows from body 1 with the higher initial temperature to body 2 which has a lower temperature. The transient conduction process decribed here serves as a model for the description of short-time contact between two (finite) bodies at different temperatures. Examples of this include the touching of different objects with a hand or foot and the short-time interaction of a heated metal body with a cooled object in reforming processes.

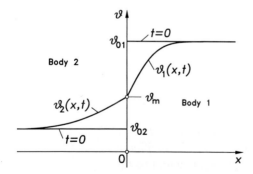

Fig. 2.24: Temperature pattern in two semi-infinite bodies with initial temperatures ϑ_{01} and ϑ_{02} in contact with each other along the plane $x = 0$

The heat conduction equation holds for both bodies and is

$$\frac{\partial \vartheta_1}{\partial t} = a_1 \frac{\partial^2 \vartheta_1}{\partial x^2}, \quad x \geq 0 ,$$

as well as

$$\frac{\partial \vartheta_2}{\partial t} = a_2 \frac{\partial^2 \vartheta_2}{\partial x^2} , \quad x \leq 0 .$$

At the interface the conditions, according to section 2.1.3

$$(\vartheta_1)_{+0} = (\vartheta_2)_{-0} = \vartheta_{\mathrm{m}}$$

and

$$\lambda_1 \left(\frac{\partial \vartheta_1}{\partial x} \right)_{+0} = \lambda_2 \left(\frac{\partial \vartheta_2}{\partial x} \right)_{-0} \qquad (2.139)$$

must be fulfilled. In addition to this the initial conditions already mentioned are valid

For the solution, we assume the contact temperature ϑ_{m} at $x = 0$ to be independent of t. As will be shown this assumption is valid and enables a simple solution to the problem to be found. The problem is split into two parts whose solutions have already been discussed in section 2.3.3.1: A temperature distribution appears in each body, which develops beginning with the constant initial temperatures ϑ_{01} and ϑ_{02}, respectively, when the surface temperature jumps to the constant value ϑ_{m}.

For body 1, with the following variable combination

$$\xi_1 = x/\sqrt{4a_1 t} \qquad (2.140)$$

we obtain, according to (2.132), the temperature

$$\vartheta_1 = \vartheta_{\mathrm{m}} + (\vartheta_{01} - \vartheta_{\mathrm{m}}) \operatorname{erf} \xi_1 \quad , \quad x \geq 0 , \qquad (2.141)$$

with a derivative at $x = +0$ of

$$\left(\frac{\partial \vartheta_1}{\partial x} \right)_{+0} = \frac{\vartheta_{01} - \vartheta_{\mathrm{m}}}{\sqrt{\pi a_1 t}} . \qquad (2.142)$$

Analogous to this, with

$$\xi_2 = x/\sqrt{4a_2 t} \qquad (2.143)$$

it holds for body 2 according to (2.124)

$$\vartheta_2 = \vartheta_{02} + (\vartheta_{\mathrm{m}} - \vartheta_{02}) \operatorname{erfc}(-\xi_2) \quad , \quad x \leq 0 , \qquad (2.144)$$

and at the boundary

$$\left(\frac{\partial \vartheta_2}{\partial x} \right)_{-0} = \frac{\vartheta_{\mathrm{m}} - \vartheta_{02}}{\sqrt{\pi a_2 t}} . \qquad (2.145)$$

Putting (2.142) and (2.145) in the boundary condition (2.139), a relationship *independent* of t follows:

$$\frac{\lambda_1}{\sqrt{a_1}} (\vartheta_{01} - \vartheta_{\mathrm{m}}) = \frac{\lambda_2}{\sqrt{a_2}} (\vartheta_{\mathrm{m}} - \vartheta_{02})$$

or

$$b_1 (\vartheta_{01} - \vartheta_{\mathrm{m}}) = b_2 (\vartheta_{\mathrm{m}} - \vartheta_{02}) . \qquad (2.146)$$

The assumption of a time independent contact temperature ϑ_{m} was therefore appropriate. Its position depends on the thermal penetration coefficients

$$b_1 = \sqrt{\lambda_1 c_1 \varrho_1} \quad \text{and} \quad b_2 = \sqrt{\lambda_2 c_2 \varrho_2}$$

of the two bodies. It follows from (2.146) that

$$\vartheta_{\mathrm{m}} = \vartheta_{02} + \frac{b_1}{b_1 + b_2} (\vartheta_{01} - \vartheta_{02}) . \qquad (2.147)$$

The time independent contact temperature lies in the vicinity of the initial temperature of the body with the larger thermal penetration coefficient. Equation (2.147) explains why different solids at the same temperature feel as if they were warmed to differing degrees when touched by the hand or foot.

2.3.3.3   Periodic temperature variations

Periodic temperature variations frequently appear in both nature and technology. These include the daily temperature changes which occur in building walls and the daily or seasonal temperature variations in the earth's crust. The cylinders in combustion engines undergo large and high frequency temperature changes, which penetrate the inner walls of the cylinders and can have some influence on the strength of the material. In the following we will consider a simple model that reproduces these temperature variations in an approximate manner and indicates their most significant properties.

We will look at a semi-infinite solid with constant material properties λ and a. It is bounded by a fluid at the free surface $x = 0$, which has a temperature ϑ_F, which changes according to the time law

$$\vartheta_F(t) = \vartheta_m + \Delta\vartheta \cos\omega t = \vartheta_m + \Delta\vartheta \cos\left(2\pi\frac{t}{t_0}\right) . \qquad (2.148)$$

This temperature oscillation is harmonic with a periodic time of t_0 and an amplitude of $\Delta\vartheta$ around the mean value ϑ_m. At the surface $x = 0$ the heat transfer condition is

$$-\lambda\left(\frac{\partial\vartheta}{\partial x}\right)_{x=0} = \alpha[\vartheta_F - \vartheta(0,t)] \qquad (2.149)$$

with a constant value for α stipulated. We are looking for the temperature field $\vartheta = \vartheta(x,t)$, which appears after a long period of time as a quasi-steady end state in the body, when the disturbances caused by the initial temperature distribution have faded away.

A solution to this heat conduction problem is once again possible applying the Laplace transformation, cf. [2.18], [2.26] or [2.1], p. 317–319, where the part originating from a constant initial temperature was calculated, which fades away to zero after a sufficiently long time interval. As this solution method is fairly complicated we will choose an alternative. It can be expected that the temperature in the interior of the body also undergoes an harmonic oscillation, which with

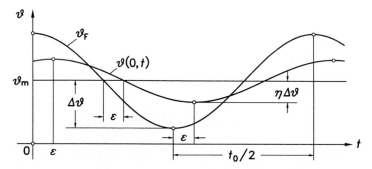

Fig. 2.25: Fluid temperature over time ϑ_F from (2.148) and surface temperature $\vartheta(0,t)$ from (2.153) with $k = (b/\alpha)\sqrt{\pi/t_0} = 1.0$

increasing depth x will be damped more and more, and in addition it will show a phase shift. The corresponding formulation for this

$$\vartheta\left(x,t\right)=\vartheta_{\mathrm{m}}+\Delta\vartheta\,\eta\,e^{-mx}\cos\left(\omega t-mx-\varepsilon\right) \qquad (2.150)$$

satisfies the heat conduction equation (2.120), when

$$m^{2}=\frac{\omega}{2a}=\frac{\pi}{at_{0}}\ . \qquad (2.151)$$

The constants η and ε are yielded from the boundary condition (2.149) to

$$\frac{1}{\eta}=\sqrt{1+2k+2k^{2}} \qquad (2.152\mathrm{a})$$

and

$$\varepsilon=\arctan\frac{k}{1+k} \qquad (2.152\mathrm{b})$$

with

$$k=\frac{m\lambda}{\alpha}=\frac{\lambda}{\alpha}\sqrt{\frac{\pi}{at_{0}}}=\frac{b}{\alpha}\sqrt{\frac{\pi}{t_{0}}}\ . \qquad (2.152\mathrm{c})$$

Their meaning is recognised when the surface temperature is calculated. From (2.150) with $x=0$ it follows that

$$\vartheta\left(0,t\right)=\vartheta_{\mathrm{m}}+\Delta\vartheta\,\eta\cos\left(\omega t-\varepsilon\right)\ . \qquad (2.153)$$

The amplitude of this oscillation is reduced by the factor $\eta<1$ against the oscillation amplitude of the fluid temperature. In addition the surface temperature has a phase shift of ε against the fluid temperature, Fig. 2.25.

The temperature at any particular depth x behaves in the same manner as the surface temperature. It oscillates harmonically with the same angular frequency ω, an increasingly damped amplitude at greater depths and a phase shift that grows with x. At each point in time (2.150) represents a temperature wave which is rapidly fading away with increasing depth inside the body, Fig. 2.26. Its wavelength is the distance between two points whose phase angles differ by 2π. If x_{1} and x_{2} are such points then from

$$\omega t-mx_{1}-\varepsilon=\omega t-mx_{2}-\varepsilon+2\pi$$

the wavelength follows as

$$\Lambda=x_{2}-x_{1}=\frac{2\pi}{m}=2\sqrt{\pi at_{0}}\ . \qquad (2.154)$$

This gives for the temperature distribution

$$\vartheta\left(x,t\right)=\vartheta_{\mathrm{m}}+\Delta\vartheta\,\eta\,e^{-2\pi x/\Lambda}\cos\left[2\pi\left(\frac{t}{t_{0}}-\frac{x}{\Lambda}\right)-\varepsilon\right]\ . \qquad (2.155)$$

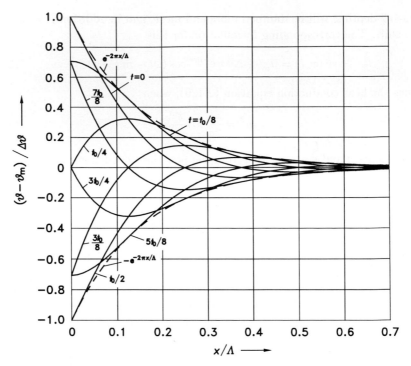

Fig. 2.26: Temperature waves according to (2.155) in a semi-infinite solid with harmonic oscillation of the surface temperature with $\eta = 1$ and $\varepsilon = 0$

In order to find the penetration depth of the temperature fluctuation we calculate the depth x_n, at which the amplitude has fallen to an n-th part of its value at the surface $x = 0$. From

$$e^{-2\pi x_n / \Lambda} = 1/n$$

we find

$$x_n = \frac{\Lambda}{2\pi} \ln n = \sqrt{\frac{a t_0}{\pi}} \ln n \ . \tag{2.156}$$

Penetration depth and wavelength increase with the thermal diffusivity of the body and the periodic time. High frequency oscillations, which appear in fast running combustion engines, have a much lower penetration depth in the cylinder wall than that in the ground caused by the daily or even seasonal temperature variations in the air temperature. A reduction in the amplitude to only 1% of the value at the surface ($n = 100$), takes place, according to (2.156) at a depth of $x_{100} = 0.733\Lambda$. Already at this depth the temperature oscillations are of little practical importance.

2.3.4 Cooling or heating of simple bodies in one-dimensional heat flow

The time dependent temperature field $\vartheta = \vartheta(r,t)$ in a body is described by the differential equation

$$\frac{\partial \vartheta}{\partial t} = a\left(\frac{\partial^2 \vartheta}{\partial r^2} + \frac{n}{r}\frac{\partial \vartheta}{\partial r}\right) \qquad (2.157)$$

with $n = 0, 1$ and 2 for a plate, a cylinder and a sphere, when geometrical one-dimensional heat flow, i.e. heat flow in only the r-direction is assumed. In a cylinder or sphere r is the radial coordinate. The cylinder must be very long compared to its diameter so that the heat flow in the axial direction can be neglected. In addition the temperature ϑ must not depend on the angular coordinates, this condition is also perscribed for the sphere. For a plate, as already done in section 1.1.2, the x-coordinate is indicated by r. It is the coordinate perpendicular to the two very large boundary planes, and is taken from the middle of the plate.

2.3.4.1 Formulation of the problem

The three bodies — plate, very long cylinder and sphere — shall have a constant initial temperature ϑ_0 at time $t = 0$. For $t > 0$ the surface of the body is brought into contact with a fluid whose temperature $\vartheta_S \neq \vartheta_0$ is constant with time. Heat is then transfered between the body and the fluid. If $\vartheta_S < \vartheta_0$, the body is cooled and if $\vartheta_S > \vartheta_0$ it is heated. This transient heat conduction process runs until the body assumes the temperature ϑ_S of the fluid. This is the steady end-state. The heat transfer coefficient α is assumed to be equal on both sides of the plate, and for the cylinder or sphere it is constant over the whole of the surface in contact with the fluid. It is independent of time for all three cases. If only half of the plate is considered, the heat conduction problem corresponds to the unidirectional heating or cooling of a plate whose other surface is insulated (adiabatic).

Under the assumptions mentioned, we obtain the boundary conditions

$$r = 0: \qquad \frac{\partial \vartheta}{\partial r} = 0$$

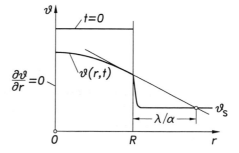

Fig. 2.27: Diagram explaining the initial and boundary conditions for the cooling of a plate of thickness $2R$, a long cylinder and a sphere, each of radius R

due to symmetry and

$$r = R : \qquad -\lambda \frac{\partial \vartheta}{\partial r} = \alpha(\vartheta - \vartheta_\mathrm{S}) \ .$$

Additionally the initial condition

$$t = 0 : \qquad \vartheta(r) = \vartheta_0, \qquad 0 \leq r \leq R \ ,$$

is valid, cf. Fig. 2.27. It is useful to introduce dimensionless variables. We put

$$r^+ := r/R \quad \text{and} \quad t^+ := at/R^2 \tag{2.158}$$

with R as half the thickness of the plate or cylinder or sphere radius, along with

$$\vartheta^+ := \frac{\vartheta - \vartheta_\mathrm{S}}{\vartheta_0 - \vartheta_\mathrm{S}} \ . \tag{2.159}$$

With these quantites the differential equation (2.157) is transformed into

$$\frac{\partial \vartheta^+}{\partial t^+} = \frac{\partial^2 \vartheta^+}{\partial r^{+2}} + \frac{n}{r^+} \frac{\partial \vartheta^+}{\partial r^+} \ . \tag{2.160}$$

The boundary conditions are

$$\frac{\partial \vartheta^+}{\partial r^+} = 0 \quad \text{for} \quad r^+ = 0 \tag{2.161}$$

and

$$-\frac{\partial \vartheta^+}{\partial r^+} = Bi\, \vartheta^+ \quad \text{for} \quad r^+ = 1 \tag{2.162}$$

with the Biot number

$$Bi = \alpha R/\lambda \ ,$$

cf. section 2.1.5. The initial condition is

$$\vartheta^+ = 1 \quad \text{for} \quad t^+ = 0 \ . \tag{2.163}$$

The desired dimensionless temperature profile in the three bodies has the form

$$\vartheta^+ = f_n(r^+, t^+, Bi) \ ,$$

where different functions f_n are yielded for the plate ($n = 0$), the cylinder ($n = 1$) and the sphere ($n = 2$), because in each case the differential equation is different. The dimensionless temperature can only assume values between $\vartheta^+ = 0$ (for $t^+ \to \infty$) and $\vartheta^+ = 1$ (for $t^+ = 0$). The temperature variations during cooling and heating are described in the same manner by ϑ^+, cf. Fig. 2.28. In cooling ($\vartheta_\mathrm{S} < \vartheta_0$) according to (2.159)

$$\vartheta(x, t) = \vartheta_\mathrm{S} + (\vartheta_0 - \vartheta_\mathrm{S})\, \vartheta^+ \left(r^+, t^+\right)$$

holds, and for heating ($\vartheta_\mathrm{S} > \vartheta_0$) it follows that

$$\vartheta(x, t) = \vartheta_0 + (\vartheta_\mathrm{S} - \vartheta_0) \left[1 - \vartheta^+ \left(r^+, t^+\right)\right] \ .$$

For $Bi \to \infty$ we obtain the special case of the surface of the bodies being kept at the constant temperature ϑ_S of the fluid, correspondingly $\vartheta^+ = 0$. This gives for the same temperature difference $\vartheta_0 - \vartheta_\mathrm{S}$ the fastest possible cooling or heating rate ($\alpha \to \infty$).

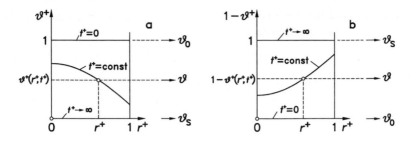

Fig. 2.28: Temperature ratio $\vartheta^+ = (\vartheta - \vartheta_S)/(\vartheta_0 - \vartheta_S)$. **a** in cooling and **b** in heating

2.3.4.2 Separating the variables

In order to find a solution to the differential equation (2.160) whilst accounting for the initial and boundary conditions, the method of separating the variables or product solution is used[3]

$$\vartheta^+(r^+, t^+) = F(r^+) \cdot G(t^+) \ .$$

The functions F and G each depend on only one variable, and satisfy the following differential equation from (2.160)

$$F(r^+)\frac{\mathrm{d}G}{\mathrm{d}t^+} = \left(\frac{\mathrm{d}^2 F}{\mathrm{d}r^{+2}} + \frac{n}{r^+}\frac{\mathrm{d}F}{\mathrm{d}r^+}\right)G(t^+)$$

or

$$\frac{1}{G(t^+)}\frac{\mathrm{d}G}{\mathrm{d}t^+} = \frac{1}{F(r^+)}\left(\frac{\mathrm{d}^2 F}{\mathrm{d}r^{+2}} + \frac{n}{r^+}\frac{\mathrm{d}F}{\mathrm{d}r^+}\right) \ . \tag{2.164}$$

The left hand side of (2.164) depends on the (dimensionless) time t^+, the right hand side on the position coordinate r^+: the variables are separated. The equality demanded by (2.164) is only possible if both sides of (2.164) are equal to a constant $-\mu^2$. This constant μ is known as the separation parameter. With this the following ordinary differential equations are produced from (2.164)

$$\frac{\mathrm{d}G}{\mathrm{d}t^+} + \mu^2 G = 0 \tag{2.165}$$

and

$$\frac{\mathrm{d}^2 F}{\mathrm{d}r^{+2}} + \frac{n}{r^+}\frac{\mathrm{d}F}{\mathrm{d}r^+} + \mu^2 F = 0 \ . \tag{2.166}$$

Products of their solutions with the same value of the separation parameter μ give solutions to the heat conduction equation (2.160).

---

[3]The application of the Laplace transformation delivers the same result. The inverse transformation from the frequency to the time region requires the use of the inversion theorem, see 2.3.2. In order to avoid this in this case the simple, classical product solution is applied.

The solution of the differential equation (2.165) for the time function $G(t^+)$ is the decaying exponential function

$$G(t^+) = C \exp\left(-\mu^2 t^+\right) \ . \tag{2.167}$$

This is true for all three bodies. The position function F is different for plates, cylinders and spheres. However, the solution functions F have to satisfy the same boundary conditions in all three cases. It follows from (2.161) and (2.162) that

$$\frac{\mathrm{d}F}{\mathrm{d}r^+} = 0 \quad \text{for} \quad r^+ = 0 \tag{2.168}$$

and

$$-\frac{\mathrm{d}F}{\mathrm{d}r^+} = Bi\, F \quad \text{for} \quad r^+ = 1 \ . \tag{2.169}$$

The boundary value problem posed by the differential equation (2.166) and the two boundary conditions (2.168) and (2.169) leads to the class of Sturm-Liouville eigenvalue problems for which a series of general theorems are valid. As we will soon show the solution function F only satisfies the boundary conditions with certain discrete values μ_i of the separation parameter. These special values μ_i are called *eigenvalues* of the boundary value problem, and the accompanying solution functions F_i are known as *eigenfunctions*. The most important rules from the theory of Sturm-Liouville eigenvalue problems are, cf. e.g. K. Jänich [2.33]:

1. All eigenvalues are real.
2. The eigenvalues form a monotonically increasing infinite series

$$\mu_1 < \mu_2 < \mu_3 \ldots \quad \text{with} \quad \lim_{i \to \infty} \mu_i = +\infty \ .$$

3. The associated eigenfunctions $F_1, F_2, \ldots$ are orthogonal, i.e. it holds that

$$\int_a^b F_i F_j \, \mathrm{d}r^+ = \begin{cases} 0 & \text{for} \quad i \neq j \\ A_i & \text{for} \quad i = j \end{cases}$$

with A_i as a positive constant. Here a and b are the two points at which the boundary conditions are stipulated. In our problem $a = 0$ and $b = 1$.
4. The eigenfunction F_i associated with the eigenvalue μ_i has exactly $(i - 1)$ zero points between the boundaries, or in other words in the interval (a, b).

These properties will now be used in the following solutions of the boundary value problems for a plate, a cylinder and a sphere.

2.3.4.3   Results for the plate

The function $F(r^+)$ satisfies the differential equation (2.166) for the plate with $n = 0$, which is known as the differential equation governing harmonic oscillations. It has the general solution

$$F = c_1 \cos\left(\mu r^+\right) + c_2 \sin\left(\mu r^+\right) \ .$$

It follows from the boundary condition (2.168) that $c_2 = 0$. The heat transfer condition (2.169) leads to a transcendental equation for the separation parameter μ, namely

$$\tan \mu = Bi/\mu \ . \tag{2.170}$$

The roots $\mu = \mu_i$ of this equation are the eigenvalues of the problem, which depend on the Biot number. As Fig. 2.29 shows, there is an infinite series of eigenvalues $\mu_1 < \mu_2 < \mu_3 \ldots$ which is in full agreement with the Sturm-Liouville theory. Only the following eigenfunctions

$$F_i = \cos\left(\mu_i r^+\right)$$

associated with the eigenvalues μ_i satisfy both the boundary conditions (2.168) and (2.169). Therefore, the infinite series shown below, made up of the infinite number of eigenfunctions and the time dependent function $G(t^+)$ according to (2.167),

$$\vartheta^+\left(r^+, t^+\right) = \sum_{i=1}^{\infty} C_i \cos\left(\mu_i r^+\right) \exp(-\mu_i^2 t^+) \tag{2.171}$$

is the general solution to the heat conduction problem in a plate. It still has to fit the initial condition (2.163). It must therefore hold that

$$1 = \sum_{i=1}^{\infty} C_i \cos\left(\mu_i r^+\right) \quad , \qquad 0 \le r^+ \le 1 \ .$$

A given function, in this case the number 1, is to be represented by the infinite sum of eigenfunctions in the interval $[0, 1]$.

The coefficient C_i is obtained by multiplying with an eigenfunction $\cos\left(\mu_j r^+\right)$ and integrating from $r^+ = 0$ to $r^+ = 1$. As the eigenfunctions are orthogonal all the terms in which $j \neq i$ vanish. The only term which remains is that with $j = i$,

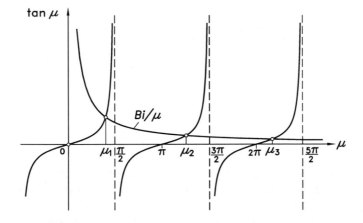

Fig. 2.29: Graphical determination of the eigenvalues according to (2.170)

so that

$$\int_0^1 \cos\left(\mu_i r^+\right)\, dr^+ = C_i \int_0^1 \cos^2\left(\mu_i r^+\right)\, dr^+$$

delivers the coefficient C_i with the result

$$C_i = \frac{2\sin\mu_i}{\mu_i + \sin\mu_i \cos\mu_i} = \frac{2Bi}{Bi^2 + Bi + \mu_i^2}\frac{1}{\cos\mu_i}. \tag{2.172}$$

With this the desired temperature distribution for cooling or heating a plate is found.

The eigenvalues μ_i and the coefficients C_i are dependent on the Biot number. Table 2.6 contains values for μ_1 and C_1. Values of the first six eigenvalues are given in [2.1]. For $Bi \to \infty$ the boundary condition $\vartheta^+(1,t^+) = 0$, corresponding to $\vartheta(R) = \vartheta_S$, is obtained with eigenvalues $\mu_1 = \pi/2$, $\mu_2 = 3\pi/2$, $\mu_3 = 5\pi/2$ ... For $Bi = 0$, the limiting case of an insulated wall with finite thermal conductivity λ, but with $\alpha = 0$, we obtain $\mu_1 = 0$, $\mu_2 = \pi$, $\mu_3 = 2\pi$, ... The temperature profile is given by $\vartheta^+ = 1$ independent of t^+, which is plausible for physical reasons.

In addition to the temperatures calculated according to (2.171) the energy, which the plate absorbs or releases as heat during a specific time interval, is often of interest. If a plate of volume V is cooled from its initial temperature ϑ_0 to the temperature of the surroundings ϑ_S (for $t \to \infty$), it releases the energy

$$Q_0 = \varrho V c(\vartheta_0 - \vartheta_S)$$

as heat to the surroundings. After a certain time t has passed the plate has released

$$Q(t) = \varrho V c\left[\vartheta_0 - \vartheta_m(t)\right]$$

as heat to the environment. In which

$$\vartheta_m(t) = \frac{1}{R}\int_0^R \vartheta(r,t)\, dr$$

is the *average temperature* of the plate at time t. The ratio

$$\frac{Q(t)}{Q_0} = \frac{\vartheta_0 - \vartheta_m(t)}{\vartheta_0 - \vartheta_S} = 1 - \vartheta_m^+(t^+) \tag{2.173}$$

is determined by the dimensionless average temperature $\vartheta_m^+(t^+)$. This is found by integrating (2.171)

$$\vartheta_m^+(t^+) = \sum_{i=1}^\infty C_i \frac{\sin\mu_i}{\mu_i}\exp\left(-\mu_i^2 t^+\right) = 2Bi^2 \sum_{i=1}^\infty \frac{\exp\left(-\mu_i^2 t^+\right)}{\mu_i^2\left(Bi^2 + Bi + \mu_i^2\right)}. \tag{2.174}$$

The equations, (2.171) for the temperature distribution in the plate as well as (2.173) and (2.174) for the released heat have been repeatedly evaluated and illustrated in diagrams, cf. [2.34] and [2.35]. In view of the computing technology available today the direct evaluation of the relationships given above is advantageous, particularly in simulation programs in which these transient heat conduction processes appear. The applications of the relationships developed here is made easier, as for large values of t^+ only the first term of the infinite series is required, cf. section 2.3.4.5. Special equations for very small t^+ will be derived in

Table 2.6: Smallest eigenvalue μ_1 from (2.170), (2.177) or (2.182), associated expansion coefficient C_1 in the series for the temperature distribution $\vartheta^+(r^+, t^+)$ as well as coefficient D_1 in the series for the average temperature $\vartheta_m^+(t^+)$ of a plate, a very long cylinder and a sphere, cf. also (2.185) and (2.188) in section 2.3.4.5

	Plate			Cylinder			Sphere		
Bi	μ_1	C_1	D_1	μ_1	C_1	D_1	μ_1	C_1	D_1
0.01	0.09983	1.0017	1.0000	0.14124	1.0025	1.0000	0.17303	1.0030	1.0000
0.02	0.14095	1.0033	1.0000	0.19950	1.0050	1.0000	0.24446	1.0060	1.0000
0.03	0.17234	1.0049	1.0000	0.24403	1.0075	1.0000	0.29910	1.0090	1.0000
0.04	0.19868	1.0066	1.0000	0.28143	1.0099	1.0000	0.34503	1.0120	1.0000
0.05	0.22176	1.0082	0.9999	0.31426	1.0124	0.9999	0.38537	1.0150	1.0000
0.06	0.24253	1.0098	0.9999	0.34383	1.0148	0.9999	0.42173	1.0179	0.9999
0.07	0.26153	1.0114	0.9999	0.37092	1.0173	0.9999	0.45506	1.0209	0.9999
0.08	0.27913	1.0130	0.9999	0.39603	1.0197	0.9999	0.48600	1.0239	0.9999
0.09	0.29557	1.0145	0.9998	0.41954	1.0222	0.9998	0.51497	1.0268	0.9999
0.10	0.31105	1.0161	0.9998	0.44168	1.0246	0.9998	0.54228	1.0298	0.9998
0.15	0.37788	1.0237	0.9995	0.53761	1.0365	0.9995	0.66086	1.0445	0.9996
0.20	0.43284	1.0311	0.9992	0.61697	1.0483	0.9992	0.75931	1.0592	0.9993
0.25	0.48009	1.0382	0.9988	0.68559	1.0598	0.9988	0.84473	1.0737	0.9990
0.30	0.52179	1.0450	0.9983	0.74646	1.0712	0.9983	0.92079	1.0880	0.9985
0.40	0.59324	1.0580	0.9971	0.85158	1.0931	0.9970	1.05279	1.1164	0.9974
0.50	0.65327	1.0701	0.9956	0.94077	1.1143	0.9954	1.16556	1.1441	0.9960
0.60	0.70507	1.0814	0.9940	1.01844	1.1345	0.9936	1.26440	1.1713	0.9944
0.70	0.75056	1.0918	0.9922	1.08725	1.1539	0.9916	1.35252	1.1978	0.9925
0.80	0.79103	1.1016	0.9903	1.14897	1.1724	0.9893	1.43203	1.2236	0.9904
0.90	0.82740	1.1107	0.9882	1.20484	1.1902	0.9869	1.50442	1.2488	0.9880
1.00	0.86033	1.1191	0.9861	1.25578	1.2071	0.9843	1.57080	1.2732	0.9855
1.10	0.89035	1.1270	0.9839	1.30251	1.2232	0.9815	1.63199	1.2970	0.9828
1.20	0.91785	1.1344	0.9817	1.34558	1.2387	0.9787	1.68868	1.3201	0.9800
1.30	0.94316	1.1412	0.9794	1.38543	1.2533	0.9757	1.74140	1.3424	0.9770
1.40	0.96655	1.1477	0.9771	1.42246	1.2673	0.9727	1.79058	1.3640	0.9739
1.50	0.98824	1.1537	0.9748	1.45695	1.2807	0.9696	1.83660	1.3850	0.9707
1.60	1.00842	1.1593	0.9726	1.48917	1.2934	0.9665	1.87976	1.4052	0.9674
1.70	1.02725	1.1645	0.9703	1.51936	1.3055	0.9633	1.92035	1.4247	0.9640
1.80	1.04486	1.1695	0.9680	1.54769	1.3170	0.9601	1.95857	1.4436	0.9605
1.90	1.06136	1.1741	0.9658	1.57434	1.3279	0.9569	1.99465	1.4618	0.9570
2.00	1.07687	1.1785	0.9635	1.59945	1.3384	0.9537	2.02876	1.4793	0.9534
2.20	1.10524	1.1864	0.9592	1.64557	1.3578	0.9472	2.09166	1.5125	0.9462
2.40	1.13056	1.1934	0.9549	1.68691	1.3754	0.9408	2.14834	1.5433	0.9389
2.60	1.15330	1.1997	0.9509	1.72418	1.3914	0.9345	2.19967	1.5718	0.9316
2.80	1.17383	1.2052	0.9469	1.75794	1.4059	0.9284	2.24633	1.5982	0.9243

Table 2.6: continued

Bi	Plate			Cylinder			Sphere		
	μ_1	C_1	D_1	μ_1	C_1	D_1	μ_1	C_1	D_1
3.00	1.19246	1.2102	0.9431	1.78866	1.4191	0.9224	2.28893	1.6227	0.9171
3.50	1.23227	1.2206	0.9343	1.85449	1.4473	0.9081	2.38064	1.6761	0.8995
4.00	1.26459	1.2287	0.9264	1.90808	1.4698	0.8950	2.45564	1.7202	0.8830
4.50	1.29134	1.2351	0.9193	1.95248	1.4880	0.8830	2.51795	1.7567	0.8675
5.00	1.31384	1.2402	0.9130	1.98981	1.5029	0.8721	2.57043	1.7870	0.8533
6.00	1.34955	1.2479	0.9021	2.04901	1.5253	0.8532	2.65366	1.8338	0.8281
7.00	1.37662	1.2532	0.8932	2.09373	1.5411	0.8375	2.71646	1.8673	0.8069
8.00	1.39782	1.2570	0.8858	2.12864	1.5526	0.8244	2.76536	1.8920	0.7889
9.00	1.41487	1.2598	0.8796	2.15661	1.5611	0.8133	2.80443	1.9106	0.7737
10.00	1.42887	1.2620	0.8743	2.17950	1.5677	0.8039	2.83630	1.9249	0.7607
12.00	1.45050	1.2650	0.8658	2.21468	1.5769	0.7887	2.88509	1.9450	0.7397
14.00	1.46643	1.2669	0.8592	2.24044	1.5828	0.7770	2.92060	1.9581	0.7236
16.00	1.47864	1.2683	0.8541	2.26008	1.5869	0.7678	2.94756	1.9670	0.7109
18.00	1.48830	1.2692	0.8499	2.27555	1.5898	0.7603	2.96871	1.9734	0.7007
20.00	1.49613	1.2699	0.8464	2.28805	1.5919	0.7542	2.98572	1.9781	0.6922
25.00	1.51045	1.2710	0.8400	2.31080	1.5954	0.7427	3.01656	1.9856	0.6766
30.00	1.52017	1.2717	0.8355	2.32614	1.5973	0.7348	3.03724	1.9898	0.6658
35.00	1.52719	1.2721	0.8322	2.33719	1.5985	0.7290	3.05207	1.9924	0.6579
40.00	1.53250	1.2723	0.8296	2.34552	1.5993	0.7246	3.06321	1.9942	0.6519
50.00	1.54001	1.2727	0.8260	2.35724	1.6002	0.7183	3.07884	1.9962	0.6434
60.00	1.54505	1.2728	0.8235	2.36510	1.6007	0.7140	3.08928	1.9974	0.6376
80.00	1.55141	1.2730	0.8204	2.37496	1.6013	0.7085	3.10234	1.9985	0.6303
100.00	1.55525	1.2731	0.8185	2.38090	1.6015	0.7052	3.11019	1.9990	0.6259
200.00	1.56298	1.2732	0.8146	2.39283	1.6019	0.6985	3.12589	1.9998	0.6170
∞	1.57080	1.2732	0.8106	2.40483	1.6020	0.6917	3.14159	2.0000	0.6079

section 2.3.4.6. In addition to these there are also approximation equations which are numerically more simple than the relationships derived here, see [2.74].

Example 2.4: In example 2.3 of section 2.3.2, the temperature $\vartheta\,(0,t^+)$ at $x = 0$ of the adiabatic surface of a flat wall (plate) was calculated. $\vartheta\,(0,t^+)$ changed, as a result of the stepwise increase in the temperature from ϑ_0 to ϑ_S at the other wall surface. For $\vartheta\,(0,t^+)$ in example 2.3 we found

$$\frac{\vartheta\,(0,t^+) - \vartheta_0}{\vartheta_S - \vartheta_0} = 2\,\mathrm{erfc}\left(\frac{1}{2\sqrt{t^+}}\right) - 2\,\mathrm{erfc}\left(\frac{3}{2\sqrt{t^+}}\right) + 2\,\mathrm{erfc}\left(\frac{5}{2\sqrt{t^+}}\right) - \dots \quad (2.175)$$

This equation will now be confronted with the relationships from (2.171).

The rapid change in the surface temperature at $r = R$ (corresponding to $x = \delta$ in example 2.3) means that $\alpha \to \infty$, and therefore $Bi \to \infty$. This gives, from (2.170), the eigenvalues

$$\mu_1 = \frac{\pi}{2} \quad , \quad \mu_2 = \frac{3\pi}{2} \quad , \quad \mu_3 = \frac{5\pi}{2} \quad , \quad \dots$$

and from (2.172) the coefficients

$$C_1 = \frac{4}{\pi} \quad , \quad C_2 = -\frac{4}{3\pi} \quad , \quad C_3 = \frac{4}{5\pi} \quad , \quad \dots$$

Then from (2.171) it follows that

$$\frac{\vartheta\left(0, t^+\right) - \vartheta_\mathrm{S}}{\vartheta_0 - \vartheta_\mathrm{S}} = \frac{4}{\pi} \sum_{i=1}^{\infty} \frac{(-1)^{i+1}}{2i - 1} \exp\left[-\frac{\pi^2}{4}(2i - 1)^2 t^+\right] \quad .$$

The corresponding equation to (2.175) is found to be

$$\frac{\vartheta\left(0, t^+\right) - \vartheta_0}{\vartheta_\mathrm{S} - \vartheta_0} = 1 - \frac{4}{\pi} \exp\left(-\frac{\pi^2}{4} t^+\right) + \frac{4}{3\pi} \exp\left(-\frac{9\pi^2}{4} t^+\right) - \frac{4}{5\pi} \exp\left(-\frac{25\pi^2}{4} t^+\right) + \dots \quad (2.176)$$

The equations (2.175) and (2.176) describe the same temperature change. While (2.175) rapidly converges for small times t^+, (2.176) is particularly well suited to large times t^+. This is shown by the following comparison of two times $t^+ = 0.1$ and $t^+ = 1.0$. For $t^+ = 0.1$ we obtain with only the *first* term from (2.175) the value 0.05070 to five decimal places. In (2.176), however, it is necessary to use three terms of the infinite series

$$1 - 0.99484 + 0.04607 - 0.00053 + \dots = 0.05070 \quad .$$

For the time $t^+ = 1.0$ the first term in the series in (2.176) is sufficient to give

$$1 - 0.10798 + \dots = 0.89202 \quad .$$

This result is only possible with the first three terms when (2.175) is used

$$0.95900 - 0.06779 + 0.00081 - \dots = 0.89202 \quad .$$

2.3.4.4   Results for the cylinder and the sphere

For an *infinitely long cylinder* the function $F(r^+)$ is determined by (2.166) with $n = 1$. This is the zero-order Bessel differential equation and its solutions are the Bessel function J_0 and the Neumann function N_0 both of zero-order:

$$F\left(r^+\right) = c_1 J_0\left(\mu r^+\right) + c_2 N_0\left(\mu r^+\right) \quad .$$

The symmetry condition (2.168) rules N_0 out ($c_2 = 0$). In the same manner as for the plate the heat transfer condition (2.169) delivers a defining equation for the eigenvalues μ_i:

$$-\frac{\mathrm{d}J_0(\mu r^+)}{\mathrm{d}r^+} = Bi\, J_0\left(\mu r^+\right) \quad \text{for} \quad r^+ = 1$$

or

$$\mu J_1(\mu) = Bi\, J_0(\mu) \tag{2.177}$$

with J_1 as the first-order Bessel function. Tables of the Bessel functions J_0 and J_1 can be found in [2.30] and [2.32]; information about their properties and equations used in their calculation are available in [2.28].

Once again an infinite number of eigenvalues μ_i is obtained as a solution of (2.177), which the eigenfunctions $F_i = J_0(\mu_i r^+)$ belong to. The sum

$$\vartheta^+\left(r^+, t^+\right) = \sum_{i=1}^{\infty} C_i J_0\left(\mu_i r^+\right) \exp\left(-\mu_i^2 t^+\right) \tag{2.178}$$

is a solution of the heat conduction equation for the cylinder which satisfies the boundary conditions. The initial condition (2.163), namely $\vartheta^+ (r^+, 0) = 1$, provides us with the expansion coefficients

$$C_i = \frac{2 J_1(\mu_i)}{\mu_i \left[J_0^2(\mu_i) + J_1^2(\mu_i) \right]} = \frac{2 Bi}{\left(Bi^2 + \mu_i^2 \right) J_0(\mu_i)} \quad . \tag{2.179}$$

Table 2.6 gives the first eigenvalue μ_1 and the associated C_1 as functions of the Biot number. Values for the first six eigenvalues can be found in [2.1].

The ratio of the heat $Q(t)$ released in time t to the maximum value Q_0 for $t \to \infty$ is given by (2.173). The dimensionless average temperature ϑ_m^+ is calculated for the cylinder from

$$\vartheta_m^+ = 2 \int_0^1 \vartheta^+ \left(r^+, t^+ \right) r^+ \, dr^+ \quad .$$

Intregration of (2.178) yields

$$\vartheta_m^+ \left(t^+ \right) = 4 Bi^2 \sum_{i=1}^{\infty} \frac{\exp \left(-\mu_i^2 t^+ \right)}{\mu_i^2 \left(Bi^2 + \mu_i^2 \right)} \quad . \tag{2.180}$$

The temperature distribution in the *sphere* is obtained in the same manner, such that

$$\vartheta^+ \left(r^+, t^+ \right) = \sum_{i=1}^{\infty} C_i \frac{\sin \left(\mu_i r^+ \right)}{\mu_i r^+} \exp \left(-\mu_i^2 t^+ \right) \quad . \tag{2.181}$$

The eigenvalues are the roots of the transcendental equation

$$\mu \cot \mu = 1 - Bi \quad . \tag{2.182}$$

The coefficients C_i are found to be

$$C_i = 2 \frac{\sin \mu_i - \mu_i \cos \mu_i}{\mu_i - \sin \mu_i \cos \mu_i} = 2 \, Bi \frac{\mu_i^2 + (Bi - 1)^2}{\mu_i^2 + Bi \, (Bi - 1)} \frac{\sin \mu_i}{\mu_i} \quad . \tag{2.183}$$

This then gives the dimensionless average temperature

$$\vartheta_m^+ \left(t^+ \right) = 3 \int_0^1 \vartheta^+ \left(r^+, t^+ \right) r^{+2} \, dr^+$$

for a sphere, as the series

$$\vartheta_m^+ \left(t^+ \right) = 6 \, Bi^2 \sum_{i=1}^{\infty} \frac{\exp \left(-\mu_i^2 t^+ \right)}{\mu_i^2 \left[\mu_i^2 + Bi \, (Bi - 1) \right]} \quad . \tag{2.184}$$

This then leaves the heat $Q(t^+)$ relative to the maximum value Q_0 to be calculated according to (2.173).

Table 2.6 contains the first eigenvalue μ_1 as the solution of (2.182) and the accompanying coefficients C_1 from (2.183) as functions of Bi. Values for the first six eigenvalues are available in [2.1]. Graphical representations of the temperature profiles in very long cylinders and spheres can be found in the publications [2.34] and [2.35] which have already been mentioned in this chapter, and the associated approximation equations are available in [2.74]. The change in the dimensionless temperature $\vartheta^+ (0, t^+)$ in the centre $r^+ = 0$ of the three bodies, plate (centre plane), very long cylinder (axis) and the sphere (centre) is shown in Fig. 2.30.

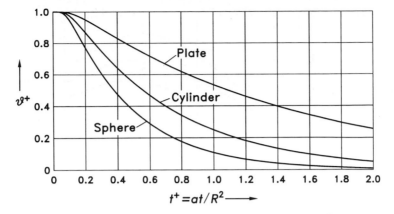

Fig. 2.30: Temperature change with time $\vartheta^+(t^+)$ in the centre ($r^+ = 0$) of a plate, a very long cylinder and a sphere for $Bi = \alpha R/\lambda = 1.0$

2.3.4.5 Approximation for large times: Restriction to the first term in the series

For sufficiently large times t^+ the first term of the series (2.171), (2.178) and (2.181), is sufficient to calculate the temperature profile for a plate, an infinitely long cylinder and a sphere. The simple result

$$\vartheta_1^+\left(r^+, t^+\right) = C \exp\left(-\mu^2 t^+\right) \begin{cases} \cos\left(\mu r^+\right) & \text{plate} \\ J_0\left(\mu r^+\right) & \text{cylinder} \\ \sin\left(\mu r^+\right)/\left(\mu r^+\right) & \text{sphere} \end{cases} \qquad (2.185)$$

is obtained. The first eigenvalue $\mu = \mu_1$ and its associated expansion coefficient $C = C_1$ depend on the Biot number $Bi = \alpha R/\lambda$. They have different values for the plate, cylinder and sphere. These values are displayed in Table 2.6.

U. Grigull and others [2.36] have investigated the errors which develop when the higher terms in the series are neglected. If a certain deviation

$$\Delta\vartheta^+ := \vartheta_1^+\left(r^+, t^+\right) - \vartheta^+\left(r^+, t^+\right) \qquad (2.186)$$

is allowed, where ϑ_1^+ is the first term from (2.185) and ϑ^+ the complete series, a dimensionless time $t_{\min}^+$ can be calculated for which this error just appears. $t_{\min}^+$ depends on r^+ and the Biot number. For all times $t^+ > t_{\min}^+$ the first term ϑ_1^+ from (2.185) can be used without an error greater than $\Delta\vartheta^+$ being incurred. Fig. 2.31 shows the results found by U. Grigull [2.36] for $r^+ = 0$ (central plane, central axis or centre) and $\Delta\vartheta^+ = 0.001$ as a function of Bi. At small Bi numbers the first term is already sufficient for rather small times $t^+ > t_{\min}^+$.

Equation (2.185) enables an explicit calculation for the heating or cooling time to be carried out. This time t_k is defined such that once the time has passed a given temperature ϑ_k is reached in the centre of the thermally conductive body

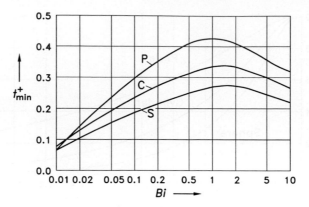

Fig. 2.31: Minimum time t_{min}^+ for which the error is less than $\Delta\vartheta^+ = 0.001$, according to (2.186), neglegting the higher series terms at $r^+ = 0$. P plate, C very long cylinder, S sphere

$(r^+ = 0)$. This temperature ϑ_k corresponds to a particular value

$$\vartheta_k^+ = (\vartheta_k - \vartheta_S) / (\vartheta_0 - \vartheta_S) \ .$$

With $r^+ = 0$ we obtain from (2.185)

$$\vartheta_k^+ = C \exp\left(-\mu^2 t_k^+\right)$$

and from that we get

$$t_k^+ = \frac{at_k}{R^2} = \frac{1}{\mu^2}\left(\ln C - \ln \vartheta_k^+\right) \ . \tag{2.187}$$

Therefore t_k^+ increases, because of $\vartheta_k^+ < 1$, the smaller the value of ϑ_k^+ that is chosen. Equation (2.187) can be easily evaluated with the assistance of Table 2.6.

The dimensionless average temperature ϑ_m^+, which is required for the calculation of the heat absorbed or released by the body according to (2.173), can also be easily calculated when only the first term of the series is used. From (2.174), (2.180) and (2.184) we obtain

$$\vartheta_m^+ = D \exp\left(-\mu^2 t^+\right) \ . \tag{2.188}$$

The coefficient $D = D_1$, which is dependent on the Biot number is given in Table 2.6.

2.3.4.6   A solution for small times

At the beginning of a heating or cooling process (t^+ small) many terms of the infinite series (2.171), (2.178) or (2.181), are required to calculate the temperature profile in a plate, cylinder or sphere exactly. This implies that an alternative

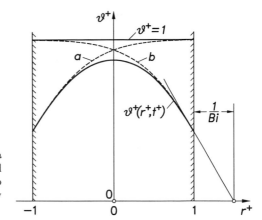

Fig. 2.32: Temperature profile in a plate from (2.189) for $Bi = 2.0$ and $t^+ = 0.25$. The dotted lines a and b correspond to the terms in the curly brackets of (2.189)

solution method for these types of thermal conduction problems needs to be found which converges well for small times. We find it with the aid of the Laplace transformation, where the Laplace transform of the temperature is expanded into a series which converges for large values of s, cf. section 2.3.2. Term by term back transformation using the correspondence table, produces expressions which are particularly suitable for the calculation of the temperature fields at small times. This method was first applied by S. Goldstein [2.37], cf. also [2.36]. We will restrict ourselves to the results for the plate. The solutions for cylinders and spheres lead to more complicated relationships. These can be found in [2.36] and [2.37].

For the heating or cooling of a plate dealt with in section 2.3.4.3 and written in dimensionless variables according to (2.158) and (2.159), we obtain

$$\vartheta^+ (r^+, t^+) = 1 - \left\{ \operatorname{erfc} \frac{1+r^+}{2\sqrt{t^+}} - \exp\left[Bi\left(1 + r^+\right) + Bi^2 t^+\right] \operatorname{erfc}\left(\frac{1+r^+}{2\sqrt{t^+}} + Bi\sqrt{t^+}\right) \right\}$$
$$- \left\{ \operatorname{erfc} \frac{1-r^+}{2\sqrt{t^+}} + \exp\left[Bi\left(1 - r^+\right) + Bi^2 t^+\right] \operatorname{erfc}\left(\frac{1-r^+}{2\sqrt{t^+}} + Bi\sqrt{t^+}\right) \right\}$$
$$+ \dots$$

$$(2.189)$$

This solution allows for a clear interpretation. The terms in the curly brackets correspond to the temperature pattern in a semi-infinite solid. Its surface lies at $r^+ = -1$ for the first term, and it stretches itself out in the positive r^+-direction. The second term belongs to a semi-infinite body which stretches out from its surface at $r^+ = +1$ in the negative r^+-direction. The subtraction of these two temperature profiles from the initial temperature $\vartheta^+ = 1$ yields the symmetrical temperature profile for the plate shown in Fig. 2.32, which satisfies the condition $\partial \vartheta^+ / \partial r^+ = 0$ at $r^+ = 0$.

The solution (2.189) is only valid for times, for which the contribution of the first curly bracket to the surface temperature at $r^+ = +1$ is negligibly small. Therefore

$$\Delta \vartheta^+ \left(r^+ = 1, t^+\right) = \operatorname{erfc} \frac{1}{\sqrt{t^+}} - \exp\left(2Bi + Bi^2 t^+\right) \operatorname{erfc}\left(\frac{1}{\sqrt{t^+}} + Bi\sqrt{t^+}\right)$$

has to be sufficiently small. An asymptotic expansion of this expression for $t^+ \to 0$ gives

$$\Delta\vartheta^+ \left(1, t^+\right) = \frac{e^{-1/t^+}\sqrt{t^+}}{\sqrt{\pi}} \left[(1-y) - \frac{t^+}{2}\left(1-y^3\right) + \frac{3}{4}t^{+2}\left(1-y^5\right) - \dots\right]$$

with $y = (1 + t^+ Bi)^{-1}$. The worst case corresponds to $Bi \to \infty$, therefore $y = 0$. This yields, for various values of the dimensionless time, the following values for $\Delta\vartheta^+$:

$$
\begin{array}{llllll}
t^+ & = & 0.05 & 0.10 & 0.15 & 0.20 & 0.25 \\
\Delta\vartheta^+ & = & 2.5 \cdot 10^{-10} & 7.7 \cdot 10^{-6} & 2.6 \cdot 10^{-4} & 1.6 \cdot 10^{-3} & 4.6 \cdot 10^{-3}
\end{array}
$$

This means that the solution given in (2.189) is generally applicable to situations where $t^+ < 0.2$.

2.3.5 Cooling and heating in multi-dimensional heat flow

In geometric multi-dimensional heat flow the temperature fields that have to be calculated depend on two or three spatial coordinates and must satisfy the heat conduction equation

$$\frac{\partial\vartheta}{\partial t} = a\nabla^2\vartheta \ . \tag{2.190}$$

Here the Laplace operator $\nabla^2\vartheta$ has the form given in 2.1.2 for cartesian and cylindrical coordinate systems. We will once again consider the transient heat conduction problem solved for the plate, the infinitely long cylinder and the sphere in section 2.3.4: A body with a constant initial temperature ϑ_0 is brought into contact with a fluid of constant temperature $\vartheta_S \neq \vartheta_0$, so that heat transfer takes place between the fluid and the body, whereby the constant heat transfer coefficient α is decisive.

In the following section we will consider two solution procedures:
- an analytical solution for special body shapes, which is a product of the temperature distributions already calculated in 2.3.4.
- an approximation method for any body shape, which is only adequate for small Biot numbers.

The latter of the two methods offers a practical, simple applicable solution to transient heat conduction problems and should always be applied for sufficiently small Biot numbers.

2.3.5.1 Product solutions

A cylinder of a finite length is formed by the perpendicular penetration of an infinitely long cylinder with a plate. In the same manner, a very long prism with a rectangular cross section can be formed by the penetration of two plates and a rectangular parallelepiped, by the penetration of three orthogonal plates. The spatial temperature distributions, for heating and cooling of bodies of this type, are products of the geometric one-dimensional temperature distributions

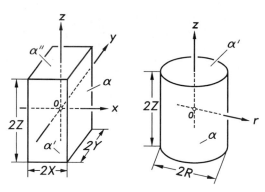 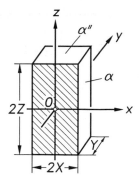

Fig. 2.33: Dimensions of a parallelepiped and a finite cylinder, with the positions of the coordinate systems for the equations (2.191) and (2.192)

Fig. 2.34: Parallelepiped with adiabatic surface at $y = 0$ (hatched). On the opposite surface heat transfer with α'

in the simple bodies which penetrate perpendicular to each other. So for the *parallelepiped* with side lengths $2X$, $2Y$ and $2Z$, see Fig. 2.33,

$$\vartheta^+ = \frac{\vartheta - \vartheta_S}{\vartheta_0 - \vartheta_S} = \vartheta^+_{Pl}\left(\frac{x}{X}, \frac{at}{X^2}, \frac{\alpha X}{\lambda}\right) \cdot \vartheta^+_{Pl}\left(\frac{y}{Y}, \frac{at}{Y^2}, \frac{\alpha' Y}{\lambda}\right) \cdot \vartheta^+_{Pl}\left(\frac{z}{Z}, \frac{at}{Z^2}, \frac{\alpha'' Z}{\lambda}\right) .$$
(2.191)

The solution for the plate ϑ^+_{Pl} is given by (2.171). Here r^+ is replaced by x/X, y/Y or z/Z and equally t^+ by at/X^2, at/Y^2 and at/Z^2 respectively. The Biot number $\alpha R/\lambda$ has the significance as indicated in (2.191). It should be noted that the heat transfer coefficient only has to have the same value on the two surfaces which lie opposite each other. It can therefore assume different values α, α' and α'' on the pairs of surfaces perpendicular to the x-, y- or z-directions. The last factor in (2.191) has to be omitted for a *prism*, which is very long in the z-direction. The temperature distribution for a *cylinder* of height $2Z$ is yielded from the products

$$\vartheta^+ = \vartheta^+_{Cy}\left(\frac{r}{R}, \frac{at}{R^2}, \frac{\alpha R}{\lambda}\right) \cdot \vartheta^+_{Pl}\left(\frac{z}{Z}, \frac{at}{Z^2}, \frac{\alpha' Z}{\lambda}\right) ,$$
(2.192)

where the temperature ϑ^+_{Cy} of the infinitely long cylinder is given by (2.178). The heat transfer coefficient α at the shell can also be different from the heat transfer coefficient α' on the two ends.

If one or two of the parallel flat surfaces of the parallelepiped, prism or cylinder are adiabatically insulated, whilst at the other heat is transferred, the equations given above can also be applied. This entails halving the dimension perpendicular to the adiabatic surface, therefore putting the zero point of the coordinate system into the adiabatic surface. An example of this for a parallelepiped is illustrated in Fig. 2.34.

The proof of the correctness of the relationships (2.191) and (2.192) can be found in the book by H.S. Carslaw and J.C. Jaeger [2.1], pp. 33–35. Other product solutions for multi-dimensional temperature fields using the solutions for the semi-infinite solid are possible. References [2.1] and [2.18] give further details.

The heat released in multi-dimensional heat conduction can also be determined from of the product solutions. According to section 2.3.4

$$\frac{Q(t)}{Q_0} = \frac{\vartheta_0 - \vartheta_m(t)}{\vartheta_0 - \vartheta_S} = 1 - \vartheta_m^+(t^+) \ , \tag{2.193}$$

holds for the heat released up until time t, cf. (2.173). The dimensionless average temperature ϑ_m^+ is yielded as the product of the average temperatures calculated in 2.3.4.3 for the plate and the infinitely long cylinder. For the *parallelepiped* we obtain

$$\vartheta_m^+ = \vartheta_{mPl}^+\left(\frac{at}{X^2}, \frac{\alpha X}{\lambda}\right) \cdot \vartheta_{mPl}^+\left(\frac{at}{Y^2}, \frac{\alpha' Y}{\lambda}\right) \cdot \vartheta_{mPl}^+\left(\frac{at}{Z^2}, \frac{\alpha'' Z}{\lambda}\right) \tag{2.194}$$

with ϑ_{mPl}^+ from (2.174). In the same way we have for the *cylinder of finite length*

$$\vartheta_m^+ = \vartheta_{mCy}^+\left(\frac{at}{R^2}, \frac{\alpha R}{\lambda}\right) \cdot \vartheta_{mPl}^+\left(\frac{at}{Z^2}, \frac{\alpha' Z}{\lambda}\right) \ , \tag{2.195}$$

where ϑ_{mCy}^+ is given by (2.180).

The equations given here for the temperature distribution and the average temperature are especially easy to evaluate if the dimensionless time t^+ is so large, that the solution can be restricted to the first term in the infinite series, which represent the temperature profile ϑ_{Pl}^+ in the plate and ϑ_{Cy}^+ in the long cylinder. The equations introduced in section 2.3.4.5 and Table 2.6 can then be used. The heating or cooling times required to reach a preset temperature ϑ_k in the centre of the thermally conductive solids handled here can be explicitly calculated when the series are restricted to their first terms. Further information is available in [2.37].

Example 2.5: A cylinder of Chromium-Nickel Steel ($\lambda = 15.0\,\text{W/K m}$, $c = 510\,\text{J/kgK}$, $\varrho = 7800\,\text{kg/m}^3$) with diameter $d = 60\,\text{mm}$ and length $L = 100\,\text{mm}$, whose initial temperature is $\vartheta_0 = 320\,°\text{C}$, is submerged in an oil bath at temperature $\vartheta_S = 30\,°\text{C}$. The heat transfer coefficient is $\alpha = 450\,\text{W/m}^2\text{K}$. How long must the cylinder remain in the oil bath for the temperature at its centre to reach the value $\vartheta_k = 70\,°\text{C}$? At this time what is the highest surface temperature of the cylinder?

The temperature distribution in a cylinder of finite length is given by (2.192). In order to explicitly calculate the cooling time t_k, we will limit ourselves to the first term in each infinite series and then later we will check whether this simplification was correct. We therefore put

$$\vartheta^+ = C_{Cy} J_0\left(\mu_{Cy} r/R\right) \exp\left(-\mu_{Cy}^2 at/R^2\right) \cdot C_{Pl} \cos\left(\mu_{Pl} z/Z\right) \exp\left(-\mu_{Pl}^2 at/Z^2\right) \tag{2.196}$$

with $R = d/2 = 30\,\text{mm}$ and $Z = L/2 = 50\,\text{mm}$. The eigenvalues μ_{Cy}, μ_{Pl} and the coefficients C_{Cy}, C_{Pl} are dependent on the Biot number. It holds for the cylinder that

$$Bi_{Cy} = \alpha R/\lambda = 0.900 \ ,$$

out of which we obtain from Table 2.6 (page 165) $\mu_{Cy} = 1.20484$ and $C_{Cy} = 1.1902$. For the plate

$$Bi_{Pl} = \alpha Z/\lambda = 1.500$$

is obtained and from Table 2.6 $\mu_{Pl} = 0.98824$ and $C_{Pl} = 1.1537$.

The temperature at the centre $(r = 0, z = 0)$ of the cylinder follows from (2.196) as

$$\vartheta_k^+ = C_{Cy} C_{Pl} \exp \left\{ - \left[(\mu_{Cy}/R)^2 + (\mu_{Pl}/Z)^2 \right] a t_k \right\} , \qquad (2.197)$$

where ϑ_k^+ can be calculated from the given temperatures:

$$\vartheta_k^+ = \frac{\vartheta_k - \vartheta_S}{\vartheta_0 - \vartheta_S} = \frac{70\,°C - 30\,°C}{320\,°C - 30\,°C} = 0.1379 .$$

The cooling time t_k we are looking for is found from (2.197) to be

$$t_k = \frac{1}{a} \frac{\ln C_{Cy} + \ln C_{Pl} - \ln \vartheta_k^+}{(\mu_{Cy}/R)^2 + (\mu_{Pl}/Z)^2} .$$

With the thermal diffusivity $a = \lambda/c\varrho = 3.77 \cdot 10^{-6}\,\mathrm{m^2/s}$ this yields

$$t_k = 304\,\mathrm{s} = 5.07\,\mathrm{min} .$$

In order to check whether it is permissible to truncate the series after the first term, we calculate the Fourier numbers using this value of t_k:

$$\left(t_k^+ \right)_{Cy} = a t_k / R^2 = 1.274$$

and

$$\left(t_k^+ \right)_{Pl} = a t_k / Z^2 = 0.459 .$$

According to Fig. 2.31 on page 170 these dimensionless times are so large that the error caused by neglecting the higher terms in the series is insignificant.

The highest surface temperatures appear either at the cylindrical surface $(r = R)$ for $z = 0$ or in the middle of the circular end surfaces, i.e. at $z = Z$ and $r = 0$. For the middle of the cylindrical surface, we get from (2.196) and (2.197)

$$\vartheta^+ (r = R, z = 0, t = t_k) = J_0 (\mu_{Cy}) \vartheta_k^+ = 0.6687 \cdot 0.1379 = 0.0922 ,$$

which gives $\vartheta (r = R, z = 0, t = t_k) = 56.7\,°C$. It holds for the middle of the end surfaces that

$$\vartheta^+ (r = 0, z = Z, t = t_k) = \cos (\mu_{Pl}) \vartheta_k^+ = 0.5502 \cdot 0.1379 = 0.0759 ,$$

from which we obtain $\vartheta (r = 0, z = Z, t = t_k) = 52.0\,°C$. The highest surface temperature therefore appears on the cylindrical surface at $z = 0$.

2.3.5.2   Approximation for small Biot numbers

A simple calculation for the heating or cooling of a body of any shape is possible for the limiting case of small Biot numbers $(Bi \to 0)$. This condition is satisfied when the resistance to heat conduction in the body is much smaller then the heat transfer resistance at its surface, cf. section 2.1.5. At a fixed time, only small temperature differences appear inside the thermally conductive body, whilst between the surface temperature and that of the surroundings a much larger difference exists, cf. Fig. 2.5 on p. 118.

We simply assume that the temperature of the body is only dependend on time and not on the spatial coordinates. This assumption corresponds to $Bi = 0$

because $\lambda \to \infty$, whilst the heat transfer coefficient $\alpha \neq 0$. We apply the first law of thermodynamics to the body being considered in order to determine the variation of temperature with time. The change in its internal energy is equal to the heat flow across its surface

$$\frac{dU}{dt} = \dot{Q} .$$ (2.198)

If a body of volume V has constant material properties ϱ and c, then it holds that

$$\frac{dU}{dt} = V\varrho\frac{du}{dt} = V\varrho c\frac{d\vartheta}{dt} .$$

With A as the surface area of the body we obtain

$$\dot{Q} = \alpha A(\vartheta - \vartheta_S)$$

for the heat flow. The differential equation now follows from (2.198) as

$$\frac{d\vartheta}{dt} = \frac{\alpha A}{c\varrho V}(\vartheta - \vartheta_S) ,$$

with the solution

$$\vartheta^+ = \frac{\vartheta - \vartheta_S}{\vartheta_0 - \vartheta_S} = \exp\left(-\frac{\alpha A}{c\varrho V}t\right)$$ (2.199)

which satisfies the initial condition $\vartheta(t = 0) = \vartheta_0$. This relationship for the temperature change with time is far easier to handle than a series expansion. All the influencing quantities, such as the heat transfer coefficient, material properties and geometry of the body are combined in one single quantity, the decay time

$$t_0 = \frac{c\varrho}{\alpha}\frac{V}{A} .$$

For the example of the plate we show that (2.199) corresponds to the exact solution for $Bi = 0$ with $\alpha \neq 0$. We investigate further, for which Biot numbers not equal to zero this simplified calculation of the temperature change according to (2.199) can be applied to, whilst still obtaining sufficiently high accuracy.

The temperature distribution in a plate is given by (2.171). For $Bi = 0$ we obtain, from (2.170), the eigenvalues $\mu_1 = 0$, $\mu_2 = \pi$, $\mu_3 = 2\pi$, ... Thereby for the first eigenvalue it holds that

$$\mu_1^2 = Bi - Bi^2/3 + \dots ,$$ (2.200)

which can easily be confirmed by series expansion of $\tan\mu$. According to (2.172) all expansion coefficents C_i with $i \geq 2$ will be equal to zero, and it holds that $C_1 = 1$. As $\cos(\mu_1 r^+) \equiv 1$, under consideration of (2.200), we obtain, for the only non-zero term in the infinite series (2.171)

$$\vartheta^+ = \frac{\vartheta - \vartheta_S}{\vartheta_0 - \vartheta_S} = \exp\left(-Bi\, t^+\right) = \exp\left(-\frac{\alpha t}{c\varrho R}\right) .$$

This agrees with (2.199), because for the plate of thickness $2R$, $V/A = R$. The same method can be used to show that for other shapes of solids (2.199) agrees with the exact solution for $Bi = 0$, corresponding to $\lambda \to \infty$, with $\alpha \neq 0$. If however, with a finite value for λ the Biot number would be zero when $\alpha = 0$, it follows from (2.171) that $\vartheta^+ = 1$. The insulated plate keeps its initial temperature ϑ_0, or in other words no equalization of the temperatures ϑ_0 and ϑ_S takes place.

The *accuracy of the approximation equation* (2.199) can be evaluated by comparison with the exact solution for parallelepipeds and prisms with different side to length ratios, for the cylinder with various diameter to height ratios as well as for a sphere. Here we can compare the true average temperature ϑ_m^+ with the temperature ϑ^+ from (2.199). The difference

$$\Delta\vartheta^+ := \vartheta_m^+ - \vartheta^+ \tag{2.201}$$

still depends on the time t or t^+. It is however, always positive; due to the assumption of a spatially constant temperature ϑ^+ we obtain too high a value for the surface temperature which is decisive for heat transfer. This then causes the calculated heat flow released to the surroundings to be too high and then ϑ^+, the average temperature from the approximate solution (2.199), is less than the true average temperature ϑ_m^+ for all times.

The time dependent difference $\Delta\vartheta^+$ from (2.201) assumes a maximum value $\Delta\vartheta_{max}^+$ for each body for a given Biot number, and this maximum appears at a certain time. This maximum deviation of the approximate solution has been calculated for different solid shapes. For $Bi = 0.1$ the maximum difference lies without exception at less than 2% of the characteristic temperature difference $\vartheta_0 - \vartheta_S$. The Biot number in this case was evaluated with the cylinder or sphere radius R as its characteristic length ($Bi = \alpha R/\lambda$). In parallelepipeds and prisms the characteristic length was taken to be half of the *smallest* side length $2X$. In bodies whose dimensions X, Y, Z or R and Z do not differ greatly, the maximum error from the approximation solution lies around 1% of $\vartheta_0 - \vartheta_S$. The error increases rapidly with larger Biot numbers. The error of less than 2% of $\vartheta_0 - \vartheta_S$ calculated for $Bi = 0.10$ could presumedly be tolerated.

The approximate solution from (2.199) offers a simple and recommended procedure for the calculation of the heating or cooling processes in any solid shape. However it is only sufficiently accurate for Biot numbers $Bi < 0.1$. This condition should be checked while choosing the characteristic length for the Biot number to be that of half the shortest length dimension of the body under consideration.

2.3.6 Solidification of geometrically simple bodies

Pure substances and eutectic mixtures solidify and melt at definite temperatures ϑ_E, which differ from substance to substance and are barely dependent on the pressure. The best known example of this is water which at atmospheric pressure, freezes at $\vartheta_E = 0$ °C. This releases the fusion enthalpy of $h_E = 333\,\text{kJ/kg}$. When a solid body melts the enthalpy of fusion must be supplied to the solid as heat.

Solidifying processes are important in cryogenics, food and process industries and also in metallurgy. A main point of interest is the speed at which the boundary between the solid and the liquid phase moves. From this the time required for solidifying layers of a given thickness can be calculated. The modelling of these processes belongs to the field of transient heat conduction, as the enthalpy of fusion released at the phase boundary has to be conducted as heat through the solid.

A general mathematical solution for this type of thermal conduction problems does not exist. Special explicit solutions have been found by F. Neumann[4] in 1865 and J. Stefan in 1891 [2.39]. In the first section we will discuss the solution given by J. Stefan. Quasi-steady solutions will be derived in the section following that. These solutions assume that any heat storage in the solidified body can be neglected. In the final section we will look at the improvements for this quasi-steady solution in which the heat stored in the body is considered approximately.

2.3.6.1 The solidification of flat layers (Stefan problem)

A solidified body is maintained by cooling at $x = 0$ at the constant temperature ϑ_0, which is lower than the solidification temperature ϑ_E, Fig. 2.35. Only one-dimensional heat conduction in the x-direction will be assumed. At the phase boundary $x = s$, which is moving to the right, the solid is touching the liquid which has already been cooled to the solidification temperature. By advance of the phase boundary, or in other words by solidifying a layer of thickness ds, the enthalpy of fusion is released and must be conducted as heat to the cooled surface of the solid at $x = 0$.

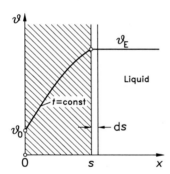

Fig. 2.35: Temperature profile (for $t = $ const) for the solidifying of a plane solid. s is the distance between the phase boundary and the cooled surface $x = 0$

The temperature $\vartheta = \vartheta(x, t)$ in the solidified body satisfies the heat conduction equation

$$\frac{\partial \vartheta}{\partial t} = a \frac{\partial^2 \vartheta}{\partial x^2} \tag{2.202}$$

with the boundary conditions

$$\vartheta = \vartheta_0 \quad \text{for} \quad x = 0 \; , \; t > 0 \; , \tag{2.203}$$

and

$$\vartheta = \vartheta_E \quad \text{for} \quad x = s \; , \; t > 0 \; , \tag{2.204}$$

---

[4]F. Neumann presented this solution in his lectures at the University of Königsberg. The first time it was published was in a German book: Die partiellen Differentialgleichungen der Physik, Editors: B. Riemann and H. Weber, Vol.2, pp. 117–121. Braunschweig: F. Vieweg 1912.

as well as the initial condition

$$s = 0 \quad \text{for} \quad t = 0 \ . \tag{2.205}$$

At the phase interface, the energy balance

$$\lambda \frac{\partial \vartheta}{\partial x} \, dt = h_{\mathrm{E}} \varrho \, ds$$

has to be satisfied, in which h_{E} is the specific enthalpy of fusion. From this we obtain the advance of the phase boundary with time (solidification speed)

$$\frac{ds}{dt} = \frac{\lambda}{h_{\mathrm{E}} \varrho} \left(\frac{\partial \vartheta}{\partial x} \right)_{x=s} \ . \tag{2.206}$$

A solution of the heat conduction equation (2.202) is the error function

$$\vartheta = \vartheta_0 + C \, \mathrm{erf} \left(\frac{x}{2\sqrt{at}} \right) \ ,$$

cf. 2.3.3.1; it satisfies the boundary condition (2.203). The condition (2.204) requires that

$$\vartheta_{\mathrm{E}} = \vartheta_0 + C \, \mathrm{erf} \left(\frac{s}{2\sqrt{at}} \right) \ .$$

The argument of the error function must therefore be a constant γ independent of t. The thickness of the solidified layer increases proportionally with $\sqrt{t}$,

$$s = \gamma \, 2\sqrt{at} \ , \tag{2.207}$$

by which the initial condition (2.205) is also satisfied. We obtain with

$$\vartheta_{\mathrm{E}} - \vartheta_0 = C \, \mathrm{erf} \, \gamma$$

the temperature profile in the solidified layer ($x \leq s$) as

$$\vartheta^+ := \frac{\vartheta - \vartheta_0}{\vartheta_{\mathrm{E}} - \vartheta_0} = \frac{\mathrm{erf} \left(x/\sqrt{4at} \right)}{\mathrm{erf} \, \gamma} = \frac{\mathrm{erf} \left(\gamma x / s \right)}{\mathrm{erf} \, \gamma} \ . \tag{2.208}$$

The still unknown constant γ is yielded from the condition (2.206) for the solidification speed. It follows from (2.207) that

$$\frac{ds}{dt} = \gamma \sqrt{\frac{a}{t}}$$

and from (2.206) and (2.208)

$$\frac{ds}{dt} = \frac{\lambda}{h_{\mathrm{E}} \varrho} \frac{\vartheta_{\mathrm{E}} - \vartheta_0}{\mathrm{erf} \, \gamma} \frac{e^{-\gamma^2}}{\sqrt{\pi} \sqrt{at}} \ .$$

This gives us the transcendental equation independent of t

$$\sqrt{\pi}\gamma e^{\gamma^2} \operatorname{erf} \gamma = \frac{c(\vartheta_E - \vartheta_0)}{h_E} = \frac{1}{Ph} \qquad (2.209)$$

for the determination of γ. This constant is only dependent on the *phase transition number*

$$Ph := \frac{h_E}{c(\vartheta_E - \vartheta_0)} = \frac{1}{St} \quad , \qquad (2.210)$$

which is the ratio of two specific energies, the specific enthalpy of fusion h_E and the difference between the internal energy of the solid at ϑ_E and at ϑ_0. The reciprocal of Ph is also known as the Stefan number St.

Using a series expansion of the error function, see, for example [2.28], [2.30], enables us to expand the left hand side of (2.209) into a series which rapidly converges for small values of γ:

$$\sum_{n=0}^{\infty} \frac{(2\gamma^2)^{n+1}}{1 \cdot 3 \cdot 5 \cdot \ldots (2n+1)} = \frac{1}{Ph} \quad . \qquad (2.211)$$

From this equation we get the series

$$\gamma^2 = \frac{1}{2}Ph^{-1} - \frac{1}{6}Ph^{-2} + \frac{7}{90}Ph^{-3} - \frac{79}{1890}Ph^{-4} + \ldots \quad , \qquad (2.212)$$

which for large values of the phase transition number allows us to directly calculate γ. Finally the time t required to solidify a plane layer of thickness s is yielded from (2.207) and (2.212) as

$$t = \frac{s^2}{4a\gamma^2} = \frac{h_E \varrho s^2}{2\lambda(\vartheta_E - \vartheta_0)}\left(1 + \frac{1}{3}Ph^{-1} - \frac{2}{45}Ph^{-2} + \frac{16}{945}Ph^{-3} - \ldots\right) \quad . \qquad (2.213)$$

The solidification time increases with the square of the layer thickness and is larger the smaller the phase transition number Ph.

Neglecting the heat stored in the solidified body corresponds, due to $c = 0$, to the limiting case of $Ph \to \infty$. This is the so-called quasi-steady approximation, which gives from (2.213), a solidification time

$$t^* = \frac{h_E \varrho s^2}{2\lambda(\vartheta_E - \vartheta_0)} \qquad (2.214)$$

which is always too small. The relative error

$$\frac{t - t^*}{t} = 1 - 2Ph\gamma^2 = \frac{1}{3}Ph^{-1} - \frac{7}{45}Ph^{-2} + \frac{79}{945}Ph^{-3} \ldots \qquad (2.215)$$

is shown in Fig. 2.36. For $Ph > 6.2$ the error is less than 5%. It increases strongly for smaller phase transition numbers.

The problem discussed here was first solved by J. Stefan [2.39] in 1891. The solution of the more general problem first given by F. Neumann has been discussed

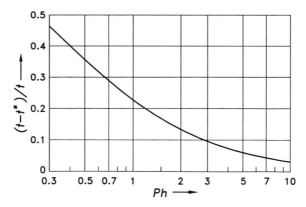

Fig. 2.36: Relative error of the solidification time t^* calculated with the quasi-steady approximation (2.214)

at some length by U. Grigull and H. Sandner [2.18]. An initial temperature in the liquid $\vartheta_{0F} > \vartheta_E$ is assumed in this case, so that the heat conduction equation must also be solved for the liquid temperature $\vartheta_F(x,t)$. It is assumed that no convection takes place. The solution also contains error functions, whereby the material properties λ_F, c_F and ϱ_F of the liquid have to be considered. Further analytical solutions of variations of the Stefan-Neumann Problem can be found in H.S. Carslaw and J.C. Jaeger [2.1]. Generalised mathematical formulations and a discussion of further solution methods are contained in the contributions from A.B. Tayler and J.R. Ockendon in [2.40], cf. also [2.41].

2.3.6.2   The quasi-steady approximation

For sufficiently large values of the phase transition number Ph defined by (2.210), approximately for $Ph > 7$, it is permissible to neglect the storage capability of the solidified layer. A temperature profile similar to one that would appear in steady-state heat conduction is assumed to exist. This quasi-steady approximation allows boundary conditions different from those of the exact solution from F. Neumann and J. Stefan to be assumed at the cooled end of the solid layer. In addition this approach also allows us to investigate solidifying processes in cylindrical and spherical geometries, for which no exact analytical solutions exist. The equations which follow for the solidification-times using the quasi-steady approximation were derived by R. Plank [2.42] and K. Nesselmann [2.43], [2.44]. They are valid due to the large phase transition number for the freezing of ice, and substances containing water, because h_E is particularly large for water.

We will first look at the solidifying process on a flat, cooled wall, as shown in Fig. 2.37. The wall, with thickness δ_W and thermal conductivity λ_W will be cooled by a fluid having a temperature ϑ_0, whereby the heat transfer coefficient α is decisive. On the other side of the wall a solidified layer develops, which at

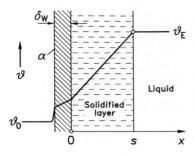

Fig. 2.37: Temperature profile in the solidification of a flat layer under the assumption of the quasi-steady approximation

time t has a thickness of s. The liquid is assumed to already be at the solidifying temperature ϑ_{E}.

During the time interval $\mathrm{d}t$ the phase interface moves a distance of $\mathrm{d}s$. This releases the fusion enthalpy

$$\mathrm{d}Q = h_{\mathrm{E}}\varrho A \, \mathrm{d}s$$

which must be transferred as heat through the layer that is already solidified. In the sense of the quasi-steady approximation, at each time overall heat transfer between the interface ($\vartheta = \vartheta_{\mathrm{E}}$) and the cooling medium ($\vartheta = \vartheta_0$) takes place. Then according to section 1.2.1 it holds that

$$\mathrm{d}Q = \dot{Q} \, \mathrm{d}t = \frac{(\vartheta_{\mathrm{E}} - \vartheta_0)A}{\dfrac{s}{\lambda} + \dfrac{\delta_{\mathrm{W}}}{\lambda_{\mathrm{W}}} + \dfrac{1}{\alpha}} \mathrm{d}t \quad,$$

where λ is the thermal conductivity of the solidified body. It follows from both equations for $\mathrm{d}Q$ that

$$\mathrm{d}t = \frac{h_{\mathrm{E}}\varrho}{\lambda(\vartheta_{\mathrm{E}} - \vartheta_0)} \left(s + \frac{\lambda}{k} \right) \mathrm{d}s \tag{2.216}$$

with

$$\frac{1}{k} = \frac{\delta_{\mathrm{W}}}{\lambda_{\mathrm{W}}} + \frac{1}{\alpha} \quad. \tag{2.217}$$

The solidification-time for a flat layer of thickness s is found by integrating (2.216) to be

$$t = \frac{h_{\mathrm{E}}\varrho s^2}{2\lambda(\vartheta_{\mathrm{E}} - \vartheta_0)} \left(1 + 2\frac{\lambda}{ks} \right) \quad. \tag{2.218}$$

For $k \to \infty$ the temperature at $x = 0$ has the value $\vartheta = \vartheta_0$. This is the boundary condition in the Stefan problem discussed in 2.3.6.1. From (2.218) we obtain the first term of the exact solution (2.213), correponding to $Ph \to \infty$, and therefore the time t^* according to (2.214). With finite heat transfer resistance ($1/k$) the solidification-time is greater than t^*; it no longer increases proportionally to s^2.

In the same manner the solidification-times for layers on cylindrical (tubes) and spherical surfaces can be calculated. We will derive the result for a layer which develops on the outside

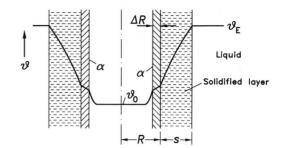

Fig. 2.38: Solidification on the outside of a tube cooled from the inside, which has an outer radius R (quasi-steady approximation)

of a tube of outer radius R being cooled from the inside, Fig. 2.38. By solidifying a layer of thickness ds, heat

$$dQ = h_E \varrho 2\pi (R + s) L ds$$

is released. In this equation L is the length of the tube. For the overall heat transfer

$$dQ = \dot{Q} dt = \frac{\vartheta_E - \vartheta_0}{\dfrac{1}{2\pi L\lambda} \ln \dfrac{R + s}{R} + \dfrac{1}{kA}} dt$$

is valid with

$$\frac{1}{kA} = \frac{1}{2\pi L\lambda_W} \ln \frac{R}{R - \Delta R} + \frac{1}{2\pi L(R - \Delta R)\alpha} ,$$

where ΔR is the thickness of the tube wall. It follows that

$$dt = \frac{h_E \varrho}{\lambda (\vartheta_E - \vartheta_0)} \left[(R + s) \ln \frac{R + s}{R} + (R + s)\beta \right] ds \qquad (2.219)$$

with the abbreviation

$$\beta = \frac{\lambda}{\lambda_W} \ln \frac{R}{R - \Delta R} + \frac{\lambda}{(R - \Delta R)\alpha} . \qquad (2.220)$$

Integration of (2.219) taking account of the initial condition $s = 0$ for $t = 0$ yields with

$$s^+ = s/R$$

the soldification-time

$$t = \frac{h_E \varrho s^2}{2\lambda (\vartheta_E - \vartheta_0)} \left[\left(1 + \frac{1}{s^+} \right)^2 \ln(1 + s^+) - \left(1 + \frac{2}{s^+} \right) \left(\frac{1}{2} - \beta \right) \right] . \qquad (2.221)$$

The relationships in Table 2.7 were found in the same manner. It should be noted here that R is the radius of the cylindrical or spherical surface on which the solidifying layer develops, cf. Fig. 2.39. At $\beta = 0$ the functions $f(s^+, \beta)$ from Table 2.7 for $s^+ \to 0$ assume the limiting value of one. With small layer thicknesses and $\beta = 0$ the solidification-time increases proportionally to s^2, irrespective of the geometrical shape, which according to (2.218) also holds for a flat layer.

Fig. 2.39: Dimensions for solidification on the outside (left hand section) and inside (right hand section) of a tube or a hollow sphere

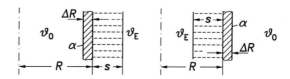

Table 2.7: Equations for the calculation of the solidifying time t for cylindrical and spherical layers of thickness s, see Fig. 2.39. R always indicates the radius of the cylinder or sphere at the side where the solidified layer develops. ΔR is the thickness of the wall of the cylinder or sphere.

Solid-ified Layer	Solidifying time $t = \dfrac{h_E \varrho s^2}{2\lambda(\vartheta_E - \vartheta_0)} f(s^+, \beta)$ with $s^+ = s/R$ and	
	$f(s^+, \beta) =$	$\beta =$
Cylinder outside	$\left(1 + \dfrac{1}{s^+}\right)^2 \ln(1 + s^+) - \left(1 + \dfrac{2}{s^+}\right)\left(\dfrac{1}{2} - \beta\right)$	$\dfrac{\lambda}{\lambda_W} \ln \dfrac{R}{R - \Delta R} + \dfrac{\lambda}{\alpha(R - \Delta R)}$
Cylinder inside	$\left(1 - \dfrac{1}{s^+}\right)^2 \ln(1 - s^+) - \left(1 - \dfrac{2}{s^+}\right)\left(\dfrac{1}{2} + \beta\right)$	$\dfrac{\lambda}{\lambda_W} \ln \dfrac{R + \Delta R}{R} + \dfrac{\lambda}{\alpha(R + \Delta R)}$
Sphere outside	$1 + \dfrac{2}{3}s^+ + \dfrac{2\beta}{s^+}\left(1 + s^+ + \dfrac{s^{+2}}{3}\right)$	$\dfrac{\lambda}{\lambda_W} \dfrac{\Delta R}{R - \Delta R} + \dfrac{\lambda}{\alpha(R - \Delta R)} \dfrac{R}{R - \Delta R}$
Sphere inside	$1 - \dfrac{2}{3}s^+ + \dfrac{2\beta}{s^+}\left(1 - s^+ + \dfrac{s^{+2}}{3}\right)$	$\dfrac{\lambda}{\lambda_W} \dfrac{\Delta R}{R + \Delta R} + \dfrac{\lambda}{\alpha(R + \Delta R)} \dfrac{R}{(R + \Delta R)}$

2.3.6.3   Improved approximations

As a comparison with the exact solution of the Stefan problem shows, the quasi-steady approximation discussed in the last section only holds for sufficiently large values of the phase transition number, around $Ph > 7$. There are no exact solutions for solidification problems with finite overall heat transfer resistances to the cooling liquid or for problems involving cylindrical or spherical geometry, and therefore we have to rely on the quasi-steady approximation. An improvement to this approach in which the heat stored in the solidifed layer is at least approximately considered, is desired and was given in different investigations.

In many cases the temperature profile is replaced by an approximate function, e.g. a second or third degree polynomial. The time dependent coefficient functions of these types of expressions are then fitted to the boundary conditions. The heat conduction equation cannot be exactly, merely approximately satisfied. This can occur, according to T.R. Goodman [2.45], by using a heat-balance integral, see also with reference to this subject [2.46] and [2.47]. F. Megerlin [2.48] however recommends that the heat conduction equation be satisfied pointwise, namely at the interface $x = s$. This produces fairly complicated equations for the solidification speed ds/dt, the integration of which delivers the solidification-time. This integration can often only be carried out numerically, cf. [2.48] and [2.49].

In the procedure of asymptotic approximation no arbitrary approximation functions are used, instead a series of functions

$$\vartheta(x, t) = \vartheta_0(x, t) + \vartheta_1(x, t)Ph^{-1} + \vartheta_2(x, t)Ph^{-2} + \ldots$$

is introduced for the temperature in the solidified layer. This series converges all the better the larger the phase transition number Ph. A similar series is used for the solidification speed:

$$\frac{\mathrm{d}s}{\mathrm{d}t} = s_0(t) + s_1(t)Ph^{-1} + s_2(t)Ph^{-2} + \ldots$$

The functions $\vartheta_i(x,t)$ and $s_i(t)$ can be recursively determined from the exact formulation of the problem with the heat conduction equation and its associated boundary conditions. Thereby $\vartheta_0(x,t)$ and $s_0(t)$ correspond to the quasi-steady approximation with $Ph \to \infty$.

K. Stephan and B. Holzknecht [2.50] have solved the solidification problems dealt with in 2.3.6.2 in this way. Unfortunately the expressions yielded for terms with $i \geq 1$ were very complex and this made it very difficult to calculate the solidification-time explicitly. K. Stephan and B. Holzknecht therefore derived simpler and rather accurate approximation equations for the solidification speed.

Finally the numerical solution of solidification problems should be mentioned, which due to the moving phase boundary contains additional difficulties. As we will not be looking at these solutions in section 2.4, at this point we would suggest the work by K. Stephan and B. Holzknecht [2.51] as well as the contributions from D.R. Atthey, J. Crank and L. Fox in [2.40] as further reading.

2.3.7   Heat sources

Heat sources appear within a heat conducting body as a result of dissipative processes and chemical or nuclear reactions. According to 2.1.2 the differential equation

$$\frac{\partial \vartheta}{\partial t} = a\nabla^2\vartheta + \frac{\dot{W}(\boldsymbol{x},t,\vartheta)}{c\varrho} \tag{2.222}$$

is valid for the temperature field under the assumption that the material properties are independent of both temperature and concentration. The function $\dot{W}(\boldsymbol{x},t,\vartheta)$ introduced in section 2.1.1 is the thermal power per volume, originating inside the body due to the processes mentioned above. The solution of the heat conduction equation (2.222) is, in general, possible if $\dot{W}$ is either independent or only linearly dependent on the temperature ϑ. The Laplace transformation is the preferred solution method for this linear problem. Alternative methods are given in [2.1], where a large number of solutions for different geometries and boundary conditions will be found. In addition the book by H. Tautz [2.26] contains several cases in which the Laplace transformation is used to solve the problem.

In the following section we will deal with an example of homogeneous heat sources. The internal heat development is continuously distributed over the whole body. In the section after that we will discuss local heat sources where the heat development is concentrated at a point or a line in the heat conducting body.

2.3.7.1   Homogeneous heat sources

Assuming geometric one-dimensional heat flow in the x-direction and a power density $\dot{W}$ which is independent of temperature, the heat conduction equation is then

$$\frac{\partial \vartheta}{\partial t} = a\frac{\partial^2 \vartheta}{\partial x^2} + \frac{\dot{W}(x,t)}{c\varrho} \ . \tag{2.223}$$

This linear non-homogeneous differential equation can easily be solved using the Laplace transformation. The steps for the solution of such a problem are shown in the following example.

In a semi-infinite body ($x \geq 0$), a spatially constant, but time dependent power density is supposed to exist:

$$\dot{W}(t) = \dot{W}_0\sqrt{t_0/t} \ , \quad t > 0 \ . \tag{2.224}$$

Here $\dot{W}_0$ represents the power density at time t_0. By using (2.224), a heat release is modelled as it occurs, for example when concrete sets. Then we have an initial large release of heat which rapidly decreases with advancing time. The semi-infinite body initially has the constant (over) temperature

$$\vartheta = 0 \quad \text{for} \quad t = 0 \tag{2.225}$$

at which the surface $x = 0$ should always be kept:

$$\vartheta(0,t) = 0 \quad \text{for} \quad x = 0 \ . \tag{2.226}$$

By applying the Laplace transformation to (2.223) and (2.224) we obtain the non-homogeneous ordinary differential equation

$$\frac{\mathrm{d}^2 u}{\mathrm{d}x^2} - p^2 u = -\frac{\dot{W}_0}{\lambda}\frac{\sqrt{\pi t_0}}{\sqrt{s}}$$

with $p^2 = s/a$ under consideration of the initial condition (2.225). The solution of the homogeneous differential equation limited for $x \to \infty$ is $u_{\text{hom}} = C\exp(-px)$. A particular solution for the non-homogeneous equation is

$$u_{\text{inh}} = \frac{\dot{W}_0}{\lambda}\frac{\sqrt{\pi t_0}}{\sqrt{s}p^2} = \frac{\dot{W}_0}{c\varrho}\frac{\sqrt{\pi t_0}}{s^{3/2}} \ .$$

The transformed function composed of these two parts

$$u(x,s) = Ce^{-px} + \frac{\dot{W}_0\sqrt{\pi t_0}}{c\varrho s^{3/2}}$$

still has to be fitted to the boundary condition (2.226). This gives

$$u(x,s) = \frac{\dot{W}_0}{c\varrho}\sqrt{\pi t_0}\left(\frac{1}{s^{3/2}} - \frac{1}{\sqrt{a}}\frac{e^{-px}}{sp}\right) \ .$$

The reverse transformation, with the assistance of a correspondence table, Table 2.3, provides us with

$$\vartheta(x,t) = \frac{\dot{W}_0\sqrt{t_0}}{c\varrho}2\sqrt{t}\left(1 - \sqrt{\pi}\,\text{ierfc}\,\frac{x}{2\sqrt{at}}\right) \tag{2.227}$$

with the integrated error function from section 2.3.3.1. Introducing into (2.227) the following dimensionless variables

$$t^+ := t/t_0, \quad x^+ := x/\sqrt{at_0}, \quad \vartheta^+ := \vartheta c\varrho/(\dot{W}_0 t_0) \ ,$$

yields

$$\vartheta^+\left(x^+,t^+\right) = 2\sqrt{t^+}\left(1 - \sqrt{\pi}\,\text{ierfc}\,\frac{x^+}{2\sqrt{t^+}}\right) \ . \tag{2.228}$$

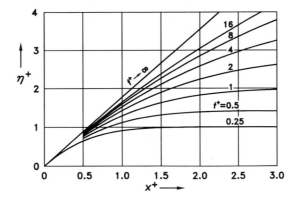

Fig. 2.40: Temperature field $\vartheta^+(x^+, t^+)$ according to (2.228) in a semi-infinite body with time dependent homogeneous heat sources according to (2.224)

This temperature distribution is illustrated in Fig. 2.40. For very small values of $x^+/(2\sqrt{t^+})$ it holds that

$$\sqrt{\pi}\,\text{ierfc}\left[x^+/\left(2\sqrt{t^+}\right)\right] = 1 - \sqrt{\pi}x^+/\left(2\sqrt{t^+}\right) + \cdots .$$

Therefore we obtain for $t^+ \to \infty$ the straight line

$$\vartheta_\infty^+ = \sqrt{\pi}x^+$$

as the steady end temperature. This straight line is simultaneously the tangent to *all* temperature curves $t^+ = \text{const}$ at the free surface $x^+ = 0$. This is where the heat flux, constant with respect to time

$$\dot{q}_0 = -\lambda\left(\frac{\partial\vartheta}{\partial x}\right)_{x=0} = -\dot{W}_0\sqrt{at_0}\left(\frac{\partial\vartheta^+}{\partial x^+}\right)_{x^+=0} = -\dot{W}_0\sqrt{\pi at_0}$$

has to be removed in order to maintain the temperature at the value $\vartheta^+ = 0$.

Solutions of (2.223) for other power densities $\dot{W}(x,t)$ and boundary conditions can be found in the same way. However, for bodies of finite thickness, the inverse transformation of $u(x,s)$ can normally only be carried out using the inversion theorem mentioned in section 2.3.2. For cases such as these see [2.1] and the detailed examples discussed in [2.26].

2.3.7.2 Point and linear heat sources

If the generation of heat is concentrated in a small, limited space we speak of local heat sources, which can be idealised as point, line or sheet singularities. For example, an electrically heated thin wire can be treated as a linear heat source. Alongside such technical applications these singularities have significant theoretical importance in the calculation of temperature fields, cf. [2.1].

The calculation of the temperature field produced when a point heat source exists will now be explained. We will consider a body which extends infinitely in all directions, with a spherical hollow space of radius R inside it, Fig. 2.41. At the surface of the sphere the boundary condition of a spatially constant heat flux $\dot{q}_0 = \dot{q}_0(t)$ is stipulated. This corresponds to a heat source within the hollow space with a thermal power (yield) of

$$\dot{Q} = 4\pi R^2 \dot{q}_0 \ .$$

The limit $R \to 0$ at a given time t, keeping $\dot{Q}$ constant, leads to a point heat source of strength $\dot{Q}(t)$, which is located at the centre of the sphere which is taken to be the origin $r = 0$ of the radial (spherical) coordinate.

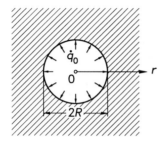

Fig. 2.41: Infinite body with respect to all dimensions with a spherical hollow space at whose surface the heat flux $\dot{q}_0$ is prescribed

The temperature field outside the spherical hollow satisfies the differential equation

$$\frac{\partial \vartheta}{\partial t} = a \left(\frac{\partial^2 \vartheta}{\partial r^2} + \frac{2}{r} \frac{\partial \vartheta}{\partial r} \right)$$

with boundary conditions

$$-\lambda \frac{\partial \vartheta}{\partial r} = \dot{q}_0 = \frac{\dot{Q}(t)}{4\pi R^2} \quad \text{for} \quad r = R$$

and

$$\vartheta = 0 \quad \text{for} \quad r \to \infty \ ,$$

if we assume the initial temperature to be $\vartheta = 0$. Applying the Laplace transformation leads to

$$\frac{d^2 u}{dr^2} + \frac{2}{r} \frac{du}{dr} - p^2 u = 0$$

with $p^2 = s/a$. The boundary conditions are

$$\frac{du}{dr} = -\frac{1}{4\pi R^2 \lambda} \mathcal{L}\left\{ \dot{Q}(t) \right\} \quad \text{for} \quad r = R \tag{2.229}$$

and $u \to 0$ for $r \to \infty$. A solution which satisfies the last condition is

$$u = \frac{B}{r} e^{-pr} \ .$$

The constant B is obtained from (2.229), so that

$$u = \frac{\mathcal{L}\left\{ \dot{Q}(t) \right\}}{4\pi \lambda r} \frac{e^{-p(r-R)}}{1 + pR}$$

is the desired transformed function. Letting $R \to 0$ we obtain for the point heat source at $r = 0$

$$u = \frac{\mathcal{L}\left\{\dot{Q}(t)\right\}}{4\pi\lambda r}e^{-pr} \quad .$$

From No. 4 in Table 2.3 on page 145

$$e^{-pr} = \mathcal{L}\left\{\frac{r}{2\sqrt{a\pi t^{3/2}}}e^{-r^2/4at}\right\} \quad ,$$

such that u is obtained as the product of two Laplace transforms

$$u = \frac{1}{4\pi\lambda}\mathcal{L}\left\{\dot{Q}(t)\right\} \cdot \mathcal{L}\left\{\frac{e^{-r^2/4at}}{2\sqrt{a\pi t^{2/3}}}\right\} \quad .$$

Which, according to the convolution theorem, No. 6 in Table 2.2 (page 143), is

$$u = \frac{1}{4\pi\lambda}\mathcal{L}\left\{\int_{\tau=0}^{\tau=t}\dot{Q}(\tau)\frac{\exp\left(-r^2/4a(t-\tau)\right)}{2\sqrt{a\pi}(t-\tau)^{3/2}}\,\mathrm{d}\tau\right\} \quad .$$

So the temperature distribution around a point heat source at $r = 0$, which, at time $t = 0$ is "switched on" with a thermal power of $\dot{Q}(t)$, is yielded to be

$$\vartheta(r,t) = \frac{1}{(4\pi a)^{3/2}c\varrho}\int_0^t \dot{Q}(\tau)\exp\left(\frac{-r^2}{4a\,(t-\tau)}\right)\frac{\mathrm{d}\tau}{(t-\tau)^{3/2}} \quad . \tag{2.230}$$

This general solution for any time function $\dot{Q}(t)$ contains special simple cases. The temperature field for a source of constant thermal power $\dot{Q}(t) = \dot{Q}_0$ is found to be

$$\vartheta(r,t) = \frac{\dot{Q}_0}{4\pi\lambda r}\,\mathrm{erfc}\,\frac{r}{\sqrt{4at}} \quad . \tag{2.231}$$

For $t \to \infty$ we obtain, with $\mathrm{erfc}(0) = 1$, the steady-state temperature field around a point heat source as

$$\vartheta(r) = \frac{\dot{Q}_0}{4\pi\lambda r} \quad .$$

A very fast reaction or an electrical short can cause a sudden release of energy Q_0 at time $t = 0$ in a small space at $r = 0$. For this "heat explosion" the limit in (2.230) is set as $\tau \to 0$ and the heat released is given by

$$Q_0 = \lim_{\tau\to 0}\int_0^\tau \dot{Q}(\tau)\,\mathrm{d}\tau \quad .$$

This then gives

$$\vartheta(r,t) = \frac{Q_0}{(4\pi at)^{3/2}c\varrho}\exp\left(-\frac{r^2}{4at}\right) \quad . \tag{2.232}$$

as the temperature distribution. It has the boundary value of $\vartheta \to \infty$ for $t = 0$ and $r = 0$. At a fixed point $r = \mathrm{const}$, where $r \neq 0$ the temperature change with

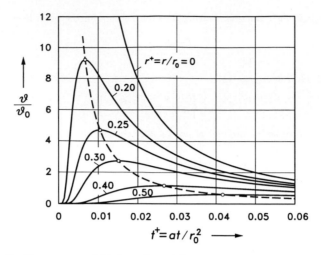

Fig. 2.42: Temperature field from (2.232) in an infinite body after a "heat explosion" at $r = 0$. The reference temperature is $\vartheta_0 = Q_0/(c\varrho r_0^3)$ with r_0 as an arbitrarily selected distance from $r = 0$

time begins with $\vartheta = 0$, goes through a maximum at time $t_{max} = r^2/(6a)$ and falls once again back to $\vartheta = 0$. The temperature profile over time for different ratios of r/r_0, according to (2.232) is shown in Fig. 2.42

We will also start from this "heat explosion" for the calculation of the temperature field around a *linear heat source* at $r = 0$. At time $t = 0$, heat Q_0 is released by a linear heat source of length L (perpendicular to the r, φ–plane of the polar coordinate system). As no other heat sources are present, at every later time the heat Q_0 has to be found as an increase in the internal energy of the environment of the heat source. Therefore the balance

$$Q_0 = \int_0^\infty 2\pi r L c\varrho\, \vartheta(r,t)\, \mathrm{d}r = 2\pi L c\varrho \int_0^\infty r\vartheta(r,t)\, \mathrm{d}r$$

holds, independent of time t. The desired temperature distribution $\vartheta(r,t)$, must satisfy this equation.

In analogy to the point "heat explosion" we introduce the function

$$\vartheta(r,t) = f(t) \exp\left(-\frac{r^2}{4at}\right)$$

corresponding to (2.232) and obtain

$$Q_0 = 2\pi L c\varrho f(t) \int_0^\infty r \exp\left(-\frac{r^2}{4at}\right) \mathrm{d}r \ .$$

With $\xi = r/\sqrt{4at}$ this becomes

$$Q_0 = 2\pi Lc\varrho f(t)4at \int_0^\infty \xi e^{-\xi^2}\, d\xi \ .$$

The definite integral which appears here has the value of $1/2$, such that the time function $f(t)$ is yielded to be

$$f(t) = \frac{Q_0/L}{4\pi c\varrho at} = \frac{Q_0/L}{4\pi\lambda t} \ .$$

From this we obtain the desired temperature distribution

$$\vartheta(r,t) = \frac{Q_0/L}{4\pi\lambda t} \exp\left(-\frac{r^2}{4at}\right) \ . \tag{2.233}$$

It is easy to prove that this satisfies the heat conduction equation (2.157) with $n = 1$ (cylindrical coordinates).

This result can then be generalised for a linear source at $r = 0$ with time dependent thermal power $\dot{Q}(t)$. During the time interval from $t = \tau$ to $t = \tau + d\tau$ it releases heat $\dot{Q}(\tau)\, d\tau$ thereby generating a temperature field corresponding to (2.233). By superimposing these small "heat explosions" over time the following temperature field, analogous to (2.230), is obtained:

$$\vartheta(r,t) = \frac{1}{4\pi\lambda L} \int_{\tau=0}^{\tau=t} \dot{Q}(\tau) \exp\left(-\frac{r^2}{4a(t-\tau)}\right) \frac{d\tau}{t-\tau} \ . \tag{2.234}$$

For the special case of constant thermal power $\dot{Q}_0$ this becomes

$$\vartheta(r,t) = -\frac{\dot{Q}_0}{4\pi\lambda L}\mathrm{Ei}\left(-\frac{r^2}{4at}\right) \ . \tag{2.235}$$

The function which is present here is known as the exponential integral

$$\mathrm{Ei}(-\xi) = \int_\xi^\infty \frac{e^{-u}}{u}\, du$$

with the series expansion

$$\mathrm{Ei}(-\xi) = 0.577216 + \ln\xi + \sum_{n=1}^\infty (-1)^n \frac{\xi^n}{n\cdot n!}$$

and the asymptotic expansion ($\xi \gg 1$)

$$\mathrm{Ei}(-\xi) = \frac{e^{-\xi}}{-\xi}\left(1 - \frac{1!}{\xi} + \frac{2!}{\xi^2} - \frac{3!}{\xi^3} + \cdots\right) \ ,$$

cf. [2.28] and [2.30]. Some values of $\mathrm{Ei}(-\xi)$, which is always negative, are given in Table 2.8.

Table 2.8: Values of the exponential integral $\mathrm{Ei}(-\xi)$

ξ	$-\mathrm{Ei}(-\xi)$	ξ	$-\mathrm{Ei}(-\xi)$	ξ	$-\mathrm{Ei}(-\xi)$	ξ	$-\mathrm{Ei}(-\xi)$	ξ	$-\mathrm{Ei}(-\xi)$
0	∞	0.15	1.4645	0.4	0.7024	0.9	0.2602	3.0	0.01305
0.01	4.0379	0.20	1.2227	0.5	0.5598	1.0	0.2194	3.5	0.00697
0.02	3.3547	0.25	1.0443	0.6	0.4544	1.5	0.1000	4.0	0.00378
0.05	2.4679	0.30	0.9057	0.7	0.3738	2.0	0.0489	5.0	0.00115
0.10	1.8229	0.35	0.7942	0.8	0.3106	2.5	0.0249	6.0	0.00036

2.4 Numerical solutions to heat conduction problems

Complicated heat conduction problems, for which no explicit solutions exist or can only be obtained with great effort, are preferentially solved using numerical methods. Problems with temperature dependent material properties, complex geometrical forms and those with particular boundary conditions, for example temperature dependent heat transfer coefficients, all belong in this group. However, despite the use of computers, numerical solution methods often demand a great deal of programming, memory and computing time which should not be underestimated. The decision to use a numerical solution procedure should therefore be carefully considered, particularly if simplifications of the problem are acceptable, which would then lead to an analytical solution.

Two methods are available for the numerical solution of initial-boundary-value problems, the finite difference method and the finite element method. Finite difference methods are easy to handle and require little mathematical effort. In contrast the finite element method, which is principally applied in solid and structure mechanics, has much higher mathematical demands, it is however very flexible. In particular, for complicated geometries it can be well suited to the problem, and for such cases should always be used in preference to the finite difference method. We will limit ourselves to an introductory illustration of the difference method, which can be recommended even to beginners as a good tool for solving heat conduction problems. The application of the finite element method to these problems has been described in detail by G.E. Myers [2.52]. Further information can be found in D. Marsal [2.53] and in the standard works [2.54] to [2.56].

2.4.1   The simple, explicit difference method for transient heat conduction problems

In the finite difference method, the derivatives $\partial\vartheta/\partial t$, $\partial\vartheta/\partial x$ and $\partial^2\vartheta/\partial x^2$, which appear in the heat conduction equation and the boundary conditions, are replaced by difference quotients. This *discretisation* transforms the differential equation into a finite difference equation whose solution approximates the solution of the differential equation at discrete points which form a grid in space and time. A reduction in the mesh size increases the number of grid points and therefore the accuracy of the approximation, although this does of course increase the computation demands. Applying a finite difference method one has therefore to make a compromise between accuracy and computation time.

2.4.1.1   The finite difference equation

For the introduction and explanation of the method we will discuss the case of transient, geometric one-dimensional heat conduction with constant material properties. In the region $x_0 \leq x \leq x_n$ the heat conduction equation

$$\frac{\partial\vartheta}{\partial t} = a\frac{\partial^2\vartheta}{\partial x^2} \qquad (2.236)$$

has to be solved for times $t \geq t_0$, whilst considering the boundary conditions at x_0 and x_n. The initial temperature distribution $\vartheta_0(x)$ for $t = t_0$ is given.

A grid is established along the strip $x_0 \leq x \leq x_n$, $t \geq t_0$, with mesh size Δx in the x-direction and Δt in the t-direction, Fig. 2.43. A grid point (i, k) has the coordinates

$$x_i = x_0 + i\Delta x \quad \text{with} \quad i = 0, 1, 2, \dots$$

and

$$t_k = t_0 + k\Delta t \quad \text{with} \quad k = 0, 1, 2, \dots$$

The approximation value of ϑ at grid point (i, k) is indicated by

$$\vartheta_i^k = \vartheta(x_i, t_k) \ . \qquad (2.237)$$

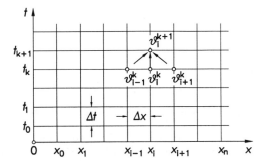

Fig. 2.43: Grid for the discretisation of the heat conduction equation (2.236) and to illustrate the finite difference equation (2.240)

We refrain from indicating the approximation value of the temperature by a different symbol to the exact temperature, e.g. Θ instead of ϑ, as is usually the case in mathematics literature. The time level t_k is indicated by the superscript k without brackets because there is no danger of it being confused with the k-th power of ϑ.

The derivatives which appear in (2.236) are replaced by difference quotients, whereby a *discretisation error* has to be taken into account

$$\text{derivative} = \text{difference quotient} + \text{discretisation error} \ .$$

The discretisation error goes to zero with a reduction in the mesh size Δx or Δt.

The second derivative in the x-direction at position x_i at time t_k is replaced by the central second difference quotient

$$\left(\frac{\partial^2 \vartheta}{\partial x^2}\right)_i^k = \frac{\vartheta_{i-1}^k - 2\vartheta_i^k + \vartheta_{i+1}^k}{\Delta x^2} + O(\Delta x^2) \ . \tag{2.238}$$

The writing of $O(\Delta x^2)$ indicates that the discretisation error is proportional to Δx^2 and therefore by reducing the mesh size the error approaches zero with the square of the mesh size. The first derivative with respect to time is replaced by the relatively inaccurate forward difference quotient

$$\left(\frac{\partial \vartheta}{\partial t}\right)_i^k = \frac{\vartheta_i^{k+1} - \vartheta_i^k}{\Delta t} + O(\Delta t) \ . \tag{2.239}$$

Its discretisation error only approaches zero proportionally to Δt. However with (2.239) a numerically simple, explicit finite difference formula is obtained. Putting (2.238) and (2.239) into the differential equation (2.236) gives, after a simple rearrangement, the finite difference equation

$$\vartheta_i^{k+1} = M\vartheta_{i-1}^k + (1 - 2M)\,\vartheta_i^k + M\vartheta_{i+1}^k \tag{2.240}$$

with

$$M := a\Delta t/\Delta x^2 \tag{2.241}$$

as the modulus or Fourier number of the difference method. The discretisation error of the difference equation is, due to $\Delta t = M\Delta x^2/a$, of order $O(\Delta x^2)$.

Equation (2.240) is an *explicit* difference formula; it allows, from three known temperatures of the time level $t = t_k$ the temperatures ϑ_i^{k+1} of the next time level $t_{k+1} = t_k + \Delta t$ to be explicitly calculated, cf. Figure 2.43. For $k = 0$, and therefore $t = t_0$, all the temperatures ϑ_i^0 are known from the given initial temperature profile. Equation (2.240) enables all ϑ_i^1 at time t_1 to be calculated and from these the values for ϑ_i^2 at time t_2, etc.

The equations (2.238) and (2.239) for the replacement of the derivatives with difference quotients can be derived using a Taylor series expansion of the temperature field around the point (x_i, t_k), cf. [2.53] and [2.57]. It is also possible to derive the finite difference formula(2.240) from an energy balance in conjunction with Fourier's law, cf. [2.58]. This requires the introduction of the heat conduction resistances $\Delta x/\lambda$ between the grid points and the consideration of the energy storage in a "block" around the position x_i. A further development of this method is the Volume Integral Method, with which various finite difference equations can be derived, cf. [2.59].

2.4.1.2 The stability condition

Many finite difference formulae have the undesirable property that small initial and rounding errors become larger as the calculation proceeds, which in the end produces a false result. This phenomenon is called (numerical) instability. In contrast a difference formula is stable when the errors become smaller during the calculation run and therefore their effect on the result declines. Most difference equations are only conditionally stable, that is they are stable for certain step or mesh sizes. The explicit equation (2.240) belongs to this group, as it is only stable when the modulus M satisfies the condition

$$M = a\Delta t/\Delta x^2 \leq \frac{1}{2} \ . \tag{2.242}$$

When Δx is given, the size of the time step Δt cannot be chosen at will, because violation of (2.242) will not just make the process inaccurate, it will be rendered completely useless.

The stability condition (2.242) can be derived in a number of ways; cf. the extensive discussion in [2.57]. A general condition for stability of explicit difference formulae is the requirement that no coefficient in such an equation is negative, cf. [2.60]. This means for (2.240)

$$1 - 2M \geq 0 \ ,$$

from which (2.242) immediately follows. The so-called ε-scheme is used to clarify stability behaviour. For $t = t_0$ we set all $\vartheta_i^0 = 0$ except for one value which is set to $\varepsilon \neq 0$. The error propagation is followed by applying the difference formula for further time steps with $k = 1, 2, \ldots$ The ε-scheme for the difference equation (2.240) has the following appearance for $M = 1/2$ (stability limit):

ε-scheme for $M = 1/2$

$k = 0$:	0	0	0	0	ε	0	0	0	0
$k = 1$:	0	0	0	0.5ε	0	0.5ε	0	0	0
$k = 2$:	0	0	0.25ε	0	0.5ε	0	0.25ε	0	0
$k = 3$:	0	0.125ε	0	0.375ε	0	0.375ε	0	0.125ε	0
$k = 4$:	0.0625ε	0	0.25ε	0	0.375ε	0	0.25ε	0	0.0625ε

The finite difference formula with $M = 1/2$ is just about stable, the error is shared between two neighbouring points and slowly declines. In addition the grid divides into two grid sections which are not connected, as in the difference formula

$$\vartheta_i^{k+1} = \frac{1}{2}\left(\vartheta_{i-1}^k + \vartheta_{i+1}^k\right) \quad , \quad \left(M = \frac{1}{2}\right) \ , \tag{2.243}$$

only *two* temperatures at time t_k, determine the new temperature ϑ_i^{k+1}; ϑ_i^k has no influence on ϑ_i^{k+1}. A loose connection of the two grid sections exists only because the temperatures at the boundaries are given. Therefore, it is not recommended that (2.240) is used with $M = 1/2$, even though this allows the largest time step Δt. The difference equation (2.243) is the basis for the previously most frequently used graphical methods from L. Binder [2.61] and E. Schmidt [2.62]. As a result of the advances in computer technology, these clear but somewhat inaccurate and time-consuming graphical procedures have lost their former importance. We will not discuss these graphical methods any further, information is available in [2.63] and [2.64].

The instability of the finite difference formula (2.240) for $M > 1/2$ is illustrated by the ε-scheme for $M = 1$:

$$\varepsilon\text{-scheme for } M = 1$$

$k = 0:$	0	0	0	0	ε	0	0	0	0
$k = 1:$	0	0	0	ε	$-\varepsilon$	ε	0	0	0
$k = 2:$	0	0	ε	-2ε	3ε	-2ε	ε	0	0
$k = 3:$	0	ε	-3ε	6ε	-7ε	6ε	-3ε	ε	0
$k = 4:$	ε	-4ε	7ε	-16ε	19ε	-16ε	7ε	-4ε	ε

An error becomes greater with every time step, such that the solution of the difference equation does not correspond to the solution of the differential equation as the stability condition has been violated.

2.4.1.3   Heat sources

The finite difference equation (2.240) can easily be extended to cases where heat sources are present. The differential equation from 2.1.2 is valid:

$$\frac{\partial \vartheta}{\partial t} = a\frac{\partial^2 \vartheta}{\partial x^2} + \frac{\dot{W}(x,t,\vartheta)}{c\varrho} \quad . \tag{2.244}$$

Its discretisation with the difference quotients according to (2.238) and (2.239) leads to the explicit finite difference formula

$$\vartheta_i^{k+1} = M\vartheta_{i-1}^k + (1 - 2M)\,\vartheta_i^k + M\vartheta_{i+1}^k + \left(M\Delta x^2/\lambda\right)\dot{W}_i^k \quad . \tag{2.245}$$

Here

$$\dot{W}_i^k = \dot{W}\left(x_i, t_k, \vartheta_i^k\right) \tag{2.246}$$

indicates the value of the power density at x_i at time t_k.

Equation (2.245) can also be used when the power density depends on the temperature. The temperature ϑ_i^k which appears in (2.246) is known from the calculation of the temperature field at time t_k. A strong temperature dependence of $\dot{W}$ can fundamentally influence the stability behaviour. Reference [2.57] shows that the stability condition (2.242) holds unaltered if $\dot{W}$ depends linearly on the temperature ϑ.

2.4.2   Discretisation of the boundary conditions

Considering the boundary conditions in finite difference methods we will distinguish three different cases:
- prescribed boundary temperatures
- preset heat flux at the boundary and
- the heat transfer condition, cf. section 2.1.3.

If the *temperature at the boundary* $x = x_R$ is given, then the grid divisions should be chosen such that the boundary coincides with a grid line $x_i = $ const. The left hand boundary is then $x_R = x_0$ with the right hand boundary $x_R = x_{n+1} = x_0 + (n + 1)\Delta x$. The given temperature $\vartheta(x_R, t_k)$ is used as the temperature value ϑ_0^k or ϑ_{n+1}^k in the difference equation (2.240).

With given *heat flux* $\dot{q}(t)$ the condition

$$-\lambda \left(\frac{\partial \vartheta}{\partial n}\right)_{x=x_R} = \dot{q}(t) \qquad (2.247)$$

has to be satisfied at the boundary. The derivative must be formed in the outward normal direction, and the heat flux $\dot{q}$ is positive if it flows in this direction. The grid will now be established such that the boundary lies between two grid lines. This gives, for the left hand boundary $x_R = x_0 + \Delta x/2$, see Fig. 2.44. This means that grid points $(0, k)$ are introduced outside the thermally conductive body. The temperatures ϑ_0^k which are present here serve only as calculated values to fulfill the boundary condition (2.247).

Fig. 2.44: Consideration of the boundary conditions (2.247) at $x_R = x_0 + \Delta x/2$ by means of the introduction of temperatures ϑ_0^k outside the body

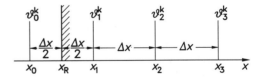

The local derivative which appears in (2.247) is replaced by the rather exact *central* difference quotient. It holds at the left hand boundary that

$$-\left(\frac{\partial \vartheta}{\partial n}\right)_{x_0+\Delta x/2} = \left(\frac{\partial \vartheta}{\partial x}\right)_{\frac{1}{2}}^k = \frac{\vartheta_1^k - \vartheta_0^k}{\Delta x} + O(\Delta x^2) \ . \qquad (2.248)$$

Then from (2.247) we obtain

$$\vartheta_0^k = \vartheta_1^k - \frac{\Delta x}{\lambda}\dot{q}(t_k) \ . \qquad (2.249)$$

At $x = x_1$ the difference equation (2.240) with $i = 1$ holds:

$$\vartheta_1^{k+1} = M\vartheta_0^k + (1 - 2M)\vartheta_1^k + M\vartheta_2^k \ . \qquad (2.250)$$

Elimination of ϑ_0^k from both these equations yields

$$\vartheta_1^{k+1} = (1 - M)\,\vartheta_1^k + M\vartheta_2^k - M\frac{\Delta x}{\lambda}\dot{q}(t_k) \ . \tag{2.251}$$

This equation replaces (2.250), when the boundary condition (2.247) is to be satisfied at the left hand boundary.

If (2.247) is to be satisfied at the right hand boundary, then the grid is chosen such that $x_R = x_n + \Delta x/2$ is valid. The elimination of ϑ_{n+1}^k from the difference equation which holds for x_n, and from the boundary condition, yields

$$\vartheta_n^{k+1} = M\vartheta_{n-1}^k + (1 - M)\,\vartheta_n^k + M\frac{\Delta x}{\lambda}\dot{q}(t_k) \ . \tag{2.252}$$

As in (2.251), $\dot{q}(t_k)$ will also be positive when the heat flux is out of the body.

The difference equations (2.251) and (2.252) are valid, with $\dot{q} \equiv 0$, for an *adiabatic boundary*. Adiabatic boundaries are also the planes of symmetry inside the body. In this case the grid is chosen such that the adiabatic plane of symmetry lies between two neighbouring grid lines. The calculation of the temperatures can then be limited to one half of the body.

If heat is transferred at the boundary to a fluid at temperature ϑ_S, where the heat transfer coefficient α is given, then the *heat transfer condition* is valid

$$-\lambda\frac{\partial\vartheta}{\partial n} = \alpha\,(\vartheta - \vartheta_S) \quad \text{for} \quad x = x_R \ . \tag{2.253}$$

The derivative is formed in the outward normal direction, and α and ϑ_S can be dependent on time. For the discretisation of (2.253) it is most convenient if the boundary coincides with a grid line, Fig. 2.45, as the boundary temperature which appears in (2.253) can immediately be used in the difference formula. The replacement of the derivative $\partial\vartheta/\partial n$ by the central difference quotient requires grid points outside the body, namely the temperatures ϑ_0^k or ϑ_{n+1}^k, which, in conjunction with the boundary condition, can be eliminated from the difference equations.

At the left hand boundary $(x_R = x_1)$ the outward normal to the surface points in the negative x-direction. Therefore from (2.253) it follows that

$$\frac{\partial\vartheta}{\partial x} = \frac{\alpha}{\lambda}\,(\vartheta - \vartheta_S) \quad \text{for} \quad x = x_1 \ .$$

Replacing with the central difference quotient,

$$\left(\frac{\partial\vartheta}{\partial x}\right)_1^k = \frac{\vartheta_2^k - \vartheta_0^k}{2\Delta x} + O(\Delta x^2) \ ,$$

gives

$$\vartheta_0^k = \vartheta_2^k - 2Bi^*\left(\vartheta_1^k - \vartheta_S^k\right) \tag{2.254}$$

with

$$Bi^* = \alpha\Delta x/\lambda \tag{2.255}$$

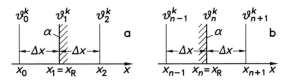

Fig. 2.45: Consideration of the heat transfer condition (2.253), **a** at the left hand boundary $x_R = x_1$, **b** at the right hand boundary $x_R = x_n$

as the Biot number of the finite difference method. Using (2.254), the temperature ϑ_0^k is eliminated from the difference equation (2.250), which assumes the form

$$\vartheta_1^{k+1} = [1 - 2M\,(1 + Bi^*)]\,\vartheta_1^k + 2M\vartheta_2^k + 2M\,Bi^*\vartheta_S^k \ . \qquad (2.256)$$

At the right hand boundary $(x_R = x_n)$, we obtain from (2.253) the boundary condition

$$-\frac{\partial \vartheta}{\partial x} = \frac{\alpha}{\lambda}\,(\vartheta - \vartheta_S) \ .$$

Its discretisation leads to an equation analogous to (2.254)

$$\vartheta_{n+1}^k = \vartheta_{n-1}^k - 2Bi^*\left(\vartheta_n^k - \vartheta_S^k\right). \qquad (2.257)$$

This is then used to eliminate ϑ_{n+1}^k from the difference equation valid for x_n, which then has the form

$$\vartheta_n^{k+1} = 2M\vartheta_{n-1}^k + [1 - 2M\,(1 + Bi^*)]\,\vartheta_n^k + 2M\,Bi^*\vartheta_S^k \ . \qquad (2.258)$$

The consideration of the heat transfer condition (2.253) detoriates the *stability behaviour* of the explicit difference formula. As the coefficients of the explicit difference equations (2.256) and (2.258) must be positive in order to guarantee stability, we obtain the stability condition

$$M \leq \frac{1}{2\,(1 + Bi^*)} \ . \qquad (2.259)$$

This tightens the condition (2.242) and leads to even smaller time steps Δt.

In the consideration of the heat transfer condition (2.253) a choice of grid different to that shown in Fig. 2.45 is frequently made. The grid is laid out according to Fig. 2.44 with the derivative $\partial \vartheta / \partial x$ at $x_R = x_0 + \Delta x/2$ being replaced by the central difference quotient from (2.248). The boundary temperature $\vartheta\,(x_R, t_k)$ is found using the approximation

$$\vartheta\,(x_R, t_k) = \vartheta_{1/2}^k = \frac{1}{2}\left(\vartheta_0^k + \vartheta_1^k\right) \ . \qquad (2.260)$$

This then gives us the relationship from (2.253)

$$\vartheta_0^k = \frac{2 - Bi^*}{2 + Bi^*}\vartheta_1^k + \frac{2Bi^*}{2 + Bi^*}\vartheta_S^k \ , \qquad (2.261)$$

which, with (2.250), leads to the difference formula

$$\vartheta_1^{k+1} = \left(1 - M\frac{2 + 3Bi^*}{2 + Bi^*}\right)\vartheta_1^k + M\vartheta_2^k + \frac{2MBi^*}{2 + Bi^*}\vartheta_S^k \ . \tag{2.262}$$

The discretisation error of this equation is $O(\Delta x)$ and not $O(\Delta x^2)$ as for (2.256), because the approximation (2.260) was used. The use of (2.262) therefore leads us to expect larger errors than those with (2.256). On the other hand, the stability behaviour is better. Instead of (2.259) the condition

$$M \le \frac{2 + Bi^*}{2 + 3Bi^*} \tag{2.263}$$

is valid, which for large values of Bi^* delivers the limiting value $M = 1/3$, against which (2.259) leads to $M \to 0$.

If the larger discretisation error of (2.262) is to be avoided, the boundary temperature $\vartheta_{1/2}^k$ has to be replaced by a more accurate expression than the simple arithmetic mean according to (2.260). A parabolic curve through the three temperatures $\vartheta_0^k, \vartheta_1^k$ and ϑ_2^k gives

$$\vartheta_{1/2}^k = \frac{1}{8}\left(3\vartheta_0^k + 6\vartheta_1^k - \vartheta_2^k\right) \ .$$

From which, instead of (2.261) we get

$$\vartheta_0^k = \frac{8 - 6Bi^*}{8 + 3Bi^*}\vartheta_1^k + \frac{Bi^*}{8 + 3Bi^*}\vartheta_2^k + \frac{8Bi^*}{8 + 3Bi^*}\vartheta_S^k$$

and from (2.250) the difference equation

$$\vartheta_1^{k+1} = \left(1 - 4M\frac{2 + 3Bi^*}{8 + 3Bi^*}\right)\vartheta_1^k + M\frac{8 + 4Bi^*}{8 + 3Bi^*}\vartheta_2^k + \frac{8MBi^*}{8 + 3Bi^*}\vartheta_S^k \tag{2.264}$$

is obtained. Although this equation for the consideration of the boundary condition (2.250) is numerically more complex it is also more exact than (2.262), because its discretisation error is $O(\Delta x^2)$. The resulting stability condition is

$$M \le \frac{1}{4}\frac{8 + 3Bi^*}{2 + 3Bi^*} \ . \tag{2.265}$$

This only leads to a tightening of the stability condition (2.242) for $Bi^* \ge 4/3$ and for very large Bi^* yields the limiting value $M \le 1/4$.

With small values of Bi^* the simple relationship (2.256) with a grid laid out according to Fig. 2.45 a delivers very accurate results. In contrast for large Bi^* values, (2.264) should be used with a grid according to that in Fig. 2.44, where the stability condition (2.265) has to be observed.

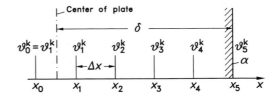

Fig. 2.46: Grid division for the calculation of the cooling of a plate of thickness 2δ

Example 2.6: A steel plate with the material properties $\lambda = 15.0\,\text{W/K}\,\text{m}$ and $a = 3.75 \cdot 10^{-6}\,\text{m}^2/\text{s}$ is $2\delta = 270\,\text{mm}$ thick and has a constant initial temperature ϑ_0. At time t_0 the plate is brought into contact with a fluid which has a temperature $\vartheta_S < \vartheta_0$ that is constant with respect to time. The heat transfer coefficient at both surfaces of the plate is $\alpha = 75\,\text{W/m}^2\text{K}$. The temperatures during the cooling of the plate are to be numerically determined. Simple initial and boundary conditions were intentionally chosen, so that the accuracy of the finite difference method could be checked when compared to the explicit solution of the case dealt with in section 2.3.3.

Due to symmetry, it is sufficient to consider only one half of the plate which is $\delta = 135\,\text{mm}$ thick. Its left hand surface can be taken to be adiabatic, whilst heat is transferred to the fluid at its right hand surface. We choose the grid from Fig. 2.46 with a mesh size of $\Delta x = 30\,\text{mm}$. The left boundary of the plate lies in the middle of the two grid lines x_0 and $x_1 = x_0 + \Delta x$; the right boundary coincides with x_5. The Biot number of the difference method is

$$Bi^* = \frac{\alpha \Delta x}{\lambda} = \frac{75\,\left(\text{W/m}^2\text{K}\right)\,0.030\,\text{m}}{15.0\,\text{W/K}\,\text{m}} = 0.15 \ .$$

For the modulus we choose

$$M = a\Delta t/\Delta x^2 = 1/3 \ ,$$

which satisfies the stability condition (2.259). The time step will be

$$\Delta t = \frac{\Delta x^2}{a} M = 80\,\text{s} \ .$$

We set $\vartheta_S = 0$ and $\vartheta_0 = 1.0000$. The temperatures ϑ_i^k calculated agree with the dimensionless temperatures $\vartheta^+(x_i, t_k)$ in the explicit solution from (2.171).

With these arrangements the five following difference equations are valid:

$$\vartheta_1^{k+1} = \frac{2}{3}\vartheta_1^k + \frac{1}{3}\vartheta_2^k$$

according to (2.251) with $\dot{q}(t_k) = 0$,

$$\vartheta_i^{k+1} = \frac{1}{3}\left(\vartheta_{i-1}^k + \vartheta_i^k + \vartheta_{i+1}^k\right)$$

for $i = 2, 3, 4$ according to (2.240) as well as

$$\vartheta_5^{k+1} = \frac{2}{3}\vartheta_4^k + 0.2333\,\vartheta_5^k$$

from (2.258). The results for the first 12 time steps are shown in Table 2.9.

In Table 2.10 the surface temperatures ϑ_5^k and the temperature profile at time $t_{12} = 960\,\text{s}$ are compared with the exact solution. Although the grid chosen was coarse the agreement with the explicit solution from (2.171) is satisfactory. However the first 12 time steps record only a small part of the cooling process. This is due to the restriction of the time step Δt in the stability condition. This can only be overcome by transferring to an implicit difference method.

Table 2.9: Temperatures in the cooling of a steel plate, calculated with the explicit difference method.

k	t_k/s	ϑ_1^k	ϑ_2^k	ϑ_3^k	ϑ_4^k	ϑ_5^k
0	0	1.0000	1.0000	1.0000	1.0000	1.0000
1	80	1.0000	1.0000	1.0000	1.0000	0.9000
2	160	1.0000	1.0000	1.0000	0.9667	0.8767
3	240	1.0000	1.0000	0.9889	0.9478	0.8490
4	320	1.0000	0.9963	0.9789	0.9286	0.8300
5	400	0.9988	0.9917	0.9679	0.9125	0.8127
6	480	0.9964	0.9861	0.9574	0.8977	0.7980
7	560	0.9930	0.9800	0.9471	0.8844	0.7847
8	640	0.9887	0.9734	0.9372	0.8721	0.7727
9	720	0.9836	0.9664	0.9276	0.8607	0.7617
10	800	0.9779	0.9592	0.9182	0.8500	0.7515
11	880	0.9717	0.9518	0.9091	0.8399	0.7420
12	960	0.9651	0.9442	0.9003	0.8303	0.7331

Table 2.10: Comparison of the surface temperature ϑ_5^k and the temperatures ϑ_i^{12} calculated using the difference method with the values of ϑ^+ from the exact solution according to (2.171).

k	ϑ_5^k	$\vartheta^+(x_5,t_k)$	k	ϑ_5^k	$\vartheta^+(x_5,t_k)$	k	ϑ_5^k	$\vartheta^+(x_5,t_k)$
1	0.9000	0.9093	5	0.8127	0.8142	9	0.7617	0.7631
2	0.8767	0.8755	6	0.7980	0.7994	10	0.7515	0.7529
3	0.8490	0.8509	7	0.7847	0.7861	11	0.7420	0.7433
4	0.8300	0.8311	8	0.7727	0.7741	12	0.7331	0.7343

i	1	2	3	4	5
ϑ_i^{12}	0.9651	0.9442	0.9003	0.8303	0.7331
$\vartheta^+(x_i,t_{12})$	0.9629	0.9427	0.8998	0.8309	0.7343

2.4.3 The implicit difference method from J. Crank and P. Nicolson

The explicit difference method discussed in 2.4.1 has the disadvantage that the time step Δt is limited by the stability conditions (2.242) and (2.259). Therefore obtaining a temperature profile at a given time usually requires a lot of time steps. This step size restriction can be avoided by choosing an implicit difference method. It requires a system of linear equations to be solved for each time step. This system has a very simple form, it is a tridiagonal sytem whose coefficient matrix is only occupied along the main diagonal and both its neighbours. Simple solution algorithms for tridiagonal systems can be found in D. Marsal [2.53] and also in the standard works on numerical mathematics, e.g. [2.66] and [2.67].

A particularly accurate implicit difference method, which is always stable, has been presented by J. Crank and P. Nicolson [2.65]. In this method the temperatures at the time levels t_k and t_{k+1} are used. However the differential equation (2.236) is discretised for the time lying between these two levels: $t_k + \Delta t/2$. This makes it possible to approximate the derivative $(\partial \vartheta/\partial t)_i^{k+1/2}$ by means of the accurate *central* difference quotient

$$\left(\frac{\partial \vartheta}{\partial t}\right)_i^{k+\frac{1}{2}} = \frac{\vartheta_i^{k+1} - \vartheta_i^k}{\Delta t} + O(\Delta t^2) \ . \tag{2.266}$$

This is advantageous because the choice of an implicit difference method allows larger time steps to be used and therefore requires a more exact approximation of the derivative with respect of time.

The second derivative $(\partial^2 \vartheta/\partial x^2)_i^{k+1/2}$ at time $t_k + \Delta t/2$ is replaced by the arithmetic mean of the second central difference quotients at times t_k and t_{k+1}. This produces

$$\left(\frac{\partial^2 \vartheta}{\partial x^2}\right)_i^{k+\frac{1}{2}} = \frac{1}{2}\left(\frac{\vartheta_{i-1}^{k+1} - 2\vartheta_i^{k+1} + \vartheta_{i+1}^{k+1}}{\Delta x^2} + \frac{\vartheta_{i-1}^k - 2\vartheta_i^k + \vartheta_{i+1}^k}{\Delta x^2}\right) + O(\Delta x^2) \ . \tag{2.267}$$

With (2.266) and (2.267) the implicit difference equation is obtained

$$- M\vartheta_{i-1}^{k+1} + (2 + 2M)\vartheta_i^{k+1} - M\vartheta_{i+1}^{k+1} = M\vartheta_{i-1}^k + (2 - 2M)\vartheta_i^k + M\vartheta_{i+1}^k \tag{2.268}$$

with $M = a\Delta t/\Delta x^2$.

The temperatures at time t_k on the right hand side of (2.268) are known; the three unknown temperatures at time t_{k+1} on the left hand side have to be calculated. The difference equation (2.268) yields a system of linear equations with $i = 1, 2, \ldots n$. The main diagonal of the coefficient matrix contains the elements $(2 + 2M)$; the sub- and superdiagonals are made up of the elements $(-M)$; all other coefficients are zero. In this tridiagonal system, the first equation $(i = 1)$ cannot contain the term $-M\vartheta_0^{k+1}$ and likewise, in the last equation $(i = n)$ the term $-M\vartheta_{n+1}^{k+1}$ may not appear. These terms are eliminated by taking the *boundary conditions* into consideration.

If the *temperatures at the boundaries* are given, the grid is chosen such that $x = x_0$ and $x = x_{n+1}$ coincide with the two boundaries. This means that $\vartheta_0^k, \vartheta_0^{k+1}$ and $\vartheta_{n+1}^k, \vartheta_{n+1}^{k+1}$ are always known, and the first equation of the tridiagonal system looks like

$$(2 + 2M)\vartheta_1^{k+1} - M\vartheta_2^{k+1} = M\left(\vartheta_0^k + \vartheta_0^{k+1}\right) + (2 - 2M)\vartheta_1^k + M\vartheta_2^k \ , \tag{2.269}$$

whilst the last equation will be

$$- M\vartheta_{n-1}^{k+1} + (2 + 2M)\vartheta_n^{k+1} = M\vartheta_{n-1}^k + (2 - 2M)\vartheta_n^k + M\left(\vartheta_{n+1}^k + \vartheta_{n+1}^{k+1}\right) \ . \tag{2.270}$$

If the *heat flux* $\dot{q}(t)$ at the left boundary is given, cf. (2.247), the grid is chosen to be the same as in Fig. 2.44, so that the boundary lies in the middle between x_0 and x_1. ϑ_0^k and ϑ_0^{k+1} will be eliminated using (2.249). The first equation now looks like this

$$(2+M)\vartheta_1^{k+1} - M\vartheta_2^{k+1} = (2-M)\vartheta_1^k + M\vartheta_2^k + M\frac{\Delta x}{\lambda}\left[\dot{q}(t_k) + \dot{q}(t_{k+1})\right] \quad . \quad (2.271)$$

The adiabatic wall is a special case with $\dot{q} \equiv 0$. When the heat flux at the right edge is given, ϑ_{n+1}^k and ϑ_{n+1}^{k+1} are eliminated in the same manner as above.

When considering the *heat transfer condition* (2.253) we lay the grid out as in Fig. 2.45, namely so that the boundary coincides with x_1 or x_n, when (2.253) is stipulated at the left or right boundary. Once again ϑ_0^k and ϑ_0^{k+1} are eliminated from the first equation of the tridiagonal system (2.268), this time using (2.254). This yields

$$[1 + M(1 + Bi^*)]\vartheta_1^{k+1} - M\vartheta_2^{k+1} = [1 - M(1 + Bi^*)]\vartheta_1^k \\ + M\vartheta_2^k + MBi^*\left(\vartheta_S^k + \vartheta_S^{k+1}\right) \quad (2.272)$$

for the first equation. We eliminate ϑ_{n+1}^k and ϑ_{n+1}^{k+1} using (2.257) giving us as the last equation

$$-M\vartheta_{n-1}^{k+1} + [1 + M(1 + Bi^*)]\vartheta_n^{k+1} = M\vartheta_{n-1}^k + [1 - M(1 + Bi^*)]\vartheta_n^k \\ + MBi^*\left(\vartheta_S^k + \vartheta_S^{k+1}\right) \quad . \quad (2.273)$$

The difference method of Crank-Nicolson is stable for all M. The size of the time steps is limited by the accuracy requirements. Very large values of M lead to finite oscillations in the numerical solution which only slowly decay with increasing k, cf. [2.57].

Example 2.7: The cooling problem discussed in Example 2.6 will now be solved using the Crank-Nicolson implicit difference method. The grid divisions will be kept as that in Fig. 2.46.

The tridiagonal system for each time step, which has to be solved consists of five equations and has the form

$$\mathbf{A}\,\vartheta^{k+1} = \mathbf{b}$$

with the coefficient matrix

$$\mathbf{A} = \begin{bmatrix} 2+M & -M & 0 & 0 & 0 \\ -M & 2+2M & -M & 0 & 0 \\ 0 & -M & 2+2M & -M & 0 \\ 0 & 0 & -M & 2+2M & -M \\ 0 & 0 & 0 & -M & 1+1{,}15M \end{bmatrix} ,$$

the solution vector

$$\vartheta^{k+1} = \left[\vartheta_1^{k+1}, \vartheta_2^{k+1}, \vartheta_3^{k+1}, \vartheta_4^{k+1}, \vartheta_5^{k+1}\right]^{\mathrm{T}}$$

and the right side

$$\mathbf{b} = \begin{bmatrix} (2-M)\vartheta_1^k + M\vartheta_2^k \\ M\vartheta_1^k + (2-2M)\vartheta_2^k + M\vartheta_3^k \\ M\vartheta_2^k + (2-2M)\vartheta_3^k + M\vartheta_4^k \\ M\vartheta_3^k + (2-2M)\vartheta_4^k + M\vartheta_5^k \\ M\vartheta_4^k + (1-1{,}15M)\vartheta_5^k \end{bmatrix} .$$

Table 2.11: Temperatures in the cooling of a steel plate, calculated using the Crank-Nicolson method with $M = a\Delta t/\Delta x^2 = 1$.

t_k/s	ϑ_1^k	ϑ_2^k	ϑ_3^k	ϑ_4^k	ϑ_5^k
240	0.9990	0.9969	0.9885	0.9573	0.8406
480	0.9937	0.9851	0.9594	0.8982	0.8044
720	0.9815	0.9657	0.9282	0.8639	0.7635
960	0.9636	0.9437	0.9013	0.8322	0.7356
1200	0.9427	0.9208	0.8756	0.8058	0.7105
1440	0.9203	0.8975	0.8514	0.7816	0.6887
1680	0.8974	0.8743	0.8282	0.7592	0.6686
1920	0.8744	0.8514	0.8058	0.7381	0.6498
2160	0.8516	0.8289	0.7841	0.7179	0.6318
2400	0.8292	0.8070	0.7630	0.6984	0.6146

Table 2.12: Comparison of the temperatures for $t = 9600\,\mathrm{s}$, calculated with different modulus values M.

M	ϑ_1	ϑ_2	ϑ_3	ϑ_4	ϑ_5
1	0.3684	0.3584	0.3388	0.3099	0.2727
2	0.3683	0.3584	0.3387	0.3099	0.2727
5	0.3677	0.3584	0.3386	0.3051	0.2801
10	0.3683	0.3532	0.3259	0.2995	0.3180
Analyt. Solution	0,3684	0.3584	0.3389	0.3101	0.2730

In the first step of the process ($k = 0$) all temperatures are set to $\vartheta_i^0 = 1.000$, corresponding to the initial condition $\vartheta(x_i, t_0) = 1$.

We chose $M = 1$, a time step three times larger than that in the explicit method from Example 2.6, namely $\Delta t = 240\,\mathrm{s} = 4.0\,\mathrm{min}$. The temperatures for the first 10 time steps are given in Table 2.11. A comparison of the temperatures at time $t_4 = 960\,\mathrm{s}$ with the values shown in Table 2.10, indicates that the Crank-Nicolson method delivers better results than the explicit method, despite the time step being three times larger.

In order to investigate the influence of the size of the time step on the accuracy we also used alongside $M = 1$, $M = 2$, 5 and 10. The temperature distributions for $t = 9600\,\mathrm{s} = 160\,\mathrm{min}$ are shown in Table 2.12. This time is reached after 40 steps with $M = 1$, 20 steps for $M = 2$, 8 steps for $M = 5$ and finally 4 steps for $M = 10$. The temperatures for $M = 1$ and $M = 2$ agree very well with each other and with the analytical solution. The values for $M = 5$ yield somewhat larger deviations, while the result for $M = 10$ is useless. This large step produces temperature oscillations which are physically impossible. In [2.57], pg. 122, a condition for the restriction of the step size, so that oscillations can be avoided, is given for a transient heat conduction problem with boundary conditions different from our example. The transfer of this condition to the present task delivers the limit

$$M < \frac{2}{\pi} \frac{\delta}{\Delta x} = 5.7 \ .$$

It confirms the result of our numerical test calculations.

2.4.4   Noncartesian coordinates. Temperature dependent material properties

In the following we will discuss the difference method with consideration for temperature dependent material properties as well as for cylindrical and spherical coordinates, whereby geometric one-dimensional heat flow is assumed in the radial direction. The decisive differential equation for the temperature field is then

$$c\varrho \frac{\partial \vartheta}{\partial t} = \frac{1}{r^m} \frac{\partial}{\partial r} \left(r^m \lambda \frac{\partial \vartheta}{\partial r} \right) \tag{2.274}$$

with the exponents

$$m = \begin{cases} 0 & \text{for the plate } (r = x), \\ 1 & \text{for the cylinder,} \\ 2 & \text{for the sphere.} \end{cases}$$

The establishment of difference equations is based on the discretisation of the self-adjoint differential operator

$$D := \frac{\partial}{\partial r} \left(f \frac{\partial \vartheta}{\partial r} \right) \tag{2.275}$$

with

$$f := r^m \lambda(\vartheta) \quad , \quad m = 0, 1, 2 \ . \tag{2.276}$$

We will now discuss this and derive the difference equations for different coordinates systems with and without consideration of the temperature dependence of the properties of the material.

2.4.4.1   The discretisation of the self-adjoint differential operator

It is often advantageous to use a grid which does not have regular divisions in the r-direction for the discretisation of the operator D from (2.275). The use of smaller separations between the grid lines allows the stronger curvature of the temperature profile for small r-values to be more acccurately considered. Therefore we will implement the use of a so-called centred grid with any (dependent on i) mesh size Δr_i for the discretisation of D, as shown in Fig. 2.47. Using central difference quotients we then obtain

$$D_i^k = \frac{1}{\Delta r_i} \left[\left(f \frac{\partial \vartheta}{\partial r} \right)_{i+\frac{1}{2}}^k - \left(f \frac{\partial \vartheta}{\partial r} \right)_{i-\frac{1}{2}}^k \right] \ .$$

Replacing the first derivatives with central difference quotients gives

$$D_i^k = \frac{1}{\Delta r_i} \left[\frac{2 f_{i+1/2}^k}{\Delta r_i + \Delta r_{i+1}} \left(\vartheta_{i+1}^k - \vartheta_i^k \right) - \frac{2 f_{i-1/2}^k}{\Delta r_i + \Delta r_{i-1}} \left(\vartheta_i^k - \vartheta_{i-1}^k \right) \right] \ . \tag{2.277}$$

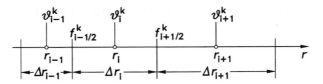

Fig. 2.47: Centred grid for the discretisation of the self-adjoint differential operator D according to (2.275)

According to (2.276)

$$f^k_{i\pm\frac{1}{2}} = \left(r_i \pm \frac{\Delta r_i}{2}\right)^m \lambda\left(\vartheta_{i\pm\frac{1}{2}}\right) \quad , \quad m = 0, 1, 2 , \qquad (2.278)$$

is valid. The thermal conductivities which appear here at the temperatures $\vartheta^k_{i+\frac{1}{2}}$ or $\vartheta^k_{i-\frac{1}{2}}$ have to be calculated by a suitable mean value formulation of the thermal conductivities at the known temperatures ϑ^k_i, ϑ^k_{i+1}, ϑ^k_{i-1} at the grid points. We will come back to this in 2.4.4.3. Next we wish to concentrate on the simple case of constant properties.

2.4.4.2 Constant material properties. Cylindrical coordinates

Equation (2.278) simplifies to

$$f_{i\pm\frac{1}{2}} = \lambda\left(r_i \pm \frac{\Delta r_i}{2}\right)^m \qquad (2.279)$$

for constant thermal conductivity. Using this assumption the difference equation for the cylinder ($m = 1$) will now be derived. The case for the sphere $m = 2$ is left to the reader to solve. The simple difference equations for a plate ($m = 0$) are available in 2.4.1 and 2.4.3, when a constant mesh size $\Delta r_i = \Delta x$ is assumed.

For a cylinder with $\lambda = $ const we obtain, from (2.277) and (2.279), the expression

$$D^k_i = \lambda \frac{r_i}{\Delta r_i^2}\left[g^+_i\left(\vartheta^k_{i+1} - \vartheta^k_i\right) - g^-_i\left(\vartheta^k_i - \vartheta^k_{i-1}\right)\right] , \qquad (2.280)$$

where we have introduced the abbreviations

$$g^+_i := \frac{2 + \Delta r_i/r_i}{1 + \Delta r_{i+1}/\Delta r_i} \quad \text{and} \quad g^-_i := \frac{2 - \Delta r_i/r_i}{1 + \Delta r_{i-1}/\Delta r_i} . \qquad (2.281)$$

A further simplification is possible if the *grid spacing* Δr is assumed to be *constant*. With

$$g^+_i = 1 + \frac{\Delta r}{2r_i} \quad \text{and} \quad g^-_i = 1 - \frac{\Delta r}{2r_i} \qquad (2.282)$$

we then obtain from (2.280)

$$D^k_i = \lambda \frac{r_i}{\Delta r^2}\left[\left(1 + \frac{\Delta r}{2r_i}\right)\vartheta^k_{i+1} - 2\vartheta^k_i + \left(1 - \frac{\Delta r}{2r_i}\right)\vartheta^k_{i-1}\right] . \qquad (2.283)$$

If an *explicit* difference equation is desired, then the time based derivative in

$$\left(\frac{\partial\vartheta}{\partial t}\right)_i^k = \frac{a}{r_i}\frac{\partial}{\partial r}\left(r\frac{\partial\vartheta}{\partial r}\right)_i^k = \frac{1}{c\varrho r_i}D_i^k$$

has to be replaced according to (2.239). This produces the explicit difference equation

$$\vartheta_i^{k+1} = M\left(1 - \frac{\Delta r}{2r_i}\right)\vartheta_{i-1}^k + (1-2M)\vartheta_i^k + M\left(1 + \frac{\Delta r}{2r_i}\right)\vartheta_{i+1}^k , \qquad (2.284)$$

whose modulus M has to satisfy the stability criterion mentioned in 2.4.1.2

$$M = a\Delta t/\Delta r^2 \le \frac{1}{2} . \qquad (2.285)$$

The boundary conditions are treated in the same manner as in section 2.4.2.

To transfer the Crank-Nicolson [2.65] *implicit difference method*, which is always stable, over to cylindrical coordinates requires the discretisation of the equation

$$\left(\frac{\partial\vartheta}{\partial t}\right)_i^{k+\frac{1}{2}} = \frac{a}{r_i}\frac{\partial}{\partial r}\left(r\frac{\partial\vartheta}{\partial r}\right)_i^{k+\frac{1}{2}} \approx \frac{1}{c\varrho r_i}\frac{1}{2}\left(D_i^{k+1} + D_i^k\right) .$$

With the time derivative from (2.266) and D_i^k or D_i^{k+1} according to (2.283) we obtain the following difference equation

$$-M\left(1 - \frac{\Delta r}{2r_i}\right)\vartheta_{i-1}^{k+1} + (2+2M)\vartheta_i^{k+1} - M\left(1 + \frac{\Delta r}{2r_i}\right)\vartheta_{i+1}^{k+1} = C_i^k , \qquad (2.286)$$

whose right hand side

$$C_i^k = M\left(1 - \frac{\Delta r}{2r_i}\right)\vartheta_{i-1}^k + (2-2M)\vartheta_i^k + M\left(1 + \frac{\Delta r}{2r_i}\right)\vartheta_{i+1}^k \qquad (2.287)$$

contains the known temperatures at time t_k. This equation leads to a tridiagonal system of linear equations which have to be solved for each time step, cf. 2.4.3. The temperatures ϑ_0^k and ϑ_0^{k+1} which appear in the first equation and those from the last equation ϑ_{n+1}^k and ϑ_{n+1}^{k+1}, can be eliminated with the help of the boundary conditions, which was extensively discussed in 2.4.3.

A non-equidistant grid, cf. Fig. 2.47, requires the use of the discretised differential operator D_i^k from (2.280) with the functions g_i^+ and g_i^- according to (2.281). The *explicit differential equation* analogous to (2.284) has the form

$$\vartheta_i^{k+1} = M_i g_i^- \vartheta_{i-1}^k + \left[1 - M_i\left(g_i^+ + g_i^-\right)\right]\vartheta_i^k + M_i g_i^+ \vartheta_{i+1}^k \qquad (2.288)$$

with the i-dependent modulus

$$M_i = a\Delta t/\Delta r_i^2 . \qquad (2.289)$$

The difference method is only stable if none of the coefficients in (2.288) is negative. This leads to the stability condition $M_i \le (g_i^+ + g_i^-)^{-1}$, and therefore

$$\Delta t \le \min\left\{\frac{1}{a}\frac{\Delta r_i^2}{g_i^+ + g_i^-}\right\} . \qquad (2.290)$$

The stable, *implicit method from Crank and Nicolson* can be used without this restriction. A generalisation of (2.286) delivers the tridiagonal linear equation system

$$- M_i g_i^- \vartheta_{i-1}^{k+1} + \left[2 + M_i \left(g_i^+ + g_i^-\right)\right] \vartheta_i^{k+1} - M_i g_i^+ \vartheta_i^{k+1} = C_i^k \qquad (2.291)$$

with the right hand side

$$C_i^k = M_i g_i^- \vartheta_{i-1}^k + \left[2 - M_i \left(g_i^+ + g_i^-\right)\right] \vartheta_i^k + M_i g_i^+ \vartheta_{i+1}^k \ , \qquad (2.292)$$

which has to be solved for each time step. The modulus M_i is given by (2.289).

2.4.4.3 Temperature dependent material properties

If λ and $c\varrho$ change with temperature, cf. section 2.1.4, a closed solution to the heat conduction equation cannot generally be found, which only leaves the possibiltiy of using a numerical solution method. We will show how temperature dependent properties are accounted for by using the example of the plate, $m = 0$ in (2.274). The transfer of the solution to a cylinder or sphere ($m = 1$ or 2 respectively) is left to the reader. It is fairly simple to carry out with the help of the general discretisation equation (2.277) for the differential operator D from (2.275).

Using an equidistant grid with spacing Δx, with $\Delta r_i = \Delta r_{i+1} = \Delta r_{i-1} = \Delta x$ and $m = 0$ we obtain from (2.277) and (2.278)

$$D_i^k = \frac{1}{\Delta x^2} \left[\lambda_{i+\frac{1}{2}}^k \left(\vartheta_{i+1}^k - \vartheta_i^k \right) - \lambda_{i-\frac{1}{2}}^k \left(\vartheta_i^k - \vartheta_{i-1}^k \right) \right] \ . \qquad (2.293)$$

In which $\lambda_{i\pm1/2}^k$ is the thermal conductivity at temperature $\vartheta_{i\pm1/2}^k$. This requires a suitable mean value to be chosen, the arithmetic, geometric or harmonic mean of the thermal conductivities at the known temperatures ϑ_i^k and ϑ_{i+1}^k or ϑ_i^k nd ϑ_{i-1}^k. The type of mean value formation does not play a decisive role if λ is only weakly dependent on ϑ or if the step size Δx is chosen to be very small. D. Marsal [2.53] recommends the use of the harmonic mean, so

$$\lambda_{i\pm\frac{1}{2}}^k = \frac{2\lambda_i^k \lambda_{i\pm1}^k}{\lambda_i^k + \lambda_{i\pm1}^k} \ , \qquad (2.294)$$

where $\lambda_i^k = \lambda(\vartheta_i^k) = \lambda[\vartheta(x_i, t_k)]$. Using the harmonic mean we obtain, from (2.294),

$$D_i^k = \frac{2\lambda_i^k}{\Delta x^2} \left(\frac{\vartheta_{i+1}^k - \vartheta_i^k}{1 + \lambda_i^k/\lambda_{i+1}^k} - \frac{\vartheta_i^k - \vartheta_{i-1}^k}{1 + \lambda_i^k/\lambda_{i-1}^k} \right) \ . \qquad (2.295)$$

In order to avoid complicated iterations it is recommended that an explicit difference method is applied to this case. We replace the time derivative with the first difference quotient according to (2.239) and obtain from (2.274) with (2.295) the difference equation

$$\vartheta_i^{k+1} = \vartheta_i^k + \frac{2a_i^k \Delta t}{\Delta x^2} \left(\frac{\vartheta_{i+1}^k - \vartheta_i^k}{1 + \lambda_i^k/\lambda_{i+1}^k} - \frac{\vartheta_i^k - \vartheta_{i-1}^k}{1 + \lambda_i^k/\lambda_{i-1}^k} \right) \ . \qquad (2.296)$$

Here $a_i^k = a(\vartheta_i^k)$ is the thermal diffusivity at the temperature ϑ_i^k.

The stability of the difference method is guaranteed by choosing a small enough time step Δt such that the coefficient of ϑ_i^k in (2.296) is positive. Therefore it always has to be

$$1 - \frac{2a_i^k \Delta t}{\Delta x^2}\left(\frac{1}{1 + \lambda_i^k/\lambda_{i+1}^k} + \frac{1}{1 + \lambda_i^k/\lambda_{i-1}^k}\right) > 0 \ . \qquad (2.297)$$

As a and λ change with each time step, a control for this inequality should be built into the computer program and should the situation arise Δt should be reduced stepwise.

The temperature dependence of λ must also be considered in the heat transfer condition (2.253) and for the boundary condition of prescribed heat flux. The Biot number Bi^* will be temperature dependent and this must be noted in the elimination of the temperatures ϑ_0^k and ϑ_{n+1}^k from (2.254) and (2.257) respectively and in (2.297) with $i = 1$ or $i = n$.

2.4.5 Transient two- and three-dimensional temperature fields

If planar or spatial temperature fields are to be determined the computation time and storage capacity increase significantly in comparison to geometrically one-dimensional problems. In the following we will restrict ourselves to rectangular regions (cartesian coordinates); cylindrical and spherical problems can be solved by discretisation of the corresponding differential equations. The finite difference method is not well suited to complicated geometries, especially when boundary conditions of the second or third type are stipulated at boundaries of any particular shape. In these cases it is more convenient to use a finite element method.

We will assume that the material properties are constant. The heat conduction equation for planar, transient temperature fields with heat sources has the form

$$\frac{\partial \vartheta}{\partial t} = a\left(\frac{\partial^2 \vartheta}{\partial x^2} + \frac{\partial^2 \vartheta}{\partial y^2}\right) + \frac{\dot{W}(x,y,t,\vartheta)}{c\varrho} \ . \qquad (2.298)$$

The temperature $\vartheta = \vartheta(x,y,t)$ is to be determined in a rectangle parallel to the x- and y-axes. We discretise the coordinates by

$$x_i = x_0 + i\Delta x \quad , \quad y_j = y_0 + j\Delta y \quad \text{and} \quad t_k = t_0 + k\Delta t \ .$$

The temperature at an intersection point of the planar grid in Fig. 2.48 will be indicated by

$$\vartheta_{i,j}^k = \vartheta\left(x_i, y_j, t_k\right) \ .$$

The corresponding equation for the thermal power density is

$$\dot{W}_{i,j}^k = \dot{W}\left(x_i, y_j, t_k\right) \quad \text{or} \quad \dot{W}_{i,j}^k = \dot{W}\left(\vartheta_{i,j}^k\right) \ ,$$

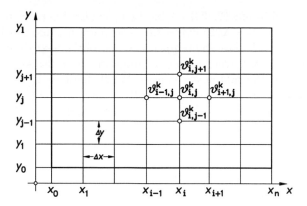

Fig. 2.48: Planar grid for the discretisation of the heat conduction equation (2.298) in the rectangular region $x_0 \le x \le x_n$, $y_0 \le y \le y_l$

in the case where $\dot{W}$ is only dependent on the temperature

The two second derivatives in the x- or y-directions are approximated by the central difference quotient, so that

$$\left(\frac{\partial^2 \vartheta}{\partial x^2}\right)^k_{i,j} = \frac{1}{\Delta x^2}\left(\vartheta^k_{i-1,j} - 2\vartheta^k_{i,j} + \vartheta_{i+1,j}\right) + O(\Delta x^2) \qquad (2.299)$$

and

$$\left(\frac{\partial^2 \vartheta}{\partial y^2}\right)^k_{i,j} = \frac{1}{\Delta y^2}\left(\vartheta^k_{i,j-1} - 2\vartheta_{i,j} + \vartheta_{i,j+1}\right) + O(\Delta y^2) \qquad (2.300)$$

are valid. In order to obtain an explicit difference equation, we use the forward difference quotient for the time derivative

$$\left(\frac{\partial \vartheta}{\partial t}\right)^k_{i,j} = \frac{1}{\Delta t}\left(\vartheta^{k+1}_{i,j} - \vartheta^k_{i,j}\right) + O(\Delta t) \qquad (2.301)$$

Using these difference quotients, we obtain, from (2.298), the explicit finite difference formula

$$\begin{aligned}
\vartheta^{k+1}_{i,j} &= \left(1 - 2M_x - 2M_y\right)\vartheta^k_{i,j} + M_x\left(\vartheta^k_{i-1,j} + \vartheta^k_{i+1,j}\right) \\
&\quad + M_y\left(\vartheta^k_{i,j-1} + \vartheta^k_{i,j+1}\right) + \frac{\Delta t}{c\varrho}\dot{W}^k_{i,j} \ .
\end{aligned} \qquad (2.302)$$

Two moduli appear here

$$M_x := \frac{a\Delta t}{\Delta x^2} \quad \text{and} \quad M_y := \frac{a\Delta t}{\Delta y^2} \ .$$

For a square grid ($\Delta x = \Delta y$) we obtain $M_x = M_y = M = a\Delta t/\Delta x^2$, and (2.302) is simplified to

$$\vartheta^{k+1}_{i,j} = (1 - 4M)\vartheta^k_{i,j} + M\left(\vartheta^k_{i-1,j} + \vartheta^k_{i+1,j} + \vartheta^k_{i,j-1} + \vartheta^k_{i,j+1}\right) + \frac{\Delta t}{c\varrho}\dot{W}^k_{i,j} \ . \qquad (2.303)$$

Equations (2.302) and (2.303) enable us to explicitly calculate the temperatures $\vartheta_{i,j}^1$ at time $t_1 = t_0 + \Delta t$ from the initial temperature distribution $\vartheta_{i,j}^0$. These can then be used to calculate the temperature $\vartheta_{i,j}^2$ of the next time level etc. This difference method is relatively simple to program. It is also suitable for a temperature dependent thermal power density $\dot{W}(\vartheta)$, because $\dot{W}_{i,j}^k$ is calculated for the already known temperature $\vartheta_{i,j}^k$. The disadvantage of this explicit difference method lies in its limited stability, cf. 2.4.1.2. The *stability condition*, from which none of the coefficients on the right hand side of (2.302) are allowed to be negative, limits the time step

$$\Delta t \leq \frac{\Delta x^2}{2a \left[1 + (\Delta x/\Delta y)^2\right]} \ . \tag{2.304}$$

This leads to an even smaller time step than in the condition (2.242) for geometrical one-dimensional heat conduction.

The restriction on the step size (2.304) due to the stability condition for the explicit difference method can be avoided by using an implicit method. This means that (2.298) is discretised at time t_{k+1} and the backward difference quotient is used to replace the time derivative. With

$$\left(\frac{\partial \vartheta}{\partial t}\right)_{i,j}^{k+1} = \frac{1}{\Delta t} \left(\vartheta_{i,j}^{k+1} - \vartheta_{i,j}^k\right) + O(\Delta t)$$

and the difference quotients (2.299) and (2.300) formulated for t_{k+1} we obtain from (2.298) , with the simplification that $\dot{W} \equiv 0$, the implicit difference equation

$$\frac{a}{\Delta x^2} \left(\vartheta_{i-1,j}^{k+1} - 2\vartheta_{i,j}^{k+1} + \vartheta_{i+1,j}^{k+1}\right) + \frac{a}{\Delta y^2} \left(\vartheta_{i,j-1}^{k+1} - 2\vartheta_{i,j}^{k+1} + \vartheta_{i,j+1}^{k+1}\right) = \frac{1}{\Delta t} \left(\vartheta_{i,j}^{k+1} - \vartheta_{i,j}^k\right) \ . \tag{2.305}$$

This equation is to be formulated for all grid points (i, j). A system of linear equations for the unknown temperatures at time t_{k+1}, that has to be solved for every time step, is obtained. Each equation contains five unknowns, only the temperature $\vartheta_{i,j}^k$ at the previous time t_k is known. A good solution method has been presented by P.W. Peaceman and H.H. Rachford [2.69]. It is known as the 'alternating-direction implicit procedure' (ADIP). Here, instead of the equation system (2.305) two tridiagonal systems are solved, through which the computation time is reduced, see also [2.53].

The boundary conditions for two dimensional temperature fields are easily fulfilled as long as the boundaries run parallel to the coordinate axes. In addition to the grid points inside the region a series of points outside the region are used. These extra grid points enable the three types of boundary condition from section 2.4.2 to be considered without any difficulties. Significant complications can appear for boundaries which do not run parallel with the coordinate axes. The simplest condition to satisfy in this case is the prescribed temperature. Here it is sufficient to approximate the curved wall with straight lines parallel to x and y, with suitably small grid spacing Δx and Δy. However the discretisation of the derivative $\partial \vartheta/\partial n$ in the normal direction to the edges leads to complicated expressions which are difficult to use.

The discretisation of the heat conduction equation can also be undertaken for three-dimensional temperature fields, and this is left to the reader to attempt. The stability condition (2.304) is tightened for the explicit difference formula which means time steps even smaller than those for planar problems. The system of equations of the implicit difference method cannot be solved by applying the ADIP-method, because it is unstable in three dimensions. Instead a similar method introduced by J. Douglas and H.H. Rachford [2.71], [2.72], is used, that is stable and still leads to tridiagonal systems. Unfortunately the discretisation error using this method is greater than that from ADIP, see also [2.53].

2.4.6 Steady-state temperature fields

The application of the difference method to steady-state heat conduction problems involves the discretisation of the two- or three-dimensional region, in which the temperature field has to be determined, by the choice of a suitable grid. The temperatures at the grid points are determined by difference equations which link each temperature to the temperatures at the neighbouring grid points. The difference equations form a system of linear equations which replaces the heat conduction equation with its associated boundary conditions, and whose solution delivers approximate values for the temperatures at each grid point. Reducing the grid spacing (mesh size) increases the number of grid points and linear difference equations, from which the temperatures calculated are better approximations to the true values. We will limit ourselves to planar temperature fields; three-dimensional fields are calculated in a similar manner. Some well suited methods and formulated algorithms for these temperature fields can be found, in particular, in the work by D. Marsal [2.53]. This also includes methods which go beyond the simple difference method which we will discuss in the following.

2.4.6.1 A simple finite difference method for plane, steady-state temperature fields

Plane, steady-state temperature fields $\vartheta = \vartheta(x, y)$ with heat sources of thermal power density $\dot{W}$ are described by the differential equation

$$\frac{\partial^2 \vartheta}{\partial x^2} + \frac{\partial^2 \vartheta}{\partial y^2} + \frac{\dot{W}(x, y, \vartheta)}{\lambda} = 0 \ . \tag{2.306}$$

A *square* grid with mesh size $\Delta x = \Delta y$ is chosen for the discretisation, such that

$$x_i = x_0 + i\Delta x \ , \qquad i = 0, 1, 2, \ldots$$

and

$$y_j = y_0 + j\Delta x \ , \qquad j = 0, 1, 2, \ldots$$

are valid. The temperature at grid point (x_i, y_j) is indicated by $\vartheta_{i,j} = \vartheta(x_i, y_j)$, and equally the thermal power density by $\dot{W}_{i,j} = \dot{W}(x_i, y_j, \vartheta_{i,j})$.

In contrast to the previous sections, we will now derive the difference equation associated with (2.306) through an energy balance in an illustrative manner. To do this we interpret each grid point as the centre of a small block, which is cut out of the thermally conductive material, Fig. 2.49. The block has a square base of side length Δx and a height b perpendicular to the x,y-plane. The temperature $\vartheta_{i,j}$ at the grid point (i, j) is taken to be the characteristic mean temperature of the entire block. Heat is conducted to the block from its four immediate neighbours with the mean temperatures $\vartheta_{i+1,j}$, $\vartheta_{i,j+1}$, $\vartheta_{i-1,j}$ and $\vartheta_{i,j-1}$. Therefore, the energy balance for this block contains the four heat flows from Fig. 2.49 and the thermal power arising from the internal heat sources $\dot{W}_{i,j}\Delta V$, in which $\Delta V = \Delta x^2 b$ is the volume of the block :

$$\dot{Q}_{i+1} + \dot{Q}_{j+1} + \dot{Q}_{i-1} + \dot{Q}_{j-1} + \dot{W}_{i,j}\Delta V = 0 \ . \tag{2.307}$$

The heat flow $\dot{Q}_{i+1}$, from block $(i+1, j)$ to block (i, j) is given by

$$\dot{Q}_{i+1} = \dot{q}_{i+1}\Delta x\, b = \frac{\lambda}{\Delta x}\left(\vartheta_{i+1,j} - \vartheta_{i,j}\right)\Delta x\, b = \lambda\left(\vartheta_{i+1,j} - \vartheta_{i,j}\right)b \ .$$

Correspondingly we get

$$\dot{Q}_{j+1} = \lambda\left(\vartheta_{i,j+1} - \vartheta_{i,j}\right)b \ ,$$
$$\dot{Q}_{i-1} = \lambda\left(\vartheta_{i-1,j} - \vartheta_{i,j}\right)b$$

and

$$\dot{Q}_{j-1} = \lambda\left(\vartheta_{i,j-1} - \vartheta_{i,j}\right)b \ .$$

In the sense of the sign agreement in thermodynamics, the four heat flows will be taken as positive quantities if they flow into the block (i, j). With that we obtain from the energy balance (2.307) the desired difference equation

$$\vartheta_{i+1,j} + \vartheta_{i,j+1} + \vartheta_{i-1,j} + \vartheta_{i,j-1} - 4\,\vartheta_{i,j} = -\dot{W}_{i,j}\Delta x^2/\lambda \ . \tag{2.308}$$

This is an approximate equation as each small, but finite block has only one discrete temperature associated with it and because only the heat conduction between immediate neighbours has been considered. As we can show with the use of the discretisation equations (2.299) and (2.300) for the second derivatives, we also obtain (2.308) by the usual discretisation of the differential equation (2.306). The dicretisation error in this case is $O(\Delta x^2)$; it decreases to zero with the square of the mesh size (= block width).

The difference equation (2.308) has to be formulated for each grid point inside the region. The corresponding difference equations for the points on the edges will be derived in the next section. A linear equation system is obtained which, with fine grid division, has many equations. Numerical mathematics offers methods which can be used to solve very large systems of equations [2.66], [2.67]. Advantages and disadvantages of these solution methods are discussed by D. Marsal [2.53] and G.D. Smith [2.57]. In general iterative methods, described in detail by D.M. Young [2.73], are used, in particular the Gauss-Seidel-method and the method of successive over-relaxation (SOR-method).

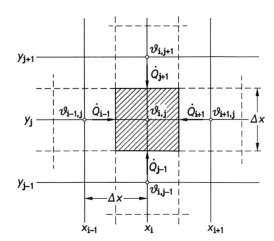

Fig. 2.49: Block with square cross section around the point (i, j) for the derivation of the energy balance (2.307)

Example 2.8: A wall of thickness δ surrounds a room with a square base; the length of the internal side of the square is $2.5\,\delta$. The wall has constant surface temperatures ϑ_i and $\vartheta_o < \vartheta_i$ respectively. Calculate the heat flow out of the room due to the temperature difference $\vartheta_i - \vartheta_o$.

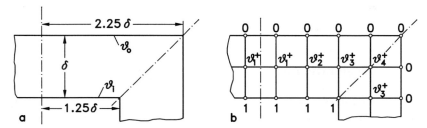

Fig. 2.50: Wall of a room with square base. **a** Dimensions, **b** Coarse grid ($\Delta x = \delta/2$) with four unknown (normalised) temperatures ϑ_1^+ to ϑ_4^+

For reasons of symmetry it is sufficient to deal with only an eigth of the base, Fig. 2.50a. The desired heat flow rate lies between two limits. They can be calculated, with the inner wall surface to be

$$\dot{Q}_i = 8 \cdot 1{,}25 \cdot \delta \cdot b \frac{\lambda}{\delta} \, (\vartheta_i - \vartheta_o) = 10\, b\lambda \, (\vartheta_i - \vartheta_o)$$

and with the outer surface of the wall

$$\dot{Q}_o = 8 \cdot 2{,}25 \cdot \delta \cdot b \frac{\lambda}{\delta} \, (\vartheta_i - \vartheta_o) = 18\, b\lambda \, (\vartheta_i - \vartheta_o)$$

where b is the dimension of the wall perpendicular to the drawing plane in Fig. 2.50a. The correct value between 10 and 18 for the shape factor

$$S_b = \frac{\dot{Q}}{\lambda b \, (\vartheta_i - \vartheta_o)} \quad ,$$

see section 2.2.5.2, is obtained by calculating the temperature field in the wall.

We use the dimensionless temperature $\vartheta^+ = (\vartheta - \vartheta_o)/(\vartheta_i - \vartheta_o)$, which lies in the region $0 \le \vartheta^+ \le 1$. A very coarse grid with $\Delta x = \delta/2$ delivers four grid points with the unknown temperatures ϑ_1^+ to ϑ_4^+, cf. Fig. 2.50b. They are calculated using the difference equations according to (2.308) with $\dot{W}_{i,j} \equiv 0$

$$\vartheta_2^+ + 0 + \vartheta_1^+ + 1 - 4\,\vartheta_1^+ = 0 ,$$

$$\vartheta_3^+ + 0 + \vartheta_1^+ + 1 - 4\,\vartheta_2^+ = 0 ,$$

$$\vartheta_4^+ + 0 + \vartheta_2^+ + 1 - 4\,\vartheta_3^+ = 0 ,$$

$$0 + 0 + \vartheta_3^+ + \vartheta_3^+ - 4\,\vartheta_4^+ = 0 .$$

The first and fourth equations concern the symmetry of the temperature field along the dotted line of symmetry in Fig. 2.50b.

The solution of the four equations yields the temperatures

$$\vartheta_1^+ = 0.4930, \quad \vartheta_2^+ = 0.4789, \quad \vartheta_3^+ = 0.4225, \quad \vartheta_4^+ = 0.2113 .$$

With these results the heat flow conducted to the outer surface of the wall is determined to be

$$\dot{Q} = 8\,b\lambda\,(\vartheta_i - \vartheta_o)\,(\vartheta_1^+ + \vartheta_2^+ + \vartheta_3^+ + \vartheta_4^+) = 12.85\,b\lambda\,(\vartheta_i - \vartheta_o) .$$

The shape factor has a value of $S_b = 12.85$. The corners increase the heat flow by 28.5% compared with $\dot{Q}_i$, the heat flow calculated with the internal wall area. These are inaccurate approximations due to the coarseness of the grid. A refined grid would deliver more accurate temperatures, but increase the number of difference equations. Halving the mesh size ($\Delta x = \delta/4$) already leads to a system of 24 linear equations.

2.4.6.2 Consideration of the boundary conditions

The system of linear equations originating from the difference equation (2.308) has to be supplemented by the difference equations for the points around the boundaries where the decisive boundary conditions are taken into account. As a simplification we will asume that the boundaries run parallel to the x- and y-directions. Curved boundaries can be replaced by a series of straight lines parallel to the x- and y-axes. However a sufficient degree of accuracy can only be reached in this case by having a very small mesh size Δx. If the boundaries are coordinate lines of a polar coordinate system (r, φ), it is recommended that the differential equation and its boundary conditions are formulated in polar coordinates and then the corresponding finite difference equations are derived.

The simplest case to consider is the first kind of boundary condition. For this the boundaries must coincide with grid lines, as in Example 2.8, and the stipulated temperatures must be used in the difference equation (2.308) for the grid points lying on the boundaries. This is achieved for boundaries which run parallel to the x- and y-axes, by the suitable choice of mesh size Δx. If necessary the square grid used up until now must be replaced by a rectangular grid with different grid spacings Δx and Δy. The difference equation (2.308) has to be altered, which can be carried out without difficulties using the explanations in 2.4.6.1.

The boundary condition of a given temperature can also be satisfied in a good approximation for curved boundaries. Fig. 2.51 shows a grid point (i, j) close to

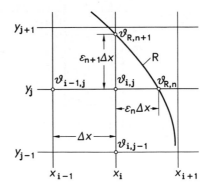

Fig. 2.51: Grid point (i,j) near the curved boundary R, at which the temperature ϑ_R is perscribed

the boundary. The straight lines $x = x_i$ and $y = y_j$ running through the point cut the curved boundary, where the temperatures $\vartheta_{R,n}$ and $\vartheta_{R,n+1}$ at the intersections are given. Instead of (2.308) the difference equation for grid point (i,j) is

$$
\frac{2}{\varepsilon_n\left(1+\varepsilon_n\right)}\vartheta_{R,n} + \frac{2}{\varepsilon_{n+1}\left(1+\varepsilon_{n+1}\right)}\vartheta_{R,n+1} + \frac{2}{1+\varepsilon_n}\vartheta_{i-1,j}
$$
$$
+\frac{2}{1+\varepsilon_{n+1}}\vartheta_{i,j-1} - 2\left(\frac{1}{\varepsilon_n}+\frac{1}{\varepsilon_{n+1}}\right)\vartheta_{i,j} = -\frac{\Delta x^2}{\lambda}\dot{W}_{i,j} \ . \tag{2.309}
$$

Its derivation can be found in G.D. Smith [2.57].

The *heat transfer condition* can only be satisfied simply if the boundary consists of straight lines which run parallel to the x- and y-axes. The boundary has to coincide with a grid line, and the difference equation for this sort of boundary point is acquired from an energy balance. This is shown for a grid point on an boundary which coincides with the grid line $y = y_j$, Fig. 2.52. The heat flows transferred by conduction from the neighbouring blocks into the shaded block are

$$
\dot{Q}_{i-1} = \frac{\lambda}{\Delta x}\left(\vartheta_{i-1,j}-\vartheta_{i,j}\right)\frac{\Delta x}{2}b = \frac{\lambda}{2}\left(\vartheta_{i-1,j}-\vartheta_{i,j}\right)b \ ,
$$

$$
\dot{Q}_{j-1} = \frac{\lambda}{\Delta x}\left(\vartheta_{i,j-1}-\vartheta_{i,j}\right)\Delta x\,b = \lambda\left(\vartheta_{i,j-1}-\vartheta_{i,j}\right)b
$$

and

$$
\dot{Q}_{i+1} = \frac{\lambda}{\Delta x}\left(\vartheta_{i+1,j}-\vartheta_{i,j}\right)\frac{\Delta x}{2}b = \frac{\lambda}{2}\left(\vartheta_{i+1,j}-\vartheta_{i,j}\right)b \ .
$$

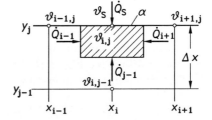

Fig. 2.52: Derivation of the finite difference equation for the boundary point (x_i, y_j) with heat transfer to a fluid at temperature ϑ_S

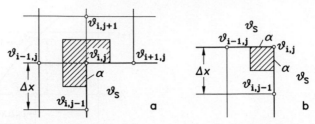

Fig. 2.53: Explanation of the difference equation for the boundary point **a** in the reflex corner, **b** at the outer corner with heat transfer to a fluid at temperature ϑ_S

The heat flow transferred from the fluid at temperature ϑ_S is

$$\dot{Q}_S = \alpha_i \left(\vartheta_S - \vartheta_{i,j}\right) \Delta x \, b$$

where α_i is the local heat transfer coefficient at the point x_i. Putting the four heat flows into the balance equation

$$\dot{Q}_{i-1} + \dot{Q}_{j-1} + \dot{Q}_{i+1} + \dot{Q}_S + \dot{W}_{i,j} \Delta x^2 b/2 = 0$$

produces the difference equation

$$\vartheta_{i,j-1} + \frac{1}{2}\left(\vartheta_{i-1,j} + \vartheta_{i+1,j}\right) + Bi_i^*\vartheta_S - (2 + Bi_i^*)\,\vartheta_{i,j} = -\dot{W}_{i,j}\Delta x^2/2\lambda \qquad (2.310)$$

with $Bi_i^* = \alpha_i \Delta x/\lambda$ as the local Biot number. This number allows us to easily account for changes in the heat transfer coefficient along the boundary.

The difference equations which hold for heat transfer at the internal (reflex) corner according to Fig. 2.53a, and at the outer corner according to Fig. 2.53b, are found in the same manner. We obtain for the internal corner

$$\vartheta_{i,j+1} + \vartheta_{i-1,j} + \frac{1}{2}\left(\vartheta_{i,j-1} + \vartheta_{i+1,j}\right) + Bi_i^*\vartheta_S - (3 + Bi_i^*)\,\vartheta_{i,j} = -3\dot{W}_{i,j}\Delta x^2/4\lambda$$
$$(2.311)$$

and for the corner in Fig. 2.53b

$$\frac{1}{2}\left(\vartheta_{i-1,j} + \vartheta_{i,j-1}\right) + Bi_i^*\vartheta_S - (1 + Bi_i^*)\,\vartheta_{i,j} = -\dot{W}_{i,j}\Delta x^2/4\lambda \ . \qquad (2.312)$$

If no internal heat sources are present, $\dot{W}_{i,j} = 0$ has to be put into the relationships (2.309) to (2.312).

Putting, $Bi_i^* = 0$, therefore $\alpha_i = 0$, into (2.310) to (2.312), makes these relationships valid for an *adiabatic* surface. This corresponds to the special case $\dot{q} = 0$, of the boundary condition of prescribed heat flux $\dot{q}$. According to section 2.4.2 this boundary condition is more accurately accounted for when the grid is laid out such that the boundary with *stipulated heat flux* has a distance of $\Delta x/2$ from the grid line, see Fig. 2.54. The energy balance for the block highlighted in Fig. 2.54 which has a temperature $\vartheta_{i,j}$ yields

$$\dot{Q}_{i+1} + \dot{Q}_{i-1} + \dot{Q}_{j-1} + \dot{q}_i\Delta x\, b + \dot{W}_{i,j}\Delta x^2 b = 0 \ ,$$

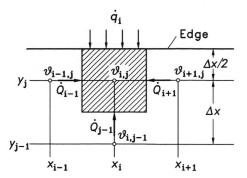

Fig. 2.54: Derivation of the difference equation (2.313) with given heat flux $\dot q_i$ at the boundary

where $\dot q_i = \dot q(x_i)$ is the given heat flux at $x = x_i$. The difference equation follows on from the energy balance:

$$\vartheta_{i+1,j} + \vartheta_{i-1,j} + \vartheta_{i,j-1} - 3\vartheta_{i,j} = -\dot q_i \frac{\Delta x}{\lambda} - \dot W_{i,j}\frac{\Delta x^2}{\lambda} \ . \tag{2.313}$$

Similar difference equations are obtained when the heat flux is stipulated on a boundary which runs parallel to the y-axis.

Example 2.9: The shaped brick shown in Fig. 2.55a of an oven wall ($\lambda = 0.80\,\mathrm{W/K\,m}$) with trapezoidal cross section ($\delta = 0.30\,\mathrm{m}$) is insulated on the perpendicular surfaces. At the upper surface heat is transferred to the air which is at constant temperature ϑ_S; the heat transfer coefficient is $\alpha = 10\,\mathrm{W/m^2K}$. The lower surface has a constant temperature $\vartheta_R > \vartheta_S$. The heat transferred through the brick to the air is to be calculated along with the temperatures of the surface which releases the heat.

In order to approximately determine the temperature field we will use the grid illustrated in Fig. 2.55b with the coarse mesh size $\Delta x = \Delta y = \delta/3 = 0.10$ m. As in the last example, we introduce the dimensionless temperature

$$\vartheta^+ = \frac{\vartheta - \vartheta_S}{\vartheta_R - \vartheta_S} \ , \qquad 0 \le \vartheta^+ \le 1$$

and compile the difference equations for the eight unknown temperatures ϑ_1^+ to ϑ_8^+. Equation (2.310) with $\dot W_{i,j} \equiv 0$ is decisive for the three surface temperatures, which results in the three equations

$$\vartheta_4^+ + \tfrac{1}{2}\left(\vartheta_1^+ + \vartheta_2^+\right) - (2 + Bi^*)\,\vartheta_1^+ \ = \ 0 \ ,$$

$$\vartheta_5^+ + \tfrac{1}{2}\left(\vartheta_1^+ + \vartheta_3^+\right) - (2 + Bi^*)\,\vartheta_2^+ \ = \ 0 \ ,$$

$$\vartheta_6^+ + \tfrac{1}{2}\left(\vartheta_2^+ + \vartheta_3^+\right) - (2 + Bi^*)\,\vartheta_3^+ \ = \ 0 \ .$$

It should be noted here that $\vartheta_S^+ = 0$. The perpendicular adiabatic walls are considered to be planes of symmetry with a reflection of the temperatures ϑ_1^+ and ϑ_3^+ in the first and last equations, respectively.

For the two inner grid points with temperatures ϑ_4^+ and ϑ_5^+ the difference equation (2.308) is applied:

$$\vartheta_5^+ + \vartheta_1^+ + \vartheta_4^+ + \vartheta_7^+ - 4\,\vartheta_4^+ \ = \ 0 \ ,$$

$$\vartheta_6^+ + \vartheta_2^+ + \vartheta_4^+ + \vartheta_8^+ - 4\,\vartheta_5^+ \ = \ 0 \ .$$

The grid points close to the lower boundary, with ϑ_6^+ to ϑ_8^+ have "arms", which are intersected by the oblique boundary along which $\vartheta_R^+ = 1$. Here we use (2.309) and imagine that Fig. 2.55b

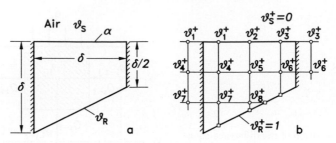

Fig. 2.55: Heat conduction in a shaped brick. **a** Dimensions and given temperatures, **b** square grid for the calculation of the temperatures ϑ_1^+ to ϑ_8^+

is rotated anticlockwise by $90°$, such that agreement with Fig. 2.51 exists. It follows for point 6 with $\varepsilon_n = 0.75$ and $\varepsilon_{n+1} = 1$ as well as $\vartheta_{R,n} = \vartheta_R^+$ and $\vartheta_{R,n+1} = \vartheta_6^+$ that

$$\frac{2}{0.75\,(1+0.75)}\vartheta_R^+ + \vartheta_6^+ + \frac{2}{1+0.75}\vartheta_3^+ + \vartheta_5^+ - 2\left(\frac{1}{0.75}+1\right)\vartheta_6^+ = 0 \ .$$

The result for the grid point ϑ_7^+ with $\vartheta_{R,n} = \vartheta_R^+$ and $\vartheta_{R,n+1} = \vartheta_8^+$ is analogous, i.e.

$$\frac{2}{0.75\,(1+0.75)}\vartheta_R^+ + \vartheta_8^+ + \frac{2}{1+0.75}\vartheta_4^+ + \vartheta_7^+ - 2\left(\frac{1}{0.75}+1\right)\vartheta_7^+ = 0 \ .$$

The difference equation for the last, closest to the boundary grid point 8 with $\varepsilon_n = 0.25$, $\varepsilon_{n+1} = 0.50$, $\vartheta_{R,n} = \vartheta_R^+$ and $\vartheta_{R,n+1} = \vartheta_R^+$ looks like

$$\frac{2}{0.25\,(1+0.25)}\vartheta_R^+ + \frac{2}{0.50\,(1+0.50)}\vartheta_R^+ + \frac{2}{1+0.25}\vartheta_5^+ + \frac{2}{1+0.50}\vartheta_7^+ - 2\left(\frac{1}{0.25}+\frac{1}{0.50}\right)\vartheta_8^+ = 0 \ .$$

With $Bi^* = \alpha\Delta x/\lambda = 10\,(\mathrm{W/m^2K})0.10\,\mathrm{m}/0.80\,(\mathrm{W/Km}) = 1.25$ and $\vartheta_R^+ = 1$ we obtain from these equations the system

$$\vartheta_1^+ = 0.1818\,\vartheta_2^+ + .3636\,\vartheta_4^+ \ , \qquad\qquad \vartheta_5^+ = 0.2500\,(\vartheta_2^+ + \vartheta_4^+ + \vartheta_6^+ + \vartheta_8^+) \ ,$$

$$\vartheta_2^+ = 0.1538\,(\vartheta_1^+ + \vartheta_3^+) + 0.3077\,\vartheta_5^+ \ , \qquad \vartheta_6^+ = 0.3117\,\vartheta_3^+ + 0.2727\,\vartheta_5^+ + 0.4156 \ ,$$

$$\vartheta_3^+ = 0.1818\,\vartheta_2^+ + 0.3636\,\vartheta_6^+ \ , \qquad\qquad \vartheta_7^+ = 0.3117\,\vartheta_4^+ + 0.2727\,\vartheta_8^+ + 0{,}4156 \ ,$$

$$\vartheta_4^+ = 0.2500\,(\vartheta_1^+ + \vartheta_4^+ + \vartheta_5^+ + \vartheta_7^+) \ , \qquad \vartheta_8^+ = 0.1333\,\vartheta_5^+ + 0.1111\,\vartheta_7^+ + 0.7556 \ .$$

Its solution, found according to the iteration method of Gauss-Seidel, delivers the temperatures

$$\vartheta_1^+ = 0.258 \ , \quad \vartheta_2^+ = 0.274 \ , \quad \vartheta_3^+ = 0.295 \ , \quad \vartheta_4^+ = 0.573 \ ,$$

$$\vartheta_5^+ = 0.613 \ , \quad \vartheta_6^+ = 0.675 \ , \quad \vartheta_7^+ = 0.848 \ , \quad \vartheta_8^+ = 0.932 \ .$$

The temperature of the surface bounded by air is not constant; it increases from the left hand corner where the brick is thickest to the right hand corner. Its mean value is

$$\vartheta_m^+ = \frac{1}{3}\left(\vartheta_1^+ + \vartheta_2^+ + \vartheta_3^+\right) = 0.276 \ .$$

This allows the heat flow transferred to the air to be calculated as

$$\dot{Q} = \alpha\,\delta b\,(\vartheta_m - \vartheta_S) = \alpha\delta b\vartheta_m^+\,(\vartheta_R - \vartheta_S)$$

where b is the brick dimension perpendicular to the drawing plane of Fig. 2.55a. We obtain

$$\dot{Q}/b\,(\vartheta_R - \vartheta_S) = 0.828\,\mathrm{W/K\,m} \ .$$

2.5   Mass diffusion

The basis for the solution of mass diffusion problems, which go beyond the simple case of steady-state and one-dimensional diffusion, sections 1.4.1 and 1.4.2, is the differential equation for the concentration field in a *quiescent* medium. It is known as the mass diffusion equation. As mass diffusion means the movement of particles, a quiescent medium may only be presumed for special cases which we will discuss first in the following sections. In a similar way to the heat conduction in section 2.1, we will discuss the derivation of the mass diffusion equation in general terms in which the concentration dependence of the material properties and chemical reactions will be considered. This will show that a large number of mass diffusion problems can be described by differential equations and boundary conditions, just like in heat conduction. Therefore, we do not need to solve many new mass diffusion problems, we can merely transfer the results from heat conduction to the analogue mass diffusion problem. This means that mass diffusion problem solutions can be illustrated in a short section. At the end of the section a more detailed discussion of steady-state and transient mass diffusion with chemical reactions is included.

2.5.1   Remarks on quiescent systems

The following observations show that in contrast with heat conduction, mass diffusion processes seldom occur in quiescent systems. Therefore mass diffusion in quiescent systems has less practical meaning compared to heat conduction. We understand mass diffusion to be mass transport as a result of the natural movement of molecules from one region of a system to another. Correspondingly, heat conduction can be described as the energy transport due to the statistical movement of elementary particles caused by an irregular temperature distribution. In this respect a close relationship between mass diffusion and heat conduction exists.

In contrast to heat conduction, in mass diffusion the average velocity of the particles of the individual materials in a volume element can be different from each other, so that a relative movement of the individual particles to each other is macroscopically perceptible. In general this results in a macroscopic movement of all particles in a volume element and therefore convection. As these considerations show, in contrast to heat conduction quiescent systems cannot always be assumed for mass diffusion. This can only be assumed under certain conditions, which we will now discuss.

We will limit ourselves to a discussion of systems where the reference velocity for the determination of the mass diffusional flux disappears, cf. (1.153). As a reference velocity we chose the gravitational velocity according to (1.154),

$$\varrho\, \boldsymbol{w} = \sum_K \varrho_K \boldsymbol{w}_K \;, \tag{2.314}$$

and the average molar velocity $\boldsymbol{u}$ according to (1.157), which we will write in the following form

$$c\, \boldsymbol{u} = \sum_K c_K \boldsymbol{w}_K \;. \tag{2.315}$$

Vanishing gravitational velocity $\boldsymbol{w} = 0$ at every position has

$$\sum_K \varrho_K \boldsymbol{w}_K = 0 \tag{2.316}$$

as a result overall in the observed system. The mass flow into the volume element has to be exactly the same size as the flow rate out of the element, so that (2.316) is satisfied. The mass in the volume element, therefore remains unchanged. The density ϱ of a volume element in a particular position is also constant. It can, however, be different for an element in a different position $\boldsymbol{x}$. This means that $\partial\varrho/\partial\boldsymbol{x} \neq 0$ is possible but $\partial\varrho/\partial t = 0$. Vanishing gravitational velocity $\boldsymbol{w} = 0$ merely has the result $\partial\varrho/\partial t = 0$, but not $\partial\varrho/\partial\boldsymbol{x} = 0$.

The model considered here is that of an incompressible body. This is defined by the fact that the density of a volume element in the material does not change during its movement, i.e. $\varrho = \varrho(\boldsymbol{x}, t) = \text{const}$ and therefore $d\varrho/dt = 0$, which is fulfiled in our case, because due to $\boldsymbol{w} = 0$ and $\partial\varrho/\partial t = 0$

$$\frac{d\varrho}{dt} = \frac{\partial\varrho}{\partial t} + \boldsymbol{w}\frac{\partial\varrho}{\partial\boldsymbol{x}} = 0 \;.$$

As an example we will take a porous body into which hydrogen diffuses. A volume element in the porous body has a certain mass and therefore a certain density, e.g. $\varrho = 7.8 \cdot 10^3 \,\text{kg/m}^3$. According to this a volume element of $1\,\text{mm}^3$ has a mass of $7.8 \cdot 10^{-3}\,\text{g}$. We assume that the volume element is at the ambient temperature $298\,\text{K}$ with a partial pressure of hydrogen of $p = 1\,\text{hPa}$. The equation of state for ideal gases is used to find the hydrogen absorbed by the volume element, which is

$$M_{\text{H}_2} = \frac{p_{\text{H}_2}V}{R_{\text{H}_2}T} = \frac{10^2\text{Pa} \cdot 10^{-9}\,\text{m}^3}{4,1245\,(\text{J/gK}) \cdot 298\,\text{K}}$$

$$M_{\text{H}_2} = 8,136 \cdot 10^{-11}\,\text{g} \;.$$

The mass of the element is hardly changed due to the added hydrogen, and so any mass change can be neglected. The centre of mass remains in the same position in space, a good approximation is $d\boldsymbol{x}/dt = \boldsymbol{w} = 0$.

Similar results are obtained for mass diffusion in very dilute solutions, for example the absorption of a gas in a liquid. The density of the liquid, away from the critical region, is much larger than that of the gas, and the mass of the gas is extraordinarily low. This means that the mass of a volume element is practically unchanged by the absorption of gas. Once again, a good approximation is $d\boldsymbol{x}/dt = \boldsymbol{w} = 0$.

the region, whilst the normal vector $\boldsymbol{n}$ points outwards, so the scalar product will be negative. The diffusional flux, according to (1.155), is defined by

$$\boldsymbol{j}_A^* = \varrho_A(\boldsymbol{w}_A - \boldsymbol{w}) \ .$$

We will deal with a quiescent medium relative to the gravitational velocity, $\boldsymbol{w} = 0$, so that the diffusional flux is

$$\boldsymbol{j}_A^* = \varrho_A \, \boldsymbol{w}_A \ . \tag{2.321}$$

It follows from (2.320) and (2.321) that

$$\mathrm{d}\dot{M}_A = -\boldsymbol{j}_A^* \, \boldsymbol{n} \, \mathrm{d}A \ .$$

Integration gives the total inlet mass flow of the component

$$\dot{M}_A = -\int\limits_{(A)} \boldsymbol{j}_A^* \, \boldsymbol{n} \, \mathrm{d}A = -\int\limits_{(V)} \mathrm{div} \, \boldsymbol{j}_A^* \, \mathrm{d}V \tag{2.322}$$

This will then be converted from the integral over the whole surface area of the region to the volume integral of the divergence $\boldsymbol{j}_A^*$ according to Gauss' integral theorem.

At the same time a certain amount of substance A is generated or absorbed by a chemical reaction inside the region. The production rate in a volume element is indicated by $\dot{\Gamma}_A$ (SI units kg/m^{3}s). It is positive when subtance A is generated, and negative when it is consumed. The mass flow

$$\int\limits_{(V)} \dot{\Gamma}_A \, \mathrm{d}V \tag{2.323}$$

of substance A is generated inside the region. The mass flow over the surface area of the region due to diffusion and the flow generated by the chemical reaction cause an increase over time of the mass stored inside the region by

$$\frac{\mathrm{d}}{\mathrm{d}t} \int\limits_{(V)} \varrho_A \, \mathrm{d}V = \int\limits_{(V)} \frac{\partial \varrho_A}{\partial t} \, \mathrm{d}V \ . \tag{2.324}$$

Taking into account (2.322) and (2.323) the mass balance for substance A is

$$\int\limits_{(V)} \frac{\partial \varrho_A}{\partial t} \, \mathrm{d}V = -\int\limits_{(V)} \mathrm{div} \, \boldsymbol{j}_A^* \, \mathrm{d}V + \int\limits_{(V)} \dot{\Gamma}_A \, \mathrm{d}V \ .$$

As the balance region can be infinitely small, it holds that

$$\frac{\partial \varrho_A}{\partial t} = -\mathrm{div} \, \boldsymbol{j}_A^* + \dot{\Gamma}_A \ . \tag{2.325}$$

A mass balance of this type is valid for each component in the mixture. Therefore, there are as many mass balances as substances. Addition of all the components leads to

$$\sum_{K=1}^{N} \frac{\partial \varrho_K}{\partial t} = -\operatorname{div} \sum_{K=1}^{N} \boldsymbol{j}_K^* + \sum_{K=1}^{N} \dot{\Gamma}_K \ . \tag{2.326}$$

In which $\sum_{K=1}^{N} \varrho_K = \varrho$ is the density of a mixture of N substances. The sum of all the diffusional fluxes is given by (1.156), i.e. $\sum_{K=1}^{N} \boldsymbol{j}_K^* = 0$. As the mass of a certain substance generated per unit time in the chemical reaction can only be as much as is consumed of another substance, it follows that

$$\sum_{K=1}^{N} \dot{\Gamma}_K = 0 \ .$$

The summation over all substances, according to (2.326), leads to

$$\frac{\partial \varrho}{\partial t} = 0 \ .$$

The density at a given position does not change with time. This is stipulated because we have presumed a quiescent medium $\boldsymbol{w} = 0$. However the density is only contant at a fixed position. It can change locally due to a different composition, $\varrho = \varrho(\boldsymbol{x})$.

As the density ϱ is independent of time, (2.325) may, because $\varrho_A = \varrho\,\xi_A$, also be written as

$$\varrho \frac{\partial \xi_A}{\partial t} = -\operatorname{div} \boldsymbol{j}_A^* + \dot{\Gamma}_A \ . \tag{2.327}$$

Introducing Fick's law (2.318) gives

$$\varrho \frac{\partial \xi_A}{\partial t} = \operatorname{div}(\varrho\, D \operatorname{grad} \xi_A) + \dot{\Gamma}_A \tag{2.328}$$

as the desired differential equation for the mass fraction $\xi_A(\boldsymbol{x}, t)$ in a quiescent, relative to the gravitational velocity, isotropic binary mixture of subtances A and B. A corresponding equation also exists for B. However this does not need to be solved, because when the mass fractiom ξ_A is known, the other mass fraction in a binary mixture is $\xi_B = 1 - \xi_A$.

Putting Fick' law (2.319) into (2.327) provides us with the corresponding equation to (2.328) for a mixture of N components

$$\varrho \frac{\partial \xi_A}{\partial t} = -\operatorname{div}(\varrho \sum_{\substack{K=1 \\ K \neq A}}^{N} \frac{\tilde{M}_A \tilde{M}_K}{\tilde{M}^2} D_{AK} \operatorname{grad} \tilde{x}_K) + \dot{\Gamma}_A \ . \tag{2.329}$$

As we have presumed a quiescent body relative to the gravitational velocity, $\boldsymbol{w} = 0$ and therefore $\partial \varrho / \partial t = 0$, the quantities D and $\dot{\Gamma}_A$ in (2.328) depend on the density

$\varrho(\boldsymbol{x})$, the temperature ϑ and the composition ξ_A. In (2.329) these quantities do not only depend on the mass fraction ξ_A of component A, but also on the mass fractions of the other substances. In addition grad $\tilde{x}_A$ also depends on the gradients of the mass fractions of the other substances, which follow from $\tilde{x}_A = \tilde{M}\xi_A/\tilde{M}_A$ with $1/\tilde{M} = \sum_K \xi_K/\tilde{M}_K$.

Therefore the diffusion equations (2.328) and (2.329) are, in general, non-linear. The general equation (2.329), contains (2.328) as a special case with $N = 2$, which can be easily checked.

We will now presuppose a quiescent system relative to the average molar velocity $\boldsymbol{u}$, with constant temperature and pressure, and derive the diffusion equation. The molar flow of component A into the surface element dA, as in Fig. 2.57b is

$$d\dot{N}_A = -c_A \, \boldsymbol{w}_A \, \boldsymbol{n} \, dA \ . \tag{2.330}$$

The minus sign appears because the normal vector $\boldsymbol{n}$ points outwards from the area dA, and the molar flow into the area should be positive. The diffusional flux is defined by (1.158) as

$$_u\boldsymbol{j}_A = c_A(\boldsymbol{w}_A - \boldsymbol{u})$$

and is equal to the molar flux

$$_u\boldsymbol{j}_A = c_A \boldsymbol{w}_A \ , \tag{2.331}$$

because we have presumed a vanishing average molar velocity $\boldsymbol{u}$. The molar flow into the total area is

$$\dot{N}_A = -\int_{(A)} c_A \, \boldsymbol{w}_A \, \boldsymbol{n} \, dA = -\int_{(A)} {_u\boldsymbol{j}_A} \, \boldsymbol{n} \, dA = -\int_{(V)} \operatorname{div} {_u\boldsymbol{j}_A} \, dV \ . \tag{2.332}$$

The amount of substance A generated by chemical reaction in the region is indicated by the molar production flux $\dot{\gamma}_A$ (SI units mol/m^{3}s). It is

$$\int_{(V)} \dot{\gamma}_A \, dV \ . \tag{2.333}$$

The molar flow into the region due to diffusion and that generated by the chemical reaction cause an increase over time in the amount of substance stored in the region of

$$\frac{d}{dt} \int_{(V)} c_A \, dV = \int_{(V)} \frac{\partial c_A}{\partial t} \, dV \ . \tag{2.334}$$

The material balance for substance A is then

$$\int_{(V)} \frac{\partial c_A}{\partial t} \, dV = -\int_{(V)} \operatorname{div} {_u\boldsymbol{j}_A} \, dV + \int_{(V)} \dot{\gamma}_A \, dV \ .$$

As the balance region can be chosen to be infinitely small, it holds that

$$\frac{\partial c_A}{\partial t} = -\mathrm{div}\,_u\boldsymbol{j}_A + \dot{\gamma}_A \ . \tag{2.335}$$

A balance of this type exists for all the substances involved in the diffusion, so they can be added together giving

$$\sum_{K=1}^{N} \frac{\partial c_K}{\partial t} = -\mathrm{div} \sum_{K=1}^{N} {}_u\boldsymbol{j}_K + \sum_{K=1}^{N} \dot{\gamma}_K \ . \tag{2.336}$$

Here $\sum_{K=1}^{N} c_K = c$ is the molar density of a mixture consisting of N substances. Furthermore if $\sum_{K=1}^{N} {}_u\boldsymbol{j}_K = 0$, cf. section 1.4.1, page 72, (2.336) is transformed into the following equation valid for the sum over all components

$$\frac{\partial c}{\partial t} = \sum_{K=1}^{N} \dot{\gamma}_K \ . \tag{2.337}$$

As the sum of the number of moles does not generally remain the same in a chemical reaction, the right hand side of (2.337) does not disappear unless exactly the same number of moles are generated as disappear. Inserting Fick's law (1.160), in which D_{AB} is replaced by D, we obtain from (2.335)

$$\frac{\partial c_A}{\partial t} = \mathrm{div}(c\,D\,\mathrm{grad}\,\tilde{x}_A) + \dot{\gamma}_A \tag{2.338}$$

as the differential equation for the molar concentration $c_A(\boldsymbol{x}, t)$ in a quiescent, relative to the average molar velocity $\boldsymbol{u}$, binary mixture of components A and B. A corresponding equation exists for substance B. It is also possible to solve (2.338) and then determine c_B, because $c_B = c - c_A$.

For a multicomponent mixture of ideal gases at constant temperature and pressure, Fick's law in the particle reference system [2.75], can be written as

$$_u\boldsymbol{j}_A = \sum_{\substack{K=1 \\ K \neq A}}^{N} D_{AK}\,\mathrm{grad}\,c_K \ . \tag{2.339}$$

Putting this into (2.335) we obtain for a multicomponent mixture at constant pressure and temperature, the diffusion equation

$$\frac{\partial c_A}{\partial t} = -\mathrm{div}\Big(\sum_{\substack{K=1 \\ K \neq A}}^{N} D_{AK}\,\mathrm{grad}\,c_K\Big) + \dot{\gamma}_A \ . \tag{2.340}$$

In (2.338) $D = D(p, \vartheta, \tilde{x}_A)$ is the diffusion coefficient and the molar production density is $\dot{\gamma}_A = \dot{\gamma}_A(p, \vartheta, \tilde{x}_A)$ with $\tilde{x}_A = c_A/c$. In (2.340) these quantities depend not only on the mole fraction $\tilde{x}_A$, but also on the mole fractions of the other substances.

2.5.3 Simplifications

We will still only deal with binary mixtures in the following, as the diffusion coefficients D_{AK} of mixtures of more than two components are mostly unknown, and therefore the diffusion in these mixtures cannot be quantitatively calculated. The diffusion equations (2.328) and (2.338) for binary mixtures can often be simplified considerably.

We have already ascertained that in the diffusion of a gas A into a solid or liquid B, the density of a volume element is practically unchanged, $\partial \varrho / \partial t = 0$, because the mass of the gas absorbed is low in comparison with the mass of the volume element. If substance B was initially homogeneous, $\varrho = \varrho(\boldsymbol{x}) = \text{const}$, the density will also be unchanged locally during the diffusion process. We can therefore say a good approximation is that the density is constant, independently of position and time. Furthermore, measurements [2.76] have shown that the diffusion coefficient in dilute liquid solutions at constant temperature may be taken as approximately constant. Equally in diffusion of a gas into a homogenous, porous solid at constant temperature, the diffusion coefficient is taken to be approximately constant, as the concentration only changes within very narrow limits. In these cases, in which $\varrho = \text{const}$ and $D_{AB} = D = \text{const}$ can be assumed, (2.328) simplifies to

$$\frac{\partial \xi_A}{\partial t} = D\nabla^2 \xi_A + \dot{\Gamma}_A / \varrho \tag{2.341}$$

with $\dot{\Gamma}_A = \dot{\Gamma}_A(\vartheta, \xi_A)$.

The equation (2.338) for vanishing average molar velocity $\boldsymbol{u} = 0$, can also be simplified, when it is applied to binary mixtures of ideal gases. As a good approximation, at low pressures generally up to about 10 bar, the diffusion coefficient is independent of the composition. It increases with temperature and is inversely proportional to pressure. The diffusion coefficients in isobaric, isothermal mixtures are constant. In this case (2.338) is transformed into the equation for $c = \text{const}$

$$\frac{\partial c_A}{\partial t} = D\nabla^2 c_A + \dot{\gamma}_A \tag{2.342}$$

with $\dot{\gamma}_A = \dot{\gamma}_A(\vartheta, \tilde{x}_A)$.

The equations (2.341) and (2.342) are equivalent to each other, because putting $c_A = \varrho_A / \tilde{M}_A$ into (2.342) results in

$$\frac{\partial \varrho_A}{\partial t} = D\nabla^2 \varrho_A + \dot{\gamma}_A \tilde{M}_A \ .$$

On the other hand, because $\varrho = \text{const}$ we can also write

$$\frac{\partial \varrho_A}{\partial t} = D\nabla^2 \varrho_A + \dot{\Gamma}_A \ .$$

By definition $\dot{\gamma}_A \tilde{M}_A = \dot{\Gamma}_A$, which shows that the two equations are equivalent. For the case where no chemical reaction occurs, $\dot{\gamma}_A = 0$, (2.342) was first presented by A. Fick in 1855 [2.76], and

$$\frac{\partial c_A}{\partial t} = D\nabla^2 c_A \qquad (2.343)$$

is therefore known as Fick's second law.

2.5.4 Boundary conditions

The equations (2.342) and (2.341) are of the same type as the heat conduction equation (2.11). The following correspondences hold

$$c_A \hat{=} \vartheta \quad , \quad D \hat{=} a \quad \text{and} \quad \dot{\gamma}_A \hat{=} \dot{W}/c\varrho \ .$$

As a result of this many solutions to the heat conduction equation can be transferred to the analogous mass diffusion problems, provided that not only the differential equations but also the initial and boundary conditions agree. Numerous solutions of the differential equation (2.342) can be found in Crank's book [2.78]. Analogous to heat conduction, the *initial condition* prescribes a concentration at every position in the body at a certain time. Timekeeping begins with this time, such that

$$c_A(x, y, z, t = 0) = c_{A\alpha}(x, y, z) \ . \qquad (2.344)$$

The local boundary conditions can be split into three groups, just as in heat conduction processes.

  1. The concentration can be set as a function of time and the position x_0, y_0, z_0 on the surface of the body (boundary condition of the 1st kind)

$$c_A(x_0, y_0, z_0, t) = c_{A0}(t) \ . \qquad (2.345)$$

Examples of this are the drying of porous substances in the first drying period, when the surface is covered by a liquid film, or the evaporation of water in dry air. In both cases the saturation pressure $p_{AS}(\vartheta)$, associated with the temperature ϑ, becomes apparent at the interface between the liquid and gas. This then gives $c_{A0} = p_{AS}(\vartheta)/R_m T$.

The solubility of a gas A in an adjacent liquid or solid B is also often of interest. If the gas is only weakly soluble in the liquid, the mole fraction $\tilde{x}_A$ of the dissolved gas is obtained from Henry's law (1.217) as

$$\tilde{x}_{A0} = p_{AS}(\vartheta)/k_H \ ,$$

where k_H is the Henry coefficient. It is a function of temperature and pressure in binary mixtures. The pressure dependence can, however, be neglected if the total pressure of the gas is so low that it can be considered to be ideal. Numerical values of $k_H(\vartheta)$ for this case can be found in the tables from Landolt-Börnstein [2.79].

2. The diffusional flux normal to the surface can be prescribed as a function of time and position (boundary condition of the 2nd kind)

$$\dot{}_{u}j_{\mathrm{A}} = -c\,D\,\frac{\partial \tilde{x}_{\mathrm{A}}}{\partial n} \;, \tag{2.346}$$

whereby for surfaces impermeable to the material we get

$$\partial \tilde{x}_{\mathrm{A}}/\partial n = 0 \;. \tag{2.347}$$

3. Contact between two quiescent bodies (1) and (2) at the interface can exist (boundary condition of the 3rd kind)

$$(c\,D)_0^{(1)} \left(\frac{\partial \tilde{x}_{\mathrm{A}}}{\partial n}\right)_0^{(1)} = (c\,D)_0^{(2)} \left(\frac{\partial \tilde{x}_{\mathrm{A}}}{\partial n}\right)_0^{(2)} \;. \tag{2.348}$$

If a quiescent body (1) is bounded by a moving fluid (2) a diffusion boundary layer develops in the fluid. Instead of (2.348), the boundary condition

$$- (c\,D)_0^{(1)} \left(\frac{\partial \tilde{x}_{\mathrm{A}}}{\partial n}\right)_0^{(1)} = \beta\,\Delta\,c_{\mathrm{A}}^{(2)} \tag{2.349}$$

is used, with concentration difference in the fluid $\Delta c_{\mathrm{A}}^{(2)} = c_{\mathrm{A}0}^{(2)} - c_{\mathrm{A}\delta}^{(2)}$ between the interface 0 and the edge δ of the boundary layer of the fluid. The mole fraction $\tilde{x}_{\mathrm{A}}$ and concentration c_{A} are transferable into each other by $c_{\mathrm{A}} = \tilde{x}_{\mathrm{A}}\,c$. The concentration $c = N/V$ is given by $c = p/R_{\mathrm{m}}T$ for ideal gases, whilst for real gases it has to be taken from the thermal equation of state.

If heat is transferred by conduction between two bodies (1) and (2) which are in contact with each other (boundary condition of the 3rd kind), then not only the heat fluxes, but also the temperatures $\vartheta_0^{(1)} = \vartheta_0^{(2)}$ will be equal. In contrast to this, in diffusion at the phase interface, a concentration jump almost always develops. This is yielded from the equilibrium conditions with regard to mass transfer (equality of the chemical potential of each component in each phase). A relationship of the form below exists for given values of temperature and pressure

$$\tilde{x}_{\mathrm{A}0}^{(1)} = f(\tilde{x}_{\mathrm{A}0}^{(2)}) \;, \tag{2.350}$$

which is provided by the thermodynamics of phase equilibria. An equation of this type is the definition equation for the equilibrium constant

$$K := \tilde{x}_{\mathrm{A}0}^{(1)}/\tilde{x}_{\mathrm{A}0}^{(2)} \;. \tag{2.351}$$

For the solubility of gases in liquids

$$K = k_{\mathrm{H}}(\vartheta)/p \;,$$

where $k_{\mathrm{H}}(\vartheta)$ is the Henry coefficient. As already mentioned in chapter 1 further information about the equilibrium constant and Henry coefficients can be found in text books on the thermodynamics of mixtures, for example [1.1]. The solubility of a gas A in a solid phase is calculated from

$$c_{A0} = L_S \, p_{A0} \tag{2.352}$$

with the solubility L_S (SI units mol/m³bar). Numerical values can be found in the tables from Landolt-Börnstein [2.79] among others.

Example 2.10: In a spherical vessel of internal diameter 1450 mm and wall thickness 4 mm made of chromium-nickel-steel, a gas containing hydrogen is stored at 85 °C under high pressure. The hydrogen has an initial partial pressure of 1 MPa. As a result of diffusion through the vessel wall, some of the hydrogen is lost into the air surrounding the container, such that the pressure in the vessel falls over time. We wish to calculate the time taken for the pressure in the vessel to fall by 10^{-3} MPa, and how much hydrogen is lost in this time interval. Hint: As the pressure in the vessel falls very slowly the change over time of the hydrogen content in the wall can be neglected, i.e. we assume steady-state diffusion. Equation (2.352) holds for the amount of hydrogen dissolved at the wall surface, with the experimentally determined solubility coefficient $L_S = 9{,}01$ mol/m³bar. The diffusion coefficient for hydrogen into the vessel wall at 85 °C is $D = 1{,}05 \cdot 10^{-13}$ m²/s. The gaseous mixture in the vessel satisfies the thermal equation of state for ideal gases. The hydrogen content in the surrounding air may be neglected.

The amount of hydrogen (substance A is hydrogen) which diffuses through the vessel wall is equal to the reduction in the amount stored:

$$\dot{j}_A A = -\frac{dN}{dt} = -\frac{dN_A}{dt} \quad .$$

Furthermore

$$\dot{j}_A A = \frac{D}{\delta} A_m (c_{Ai} - c_{Aa})$$

with wall thickness δ and the geometric mean $A_m = \sqrt{A_i \, A_o} \cong A_i$ of the internal and external surface areas. With $c_{Ao} = 0$ and $c_{Ai} = L_S p_A$ we obtain

$$\dot{j}_A A = \frac{D}{\delta} A_m L_S \, p_A \quad .$$

On the other hand, for the amount of hydrogen inside the vessel we have $N_A = p_A V / R_m T$. This then gives

$$-\frac{V}{R_m T} \frac{dp_A}{dt} = \frac{D}{\delta} A_m L_S \, p_A$$

or

$$d \ln p_A = -\frac{D}{\delta} A_m L_S \frac{R_m T}{V} dt \quad .$$

Integration yields

$$p_A(t) = p_A(t = 0) \exp\left(-\frac{D}{\delta} A_m L_S \frac{R_m T}{V} t\right)$$

Putting in the numerical values we obtain

$$p_A(t_1) = 1 \, \text{MPa} \cdot \exp\left(-2.915 \cdot 10^{-11} t_1/\text{s}\right)$$

from which we get

$$t_1 = \frac{1}{-2.915 \cdot 10^{-11}} \ln \frac{(1 - 10^{-3}) \, \text{MPa}}{1 \, \text{MPa}} \, \text{s} = 3.432 \cdot 10^7 \, \text{s} = 9534 \, \text{h} = 397 \, \text{days} \quad .$$

The loss of hydrogen follows from the equation of state for ideal gases

$$\frac{M_A}{M_A(t = 0)} = \frac{\Delta p_A}{p_A(t = 0)} = \frac{10^{-3} \, \text{MPa}}{1 \, \text{MPa}} \quad .$$

This is 0.1 % of the initial amount of hydrogen present, which was

$$M_A(t = 0) = \tilde{M}_A N_A(t = 0) = \tilde{M}_A \, p_A(t = 0) V / R_m T = 1.072 \, \text{kg} \quad .$$

2.5.5 Steady-state mass diffusion with catalytic surface reaction

We have already considered steady-state one-dimensional diffusion in the introductory sections 1.4.1 and 1.4.2. Chemical reactions were excluded from these discussions. We now want to consider the effect of chemical reactions, firstly the reactions that occur in a catalytic reactor. These are *heterogeneous* reactions, which we understand to be reactions at the contact area between a reacting medium and the catalyst. It takes place at the surface, and can therefore be formulated as a boundary condition for a mass transfer problem. In contrast *homogeneous* reactions take place inside the medium. Inside each volume element, depending on the temperature, composition and pressure, new chemical compounds are generated from those already present. Each volume element can therefore be seen to be a source for the production of material, corresponding to a heat source in heat conduction processes.

As an example we will consider a catalytic reactor, Fig. 2.58, in which by a chemical reaction between a gas A and its reaction partner R, a new reaction product P is formed. The reaction partner R and the gas A are fed into the reactor, excess gas A and reaction product P are removed from the reactor. The reaction is filled with spheres, whose surfaces are covered with a catalytic material. The reaction between gas A and reactant R occurs at the catalyst surface and is accelerated due to the presence of the catalyst. In most cases the complex reaction mechanisms at the catalyst surface are not known completely, which suggests the use of very simplified models. For this we will consider a section of the catalyst surface, Fig. 2.59. On the catalyst surface $x = 0$ at steady-state, the same amount of gas as is generated will be transported away by diffusion. The reaction rate is equal to the diffusive flux. In general the reaction rate $\dot{n}_{A0}$ of a catalytic reaction depends on the concentration of the reaction partner. In the present case we assume that the reaction rate will be predominantly determined by the concentration $c_A(x = 0) = c_{A0}$ of gas A at the surface. For a first order reaction

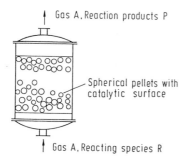

Fig. 2.58: Catalytic reactor.

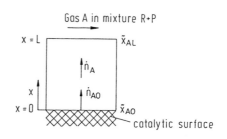

Fig. 2.59: Uni-directional diffusion with reaction at the catalyst surface

it is given by

$$\dot{n}_{A0} = -k_1'' c_{A0} \; . \tag{2.353}$$

For an n-th order reaction we have

$$\dot{n}_{A0} = -k_n'' c_{A0}^n \; . \tag{2.354}$$

Here k_n'' is the rate constant with SI units (mol / m²s) / (mol / m³)^n. The rate constant k_1'' for a first order reaction has SI units m/s. The superscript indices $''$ indicate that the reaction takes place at the surface. The minus sign makes $\dot{n}_{A0}$ negative, as substance A is consumed in the chemical reaction. If substance A had been generated (2.353) would have a positive sign.

Presuming that the reaction is first order, at the catalyst surface it holds that

$$\dot{n}_{A0} = -\left(c D \frac{\partial \tilde{x}_A}{\partial x} \right)_{x=0} = -k_1'' c_{A0} \; . \tag{2.355}$$

Above the catalyst surface, substance A will be transported by diffusion, mainly in the x-direction. In a thin layer close to the wall, the diffusion boundary layer, mass transport by convection is negligible, and from (2.338) we obtain the diffusion equation valid for steady one-dimensional diffusion without chemical reaction

$$\frac{d}{dx}\left(c D \frac{d\tilde{x}_A}{dx} \right) = 0 \; . \tag{2.356}$$

Its solution must not only satisfy the boundary condition (2.355) but also

$$\tilde{x}_A(x = L) = \tilde{x}_{AL} \tag{2.357}$$

where L is the thickness of the diffusion boundary layer. Under the assumption of $c D = $ const we obtain the solution

$$\tilde{x}_A - \tilde{x}_{AL} = \frac{k_1'' c_{A0}}{c D}(x - L) \; . \tag{2.358}$$

The mole fraction at the wall $x = 0$ is found to be

$$\tilde{x}_{A0} = -\frac{k_1'' c_{A0}}{c D} L + \tilde{x}_{AL} = -\frac{k_1'' \tilde{x}_{A0} L}{D} + \tilde{x}_{AL}$$

or

$$\tilde{x}_{A0} = \frac{\tilde{x}_{AL}}{1 + k_1'' L/D} \; . \tag{2.359}$$

Futhermore, if we take $c = $ const as an assumption, due to $c_A = \tilde{x}_A c$, we get

$$c_{A0} = \frac{c_{AL}}{1 + k_1'' L/D} \tag{2.360}$$

and therefore the reaction rate with (2.355)

$$\dot{n}_{A0} = -\frac{k_1'' c_{AL}}{1 + (k_1'' L/D)} \; . \tag{2.361}$$

In the case of convective mass transfer at the catalyst surface, the mass transfer coefficient β appears in this equation in place of D/L, $\beta = D/L$. This can be easily checked because (2.355) can be replaced by

$$\dot{n}_{A0} = -k_1'' c_{A0} = \beta(c_{A0} - c_{AL}) \ ,$$

which after elimination of c_{A0}, yields the relationship

$$\dot{n}_{A0} = -\frac{k_1'' c_{AL}}{1 + (k_1''/\beta)} \ . \tag{2.362}$$

The negative sign in (2.361) and (2.362) indicates that the mass flow of A is towards the catalyst surface.

The dimensionless quantitiy

$$Da_1 = k_1'' L/D \tag{2.363}$$

is called the *Damköhler*[5] number for a heterogeneous first order reaction (there are further Damköhler numbers). It is the ratio of the reaction rate k_1'' to the diffusion rate D/L.

In (2.361) or (2.362) two limiting case are of interest:

a) $Da_1 = k_1'' L/D$ or k_1''/β are very small, because $k_1'' \ll D/L$ or $k_1'' \ll \beta$. Then

$$\dot{n}_{A0} = -k_1'' c_{AL} \ . \tag{2.364}$$

The material conversion will be determined by the reaction rate. It is "reaction controlled". According to (2.360) the concentration of substance A perpendicular to the catalyst surface is constant, i.e. $c_{A0} = c_{AL}$.

b) $Da_1 = k_1'' L/D$ or k_1''/β is very large because $k_1 \gg D/L$ or $k_1 \gg \beta$. Then we get

$$\dot{n}_{A0} = -\frac{D}{L} c_{AL} \quad \text{or} \quad \dot{n}_{A0} = -\beta c_{AL} \ . \tag{2.365}$$

The material conversion is determined by diffusion. It is "diffusion controlled". So according to (2.360), $c_{A0} = 0$. Substance A will be completely converted in a rapid reaction at the catalyst surface.

Example 2.11: In the catalytic converter of a car, nitrogen oxide is reduced at the catalyst surface, according to the following reaction

$$NO + CO \rightarrow \frac{1}{2} N_2 + CO_2 \ .$$

The NO-reduction (substance A) is approximately a first order reaction $\dot{n}_{A0} = -k_1'' c_{A0}$. The exhaust flow of a 75 kW engine is $M = 350 \, \text{kg/h}$, the molar mass of the exhaust gases is

---

[5]Gerhard Damköhler (1908–1944) was the first to develop a similarity theory for chemical processes with consideration of both heat and mass transfer. With his work "Einfluß von Diffusion, Strömung und Wärmetransport auf die Ausbeute von chemischen Reaktionen" (Der Chemie-Ingenieur, Leipzig, 1937, 359–485), in english, 'The influence of diffusion, flow and heat transport on the yield of a chemical reaction', he paved the way for further investigation into chemical reaction technology.

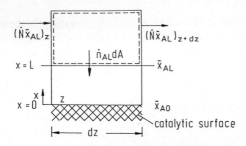

Fig. 2.60: For the material balance of an exhaust catalytic converter

$\tilde{M} = 32\,\mathrm{kg/kmol}$ and can be approximated as constant. The exhaust gases are at 480 °C and 0.12 MPa and contain NO with a mole fraction of $\tilde{x}_{AL} = 10^{-3}$. 80 % of this has to be removed. The mass transfer coefficient $\beta = 0.1\,\mathrm{m/s}$ and the rate constant $k_1'' = 0.05\,\mathrm{m/s}$ are given. How large a catalyst surface is required?

The molar flow rate of the exhaust is

$$\dot{N} = \dot{M}/\tilde{M} = \frac{350\,\mathrm{kg/h}}{3600\,\mathrm{s/h} \cdot 32\,\mathrm{kg/kmol}} = 3.038 \cdot 10^{-3}\,\mathrm{kmol/s} \ .$$

The material balance around the dotted balance region, Fig. 2.60, yields

$$(\dot{N}\,\tilde{x}_{AL})_z = (\dot{N}\,\tilde{x}_{AL})_{z+dz} + \dot{n}_{AL}\,dA$$

or, because $\dot{N} = \mathrm{const}$

$$\dot{N}\,d\tilde{x}_{AL} = -\dot{n}_{AL}\,dA \ .$$

Here $\dot{n}_{AL} = \beta\,c\,(\tilde{x}_{AL} - \tilde{x}_{A0})$, if we presume a low mass flow rate normal to the wall and we neglect the Stefan correction factor for the mass transfer coefficients. With $c = N/V = p/R_mT$ we get

$$\dot{N}\,d\tilde{x}_{AL} = \frac{-\beta p}{R_m T}\,\tilde{x}_{AL}\left(1 - \frac{\tilde{x}_{A0}}{\tilde{x}_{AL}}\right)dA \ .$$

Putting in (2.359) yields, whilst accounting for $\beta = D/L$

$$\dot{N}\,d\tilde{x}_{AL} = \frac{-p}{R_m T}\,\tilde{x}_{AL}\,\frac{k_1''}{1 + k_1''/\beta}\,dA$$

or

$$\frac{1}{\tilde{x}_{AL}}d\tilde{x}_{AL} = \frac{-p}{\dot{N}R_m T}\,\frac{k_1''}{1 + k_1''/\beta}\,dA \ .$$

Integration between the inlet cross section i and the exit cross section e of the catalytic converter of area A yields

$$\ln\frac{(\tilde{x}_{AL})_e}{(\tilde{x}_{AL})_i} = -\frac{p}{\dot{N}R_m T}\,\frac{k_1''}{1 + k_1''/\beta}\,A \ ,$$

and therefore

$$A = \frac{\dot{N}\,R_m T(1 + k_1''/\beta)}{p\,k_1''}\,\ln\frac{(\tilde{x}_{AL})_i}{(\tilde{x}_{AL})_e} \ .$$

This then becomes

$$A = \frac{3.038\,\mathrm{mol/s} \cdot 8.31451\,\mathrm{Nm/molK} \cdot 753.15\,\mathrm{K}(1 + 0.05\,\mathrm{m/s}/0.1\,\mathrm{m/s})}{0.12 \cdot 10^6\,\mathrm{N/m^2} \cdot 0.05\,\mathrm{m/s}}\,\ln(10^{-3}/2 \cdot 10^{-4})$$

$$A = 7.65\,\mathrm{m^2} \ .$$

2.5.6   Steady-state mass diffusion with homogeneous chemical reaction

We will consider the case of a substance A which diffuses into a porous or paste type solid or a quiescent fluid B, as shown in Fig. 2.61, where it reacts chemically with other reaction partners. A good example of this is the biological treatment of waste water. Here oxygen diffuses out of air or oxygen bubbles into the waste water that surrounds them, where it is converted along with the organic pollutants, e.g. hydrocarbons, into carbon dioxide and water by microorganisms. Substance B can also be a catalyst. Catalysts frequently consist of spherical or cylindrical pellets with have fine capillaries runing through them. The porous internal surface area is determined by the fine capillaries and is many times greater than the outer surface area. Chemical conversion can therefore be accelerated both at the outer and inner surfaces.

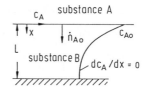

Fig. 2.61: Unidirectional mass diffusion with homogeneous reaction

According to (2.338), for steady-state, geometric one-dimensional diffusion with chemical reaction we have

$$\frac{\mathrm{d}}{\mathrm{d}x}\left(c\,D\,\frac{\mathrm{d}\tilde{x}_A}{\mathrm{d}x}\right) + \dot{\gamma}_A = 0 \ . \tag{2.366}$$

We will now assume that substance A and the products created by the chemical reaction are only present in very small amounts in substance B. So, this is approximately $c = N/V = $ const. In addition we will set $D = $ const.

If substance B is a porous solid the molecular diffusion coefficient D has to be replaced by the effective diffusion coefficient D_{eff}. This is smaller than the molecular diffusion coefficient because the movement of the molecules is impeded by the pores. It is common practice to define a *diffusion resistance factor*

$$\mu := \frac{D}{D_{\mathrm{eff}}} \ .$$

A porous solid of volume V consists of the volume V_S of the solid material and $V_G = V - V_S$ the volume of the solid-free spaces. We call $\varepsilon_p := V_G/V$ the void fraction. The resistance factor depends on the voidage of the porous body and on a *diversion* or *winding factor* μ_p such that the relationships

$$\mu = \mu_p/\varepsilon_p \ ,$$

and

$$D_{\text{eff}} = \frac{1}{\mu} D = \frac{\varepsilon_p}{\mu_p} D \ .$$

hold.

The following table contains several values for the void fraction and the diversion factor.

Table 2.13: Void fraction ε_p and diversion factor μ_p of some dry materials

Material	Density ϱ kg/m^3	Void fraction ε_p	Diversion factor μ_p
Wall bricks	1360	0.49	3.3 to 3.4
Clinker bricks	2050	0.19	73.0 to 89.0
Lime sandstone	900	0.63	5.4 to 5.9
Pumice concrete	840	0.62	4.3 to 5.5
Wood fibre	300	0.81	2.0 to 2.6
Molecular sieve 4 Å	1100	0.45	10.3
Molecular sieve 10 Å	1180	0.57	8.0
Silica gel	1090	0.46	3.6
Activated carbon	760	0.56	7.0

Substance A is converted in a first order reaction

$$\dot{\gamma}_A = -k_1 c_A \ . \tag{2.367}$$

Where k_1 is the rate constant of the chemical reaction (SI unit s^{-1}). The subscript index 1 indicates that it concerns a first order reaction. The reaction rate $\dot{\gamma}_A$ is negative, because substance A is consumed in the reaction. If A was generated, a positive sign would be present in (2.367). In porous solids the reaction is quicker the larger the inner and outer surface areas. The rate constant k_1 is therefore split into two factors

$$k_1 = a_p k_1' \ , \tag{2.368}$$

in which a_p is the specific area (SI units m^2/m^3). It is the area available for reaction divided by the volume of the porous solid. The rate constant k_1' has SI units of m/s.

With $\tilde{x}_A = c_A/c$, (2.366) and (2.367) are transformed, under the assumptions already given into

$$\frac{d^2 c_A}{dx^2} - \frac{k_1}{D} c_A = 0 \ . \tag{2.369}$$

This has to be solved under the boundary conditions shown in Fig. 2.61

$$c_A(x = 0) = c_{A0} \tag{2.370}$$

and

$$\left(\frac{d c_A}{dx}\right)_{x=L} = 0 \ . \tag{2.371}$$

D is replaced by D_{eff} for porous solids. Equation (2.369) agrees in terms of its form with (2.53) for heat conduction in a rod. In this case the temperature profile was calculated with the boundary conditions of constant temperature at the beginning of the rod and vanishing heat flow at the end of the rod. In the current problem these correspond to the boundary conditions (2.370) and (2.371). The solution of (2.369) to (2.371) therefore corresponds to the relationship (2.59) found earlier for heat conduction in rods. The solution is

$$\frac{c_A}{c_{A0}} = \frac{\cosh m(L - x)}{\cosh m L} \qquad (2.372)$$

with

$$m = \sqrt{k_1/D} \ . \qquad (2.373)$$

The amount of substance transferred is

$$\dot{N}_{A0} = A\,\dot{n}_{A0} = -A\,D\left(\frac{\mathrm{d}c_A}{\mathrm{d}x}\right)_{x=0} = A\,D\,c_{A0}\,m\,\tanh(mL) \ . \qquad (2.374)$$

The dimensionless quantity

$$m L = \sqrt{k_1 L^2/D} := Ha \qquad (2.375)$$

is known as the *Hatta*[6] number Ha. The square of Ha is equal to the ratio of the so-called relaxation time of the reaction

$$t_R = 1/k_1$$

and the relaxation time of diffusion

$$t_D := L^2/D \ ,$$

which is then

$$Ha^2 = t_D/t_R \ .$$

A large value for the Hatta number means that the chemical reaction is rapid in comparison with diffusion. Substance A cannot penetrate very far into substance B by diffusion, but will be converted by chemical reaction in a layer close to the surface.

We introduce another factor for porous bodies, it is called the *pore effectiveness factor* η_P. This is understood to be the ratio of the actual amount of substance transferred $\dot{N}_{A0}$ to the amount which would be transferred $\dot{N}_A$, if the concentration c_{A0} prevailed throughout the porous body. This basically means, if the effective diffusion coefficient D_{eff}, which is used in place of the molecular diffusion coefficient

---

[6]Shironji Hatta (1895–1973) was a professor at the Tohoku Imperial University, now the Tohoku University, in Tokyo, Japan. He undertook a number of fundamental studies about the absorption of gases in liquids, in particular absorption with simultaneous chemical reaction.

D, was very large $D_{\text{eff}} \to \infty$, then $Ha \to 0$. Then according to (2.367) the reaction rate of substance A

$$\dot{\gamma}_A = -k_1 \, c_{A0}$$

disappears and the total amount delivered would be

$$\dot{N}_A = k_1 \, c_{A0} \, A \, L \ . \tag{2.376}$$

The pore effectiveness factor would then be

$$\eta_P = \frac{\dot{N}_{A0}}{\dot{N}_A} = \frac{D_{\text{eff}}}{k_1 L} m \tanh(mL)$$

or

$$\eta_P = \frac{\tanh(mL)}{mL} \tag{2.377}$$

with $m = \sqrt{k_1/D_{\text{eff}}} = \sqrt{k_1' a_p/D_{\text{eff}}}$.

The pore effectivness factor is valid for pores with constant cross sectional area. It corresponds to the fin efficiency for a straight, rectangular fin, eq. (2.79). The pore effectiveness factor for pores of any cross sectional shape, can also be calculated, as a fairly good approximation using (2.377), as shown by Aris [2.81], if the length L is formed as a characteristic length

$$L = V/A \tag{2.378}$$

with the volume V of the porous body and its outer surface area A. A spherical pellet of radius R would, for example, have $L = (4/3)\pi R^3 / 4\pi R^2 = R/3$.

For small values $mL < 0.3$ in (2.377) we obtain $\eta_p > 0.97$, which lies close to 1. This implies that the composition of the reaction partner A hardly changes over the length of the pore. The resistance to diffusion is negligible in comparison to the other resistances. A small value of $mL = \sqrt{k_1/D_{\text{eff}}} \, L$ indicates a small pore, a slow reaction or rapid diffusion.

Example 2.12: In order to reduce the CO content in the waste gas from a furnace, the exhaust gas is passed over the porous CuO particles (pellets) of a catalytic reactor in the exhaust pipe. The CO (substance A) is oxidised with O_2 to CO_2 inside and at the surface of the pellets, according to the reaction

$$CO + \frac{1}{2}O_2 = CO_2 \ .$$

This can be approximately taken to be a first order reaction, so

$$\dot{n}_A = -k_1' \, a_P \, c_A \ .$$

a) Find the pore effectiveness factor. b) How many kg CuO are required if the mole fraction of CO has to be reduced to 1/10 of its initial value $\tilde{x}_A = 0.04$? The molar flow rate $\dot{N}$ of the exhaust gas can be assumed to be constant.
Given: Molar flow rate of the exhaust gas $\dot{N} = 3$ mol/s, mole fraction of CO in the exhaust gas $\tilde{x}_A = 0.04$, exhaust temperature 480 °C, pressure 0.12 MPa, diameter of the spherical pellets 5 mm, specific area $a_P = 5 \cdot 10^6$ m^2/m^3, effective diffusion coefficient $D_{\text{eff}} = 5 \cdot 10^{-5}$ m^2/s, reaction rate constant $k_1' = 10^{-3}$ m/s, density of CuO: $\varrho_{CuO} = 8.9 \cdot 10^3$ kg/m^3.

a) We have $mL = (k_1' a_P / D_{\text{eff}})^{1/2} L$ with $L = R/3$. Which gives

$$mL = (10^{-3} \text{m/s} \cdot 5 \cdot 10^6 \, \text{m}^2/\text{m}^3/5 \cdot 10^{-5} \, \text{m}^2/\text{s})^{1/2} \, (2.5 \cdot 10^{-3} \, \text{m}/3) = 8.33$$

and therefore

$$\eta_P = \tanh(mL)/(mL) = 0.12 \ .$$

b) The CO flowing outside the pores in the gas space is decomposed by a chemical reaction. Along an element dz the amount of CO changes by $d(\dot{N} \tilde{x}_{A0})$. This amount is converted by chemical reaction in the pores

$$d\dot{N}_{A0} = \dot{\gamma}_A \, dV_P = -k_1' \, a_P \, c_{A_0} \, \eta_P \, dV_P = -k_1' \, a_P \, c \tilde{x}_{A0} \, \eta_P \, dV_P \ .$$

V_P is the volume of the pellets. With that we get

$$d(\dot{N} \tilde{x}_{A0}) = -k_1' \, a_P \, c \, \tilde{x}_{A0} \, \eta_P \, dV_P \ .$$

From which, with $\dot{N} = \text{const}$ and $c = p/R_m T$ it follows that

$$\frac{d\tilde{x}_{A0}}{\tilde{x}_{A0}} = -\frac{k_1' a_P p \eta_P}{\dot{N} R_m T} dV_P \ .$$

After integration between mole fraction $\tilde{x}_{Ai}$ at the inlet and $\tilde{x}_{Ao}$ at the outlet of the reactor we obtain the volume V_P of the pellets

$$V_P = \frac{\dot{N} R_m T}{k_1' a_P p \, \eta_P} \ln \frac{\tilde{x}_{Ai}}{\tilde{x}_{Ao}} \ ,$$

$$V_P = \frac{3 \, \text{mol/s} \cdot 8.31451 \, \text{Nm/mol K} \cdot 753.15 \, \text{K}}{10^{-3} \, \text{m/s} \cdot 5 \cdot 10^6 \, \text{m}^2/\text{m}^3 \cdot 1.2 \cdot 10^5 \, \text{N/m}^2 \cdot 0.12} \ln 10 = 6.008 \cdot 10^{-4} \, \text{m}^3 \ .$$

Therefore we need

$$M = \varrho V_P = 8.9 \cdot 10^3 \, \text{kg/m}^3 \cdot 6.008 \cdot 10^{-4} \, \text{m}^3 = 5.35 \, \text{kg CuO} \ .$$

2.5.7 Transient mass diffusion

In section 2.5.3, page 229, it was shown that the differential equation for transient mass diffusion is of the same type as the heat conduction equation, a result of which is that many mass diffusion problems can be traced back to the corresponding heat conduction problem. We wish to discuss this in detail for transient diffusion in a semi-infinite solid and in the simple bodies like plates, cylinders and spheres.

2.5.7.1 Transient mass diffusion in a semi-infinite solid

We will consider transient diffusion of a substance in a semi-infinite body B. At time $t = 0$, substance A is stored in the body at a concentration $c_{A\alpha}$. The desired concentration profile $c_A = c_A(x,t)$ satisfies the following differential equation, under the assumption $cD = \text{const}$

$$\frac{\partial c_A}{\partial t} = D\frac{\partial^2 c_A}{\partial x^2}, \quad t \geq 0, \quad x \geq 0 \qquad (2.379)$$

and should fulfill the initial condition

$$c_A(t = 0, x) = c_{A\alpha} = \text{const} .$$

In analogy to the heat conduction problem, the following conditions are possible at the surface of the body:
- a stepwise change in the surface concentration to the value c_{A0}, which should remain constant for $t > 0$,
- input of a constant molar flux $\dot{n}_{A0}$,
- a stepwise change in the concentration in the surroundings to $c_{AU} \neq c_{A0}$, so that mass transfer takes place with a mass transfer coefficient β.

As we already know, the solutions for the heat conduction problems in section 2.3.3 can be transferred to the mass diffusion problem, due to the similarity of the differential equations, initial and boundary conditions. The corresponding quantities to heat conduction for mass diffusion are shown in the following table.

Table 2.14: Corresponding relationships between quantities in heat conduction and diffusion

Heat conduction	Diffusion
ϑ	c_A
a	D
$t^+ = at/L^2$	$t_D^+ = Dt/L^2$
$Bi = \alpha L/\lambda$	$Bi_D = \beta L/D$
$x/2\sqrt{at}$	$x/2\sqrt{Dt}$
$b = \sqrt{\lambda c\varrho}$	$b_D = \sqrt{D}$

Using these correspondences allows us to write up the solutions to the diffusion problem from the solutions already discussed for heat conduction.

In transient diffusion in a semi-infinite solid with stepwise change in the surface concentration, we find from (2.126)

$$\frac{c_A - c_{A\alpha}}{c_{A0} - c_{A\alpha}} = \text{erfc}\frac{x}{2\sqrt{Dt}} \qquad (2.380)$$

and with (2.131) the transferred molar flux

$$\dot{n}_A(t, x) = \frac{\sqrt{D}}{\sqrt{\pi t}}(c_{A0} - c_{A\alpha})\exp\left(-\frac{x^2}{4Dt}\right) . \qquad (2.381)$$

The molar flux transferred at the surface $x = 0$ is

$$\dot{n}_{\text{A}}(t, x = 0) = \frac{\sqrt{D}}{\sqrt{\pi t}}(c_{\text{A0}} - c_{\text{A}\alpha}) \ . \tag{2.382}$$

Corresponding solutions are found from (2.135), when a constant molar flux $\dot{n}_{\text{A0}}$ is fed in at the surface. In the same way the solution for mass transfer from the surface to another fluid is obtained from (2.140).

2.5.7.2 Transient mass diffusion in bodies of simple geometry with one-dimensional mass flow

The time dependent concentration field $c_{\text{A}}(x, t)$ in a body is determined by the diffusion equation which corresponds to the heat conduction equation (2.157). With $cD = \text{const}$ this equation is as follows

$$\frac{\partial c_{\text{A}}}{\partial t} = D\left(\frac{\partial^2 c_{\text{A}}}{\partial r^2} + \frac{n}{r}\frac{\partial c_{\text{A}}}{\partial r}\right) \tag{2.383}$$

with $n = 0$ for a plate, $n = 1$ for a cylinder and $n = 2$ for a sphere, when geometric one-dimensional flow in the r-direction is assumed. As in heat conduction, r is the radial coordinate for cylinders and spheres. The cylinder should be very long in comparison to its diameter. The concentrations c_{A} in the cylinder and sphere cannot be dependent on the angular coordinate.

With a plate the x-coordinate will, for now, be indicated by r. It goes from the centre of the plate outwards. With the initial condition

$$t = 0 : \quad c_{\text{A}}(r) = c_{\text{A}\alpha}, \quad 0 \leq r \leq R$$

and the boundary conditions

$$r = 0 : \quad \partial c_{\text{A}}/\partial r = 0$$

$$r = R : \quad -D\partial c_{\text{A}}/\partial r = \beta(c_{\text{A}} - c_{\text{AS}})$$

we obtain the results by applying the basics of the solution to the corresponding heat conduction problem. We put in

$$c_{\text{A}}^+ = \frac{c_{\text{A}} - c_{\text{AS}}}{c_{\text{A}\alpha} - c_{\text{AS}}} \ .$$

For the *flat plate*, from (2.171) and (2.172), as well as with Bi_{D} and t_{D}^+ according to Tab. 2.14 and $r^+ = r/R$, it follows that:

$$c_{\text{A}}^+(r^+, t_{\text{D}}^+) = \sum_{i=1}^{\infty} \frac{2\,Bi_{\text{D}}}{Bi_{\text{D}}^2 + Bi_{\text{D}} + \mu_i^2} \frac{1}{\cos \mu_i} \cos(\mu_i r^+) \exp(-\mu_i^2 t_{\text{D}}^+) \tag{2.384}$$

with the eigenvalues μ_i from

$$\tan \mu = Bi_{\text{D}}/\mu \ . \tag{2.385}$$

The average concentration c_{Am}^+ follows from (2.174)

$$c_{Am}^+(t_D^+) = 2Bi_D^2 \sum_{i=1}^{\infty} \frac{\exp(-\mu_{iD}^2 t_D^+)}{\mu_i^2(Bi_D^2 + Bi_D + \mu_i^2)} \ . \tag{2.386}$$

Analogous results are yielded with the equations from section 2.3.4.4, page 168, for the cylinder and the sphere.

Example 2.13: What is the equation for the average concentration $c_{Am}^+(t_D^+)$ of a sphere, if the concentration jumps from its initial value $c_A(r, t = 0) = c_{A\alpha}$ to a value of $c_A(r = R, t) = c_{A0}$?

The jump in the surface concentration at $r = R$ is only possible if the resistance to mass trasnfer $1/\beta A$ between the surface area A and the surroundings is small enough to be neglected. This means that $\beta \to \infty$, giving $Bi_D = \beta R/D_{AB} \to \infty$ and $c_{A0} = c_{AS}$, and also $c_{Am}^+ = (c_{Am} - c_{A0})/(c_{A\alpha} - c_{A0})$. For a *sphere* we obtain, from (2.183) and (2.185)

$$c_A^+(r^+, t_D^+) = \sum_{i=1}^{\infty} 2\,Bi_D \frac{\mu_i^2 + (Bi_D - 1)^2}{\mu_i^2 + Bi_D(Bi_D - 1)} \frac{\sin \mu_i}{\mu_i} \frac{\sin(\mu_i r^+)}{\mu_i r^+}\exp(-\mu_i^2 t_D^+) \tag{2.387}$$

with the eigenvalues μ_i from

$$\mu \cot \mu = 1 - Bi_D \ . \tag{2.388}$$

The average concentration is, in accordance with (2.184),

$$c_{Am}^+(t_D^+) = 6\,Bi_D^2 \sum_{i=1}^{\infty} \frac{\exp(-\mu_i^2 t_D^+)}{\mu_i^2[\mu_i^2 + Bi_D(Bi_D - 1)]} \ . \tag{2.389}$$

For $Bi_D \to \infty$ we obtain, from (2.388), the eigenvalues $\mu_1 = \pi$, $\mu_2 = 2\pi$, $\mu_3 = 3\pi \ldots$ and as the average concentration from (2.389)

$$c_{Am}^+(t_D^+) = 6 \sum_{i=1}^{\infty} \frac{\exp(-\mu_i^2 t_D^+)}{\mu_i^2} \ .$$

With the eigenvalues $\mu_i = i\pi$, $i = 1, 2, 3 \ldots$ we can also write

$$c_{Am}^+(t_D^+) = \frac{6}{\pi^2} \sum_{i=1}^{\infty} \frac{\exp(-i^2\pi^2 t_D^+)}{i^2} \ . \tag{2.390}$$

2.6   Exercises

2.1: Derive the differential equation for the temperature field $\vartheta = \vartheta(r, t)$, that appears in a cylinder in transient, geometric one-dimensional heat conduction in the radial direction. Start with the energy balance for a hollow cylinder of internal radius r and thickness Δr and execute this to the limit $\Delta r \to 0$. The material properties λ and c depend on ϑ; internal heat sources are not present.

2.2: One surface ($x = 0$) of a cooled plate, which has thickness δ, is insulated, whilst at the other surface heat is transferred to a fluid at a temperature of ϑ_F. Sketch the temperature profile $\vartheta = \vartheta(x, t^*)$ in the plate for a fixed time t^*. Which conditions must the curve satisfy close to the two surfaces $x = 0$ und $x = \delta$, if the Biot number is $Bi = \alpha\delta/\lambda = 1.5$? In addition sketch the temperature profile of the fluid temperature in the boundary layer on the surface of the plate, taking account of the condition $Nu = \alpha\delta/\lambda_F = 10$; λ_F is the thermal conductivity of the fluid.

2.3: The temperature profile in a steel plate of thickness $\delta = 60\,\text{mm}$ and constant thermal diffusivity $a = 12.6 \cdot 10^{-6}\,\text{m}^2/\text{s}$, at a fixed time t_0 is given by

$$\vartheta^+ := \frac{\vartheta - \vartheta_1}{\vartheta_2 - \vartheta_1} = x^+ - B(t_0) \cos\left(\pi(x^+ - \frac{1}{2})\right) \quad , \qquad 0 \le x^+ \le 1 \ ,$$

with $B(t_0) = 0.850$. Here $x^+ = x/\delta$; $\vartheta_1 = 100\,°\text{C}$ and $\vartheta_2 = 250\,°\text{C}$ are the constant surface temperatures of the plate at $x^+ = 0$ and $x^+ = 1$.

a) Draw the temperature distribution $\vartheta = \vartheta(x, t_0)$. Is the plate heated or cooled?

b) At which position x_T^+ does the temperature change the fastest with time? How big is $\partial\vartheta/\partial t$ there?

c) Does x_T^+ agree with the position $x_{\min}^+$ at which the temperature profile at time t_0 has its minimum?

d) Determine the time function $B(t)$ whilst accounting for the initial condition $B(t_0) = 0.850$. Which steady temperature pattern is yielded for $t \to \infty$?

2.4: Heat is released due to a reaction in a very long cylinder of radius R. The thermal power density increases with distance r from the cylinder axis:

$$\dot{W}(r) = \dot{W}_R \, (r/R)^m \quad , \qquad m \ge 0 \ .$$

a) How big is the heat flux $\dot{q}(R)$ released from the cylinder? How large does $\dot{W}_R$ have to be so that $\dot{q}(R)$ agrees with the heat flux released from a cylinder of the same dimensions with a spatially constant thermal power density $\dot{W}_0$?

b) For a cylinder with constant λ calculate the overtemperature $\Theta(r) = \vartheta(r) - \vartheta(R)$ using the formulation

$$\Theta(r) = A\left[1 - (r/R)^k\right] \ ,$$

which satisfies the boundary conditions

$$r = 0: \ d\Theta/dr = 0 \quad \text{and} \quad r = R: \ \Theta = 0 \ .$$

c) Compare the maximum overtemperature $\Theta_{\max}$ with the maximum overtemperature $\Theta_{\max}^0$ of a cylinder with constant thermal power density $\dot{W}_0$. How big is the ratio $\Theta_{\max}/\Theta_{\max}^0$, if both cylinders release the same amount of heat? Calculate $\Theta_{\max}/\Theta_{\max}^0$ for $m = 0, 1, 2$ and 3.

2.5: A steel bolt ($d = 20\,\text{mm}$, $\lambda = 52.0\,\text{W/K m}$) protrudes from an insulation layer, cf. Fig. 2.62. Its left hand end is kept at the constant temperature $\vartheta^* = 75.0\,°\text{C}$. The bolt is completely insulated over the length $L_{\text{is}} = 100\,\text{mm}$. The piece sticking out, $L = 200\,\text{mm}$, releases heat to the surroundings, which are at $\vartheta_S = 15.0\,°\text{C}$, where the heat transfer coefficient is $\alpha = 8.85\,\text{W/m}^2\text{K}$ at the outside of the bolt and the free head surface.

a) Find the temperatures ϑ_0 and ϑ_L.

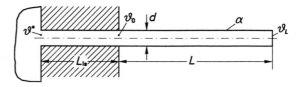

Fig. 2.62: Steel bolt sticking out of an insulation layer

b) How large is the heat flow released by the bolt to its surroundings? Determine the heat flow $\dot{Q}_L$ out of the free head end.

c) Compare the results obtained from the temperature profile according to (2.55), with the values yielded for a replacement bolt of length L_C with an adiabatic head end, according to (2.60).

2.6: The efficiency η_f of a straight fin with a rectangular cross section is determined by measuring three temperatures: the temperature ϑ_0 at the fin base, the temperature ϑ_h at the top of the fin and the temperature ϑ_S of the fluid which surrounds the fin. Calculate η_f for $\vartheta_0 = 75.0\ ^\circ\text{C}$, $\vartheta_h = 40.5\ ^\circ\text{C}$ and $\vartheta_S = 15.0\ ^\circ\text{C}$.

2.7: A brass tube with outer diameter $d = 25\,\text{mm}$ is exposed to an air stream flowing across it perpendicular to its axis, the heat transfer coefficient is $\alpha = 90\,\text{W/m}^2\text{K}$. The tube has annular fins made of brass ($\lambda_f = 126\,\text{W/K m}$) attached to it, which have a thickness of $\delta_f = 1.5\,\text{mm}$ and are $h = 20\,\text{mm}$ high. The surface temperature of the tube and the air temperature can be taken to be constant.

a) How many fins per metre must be fixed to the tube for the heat flow released to the air to be six times larger? $\alpha_f = \alpha$ can be used.

b) The height of the fins is increased to $h = 30\,\text{mm}$. By what factor does the heat flow increase in comparison to the finned tube in a)?

2.8: A tube of diameter $d = 0.30\,\text{m}$ and surface temperature $\vartheta_R = 60\ ^\circ\text{C}$ is placed in the ground ($\lambda = 1.20\,\text{W/K m}$) such that its axis lies $0.80\,\text{m}$ below the surface. The air temperature is $\vartheta_A = 10\ ^\circ\text{C}$; the heat transfer coefficient between the air and the earth's surface is $\alpha = 8.5\,\text{W/m}^2\text{K}$. How large is the heat loss $\dot{Q}/L$ per length L of the tube?

2.9: An asphalt road coating ($\lambda = 0.65\,\text{W/K m}$, $\varrho = 2120\,\text{kg/m}^3$, $c = 920\,\text{J/kg K}$) has reached the temperature $\vartheta_0 = 55\ ^\circ\text{C}$, not only at its surface but in a layer several centimetres below the surface, due to a long period of sunshine. A sudden rainstorm reduces the surface temperature over 10 min to $\vartheta_S = 22\ ^\circ\text{C}$. Find

a) the heat released per unit area by the asphalt during the rainstorm, and

b) the temperature at a depth of $3.0\,\text{cm}$ at the end of the rain shower.

2.10: A very thick concrete wall ($\lambda = 0.80\,\text{W/K m}$, $\varrho = 1950\,\text{kg/m}^3$, $c = 880\,\text{J/kg K}$) initially has as temperature throughout the wall of $\vartheta_0 = 20\ ^\circ\text{C}$. It is heated at its surface with a constant heat flux $\dot{q}_0 = 650\,\text{W/m}^2$ (sunshine). What temperature has the surface reached after $2.0\,\text{h}$? How high is the temperature at this time at a depth of $10\,\text{cm}$?

2.11: Solve exercise 2.10 with the additional condition that the concrete wall releases heat to air at a temperature $\vartheta_S = \vartheta_0 = 20\ ^\circ\text{C}$, whereby the heat transfer coefficient $\alpha = 15.0\,\text{W/m}^2\text{K}$ is decisive.

2.12: The penetration of the daily and seasonal temperature fluctuations in the ground ($a = 0.35 \cdot 10^{-6}\,\mathrm{m^2/s}$) is to be investigated. For this it is assumed that the surface temperature follows an harmonic oscillation

$$\vartheta_0(t) = 10.0\,\mathrm{°C} + \Delta\vartheta\cos(2\pi\,t/t_0)$$

with periodic time t_0.

a) Show that the amplitude of the daily temperature fluctutations ($\Delta\vartheta = 10\,\mathrm{°C}$, $t_0 = 24\,\mathrm{h}$) is already small enough to be neglected at a depth of 1 m.

b) Find, for the seasonal temperature fluctuation ($\Delta\vartheta = 25\,\mathrm{°C}$, $t_0 = 365\,\mathrm{d}$), the temperature profile at a depth of 2 m. What are the highest and lowest temperatures at this depth, and on which day of the year do they appear, in each case using the assumption that the maximum surface temperature is achieved on 1st August.

2.13: A flat wall of firestone has thickness $\delta = 0.325\,\mathrm{m}$ and material properties $\lambda = 1.15\,\mathrm{W/K\,m}$, $a = 0.613 \cdot 10^{-6}\,\mathrm{m^2/s}$. One of its surfaces is insulated, whilst heat is transferred to the surroundings ($\vartheta_S = 20.0\,\mathrm{°C}$) at the other. The wall has an initial temperature $\vartheta_0 = 180.0\,\mathrm{°C}$. At time t_1 the temperature of the non-insulated surface is $\vartheta_{W1} = 50.0\,\mathrm{°C}$; at time $t_2 = t_1 + 8.50\,\mathrm{h}$ the surface temperature has fallen to $\vartheta_{W2} = 41.9\,\mathrm{°C}$.

a) What value has the heat transfer coefficient α decisive for the cooling?

b) Find the temperatures of the insulated surface at times t_1 and t_2.

2.14: A sphere made of plastic ($\lambda = 0.35\,\mathrm{W/K\,m}$, $c = 2300\,\mathrm{J/kg\,K}$, $\varrho = 950\,\mathrm{kg/m^3}$) of diameter $d = 20\,\mathrm{mm}$, which was heated to $\vartheta_0 = 110.0\,\mathrm{°C}$, is cooled in air at $\vartheta_S = 15.0\,\mathrm{°C}$; the heat transfer coefficient is $\alpha = 8.75\,\mathrm{W/m^2K}$.

a) What temperature has the sphere reached after 20 min? Use the approximation equation for small Biot numbers, cf. section 2.3.5.2.

b) Compare the result from a) with the exact solution by calculating the mean temperature ϑ_m and the temperatures at the surface and the centre of the sphere.

2.15: A very long steel tape with rectangular cross section ($h = 15\,\mathrm{mm}$, $b = 50\,\mathrm{mm}$) and thermal diffusivity $a = 3.8 \cdot 10^{-6}\,\mathrm{m^2/s}$ is pulled through an oil bath of length $L = 4.0\,\mathrm{m}$. The oil bath is at the constant temperature $\vartheta_F = 50\,\mathrm{°C}$; the steel tape has a temperature of $\vartheta_0 = 375\,\mathrm{°C}$ as it enters the oil bath. The heat transfer coefficient between the steel tape and the oil is so large that the surface of the steel tape assumes the temperature ϑ_F. With what speed w does the steel tape have to pulled through the oil for it to leave the oil bath with a temperature at its central axis of $\vartheta_k = 65\,\mathrm{°C}$? The axial heat conduction may be neglected.

2.16: A cast-iron pipe ($\lambda = 51.0\,\mathrm{W/K\,m}$) with internal diameter $d_i = 40\,\mathrm{mm}$ and wall thickness $\delta = 7.5\,\mathrm{mm}$ is filled with water which is at a temperature of $\vartheta_E = 0.0\,\mathrm{°C}$. The external air is at $\vartheta_0 = -8.0\,\mathrm{°C}$; the heat transfer coefficient between the pipe and air is $\alpha = 25\,\mathrm{W/m^2K}$. How long does it take for an ice layer of thickness $s = 15\,\mathrm{mm}$ to form in the pipe, such that the increase in the volume of the ice compared to the water can lead to a burst pipe? How long will it take before the contents of the pipe are completely frozen? Properties of ice at 0 °C: $\varrho = 917\,\mathrm{kg/m^3}$, $\lambda = 2.25\,\mathrm{W/K\,m}$, $h_E = 333\,\mathrm{kJ/kg}$.

2.17: A very thin platinum wire of length L is mounted into the centre of a very large, normally cylindrical sample of the material whose thermal conductivity is to be measured. Good contact between the wire and the material surrounding it can be assumed. Until time $t = 0$ this arrangement is at a constant temperature ϑ_0. The wire is then heated electrically with a constant power $\dot{Q}_0$. The temperatures ϑ_1 and ϑ_2 of the platinum wire are determined, (for example by the change in its resistance), at times t_1 and t_2. According to [2.82], pg. 442, the thermal conductivity of the material surrounding the wire is obtained from

$$\lambda = \frac{\dot{Q}_0/L}{4\pi(\vartheta_2 - \vartheta_1)} \ln \frac{t_2}{t_1} \quad .$$

Derive this relationship from eq. (2.235) from section 2.3.7.2.

2.18: A very thick wall with constant thermal diffusivity a and a constant initial temperature ϑ_0 is heated at its surface. The temperature rises there, between $t = 0$ und $t = t^*$ linearly with time t, to the value $\vartheta_1 > \vartheta_0$, which remains constant for $t > t^*$. The temperature profile in the wall at times $t = t^*$ and $t = 2\,t^*$ is to be calculated numerically. The simple explicit difference method is to be used, with $\Delta t = t^*/6$ and $M = 1/3$, and the normalised temperature $\vartheta^+ = (\vartheta - \vartheta_0)/(\vartheta_1 - \vartheta_0)$ is to be used. Compare the numerically calculated values with the exact solution

$$\vartheta^+(x,t) = \begin{cases} \dfrac{t}{t^*} F\left(\dfrac{x}{2\sqrt{at}}\right) & \text{for} \quad 0 \leq t \leq t^* \ , \\[3mm] \dfrac{t}{t^*} F\left(\dfrac{x}{2\sqrt{at}}\right) - \dfrac{t - t^*}{t^*} F\left(\dfrac{x}{2\sqrt{a(t - t^*)}}\right) & \text{for} \quad t \geq t^* \ , \end{cases}$$

where

$$F(\xi) = (1 + 2\,\xi^2)\,\mathrm{erfc}\,\xi - \frac{2}{\sqrt{\pi}}\,\xi\exp(-\xi^2) \quad .$$

2.19: A long hollow plastic cylinder ($\lambda = 0.23\,\mathrm{W/K\,m}$, $c = 1045\,\mathrm{J/kg\,K}$, $\varrho = 2200\,\mathrm{kg/m^3}$) with diameters $d_\mathrm{i} = 60\,\mathrm{mm}$ and $d_\mathrm{a} = 100\,\mathrm{mm}$ initially has a temperature of $\vartheta_0 = 20\,°\mathrm{C}$, which is the same as that of the surroundings (air). At time $t = 0$ the hollow cylinder is filled with a hot liquid and is heated such that the inner wall assumes a temperature of $\vartheta_\mathrm{i} = 80\,°\mathrm{C}$ for $t \geq 0$. At the outer wall, heat is transferred to the air with a heat transfer coefficent of $\alpha = 12.0\,\mathrm{W/m^2K}$.

a) Using the explicit difference method with $\Delta r = 4.0\,\mathrm{mm}$, show that the step size $\Delta t = 60\,\mathrm{s}$ is permissible and go on to calculate the temperature profile at time $t^* = 15\,\mathrm{min}$.

b) How large is the difference between the steady temperature profile and the temperatures calculated for time t^*?

2.20: The efficiency of a square sheet fin which sits on a tube of radius r_0 is to be determined approximately using a difference method. The fin has side length $s = 4\,r_0$, a constant thickness δ_f and thermal conductivity λ_f. The heat transfer coefficient α_f is constant over the surface area of the fin. The temperature at the fin base is ϑ_0, the ambient temperature is ϑ_S. The difference method for the calculation of the dimensionless temperature field $\vartheta^+ = (\vartheta - \vartheta_\mathrm{S})/(\vartheta_0 - \vartheta_\mathrm{S})$ will be based on the square grid with $\Delta x = 0.40\,r_0$, illustrated in Fig. 2.63. Due to symmetry it is sufficient to determine the 12 temperatures ϑ_1^+ to ϑ_{12}^+.

a) Set up the linear equations for the temperatures ϑ_1^+ to ϑ_{12}^+, using as an abbreviation the quantity $m = \sqrt{2\alpha_\mathrm{f}/\lambda_\mathrm{f}\delta_\mathrm{f}}$ according to (2.71). Note that the grid points 5, 9 and 12 are only associated with the area $\Delta x^2/2$. In addition points 10 and 11 have another part of the area Δx^2 associated with heat transfer that has to be calculated.

b) Find, for $m\Delta x = 0.40$, ϑ_1^+ to ϑ_{12}^+ and determine an approximate value for η_f. Compare this value with the result yielded from the approximation equations (2.82) and (2.83).

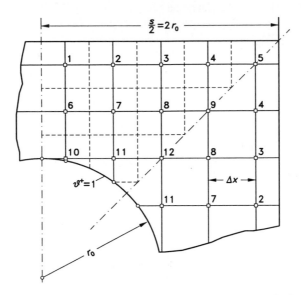

Fig. 2.63: Square sheet fin on a circular pipe, with square grid of mesh size $\Delta x = 0.40 \, r_0$. The dotted lines bound the areas associated with the temperatures ϑ_i^+.

2.21: A 75 mm thick wooden plate contains 2.8 % by mass water, relative to dry wood, which corresponds to a mass fraction of $\xi_{A_\alpha} = 0.218$. At equilibrium with the environment the mass fraction is reduced to $\xi_{A_S} = 0.065$.
How long does it take for the mass fraction of water in the middle of the plate to fall to $\xi_A(x = 0) = 0.08$? The diffusion coefficient of water in wood is $D = 1.2 \cdot 10^{-9} \, \text{m}^2/\text{s}$.

2.22: Instead of the 75 mm thick wooden plate in the previous exercise, a very long, square wooden trunk, with a side length of 75 mm is to be dried. The mass fraction of water and the diffusion coefficient are the same as in the previous question: $\xi_{A_\alpha} = 0.218, \xi_{A_S} = 0.065, D = 1.2 \cdot 10^{-9} \, \text{m}^2/\text{s}$. How long will it take for the mass fraction of water in the core of the trunk to fall to $\xi_A(x = 0, y = 0) = 0.08$?

2.23: In a spray absorber, a waste gas containing ammonia will be purified with water. The waste gas and water droplets flow countercurrent to each other as shown in Fig. 2.64. The saturated concentration of ammonia (substance A) for a temperature of 10 °C of the spray water and a pressure of 0.11 MPa appears at the surface of the falling water droplets. The mass fraction of ammonia at saturation is $\xi_{A_o} = 0.4$. The following mass flow rates and mass fractions are given: water feed $\dot{M}_W = 2.3 \, \text{kg/s}$, waste gas feed $\dot{M}_{Gi} = 4.4 \, \text{kg/s}$, mass fraction of ammonia in the waste gas feed $\xi_{Ai}^G = 0.12$; this should be reduced to a value of $\xi_{Ao}^G = 0.006$ in the waste gas exit stream; the mass fraction of ammonia in the water outlet is allowed to reach $\xi_{Ao}^L = 0.24$. The fraction of water in the waste gas is negligible.

a) What are the mass flow rates $\dot{M}_{Go}$ of the waste gas, $\dot{M}_{Lo}$ of the waste water and what mass flow rate of water $\dot{M}_W$ has to be fed into the absorber?

b) After what period of time has the average mass fraction of NH_3 in the water outlet stream increased to $\xi_{Ao}^L = 0.24$? The diameter of the water droplets is 3 mm and the diffusion coefficient of NH_3 in H_2O is $D = 1.5 \cdot 10^{-9} \, \text{m}^2/\text{s}$.

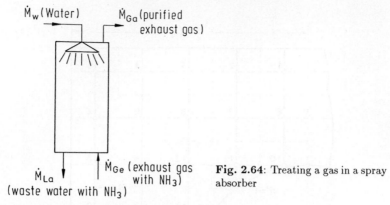

$\dot{M}_w$ (Water) $\dot{M}_{Ga}$ (purified
                      exhaust gas)

$\dot{M}_{La}$ $\dot{M}_{Ge}$ (exhaust gas   **Fig. 2.64**: Treating a gas in a spray
(waste water with NH$_3$)   with NH$_3$)   absorber

c) How high is the spray absorber if the water droplets are to fall through the rising waste gas
   with a velocity of $w = 0.1$ m/s?

2.24: Dry polystyrene balls of diameter 60 mm are sprayed with water for 1 h. How much
water does each ball take up if the diffusion coefficient for water (substance A) in polystyrene
(substance B) is $D = 1.8 \cdot 10^{-8}$ m^2/s?

3 Convective heat and mass transfer. Single phase flow

In the first chapter the heat transfer coefficient was defined by

$$\dot{q} = \alpha \Delta \vartheta$$

and the mass transfer coefficient for a substance A by

$$\dot{n}_{A} = \beta \Delta c_{A}$$

This mass transfer coefficient was valid for vanishing convective flows and had to be corrected for finite convective flows. These equations describe convective heat and mass transfer, but they are nothing more than definition equations for the heat transfer coefficient α and the mass transfer coefficient β, and should in no way be seen as laws for heat and mass transfer. Rather the laws of nature for the course of processes of heat and mass transfer are hidden in the heat and mass transfer coefficients. Generally both are not constant, but locally variable and in transient processes they also change with time. In addition they depend on the flow, the properties of the fluid and the geometry of the heat or mass transfer surfaces. Therefore, the definition equations for heat and mass transfer coefficients given above are not suitable for describing the mechanisms of heat and mass transfer. This is only possible with an in depth study of flow and will be the subject of the following exposition.

Basically we differentiate between forced and free flow. A forced flow is produced by external forces, for example when a flow is caused by a pump or blower. A free flow, on the other hand, is due to changes in density which are caused by temperature, pressure and concentration fields. We will discuss heat and mass transfer in forced flow first, moving onto free flow at the end of the section.

3.1 Preliminary remarks: Longitudinal, frictionless flow over a flat plate

In order to discover how heat and mass transfer coefficients change according to the flow we will first consider the longitudinal flow over a flat plate. We assume that the flow over the plate is at a constant average velocity w_m. Synonymous with that is the assumption that the viscosity of the fluid is vanishingly small. Real flows adhere to the wall such that the flow velocity increases asymptotically from zero at the wall to the velocity at the centre of the flow. A fluid with negligible viscosity will not adhere to the wall. These types of fluids do not exist in practice, but their introduction enables us to calculate crude local temperature and concentration fields and from them the local heat and mass trasnfer coefficients using the methods we already know. Through this the understanding of the general considerations with the decisive partial differential equations, which follow later, will be made easier.

A fluid at temperature ϑ_α flows along a flat plate with a constant average velocity w_m, Fig. 3.1. The surface temperature of the plate is constant and equal to ϑ_0. If the plate is hotter than the fluid $\vartheta_0 > \vartheta_\alpha$ a temperature profile develops, as shown in the left hand diagram in Fig. 3.1a. If however the plate is colder than the fluid, $\vartheta_0 < \vartheta_\alpha$, the temperature profile shown on the right in Fig. 3.1a, is formed. The temperature changes are limited to a thin layer close to the wall which increases along the flow, because the temperature variation always propagates into the fluid. The layer close to the wall of thickness $\delta_T(x)$, in which this temperature variation exists, is known as the *thermal boundary layer*, Fig. 3.1b. It stretches asymptotically into space, so a definition of its thickness has to be ascertained. For example, the boundary could be at the point where the temperature only very slightly deviates, by say 1 %, from the temperature of the core fluid.

In the thermal boundary the heat fed into a fluid element by conduction is stored as internal energy and leaves with the fluid element. This is $\vartheta = \vartheta(x, y)$. Outside the thermal boundary layer a constant temperature ϑ_α prevails in the outer flow. An observer moving with a fluid element, at a distance y from the wall finds himself, after time t at the position $x = w_m t$. For him the temperature $\vartheta(x, y) = \vartheta(w_m t, y)$ is only a function of time and the distance from the wall y, as the velocity w_m should be constant. The temperature $\vartheta(t, y)$ of the observer can therefore be described by Fourier's equation (2.14) for transient heat conduction,

$$\frac{\partial \vartheta}{\partial t} = a \frac{\partial^2 \vartheta}{\partial y^2} \ , \tag{3.1}$$

if we assume the thermal conductivity λ to be independent of temperature. At position $x = 0$ or at time $t = 0$ the initial temperature ϑ_α prevails. The temperature at the wall ϑ_0 is given. The boundary conditions are

$$\vartheta(t = 0, y) = \vartheta_\alpha \tag{3.2}$$

$$\vartheta(t, y = 0) = \vartheta_0 \ . \tag{3.3}$$

Together with (3.1) these boundary conditions also describe the temperature field in a semi-infinite body with the initial temperature ϑ_α, if the surface temperature suddenly assumes the constant value $\vartheta_0 \neq \vartheta_\alpha$. This problem has already been

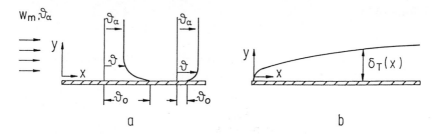

Fig. 3.1: **a** Temperature profile and **b** Thermal boundary layer in flow along a flat plate

discussed in section 2.3.3.1. The solution is, cf. (2.124),

$$\vartheta - \vartheta_\alpha = (\vartheta_0 - \vartheta_\alpha)\,\mathrm{erfc}\,\frac{y}{2\sqrt{at}}\;. \qquad (3.4)$$

The local heat transfer coefficient α is obtained from the energy balance at the wall

$$\dot{q} = \alpha(\vartheta_0 - \vartheta_\alpha) = -\lambda\left(\frac{\partial\vartheta}{\partial y}\right)_{y=0}\;. \qquad (3.5)$$

The gradient present at the wall is yielded by differentiation of (3.4) to be

$$\left(\frac{\partial\vartheta}{\partial y}\right)_{y=0} = -(\vartheta_0 - \vartheta_\alpha)\frac{1}{\sqrt{\pi a t}}\;.$$

Which then gives the heat transfer coefficient as

$$\alpha = \frac{\lambda}{\sqrt{\pi a t}}$$

or with $t = x/w_\mathrm{m}$

$$\alpha = \frac{1}{\sqrt{\pi}}\lambda\sqrt{\frac{w_\mathrm{m}}{ax}}\;. \qquad (3.6)$$

The mean heat transfer coefficient α_m is the integral mean value over the plate length L

$$\alpha_\mathrm{m} = \frac{1}{L}\int\limits_{x=0}^{L}\alpha\,\mathrm{d}x = 2\frac{1}{\sqrt{\pi}}\lambda\sqrt{\frac{w_\mathrm{m}}{aL}} = 2\alpha(x = L)\;. \qquad (3.7)$$

It is twice as big as the local heat transfer coefficient at the position L. As (3.6) indicates, the local heat transfer coefficient falls along the length by $\alpha \sim x^{-1/2}$. At the start of the plate $x \to 0$ the heat transfer coefficient is extremely large, $\alpha \to \infty$, and accordingly the transferred heat flux will also be very big, $\dot{q} \to \infty$. As the length $x \to \infty$, the heat transfer coefficient becomes negligibly small $\alpha \to 0$.

An approximation of the thickness δ_T of the thermal boundary layer is obtained by linearising the temperature increase

$$\left(\frac{\partial\vartheta}{\partial y}\right)_{y=0} \approx \frac{\vartheta_\alpha - \vartheta_0}{\delta_T}\;.$$

Then from (3.5)

$$\alpha \approx \frac{\lambda}{\delta_T} \ .$$

The heat transfer coefficient is inversely proportional to the thickness of the thermal boundary layer. The resistance to heat transfer

$$\frac{1}{\alpha} \approx \frac{\delta_T}{\lambda}$$

is therefore proportional to the thermal boundary layer thickness. It is very small at the start of the plate ($\delta_T \to 0$, $\alpha \to \infty$) and increases beyond all measure if the plate were infinitely long ($\delta_T \to \infty$, $\alpha \to 0$). Using (3.6) we find that the thermal boundary layer

$$\delta_T \approx \frac{\lambda}{\alpha} = \sqrt{\pi} \sqrt{\frac{ax}{w_m}} \tag{3.8}$$

increases with the square root of the length x.

The investigation of the problem in terms of mass transfer follows along the same lines. We will consider a flat plate coated with a substance A, for example naphthalene, that diffuses into a fluid, for example air, which is flowing along the plate. So that our precondition for the flow velocity w_m of the fluid along the plate is not altered, we have to assume that the amount of material transferred by diffusion is negligible when compared to the amount of material flowing over the plate. The convection normal to the wall does not play a role. The concentration c_{A0} of substance A on the surface of the plate is constant, and the concentration of the fluid arriving at the surface is $c_{A\alpha} < c_{A0}$. In the case of an ideal gas we get $c_A = N_A/V = p_A/R_m T$, where c_{A0} is formed with the saturation pressure $p_A(\vartheta_0)$ at the plate surface, and $c_{A\alpha}$ is formed with the partial pressure of substance A in the fluid. A concentration profile develops. In exactly the same way as the temperature variation from before, the concentration changes are limited to within a thin layer of thickness $\delta_c(x)$ close to the wall, that grows in the direction of flow. This layer is called the *concentration boundary layer*. Once again for an observer moving in a volume element, it is a function of time and position, which is described by (2.343), known to us for transient diffusion, where we presume that the diffusion coefficient is independent of concentration

$$\frac{\partial c_A}{\partial t} = D \frac{\partial^2 c_A}{\partial y^2} \ . \tag{3.9}$$

The boundary conditions are

$$c_A(t = 0, y) = c_{A\alpha} \tag{3.10}$$

$$c_A(t, y = 0) = c_{A0} \ . \tag{3.11}$$

The analogous problem for heat conduction was given by (3.1) to (3.3). Correspondingly we obtain the solution related to (3.4)

$$c_A - c_{A\alpha} = (c_{A0} - c_{A\alpha}) \operatorname{erfc} \frac{y}{2\sqrt{Dt}} \ . \tag{3.12}$$

As $c_A = \tilde{x}_A c$, the concentration c_A may be replaced here by the mole fraction $\tilde{x}_A$. The local heat transfer coefficient is obtained from a material balance at the wall

$$\dot{n}_{A0} = \beta(c_{A0} - c_{A\alpha}) = -cD\left(\frac{\partial \tilde{x}_A}{\partial y}\right)_{y=0} . \qquad (3.13)$$

The gradient at the wall is found by differentiating (3.12) to be

$$\left(\frac{\partial c_A}{\partial y}\right)_{y=0} = -(c_{A0} - c_{A\alpha})\frac{1}{\sqrt{\pi D t}} .$$

Which means the mass transfer coefficient is

$$\beta = \frac{D}{\sqrt{\pi D t}}$$

or with $t = x/w_m$

$$\beta = \frac{1}{\sqrt{\pi}}\sqrt{\frac{Dw_m}{x}} . \qquad (3.14)$$

The mean mass transfer coefficient β_m over the plate length L

$$\beta_m = \frac{1}{L}\int_{x=0}^{L} \beta \, \mathrm{d}x = 2\beta(x = L)$$

is twice as big as the local value at point L, just like for the mean heat transfer coefficient. An approximation of the thickness δ_c of the concentration boundary layer follows from (3.13) by linearising the concentration profile,

$$c\frac{\partial \tilde{x}_A}{\partial y} \approx c\frac{\tilde{x}_{A\alpha} - \tilde{x}_{A0}}{\delta_c} = \frac{c_{A\alpha} - c_{A0}}{\delta_c} ,$$

and leads to the expression corresponding to (3.8)

$$\delta_c \approx \frac{D}{\beta} = \sqrt{\pi}\sqrt{\frac{Dx}{w_m}} , \qquad (3.15)$$

according to which the thickness of the concentration boundary layer increases with the square root of the length x.

It follows from (3.8) and (3.15) that the ratio between the thicknesses of the thermal and concentration boundary layers is given approximately by the Lewis number $Le = a/D$, which has already been introduced, see Table 1.5

$$\frac{\delta_T}{\delta_c} \approx \sqrt{Le} .$$

The heat and mass transfer coefficients are also linked by the Lewis number due to (3.6) and (3.14)

$$\frac{\alpha}{\varrho c_p \beta} = \sqrt{\frac{a}{D}} = \sqrt{Le} .$$

As the Lewis number is of order one for ideal gases, the relationship (1.199): $\beta = \alpha/\varrho\, c_p$, discussed earlier, holds.

3.2   The balance equations

Real flows are not frictionless. As a result of friction, a flow field is formed with local and time related changes in the velocity. The temperature and concentration fields will not only be determined by conduction and diffusion but also by the flow itself. The form and profiles of flow, temperature and concentration fields are found by solving the mass, momentum and energy balances, which are the subject of the next section.

3.2.1   Reynolds' transport theorem

The derivation of the balance equations is simplified by using Reynolds' transport theorem. In order to derive it, we will observe a particular, infinitesimally small fluid mass $\mathrm{d}M$ and follow its movements in a flow field. The fluid mass should always consist of the same parts; is occupies a certain volume and has a certain surface area. As the form of the fluid volume generally changes during the course of its motion, its volume and the surface area can also change with time.

The mass M of the bounded fluid volume is found from the sum of all the mass elements $\mathrm{d}M$

$$M = \int_{(M)} \mathrm{d}M \ ,$$

which, with the density

$$\varrho = \lim_{\Delta V \to 0} \frac{\Delta M}{\Delta V} = \frac{\mathrm{d}M}{\mathrm{d}V} \ ,$$

as a steady function of time and the position coordinate in a continuum, can also be written as

$$M = \int_{V(t)} \varrho \, \mathrm{d}V \ .$$

Correspondingly each of the other quantities of state Z, such as internal energy, enthalpy, and entropy among others, can be formed by integration of the associated specific state quantity. For the specific quantity of state z it holds that

$$z = \lim_{\Delta M \to 0} \frac{\Delta Z}{\Delta M} = \frac{\mathrm{d}Z}{\mathrm{d}M},$$

which results in

$$Z = \int_{M} z \, \mathrm{d}M = \int_{V(t)} z \varrho \, \mathrm{d}V \ .$$

The specific quantity of state z can be a scalar, a vector or also a tensor of any rank. Just like the density, it is time and position dependent, whilst the extensive quantity of state Z is only time dependent.

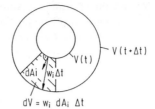

Fig. 3.2: Deformation of a closed system of volume $V(t)$ in a flow

The time derivative of Z will now be formed

$$\frac{dZ}{dt} = \frac{d}{dt} \int\limits_{V(t)} z\varrho \, dV \ .$$

As an abbreviation for the quantity of state per volume we will write $z\varrho = Z_V$ from now on. This depends on both time and position. Then by definition it is

$$\frac{dZ}{dt} = \lim_{\Delta t \to 0} \frac{Z(t + \Delta t) - Z(t)}{\Delta t}$$

$$= \lim_{\Delta t \to 0} \frac{1}{\Delta t} \left[\int\limits_{V(t+\Delta t)} Z_V(t + \Delta t) \, dV - \int\limits_{V(t)} Z_V(t) \, dV \right] ,$$

in which the position dependence of Z_V has not been written to prevent the expression becoming too confusing. The derivative can also be written as

$$\frac{dZ}{dt} = \lim_{\Delta t \to 0} \frac{1}{\Delta t} \left[\int\limits_{V(t+\Delta t)} (Z_V(t + \Delta t) - Z_V(t)) \, dV + \int\limits_{V(t)} Z_V(t) \, dV - \int\limits_{V(t)} Z_V(t) \, dV \right]$$

or

$$\frac{dZ}{dt} = \int\limits_{V(t)} \lim_{\Delta t \to 0} \frac{Z_V(t + \Delta t) - Z_V(t)}{\Delta t} \, dV + \lim_{\Delta t \to 0} \frac{1}{\Delta t} \int\limits_{\Delta V(t)} Z_V(t) \, dV \ . \qquad (3.16)$$

The volume increase with time Δt is indicated here by $\Delta V(t) = V(t+\Delta t) - V(t)$. According to Fig. 3.2 the volume element is given by $dV = w_i \, dA_i \Delta t$.[1] Which transforms the second integral into a surface integral,

$$\frac{1}{\Delta t} \int\limits_{A(t)} Z_V(t) w_i \, dA_i \Delta t = \int\limits_{A(t)} Z_V(t) w_i \, dA_i \ ,$$

---

[1]In this and the following sections 3.3 and 3.4 as well as 3.6 and 3.10 we will use the tensor notation, because it allows the balance equations to be written very clearly. An overview of tensor notations can be found in Appendix A1.

and the formation of limits is superfluous, as the time interval Δt is no longer present. The integrand of the first integral goes to $\partial Z_V / \partial t$ according to the limits. Therefore, the *Reynolds transport theorem* is obtained from (3.16) for volumina, in which we put once again $Z_V = z\varrho$

$$\frac{\mathrm{d}Z}{\mathrm{d}t} = \int\limits_{V(t)} \frac{\partial(z\varrho)}{\partial t}\,\mathrm{d}V + \int\limits_{A(t)} z\varrho w_i\,\mathrm{d}A_i \ . \tag{3.17}$$

The first term on the right hand side of this equation shows how much the observed quantity of state Z has increased by in the volume V, at time t. The second term indicates which part of the state quantity flows out with the material. The surface element $\mathrm{d}A_i$ is, as shown in Fig. 3.2, by definition in the outward direction so that positive $w_i\,\mathrm{d}A_i$ indicates a flow out of the volume. Clearly, according to (3.17), the change in the extensive quantity of state Z of a substance amount M with a volume $V(t)$ that changes with time, is equal to the increase in the extensive quantity of state inside the volume V at time t and the fraction of the quantity of state that flows out of the volume with the material at the same time.

With the help of Gauss' theorem the surface integral can be converted into a volume integral, giving

$$\frac{\mathrm{d}Z}{\mathrm{d}t} = \int\limits_{V(t)} \frac{\partial(z\varrho)}{\partial t}\,\mathrm{d}V + \int\limits_{V(t)} \frac{\partial(z\varrho w_i)}{\partial x_i}\,\mathrm{d}V \ . \tag{3.18}$$

3.2.2 The mass balance

3.2.2.1 Pure substances

We will now apply (3.18) to the mass as an extensive quantity of state, i.e. $Z = M$ and $z = M/M = 1$. Equation (3.18) then becomes

$$\frac{\mathrm{d}M}{\mathrm{d}t} = \int\limits_{V(t)} \frac{\partial\varrho}{\partial t}\,\mathrm{d}V + \int\limits_{V(t)} \frac{\partial\varrho w_i}{\partial x_i}\,\mathrm{d}V \ . \tag{3.19}$$

As by definition the mass $\mathrm{d}M$ of the infinitesimally small subsystem being observed is constant, the mass of the whole system must also be constant, within the volume $V(t)$ which changes with time. We are therefore considering a closed system, with $\mathrm{d}M/\mathrm{d}t = 0$. Due to the precondition of constant mass of the infinitesimally small subsystem, (3.19) also holds for $V(t) \to 0$. Therefore the sum of the two integrands must disappear. Which means we obtain

$$\frac{\partial\varrho}{\partial t} + \frac{\partial(\varrho w_i)}{\partial x_i} = 0 \ . \tag{3.20}$$

By differentiation

$$\frac{\partial \varrho}{\partial t} + w_i \frac{\partial \varrho}{\partial x_i} + \varrho \frac{\partial w_i}{\partial x_i} = 0$$

the relationship equivalent to equation (3.20) is found to be

$$\frac{d\varrho}{dt} + \varrho \frac{\partial w_i}{\partial x_i} = 0 \ , \qquad (3.21)$$

because we have

$$\frac{d\varrho}{dt} = \frac{\partial \varrho}{\partial t} + w_i \frac{\partial \varrho}{\partial x_i} \ .$$

The equations (3.20) and (3.21) say that the mass is conserved in an infinitesimally small volume element, and they are called *continuity equations*.

In an incompressible fluid the density remains constant, $\varrho = \text{const}$. The continuity equation is simplified to

$$\frac{\partial w_i}{\partial x_i} = 0 \ .$$

Example 3.1: Show, with help from (3.18) and (3.20), that the following equation is valid:

$$\frac{dZ}{dt} = \frac{d}{dt} \int_{V(t)} z \varrho \, dV = \int_{V(t)} \varrho \frac{dz}{dt} \, dV \ .$$

Differentiation of the integrands on the right hand side also allows (3.18) to be written

$$\frac{dZ}{dt} = \frac{d}{dt} \int_{V(t)} z \varrho \, dV = \int_{V(t)} \left(z \frac{\partial \varrho}{\partial t} + \varrho \frac{\partial z}{\partial t} \right) dV + \int_{V(t)} \left(z \frac{\partial(\varrho \, w_i)}{\partial x_i} + \varrho \, w_i \frac{\partial z}{\partial x_i} \right) dV$$

$$= \int_{V(t)} z \left(\frac{\partial \varrho}{\partial t} + \frac{\partial(\varrho \, w_i)}{\partial x_i} \right) dV + \int_{V(t)} \varrho \left(\frac{\partial z}{\partial t} + w_i \frac{\partial z}{\partial x_i} \right) dV \ .$$

The use of the continuity equation (3.20) makes the first integral on the right hand side disappear. The integrand in the brackets of the second integral is equal to the total differential

$$\frac{dz}{dt} = \frac{\partial z}{\partial t} + w_i \frac{\partial z}{\partial x_i} \ .$$

Whereby the relationship

$$\frac{dZ}{dt} = \frac{d}{dt} \int_{V(t)} z \varrho \, dV = \int_{V(t)} \varrho \frac{dz}{dt} \, dV$$

is proved.

3.2.2.2   Multicomponent mixtures

The continuity equation for any component A of a mixture consisting of N components will now be set up. We will only consider substance A. At time t it will enter the volume $V(t)$ and have a surface area $A(t)$. The discharge of substance A enlarges the volume represented in Fig. 3.2 by $dV = w_{Ai}\, dA_i \Delta t$, where w_{Ai} is the flow velocity of substance A. Transferring (3.16) to (3.17) and (3.18) means the gravitational velocity w_i is replaced by the velocity w_{Ai} of substance A. Furthermore, for the extensive quantity of state Z in (3.17) we put the mass M_A, $z = M_A/M = \xi_A$ and $z\varrho = M_A/V = \varrho_A$ and with that we obtain

$$\frac{dM_A}{dt} = \int\limits_{V(t)} \frac{\partial \varrho_A}{\partial t}\, dV + \int\limits_{A(t)} \varrho_A w_{Ai}\, dA_i \; . \tag{3.22}$$

According to this equation the increase in substance A is made up of two parts: from the increase inside the system and from the amount of substance A which flows out of the system.

$\dot{\Gamma}_A$ is the production density (SI-units kg/m³s) of component A in a volume element, so

$$\frac{dM_A}{dt} = \int\limits_{V(t)} \dot{\Gamma}_A\, dV \; .$$

The production density $\dot{\Gamma}_A$ changes, in general, with time and position. It is determined by the course of the chemical reaction in the system, and it is a task of reaction kinetics to investigate it. Converting the surface integral in (3.22) to a volume integral, using Gauss' theorem, gives

$$\int\limits_{V(t)} \dot{\Gamma}_A\, dV = \int\limits_{V(t)} \frac{\partial \varrho_A}{\partial t}\, dV + \int\limits_{V(t)} \frac{\partial \varrho_A w_{Ai}}{\partial x_i}\, dV \; .$$

As this relationship also holds for $V(t) \to 0$, the integrands have to agree. Therefore we obtain

$$\dot{\Gamma}_A = \frac{\partial \varrho_A}{\partial t} + \frac{\partial(\varrho_A w_{Ai})}{\partial x_i} \; . \tag{3.23}$$

This type of mass balance is known as a *component continuity equation*. It can be set up for each component. This means that there are as many of these equations as there are components. The summation over all the components leads to a continuity equation for the total mass, due to $\sum \dot{\Gamma}_K = 0$, $\sum \varrho_K = \varrho$ and $\sum \varrho_K w_{Ki} = \varrho w_i$. In place of the N component continuity equations for a system of N components, $N - 1$ component continuity equations along with the continuity equation for the total mass can be used.

The mass flux $\varrho_A w_{Ai}$ in (3.23) can be found from the diffusional flux

$$\varrho_A w_{Ai} = j_{Ai}^* + \varrho_A w_i \; .$$

This transforms (3.23) into

$$\dot{\Gamma}_A = \frac{\partial \varrho_A}{\partial t} + \frac{\partial}{\partial x_i}(j^*_{Ai} + \varrho_A w_i) \qquad (3.24)$$

or

$$\frac{\partial \varrho_A}{\partial t} + \frac{\partial}{\partial x_i}(\varrho_A w_i) = -\frac{\partial j^*_{Ai}}{\partial x_i} + \dot{\Gamma}_A \ .$$

Here the partial density ϱ_A can be expressed in terms of the mass fraction ξ_A and the density $\varrho_A = \xi_A \varrho$, so that the left hand side of the equation is rearranged into

$$\frac{\partial(\xi_A \varrho)}{\partial t} + \frac{\partial}{\partial x_i}(\xi_A \varrho w_i) = \xi_A \left[\frac{\partial \varrho}{\partial t} + \frac{\partial(\varrho w_i)}{\partial x_i} \right] + \varrho \frac{\partial \xi_A}{\partial t} + \varrho w_i \frac{\partial \xi_A}{\partial x_i} \ .$$

The term in the square brackets disappears due to the continuity equation (3.20) for the total mass. The component continuity equation is then

$$\varrho \frac{\partial \xi_A}{\partial t} + \varrho w_i \frac{\partial \xi_A}{\partial x_i} = -\frac{\partial j^*_{Ai}}{\partial x_i} + \dot{\Gamma}_A$$

or

$$\varrho \frac{d\xi_A}{dt} = -\frac{\partial j^*_{Ai}}{\partial x_i} + \dot{\Gamma}_A \ . \qquad (3.25)$$

For a binary mixture, with the introduction of the diffusional flow j^*_{Ai} according to (1.161) with $D_{AB} = D_{BA} = D$, this is transformed into

$$\varrho \frac{\partial \xi_A}{\partial t} + \varrho w_i \frac{\partial \xi_A}{\partial x_i} = \frac{\partial}{\partial x_i} \left(\varrho D \frac{\partial \xi_A}{\partial x_i} \right) + \dot{\Gamma}_A \ . \qquad (3.26)$$

In the case of quiescent systems, $w_i = 0$, this yields the relationship already known for transient diffusion, (2.341). If constant density is presumed, it follows from (3.26) that

$$\frac{\partial \xi_A}{\partial t} + w_i \frac{\partial \xi_A}{\partial x_i} = \frac{\partial}{\partial x_i} \left(D \frac{\partial \xi_A}{\partial x_i} \right) + \dot{\Gamma}_A/\varrho \qquad (3.27)$$

or with $\xi_A = \tilde{M}_A c_A/\varrho$, in which c_A is the molar concentration $c_A = N_A/V$,

$$\frac{\partial c_A}{\partial t} + w_i \frac{\partial c_A}{\partial x_i} = \frac{\partial}{\partial x_i} \left(D \frac{\partial c_A}{\partial x_i} \right) + \dot{r}_A \qquad (3.28)$$

with the reaction rate $\dot{r}_A = \dot{\Gamma}_A/\tilde{M}_A$ (SI-units kmol/m³s), the amount of substance A generated by chemical reaction.

Example 3.2: Show that, under the presumption of film theory — steady-state mass transfer only in the direction of the coordinate axis adjacent to the wall, vanishing production density — that the component continuity equation (3.25) transforms into (1.186) of film theory.

Indicating the coordinate adjacent to the wall with y, under the given preconditions, (3.25) is transformed into

$$\varrho w \frac{\partial \xi_A}{\partial y} = -\frac{\partial j^*_A}{\partial y} \ ,$$

wherein j_A^* is the diffusional flow and w is the velocity in the y-direction. On the other hand, as a result of the continuity equation (3.20)

$$\frac{\partial(\varrho\,w)}{\partial y} = 0 \ .$$

With that $\varrho\,w = \text{const}$, and (3.25) can also be written as

$$\frac{\partial}{\partial y}(j_A^* + \varrho\,w\,\xi_A) = \frac{\partial}{\partial y}(j_A^* + \varrho_A\,w) = 0 \ .$$

Now, by definition from (1.155) $j_A^* = \varrho_A\,(w_A - w)$, and therefore $j_A^* + \varrho_A\,w = \varrho_A\,w_A$ which means that $\partial(\varrho_A\,w_A)/\partial y = 0$ or $d\dot{M}_A/dy = 0$, from which, because $\dot{M}_A = \tilde{M}_A\,\dot{N}_A$, eq. (1.186) $d\dot{N}_A/dy = 0$ follows. This result can be found directly from (3.23), if $\Gamma_A = 0$, $d\varrho_A/dt = 0$ are put in, and one-dimensional mass flow is assumed.

3.2.3 The momentum balance

The mass elements of a flowing fluid transfer momentum. This is understood to be the product of the mass and velocity. A mass element dM, which flows at a velocity of w_j transports a momentum $w_j\,dM = w_j\varrho\,dV$. The total momentum I_j transported in a fluid of volume $V(t)$ is therefore

$$I_j = \int_{V(t)} \varrho\,w_j\,dV \ . \tag{3.29}$$

According to Newton's second law of mechanics the change in momentum of a body with time is equal to the resultant of all the forces acting on the body

$$\frac{dI_j}{dt} = F_j \tag{3.30}$$

and therefore

$$\frac{d}{dt}\int_{V(t)} \varrho w_j\,dV = F_j \ . \tag{3.31}$$

Applying the transport theorem gives, if we put $z = w_j$ into (3.17),

$$\int_{V(t)} \frac{\partial(\varrho w_j)}{\partial t}\,dV + \int_{A(t)} \varrho w_j w_i\,dA_i = F_j \tag{3.32}$$

or taking into account the continuity equation (3.20) (see also example 3.1)

$$\int_{V(t)} \varrho\frac{dw_j}{dt}\,dV = F_j \tag{3.33}$$

with

$$\frac{\mathrm{d}w_j}{\mathrm{d}t} = \frac{\partial w_j}{\partial t} + w_i \frac{\partial w_j}{\partial x_i} \ . \tag{3.34}$$

The forces F_j attacking the body can be split into two classes: in body forces which are proportional to the mass and in surface forces which are proportional to the surface area.

The *body forces* have an effect on all the particles in the body. They are far ranging forces and are caused by force fields. An example is the earth's gravitational field. The acceleration due to gravity g_j acts on each molecule, so that the force of gravity on a fluid element of mass ΔM is

$$\Delta F_j = g_j \Delta M \ .$$

It is proportional to the mass of the fluid element. The body force is defined by

$$k_j := \lim_{\Delta M \to 0} \frac{\Delta F_j}{\Delta M} = \frac{\mathrm{d}F_j}{\mathrm{d}M} \tag{3.35}$$

and the surface force by

$$f_j := \lim_{\Delta V \to 0} \frac{\Delta F_j}{\Delta V} = \frac{\mathrm{d}F_j}{\mathrm{d}V} \ . \tag{3.36}$$

In the case of gravitational force with $k_j = g_j$ and $f_j = \varrho g_j$, in general it holds that $f_j = \varrho k_j$. Other body forces are centrifugal forces or forces created by electromagnetic fields.

In the case of a multicomponent mixture, the different effects of the body forces on the individual components must also be considered. The body force k_{Aj} acting on component A will be defined by

$$k_{Aj} := \lim_{\Delta M_A \to 0} \frac{\Delta F_{Aj}}{\Delta M_A} = \frac{\mathrm{d}F_{Aj}}{\mathrm{d}M_A} \ .$$

Then $\mathrm{d}F_{Aj} = k_{Aj}\,\mathrm{d}M_A = k_{Aj}\varrho_A\,\mathrm{d}V$ and $\mathrm{d}F_j = \sum \mathrm{d}F_{Kj} = \mathrm{d}V\sum k_{Kj}\varrho_K$, where the summation is formed over all the substances K. On the other hand, due to (3.35), $\mathrm{d}F_j = k_j\,\mathrm{d}M = k_j\varrho\,\mathrm{d}V$ is valid and therefore

$$k_j\varrho = \sum k_{Kj}\varrho_K \ . \tag{3.37}$$

If the gravitational force is the only mass force acting on the body then $g_j = k_j = k_{Kj}$ and $\varrho = \sum \varrho_K$.

The *surface forces* are short-range forces. They exist in the area immediately adjacent to the fluid being considered and act on its surface. If ΔA is a surface element of a body with a force ΔF_j acting on it, Fig. 3.3, then

$$t_j := \lim_{\Delta A \to 0} \frac{\Delta F_j}{\Delta A} = \frac{\mathrm{d}F_j}{\mathrm{d}A} \tag{3.38}$$

is called the stress vector. It is not only dependent on the position and time but also on the orientation, and is therefore dependent on the normal vector to the surface element. In order to see this dependence we will consider a flow along a

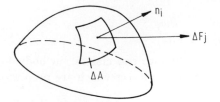
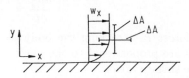
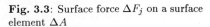

Fig. 3.3: Surface force ΔF_j on a surface element ΔA

Fig. 3.4: Dependence of the surface forces on the orientation of the surface element ΔA

flat plate, Fig. 3.4. A normal force acts on a surface element ΔA perpendicular to the plate. If we imagine that the surface element is rotated at the same place to a position parallel to the wall, the element is only subjected to shear stress. The two forces are, in general, of different magnitude.

The total force acting at time t on a fluid of volume $V(t)$ and surface area $A(t)$ is found by integrating the body and surface forces to be

$$F_j = \int_{V(t)} \varrho k_j \, \mathrm{d}V + \int_{A(t)} t_j \, \mathrm{d}A \ . \tag{3.39}$$

With this the momentum equation (3.33) can be written as

$$\int_{V(t)} \varrho \frac{\mathrm{d}w_j}{\mathrm{d}t} \, \mathrm{d}V = \int_{V(t)} \varrho k_j \, \mathrm{d}V + \int_{A(t)} t_j \, \mathrm{d}A \ . \tag{3.40}$$

It means that the change in momentum over time of a fluid of volume V at time t is effected by the body and the surface forces.

3.2.3.1   The stress tensor

For the calculation of the stress vector t_j we will consider a fluid element in the shape of an infinitesimally small tetrahedron, Fig. 3.5, which has one surface in any direction whilst the others stretch along the coordinate axes. The oblique surface has a normal unit vector n_i in the outward direction and an area $\mathrm{d}A$. The stress vector t acts on this area. Correspondingly, the stress vector t_1 acts at the surface $\mathrm{d}A_1$, perpendicular to the x-axis, and on the other two areas $\mathrm{d}A_2$ and $\mathrm{d}A_3$ we have the stress vectors t_2 and t_3. The forces acting on the four surface elements must be in equilibrium with each other independent of the momentary movement of the fluid element. This immediately follows from the momentum equation (3.40), when it is applied to an infinitesimally small volume element. At the limit $V \to 0$ the volume integral disappears more rapidly than the surface integral, and therefore the surface forces are in equilibrium locally. Applying this to the tetrahedron gives

$$t \, \mathrm{d}A = t_1 \, \mathrm{d}A_1 + t_2 \, \mathrm{d}A_2 + t_3 \, \mathrm{d}A_3 \ .$$

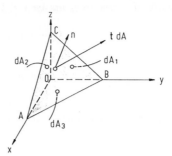

Fig. 3.5: Equilibrium of surface forces

The surface elements can be still be eliminated with the relationships from Fig. 3.6 of

$$dA = \frac{1}{2}\overline{AD} \cdot \overline{BC} \quad , \qquad dA_1 = \frac{1}{2}\overline{OD} \cdot \overline{BC}$$

and therefore

$$dA_1 = dA\frac{\overline{OD}}{\overline{AD}} = dA \cos \alpha \ .$$

Then, because $\cos \alpha = n_1/n = n_1$, we can also write $dA_1 = dA\, n_1$. Correspondingly, it holds that $dA_2 = dA\, n_2$, $dA_3 = dA\, n_3$.

The following relationship exists between the stress vectors

$$\boldsymbol{t} = \boldsymbol{t}_1 n_1 + \boldsymbol{t}_2 n_2 + \boldsymbol{t}_3 n_3 \ . \tag{3.41}$$

Each of the stress vectors $\boldsymbol{t}_1$, $\boldsymbol{t}_2$ and $\boldsymbol{t}_3$ can be represented by the three components of stress, which are in any case forces over the unit area. Two indices are required to indicate the components of stress. By agreement the first index relates to the area or surface on which the stress acts, and is identical to the index of the coordinate axis normal to this surface. The second index indicates in which direction the stress component is acting. As an example, Fig. 3.7 shows the three

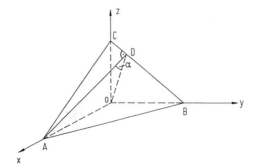

Fig. 3.6: Relationship between the areas.

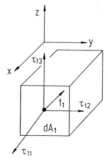

Fig. 3.7: Stress components on the surface dA_1

components of stress over an area dA_1 perpendicular to the x-axis. This is

$$t_1 = \tau_{11}e_1 + \tau_{12}e_2 + \tau_{13}e_3$$

and correspondingly

$$t_2 = \tau_{21}e_1 + \tau_{22}e_2 + \tau_{23}e_3 \ , \tag{3.42}$$
$$t_3 = \tau_{31}e_1 + \tau_{32}e_2 + \tau_{33}e_3 \ .$$

The stress components form a tensor which is made up of nine components:

$$\tau_{ij} = \begin{pmatrix} \tau_{11} & \tau_{12} & \tau_{13} \\ \tau_{21} & \tau_{22} & \tau_{23} \\ \tau_{31} & \tau_{32} & \tau_{33} \end{pmatrix} \ .$$

The stresses acting normal to the surface τ_{11},τ_{22}, τ_{33} or τ_{ij} with $i = j$ are known as normal stresses, the stresses tangential to the surface τ_{ij} with $i \neq j$ are called tangential or shear stresses. Due to the rule that shear stresses associated with each other are equal we can write $\tau_{ij} = \tau_{ji}$: the stress tensor is symmetrical. Putting the stress vectors from (3.42) into (3.41), gives

$$\begin{aligned} t &= (\tau_{11}e_1 + \tau_{12}e_2 + \tau_{13}e_3)\,n_1 \\ &+ (\tau_{21}e_1 + \tau_{22}e_2 + \tau_{23}e_3)\,n_2 \\ &+ (\tau_{31}e_1 + \tau_{32}e_2 + \tau_{33}e_3)\,n_3 \ . \end{aligned}$$

On the other hand the stress vector t acting on the oblique surface of the tetrahedron in Fig. 3.5, can be split into its three components t_1, t_2, t_3 in the direction of the coordinate axes

$$t = t_1e_1 + t_2e_2 + t_3e_3 \ .$$

As the comparison with the previous relationship shows, the following holds for the components of the stress vector

$$\begin{aligned} t_1 &= \tau_{11}n_1 + \tau_{21}n_2 + \tau_{31}n_3 \\ t_2 &= \tau_{12}n_1 + \tau_{22}n_2 + \tau_{32}n_3 \\ t_3 &= \tau_{13}n_1 + \tau_{23}n_2 + \tau_{33}n_3 \end{aligned}$$

or

$$t_j = \tau_{ji}n_i \qquad (i, j = 1, 2, 3) \ . \tag{3.43}$$

Theoretically the stress tensor τ_{ji} can be written as the sum of two tensors

$$\tau_{ji} = \begin{pmatrix} \tau_{11} - \frac{1}{3}\tau_{kk} & \tau_{12} & \tau_{13} \\ \tau_{21} & \tau_{22} - \frac{1}{3}\tau_{kk} & \tau_{23} \\ \tau_{31} & \tau_{32} & \tau_{33} - \frac{1}{3}\tau_{kk} \end{pmatrix} + \begin{pmatrix} \frac{1}{3}\tau_{kk} & 0 & 0 \\ 0 & \frac{1}{3}\tau_{kk} & 0 \\ 0 & 0 & \frac{1}{3}\tau_{kk} \end{pmatrix}$$

or

$$\tau_{ji} = \hat{\tau}_{ji} + \frac{1}{3}\delta_{ji}\tau_{kk} \tag{3.44}$$

with the *unit tensor* δ_{ji}, also called the *Kronecker-δ*,

$$\delta_{ji} = \begin{cases} 1 & \text{for} \quad i = j \\ 0 & \text{for} \quad i \neq j \end{cases} \quad \text{i.e.} \quad \delta_{ji} = \begin{pmatrix} 1 & 0 & 0 \\ 0 & 1 & 0 \\ 0 & 0 & 1 \end{pmatrix} . \tag{3.45}$$

From this definition it immediately follows that $\delta_{ji} = \delta_{11} + \delta_{22} + \delta_{33} = 3$.

We call $\hat{\tau}_{ji}$ the *deviator* of the tensor τ_{ji}. It is thereby characterised such that the diagonal elements, the so-called *trace of the tensor*, disappear, which then gives

$$(\tau_{11} - \frac{1}{3}\tau_{kk}) + (\tau_{22} - \frac{1}{3}\tau_{kk}) + (\tau_{33} - \frac{1}{3}\tau_{kk}) = \tau_{kk} - 3 \cdot \frac{1}{3}\tau_{kk} = 0 .$$

In general a deviator is a tensor with a zero trace.

As the diagonal elements characterise a normal stress and cancel each other out, the deviator is decisive for the shearing of a fluid element, whilst the term $1/3\,\delta_{ij}\tau_{kk}$ contains normal stresses equal in all directions, or so-called hydrostatic stresses. The arithmetic mean value of the three normal stresses τ_{kk} are known as the *average pressure*

$$-\bar{p} = \frac{1}{3}\tau_{kk} .$$

The minus sign stems from the fact that liquids can take practically no tensile stresses in, and therefore τ_{kk} is generally negative, whereas the average pressure is positive. With this (3.44) can also be written as

$$\tau_{ji} = \hat{\tau}_{ji} - \delta_{ji}\bar{p} . \tag{3.46}$$

The average pressure $\bar{p}$ is not identical to the *thermodynamic pressure*, which is obtained, for quiescent fluids, from the thermal equation of state $p = p(v, T)$. However, we can show, see Appendix A2, that for fairly slow changes in volume the average and thermodynamic pressures are linked by

$$\bar{p} - p := -\zeta \frac{\partial w_k}{\partial x_k} , \tag{3.47}$$

in which, the factor $\zeta > 0$, defined in this equation, is the so-called *bulk viscosity* (SI units kg/sm; $0.1\,\text{kg/sm} = 1\,\text{Poise}$). Methods of statistical mechanics have shown that this bulk viscosity disappears in low density gases, and that it is very small in dense gases and liquids. Therefore in fluid mechanics and heat and mass transfer we always presume $\zeta = 0$, thereby setting the average pressure equal to the thermodynamic pressure. Furthermore in an incompressible fluid, because $\partial w_k/\partial x_k = 0$, the average pressure is always equal to the thermodynamic pressure.

3.2.3.2 Cauchy's equation of motion

By introducing the stress vector t_j according to (3.43) into the momentum equation (3.40) we obtain

$$\int\limits_{V(t)} \varrho \frac{\mathrm{d}w_j}{\mathrm{d}t}\,\mathrm{d}V = \int\limits_{V(t)} \varrho k_j\,\mathrm{d}V + \int\limits_{A(t)} \tau_{ji} n_i\,\mathrm{d}A\ .$$

With the help of Gauss' law we rearrange the surface integral

$$\int\limits_{A(t)} \tau_{ji} n_i\,\mathrm{d}A = \int\limits_{V(t)} \frac{\partial \tau_{ji}}{\partial x_i}\,\mathrm{d}V\ ,$$

and obtain

$$\int\limits_{V(t)} \varrho \frac{\mathrm{d}w_j}{\mathrm{d}t}\,\mathrm{d}V = \int\limits_{V(t)} \varrho\,k_j\,\mathrm{d}V + \int\limits_{V(t)} \frac{\partial \tau_{ji}}{\partial x_i}\,\mathrm{d}V\ .$$

This equation holds for any volume $V(t)$ and therefore also in the limiting case of $V(t) \to 0$. From this follows *Cauchy's equation of motion*[2]:

$$\varrho \frac{\mathrm{d}w_j}{\mathrm{d}t} = \varrho\,k_j + \frac{\partial \tau_{ji}}{\partial x_i} \qquad (i,j = 1,2,3)\ . \tag{3.48}$$

It is valid for each continuum independent of the individual material properties and is therefore one of the fundamental equations in fluid mechanics and subsequently also in heat and mass transfer. The movement of a particular substance can only be described by introducing a so-called constitutive equation which links the stress tensor with the movement of a substance. Generally speaking, constitutive equations relate stresses, heat fluxes and diffusion velocities to macroscopic variables such as density, velocity and temperature. These equations also depend on the properties of the substances under consideration. For example, Fourier's law of heat conduction is invoked to relate the heat flux to the temperature gradient using the thermal conductivity. An understanding of the strain tensor is useful for the derivation of the consitutive law for the shear stress. This strain tensor is introduced in the next section.

---

[2] Auguste-Louis Cauchy (1789–1857) was, as a contemporary of Leonhard Euler and Carl-Friedrich Gauss, one of the most important mathematicians of the first half of the 19th century. His most famous publications are "Traité des Fonctions" and " Méchanique Analytique". As he refused to take the oath to the new regime after the revolution in 1830, his positions as professor at the Ecole Polytechnique and at the Collège de France were removed and he was dismissed from the Académie Française. He spent several years in exile in Switzerland, Turin and Prague. He was permitted to return to France in 1838, and it was there that he was reinstated as a professor at the Sorbonne after the revolution in 1848.

3.2.3.3   The strain tensor

The individual fluid elements of a flowing fluid are not only displaced in terms of their position but are also deformed under the influence of the normal stresses τ_{ii} and the shear stresses $\tau_{ij}(i \neq j)$. The deformation velocity depends on the relative movement of the individual points of mass to each other. It is only in the case when the points of mass in a fluid element do not move relatively to each other that the fluid element behaves like a rigid solid and will not be deformed. Therefore a relationship between the velocity field and the deformation, and with that also between the velocity field and the stress tensor τ_{ij} must exist. This relationship is required if we wish to express the stress tensor in terms of the velocities in Cauchy's equation of motion.

Normal stresses change the magnitude of a fluid element of given mass. If they are different, for example $\tau_{11} \neq \tau_{22}$, the shape of the fluid element will also be altered. As we can see in Fig. 3.8, a rectangle would be transformed into a prism, a spherical fluid element could be deformed into an ellipsoid. Fig. 3.9 shows the front view of a cube which will be stretched by a normal stress τ_{11}. We recognise that the volume increases by

$$\frac{\partial w_1}{\partial x_1}\, \mathrm{d}x_1\, \mathrm{d}t\, \mathrm{d}x_2\, \mathrm{d}x_3 = \frac{\partial w_1}{\partial x_1}\, \mathrm{d}t\, \mathrm{d}V \ .$$

Corresponding expressions for the volume increase due to normal stresses τ_{22} and τ_{33} can be obtained, so that the total volumetric increase is given by

$$\left(\frac{\partial w_1}{\partial x_1} + \frac{\partial w_2}{\partial x_2} + \frac{\partial w_3}{\partial x_3} \right) \mathrm{d}t\, \mathrm{d}V \ .$$

The volume increase at each unit of time relative to the original volume is known as the *dilatation*. It is

$$\frac{\partial w_1}{\partial x_1} + \frac{\partial w_2}{\partial x_2} + \frac{\partial w_3}{\partial x_3} = \frac{\partial w_i}{\partial x_i} = \dot{\varepsilon}_{ii} \ . \tag{3.49}$$

In an incompressible fluid, $\varrho = \mathrm{const}$, there is no dilatation, as we have seen from the continuity equation. The volume of a fluid element of a given mass remains constant.

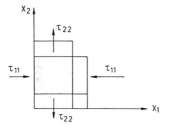

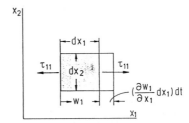

Fig. 3.8: Deformation of a fluid element due to normal stress

Fig. 3.9: For the relationship between deformation and velocity

The shear stresses $\tau_{ij}(i \neq j)$ cause an originally cubic volume element to be deformed into a rhomboid, as shown in the view in Fig. 3.10. The original right angle at A changes by the angle

$$d\gamma_{12} = \frac{\partial w_1}{\partial x_2} dt \quad \text{and} \quad d\gamma_{21} = \frac{\partial w_2}{\partial x_1} dt \ .$$

The arithmetic mean of both angular velocities is called the *strain tensor*, often also called the deformation velocity tensor

$$\dot{\varepsilon}_{12} := \frac{1}{2} \left(\frac{\partial w_1}{\partial x_2} + \frac{\partial w_2}{\partial x_1} \right)$$

or in general

$$\dot{\varepsilon}_{ji} := \frac{1}{2} \left(\frac{\partial w_j}{\partial x_i} + \frac{\partial w_i}{\partial x_j} \right) \ . \tag{3.50}$$

It is a symmetrical tensor, as $\dot{\varepsilon}_{ji} = \dot{\varepsilon}_{ij}$, and transforms for $i = j$ back to the dilatation. The elements $\dot{\varepsilon}_{ii}$ form in this case the diagonal of the strain tensor. If the sum of these terms disappears, the volume of a fluid element of given mass does not change, as already mentioned. If the tensor is split into two parts, from which the sum of the diagonal of one of the tensors, the *deviator*, disappears, this describes the change in shape at constant volume. The shape change at constant volume will, thus, be described by the deviator of the strain tensor. In order to split up the deviator we can form

$$\dot{\varepsilon}_{ji} = \hat{\dot{\varepsilon}}_{ji} + \frac{1}{3}\delta_{ji}\dot{\varepsilon}_{kk} \ . \tag{3.51}$$

As follows with (3.49) and (3.50), we have

$$\hat{\dot{\varepsilon}}_{ji} = \frac{1}{2} \left[\left(\frac{\partial w_j}{\partial x_i} + \frac{\partial w_i}{\partial x_j} \right) - \frac{2}{3}\delta_{ji}\frac{\partial w_k}{\partial x_k} \right] \ . \tag{3.52}$$

When written out, (3.51) means

$$\dot{\varepsilon}_{ji} = \begin{pmatrix} \dot{\varepsilon}_{11} - \frac{1}{3}\dot{\varepsilon}_{kk} & \dot{\varepsilon}_{12} & \dot{\varepsilon}_{13} \\ \dot{\varepsilon}_{21} & \dot{\varepsilon}_{22} - \frac{1}{3}\dot{\varepsilon}_{kk} & \dot{\varepsilon}_{23} \\ \dot{\varepsilon}_{31} & \dot{\varepsilon}_{32} & \dot{\varepsilon}_{33} - \frac{1}{3}\dot{\varepsilon}_{kk} \end{pmatrix} + \begin{pmatrix} \frac{1}{3}\dot{\varepsilon}_{kk} & 0 & 0 \\ 0 & \frac{1}{3}\dot{\varepsilon}_{kk} & 0 \\ 0 & 0 & \frac{1}{3}\dot{\varepsilon}_{kk} \end{pmatrix} \ .$$

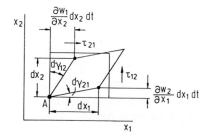

Fig. 3.10: The deformation of a cubic volume element into a rhomboid

The first term is already the desired deviator, as its trace disappears, this is then

$$\left(\dot{\varepsilon}_{11} - \frac{1}{3}\dot{\varepsilon}_{kk}\right) + \left(\dot{\varepsilon}_{22} - \frac{1}{3}\dot{\varepsilon}_{kk}\right) + \left(\dot{\varepsilon}_{33} - \frac{1}{3}\dot{\varepsilon}_{kk}\right) = \dot{\varepsilon}_{kk} - 3\frac{1}{3}\dot{\varepsilon}_{kk} = 0 \ .$$

Whilst the deviator describes the change in shape at constant volume, the (isotropic) tensor $1/3\,\delta_{ij}\dot{\varepsilon}_{kk} = 1/3\,\delta_{ij}\partial w_k/\partial x_k$ is decisive for the volume change for a constant shape. This follows from the fact that it only contains equal diagonals, so that the deformation in all coordinate directions is the same.

3.2.3.4 Constitutive equations for the solution of the momentum equation

In order to solve Cauchy's equation of motion, which is valid for any substance, a further relationship between the stress and strain tensors, or between the stress tensor and the velocity field is required. This type of equation, a so-called constitutive equation, is material specific and similar to the equations of state in thermodynamic characteristics for a certain substance. As a result of (3.46) and (3.47) it still generally holds that

$$\tau_{ji} = \hat{\tau}_{ji} - \delta_{ji}p + \zeta\frac{\partial w_k}{\partial x_k} \ . \tag{3.53}$$

The elements present in the deviator $\hat{\tau}_{ji}$ of the stress tensor cause the stretching described by the elements in the deviator $\hat{\dot{\varepsilon}}_{ji}$ of the stress tensor, which can be seen by a comparison of the two deviators. Therefore a substance specific relationship

$$\hat{\tau}_{ji} := f(\hat{\dot{\varepsilon}}_{ji})$$

must exist.

For low strain velocities a linear approach suggests itself

$$\hat{\tau}_{ji} = \eta\,2\hat{\dot{\varepsilon}}_{ji} \ . \tag{3.54}$$

The factor of 2 is introduced, so that, for a one-dimensional incompressible flow ($j = 1$, $i = 2$; $\partial w_k/\partial x_k = 0$) with $\hat{\dot{\varepsilon}} = (1/2)\,\partial w_1/\partial x_2$ from (3.52), Newton's law $\tau_{12} = \eta\partial w_1/\partial x_2$ is obtained.

A fluid in which the shear stress is proportional to the shear velocity, corresponding to this law, is called an *ideal viscous* or *Newtonian fluid*. Many gases and liquids follow this law so exactly that they can be called Newtonian fluids. They corrspond to ideal Hookeian bodies in elastomechanics, in which the shear strain is proportional to the shear. A series of materials cannot be described accurately by either Newtonian or Hookeian behaviour. The relationship between shear stress and strain can no longer be described by the simple linear rule given above. The study of these types of material is a subject of rheology.

After introducing $\hat{\dot{\varepsilon}}_{ji}$ according to (3.52), (3.54) is transformed into

$$\hat{\tau}_{ji} = \eta \left[\left(\frac{\partial w_j}{\partial x_i} + \frac{\partial w_i}{\partial x_j} \right) - \frac{2}{3} \delta_{ji} \frac{\partial w_k}{\partial x_k} \right] . \tag{3.55}$$

The factor η defined in (3.54) and (3.55) is the dynamic viscosity (SI units kg/sm = $1 \, \text{Pa} \cdot \text{s}$; $0.1 \, \text{kg/sm} = 1 \, \text{Poise}$). With the help of (3.53) we obtain, for the stress tensor

$$\tau_{ji} = \eta \left[\left(\frac{\partial w_j}{\partial x_i} + \frac{\partial w_i}{\partial x_j} \right) - \frac{2}{3} \delta_{ji} \frac{\partial w_k}{\partial x_k} \right] - \delta_{ji} p + \zeta \frac{\partial w_k}{\partial x_k} . \tag{3.56}$$

3.2.3.5 The Navier-Stokes equations

We will presume a Newtonian fluid and neglect the normally very small bulk viscosity. The statement (3.56) for the stress tensor is then transformed into Stokes' formulation, the so-called *Stokes hypothesis*:

$$\tau_{ji} = \eta \left[\left(\frac{\partial w_j}{\partial x_i} + \frac{\partial w_i}{\partial x_j} \right) - \frac{2}{3} \delta_{ji} \frac{\partial w_k}{\partial x_k} \right] - \delta_{ji} p . \tag{3.57}$$

Putting this expression into Cauchy's equation of motion, (3.48), yields the so-called *Navier-Stokes equation*

$$\varrho \frac{dw_j}{dt} = \varrho k_j - \frac{\partial p}{\partial x_j} + \frac{\partial}{\partial x_i} \eta \left[\left(\frac{\partial w_j}{\partial x_i} + \frac{\partial w_i}{\partial x_j} \right) - \frac{2}{3} \delta_{ji} \frac{\partial w_k}{\partial x_k} \right] . \tag{3.58}$$

For an incompressible fluid we have $\partial w_k / \partial x_k = \partial w_i / \partial x_i = 0$. Furthermore, if we presume a constant viscosity, the equation is simplified to

$$\varrho \frac{dw_j}{dt} = \varrho k_j - \frac{\partial p}{\partial x_j} + \eta \frac{\partial^2 w_j}{\partial x_i^2} . \tag{3.59}$$

The equation clearly shows that the forces acting on a fluid element are made up of body, pressure and viscosity forces. A momentum balance exists for each of the three coordinate directions $j = 1, 2, 3$, so that (3.58) and (3.59) each represent three equations independent of each other. (3.59) can be seen written out in Appendix A3. Its formulation for cylindrical coordinates is also given there.

3.2.4 The energy balance

Pure substances
According to the first law of thermodynamics, the internal energy U of a closed system changes due to the addition of heat Q_{12} and work W_{12} into the system

$$U_2 - U_1 = Q_{12} + W_{12}$$

or in differential form

$$dU = dQ + dW \tag{3.60}$$

or

$$\frac{dU}{dt} = \frac{dQ}{dt} + \frac{dW}{dt} = \dot{Q} + P \; . \tag{3.61}$$

We will apply these equations to a fluid element of given mass. As this will be considered to be a closed system we can immediately exclude mass transport over the system boundary and with that diffusion.

With the help of the transport theorem (3.17), by putting in $z = u$, we obtain the following temporal change in the internal energy of a flowing fluid

$$\frac{dU}{dt} = \frac{d}{dt} \int_{V(t)} u\varrho \, dV = \int_{V(t)} \frac{\partial u\varrho}{\partial t} \, dV + \int_{A(t)} u\varrho w_i \, dA_i \; . \tag{3.62}$$

According to this, the change in internal energy over time in a flowing fluid is equal to internal energy stored inside the fluid volume $V(t)$ at time t, and the internal energy flowing out over the surface $A(t)$ of the fluid volume.

Heat is transferred, by definition, between a system and its surroundings, and therefore the heat fed into the system via its surface is

$$\dot{Q} = \int_{A(t)} \dot{q} \, dA \; , \tag{3.63}$$

in which $\dot{q}$ is the heat flux. It is a scalar. (3.63), with the normal vector $\boldsymbol{n}$, which indicates the orientation of the surface elements in space, can also be written as

$$\dot{Q} = \int_{A(t)} (\dot{q} n_i)(dA \, n_i)$$

because $n_i n_i = 1$. The products obtained

$$dA \, n_i = dA_i \quad (i = 1, 2, 3)$$

are, as shown in section 3.2.3.1, the projections of the surface dA on the plane formed by the coordinate axes. It is known that dA_i are the components of the surface vector $\boldsymbol{n} \, dA$. Correspondingly,

$$\dot{q} n_i = -\dot{q}_i \tag{3.64}$$

can be interpreted as the components of a vector $\dot{\boldsymbol{q}}$ of the heat flux. $\dot{q}_1$ is the heat flux over the area dA_1, $\dot{q}_2$ is that over dA_2 and $\dot{q}_3$ that over dA_3. The minus sign in (3.64) stems from the fact that a heat flux fed into a system is counted as positive, but the normal vector n_i of a closed area points outwards. With that we get

$$\dot{Q} = - \int_{A(t)} \dot{q}_i n_i \, dA \; . \tag{3.65}$$

The work done to change the internal energy of a system is exchanged between the system and its surroundings and 'flows' over the surface of the system. The body forces displace each element as a whole, and therefore also contribute to the change in the kinetic and potential energy, but not to the change in the internal energy, as long as the mass of the fluid element is not altered by mass exchange with its neighbours. In this case the individual particles will move at different speeds and the body forces acting on them will contribute to the total power. This contribution plays an important role in multicomponent mixtures and must be taken into account there.

In addition the surface forces contain parts which cause a change in the internal energy, and some that displace the fluid element as a whole without changing its internal energy. The fraction of the power, which contributes to a change in the internal energy, fed into the system via its surface can also be written as

$$P := \int_{A(t)} \dot{\omega} \, dA \ . \tag{3.66}$$

The *power density* $\dot{\omega}$ (SI units W/m^2) is defined by this equation. We choose the letter $\dot{\omega}$ for the power density, rather than $\dot{w}$, otherwise it would be easy to confuse it with velocity w. Similar to the considerations for heat flux, (3.66) can be rearranged into

$$P = \int_{A(t)} (\dot{\omega} n_i)(\, dA n_i)$$

and here the quantities

$$\dot{\omega} n_i = -\dot{\omega}_i \tag{3.67}$$

can be interpreted as the the components of the vector $\dot{\boldsymbol{\omega}}$ of the power density; $\dot{\omega}_1$ is the power density across the area dA_1, $\dot{\omega}_2$ that across the area dA_2 and $\dot{\omega}_3$ that across the area dA_3. The minus sign appears once more because the power being fed into the system is taken to be positive. It holds, therefore, that

$$P = - \int_{A(t)} \dot{\omega}_i n_i \, dA \ . \tag{3.68}$$

The first law (3.61), in conjunction with the expressions for internal energy (3.62), the heat flow (3.65) and the power (3.68), can also be written as

$$\int_{V(t)} \frac{\partial(u\varrho)}{\partial t} \, dV + \int_{A(t)} u\varrho w_i \, dA_i = - \int_{A(t)} \dot{q}_i n_i \, dA - \int_{A(t)} \dot{\omega}_i n_i \, dA \ .$$

Converting the surface into the volume integral using Gauss' law and then transferring to a small volume element $V(t) \to 0$ yields

$$\frac{\partial(u\varrho)}{\partial t} + \frac{\partial(u\varrho w_i)}{\partial x_i} = - \frac{\partial \dot{q}_i}{\partial x_i} - \frac{\partial \dot{\omega}_i}{\partial x_i} \ .$$

The left hand side can also be written as

$$\varrho\frac{\partial u}{\partial t} + u\frac{\partial \varrho}{\partial t} + u\frac{\partial(\varrho w_i)}{\partial x_i} + \varrho w_i\frac{\partial u}{\partial x_i} \ .$$

By taking into account the continuity equation (3.20), the equation above is simplified to

$$\varrho\frac{\partial u}{\partial t} + \varrho w_i\frac{\partial u}{\partial x_i} = \varrho\frac{du}{dt} \ .$$

Which then gives the energy equation

$$\varrho\frac{du}{dt} = -\frac{\partial \dot{q}_i}{\partial x_i} - \frac{\partial \dot{w}_i}{\partial x_i} \ . \tag{3.69}$$

In order to calculate the power density $\dot{w}_i$, we will now consider the total power P_{tot} produced by the surface forces and separate from that the part known as the *drag*, which causes a displacement of the fluid element. This then leaves only the power which contributes to a change in the internal energy.

Using the stress tensor $t_i = \tau_{ji}n_j$ the total power from the surface forces is obtained as

$$P_{\text{tot}} = \int\limits_{A(t)} w_i t_i \, dA = \int\limits_{A(t)} w_i \tau_{ji} n_j \, dA = \int\limits_{V(t)} \frac{\partial(w_i\tau_{ji})}{\partial x_j} \, dV \ .$$

On the other hand, the surface forces acting on the volume element

$$F_i = \int\limits_{A(t)} t_i \, dA = \int\limits_{A(t)} \tau_{ji} n_j \, dA = \int\limits_{V(t)} \frac{\partial \tau_{ji}}{\partial x_j} \, dV$$

cause a movement of the system. The force acting on a volume element dV

$$dF_i = \frac{\partial \tau_{ji}}{\partial x_j} \, dV$$

displaces it during a short time dt along a path dx_i, such that the drag is

$$dF_i\frac{dx_i}{dt} = \frac{\partial \tau_{ji}}{\partial x_j} \, dV \frac{dx_i}{dt} = \frac{\partial \tau_{ji}}{\partial x_j} \, dV w_i \ .$$

The total drag is therefore

$$P_S = \int\limits_{V(t)} \frac{\partial \tau_{ji}}{\partial x_j} w_i \, dV \ ,$$

and the part of the surface forces contributing to the change in internal energy will be

$$P = P_{\text{tot}} - P_S = \int\limits_{V(t)} \left(\frac{\partial(w_i\tau_{ji})}{\partial x_j} - \frac{\partial \tau_{ji}}{\partial x_j}w_i\right) \, dV \ ,$$

out of which

$$P = \int_{V(t)} \tau_{ji} \frac{\partial w_i}{\partial x_j} \, dV$$

follows. As (3.68) can be rearraged using Gauss' law into

$$P = -\int_{V(t)} \frac{\partial \dot{\omega}_i}{\partial x_i} \, dV$$

then

$$-\frac{\partial \dot{\omega}_i}{\partial x_i} = \tau_{ji} \frac{\partial w_i}{\partial x_j} \ . \tag{3.70}$$

The energy balance (3.69) is given by

$$\varrho \frac{du}{dt} = -\frac{\partial \dot{q}_i}{\partial x_i} + \tau_{ji} \frac{\partial w_i}{\partial x_j} \ . \tag{3.71}$$

As we have taken the fluid element to be a closed system, this equation is not valid for multicomponent mixtures and diffusion.

Multicomponent mixtures

Diffusion appears in multicomponent mixtures when they are not at equilibrium. Individual material flows cross over the system boundaries. Therefore a closed system cannot be presupposed. The first law now becomes [3.1]

$$dU = dQ + dW + \sum_K h_K \, d(_e M_K)$$

or

$$\frac{dU}{dt} = \dot{Q} + P + \sum_K h_K {_e}\dot{M}_K \ . \tag{3.72}$$

Here, for a particular substance A, the quantity $h_A = (\partial H/\partial M_A)_{T,p,\,M_{K \neq A}}$ is the partial specific enthalpy and $_e\dot{M}_A$ is the mass flow rate of component A fed into the system from outside. The sum is made over all the material flows. The heat flow $\dot{Q}$ and the power P are calculated in the same way as before from

$$\dot{Q} = -\int_{A(t)} \dot{q}_i n_i \, dA = -\int_{V(t)} \frac{\partial \dot{q}_i}{\partial x_i} \, dV$$

and

$$P = -\int_{A(t)} \dot{\omega}_i n_i \, dA = -\int_{V(t)} \frac{\partial \dot{\omega}_i}{\partial x_i} \, dV \ .$$

Substance A flowing at a velocity w_{Ai} across the surface element is an energy carrier. It increases the internal energy of a fluid element moving with the centre of mass velocity w_i by

$$-h_A \varrho_A (w_{A_i} - w_i) n_i \, dA = -h_A j^*_{A_i} n_i \, dA \ .$$

Once again the minus sign is necessary because energy flowing into a system is counted as positive, but the surface vector has the opposite sign. The energy from all the feed streams flowing over the total surface is

$$\sum_K h_K {_e}\dot{M}_K = -\sum_K \int_{A(t)} h_K j^*_{Ki} n_i \, dA = -\sum_K \int_{V(t)} \frac{\partial (h_K j^*_{Ki})}{\partial x_i} \, dV \ .$$

When this is applied to a fluid element, the energy equation (3.72) for a multicomponent mixture can be written as

$$\varrho \frac{du}{dt} = -\frac{\partial \dot{q}_i}{\partial x_i} - \frac{\partial \dot{\omega}_i}{\partial x_i} - \sum_K \frac{\partial (h_K j^*_{Ki})}{\partial x_i} \ . \tag{3.73}$$

The power density $\dot{\omega}_i$ which appears here consists of the power from the surface forces minus the drag forces, which displace the fluid element as a whole and therefore do not contribute to a change in the internal energy. In contrast to the equation for pure substances this also contains an additional term because the different particle types move with different velocities w_{A_i}, such that the associated body forces k_{A_i} on substance A in a volume element dV deliver a contribution of

$$\varrho_A (w_{Ai} - w_i) k_{Ai}\, dV = j^*_{Ai} k_{Ai}\, dV \ .$$

For all the substances, the contribution on a system of volume $V(t)$ would be

$$\sum_K \int_{V(t)} j^*_{Ki} k_{Ki}\, dV \ .$$

The total power P for the change in the internal energy is made up of the power of the surface forces

$$\int_{A(t)} w_i t_i\, dA = \int_{A(t)} w_i \tau_{ji} n_j\, dA = \int_{V(t)} \frac{\partial (w_i \tau_{ji})}{\partial x_i}\, dV$$

and the body forces

$$\sum_K \int_{V(t)} j^*_{Ki} k_{Ki}\, dV$$

minus the drag force

$$\int_{V(t)} \frac{\partial \tau_{ji}}{\partial x_j} w_i\, dV \ .$$

$$P = \int_{V(t)} \left(\tau_{ji} \frac{\partial w_i}{\partial x_j} + \sum_K j^*_{Ki} k_{Ki} \right) dV \ .$$

Then for the *power density*, it follows, due to

$$P = -\int_{A(t)} \dot{\omega}_i n_i\, dA = -\int_{V(t)} \frac{\partial \dot{\omega}_i}{\partial x_i}\, dV \ ,$$

that

$$-\frac{\partial \dot{\omega}_i}{\partial x_i} = \tau_{ji} \frac{\partial w_i}{\partial x_j} + \sum_K j^*_{Ki} k_{Ki} \ . \tag{3.74}$$

It is different to the power density (3.70) for pure substances due to the presence of the term $\sum_K j^*_{Ki} k_{Ki}$ for the power exerted by the body forces on the individual substances. The energy equation (3.73) is transformed to the following for multicomponent mixtures

$$\varrho \frac{du}{dt} = -\frac{\partial \dot{q}'_i}{\partial x_i} + \tau_{ji} \frac{\partial w_i}{\partial x_j} + \sum_K j^*_{Ki} k_{Ki} \tag{3.75}$$

with

$$\dot{q}'_i = \dot{q}_i + \sum_K h_K j^*_{Ki} \ , \tag{3.76}$$

which is transformed into the energy balance (3.71) for pure substances when a vanishing diffusion flow $j^*_{Ki} = 0$ exists.

3.2.4.1 Dissipated energy and entropy

The energy equation (3.71), clearly shows that the internal energy in a fluid element is changed by the influx of heat and the performance of work. In the case of multicomponent mixtures, as shown in (3.75), a further term for the energy fed into the system with the material is added.

The work done consists of the reversible part and the dissipated work. We want to calculate these contributions individually with the help of an entropy balance. For the sake of simplicity we will only consider pure substances. According to Gibb's fundamental equation, we have

$$T\frac{ds}{dt} = \frac{du}{dt} + p\frac{dv}{dt} = \frac{du}{dt} - \frac{p}{\varrho^2}\frac{d\varrho}{dt} \quad . \tag{3.77}$$

Inserting the energy equation (3.71) and the continuity equation (3.21) yields

$$T\frac{ds}{dt} = -\frac{1}{\varrho}\frac{\partial \dot{q}_i}{\partial x_i} + \frac{1}{\varrho}\tau_{ji}\frac{\partial w_i}{\partial x_j} + \frac{p}{\varrho}\frac{\partial w_i}{\partial x_i} \quad .$$

Now we use

$$\frac{\partial(\dot{q}_i/T)}{\partial x_i} = \frac{1}{T}\frac{\partial \dot{q}_i}{\partial x_i} - \frac{\dot{q}_i}{T^2}\frac{\partial T}{\partial x_i}$$

or

$$\frac{\partial \dot{q}_i}{\partial x_i} = T\frac{\partial(\dot{q}_i/T)}{\partial x_i} + \frac{\dot{q}_i}{T}\frac{\partial T}{\partial x_i}$$

and according to (3.53)

$$\tau_{ji} = \hat{\tau}_{ji} - \delta_{ji}p \quad ,$$

if we presume that the bulk viscosity disappears. With that we can write the entropy balance (3.77) as

$$\varrho\frac{ds}{dt} = -\frac{\partial(\dot{q}_i/T)}{\partial x_i} - \frac{\dot{q}_i}{T^2}\frac{\partial T}{\partial x_i} + \frac{1}{T}\hat{\tau}_{ji}\frac{\partial w_i}{\partial x_j} + \frac{1}{T}\left(-\delta_{ji}p\frac{\partial w_i}{\partial x_j}\right) + \frac{p}{T}\frac{\partial w_i}{\partial x_i} \quad .$$

Because $\delta_{ji}\partial w_i/\partial x_j = \partial w_j/\partial x_j = \partial w_i/\partial x_i$ this equation simplifies to

$$\varrho\frac{ds}{dt} = -\frac{\partial(\dot{q}_i/T)}{\partial x_i} - \frac{\dot{q}_i}{T^2}\frac{\partial T}{\partial x_i} + \frac{1}{T}\hat{\tau}_{ji}\frac{\partial w_i}{\partial x_j} \quad . \tag{3.78}$$

The first term on the right hand side of (3.78) represents the entropy fed in with the heat. This is known as the *entropy flux*. The two remaining terms represent the *production of entropy*. The second term stems from the finite temperature differences in thermal conduction, the third from the mechanical energy. We call

$$\phi := \hat{\tau}_{ji}\frac{\partial w_i}{\partial x_j} \tag{3.79}$$

the *viscous dissipation*. This is the mechanically dissipated energy (SI units W/m^3) per unit volume. Taking (3.53) into consideration

$$\tau_{ji} = \hat{\tau}_{ji} - p\delta_{ji}$$

the first law (3.71) can also be written

$$\varrho\frac{du}{dt} = -\frac{\partial \dot{q}_i}{\partial x_i} - p\frac{\partial w_i}{\partial x_i} + \phi \quad . \tag{3.80}$$

Vanishing bulk viscosity is also presupposed here. Introducing the viscous dissipation into the energy equation (3.75) for multicomponent mixtures, gives

$$\varrho\frac{du}{dt} = -\frac{\partial \dot{q}'_i}{\partial x_i} - p\frac{\partial w_i}{\partial x_i} + \phi + \sum_K j^*_{Ki}k_{Ki} \quad . \tag{3.81}$$

Likewise (3.81) presupposes vanishing bulk viscosity. The derivation of the entropy balance for mixtures can be found in Appendix A 5.

Example 3.3: Calculate the viscous dissipation for a Newtonian fluid. How large is the viscous dissipation for the special case of one dimensional flow $w_1 = w_1(x_2)$?

For a Newtonian fluid, according to (3.55) we have

$$\hat{\tau}_{ji} = \eta \left[\left(\frac{\partial w_j}{\partial x_i} + \frac{\partial w_i}{\partial x_j} \right) - \frac{2}{3} \delta_{ji} \frac{\partial w_k}{\partial x_k} \right]$$

and therefore

$$\Phi = \hat{\tau}_{ji} \frac{\partial w_i}{\partial x_j} = \eta \frac{\partial w_i}{\partial x_j} \left[\left(\frac{\partial w_j}{\partial x_i} + \frac{\partial w_i}{\partial x_j} \right) - \frac{2}{3} \delta_{ji} \frac{\partial w_k}{\partial x_k} \right] \ .$$

In the case of $w_1 = w_1(x_2)$ the expression is reduced to

$$\Phi = \eta \frac{\partial w_1}{\partial x_2} \frac{\partial w_1}{\partial x_2} = \eta \left(\frac{\partial w_1}{\partial x_2} \right)^2 \ .$$

3.2.4.2 Constitutive equations for the solution of the energy equation

In order to be able to solve the energy equation (3.71) or (3.75), some constitutive equations are required. We will now consider equation (3.71) for pure substances. This necessitates the introduction of the caloric equation of state $u = u(\vartheta, v)$. By differentiation we obtain

$$du = \left(\frac{\partial u}{\partial \vartheta} \right)_v d\vartheta + \left(\frac{\partial u}{\partial v} \right)_\vartheta dv = c_v \, d\vartheta + \left(\frac{\partial u}{\partial v} \right)_\vartheta dv$$

In thermodynamics, as shown, for example, in [3.2],

$$\left(\frac{\partial u}{\partial v} \right)_\vartheta = T \left(\frac{\partial p}{\partial \vartheta} \right)_v - p \ .$$

Furthermore $dv = -d\varrho/\varrho^2$. This leads to

$$\varrho \frac{du}{dt} = \varrho c_v \frac{d\vartheta}{dt} - \left[T \left(\frac{\partial p}{\partial \vartheta} \right)_v - p \right] \frac{1}{\varrho} \frac{d\varrho}{dt} \ . \tag{3.82}$$

With the continuity equation (3.21) we obtain for this

$$\varrho \frac{du}{dt} = \varrho c_v \frac{d\vartheta}{dt} + \left[T \left(\frac{\partial p}{\partial \vartheta} \right)_v - p \right] \frac{\partial w_i}{\partial x_i} \ .$$

The expression in the square brackets disappears for ideal gases. For incompressible fluids, $\varrho = $ const, $d\varrho/dt = 0$ and $\partial w_i/\partial x_i = 0$. The expression is simplified in both cases to

$$\varrho \frac{du}{dt} = \varrho c_v \frac{d\vartheta}{dt} \ .$$

In incompressible fluids it is not necessary to differentiate between c_p and c_v, $c_p = c_v = c$. While in an isotropic body, the heat flux is given by Fourier's law

$$\dot{q}_i = -\lambda \frac{\partial \vartheta}{\partial x_i} \ ,$$

in nonisotropic materials, for example in crystals, there are preferred directions for heat flow. The heat flux $\dot{q}_i$ is no longer dependent on just the gradients $\partial \vartheta / \partial x_i$ but also in the most general case on all three components of the temperature gradients. For example, for the heat flux $\dot{q}_1$ in the direction of the x-axis, it holds that

$$-\dot{q}_1 = \lambda_{11} \frac{\partial \vartheta}{\partial x_1} + \lambda_{21} \frac{\partial \vartheta}{\partial x_2} + \lambda_{31} \frac{\partial \vartheta}{\partial x_3} = \lambda_{i1} \frac{\partial \vartheta}{\partial x_i} \; .$$

There are corresponding expressions for the other two heat fluxes, so that in general, we can write

$$\dot{q}_i = -\lambda_{ji} \frac{\partial \vartheta}{\partial x_i} \quad (i, j = 1, 2, 3)$$

The heat flux in the direction of a coordinate axis depends on the temperature gradients in the direction of all the coordinate axes. The thermal conductivity λ_{ji} in Fourier's law is a tensor. As we can prove with methods of thermodynamics for irreversible processes [3.3], the thermal conductivity tensor is symmetrical $\lambda_{ij} = \lambda_{ji}$. It consists of six components. Out of these, for certain crystals a few or several agree with each other or disappear.

Example 3.4: Quartz crystals have different thermal conductivities along the directions of the individual coordinate axes. The thermal conductivity tensor is given by

$$\lambda_{ji} = \begin{pmatrix} \lambda_{11} & 0 & 0 \\ 0 & \lambda_{22} & 0 \\ 0 & 0 & \lambda_{33} \end{pmatrix} \; .$$

What is the differential equation for steady-state heat conduction, if λ_{ii} is presumed to be independent of temperature?

What is the ratio of heat flows by conduction in a cube of side length a in the three coordinate directions? As an approximate solution of this part of the question, presume that the heat flows in the three coordinate directions are independent of each other.

As no flow is present, the term $\tau_{ji} \, \partial w_i / \partial x_j$ vanishes from the energy equation (3.71). Steady-state conduction is presumed, $du/dt = 0$, which means the energy equation is simplified to

$$\frac{\partial \dot{q}_i}{\partial x_i} = 0 = \frac{\partial \dot{q}_1}{\partial x_1} + \frac{\partial \dot{q}_2}{\partial x_2} + \frac{\partial \dot{q}_3}{\partial x_3} \; .$$

For the Quartz crystal $\dot{q}_1 = -\lambda_{11} \, \partial \vartheta / \partial x_1$, $\dot{q}_2 = -\lambda_{22} \, \partial \vartheta / \partial x_2$, $\dot{q}_3 = -\lambda_{33} \, \partial \vartheta / \partial x_3$. The energy equation is thereby transformed into

$$0 = \lambda_{11} \frac{\partial^2 \vartheta}{\partial x_1{}^2} + \lambda_{22} \frac{\partial^2 \vartheta}{\partial x_2{}^2} + \lambda_{33} \frac{\partial^2 \vartheta}{\partial x_3{}^2} \; .$$

Under the assumption that the heat flows in the three coordinate directions are independent of each other, we have

$$0 = \lambda_{11} \frac{\partial^2 \vartheta_1}{\partial x_1{}^2} = \lambda_{22} \frac{\partial^2 \vartheta_2}{\partial x_2{}^2} = \lambda_{33} \frac{\partial^2 \vartheta_3}{\partial x_3{}^2} \; .$$

The temperature profile between the cube surfaces opposite each other is linear. The heat fluxes are then $\dot{q}_1 = \lambda_{11} \, \Delta \vartheta_1 / a$, $\dot{q}_2 = \lambda_{22} \, \Delta \vartheta_2 / a$, $\dot{q}_3 = \lambda_{33} \, \Delta \vartheta_3 / a$, and

$$\dot{q}_1 : \dot{q}_2 : \dot{q}_3 = \lambda_{11} \, \Delta \vartheta_1 : \lambda_{22} \, \Delta \vartheta_2 : \lambda_{33} \, \Delta \vartheta_3 \; .$$

In the case of equal temperature differences we get

$$\dot{q}_1 : \dot{q}_2 : \dot{q}_3 = \lambda_{11} : \lambda_{22} : \lambda_{33} \; .$$

3.2.4.3 Some other formulations of the energy equation

For practical calculations it is often advantageous to use other dependent variables instead of the internal energy in the energy equation. Especially useful are the "*enthalpy*" and the "*temperature form*" of the *energy equation*.

The *enthalpy form* is obtained, due to $h = u + pv$, by the addition of $\varrho\,d(pv)/dt$ to both sides of the energy equation (3.80), giving

$$\varrho\frac{dh}{dt} = -\frac{\partial \dot{q}_i}{\partial x_i} - p\frac{\partial w_i}{\partial x_i} + \varrho\frac{d(pv)}{dt} + \phi \ .$$

The third term on the right hand side is

$$\varrho\frac{d(pv)}{dt} = \varrho v\frac{dp}{dt} + \varrho p\frac{dv}{dt} = \frac{dp}{dt} - \frac{1}{\varrho}p\frac{d\varrho}{dt}$$

from which, using the continuity equation (3.21)

$$\varrho\frac{d(pv)}{dt} = \frac{dp}{dt} + p\frac{\partial w_i}{\partial x_i}$$

is obtained. This yields a form of the energy equation valid for both compressible and incompressible flow

$$\varrho\frac{dh}{dt} = -\frac{\partial \dot{q}_i}{\partial x_i} + \frac{dp}{dt} + \phi \ . \tag{3.83}$$

Correspondingly, for multicomponent mixtures, we find from (3.81)

$$\varrho\frac{dh}{dt} = -\frac{\partial \dot{q}_i'}{\partial x_i} + \frac{dp}{dt} + \sum_K j_{Ki}^* k_{Ki} + \phi \tag{3.84}$$

with $\dot{q}_i' = \dot{q}_i + \sum_K h_K j_{Ki}^*$.

The *temperature form of the energy equation* is especially important, as its solution describes the temperature field in terms of time and space. For *pure substances*, this equation is obtained, if we make use of the well known relationship from thermodynamics [3.2]

$$dh = c_p\,d\vartheta - \left[T\left(\frac{\partial v}{\partial \vartheta}\right)_p - v\right] dp \ .$$

and with that eliminate the enthalpy differential in (3.83). In addition $\dot{q}_i = -\lambda\partial\vartheta/\partial x_i$. This yields

$$\varrho c_p\frac{d\vartheta}{dt} = \frac{\partial}{\partial x_i}\left(\lambda\frac{\partial\vartheta}{\partial x_i}\right) + \frac{T}{v}\left(\frac{\partial v}{\partial \vartheta}\right)_p \frac{dp}{dt} + \phi \ . \tag{3.85}$$

For ideal gases, because of $pv = RT$, the expression will be

$$\frac{T}{v}\left(\frac{\partial v}{\partial \vartheta}\right)_p = \frac{T}{v}\left(\frac{\partial v}{\partial T}\right)_p = 1 \ .$$

The enthalpy of *mixtures* is dependent on temperature, pressure and composition. Its differential is given by

$$dh = c_p\, d\vartheta - \left[T\left(\frac{\partial v}{\partial \vartheta}\right)_p - v \right] dp + \sum_K h_K\, d\xi_K \ , \tag{3.86}$$

when h_A is the partial specific enthaply of substance A, defined by

$$h_A = \left(\frac{\partial H}{\partial M_A}\right)_{T,p,M_{K\neq A}} .$$

Therefore, taking into consideration the continuity equation (3.25), this becomes

$$\varrho\frac{dh}{dt} = \varrho c_p\frac{d\vartheta}{dt} + \left[1 - \frac{T}{v}\left(\frac{\partial v}{\partial \vartheta}\right)_p\right]\frac{dp}{dt} - \sum_K h_K\frac{\partial j_{Ki}^*}{\partial x_i} + \sum_K h_K\dot{\Gamma}_K \ .$$

After inserting this expression into the energy equation (3.84), and some rearranging, we get

$$\varrho c_p\frac{d\vartheta}{dt} = \frac{\partial}{\partial x_i}\left(\lambda\frac{\partial \vartheta}{\partial x_i}\right) + \frac{T}{v}\left(\frac{\partial v}{\partial T}\right)_p\frac{dp}{dt} + \phi$$
$$+ \sum_K j_{Ki}^*\left(k_{Ki} - \frac{\partial h_K}{\partial x_i}\right) - \sum_K h_K\dot{\Gamma}_K$$

As shown in Apendix A5, the partial specific enthalpy h_K in the penultimate sum can be eliminated and replaced by the specific enthalpy h of the mixture, because

$$\sum_{K=1}^N j_{Ki}^*\frac{\partial h_K}{\partial x_i} = \sum_{K=1}^{N-1} j_{Ki}^*\frac{\partial}{\partial x_i}\left(\frac{\partial h}{\partial \xi_K}\right)$$

holds, where the derivative $\partial h/\partial \xi_K$ is formed at fixed temperature, pressure and all mass fractions except ξ_K.

The *temperature form of the energy equation for mixtures* is now

$$\varrho c_p\frac{d\vartheta}{dt} = \frac{\partial}{\partial x_i}\left(\lambda\frac{\partial \vartheta}{\partial x_i}\right) + \frac{T}{v}\left(\frac{\partial v}{\partial \vartheta}\right)_p\frac{dp}{dt} + \phi$$
$$+ \sum_{K=1}^{N-1} j_{Ki}^*\left[k_{Ki} - \frac{\partial}{\partial x_i}\left(\frac{\partial h}{\partial \xi_K}\right)\right] - \sum_K h_K\dot{\Gamma}_K \ . \tag{3.87}$$

Equations (3.85) and (3.87) are valid irrespective of whether the fluid is compressible or incompressible.

It is known that incompressible fluids represent a useful model for real fluids in fluid mechanics and heat and mass transfer. Their thermal equation of state is $v = v_0 = \text{const}$. For pure substances and also for mixtures, isobaric and isochoric specific heat capacities agree with each other, $c_p = c_v = c$.

The *temperature form of the energy equation for incompressible pure substances* is yielded from (3.85) under the assumption of constant thermal conductivity

$$\varrho c\frac{d\vartheta}{dt} = \lambda\frac{\partial^2\vartheta}{\partial x_i^2} + \phi \ . \tag{3.88}$$

If the flow is isobaric, the second term on the right hand side of (3.85) disappears. In the case of an ideal gas $(T/v)(\partial v/\partial \vartheta)_p = 1$ has to be put into (3.85).

In *incompressible mixtures* we find from (3.87), presuming constant thermal conductivity,

$$\varrho c \frac{d\vartheta}{dt} = \lambda \frac{\partial^2 \vartheta}{\partial x_i{}^2} + \phi + \sum_K j_{Ki}^* \left[k_{Ki} - \frac{\partial}{\partial x_i} \left(\frac{\partial h}{\partial \xi_K} \right) \right] - \sum_K h_K \dot{\Gamma}_K \ . \tag{3.89}$$

The energy balance for mixtures is different from that for pure substances as it contains two additional expressions, one for the energy transport by mass diffusion and one for the enthalpy or temperature changes produced by chemical reactions.

3.2.5 Summary

The most important balance equations derived in this chapter shall be summarised here. In this summary we will use the abbreviation

$$\frac{d}{dt} = \frac{\partial}{\partial t} + w_i \frac{\partial}{\partial x_i} \ .$$

The following balance equations are valid for *pure substances*:

$$\frac{d\varrho}{dt} = -\varrho \frac{\partial w_i}{\partial x_i} \ , \tag{3.90}$$

$$\varrho \frac{dw_j}{dt} = \varrho k_j + \frac{\partial \tau_{ji}}{\partial x_i} \ , \tag{3.91}$$

$$\varrho \frac{du}{dt} = -\frac{\partial \dot{q}_i}{\partial x_i} + \tau_{ji} \frac{\partial w_i}{\partial x_j} \ . \tag{3.92}$$

Equation (3.90) is the mass balance or continuity equation, (3.91) the momentum balance or Cauchy's equation of motion and (3.92) is the energy balance. As a momentum balance exists for each of the three coordinate directions, $j = 1, 2, 3$, there are five balance equations in total. The enthalpy form (3.83) is equivalent to the energy balance (3.92).

In *multicomponent mixtures* $N - 1$ mass balances for the components have to be added to the continuity equation, see (3.25), whilst the energy balance is replaced by (3.81) or (3.87).

The system of equations still has to be supplemented by the so-called constitutive equations or material laws which describe the behaviour of the materials being investigated.

If the frequently used case of an *incompressible Newtonian fluid*, $\varrho = $ const, $d\varrho/dt = 0$, of constant viscosity is presumed, the continuity equation, momentum and energy balances are transformed into

$$\frac{\partial w_i}{\partial x_i} = 0 \ , \tag{3.93}$$

$$\varrho \frac{dw_j}{dt} = \varrho k_j - \frac{\partial p}{\partial x_j} + \eta \frac{\partial^2 w_j}{\partial x_i{}^2} \ , \tag{3.94}$$

$$\varrho c \frac{d\vartheta}{dt} = \lambda \frac{\partial^2 \vartheta}{\partial x_i^2} + \phi \tag{3.95}$$

with

$$\phi = \eta \frac{\partial w_i}{\partial x_j} \left(\frac{\partial w_j}{\partial x_i} + \frac{\partial w_i}{\partial x_j} \right)$$

Equation (3.94) is the Navier-Stokes equation for an incompressible fluid.

For *multicomponent mixtures* of N components, in addition to the continuity equations for the total mass (3.93), we also need continuity equations for $N-1$ components

$$\frac{d\varrho_A}{dt} = -\frac{\partial j^*_{Ai}}{\partial x_i} \dot{\Gamma}_A . \tag{3.96}$$

The momentum balance (3.94) remains unchanged, whilst for the energy equation we obtain (3.89):

$$\varrho c \frac{d\vartheta}{dt} = \lambda \frac{\partial^2 \vartheta}{\partial x_i^2} + \phi + \sum_{K=1}^{N-1} j^*_{Ki} \left[k_{Ki} - \frac{\partial}{\partial x_i} \left(\frac{\partial h}{\partial \xi_K} \right) \right] - \sum_K h_K \dot{\Gamma}_K . \tag{3.97}$$

Example 3.5: A so-called Couette flow consists of a fluid between two parallel, infinitely large plates, where one plate — the top one in Fig. 3.11 — moves with constant velocity w_L whilst the other plate is kept still.

A good example of this type of flow is oil in a friction bearing. Here we will consider an incompressible steady-state flow, whose velocity profile is given by $w_1(x_2)$. Furthermore $\partial/\partial t = 0$, $w_2 = w_3 = 0$.

Find the velocity profile $w_1(x_2)$.

The temperature of the upper plate is ϑ_L, and that of the lower is $\vartheta_0 < \vartheta_L$. Calculate and discuss the pattern of the temperature profile $\vartheta(x_2)$, under the assumption that the viscous dissipation cannot be neglected.

Velocities are only present along the x_1-axis. This means that only the momentum equation (3.94), for $j = 1$ has to be considered. In this equation the left hand side disappears, as $\partial/\partial t = 0$, $\partial w/\partial x_1 = 0$, $w_2 = w_3 = 0$. Futhermore the mass force is $k_1 = 0$, and also $\partial p/\partial x_1 = 0$, as the flow is caused by the movement of the upper plate and not because of a pressure difference. The momentum equation is reduced to

$$0 = \eta \frac{\partial^2 w_1}{\partial x_2^2} , \quad \text{and therefore} \quad w_1 = \frac{x_2}{L} w_L .$$

Likewise, in the energy equation (3.95), all the terms on the left hand side disappear because $\partial/\partial t = 0$, $\partial \vartheta/\partial x_1 = 0$, $w_2 = 0$. This then becomes

$$\lambda \frac{\partial^2 \vartheta}{\partial x_2^2} + \phi = \quad \text{with} \quad \phi = \eta \left(\frac{\partial w_1}{\partial x_2} \right)^2 = \eta \left(\frac{w_L}{L} \right)^2 .$$

Integration of the energy equation yields a parabolic temperature profile in x_2

$$\vartheta(x_2) = -\frac{\eta}{2\lambda} \left(\frac{w_L}{L} \right)^2 x_2^2 + a_1 x_2 + a_0 .$$

The constants a_0 and a_1 follow out of the boundary conditions

$$\vartheta(x_2 = 0) = \vartheta_0 \quad \text{und} \quad \vartheta(x_2 = L) = \vartheta_L .$$

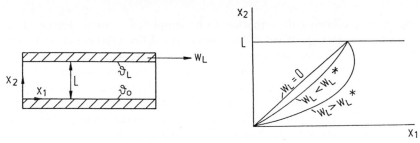

Fig. 3.11: Couette flow

With this the temperature profile is found to be

$$\vartheta = \vartheta_0 + \frac{\eta\, w_L^2}{2\,\lambda} \left[\frac{x_2}{L} - \left(\frac{x_2}{L}\right)^2 \right] + (\vartheta_L - \vartheta_0) \frac{x_2}{L} \ .$$

The pattern of the temperature profile over x_2 is shown on the right in Fig. 3.11. If the upper plate does not move, $w_L = 0$, the temperature profile is the same as that for pure heat conduction between two flat plates.

For sufficiently large values of w_L the temperature profile has a maximum, the position of which is calculated from $\mathrm{d}\vartheta/\mathrm{d}x_2 = 0$ to be

$$(x_2)_{\max} = \left[\frac{\lambda}{\eta\, w_L^2} (\vartheta_L - \vartheta_0) + \frac{1}{2} \right] L \ .$$

It has to be that $(x_2)_{\max} \leq L$. This criterium is met, if

$$w_L \geq \left[\frac{2\,\lambda}{\eta} (\vartheta_L - \vartheta_0) \right]^{1/2} = w_L^* \ .$$

At velocities less than w_L^* very little energy will be dissipated, the maximum temperature will coincide with the temperature of the upper plate. At velocities greater than w_L^*, so much energy will be dissipated that at a certain region between the plates the fluid will be heated to a temperature above that of the upper plate.

3.3 Influence of the Reynolds number on the flow

A general solution of the Navier-Stokes equations has not been possible until now. The main cause of these difficulties is the non-linear character of the differential equations by the product of the inertia terms

$$\varrho \frac{\mathrm{d}w_j}{\mathrm{d}t} = \varrho \frac{\partial w_j}{\partial t} + \varrho w_i \frac{\partial w_j}{\partial x_i}$$

on the left hand side of the Navier-Stokes equation. Solutions are only known for certain special cases. These are yielded because it is possible to determine under what conditions individual terms in the equation can be neglected compared to others, thereby allowing the equation to be simplified to some degree. In order to assess the magnitude of individual terms, the differential equation is sensibly rearranged so that it contains dimensionless quantities. This means that the magnitude of the terms is then independent of the measurement system chosen. This type of presentation of the equation by dimensionless groups is always possible, and is an expression of the general principle that the description and solution of a physical problem have to be independent of the measurement system. The use of problem orientated standard measures of the quantities is equivalent to the introduction of dimensionless groups. We will consider here the Navier-Stokes equation for incompressible flow (3.94) without body forces

$$\varrho \frac{\mathrm{d}w_j}{\mathrm{d}t} = \varrho \frac{\partial w_j}{\partial t} + \varrho w_i \frac{\partial w_j}{\partial x_i} = -\frac{\partial p}{\partial x_j} + \eta \frac{\partial^2 w_j}{\partial x_i^2} \ . \tag{3.98}$$

All velocities are divided by a reference velocity w_α given for the problem, for example the upstream velocity of a body in cross-flow. In the same way all lengths are divided by a reference length L given for the problem, an example of this is the length over which the fluid flows across a body. With this we can then form dimensionless quantities

$$w_i^+ = w_i/w_\alpha \quad , \qquad p^+ = p/\varrho w_\alpha^2$$
$$x_i^+ = x_i/L \quad , \qquad t^+ = t\,w_\alpha/L$$

and then introduce them into the Navier-Stokes equation (3.98). The viscosity η will be replaced by the kinematic viscosity ν according to $\eta = \nu \varrho$. This then gives

$$\frac{\partial w_j^+}{\partial t^+} + w_i^+ \frac{\partial w_j^+}{\partial x_i^+} = -\frac{\partial p^+}{\partial x_j^+} + \frac{1}{Re} \frac{\partial^2 w_j^+}{\partial x_i^{+2}} \tag{3.99}$$

with the already well known Reynolds number

$$Re = \frac{w_\alpha L}{\nu} \ .$$

In addition we have the dimensionless form of the continuity equation for incompressible flow

$$\frac{\partial w_i^+}{\partial x_i^+} = 0 \ , \tag{3.100}$$

which emerges from (3.93) after the inmtroduction of the dimensionless groups together with the boundary conditions, (3.99) and (3.100) completely describe the flow. As a solution we obtain the velocity field $w_j^+(t^+, x_i^+)$ and the presure field $p^+(t^+, x_i^+)$.

Obviously the solution still depends on the Reynolds number. This can be interpreted as the ratio of the inertia to friction forces. The inertia force $\varrho \mathrm{d}w_1/\mathrm{d}t$ is of magnitude $\varrho w_\alpha/t$, in which the time t is of magnitude L/w_α and therefore, the inertia force is of magnitude

$$\varrho w_\alpha^2/L$$

The friction force $\eta \partial^2 w_1 / \partial x_1^2$ is of magnitude

$$\eta w_\alpha / L^2 \ .$$

The ratio of the two terms is of the magnitude of

$$\frac{\varrho w_\alpha^2 / L}{\eta w_\alpha / L^2} = \frac{\varrho w_\alpha L}{\eta} = \frac{w_\alpha L}{\nu} = Re$$

and will just be described by the Reynolds number Re.

As we can see the inertia forces increase with the square of the velocity. They increase far more rapidly with the velocity than the friction forces which are only linearly dependent on the velocity. At large Reynolds numbers the disturbances to the velocity that are always present cannot be dampenend by the comparatively small friction forces. This manifests itself in the fact that flows at a certain Reynolds number, the so-called critical Reynolds number, change their pattern.

Whilst below the critical Reynolds number the fluid particles move along distinct stream lines and disturbances in the velocity rapidly disappear again, above this critical Reynolds number disturbances in the velocity are no longer dampened but intensified. This type of flow is called *turbulent*. The flow along distinct streamlines is known as *laminar*. Turbulent flow is always three-dimensional, unsteady and exhibits an irregular vortex pattern. The velocity at a fixed point fluctuates irregularly around a mean value. The momentary values of velocity, pressure, temperature and concentration are random quantities.

Solutions of the Navier-Stokes equations for laminar flows at large Reynolds numbers only describe real flows if the solutions are stable against small disturbances. A momentarily small disturbance has to disappear again. This is no longer the case above the critical Reynolds number. A momentarily small disturbance does not disappear, rather it grows. The flow form changes from laminar to turbulent flow.

These occurences were first observed by Osborne Reynolds (1842–1912), when he added a dye to the flow along the axis of a glass tube. In laminar flow a thin thread of colour forms in the axis of the tube, which, due to the low rate of molecular diffusion, hardly gets wider. Increasing the velocity so that the critical Reynolds number is sufficiently surpassed, means that the thread is rapidly mixed up into the flow. In tube flow it can be shown that at Reynolds numbers

$$Re_{\mathrm{crit}} = w_{\mathrm{m}} d / \nu \leq 2300 \ ,$$

formed with the velocity w_{m} averaged over the cross section and the tube diameter d, even in highly disturbed inlet flow, the flow always remains laminar. With a flow that is especially free of disturbances critical Reynolds numbers have been measured up to $40\,000$. However because in most technical applications it is impossible to have a flow completely free of disturbances, flow is only laminar below $Re_{\mathrm{crit}} = 2\,300$. At Reynolds numbers up to about $2\,600$ the flow has an intermittent character. It varies periodically with time, partly laminar, partly turbulent. Above $Re = 2\,600$ flow is completely turbulent.

In transverse flow along a flat plate the transition from laminar to turbulent flow occurs at Reynolds numbers $w_\alpha L/\nu$ between $3 \cdot 10^5$ and $5 \cdot 10^5$; w_α is the initial flow velocity, L is the length of the plate over which the fluid is flowing. The heat and mass transfer in turbulent flows is more intensive than in laminar. In general, at the same time there is also an increase in the pressure drop.

3.4   Simplifications to the Navier-Stokes equations

Simplifications to the Navier-Stokes equations are produced when the Reynolds number is very small or very large, $Re \to 0$ or $Re \to \infty$. These limiting cases are never reached in reality but they represent asymptotic solutions and are better approximations the larger or smaller the Reynolds number is. We will investigate these limiting cases in the following.

3.4.1   Creeping flows

The limiting case $Re \to 0$ is known as creeping flow. It can be realised if the velocity w_α is very small, the density ϱ is very low, for example in very dilute gases, and the viscosity η is very large, i.e. in highly viscous fluids, or if the typical body dimension L is very small as in flow around dust particles or fog droplets. The viscous forces are far greater than the inertia, so the left hand side of (3.99) is negligible in comparison to the term $(1/Re)(\partial^2 w_j^+/\partial x_i^{+2})$. In contrast to this the pressure gradient $\partial p^+/\partial x_j^+$ cannot be neglected, it is of magnitude

$$\frac{\Delta p}{\varrho w_\alpha^2} = \frac{\Delta p/L}{\varrho w_\alpha^2/L} \quad,$$

and contains the inertia forces as the small quantity in the denominator. The pressure term can only be neglected if the pressure drop is significantly smaller than the inertia forces. We can decide whether this applies, by first solving the momentum and continuity equation. By neglecting the inertia forces in (3.99) the non-linear terms drop out, and after transforming the equation back into dimensioned quantities we are left with the linear differential equation

$$\frac{\partial p}{\partial x_j} = \eta \frac{\partial^2 w_j}{\partial x_i^2} \quad. \tag{3.101}$$

A further differentiation provides

$$\frac{\partial^2 p}{\partial x_j{}^2} = \eta \frac{\partial}{\partial x_j} \frac{\partial^2 w_j}{\partial x_i{}^2} = \eta \frac{\partial^2}{\partial x_i{}^2} \left(\frac{\partial w_j}{\partial x_j} \right) = 0 \ . \tag{3.102}$$

As the flow was presumed to be incompressible, as a result of the continuity equation, $\partial w_j / \partial x_j = 0$ holds. The pressure satisfies the potential equation. Equations (3.101) and (3.102) are the basis of the hydrodynamic theory of lubrication, which encompasses, among others, the oil flow in bearings.

3.4.2   Frictionless flows

If we presume that the flow is completely free of friction $\eta = 0$, then $1/Re = 0$. The friction term on the right hand side of (3.99) disappears and after transformation to dimensioned quantities, we obtain

$$\varrho \frac{\partial w_j}{\partial t} + \varrho w_i \frac{\partial w_j}{\partial x_i} = -\frac{\partial p}{\partial x_j} \ . \tag{3.103}$$

This is *Euler's equation*. It contains, in the special case of one-dimensional steady-state flow, the relationship

$$\varrho w_1 \frac{\mathrm{d}w_1}{\mathrm{d}x_1} = -\frac{\mathrm{d}p}{\mathrm{d}x_1}$$

which when intregrated

$$\varrho \frac{w_1^2}{2} + p = \mathrm{const}$$

is known as *Bernoulli's equation*. The influence of gravity here is neglected.

Whilst the Navier-Stokes equation (3.98) is of second order, Euler's equation (3.103) only contains first order terms. As its order is one lower than the Navier-Stokes equation, after integration one less boundary condition can be satisfied. As a result of this, the no-slip condition, zero velocity at the wall, cannot be satisfied. Rather a finite velocity at the wall is obtained, as the absence of friction was presumed, whilst in real flows the velocity is zero at the wall.

3.4.3   Boundary layer flows

As we have seen in the previous sections, the friction term in the Navier-Stokes equation (3.98) may not be neglected for large Reynolds numbers $Re \to \infty$, if we wish to correctly describe the flow close to the wall, and satisfy the no-slip condition. The region in which the friction forces may not be neglected compared to the inertia forces is generally bounded by a very thin zone close to the wall, as

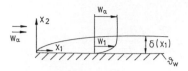

Fig. 3.12: Velocities in a boundary layer

the velocity increase and therefore also the shear stress are very large, as is clearly illustrated in Fig. 3.12. The shear stress

$$\tau_{21} = \eta \frac{\partial w_1}{\partial x_2}$$

is especially large close to the wall. In contrast to this it is negligible a large distance away from the wall. The layer close to the wall, in which the inertia and friction forces are of the same magnitude, is called the *velocity boundary layer* or simply the *boundary layer*. The calculation of the momentum, heat and mass transfer in these types of flows is the object of boundary layer theory and will be dealt with in the following section. The thickness of the boundary layer $\delta(x_1)$ will merely be guessed at here. This will only involve consideration of the inertia and friction force acting in the direction of the x_1-axis. The characteristic inertia term is of the magnitude

$$\varrho w_\alpha^2 / x_1 \ ,$$

if we replace the characteristic length by the distance from the leading edge. In the x_1-direction, the following friction terms appear in (3.98) for the flow being considered here, which has a velocity $w_1(x_1, x_2)$

$$\eta \frac{\partial^2 w_1}{\partial x_i^2} = \eta \frac{\partial^2 w_1}{\partial x_1^2} + \eta \frac{\partial^2 w_1}{\partial x_2^2} \ .$$

Whilst the first term on the right hand side is of magnitude $\eta w_\alpha / x_1^2$, the order of magnitude of the second term is $\eta w_\alpha / \delta^2$.

As the boundary layer is a great deal thinner when compared to the plate length, at a sufficient distance from the front of the plate, $\delta \ll x_1$, the second term for the friction inside the boundary layer is decisive.

Equating the orders of magnitude of the inertia and friction forces yields

$$\frac{\varrho w_\alpha^2}{x_1} \approx \frac{\eta w_\alpha^2}{\delta^2}$$

or

$$\delta \approx x_1 / \sqrt{Re_{x_1}} \tag{3.104}$$

with $Re_{x_1} = w_\alpha x_1 / \nu$. The boundary layer increases with $\sqrt{x_1}$. It is thinner the higher the Reynolds number. Within the narrow region of the boundary layer the friction forces can clearly not be neglected, whilst outside the boundary layer they have little significance, so that Euler's equation holds for the external region.

3.5 The boundary layer equations

3.5.1 The velocity boundary layer

At large Reynolds numbers, as has already been discussed, the flow area can be split into two regions, the outer, frictionless flow which is described by Euler's equation, and the flow inside the boundary layer, which is characterised by the fact that the friction forces are no longer negligible when compared to the inertia forces. From Euler's equation, for a particular pressure field, we obtain a velocity profile in the outer space. However we are still not in a position to investigate the resistances to flow. To do this we also need to know the velocity profile in the boundary layer. Inside the boundary layer the velocity profile changes from a value of zero at the wall to the asymptotic value of the velocity in the outer, frictionless region. In the same way the values for the temperature and concentration change from the values at the wall to the asymptotic values of the outer region.

In the following we will assume the velocities, temperatures and concentrations in the outer region to be known, and consider a steady-state, two-dimensional flow. The body forces are negligible. Flow along a curved wall can be taken to be two-dimensional as long as the radius of curvature of the wall is much bigger compared to the thickness of the boundary layer. The curvature is then insignificant for the thin boundary layer, and it develops just as if it was on a flat wall. The curvature of the wall is merely of influence on the outer flow and its pressure distribution.

We will introduce so-called boundary layer coordinates, Fig. 3.13, in which the coordinate $x_1 = x$ is chosen to be along the surface of the body and $x_2 = y$ as perpendicular to it. We will presume an initial velocity $w_\alpha(y)$; its integral mean value will be w_m.

Density changes in flowing liquids are very small, if we disregard the extreme cases, for example density changes caused by the very fast opening and closing of valves in piping systems. So, in general, for flowing liquids, $d\varrho/dt \to 0$ holds, and we can apply the continuity and momentum equations for incompressible flow. The density changes in gases are also small, if the velocities ar so low that the *Mach number* formed with the velocity of sound is $Ma = w_\mathrm{m}/w_\mathrm{S} \ll 1$. We will illustrate this for the example of a reversible, adiabatic flow. It follows from the

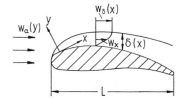

Fig. 3.13: Boundary layer in transverse flow along a body

equation of state $p = p(\varrho, s)$ that

$$dp = \left(\frac{\partial p}{\partial \varrho}\right)_s d\varrho \ .$$

The derivative here is

$$\left(\frac{\partial p}{\partial \varrho}\right)_s := w_S^2$$

where w_S is the velocity of sound. So, for a reversible, adiabatic flow, it holds that

$$dp = w_S^2 \, d\varrho \ .$$

As no technical work is executed, we have

$$dw_t = 0 = v \, dp + d\left(\frac{w^2}{2}\right) \ ,$$

when we neglect changes in the potential energy. With that it follows that

$$v \, dp = v w_S^2 \, d\varrho = w_S^2 \, d\varrho/\varrho$$

and

$$\frac{d\varrho}{\varrho} = \frac{1}{w_S^2} v \, dp = -\frac{1}{w_S^2} d\left(\frac{w^2}{2}\right) = -\frac{w}{w_S^2} dw$$

or

$$\frac{d\varrho}{\varrho} = -Ma^2 \frac{dw}{w} \ .$$

$dw/w \leq 1$ is valid for the relative velocity change. The relative density variation $d\varrho/\varrho$ is small for $Ma < 1$, e. g. for $Ma = 0.3$, which corresponds to a velocity of $100 \, \text{m/s}$ of air at ambient state, we obtain

$$\frac{d\varrho}{\varrho} \leq 0.09 \ .$$

Therefore, gases at moderate velocities, without large changes in their temperature, may be assumed to be incompressible, $d\varrho/dt \to 0$, and their flow can be approximately described by the continuity and momentum equations for incompressible flow.

Under these presumptions, the continuity and Navier-Stokes equations (3.93) and (3.94), with $w_1 = w_x$ and $w_2 = w_y$, i.e. leaving the tensor system out, may be written as

$$\frac{\partial w_x}{\partial x} + \frac{\partial w_y}{\partial y} = 0 \tag{3.105}$$

$$\varrho w_x \frac{\partial w_x}{\partial x} + \varrho w_y \frac{\partial w_x}{\partial y} = -\frac{\partial p}{\partial x} + \eta \left(\frac{\partial^2 w_x}{\partial x^2} + \frac{\partial^2 w_x}{\partial y^2}\right) \tag{3.106}$$

$$\varrho w_x \frac{\partial w_y}{\partial y} + \varrho w_y \frac{\partial w_y}{\partial y} = -\frac{\partial p}{\partial y} + \eta \left(\frac{\partial^2 w_y}{\partial x^2} + \frac{\partial^2 w_y}{\partial y^2}\right) \ . \tag{3.107}$$

In order to estimate the order of magnitude of the individual terms in these equations, dimensionless quantities are introduced. It is useful to measure the distance y from the wall relattive to a "mean boundary layer thickness"

$$\delta_m \sim L/Re^{1/2}$$

with $Re = w_m L/\nu$, whilst the length L is chosen as the length scale for the flow direction x, and the velocity w_x is divided by the mean velocity w_m in the initial flow direction. Out of the continuity equation follows

$$w_y \sim \frac{w_m}{L}\delta_m$$

and therefore

$$w_y \sim w_m/Re^{1/2} \ .$$

As a standard for the velocity w_y, $w_m/Re^{1/2}$ is introduced. In this manner the following dimensionless quantities are yielded:

$$w_x{}^+ = \frac{w_x}{w_m} \ , \quad w_y{}^+ = \frac{w_y}{w_m}Re^{1/2} \ , \quad x^+ = \frac{x}{L} \ , \quad y^+ = \frac{y}{\delta_m} = \frac{y}{L}Re^{1/2} \ , \quad p^+ = \frac{p}{\varrho w_m^2} \ .$$

With these quantities (3.105) to (3.107) assume the following forms:

$$\frac{\partial w_x^+}{\partial x^+} + \frac{\partial w_y^+}{\partial y^+} = 0 \ ,$$

$$w_x^+\frac{\partial w_x^+}{\partial x^+} + w_y^+\frac{\partial w_x^+}{\partial y^+} = -\frac{\partial p^+}{\partial x^+} + \frac{1}{Re}\frac{\partial^2 w_x^+}{\partial x^{+2}} + \frac{\partial^2 w_x^+}{\partial y^{+2}} \ ,$$

$$\frac{1}{Re}\left(w_x^+\frac{\partial w_y^+}{\partial x^+} + w_y^+\frac{\partial w_y^+}{\partial y^+}\right) = -\frac{\partial p^+}{\partial y^+} + \frac{1}{Re^2}\frac{\partial^2 w_y^+}{\partial x^{+2}} + \frac{1}{Re}\frac{\partial^2 w_y^+}{\partial y^{+2}} \ .$$

For large Reynolds numbers $Re \to \infty$ all terms with the factor $1/Re$ and even more those with $1/Re^2$ will be vanishingly small. After neglecting these terms and returning to equations with dimensions, we obtain the boundary layer equations, which were first given in this form by L. Prandtl (1875–1953) [3.4] in 1904:

$$\frac{\partial w_x}{\partial x} + \frac{\partial w_y}{\partial y} = 0 \ , \tag{3.108}$$

$$\varrho w_x\frac{\partial w_x}{\partial x} + \varrho w_y\frac{\partial w_x}{\partial y} = -\frac{\partial p}{\partial x} + \eta\frac{\partial^2 w_x}{\partial y^2} \ , \tag{3.109}$$

$$\frac{\partial p}{\partial y} = 0 \ . \tag{3.110}$$

The two unknown velocities w_x and w_y are obtained from (3.108) and (3.109). The pressure is no longer an unknown, as according to (3.110) the pressure $p = p(x)$ is not a function of the coordinate y normal to the wall. It is determined by the outer flow region, that is the shape of the body around which the fluid is flowing or the channel through which the fluid flows. The pressure has the same value in the boundary layer for any position x as in the outer flow region and can, with the assistance of Euler's equation (3.103), be replaced by the velocity $w_\delta(x)$ of the outer flow region:

$$\varrho w_\delta \frac{\partial w_\delta}{\partial x} = -\frac{\mathrm{d}p}{\mathrm{d}x} . \tag{3.111}$$

The following boundary conditions are available for the solution of this system of equations:

The zero velocity condition at the wall

$$y = 0: \quad w_x = w_y = 0 , \tag{3.112}$$

the condition at the outer edge of the boundary layer, according to which the velocity w_y changes asymptotically to that of the outer flow region,

$$y \to \infty: \quad w_x = w_\delta \tag{3.113}$$

and the initial condition is

$$x = 0: \quad w_x = w_\alpha(y) . \tag{3.114}$$

The boundary layer equations are non-linear and with the exception of a few special cases, cannot be solved analytically.

3.5.2 The thermal boundary layer

In a similar way to the simplification of the Navier-Stokes equation for the calculations in the velocity boundary layer, the energy equation can also be simplified to allow calculation of the temperatures in the thermal boundary layer. To explain this further we we once again consider a body, as in Fig. 3.12, whose surface has a certain wall temperature ϑ_0, which is different from the fluid temperature ϑ_α at the leading edge. As before we have presumed steady-state, two-dimensional flow. Large temperature and pressure changes are to be excluded, so that we can assume constant thermal conductivity. Liquids are, as already mentioned, on the whole incompressible, $\mathrm{d}\varrho/\mathrm{d}t = 0$, whilst for gases, the contribution to the energy balance from the density changes, even if they are small, still has to be estimated. Therefore we will use the temperature form (3.85) of the energy equation, which is valid for both incompressible and compressible flow

$$\varrho c_p \left(w_x \frac{\partial \vartheta}{\partial x} + w_y \frac{\partial \vartheta}{\partial y} \right) = \lambda \left(\frac{\partial^2 \vartheta}{\partial x^2} + \frac{\partial^2 \vartheta}{\partial y^2} \right) + \frac{T}{v} \left(\frac{\partial v}{\partial \vartheta} \right)_p \frac{\mathrm{d}p}{\mathrm{d}t} + \phi \tag{3.115}$$

and estimate the magnitude of the individual terms as before, in which we will use some of the dimensionless quantites that have already been defined:

$$w_x^+ = \frac{w_x}{w_m}; \quad w_y^+ = \frac{w_y}{w_m} Re^{1/2}; \quad x^+ = \frac{x}{L}; \quad p^+ = \frac{p}{\varrho w_m^2} .$$

As a standard for the y coordinate the mean thickness of the thermal boundary layer is introduced. An approximation of the thermal boundary layer has already been given in (3.8)

$$\delta_T \sim x \left(\frac{a}{w_m x} \right)^{1/2}$$

where w_m was presumed to be a constant velocity.

The thermal diffusivity is defined here by $a := \lambda/\varrho c_p$. The mean thickness δ_{Tm} of the thermal boundary layer

$$\delta_{Tm} \sim L \left(\frac{a}{w_m L} \right)^{1/2} \quad ,$$

is determined by this equation, except for the still unknown proportionality factor. The quantity which exists here

$$\frac{w_m L}{a}$$

is the already well known *Péclet number Pe*. It is linked with the Reynolds number by

$$Pe = \frac{w_m L}{a} = \frac{w_m L}{\nu} \frac{\nu}{a} = Re \, Pr \ .$$

$Pr = \nu/a$ is the Prandtl number. With δ_{Tm} we form the dimensionless coordinate

$$y^+ = \frac{y}{\delta_{Tm}} = \frac{y}{L} Pe^{1/2} \ .$$

In addition a dimensionless time is introduced

$$t^+ = \frac{t}{L} w_m$$

and as the dimensionless temperature we have

$$\vartheta^+ = \frac{\vartheta - \vartheta_0}{\vartheta_\alpha - \vartheta_0} \ .$$

By this normalisation the temperature in the thermal boundary layer varies between the values $0 \leq \vartheta^+ \leq 1$. With these dimensionless quantities the energy equation (3.115) can be rearranged into

$$w_x^+ \frac{\partial \vartheta^+}{\partial x^+} + Pr^{1/2} w_y^+ \frac{\partial \vartheta^+}{\partial y^+} = \frac{1}{Pe} \frac{\partial^2 \vartheta^+}{\partial x^{+2}} + \frac{\partial^2 \vartheta^+}{\partial y^{+2}}$$

$$+ \frac{w_m^2}{c_p \Delta \vartheta} \frac{T}{v} \left(\frac{\partial v}{\partial \vartheta} \right)_p \frac{dp^+}{dt^+} + \frac{\phi L}{w_m \varrho c_p \Delta \vartheta} \quad (3.116)$$

with $\Delta \vartheta = \vartheta_\alpha - \vartheta_0$. The viscous dissipation ϕ which appears in this is of the magnitude (see also Example 3.3)

$$\phi \sim \eta \left(\frac{\partial w_x}{\partial y} \right)^2 \sim \eta \frac{w_m^2}{\delta_m^2} \quad \text{or with} \quad \delta_m \sim L/Re^{1/2}$$

$$\phi \sim \eta \frac{w_{\mathrm{m}}^2}{L^2} Re = \eta \frac{w_{\mathrm{m}}^2}{L^2} \frac{w_{\mathrm{m}} L \varrho}{\eta} = \frac{w_{\mathrm{m}}^3 \varrho}{L} \quad .$$

With that the magnitude of the last term in (3.116) is

$$\frac{\phi L}{w_{\mathrm{m}} \varrho c_p \Delta \vartheta} \sim \frac{w_{\mathrm{m}}^2}{c_p \Delta \vartheta} \quad .$$

The quantity

$$\frac{w_{\mathrm{m}}^2}{c_p \Delta \vartheta} = Ec$$

is the already well known *Eckert number*. It is the ratio of the average kinetic
energy to the average internal energy in the boundary layer. In heat and mass
transfer problems it is very small, except when the flow is at high velocity or above
the velocity of sound. These types of flows were however discounted here. This
means that the contribution of the viscous dissipation to the energy equation can
be neglected. In the penultimate expression of (3.116)

$$\frac{w_{\mathrm{m}}^2}{c_p \Delta \vartheta} \frac{T}{v} \left(\frac{\partial v}{\partial \vartheta}\right)_p \frac{\mathrm{d}p^+}{\mathrm{d}t^+} = Ec \frac{T}{v} \left(\frac{\partial v}{\partial \vartheta}\right)_p \frac{\mathrm{d}p^+}{\mathrm{d}t^+}$$

for ideal gases we have

$$\frac{T}{v} \left(\frac{\partial v}{\partial \vartheta}\right)_p = 1 \quad ,$$

whilst for liquids, because $v = v_0$ =const the expression disappears. As the
Eckert number Ec is very small, the penultimate expression can also be neglected
for gases. We have presumed large Reynolds numbers $Re \to \infty$, and the Prandtl
number is of order 1, and so the first term on the left hand side of (3.116) can
also be neglected. The heat conduction in the direction of flow x is negligible in
comparison with that through the thin boundary layer. After transforming this all
back into dimensioned coordinates we get the following equation for the thermal
boundary layer

$$\varrho c_p w_x \frac{\partial \vartheta}{\partial x} + \varrho c_p w_y \frac{\partial \vartheta}{\partial y} = \lambda \frac{\partial^2 \vartheta}{\partial y^2} \quad . \tag{3.117}$$

Corresponding considerations are also valid for the *thermal boundary layer in multicomponent mixtures*. The energy transport through conduction and diffusion in the direction of the
transverse coordinate x is negligible in comparison to that through the boundary layer. The energy equation for the boundary layer follows from (3.97), in which we will presuppose vanishing
mass forces k_{Ki}:

$$\varrho c_p w_x \frac{\partial \vartheta}{\partial x} + \varrho c_p w_y \frac{\partial \vartheta}{\partial y} = \lambda \frac{\partial^2 \vartheta}{\partial y^2} - \sum_{K=1}^{N-1} j_{Ky}^* \frac{\partial}{\partial y}\left(\frac{\partial h}{\partial \xi_K}\right) - \sum_K h_K \dot{\Gamma}_K \quad . \tag{3.118}$$

Equation (3.118) also holds in this form for multicomponent mixtures consisting of more than
two components.
For a binary mixture of components A and B, the enthalpy of the mixture is

$$h = h_{0A} \xi_A + h_{0B}(1 - \xi_A) + \Delta h \quad ,$$

when Δh is the mixing enthalpy and h_{0A}, h_{0B} are the enthalpies of the pure substances A and B. The enthalpy of mixing can be neglected for gases far away from the critical state, and by definiton it disappears for ideal liquid mixtures. Then

$$\frac{\partial h}{\partial \xi_A} = h_{0A} - h_{0B} \ .$$

Furthermore for an incompressible pure fluid, due to $v = v_0 = \text{const}$

$$dh = c_p \, d\vartheta + v_0 \, dp$$

and

$$\frac{\partial h}{\partial y} = c_p \frac{\partial \vartheta}{\partial y}$$

as $\partial p / \partial y = 0$ in the boundary layer. With this we obtain

$$\frac{\partial}{\partial y} \frac{\partial h}{\partial \xi_A} = (c_{pA} - c_{pB}) \frac{\partial \vartheta}{\partial y}$$

where c_{pA} and c_{pB} are the specific heat capacities of the pure substances A and B. The energy equation of the boundary layer of an incompressible binary mixture then reads

$$\varrho c_p w_x \frac{\partial \vartheta}{\partial x} + \varrho c_p w_y \frac{\partial \vartheta}{\partial y} = \lambda \frac{\partial^2 \vartheta}{\partial y^2} - j_{Ay}^* (c_{pA} - c_{pB}) \frac{\partial \vartheta}{\partial y} - (h_A - h_B) \dot{\Gamma}_A \qquad (3.119)$$

with

$$j_{Ay}^* = -\varrho D \frac{\partial \xi_A}{\partial y} \ .$$

If both the substances have the same specific heat capacities, the term for mass transfer drops out of the equation. However in all other cases it does not assume a negligible value. In particular for substances such as water vapour and air, the specific heat capacities are so different that the mass transfer term cannot be removed. In addition it should be recognised that the energy equation agrees with that for pure substances when the specific heat capacities of the two components are equal and when no chemical reactions occur.

Example 3.6: In the energy equation (3.119) for a binary mixture consisting of water vapour and air at 100 °C estimate the magnitude of the individual terms. Chemical reactions are excluded. The specific heat capacity of the water vapour is $c_{pA} = 1.8477 \, \text{kJ/kgK}$, and that of the air $c_{pB} = 1.00258 \, \text{kJ/kgK}$. In addition the Lewis number is $Le = a/D = 1$.

Introducing the dimensionless quantities which will be used for the derivation of the energy equation for the boundary layer for pure substances, gives

$$w_x^+ = \frac{w_x}{w_m}; \quad w_y^+ = \frac{w_y}{w_m} Re^{1/2}; \quad x^+ = \frac{x}{L}; \quad y^+ = \frac{y}{L} Pe^{1/2} \ ;$$

$$\vartheta^+ = \frac{\vartheta - \vartheta_0}{\vartheta_\alpha - \vartheta_0} \quad \text{and additionally} \quad \xi_A^+ = \frac{\xi_A - \xi_{A0}}{\xi_{A\alpha} - \xi_{A0}} \ .$$

With that, (3.119) is transformed into

$$w_x^+ \frac{\partial \vartheta^+}{\partial x^+} + Pr^{1/2} w_y^+ \frac{\partial \vartheta^+}{\partial y^+} = \frac{\partial^2 \vartheta^+}{\partial y^{+2}} + \frac{(\xi_{A\alpha} - \xi_{A0}) \varrho D (c_{pA} - c_{pB}) L}{w_m \varrho c_p L^2} Pe \frac{\partial \xi_A^+}{\partial y^+} \frac{\partial \vartheta^+}{\partial y^+} \ .$$

Due to the fact that $Pe = w_m L/a$ we can then write for this

$$w_x^+ \frac{\partial \vartheta^+}{\partial x^+} + Pr^{1/2} w_y^+ \frac{\partial \vartheta^+}{\partial y^+} = \frac{\partial^2 \vartheta^+}{\partial y^{+2}} + \frac{(\xi_{A\alpha} - \xi_{A0}) (c_{pA} - c_{pB})}{c_p} \frac{D}{a} \frac{\partial \xi_A^+}{\partial y^+} \frac{\partial \vartheta^+}{\partial y^+} \ .$$

The magnitude of the last term is determined by its coefficient. This is yielded by setting for c_p, a mean value $(c_{pA} + c_{pB})/2 = 1.42514 \, \text{kJ/kgK}$ and putting $\xi_{A\alpha} - \xi_{A0} = 1$, that means that we assume the mass fraction of water vapour to be low $\xi_{A\alpha} = 0$ in the core flow, and large $\xi_{A0} = 1$ at the wall, because this is where the water vapour condenses. With $a/D = 1$ the magnitude of the coefficient is 0.593. As all the other expressions are of order 1 the mass transfer term cannot be neglected. This would only be possible if $\xi_{A\alpha} - \xi_{A0}$ was very small and of the order 10^{-2}.

3.5.3 The concentration boundary layer

The same reasoning as before, whereby the mass transfer in the direction of flow is negligible in comparison to that through the boundary layer, leads to the equations for the concentration boundary layer. These are found from the component continuity equations (3.25) by neglecting the relevant expressions, to be

$$\varrho w_x \frac{\partial \xi_A}{\partial x} + \varrho w_y \frac{\partial \xi_A}{\partial y} = -\frac{\partial j_{Ay}^*}{\partial y} + \dot{\Gamma}_A , \qquad (3.120)$$

with $j_{Ay}^* = -\varrho D \partial \xi_A / \partial y$ for a binary mixture.

3.5.4 General comments on the solution of boundary layer equations

The equations (3.109), (3.117) or (3.118) and (3.120) for the velocity, thermal and concentration boundary layers show some noticeable similarities. On the left hand side they contain "convective terms", which describe the momentum, heat or mass exchange by convection, whilst on the right hand side a "diffusive term" for the momentum, heat and mass exchange exists. In addition to this the energy equation for multicomponent mixtures (3.118) and the component continuity equation (3.25) also contain terms for the influence of chemical reactions. The remaining expressions for pressure drop in the momentum equation and mass transport in the energy equation for multicomponent mixtures cannot be compared with each other because they describe two completely different physical phenomena.

The equations are the basis for many technical applications. However, before we investigate particular solutions, it would be sensible to make some general remarks on the solution of these equations.

To this effect, the equations will once again be brought into a dimensionless form. As we are only interested in discussing the characteristic features of the solution, and not as in the derivation of the boundary layer equations, where an estimation of the magnitude of the individual terms in a particular equation was required, all the quantities in the equations will be made dimensionless in the same way. We will now introduce the following dimensionless quantities

$$x^+ = \frac{x}{L} ; \qquad y^+ = \frac{y}{L}; \qquad w_x^+ = \frac{w_x}{w_m} ; \qquad w_y^+ = \frac{w_y}{w_m} ;$$

$$\vartheta^+ = \frac{\vartheta - \vartheta_0}{\vartheta_\alpha - \vartheta_0} ; \qquad \xi_A^+ = \frac{\xi_A - \xi_{A0}}{\xi_{A\alpha} - \xi_{A0}} ; \qquad p^+ = \frac{p}{\varrho w_m^2} .$$

Chemical reactions will not play any role in this discussion. The considerations here will be restricted to pure substances or binary mixtures which have components of approximately the same specific heat capacity. The energy equation

(3.119) then agrees with that for pure subtances (3.117). After introduction of
the dimensionless quantities the continuity equation is

$$\frac{\partial w_x^+}{\partial x^+} = \frac{\partial w_y^+}{\partial y^+} = 0 \ , \tag{3.121}$$

whilst the momentum equation becomes

$$w_x^+ \frac{\partial w_x^+}{\partial x^+} + w_y^+ \frac{\partial w_x^+}{\partial y^+} = -\frac{\partial p^+}{\partial x^+} + \frac{1}{Re} \frac{\partial^2 w_x^+}{\partial y^{+2}} \ , \tag{3.122}$$

the energy equation reads

$$w_x^+ \frac{\partial \vartheta^+}{\partial x^+} + w_y^+ \frac{\partial \vartheta^+}{\partial y^+} = \frac{1}{RePr} \frac{\partial^2 \vartheta^+}{\partial y^{+2}} \tag{3.123}$$

and the continuity equation for component A is

$$w_x^+ \frac{\partial \xi_A^+}{\partial x^+} + w_y^+ \frac{\partial \xi_A^+}{\partial y^+} = \frac{1}{ReSc} \frac{\partial^2 \xi_A^+}{\partial y^{+2}} \tag{3.124}$$

with the Reynolds number $Re = w_m L/\nu$, the Prandtl number $Pr = \nu/a$ and the
Schmidt number $Sc = \nu/D$.

As solutions to the continuity and momentum equations, for a given flow and
thereby a given pressure drop $\partial p^+/\partial x^+$, we obtain

$$w_x^+ = f(x^+, y^+, Re) \quad \text{and} \quad w_y^+ = f(x^+, y^+, Re) \ .$$

Out of this the shear stress at the wall is found to be

$$\tau_0 = \eta \left(\frac{\partial w_x}{\partial y} \right)_{y=0} = \eta \left(\frac{\partial w_x^+}{\partial y^+} \right)_{y^+=0} \frac{w_m}{L}$$

and the friction factor

$$c_f = \frac{\tau_0}{\varrho w_m^2/2} = \frac{2}{Re} \left(\frac{\partial w_x^+}{\partial y^+} \right)_{y^+=0} = \frac{2}{Re} f(x^+, Re) \ . \tag{3.125}$$

For a given shape of the body, the friction factor only depends on the coordinate x^+ and the Reynolds number, not on the type of fluid.

The solution to the energy equation is of the form

$$\vartheta^+ = f(x^+, y^+, Re, Pr) \ .$$

This yields the heat flux transferred to be

$$\dot{q} = -\lambda \left(\frac{\partial \vartheta}{\partial y} \right)_{y=0} = -\lambda \left(\frac{\partial \vartheta^+}{\partial y^+} \right)_{y^+=0} \frac{\vartheta_\alpha - \vartheta_0}{L} \ .$$

On the other hand the heat transfer coefficient is defined by

$$\dot{q} = \alpha(\vartheta_0 - \vartheta_\alpha) \ .$$

Setting these expressions equal to each other gives

$$\frac{\alpha L}{\lambda} = Nu = \left(\frac{\partial \vartheta^+}{\partial y^+}\right)_{y^+=0} = f(x^+, Re, Pr) \ . \tag{3.126}$$

The Nusselt number is equal to the dimensionless temperature gradient at the wall. It is a universal function of x^+, Re and Pr for every fluid with a body of a given shape. The mean Nusselt number is independent of x^+, as it is the integral mean value over the heat transfer surface

$$Nu_{\mathrm{m}} = \frac{\alpha_{\mathrm{m}} L}{\lambda} = f(Re, Pr) \ . \tag{3.127}$$

Taking into consideration the velocity field $w_x^+(x^+, Re)$ and $w_y^+(y^+, Re)$, we obtain, from the component continuity equation (3.124), as a general solution

$$\xi_{\mathrm{A}}^+ = f(x^+, y^+, Re, Sc) \ .$$

The mass flux transferred is

$$\dot{j}_{\mathrm{A}}^* = -\varrho D \left(\frac{\partial \xi_{\mathrm{A}}}{\partial y}\right)_{y=0} = -\varrho D \left(\frac{\partial \xi_{\mathrm{A}}^+}{\partial y^+}\right)_{y^+=0} \frac{\xi_{\mathrm{A}\alpha} - \xi_{\mathrm{A}0}}{L} \ .$$

For a vanishing convective flow across the boundary layer the mass transfer coefficient is defined by

$$\dot{j}_{\mathrm{A}}^* = \beta \varrho(\xi_{\mathrm{A}0} - \xi_{\mathrm{A}\alpha}) \ .$$

It follows from the two equations that

$$\frac{\beta L}{D} = Sh = \left(\frac{\partial \xi_{\mathrm{A}}^+}{\partial y^+}\right)_{y^+=0} = f(x^+, Re, Sc) \ . \tag{3.128}$$

The Sherwood number is equal to the dimensionless gradient of the concentration profile at the wall. It is a universal function of x^+, Re and Sc independent of the type of fluid for a body of a certain shape. The mean Sherwood number, as the integral mean value over the mass transfer surface, is independent of x^+:

$$Sh_{\mathrm{m}} = \frac{\beta_{\mathrm{m}} L}{D} = f(Re, Sc) \ . \tag{3.129}$$

As soon as the functional relationships between the Nusselt, Reynolds and Prandtl numbers or the Sherwood, Reynolds and Schmidt numbers have been found, be it by measurement or calculation, the heat and mass transfer laws worked out from this hold for all fluids, velocities and length scales. It is also valid for all geometrically similar bodies. This is presuming that the assumptions which lead to the boundary layer equations apply, namely negligible viscous dissipation and

body forces and no chemical reactions. As the differential equations (3.123) and
(3.124) basically agree with each other, the solutions must also be in agreement,
presuming that the boundary conditions are of the same kind. The functions
(3.126) and (3.128) as well as (3.127) and (3.129) are therefore of the same type.
So, it holds that

$$\frac{Sh_m}{Nu_m} = \frac{f(Re, Sc)}{f(Re, Pr)} \ .$$

As already explained, section 1.5, these types of function can frequently be approximated by power products over a wide range of states

$$Nu_m = c\,Re^n\,Pr^m \quad \text{and} \quad Sh_m = c\,Re^n\,Sc^m \ .$$

Out of this follows

$$\frac{Sh_m}{Nu_m} = \frac{\lambda\beta_m}{\alpha_m D} = \left(\frac{Sc}{Pr}\right)^m = \left(\frac{a}{D}\right)^m$$

or

$$\frac{\beta_m}{\alpha_m} = \frac{D}{\lambda}\left(\frac{a}{D}\right)^m = \frac{1}{\varrho c_p} Le^{m-1}$$

the well known Lewis relationship, see also (1.198), with the Lewis number $Le = a/D$, from which the mass transfer coefficient β_m can be calculated, if the heat
transfer coefficient is known. A good approximation is $m = 1/3$.

The equations (3.123) and (3.124) agree for $Pr = Sc = 1$, as well as with the
momentum equation (3.122), if we presuppose vanishing pressure drop. Under
the assumption of the same boundary conditions, the velocity profile agrees with
the temperature and the concentration profile. Therefore

$$\left(\frac{\partial w_x^+}{\partial y^+}\right)_{y^+=0} = \frac{\tau_0}{\eta w_m} = \left(\frac{\partial \vartheta^+}{\partial y^+}\right)_{y^+=0} = Nu = \left(\frac{\partial \xi_A^+}{\partial y^+}\right)_{y^+=0} = Sh$$

or after introduction of the friction factor $c_f = \tau_0/(\varrho w_m^2/2)$ and the Reynolds
number $Re = w_m L/\nu$

$$\frac{Nu}{Re} = \frac{Sh}{Re} = \frac{c_f}{2} \quad \text{für} \quad Pr = Sc = 1 \ . \tag{3.130}$$

In this the Nusselt number $Nu = \alpha L/\lambda$ and the Sherwood number $Sh = \beta L/D$
are formed with the local heat and mass transfer coefficients. The ratio Nu/Re
and Sh/Re is independent of the characteristic length as this is also contained
in the Reynolds number $Re = w_m L/\nu$. The equation (3.130) is known as the
Reynolds analogy. The heat and mass transfer coefficients can be calculated with
this as long as the friction factor is known.

3.6   Influence of turbulence on heat and mass transfer

A turbulent flow is characterised by velocity fluctuations which overlap the main flow. The disturbed flow is basically three-dimensional and unsteady. At sufficiently high Reynolds numbers, the boundary layer is also no longer laminar but turbulent, such that the velocities, temperatures and concentrations all vary locally at a fixed position, as Fig. 3.14 shows for a velocity component w_i. At every position it can be formed as the sum of a time-mean value (T here is the integration time)

$$\bar{w}_i = \lim_{T \to \infty} \frac{1}{T} \int_0^T w_i(x_j, t)\, \mathrm{d}t = \bar{w}_i(x_j) \tag{3.131}$$

and a fluctuating component w_i':

$$w_i = \bar{w}_i + w_i' \ .$$

Fig. 3.14 shows a flow with a statistically steady time-mean velocity because the velocity at a fixed point is independent of time after a sufficiently long time has passed. It is obtained by the formation of a mean value as defined in (3.131). A flow can also be unsteady with respect to its time-mean velocity, for example if a fluid flows through a pulsating plastic duct. Then $\bar{w}_i(x_j, t)$, and the centre line in Fig. 3.14 would still change with time.

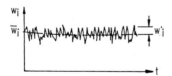

Fig. 3.14: Velocity fluctuations at a fixed position in a statistically steady flow

Only flows steady with respect to the time-mean properties will be considered here. The time-mean value of the fluctuation velocity is, by definition, equal to zero, $\bar{w}_i' = 0$. Correspondingly, pressure, temperature and concentration can also be split into mean and fluctuating values

$$p = \bar{p} + p' \ , \quad \vartheta = \bar{\vartheta} + \vartheta' \ , \quad \xi_A = \bar{\xi}_A + \xi_A' \ .$$

Incompressible flow will be presumed, so the splitting of the density into these two values is not required.

Measurements have shown that fluid particles of different sizes cause turbulent fluctuations and that they simultaneously rotate on their axes. This produces a spectrum of eddies of different size and frequency. The smallest eddies are of a magnitude from 0.1 to $1\,\mathrm{mm}$, still far greater than the mean free path of molecules, 10^{-14} mm The eddy consists of a sufficiently large number of molecules, such that the balance equations are still valid. Velocity, pressure, temperature and concentration are momentary values at each point and can be found using numerical methods from the equations for three-dimensional transient flow. In order to carry this out the time step is chosen to be exceptionally small due to the high frequency of the turbulent fluctuations of 10^4 s^{-1} to 1 s^{-1}. Even with the most powerful computers now available this would lead to a calculation time that cannot be accomplished. Anyway in heat and mass transfer the interest does not lie in the fluctuations with time, rather the mean values. However these are also not particularly easy to determine because of the non-linearity of the convective terms in the balance equations, fluctuations in the velocity, temperature and concentration also have an influence on the mean values. The balance equations formed with these terms contain additional expressions which are not yielded from the equations themselves. Phenomenological statements or statistical turbulence models have to be developed for this. In order to show this we will consider a boundary layer flow. Chemical reactions are excluded from this discussion. In addition we will only consider pure substances or binary mixtures with components that have approximately the same specific heat capacity, so that the energy equation agrees with that for pure substances.

Splitting the velocity into a mean value and a fluctuation velocity, $w_i = \bar{w}_i + w'_i$, in the continuity equation (3.93) leads to

$$\frac{\partial}{\partial x_i}(\bar{w}_i + w'_i) = \frac{\partial \bar{w}_i}{\partial x_i} + \frac{\partial w'_i}{\partial x_i} = 0 \ .$$

After averaging this over time remains

$$\frac{\partial \bar{w}_i}{\partial x_i} = 0 \ . \tag{3.132}$$

The convective term in the momentum equation (3.94) can, under consideration of the continuity equation $\partial w_i / \partial x_i = 0$, also be written as

$$w_i \frac{\partial w_j}{\partial x_i} = \frac{\partial}{\partial x_i}(w_i w_j) \ .$$

From which we get

$$\frac{\partial}{\partial x_i}[(\bar{w}_i + w'_i)(\bar{w}_j + w'_j)] = \frac{\partial}{\partial x_i}\left[\bar{w}_i \bar{w}_j + \bar{w}_i w'_j + w'_i \bar{w}_j + w'_i w'_j\right] \ .$$

After time averaging, all the terms linear in w'_j and w'_i disappear, and with (3.132), this leaves

$$\overline{w_i \frac{\partial w_j}{\partial x_i}} = \bar{w}_i \frac{\partial \bar{w}_j}{\partial x_i} + \frac{\partial}{\partial x_i}(\overline{w'_i w'_j}) \ .$$

Correspondingly, for the convective term in the energy equation (3.95), we obtain

$$\overline{w_i \frac{\partial \vartheta}{\partial x_i}} = \bar{w}_i \frac{\partial \bar{\vartheta}}{\partial x_i} + \frac{\partial}{\partial x_i}(\overline{w'_i \vartheta'})$$

and for the convective term in the component continuity equation

$$\overline{w_i \frac{\partial \xi_A}{\partial x_i}} = \bar{w}_i \frac{\partial \bar{\xi}_A}{\partial x_i} + \frac{\partial}{\partial x_i}(\overline{w_i' \vartheta'}) \ .$$

Therefore, the boundary layer equations (3.108), (3.109), (3.117) and (3.120) have the following form for turbulent flow under the assumptions mentioned above, if we ignore the tensor notation and put $w_1 = w_x$, $w_2 = w_y$, $x_1 = y$ and $x_2 = y$:

$$\frac{\partial \bar{w}_x}{\partial x} + \frac{\partial \bar{w}_y}{\partial y} = 0 \tag{3.133}$$

$$\varrho \left(\bar{w}_x \frac{\partial \bar{w}_x}{\partial x} + \bar{w}_y \frac{\partial \bar{w}_x}{\partial y} \right) = -\frac{\partial \bar{p}}{\partial x} + \frac{\partial}{\partial y} \left(\eta \frac{\partial \bar{w}_x}{\partial y} - \varrho \overline{w_x' w_y'} \right) \tag{3.134}$$

$$\varrho c_p \left(\bar{w}_x \frac{\partial \bar{\vartheta}}{\partial x} + \bar{w}_y \frac{\partial \bar{\vartheta}}{\partial y} \right) = \frac{\partial}{\partial y} \left(\lambda \frac{\partial \bar{\vartheta}}{\partial y} - \varrho c_p \overline{w_y' \vartheta'} \right) \tag{3.135}$$

$$\varrho \left(\bar{w}_x \frac{\partial \bar{\xi}_A}{\partial x} + \bar{w}_y \frac{\partial \bar{\xi}_A}{\partial y} \right) = \frac{\partial}{\partial y} \left(\varrho D \frac{\partial \bar{\xi}_A}{\partial y} - \varrho \overline{w_y' \xi_A'} \right) \ . \tag{3.136}$$

The equations are the same as those for laminar flow, except for the terms of the form $\overline{a'b'}$. They account for the effect of the turbulent fluctuations on momentum, heat and mass transfer.

The expression $-\varrho \overline{w_x' w_y'}$ (SI units N/m^2) is an averaged momentum flow per unit area, and so comparable to a shear stress: A force in the direction of the y-axis acts at a surface perpendicular to the x-axis. Terms of the general form $-\varrho \overline{w_i' w_j'}$ are called *Reynolds' stresses* or *turbulent stresses*. They are symmetrical tensors. In a corresponding manner, the energy equation (3.135), contains a 'turbulent heat flux' of the form

$$q_i = -\varrho c_p \overline{w_i' \vartheta'}$$

and in the component continuity equation a 'turbulent diffusional flux' of the form

$$\overset{*}{j}_{Ai} = -\varrho \overline{w_i' \xi_A'}$$

appears.

As a result of these equations, a total shear stress can also be defined

$$(\tau_{xy})_{\text{tot}} = \eta \frac{\partial \overline{w_x}}{\partial y} - \varrho \overline{w_x' w_y'} \tag{3.137}$$

and correspondingly a toal heat flux

$$(q_y)_{\text{tot}} = -\left(\lambda \frac{\partial \bar{\vartheta}}{\partial y} - \varrho c_p \overline{w_y' \vartheta'} \right) = -\varrho c_p \left(a \frac{\partial \bar{\vartheta}}{\partial y} - \overline{w_y' \vartheta'} \right) \tag{3.138}$$

and a total diffusional flux

$$(\overset{*}{j}_{Ay})_{\text{tot}} = -\varrho \left(D \frac{\partial \overline{\xi}_A}{\partial y} - \overline{w_y' \xi_A'} \right) \ . \tag{3.139}$$

They each consist of a molecular and a turbulent contribution. Momentum, heat and mass transfer are increased by the turbulent contribution.

The equations do not make any statement as to how the turbulent contributions are calculated. An additional assumption has to be made for this. A particularly simple rule was made by J.V. Boussinesq (1842–1929), who suggested the following for the Reynolds' stress

$$- \varrho \, \overline{w'_x w'_y} := \varepsilon_t \frac{\partial \bar{w}_x}{\partial y} \ . \tag{3.140}$$

In which ε_t is the "*turbulent viscosity*" or "*eddy diffusivity for momentum transfer*" (SI units m^2/s). This then allows the total shear stress to be expresssed by

$$(\tau_{xy})_{\text{tot}} = \varrho (\nu + \varepsilon_t) \frac{\partial \bar{w}_x}{\partial y} \ . \tag{3.141}$$

As the Reynolds stress has to disappear approaching the wall, the turbulent viscosity cannot be constant. Flows adjacent to the wall cannot be described by Boussinesq's rule, when $\varepsilon_t = $ const is presumed. However for flows like those which occur in turbulent free jets, the assumption of constant turbulent viscosity is highly suitable. Corresponding to the Boussinesq rule a "*turbulent thermal diffusivity*" a_t (SI units m^2/s) is introduced, through

$$a_t \frac{\partial \bar{\vartheta}}{\partial y} := -\overline{w'_y \vartheta'} \ . \tag{3.142}$$

It is $a_t = \lambda_t / \varrho c_p$ with the "*turbulent thermal conductivity*" or "*eddy diffusivity for heat transfer*" λ_t (SI units W/K m). The total heat flux is

$$(\dot{q}_y)_{\text{tot}} = -\varrho c_p (a + a_t) \frac{\partial \bar{\vartheta}}{\partial y} \ . \tag{3.143}$$

In addition to this a "turbulent diffusion coefficient" or "eddy diffusivity for mass transfer" D_t (SI units m^2/s) is defined by

$$D_t \frac{\partial \bar{\xi}_A}{\partial y} := \overline{w'_y \xi'_A} \ . \tag{3.144}$$

With that the total mass flux is

$$(j^*_{Ay})_{\text{tot}} = -\varrho (D + D_t) \frac{\partial \bar{\xi}_A}{\partial y} \ . \tag{3.145}$$

Just like the turbulent viscosity, the turbulent fractions of the thermal diffusivity, thermal conductivity and the diffusion coefficient have to disappear at the solid wall. In contrast, at some distance away from the wall the turbulent exchange is far more intensive than the molecular motion. This leads to good mixing of the fluid particles. The result of this is that the velocity, temperature

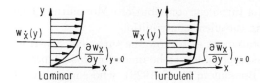

Fig. 3.15: Laminar and turbulent velocity profile in flow over a body

and concentration profiles are more uniform in the core than those in laminar flows, as shown in Fig. 3.15 for the velocity profiles in flow over a body.

The velocity at the wall increases more steeply in turbulent flow than in laminar. The shear stress and with that the resistances to flow are larger in turbulent flows than in laminar. Likewise the temperature and concentration gradients at the surface and therefore the heat and mass transfer rates are larger for turbulent flows than in laminar ones. Therefore turbulent flows are to be strived for in heat and mass transfer and for this reason they are present in most technical applications. However better heat and mass transfer has to be paid for by the increased power required for a pump or blower to overcome the resistances to flow.

3.6.1   Turbulent flows near solid walls

In many technical applications, for example flows in channels, the velocity profile close to the wall is only dependent on the distance from the wall. Indicating the velocity parallel to the wall with w_x and the coordinate normal to the wall by y, then $w_x(y)$, whilst the other velocity components disappear, $w_y = w_z = 0$. This type of flow is known as *stratified flow*. In steady-state, *laminar* flows with vanishing pressure gradients, the momentum equation (3.98) is simplified to

$$\frac{\partial}{\partial y}\left(\eta\frac{\partial w_x}{\partial y}\right) = 0 \quad \text{oder} \quad \frac{\partial \tau_{xy}}{\partial y} = 0 \; ,$$

which yields a linear velocity profile $w_x(y)$ and a constant shear stress τ_{xy}, that is $\tau_{xy} = \tau_0 = \text{const.}$

In the case of steady-state, *turbulent*, stratified flows with vanishing pressure gradients according to (3.134) we obtain

$$\frac{\partial}{\partial y}\left(\eta\frac{\partial \bar{w}_x}{\partial y} - \varrho\overline{w'_x w'_y}\right) = 0 \; ,$$

from which, by integration

$$\eta\frac{\partial \bar{w}_x}{\partial x_y} - \varrho\overline{w'_x w'_y} = \text{ const } = \tau_0 \tag{3.146}$$

follows. The constant of integration is equal to the shear stress, as at the wall, $y = 0$, the Reynolds' stresses disappear, $\varrho\overline{w'_x w'_y} = 0$. In contrast to laminar flow, the velocity $\bar{w}_x(y)$ is no longer a linear function of y.

Equation (3.146) can be rearranged into

$$\frac{\tau_0}{\varrho} = \nu \frac{\partial \bar{w}_x}{\partial y} - \overline{w_x' w_y'} \ . \tag{3.147}$$

As we can see τ_0/ϱ has the dimensions of the square of the velocity. So

$$w_\tau := \sqrt{\tau_0/\varrho} \tag{3.148}$$

is known as the *friction velocity*. With this (3.147) can also be written as the differential equation

$$1 = \frac{\mathrm{d}(\bar{w}_x/w_\tau)}{\mathrm{d}(w_\tau y/\nu)} - \frac{\overline{w_x' w_y'}}{w_\tau^2} \ . \tag{3.149}$$

The velocity profile obeys a function of the form

$$\frac{\bar{w}_x}{w_\tau} = \bar{w}_x^+ = f\left(\frac{w_\tau y}{\nu}\right) = f(y^+) \ , \tag{3.150}$$

and in addition we have

$$\frac{\overline{w_x' w_y'}}{w_\tau^2} = g\left(\frac{w_\tau y}{\nu}\right) = g(y^+) \ . \tag{3.151}$$

Eq. (3.150) represents the *wall law for turbulent flow*, first formulated by Prandtl in 1925. The functions $f(y^+)$ and $g(y^+)$ are of a universal nature, because they are independent of external dimensions such as the height of a channel and are valid for all stratified flows independent of the boundary layer thickness.

In order to calculate $\bar{w}_x/w_\tau$ by solving the differential equation (3.149), the Reynolds stress $\overline{w_x' w_y'}$ has to be known. The hypothesis introduced by Boussinesq (3.140) is unsuitable for this, as according to it, the Reynolds stress does not disappear at the wall. However, the condition $\overline{w_x' w_y'} = 0$ at the wall is satisfied by Prandtl's *mixing length theory*, which will now be explained. In order to do this we will consider a fluid element in a turbulent boundary layer, at a distance y from the wall, Fig. 3.16. It has, at a distance y, the mean velocity $\bar{w}_x(y)$ and may approach the wall over a small length l' with the velocity $w_y' < 0$. If, during this, the fluid element maintains its original velocity, at the new position it will have a velocity that is greater by $\Delta \bar{w}_x$ than that of its surroundings. The velocity difference

$$\Delta \bar{w}_x = \bar{w}_x(y) - \bar{w}_x(y - l') = l' \frac{\partial \bar{w}_x}{\partial y}$$

is a measure of the fluctuation velocity w_x'. The fluid element displaces another at this new position, thereby generating a cross velocity w_y', which with the assumption of small fluctuation velocities, is proportional to the fluctuation velocity. Therefore w_x' and w_y' are proportional to $l' \partial \bar{w}_x/\partial y$ and so

$$\tau_t = -\varrho \overline{w_x' w_y'} = \varrho k l'^2 \left| \frac{\partial \bar{w}_x}{\partial y} \right| \frac{\partial \bar{w}_x}{\partial y}$$

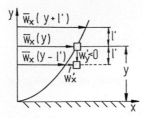

Fig. 3.16: For Prandtl's mixing length theory

or with $kl'^2 = l^2$

$$\tau_t = -\varrho\overline{w'_x w'_y} = \varrho l^2 \left|\frac{\partial \bar{w}_x}{\partial y}\right| \frac{\partial \bar{w}_x}{\partial y} \ . \tag{3.152}$$

Using the notation with the absolute lines ensures that, corresponding to Newton's law for laminar flow

$$\tau_{xy} = \eta \frac{\partial w_x}{\partial y} \ ,$$

the turbulent shear stress τ_t has the same sign as the velocity gradient $\partial \bar{w}_x/\partial y$. According to (3.152), the turbulent viscosity is

$$\varepsilon_t = l^2 \left|\frac{\partial \bar{w}_x}{\partial y}\right| \ . \tag{3.153}$$

The quantity l is known as the *mixing length*. As the Reynolds' stresses disappear at the wall, Prandtl chose the simple hypothesis

$$l = \kappa y \ . \tag{3.154}$$

This means

$$\tau_t = -\varrho\overline{w'_x w'_y} = \varrho \kappa^2 y^2 \left|\frac{\partial \bar{w}_x}{\partial y}\right| \frac{\partial \bar{w}_x}{\partial y} \ . \tag{3.155}$$

Equation (3.149), for stratified flow, is transformed into

$$1 = \frac{\mathrm{d}(\bar{w}_x/w_\tau)}{\mathrm{d}(w_\tau y/\nu)} + \kappa^2 y^2 \left(\frac{\mathrm{d}\bar{w}_x}{\mathrm{d}y}\right)^2 \frac{1}{w_\tau^2} \ . \tag{3.156}$$

As $(\mathrm{d}\bar{w}_x/\mathrm{d}y) > 0$, here $|\mathrm{d}\bar{w}_x/\mathrm{d}y|(\mathrm{d}\bar{w}_x/\mathrm{d}y)$ is written as $(\mathrm{d}w_x/\mathrm{d}y)^2$. With the abbreviations

$$\bar{w}_x^+ = \frac{\bar{w}_x}{w_\tau} \quad \text{and} \quad y^+ = \frac{w_\tau y}{\nu}$$

we obtain

$$1 = \frac{\mathrm{d}\bar{w}_x^+}{\mathrm{d}y^+} + \kappa^2 y^{+2} \left(\frac{\mathrm{d}\bar{w}_x^+}{\mathrm{d}y^+}\right)^2 \ . \tag{3.157}$$

The following solutions are yielded by integration

    a) In the laminar sublayer, $y \to 0$ and therefore $y^+ \to 0$, the second term is negligible, giving

$$\bar{w}_x^+ = y^+ \tag{3.158}$$

or taking into account the definitions for $\bar{w}_x^+, y^+$ and $w_\tau \sqrt{\tau_0/\varrho}$:

$$\tau_0 = \eta \frac{\bar{w}_x}{y} \ .$$

The velocity profile is replaced by a straight line in the laminar sublayer.

b) In the completely turbulent region, a long distance away from the wall, $y \to \infty$ and therefore $y^+ \to \infty$, the second term outweighs the first, and it holds that

$$1 = \kappa^2 y^{+2} \left(\frac{d\bar{w}_x^+}{dy^+} \right)^2$$

or

$$d\bar{w}_x^+ = \frac{1}{\kappa} \frac{dy^+}{y^+} \ .$$

Integration yields a logarithmic velocity profile

$$\bar{w}_x^+ = \frac{\bar{w}_x}{w_\tau} = \frac{1}{\kappa} \ln y^+ + c \ . \tag{3.159}$$

The constants κ and c have to be found by experiment. Values for them have been found to be $\kappa \approx 0.4$ and $c \approx 5$.

In reality the laminar sublayer is continuously transformed into the fully turbulent region. A transition region exists between the two, known as the buffer layer, so that the wall law of velocity can be split into three areas, whose boundaries are set by experimentation. The laminar sublayer extends over the region

$$0 < y^+ < 5 \ ,$$

the buffer-layer region over

$$5 < y^+ < 60$$

and the fully turbulent core over

$$y^+ > 60 \ .$$

Fig. 3.17 shows the complete pattern for the velocity profile $\bar{w}_x^+$ as a function of y^+.

3.7 External forced flow

In this section we will focus on the heat and mass transfer from or to the surface of a body with external flow. Neighbouring bodies should not be present or should be so far away that the boundary layers on the bodies over which the fluid is flowing can develop freely. Velocities, temperatures and concentrations shall only change in the boundary layer and be constant in the flow outside of the boundary

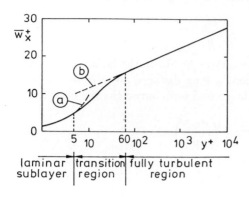

laminar | transition | fully turbulent
sublayer | region | region

Fig. 3.17: Universal velocity distribution law in the turbulent boundary layer **a**: Eq. (3.158) **b**: Eq. (3.159). The abcissa is logarithmic

layer. A forced flow, which we will consider here, is obtained from a pump or blower. Local heat and mass transfer coefficients are yielded from equations of the form

$$Nu = f(x^+, Re, Pr) \quad , \quad Sh = f(x^+, Re, Sc) \ ,$$

whilst their mean values are given by

$$Nu_m = f(Re, Pr) \quad , \quad Sh_m = f(Re, Sc) \ .$$

In many cases the shape of the body over which the fluid is flowing and with that the flow pattern are so involved that the functional relationship between the quantities can only be found experimentally. This requires the measurement of the heat and mass flows transferred at the body or a geometrically similar model and also the associated temperature and concentration differences. This then allows the calculation of the heat and mass transfer coefficients as

$$\alpha = \frac{\dot{Q}}{A\Delta\vartheta} \quad \text{and} \quad \beta = \frac{\dot{M}}{A\varrho\Delta\xi}$$

and the representation of the Nusselt and Sherwood numbers as functions of the other dimensionless numbers Re, Pr and Re, Sc respectively. The power product rules already mentioned are most suitable for the representation of these as equations. The heat and mass transfer coefficients for simple bodies can be calculated by solving the boundary layer equations. This will be discussed in the following sections by means of several characteristic examples.

3.7.1   Parallel flow along a flat plate

Flat plates with parallel flow exist, for example in plate heat exchangers. The fins on a finned tube are exposed to parallel flows. Curved surfaces like aerofoils or

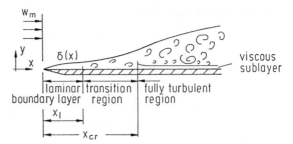

Fig. 3.18: Parallel flow on a flat plate

turbine blades can often be considered, with reference to the boundary layer, to be flat plates, because their boundary layers are almost always small in comparison with the radius of curvature. A laminar boundary layer develops right from the front of the plate, Fig. 3.18. It becomes unstable for a certain length, more exactly from a certain Reynolds number $Re = w_m x_l / \nu \gtrsim 6 \cdot 10^4$ formed with the length x_l onwards. Whilst below this Reynolds number — we talk of the indifference point of the flow — the flow is always laminar, the small disturbances in flow above this Reynolds number no longer die out. However, disturbances of very large and also those of very small wave lengths will be dampened as before. This was first proved by Tollmien [3.5]. The flow does not immediately become fully turbulent, instead a transition region, which is partly laminar and partly turbulent, follows the laminar region from the indifference point onwards. It first becomes fully turbulent flow when a sufficiently large Reynolds number $Re_{cr} = w_m x_{cr} / \nu$ of

$$Re_{cr} = 3 \cdot 10^5 \quad \text{to} \quad 5 \cdot 10^5$$

has been attained. The transition region strongly depends on the surface roughness of the plate and the inlet flow at its front. When all disturbances are kept very small the flow will be laminar up to Reynolds numbers of 10^6. The state is then metastable and the smallest disturbance leads to a transition into fully turbulent flow.

In the immediate vicinity of the solid wall, the turbulent fluctuations will be damped even in fully turbulent flow. In this thin layer adjacent to the wall, also known as the viscous sublayer, the viscous effect of the fluid outweighs that of its turbulent viscosity

3.7.1.1 Laminar boundary layer

We will now focus on heat and mas transfer in the laminar boundary layer of parallel flow on a plate. The flow is steady, disspation is negligible and we will presume constant material properties. Chemical reactions shall not occur. The boundary layer equations consist of the continuity equation

$$\frac{\partial w_x}{\partial x} + \frac{\partial w_y}{\partial y} = 0 \; , \tag{3.160}$$

the momentum equation

$$w_x \frac{\partial w_x}{\partial x} + w_y \frac{\partial w_x}{\partial y} = \nu \frac{\partial^2 w_x}{\partial y^2} \qquad (3.161)$$

and the energy equation

$$w_x \frac{\partial \vartheta}{\partial x} + w_y \frac{\partial \vartheta}{\partial y} = a \frac{\partial^2 \vartheta}{\partial y^2} \; . \qquad (3.162)$$

With regard to mass transfer we will restrict ourselves to a binary mixture with components that have approximately the same specific heat capacities, so that the energy equation remains valid in the form given above. In addition the continuity equation for a component holds

$$w_x \frac{\partial \xi_A}{\partial x} + w_y \frac{\partial \xi_A}{\partial y} = D \frac{\partial^2 \xi_A}{\partial y^2} \; . \qquad (3.163)$$

As the material properties have been presumed to be independent of the temperature and composition, the velocity field is independent of the temperature and concentration fields, so the continuity and momentum equations can be solved independently of the energy and component continuity equations.

Integral methods
We will first study a solution of the system of equations using an *integral method*, which leads to a simple and closed approximate solution. Integral methods are applied to many other boundary layer problems, in particular those involving compressible flows. However, with the introduction of electronic computers they have lost their importance. The basic idea behind the integral methods is that we do not need a complete solution of the boundary layer equation and are content instead with a solution that satisfies the equations in an integral (average) fashion over the entire boundary.

By integrating the continuity equation (3.160) over the thickness of the boundary layer δ we obtain

$$\int\limits_0^\delta \frac{\partial w_x}{\partial x} \, dy + \int\limits_0^\delta \frac{\partial w_y}{\partial y} \, dy = 0$$

or with $w_y(y = 0) = 0$:

$$w_y(\delta) = - \int\limits_0^\delta \frac{\partial w_x}{\partial x} \, dy \; . \qquad (3.164)$$

The momentum equation (3.161) can, taking into account the continuity equation (3.160), also be written in the following form

$$\frac{\partial w_x^2}{\partial x} + \frac{\partial (w_x w_y)}{\partial y} = \nu \frac{\partial^2 w_x}{\partial y^2} \; .$$

Integration between the limits $y = 0$ and $y = \delta$ gives, under consideration of

$$\int_0^\delta \frac{\partial(w_x w_y)}{\partial y}\, \mathrm{d}y = w_x(\delta)w_y(\delta) = -w_\delta \int_0^\delta \frac{\partial w_x}{\partial x}\, \mathrm{d}y \quad,$$

the integral condition for momentum

$$\frac{\mathrm{d}}{\mathrm{d}x}\int_0^\delta w_x(w_\delta - w_x)\, \mathrm{d}y = \nu \left(\frac{\partial w_x}{\partial y}\right)_{y=0} \quad. \tag{3.165}$$

The energy equation (3.162), taking into account the continuity equation (3.160), may be written as follows

$$\frac{\partial(w_x \vartheta)}{\partial x} + \frac{\partial(w_y \vartheta)}{\partial y} = a\frac{\partial^2 \vartheta}{\partial y^2} \quad.$$

Integration between the limits $y = 0$ and $y = \delta_T$ yields, with consideration of

$$\int_\delta^{\delta_T} \frac{\partial(w_y \vartheta)}{\partial y}\, \mathrm{d}y = w_y(\delta_T)\vartheta(\delta_T) = -\vartheta(\delta_T)\int_0^{\delta_T} \frac{\partial w_x}{\partial x}\, \mathrm{d}y$$

the integral condition for energy

$$\frac{\mathrm{d}}{\mathrm{d}x}\int_0^{\delta_T} w_x(\vartheta(\delta_T) - \vartheta)\, \mathrm{d}y = a\left(\frac{\partial \vartheta}{\partial y}\right)_{y=0} \quad. \tag{3.166}$$

After a corresponding integration between the limits $y = 0$ and $y = \delta_c$, the continuity equation (3.162) for a component A is transformed into the integral condition for mass transfer

$$\frac{\mathrm{d}}{\mathrm{d}x}\int_0^{\delta_c} w_x(\xi_{A\delta_c} - \xi_A\, \mathrm{d}y = D\left(\frac{\partial \xi_A}{\partial y}\right)_{y=0} \quad. \tag{3.167}$$

In the integral condition (3.165) for momentum the velocity profile is approximated by a polynomial first suggested by Pohlhausen [3.6], such that

$$\frac{w_x}{w_\delta} = a_0 + a_1\left(\frac{y}{\delta}\right) + a_2\left(\frac{y}{\delta}\right)^2 + a_3\left(\frac{y}{\delta}\right)^3 \quad, \tag{3.168}$$

in which the free coefficients are determined, so that the boundary conditions

$$w_x(y=0) = 0 \quad, \qquad w_x(y=\delta) = w_\delta \quad, \qquad \left(\frac{\partial w_x}{\partial y}\right)_{y=\delta} = 0$$

and the wall condition from the momentum equation (3.161)

$$\left(\frac{\partial^2 w_x}{\partial y^2}\right)_{y=0} = 0 \; ,$$

according to which the curvature of the velocity profile at the wall is zero, are satisfied.

We find $a_0 = a_2 = 0$, $a_1 = 3/2$ and $a_2 = -1/2$ and with this

$$\frac{w_x}{w_\delta} = \frac{3}{2}\frac{y}{\delta} - \frac{1}{2}\left(\frac{y}{\delta}\right)^3 \; . \tag{3.169}$$

This velocity profile transforms the integral condition (3.165) for momentum into an ordinary differential equation for the still unknown boundary layer thickness δ

$$\frac{\mathrm{d}}{\mathrm{d}x}\left(\frac{39}{280}w_\delta^2\delta\right) = \nu\frac{3}{2}\frac{w_\delta}{\delta}$$

or

$$\frac{\mathrm{d}(\delta^2/2)}{\mathrm{d}x} = \frac{140}{13}\frac{\nu}{w_\delta} \; .$$

Through integration, with $\delta(x = 0) = 0$ follows

$$\delta = 4{,}64\left(\frac{\nu x}{w_\delta}\right)^{1/2} = 4{,}64\frac{x}{Re_x^{1/2}} \tag{3.170}$$

with $Re_x = w_\delta x/\nu$. The boundary layer increases with the square root of the length x. The gradient of the velocity profile at the wall, according to (3.169), is

$$\left(\frac{\partial w_x}{\partial y}\right)_{y=0} = \frac{3}{2}\frac{w_\delta}{\delta} \; .$$

With the introduction of the boundary layer thickness, this yields the wall shear stress to be

$$\tau_0 = \eta\left(\frac{\partial w_x}{\partial y}\right)_{y=0} = \varrho w_\delta^2\frac{0{,}323}{(w_\delta x/\nu)^{1/2}} \; .$$

The friction factor is

$$\frac{\tau_0}{\varrho w_\delta^2/2} = \frac{0{,}646}{Re_x^{1/2}} \; . \tag{3.171}$$

Likewise, a polynomial is introduced for the temperature profile

$$\vartheta^+ = \frac{\vartheta - \vartheta_0}{\vartheta_{\delta_T} - \vartheta_0} = b_0 + b_1\frac{y}{\delta_T} + b_2\left(\frac{y}{\delta_T}\right)^2 + b_3\left(\frac{y}{\delta_T}\right)^3 \; ,$$

which should satisfy both the boundary conditions

$$\vartheta^+(y = 0) = 0 \quad ; \quad \vartheta^+(y = \delta_T) = 1 \quad ; \quad \left(\frac{\partial\vartheta^+}{\partial y}\right)_{y=\delta_T} = 0$$

and the wall condition that follows from the energy equation (3.162)

$$\left(\frac{\partial^2 \vartheta^+}{\partial y^2}\right)_{y=0} = 0 \ .$$

The same constants as those for the velocity profile (3.169) are found with this. They are

$$\vartheta^+ = \frac{2}{3}\frac{y}{\delta_T} - \frac{1}{2}\left(\frac{y}{\delta_T}\right)^3 \ . \tag{3.172}$$

Once the temperature and velocity profile (3.169) have been substituted into the integral condition for energy (3.166), the integration yields a differential equation for the boundary layer thickness δ_T

$$\frac{d}{dx}\left[\delta\left(\frac{1}{10}\frac{\delta_T^2}{\delta^2} - \frac{1}{140}\frac{\delta_T^4}{\delta^4}\right)\right] = \frac{a}{w_\delta}\frac{1}{\delta_T} \ .$$

We will now consider flows where $\delta_T/\delta < 1$. This allows us to neglect the second term in the square brackets in comparison to the first term, thereby simplifying the equation to

$$\frac{d}{dx}\left(\frac{\delta_T^2}{\delta}\right) = \frac{10a}{w_\delta}\frac{1}{\delta_T} \ .$$

By abbreviating $k = \delta_T/\delta$ and writing equation (3.170) as $\delta = c_0 x^{1/2}$ with $c_0 = (280/13)^{1/2}\left(\nu/w_\delta\right)^{1/2}$, the differential equation given above is transformed into

$$k x^{1/2}\frac{d}{dx}(k^2 x^{1/2}) = \frac{10a}{w_\delta c_0^2} = \frac{13}{28\,Pr} \ .$$

The substitution of $z = k^3$ converts this equation into the ordinary differential equation

$$z^{1/3} x^{1/2}\frac{d}{dx}(z^{2/3} x^{1/2}) = \frac{13}{28\,Pr}$$

or differentiated out

$$\frac{4}{3}x\frac{dz}{dx} + z = \frac{13}{14\,Pr} \ . \tag{3.173}$$

A particular solution is

$$z = \frac{13}{14\,Pr} \ ,$$

and as the solution of the homogeneous equation, with $z = x^m$ we find the value $m = -3/4$. The complete solution is

$$z = \frac{13}{14\,Pr} + c x^{-3/4} \ .$$

The remaining free constant c is yielded from the condition that the plate is first heated from a position x_0; so $z(x_0) = 0$ and

$$c = -\frac{13}{14\,Pr}x_0^{3/4} \ .$$

From this follows

$$k = \sqrt[3]{z} = \frac{\delta_T}{\delta} = \left(\frac{13}{14\,Pr}\right)^{1/3}\left[1 - \left(\frac{x_0}{x}\right)^{3/4}\right]^{1/3} . \qquad (3.174)$$

If the plate is heated over its entire length, then $x_0 = 0$ and the ratio of thermal to velocity boundary layer is simply a function of the Prandtl number

$$\frac{\delta_T}{\delta} = \left(\frac{13}{14\,Pr}\right)^{1/3} = \frac{0.976}{Pr^{1/3}} . \qquad (3.175)$$

The Prandtl numbers of ideal gases lie between around 0.6 and 0.9, so that their thermal boundary layer is only slightly thicker than their velocity boundary layer. Liquids have Prandtl numbers above one and viscous oils greater than 1000. The thermal boundary layer is therefore thinner than the velocity boundary layer. By the presumptions made, the solution is only valid if $\delta_T/\delta < 1$. This means that the solution is good for liquids, approximate for gases but cannot be applied to fluids with Prandtl numbers $Pr \ll 1$, such as appear in liquid metals.

Heat transfer coefficients are calculated from the transferred heat flux to be

$$\dot{q} = -\lambda\left(\frac{\partial\vartheta}{\partial y}\right)_{y=0} = \alpha(\vartheta_0 - \vartheta(\delta_T)) .$$

In which, according to (3.172),

$$\left(\frac{\partial\vartheta}{\partial y}\right)_{y=0} = \frac{3}{2\delta_T}(\vartheta(\delta_T) - \vartheta_0)$$

and therefore

$$\alpha = \frac{3}{2}\frac{\lambda}{\delta_T} .$$

The heat transfer coefficient is inversely proportional to the thickness of the thermal boundary layer. As this increases with $x^{1/2}$, $\alpha \sim x^{-1/2}$.

Putting the thickness of the thermal boundary layer according to (3.175), along with that of the velocity boundary layer (3.170) yields, after a slight rearrangement, the Nusselt number

$$Nu_x = \frac{\alpha x}{\lambda} = 0.331\,Re_x^{1/2}\,Pr^{1/3} . \qquad (3.176)$$

It will further be shown that the exact solution of the boundary layer equation yields a value of 0.332 which is only very slightly different from the value 0.331 obtained here.

The concentration boundary layer and the mass transfer coefficient can immediately be found from the equations given previously, as the integral condition (3.167) for mass transfer corresponds to that for heat transfer. The temperature ϑ is replaced by the mass fraction ξ_A, the thermal diffusivity a by the diffusion coefficient D, and instead of the thermal boundary layer δ_T the concentration boundary layer δ_c is used. This then gives us the concentration profile corresponding to (3.172):

$$\frac{\xi_A - \xi_{A0}}{\xi_{A\delta} - \xi_{A0}} = \frac{3}{2}\frac{y}{\delta_c} - \frac{1}{2}\left(\frac{y}{\delta_c}\right)^3 . \qquad (3.177)$$

The thickness of the boundary layer corresponding to (3.175), under the prerequisite that the flow and concentration boundary layers begin at the same position,

$$\frac{\delta_c}{\delta} = \frac{0.976}{Sc^{1/3}} , \qquad (3.178)$$

is valid for $Sc \geq 0.6$, and in place of (3.176) we have the equation for mass transfer

$$Sh_x = \frac{\beta x}{D} = 0.331 \, Re_x^{1/2} \, Sc^{1/3} . \qquad (3.179)$$

Example 3.7: Based on the solution (3.174) calculate the Nusselt number for parallel flow on a flat plate, if the thermal and velocity boundary layers are separated by x_0, see Fig. 3.19. Find the Sherwood number for the case of $x_0 \neq 0$.

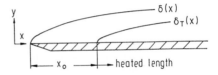

Fig. 3.19: Flow and thermal boundary layer in a laminar parallel flow on a flat plate

We have $\alpha = (3/2)\,\lambda/\delta_T$. Introducing δ_T according to (3.174), under consideration of the expression (3.170) for the velocity boundary layer, yields

$$Nu_x = \frac{\alpha\, x}{\lambda} = 0.331 \, Re_x^{1/2} \, Pr^{1/3} \left[1 - \left(\frac{x_0}{x}\right)^{3/4}\right]^{-1/3} ,$$

from which, for $x_0 = 0$, the known equation (3.176) follows. Correspondingly for the Sherwood number we obtain

$$Sh = \frac{\beta\, x}{D} = 0.331 \, Re_x^{1/2} \, Sc^{1/3} \left[1 - \left(\frac{x_0}{x}\right)^{3/4}\right]^{-1/3} .$$

Exact solution of the boundary layer equations

We presume the outer flow to be an undisturbed parallel flow of velocity w_∞. The calculation of the velocities w_x and w_y follows from the solution of the continuity equation (3.160) and the momentum equation (3.161) under the boundary conditions

$$y = 0, \quad x > 0: \qquad w_x = w_y = 0$$

$$y \to \infty: \qquad w_x = w_\infty$$

and the initial condition

$$x = 0, \quad y > 0: \qquad w_x = w_\infty .$$

Blasius [3.6] introduced in his thesis at the University of Göttingen in 1908 a stream function $\psi(x, y)$ for the solution, which has the property that

$$w_x = \frac{\partial \psi}{\partial y} \quad \text{and} \quad w_y = -\frac{\partial \psi}{\partial x} \ . \tag{3.180}$$

Through this the continuity equation will be identically satisfied. The momentum equation is transformed into a partial differential equation

$$\frac{\partial \psi}{\partial y}\frac{\partial^2 \psi}{\partial y \partial x} - \frac{\partial \psi}{\partial x}\frac{\partial^2 \psi}{\partial y^2} = \nu \frac{\partial^3 \psi}{\partial y^3} \ . \tag{3.181}$$

As according to (3.170), $\delta \sim (\nu x/w_\infty)^{1/2}$ holds for the velocity boundary layer, the distance y from the wall can usefully be related to the boundary layer thickness δ, so introducing a dimensionless variable

$$\eta^+ := y \left(\frac{w_\infty}{\nu x}\right)^{1/2} = g(x, y) \ .$$

The earlier considerations have already shown that the velocity profile can be approximately represented by statements of the form

$$\frac{w_x}{w_\delta} = \varphi \left(\frac{y}{\delta}\right) \ .$$

It is therefore appropriate to assume that the velocity profile can be represented by the variable η^+ alone. We choose the statement

$$\frac{w_x}{w_\infty} = \varphi(\eta^+) \quad \text{with} \quad \eta^+ = y \left(\frac{w_\infty}{\nu x}\right)^{1/2} \ .$$

A solution exists when the differential equation (3.181) and its associated boundary conditions can be fulfilled with this statement. In order to show that this is applicable, we will form a stream function

$$\psi = \int_0^y w_x \, \mathrm{d}y = w_\infty \delta \int_0^y \varphi(\eta^+) \, \mathrm{d}\left(\frac{y}{\delta}\right) = w_\infty \left(\frac{\nu x}{w_\infty}\right)^{1/2} \int_0^{\eta^+} \varphi(\eta^+) \, \mathrm{d}\eta^+$$

or

$$\psi = (w_\infty \nu x)^{1/2} f(\eta^+) \ .$$

The normalisation of the stream function and the coordinate y normal to the wall strongly follows from the fact that (3.181) is invariant in a transformation

$$\tilde{\psi}(\tilde{x}, \tilde{y}) = c\psi(x, y)$$

with $\tilde{x} = c^2 x$ and $\tilde{y} = cy$. This can be checked by putting this into (3.181). The variables ψ, x, y can therefore only appear in the solution in certain combinations that do not contain the factor c. These types of combinations are

$$\frac{\psi}{y} = f \left(\frac{x}{y^2}\right) \ , \quad \text{from which follows} \quad \frac{\tilde{\psi}}{\tilde{y}} = f \left(\frac{\tilde{x}}{\tilde{y}^2}\right)$$

or

$$\frac{\psi}{\sqrt{x}} = f \left(\frac{y}{\sqrt{x}}\right) \ , \quad \text{from which follows} \quad \frac{\tilde{\psi}}{\sqrt{\tilde{x}}} = f \left(\frac{\tilde{y}}{\sqrt{\tilde{x}}}\right) \ .$$

The stream function given above is just of the form $\psi/\sqrt{x} = f(y/\sqrt{x})$. With the stream function we obtain the derivatives which appear in (3.181)

$$\frac{\partial \psi}{\partial y} = w_x = w_\infty f' \quad ; \quad \frac{\partial \psi}{\partial x} = -w_y = \frac{1}{2}\sqrt{\frac{w_\infty \nu}{x}}(f - \eta^+ f') \quad ;$$

$$\frac{\partial^2 \psi}{\partial y^2} = w_\infty \sqrt{\frac{w_\infty}{\nu x}} f'' \quad ; \qquad \frac{\partial^3 \psi}{\partial y^3} = \frac{w_\infty^2}{\nu x} f''' \quad \text{and} \quad \frac{\partial^2 \psi}{\partial y \partial x} = -\frac{1}{2} \frac{w_\infty}{x} \eta^+ f'' \ . \tag{3.182}$$

Putting these derivatives into (3.181), leaves the ordinary non-linear differential equation

$$f''' + \frac{1}{2} f f'' = 0 \tag{3.183}$$

for the function $f(\eta^+)$. The boundary conditions are

$$y = 0 \quad \text{or} \quad \eta^+ = 0 : \qquad f = f' = 0$$

$$y \to 0 \quad \text{or} \quad \eta^+ \to \infty : \qquad f' = 1 \ . \tag{3.184}$$

The initial condition $x = 0$, $y > 0 : w_x = w_\infty$ is already contained in the boundary condition $f'(\eta^+ \to \infty) = 1$. Equation (3.183) was first solved numerically by Blasius [3.6] through power series laws, and later by many other authors (e.g. [3.7], [3.8]).

In the numerical integration, instead of the boundary condition $f'(\eta^+ \to \infty) = 1$ a further initial condition $f''(\eta^+ = 0) = \text{const} = c_0$ could be introduced. However this would require multiple estimations of c_0, until the condition $f'(\eta^+ \to \infty) = 1$ is satisfied. The multiple numerical solutions can be avoided if the boundary value problem is traced back to an initial value problem. This is possible in a simple manner, as (3.183) remains invariant through the transformation $\tilde{f}(\tilde{\eta}^+) = cf(\eta^+)$ with $\tilde{\eta}^+ = \eta^+/c$. This means it is transformed into an equation

$$\tilde{f}''' + \frac{1}{2} \tilde{f} \tilde{f}'' = 0$$

which is independent of the constants c. The boundary conditions

$$f(\eta^+ = 0) = f'(\eta^+ = 0) = 0$$

also remain independent of the choice of c

$$\tilde{f}(\tilde{\eta}^+ = 0) = \tilde{f}'(\tilde{\eta}^+ = 0) = 0 \ .$$

The equation above is solved with the initial conditions $\tilde{f}(\tilde{\eta}^+ = 0) = \tilde{f}'(\tilde{\eta}^+ = 0) = 0$ and $\tilde{f}''(\tilde{\eta}^+ = 0) = 1$ and then the constants c are determined from $\tilde{f}'(\tilde{\eta}^+ \to \infty) = c^2 f'(\eta^+ \to \infty) = c^2$.

Fig. 3.20 illustrates the velocity profile $w_x/w_\infty = f'(\eta^+)$. In the region close to the wall $f(\eta^+) = 0.332\,057 \eta^{+2}/2 + O(\eta^{+5})$ is valid. The coordinate transformation makes all the velocities coincide. The boundary layer approaches the core flow asymptotically and in principle stretches into infinity. The deviation of the velocity w_x from that of the core flow is, however, negligibly small at a finite distance from the wall. Therefore the boundary layer thickness can be defined as the distance from the wall at which w_x/w_∞ is slightly different from one. As an example, if we choose the value of 0.99 for w_x/w_∞, the numerical calculation yields that this value will be reached at the point $\eta^+ \approx 4.910$. The boundary layer thickness defined by this is then

$$\delta = 4.910 \frac{x}{Re_x^{1/2}} \tag{3.185}$$

which is in good agreement with the approximation (3.170). The wall shear stress is

$$\tau_0 = \eta \left(\frac{\partial w_x}{\partial x_y} \right)_{y=0} = \eta \left(\frac{\partial w_x}{\partial \eta^+} \right)_{\eta^+=0} \left(\frac{\partial \eta^+}{\partial y} \right)_{y=0}$$

$$= \eta w_\infty f''(\eta^+ = 0) \left(\frac{w_\infty}{\nu x} \right)^{1/2} \ .$$

The numerical solution delivers $f''(\eta^+ = 0) = 0.3321$. The friction factor is found to be

$$\frac{\tau_0}{\varrho w_\infty^2/2} = \frac{0.664}{Re_x^{1/2}} \ , \tag{3.186}$$

which only varies slightly from the approximate value in (3.171).

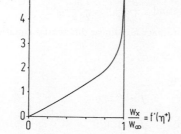

Fig. 3.20: Velocity profile on a flat plate

$\frac{W_x}{W_\infty} = f'(\eta^+)$

In the calculation of the temperature and concentration fields the velocities are replaced by the stream function in the energy equation (3.162) and the component continuity equation (3.163). In addition a dimensionless temperature

$$\vartheta^+ = \frac{\vartheta - \vartheta_0}{\vartheta_\infty - \vartheta_0}$$

and a dimensionless mass fraction

$$\xi_A^+ = \frac{\xi_A - \xi_{A0}}{\xi_{A\infty} - \xi_{A0}}$$

are introduced. The temperature will be represented by $\vartheta(\eta^+)$ and the concentration profile by $\varphi(\eta^+)$. With that, the energy equation (3.162), taking into account (3.180), is transformed into

$$\frac{\partial \psi}{\partial y}\frac{\partial \vartheta^+}{\partial \eta^+}\frac{\partial \eta^+}{\partial x} - \frac{\partial \psi}{\partial x}\frac{\partial \vartheta^+}{\partial \eta^+}\frac{\partial \eta^+}{\partial y} = a\frac{\partial^2 \vartheta^+}{\partial \eta^{+2}}\left(\frac{\partial \eta^+}{\partial y}\right)^2 \tag{3.187}$$

and the continuity equation (3.163) into

$$\frac{\partial \psi}{\partial y}\frac{\partial \xi_A^+}{\partial \eta^+}\frac{\partial \eta^+}{\partial x} - \frac{\partial \psi}{\partial x}\frac{\partial \xi_A^+}{\partial \eta^+}\frac{\partial \eta^+}{\partial y} = D\frac{\partial^2 \xi_A^+}{\partial \eta^{+2}}\left(\frac{\partial \eta^+}{\partial y}\right)^2 \; . \tag{3.188}$$

Using the derivatives (3.182) of the stream function these equations can be rearranged into ordinary differential equations

$$\vartheta^{+\prime\prime} + \frac{1}{2}Pr\,f\vartheta^{+\prime} = 0 \tag{3.189}$$

$$\xi_A^{+\prime\prime} + \frac{1}{2}Sc\,f\xi_A^{+\prime} = 0 \; . \tag{3.190}$$

The boundary conditions are

$$\eta^+ = 0 : \qquad \vartheta^+ = \xi_A^+ = 0$$

$$\eta^+ \to \infty : \qquad \vartheta^+ = \xi_A^+ = 1 \; .$$

Integrating (3.189) once gives

$$\vartheta^{+\prime} = \vartheta^{+\prime}(\eta^+ = 0)\exp\int_0^{\eta^+}\left(-\frac{1}{2}Pr\,f\right)\,\mathrm{d}\eta^+ \; . \tag{3.191}$$

On the other hand (3.183) can also be written as

$$\frac{d \ln f''}{d\eta^+} = -\frac{1}{2}f \ ,$$

from which

$$f'' = f''(\eta^+ = 0) \exp \int_0^{\eta^+} \left(-\frac{1}{2}f \, d\eta^+\right)$$

follows. This allows us to eliminate the exponential function in (3.191), producing

$$\vartheta^{+\prime} = \vartheta^{+\prime}(\eta^+ = 0) \left[\frac{f''}{f''(\eta^+ = 0)}\right]^{Pr} \ .$$

After integration, and by taking the boundary conditions $\vartheta^+(\eta^+ = 0) = 0$ and $\vartheta^+(\eta^+ \to \infty) = 1$ into account, the temperature profile follows as

$$\vartheta^+ = \frac{\vartheta - \vartheta_0}{\vartheta_\infty - \vartheta_0} = \frac{\int_0^{\eta^+} (f'')^{Pr} \, d\eta^+}{\int_0^\infty (f'')^{Pr} \, d\eta^+} \ . \tag{3.192}$$

Correspondingly the concentration profile is yielded as

$$\xi_A^+ = \frac{\xi_A - \xi_{A0}}{\xi_{A\infty} - \xi_{A0}} = \frac{\int_0^{\eta^+} (f'')^{Sc} \, d\eta^+}{\int_0^\infty (f'')^{Sc} \, d\eta^+} \ . \tag{3.193}$$

The heat flux transferred is

$$\dot{q} = -\lambda \left(\frac{\partial \vartheta}{\partial y}\right)_{y=0} = -\lambda \left(\frac{\partial \vartheta^+}{\partial \eta^+}\right)_{\eta^+=0} \frac{\partial \eta^+}{\partial y}(\vartheta_\infty - \vartheta_0)$$

with

$$\frac{\partial \eta^+}{\partial y} = \left(\frac{w_\infty}{\nu x}\right)^{1/2} \ .$$

On the other hand the heat transfer coefficient α is defined by $\dot{q} = \alpha(\vartheta_0 - \vartheta_\infty)$ and with this we obtain

$$\frac{\alpha x}{\lambda} = \left(\frac{w_\infty x}{\nu}\right)^{1/2} \frac{[f''(\eta^+ = 0)]^{Pr}}{\int_0^\infty (f'')^{Pr} \, d\eta^+}$$

with $f''(\eta^+ = 0) = 0.332$. This can also be written as

$$Nu_x = \frac{\alpha x}{\lambda} = Re_x^{1/2} F(Pr) \ . \tag{3.194}$$

Once again a corresponding equation exists for mass transfer

$$Sh_x = \frac{\beta x}{D} = Re_x^{1/2} F(Sc) \ . \tag{3.195}$$

The functions $F(Pr)$ and $F(Sc)$ are identical and are found from a numerical solution. The following relationships are good approximations

$$
\begin{aligned}
Nu_x &= 0.564 \, Re_x^{1/2} \, Pr^{1/2} &\text{for} \quad &Pr \to 0 \\
Nu_x &= 0.500 \, Re_x^{1/2} \, Pr^{1/2} &\text{for} \quad &0.005 < Pr < 0.05 \\
Nu_x &= 0.332 \, Re_x^{1/2} \, Pr^{1/3} &\text{for} \quad &0.6 < Pr < 10 \\
Nu_x &= 0.339 \, Re_x^{1/2} \, Pr^{1/3} &\text{for} \quad &Pr \ge 10 \ .
\end{aligned}
\tag{3.196}
$$

An equation valid for all Prandtl-numbers $0 \le Pr \le \infty$ is

$$Nu_x = \frac{1}{\sqrt{\pi}} Re_x^{1/2} \, \varphi_1(Pr) \tag{3.196a}$$

$$\text{with} \quad \varphi_1(Pr) = \frac{Pr^{1/2}}{(1 + 1.7 \, Pr^{1/4} + 21.36 \, Pr)^{1/6}}$$

In the range $0.25 \le Pr \le \infty$ the error in Nu_x is below 0.5%. A similar equation holds for constant heat flux $\dot{q} =$const at the wall:

$$Nu_x = \frac{\sqrt{\pi}}{2} Re_x^{1/2} \, \varphi_2(Pr) \tag{3.196b}$$

$$\text{with} \quad \varphi_2(Pr) = \frac{Pr^{1/2}}{(1 + 2.09 \, Pr^{1/4} + 48.74 \, Pr)^{1/6}}$$

In the range $0.25 \le Pr \le \infty$ the error in Nu_x is below 1%.

Corresponding equations also hold for mass transfer. The Sherwood number Sh_x appears in place of the Nusselt number Nu_x, and in the same way the Prandtl number Pr number is replaced by the Schmidt number Sc. In the region $0.6 < Pr < 10$ the agreement with the approximation equations (3.176) is excellent.

Example 3.8: Derive from the local Nusselt number (3.196), equations for the mean Nusselt numbers $Nu_m = \alpha_m L/\lambda$ with

$$\alpha_m = \frac{1}{L} \int_0^L \alpha(x) \, dx \ .$$

For a given Prandtl number, according to (3.196), $\alpha = c \, x^{-1/2}$. With that we have

$$\alpha_m = \frac{1}{L} c \int_0^L x^{-1/2} \, dx = 2 \, c \, L^{-1/2} = 2 \, \alpha(x = L) \ .$$

In addition the following is valid

$$Nu_m = \frac{\alpha_m \, L}{\lambda} = 2 \, Nu(x = L) \ .$$

Then with $Re = (w_\infty L)/\nu$

$$
\begin{aligned}
Nu_m &= 1.128\, Re^{1/2}\, Pr^{1/2} \quad \text{for} \quad Pr \to 0 \\
Nu_m &= 1.000\, Re^{1/2}\, Pr^{1/2} \quad \text{for} \quad 0.005 < Pr < 0.05 \\
Nu_m &= 0.664\, Re^{1/2}\, Pr^{1/3} \quad \text{for} \quad 0.6 < Pr < 10 \\
Nu_m &= 0.678\, Re^{1/2}\, Pr^{1/3} \quad \text{for} \quad Pr \geq 10 \; .
\end{aligned}
$$

3.7.1.2 Turbulent flow

The Reynolds analogy, which links the heat and mass transfer coefficients to the friction factor, according to

$$
\frac{Nu_x}{Re_x} = \frac{Sh_x}{Re_x} = \frac{c_f}{2} \quad \text{for} \quad Pr = Sc = 1 \; ,
$$

already delivers a simple relationship for the heat and mass transfer coefficients, as the friction factor is known from measurements [3.9]:

$$
c_f = 0.0592\, Re_x^{-1/5} \quad \text{for} \quad 5 \cdot 10^5 < Re < 10^7
$$

with $Re_x = w_m x/\nu$. With that we get

$$
\frac{Nu_x}{Re_x} = \frac{Sh_x}{Re_x} = 0.0296\, Re_x^{-1/5} \quad \text{for} \quad Pr = Sc = 1
$$

in the same range of Reynolds numbers.

According to Chilton and Colburn [3.10], [3.11] the effect of the Prandtl number on the heat transfer can be described by the empirical statement

$$
Nu_x = 0.0296\, Re_x^{4/5}\, Pr^{1/3} \; , \tag{3.197}
$$

which is valid for $0.6 < Pr < 60$ and $5 \cdot 10^5 < Re_x < 10^7$. Correspondingly for the mass transfer coefficients

$$
Sh_x = 0.0296\, Re_x^{4/5}\, Sc^{1/3} \; , \tag{3.198}
$$

valid for $0.6 < Sc < 3\,000$ and $5 \cdot 10^5 < Re_x < 10^7$. Colburn has introduced the Stanton number

$$
St = \frac{Nu_x}{Re_x Pr} = 0.0296\, Re_x^{-1/5}\, Pr^{-2/3} = \frac{c_f}{2} Pr^{-2/3} \tag{3.199}
$$

instead of the Nusselt number in (3.197), cf. section 1.1.4.

A better analytically based equation which is valid over a wide range of Prandtl or Schmidt numbers is obtained if we presume a turbulent parallel flow, i.e. a

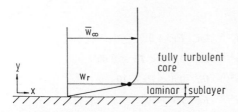

fully turbulent
core

laminar sublayer

Fig. 3.21: The two layer model of the Prandtl analogy

steady-state turbulent flow with vanishing pressure gradient, and velocity, temperature and concentration profiles which are only dependent on the coordinate y normal to the wall. Then, as follows from (3.134) to (3.139),

$$\frac{\partial(\tau_{xy})_{\text{tot}}}{\partial y} = \frac{\partial(\dot{q}_y)_{\text{tot}}}{\partial y} = \frac{\partial(\dot{j}^*_{\text{A}y})_{\text{tot}}}{\partial y} = 0 \ .$$

The total values for the shear stress, heat and diffusional fluxes are independent of the coordinate normal to the wall and therefore equal to the values at the wall. In the laminar sublayer we have

$$\tau_{xy} = \varrho\nu\frac{\partial w_x}{\partial y} = \tau_0 \quad , \qquad \dot{q}_y = -\varrho c_p a\frac{\partial\vartheta}{\partial y} = \dot{q}_0$$

and

$$\dot{j}^*_{\text{A}y} = -\varrho D\frac{\partial\xi_{\text{A}}}{\partial y} = \dot{j}^*_{\text{A}0} \ .$$

Out of these we obtain

$$\frac{\tau_0}{\dot{q}_0} = -\frac{1}{c_p}Pr\frac{\partial w_x}{\partial\vartheta}$$

and after integrating from the wall to the boundary (index r) of the laminar sublayer, Fig. 3.21

$$\vartheta_r - \vartheta_0 = -\frac{\dot{q}_0\,Pr}{\tau_0 c_p}w_r \ . \tag{3.200}$$

Correspondingly

$$\xi_r - \xi_0 = -\frac{\dot{j}^*_{\text{A}0}}{\tau_0}Sc\,w_r \ . \tag{3.201}$$

We will now assume that the fully turbulent core adjoins directly the laminar sublayer, as in Fig. 3.21, and that Boussinesq's laws (3.141), (3.143) and (3.145) are valid, which accordingly give

$$(\tau_{xy})_{\text{tot}} = \varrho(\nu + \varepsilon_t)\frac{\partial\bar{w}_x}{\partial y} \ ,$$

$$(\dot{q}_y)_{\text{tot}} = -\varrho c_p(a + a_t)\frac{\partial\bar{\vartheta}}{\partial y} \quad \text{and}$$

$$(\dot{j}^*_{\text{A}y})_{\text{tot}} = -\varrho(D + D_t)\frac{\partial\bar{\xi}_{\text{A}}}{\partial y} \ .$$

In the turbulent core $\nu \ll \varepsilon_t$, $a \ll a_t$ and $D \ll D_t$. Therefore

$$\frac{(\tau_{xy})_{\text{tot}}}{(\dot{q}_y)_{\text{tot}}} = \frac{\tau_0}{\dot{q}_0} = \frac{1}{c_p} Pr_t \frac{\partial \bar{w}_x}{\partial \bar{\vartheta}}$$

with the 'turbulent Prandtl number' $Pr_t = \varepsilon_t/a_t$. Integration from the edge r of the laminar sublayer to the turbulent core (index ∞) yields, with the assumption that the turbulent Prandtl number Pr_t is constant,

$$\bar{\vartheta}_\infty - \vartheta_r = -\frac{\dot{q}_0}{\tau_0 c_p} Pr_t(\bar{w}_\infty - w_r) \ . \tag{3.202}$$

The corresponding equation for mass transfer is

$$\bar{\xi}_{A\infty} - \xi_{Ar} = -\frac{j_{A0}^*}{\tau_0} Sc_t (\bar{w}_\infty - w_r) \ . \tag{3.203}$$

By adding (3.200) and (3.202) together, and under the assumption $Pr_t = 1$, we obtain

$$\bar{\vartheta}_\infty - \vartheta_0 = -\frac{\dot{q}_0}{\tau_0 c_p} [\bar{w}_\infty + (Pr - 1)w_r] \ .$$

Then, because $\dot{q}_0 = \alpha(\vartheta_0 - \vartheta_\infty)$ the following relationship for the heat transfer coefficient α is obtained

$$\alpha = \frac{\tau_0 c_p}{\bar{w}_\infty} \frac{1}{1 + (Pr - 1)w_r/\bar{w}_\infty} \ .$$

With $c_f = \tau_0/(\varrho \bar{w}_\infty^2/2)$ as the friction factor the equation given above can be written as

$$\frac{\alpha}{\varrho c_p \bar{w}_\infty} = St_x = \frac{c_f}{2} \frac{1}{1 + (Pr - 1)w_r/\bar{w}_\infty} \ . \tag{3.204}$$

The expression $\alpha/(\varrho c_p \bar{w}_\infty)$ is equal to the Stanton number,

$$St = \frac{Nu_x}{Re_x Pr} = \frac{\alpha x/\lambda}{\dfrac{\bar{w}_\infty}{\nu} \dfrac{\nu \varrho c_p}{\lambda}} = \frac{\alpha}{\varrho c_p \bar{w}_\infty} \ .$$

The velocity ratio $w_r/\bar{w}_\infty$ can be eliminated from (3.204), because in the laminar sublayer it holds that

$$\tau_0 = \eta \frac{w_r}{\delta_r} \ ,$$

where δ_r is its thickness. As explained in 3.6.1, the laminar sublayer extends to the point $y^+ = w_r \delta_r/\nu \approx 5$. Therefore, because $w_r = \sqrt{\tau_0/\varrho}$

$$\frac{\sqrt{\tau_0/\varrho}\, \delta_r}{\nu} \approx 5 \ .$$

Eliminating δ_r from the previous equation with this, and taking into consideration $\eta = \nu\varrho$, we obtain

$$\frac{\tau_0}{\varrho} = \frac{w_x^2}{25} \ .$$

It follows from this that

$$\frac{\tau_0}{\varrho \bar{w}_\infty^2/2} = c_f = \frac{w_r^2}{\bar{w}_\infty^2} \frac{2}{25}$$

or

$$\frac{w_r}{\bar{w}_\infty} = 5\sqrt{\frac{c_f}{2}} \ .$$

With this the relationship in (3.204) is transformed into

$$St = \frac{c_f}{2} \frac{1}{1 + (Pr - 1)5\sqrt{c_f/2}} \ . \tag{3.205}$$

A corresponding calculation for mass transfer, under the assumption of a 'turbulent Schmidt number' $Sc_t = \varepsilon_t/D_t = 1$, leads to the analogous expression

$$St' = \frac{c_f}{2} \frac{1}{1 + (Sc - 1)5\sqrt{c_f/2}} \tag{3.206}$$

with the Stanton number St' for mass transfer $St' = Sh_x/Re_x Sc$. Equations (3.205) and (3.206) for the local heat and mass transfer coefficients represent the Prandtl analogy. In the limiting case of $Pr = Sc = 1$ they are transformed into the Reynolds analogy (3.130).

In practical terms the mean Stanton number is of greatest interest. It is obtained, after introduction of the friction factor $c_f(Re_x)$, by integrating (3.205) or (3.206). However the agreement with measured values is still unsatisfactory because the splitting of the boundary layer into a laminar sublayer which adjoins a fully turbulent layer based on the Prandtl analogy is too coarse, and in addition to this the determination of the thickness of the laminar sublayer with $y^+ = 5$ is only an approximation. However the Prandtl analogy was the basis for the establishment of empirical equations which agree better with measured values. An example of this is the equation derived by Gnielinski [3.13] from a relationship given by Petukhov und Popov [3.12] for the mean Nusselt number

$$Nu_m = \frac{0.037 \, Re^{0.8} Pr}{1 + 2.443 \, Re^{-0.1}(Pr^{2/3} - 1)} \tag{3.207}$$

with $Nu_m = \alpha_m L/\lambda$, $Re = \bar{w}_\infty L/\nu$. It is valid for $5 \cdot 10^5 < Re < 10^7$ and $0.6 < Pr < 2\,000$. The material properties are calculated at the average fluid temperature $\vartheta_m = (\vartheta_0 + \vartheta_F)/2$, where ϑ_0 is the wall temperature and ϑ_F is a mean fluid temperature, defined as the arithmetic mean of the inlet and outlet temperatures. In the transition region between laminar and turbulent flow, good agreement is obtained through a quadratic superposition

$$Nu_m = \sqrt{Nu_{m,\,lam}^2 + Nu_{m,\,turb}^2} \ , \tag{3.208}$$

in which $Nu_{m,\,lam}$ is the mean Nusselt number for laminar flow $Nu_{m,\,lam} = 2Nu_x$ with Nu_x according to (3.196) and $Nu_{m,\,turb}$ is that of the turbulent flow according to (3.207). In (3.208), at low Reynolds numbers the turbulent contribution will be small and conversely at large Reynolds numbers the laminar contribution will be small. It can therefore be used over the total range of Reynolds numbers

$$10 < Re < 10^7 \; .$$

Naturally a corresponding equation for mass transfer exists for the same range of Reynolds numbers and for $0.7 < Sc < 70\,000$. In this case the Nusselt number is replaced by the Sherwood number and likewise the Prandtl by the Schmidt number.

Example 3.9: Derive a relationship for the mean heat transfer coefficient for parallel flow on a plate from the local heat transfer equations (3.196) for $0.6 < Pr < 10$ and (3.197). Within this you should take into account that the boundary layer flow is laminar to begin with and becomes turbulent above a critical Reynolds number $(Re_x)_{\text{cr}} = 5 \cdot 10^5$.

We have

$$\alpha_{\text{m}} = \frac{1}{L} \left[\int_{x=0}^{x_{\text{cr}}} \alpha_{\text{lam}} \, dx + \int_{x=x_{\text{cr}}}^{L} \alpha_{\text{turb}} \, dx \right] \; .$$

With (3.196) and (3.197) we get

$$\alpha_{\text{m}} = \frac{\lambda}{L} \left[0.332 \left(\frac{w_\infty}{\nu} \right)^{1/2} \int_{x=0}^{x_{\text{cr}}} \frac{dx}{dx^{1/2}} + 0.0296 \left(\frac{w_\infty}{\nu} \right)^{4/5} \int_{x=x_{\text{cr}}}^{L} \frac{dx}{dx^{1/5}} \right] Pr^{1/3}$$

$$Nu_{\text{m}} = \frac{\alpha_{\text{m}} L}{\lambda} = \left[0.664 \left(Re_x^{1/2} \right)_{\text{cr}} + 0.037 \left(Re^{4/5} - \left(Re_x^{4/5} \right)_{\text{cr}} \right) \right] Pr^{1/3} \; .$$

From this, with $(Re_x)_{\text{cr}} = 5 \cdot 10^5$, follows

$$Nu_{\text{m}} = (0.037 \, Re^{4/5} - 871) \, Pr^{1/3} \; ,$$

which holds for $Re = w_0 L / \nu \geq 5 \cdot 10^5$.

3.7.2 The cylinder in cross flow

The boundary layer for a cylinder in cross flow already has a finite thickness at the forward stagnation point, $x = 0$ in Fig. 3.22, whilst in parallel flow along a plate its initial thickness is zero. At the forward stagnation point the kinetic energy of the fluid in an adiabatic flow is completely converted into enthalpy, $h + w^2/2 = h_0$. A fluid element possesses a higher enthalpy at the stagnation point than upstream of it. If the flow is considered as reversible this leads to an increase in the pressure at the stagnation point because $h = h(s, p)$ with $s = \text{const}$. The pressure decreases from the stagnation point onwards. A fluid element on a streamline close to the surface of the body, accelerates again and so moves into the area of decreasing pressure. Beyond the thickest part of the body, the pressure increases again and the fluid element will be slowed down.

However real flows are not reversible, some of the kinetic energy will be dissipated and converted into internal energy due to friction in the boundary layer. The kinetic energy gained in the acceleration phase is therefore lower than in reversible flow. As a result of this the total kinetic energy of the fluid element will have been converted into enthalpy whilst it is still in the region of increasing pressure. At this point, known as the separation point, the fluid near the surface lacks sufficient momentum to overcome the pressure gradient. Under the influence of this pressure increase the element moves against the direction of flow. The boundary layer detaches from the surface and a wake is formed in the downstream region. The core flow is pushed away from the surface of the body. Beyond this separation point the pressure is practically constant, the flow is irregular and characterised by vortex formations. In this region, known as dead water, heat and mass transfer come to a virtual stand still.

In boundary layers with a pressure increase in the direction of flow, the flow can become detached. This can be seen in the momentum equation (3.109), which due to $w_x = w_y = 0$ is transformed into

$$\eta \left(\frac{\partial^2 w_x}{\partial y^2} \right)_{y=0} = \frac{\mathrm{d}p}{\mathrm{d}x} \tag{3.209}$$

at the wall. According to this the pressure drop in the outer flow determines the curvature of the velocity profile at the wall. When the core flow is at constant pressure the curvature is zero, and the velocity profile ajacent to the wall can be replaced by a straight line. In regions where the pressure is increasing, $\mathrm{d}p/\mathrm{d}x > 0$ and $\eta(\partial^2 w_x/\partial y^2)_{y=0} > 0$. This corresponds to a velocity profile like that drawn on the right in Fig. 3.22. According to Bernoulli's equation

$$\varrho w_\delta \frac{\mathrm{d}w_\delta}{\mathrm{d}x} = -\frac{\mathrm{d}p}{\mathrm{d}x}$$

the core flow decelerates. Detachment of the boundary layer is only possible in a decelerating core flow.

In turbulent flow, momentum is constantly fed into the layer adjacent to the wall because of the momentum transfer between layers at different velocities. The kinetic energy of the fluid elements close to the wall does not decrease as rapidly as in laminar flow. This means that turbulent boundary layers do not become detached as quickly as laminar boundary layers. Heat and mass trasnfer close to the wall is not only promoted by turbulence, the fluid also flows over a larger

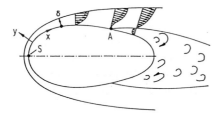

Fig. 3.22: Flow boundary layer around a cylindrical solid. S stagnation point, A detachment point of the boundary layer

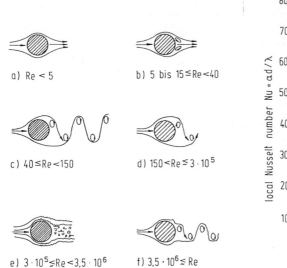

a) Re < 5   b) 5 bis 15 ≲ Re < 40

c) 40 ≲ Re < 150   d) 150 < Re ≲ 3 · 10⁵

e) 3 · 10⁵ ≲ Re < 3,5 · 10⁶   f) 3,5 · 10⁶ ≲ Re

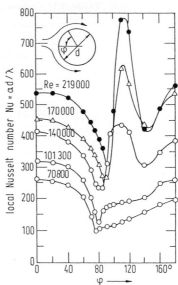

Fig. 3.23: Flow across a circular cylinder at different Reynolds numbers

Fig. 3.24: Local heat transfer coefficients in flow across a circular cylinder, according to Giedt [3.14]

surface area without detachment. At the same time the pressure resistance is lower because the fluid flow does not separate from the surface for a longer flow path.

The flow pattern *around a cylinder in cross flow* is heavily dependent on the Reynolds number, as Fig. 3.23 clearly shows for a circular cylinder. At low Reynolds numbers $Re = w_\infty d/\nu < 5$, Fig. 3.23a, the flow surrounds the cylinder; between 5 and up to $15 \leq Re \leq 40$, Fig. 3.23b, it already becomes detached, and vortices form. At higher Reynolds numbers $40 \leq Re \leq 150$, Fig. 3.23c, the vortices are periodically carried away forming a vortex pattern in which the flow is still laminar. This becomes turbulent at Reynolds numbers between $150 \leq Re \leq 300$, and is fully turbulent in the range $300 < Re < 3 \cdot 10^5$, Fig. 3.23d. Above $Re = 3 \cdot 10^5$ the boundary layer downstream of the cylinder is also turbulent. The wake is concentrated in a narrow region, is fully turbulent, Fig. 3.23e, and does not contains any large eddies. A narrow, fully turbulent vortex pattern first reforms above $Re \approx 3.5 \cdot 10^6$, Fig. 3.23 f.

The flow influences the heat and mass transfer in a complex way, as Fig. 3.24 shows, in which the local Nusselt number is plotted against the angular coordinate starting from the forward stagnation point. The Nusselt number decreases from the stagnation point onwards as the boundary layer develops. It reaches a minimum at an angle of around 80°. The flow begins to detach itself at this point, and the heat transfer increases along the perimeter where the fluid is well mixed

Table 3.1: Constants and exponents in equation (3.210)

Re	c	m	n
1 to 40	0.76	0.4	0.37
40 to 10^3	0.52	0.5	0.37
10^3 to $2 \cdot 10^5$	0.26	0.6	0.37
$2 \cdot 10^5$ to 10^7	0.023	0.8	0.4

Heating the fluid:	$p=0.25$
Cooling the fluid:	$p=0.20$

due to the vortices. At high Reynolds numbers above 10^5 two minima in the local Nusselt number appear. The stark increase between an angle of 80° and 100° is caused by the transition between the initially laminar into a turbulent boundary layer. This increases downstream, inhibiting the heat transfer, until an angle of around 140° is reached. At this point the heat transfer once more improves as a result of better mixing of fluid in the wake.

In practice, the mean heat transfer coefficient is of greatest interest. It can be described by empirical correlations of the form

$$Nu_{\mathrm{m}} = c\,Re^m\,Pr^n \left(\frac{Pr}{Pr_0} \right)^p , \qquad (3.210)$$

in which the mean Nusselt and the Reynolds number are formed with the tube diameter. All material properties should be calculated at the average fluid temperature $\vartheta = (\vartheta_0 + \vartheta_\infty)/2$, except the Prandtl number Pr_0, that should be taken at the wall temperature ϑ_0. The coefficients c, m, n, and p are taken from a paper by Žukauskas [3.15] and are exhibited in Table 3.1.

A single, empirical equation for all Reynolds numbers $1 \leq Re \leq 10^6$ and for Prandtl numbers $0.7 < Pr < 600$ has been communicated by Gnielinski [3.13]. It is

$$Nu_{\mathrm{m}} = 0.3 + \sqrt{Nu_{\mathrm{m,lam}}^2 + Nu_{\mathrm{m,turb}}^2} \qquad (3.211)$$

with $Nu_{\mathrm{m,lam}} = 2Nu_x(x = L)$ according to (3.196) and $Nu_{\mathrm{m,turb}}$ from (3.207). The length L in these equations is taken to be the circumference over which the fluid flows $L = d\pi/2$. In the case of mass transfer the Nusselt number has to be replaced by the Sherwod number, and accordingly the Prandtl by the Schmidt number. The equation is valid for $1 \leq Re \leq 10^6$ and for $0.7 < Sc < 7 \cdot 10^4$.

Example 3.10: A copper cylinder of diameter 10 mm and 1 m length has an initial temperature of $\vartheta_\alpha = 423.15\,\mathrm{K} = 150\,°\mathrm{C}$. Air at $\vartheta_A = 298.15\,\mathrm{K} = 25\,°\mathrm{C}$ then flows across the cylinder. After what period of time has the cylinder been cooled to $\vartheta_w = 308.15\,\mathrm{K} = 35\,°\mathrm{C}$?

Given properties: thermal conductivity of the cylinder $\lambda = 399\,\mathrm{W/Km}$, density $\varrho = 8933$ kg/m^3, specific heat capacity $c_A = 0.387\,\mathrm{kJ/kgK}$, thermal conductivity of the air $\lambda_A = 26.02 \cdot 10^{-3}\,\mathrm{W/Km}$, Prandtl number $Pr_A(25\,°\mathrm{C}) = 0.714$, $Pr_A(35\,°\mathrm{C}) = 0.713$, $Pr_A(92.5\,°\mathrm{C}) = 0.707$, kinematic viscosity $\nu_A(25\,°\mathrm{C}) = 158.2 \cdot 10^{-7}\,\mathrm{m}^2/\mathrm{s}$, $\nu_A(35\,°\mathrm{C}) = 167.8 \cdot 10^{-7}\,\mathrm{m}^2/\mathrm{s}$.

In the solution of this problem we first have to show that the thermal resistance of the cylinder is negligible compared to that of the air, and therefore that the temperature of the cylinder is only a function of time.

The thermal resistance of the air is yielded from the heat transfer coefficients. These are found from (3.210). The Reynolds number of the air flowing over the cylinder is

$$Re = \frac{w_\infty d}{\nu} = \frac{10\,\text{m/s} \cdot 10 \cdot 10^{-3}\,\text{m}}{158.2 \cdot 10^{-7}\,\text{m}^2/\text{s}} = 6321$$

With the values from Table 3.1, the average Nusselt number is

$$Nu_m = 0.26\,Re^{0.6}\,Pr^{0.37}\left(\frac{Pr}{Pr_0}\right)^{0.25}$$

with

$$Pr_0 = Pr\left(\frac{150 + 35}{2}\,°\text{C}\right) = Pr(92.5\,°\text{C}) = 0.707 \ .$$

This gives

$$Nu_m = 0.26 \cdot 6321^{0.6} \cdot 0.741^{0.37}\left(\frac{0.741}{0.707}\right)^{0.25} = 44.91$$

$$\alpha_m = Nu_m\,\frac{\lambda}{d} = 44.91\,\frac{26.02 \cdot 10^{-3}\,\text{W/Km}}{10 \cdot 10^{-3}\,\text{m}} = 116.9\,\text{W/m}^2\text{K}$$

The Biot number is decisive for the ratio of the thermal resistance of the cylinder to that of the air

$$Bi = \frac{\alpha_m\,d}{\lambda} = \frac{116.9\,\text{W/m}^2\text{K} \cdot 10 \cdot 10^{-3}\,\text{m}}{399\,\text{W/Km}} = 2.92 \cdot 10^{-3} \ .$$

This is well below the limit of $Bi = 0.1$ given in section 2.3.5.2. So the thermal resistance of the cylinder is negligible compared to that of the air. This allows us to associate a single temperature that only changes with time to the cylinder. The cooling time of the cylinder is yielded from (2.199) to be

$$t = \frac{\varrho\,c\,V}{\alpha_m\,A}\,\ln\frac{\vartheta_\alpha - \vartheta_A}{\vartheta_\omega - \vartheta_A} \quad \text{with} \quad \frac{V}{A} = \frac{d^2\,\pi}{4}\,L\,\frac{1}{d\,\pi\,L} = \frac{d}{4}$$

$$t = \frac{8933\,\text{kg/m}^3 \cdot 387\,\text{J/kgK} \cdot 10 \cdot 10^{-3}\,\text{m}}{116.9\,\text{W/m}^2\text{K} \cdot 4}\,\ln\frac{150 - 25}{35 - 25} = 186.7\,\text{s} = 3.1\,\text{min} \ .$$

3.7.3 Tube bundles in cross flow

In the following, the heat transfer and pressure drop in a tube bundle in cross flow will be investigated. The individual tubes in the bundle are either in alignment with each other or in a staggered arrangement, according to Fig. 3.25.

The heat transfer is greater in a staggered arrangement at the same Reynolds number. However this is paid for with a larger pressure drop. With reference to Fig. 3.25a to c, the separation between the tube centres perpendicular to the flow direction is known as *transverse pitch* s_q, tube separation in the flow direction as *longitudinal pitch* s_1 and the tube separation in the diagonal direction as *stagger* s_v. The quotient s_q/d is called the transverse pitch ratio, s_1/d the longitudinal pitch ratio, and s_v/d the staggered pitch.

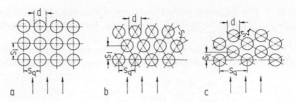

Fig. 3.25: Tube bundle in cross flow. **a** aligned tube arrangement; **b** staggered tube arrangement with the smallest cross section perpendicular to the initial flow direction; **c** staggered tube arrangement with the smallest cross section in the diagonal

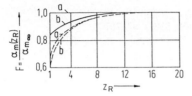

Fig. 3.26: Influence of the tube row number z_R on the heat transfer in bundles of smooth tubes from [3.16]. a) aligned, b) staggered tube arrangement. The broken lines for $Re > 10^3$, full lines for $10^2 < Re < 10^3$

The flow around and therefore the heat transfer around an individual tube within the bundle is influenced by the detachment of the boundary layer and the vortices from the previous tubes. The heat transfer on a tube in the first row is roughly the same as that on a single cylinder with a fluid in cross flow, provided the transverse pitch between the tubes is not too narrow. Further downstream the heat transfer coefficient increases because the previous tubes act as turbulence generators for those which follow. From the fourth or fifth row onwards the flow pattern hardly changes and the mean heat transfer coefficient of the tubes approach a constant end value. As a result of this the mean heat transfer coefficient over all the tubes reaches for an end value independent of the row number. It is roughly constant from about the tenth row onwards. This is illustrated in Fig. 3.26, in which the ratio F of the mean heat transfer coefficient $\alpha_m(z_R)$ up to row z_R with the end value $\alpha_m(z_R \to \infty) = \alpha_{m\infty}$ is plotted against the row number z_R.

The results for the mean heat transfer coefficients published up until now have been described by various authors [3.17] to [3.21] through correlations. A simple equation, which represents the results published hitherto, stems from Žukauskas et al. [3.16]. According to this, the mean Nusselt number for flow across a bundle of smooth tubes is

$$Nu_m = c\,Re^m\,Pr^{0,3}\left(\frac{Pr_m}{Pr_0}\right)^p F \ . \tag{3.212}$$

The equation is valid over the range

$$30 \le Re \le 1.2\cdot 10^6 \quad \text{and} \quad 0.71 \le Pr \le 500 \ .$$

The Nusselt and Reynolds numbers are formed with the outer diameter of the tube. The velocity in the Reynolds number is that in the narrowest cross section of the bundle. All the physical properties of the material are calculated at the inlet temperature of the fluid, only the Prandtl number Prandtl-Zahl Pr_{m} is formed with the mean temperature $\vartheta_{\mathrm{m}} = (\vartheta_{\mathrm{i}} + \vartheta_{\mathrm{e}})/2$ of the fluid, where ϑ_{i} is the inlet temperature of the fluid and ϑ_{e} is the temperature at the exit. The Prandtl number Pr_0 is taken at the outer wall temperature. If this is not constant the Prandtl number Pr_0 should be formed with the mean wall temperature. The factor F is taken from Fig. 3.26; for tube rows $z_{\mathrm{R}} \geq 15$, $F = 1$. The constants c, m and p are given in Table 3.2. The values c and m in the table are valid for liquids, for gases they have to be multiplied by a factor 0.88.

Table 3.2: Constants for (3.212) from Žukauskas [3.16]

| Flow region | Tube arrangement | | | |
| | Aligned | | Staggered | |
	m	c	m	c
laminar flow $200 \leq Re \leq 10^3$	0.50	0.52	0.50	0.50
transition region $10^3 \leq Re \leq 2 \cdot 10^5$	0.63	0.27	0.60	0.40
turbulent flow $Re > 2 \cdot 10^5$	0.84	0.02	0.84	0.021

Heating the fluid: $p=0.25$

Cooling the fluid: $p=0.20$

The heat transfer in a finned tube bundle is calculated using similar equations. For this we suggest [3.22] in the literature.

In the calculation of the heat transferred in a tube bundle, we have to consider that the fluid temperature between the inlet and outlet can significantly change. The heat flow from a tube surface element $\mathrm{d}A$ to the fluid amounts to

$$\mathrm{d}\dot{Q} = \alpha\,\mathrm{d}A(\vartheta_0 - \vartheta_{\mathrm{F}}) = \dot{M}c_p\,\mathrm{d}\vartheta_{\mathrm{F}} \ ,$$

where ϑ_{F} is the mean fluid temperature in a cross section between the tubes. This then gives

$$\frac{\alpha}{\dot{M}c_p}\,\mathrm{d}A = \frac{\mathrm{d}\vartheta_{\mathrm{F}}}{\vartheta_0 - \vartheta_{\mathrm{F}}} \ .$$

By integrating between the inlet cross section i, where the mean fluid temperature is ϑ_{i}, and the outlet (exit) cross section e with a mean fluid temperature ϑ_{e} we obtain

$$\frac{1}{\dot{M}c_p}\int_{\mathrm{i}}^{\mathrm{e}} \alpha\,\mathrm{d}A = \frac{\alpha_{\mathrm{m}}A_0}{\dot{c}_p} = \ln\frac{\vartheta_0 - \vartheta_{\mathrm{i}}}{\vartheta_0 - \vartheta_{\mathrm{e}}} \ .$$

Here α_m is the mean heat transfer coefficient and A_0 is the total tube surface area. With that the heat flow amounts to

$$\dot{Q} = \dot{M} c_p (\vartheta_e - \vartheta_i) = \alpha_m A_0 (\vartheta_i - \vartheta_e)/\ln \frac{\vartheta_0 - \vartheta_i}{\vartheta_0 - \vartheta_e}$$

or

$$\dot{Q} = \alpha_m A_0 \Delta\vartheta_{\log} \tag{3.213}$$

with the logarithmic mean temperature difference

$$\Delta\vartheta_{\log} = (\vartheta_e - \vartheta_i)/\ln \frac{\vartheta_0 - \vartheta_i}{\vartheta_0 - \vartheta_e} . \tag{3.214}$$

Example 3.11: Atmospheric air ($p = 0.1\,\mathrm{MPa}$) is to be heated in a tube bundle heat exchanger from 10 °C to 30 °C. The exchanger consists of 4 neighbouring rows and z_R rows of tubes aligned one behind the other. The outer diameter of the tubes is 25 mm, their length 1.5 m, the longitudinal pitch is the same as the transverse pitch: $s_l/d = s_q/d = 2$. The wall temperature of the tubes is 80 °C with an intial velocity of the air of 4 m/s. Calculate the required number z_R of tube rows.

The following properties are given: Air at $\vartheta_m = 20$ °C: viscosity $\nu = 15.11 \cdot 10^{-6}\,\mathrm{m^2/s}$, thermal conductivity $\lambda = 0.0257\,\mathrm{W/Km}$, density $\varrho = 1.293\,\mathrm{kg/m^3}$, specific heat capacity $c_p = 1.007\,\mathrm{kJ/kgK}$, Prandtl number $Pr = 0.715$. Further for air $Pr(10\,°\mathrm{C}) = 0.716$, $Pr(80\,°\mathrm{C}) = 0.708$.

The velocity w_n in the narrowest cross section is yielded from the mass balance

$$w_\infty\, 4\, s_q = w_n\, (4\, s_q - 4\, d) \quad \text{to} \quad w_n = w_\infty \frac{s_q}{s_q - d} = 4\,\frac{\mathrm{m}}{\mathrm{s}}\,\frac{2\,d}{2\,d - d} = 8\,\mathrm{m/s}\;;$$

which gives a Reynolds number of

$$Re = \frac{w_n\, d}{\nu} = \frac{8\,\mathrm{m/s} \cdot 2.5 \cdot 10^{-2}\,\mathrm{m}}{15.11 \cdot 10^{-6}\,\mathrm{m^2/s}} = 1.32 \cdot 10^4 .$$

The mean Nusselt number is calculated from (3.212) with the values for s and m from Table 3.2, which have to be multiplied by a factor of 0.88, as air is a gas. It is

$$Nu_m = 0.2376\, Re^{0.554}\, Pr^{0.3} \left(\frac{Pr_m}{Pr_0}\right)^{0.25} F ,$$

$$Nu_m = 0.2376 \cdot (1.32 \cdot 10^4)^{0.554} \cdot (0,716)^{0.3} \cdot \left(\frac{0.715}{0.708}\right)^{0.125} F .$$

We will first assume that $z_R \geq 15$, so $F = 1$. This assumption has to be corrected if $z_R < 15$ is yielded. We now have

$$Nu_m = 41.27 , \qquad \alpha_m = Nu_m \frac{\lambda}{d} = 41.27 \cdot \frac{0.0257\,\mathrm{W/Km}}{25 \cdot 10^{-3}\,\mathrm{m}} = 42.43\,\mathrm{W/(m^2K)} .$$

The heat flow is

$$\dot{Q} = \dot{M}\, c_p\,(\vartheta_i - \vartheta_e) = \varrho\, w_\infty\, 4\, s_q\, d\, L\, c_p\,(\vartheta_i - \vartheta_e)$$

$$= 1.293\,\mathrm{kg/m^3} \cdot 4\,\mathrm{m/s} \cdot 4 \cdot 2 \cdot 25 \cdot 10^{-3}\,\mathrm{m} \cdot 1.5\,\mathrm{m} \cdot 1.007\,\mathrm{kJ/kg} \cdot 20\,\mathrm{K} = 31.25\,\mathrm{kW} .$$

On the other hand, according to (3.213) $\dot{Q} = \alpha_m\, 4\, z_R\, d\, \pi\, L\, \Delta\vartheta_m$. With $\Delta\vartheta_m$ from (3.214),

$$\Delta\vartheta_m = (\vartheta_e - \vartheta_i)/\ln \frac{\vartheta_0 - \vartheta_i}{\vartheta_0 - \vartheta_e} = (30\,\mathrm{K} - 10\,\mathrm{K})/\ln\frac{80 - 10}{80 - 30} = 59.44\,\mathrm{K}$$

we get

$$\dot{Q} = 31.25 \cdot 10^3\,\mathrm{W} = 42.43\,\mathrm{W/(m^2K)} \cdot 4 \cdot z_R \cdot 25 \cdot 10^{-3}\,\mathrm{m} \cdot \pi \cdot 1.5\,\mathrm{m} \cdot 59.44\,\mathrm{K} .$$

This gives us a value of $z_R = 26.3$. The chosen value for the number of tube rows is $z_R = 27$. It is $z_R \geq 15$; it was therefore correct to put $F = 1$.

3.7.4 Some empirical equations for heat and mass transfer in external forced flow

As the previous illustrations showed, the heat and mass transfer coefficients for simple flows over a body, such as those over flat or slightly curved plates, can be calculated exactly using the boundary layer equations. In flows where detachment occurs, for example around cylinders, spheres or other bodies, the heat and mass transfer coefficients are very difficult if not impossible to calculate and so can only be determined by experiments. In terms of practical applications the calculated or measured results have been described by empirical correlations of the type $Nu = f(Re, Pr)$, some of which have already been discussed. These are summarised in the following along with some of the more frequently used correlations. All the correlations are also valid for mass transfer. This merely requires the Nusselt to be replaced by the Sherwood number and the Prandtl by the Schmidt number.

1. **Longitudinal flow along a flat plate, heated or cooled from the leading edge onwards**

   Correlation   Range of validity

   $$Nu_x = \frac{1}{\sqrt{\pi}}(Re_x Pr)^{1/2} \qquad\qquad Pr \to \infty \qquad\qquad \text{lam. flow}$$
   $$Nu_x = 0.332\, Re_x^{1/2} Pr^{1/3} \qquad\qquad 0.5 \le Pr \le 1000 \quad Re_x \le 5 \cdot 10^5$$
   $$Nu_x = 0.339\, Re_x^{1/2} Pr^{1/3} \qquad\qquad Pr \to \infty$$
   $$Nu_m = 0.664\, Re^{1/2} Pr^{1/3} \qquad\qquad 0.5 \le Pr \le 1000$$
   $$Nu_m = \frac{0.037\, Re^{0.8}\, Pr}{1 + 2.443\, Re^{-0.1}(Pr^{2/3} - 1)} \qquad 0.6 \le Pr \le 2000 \qquad \text{turb. flow:}$$
   $$5 \cdot 10^5 < Re < 10^7$$

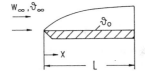

Fig. 3.27: Longitudinal flow over a flat plate, heated from its leading edge

2. **Longitudinal flow over a flat plate, heated or cooled from the point x_0**

   $$Nu_x = \frac{0.332\, Re_x^{1/2} Pr^{1/3}}{\left[1 - (x_0/x)^{3/4}\right]^{1/3}}\;, \qquad 0.5 \le Pr \le 1000\;, \qquad Re_x \le 5 \cdot 10^5$$

3. **Cylinder in cross flow**

   $$Nu_m = 0.3 + (Nu_{m,\text{lam}}{}^2 + Nu_{m,\text{turb}}{}^2)^{1/2}$$

   with

   $$Nu_{m,\text{lam}} = 0.664\, Re^{1/2} Pr^{1/3}$$

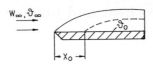

Fig. 3.28: Longitudinal flow over a flat plate heated or cooled from the point x_0

and

$$Nu_{m,\,turb} = \frac{0.037\,Re^{0.8}\,Pr}{1 + 2.443\,Re^{-0.1}(Pr^{2/3} - 1)}$$

$$10 < Re < 10^7\,, \qquad 0.6 < Pr < 1\,000\,.$$

The Nusselt and Reynolds numbers are formed with the length $L = d\pi/2$ over which the fluid flows.

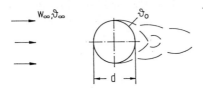

Fig. 3.29: Cylinder in cross flow

4. Cylinder of arbitrary profile in cross flow

The equation is the same as that for a cylinder in cross flow, the Nusselt and Reynolds numbers are formed with the length over which the fluid flows, in the example below $L = a + b$.

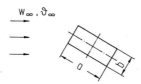

Fig. 3.30: Cylinder of arbitrary profile in cross flow

5. Flow over a sphere

The terms are the same as for a cylinder in cross flow

$$Nu_m = 2 + \left(Nu_{m,\,lam}^2 + Nu_{m,\,turb}^2\right)^{1/2}$$

$$Nu_{m,\,lam} = 0.664\,Re^{1/2}\,Pr^{1/3}$$

$$Nu_{m,\,turb} = \frac{0.037\,Re^{0.8}\,Pr}{1 + 2.443\,Re^{-0.1}(Pr^{2/3} - 1)}$$

$$Nu_m = \frac{\alpha_m d}{\lambda}\,; \qquad Re = \frac{w_\infty d}{\nu}$$

$$0.7 < Pr < 600 \qquad 1 \le Re \le 10^6$$

In the case of mass transfer, Nu_m is replaced by Sh_m and Pr by Sc, this is valid for the range $0.7 < Sc < 70000$.

6. Free falling liquid droplets

$$Nu_{\mathrm{m}} = 2 + 0.6\,Re^{1/2}Pr^{1/3}\left[25\left(\frac{x}{d}\right)^{-0.7}\right] .$$

The Nusselt and Reynolds numbers are formed with the droplet diameter d; x is the distance over which the droplet falls. With $10 \leq x/d \leq 600$ and $3\text{ mm} \leq d \leq 6\text{ mm}$.

7. Bundles of smooth tubes

$$Nu_{\mathrm{m}} = c\,Re^m\,Pr^{0.3}\left(\frac{Pr}{Pr_0}\right)^p F$$

with the constants c, m, p, which are dependent on the arrangement of the tubes and the Reynolds number, taken from Table 3.2 and the factor F which is dependent on the tube row number z_{R} F from Fig. 3.26. Pr_0 is the Prandtl number at the wall temperature. Properties of the material are those taken at the mean fluid temperature $\vartheta_{\mathrm{m}} = (\vartheta_0 + \vartheta_\infty)/2$. The equation is valid over the range

$$30 \leq Re \leq 1.2\cdot10^6 \quad \text{and} \quad 0.71 \leq Pr \leq 500 .$$

3.8 Internal forced flow

Heat and mass transfer apparatus normally consist of channels, frequently tubes, in which a fluid is heated, cooled or changes its composition. While the boundary layers in flow over bodies, for example over a flat plate, can develop freely without influence from neighbouring restrictions, in channels it is completely enclosed and so the boundary layer cannot develop freely. In the following the flow, and then the heat and mass transfer in tubes will be discussed. After this we will study flow through packed and fluidized beds.

3.8.1 Laminar flow in circular tubes

We will consider a fluid in laminar flow that enters a circular tube with constant velocity, Fig. 3.31. Friction causes the liquid adjacent to the wall to decelerate. A boundary layer develops downstream. As the same amount flows through every cross section, the core flow must accelerate in the flow direction. The driving force is the pressure drop in the flow direction. As the boundary layer grows asymptotically to the axis of the tube, the velocity profile asymptotically approaches a parabolic end profile, the so-called *Poiseuille parabola*. The deviation from this is already negligible after a finite distance. The type of flow where the velocity

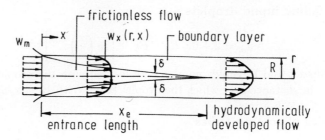

Fig. 3.31: Velocity profile and boundary layer in laminar, tubular flow

profile no longer changes is known as hydrodynamic, fully developed flow. The distance from the inlet until the deviation from the asymptotic end value is negligible is known as the *hydrodynamic entry-length*. It can be calculated exactly [3.23] for the laminar tubular flow sketched in Fig. 3.31 and is

$$x_e \approx 0.056 \, Re \, d \ , \tag{3.215}$$

when the velocity at the tube axis has a deviation of 1% from the value for Poiseuille flow.

Just as for a flat plate, whether the flow is laminar or turbulent is determined by the Reynolds number. As we have already explained, a flow with a Reynolds number below $Re = w_m d/\nu = 2300$ is always laminar and above $Re = 4000$ it is turbulent. Turbulent flows are hydrodynamically fully developed after a short distance. The entry-length lies in the range [3.24]

$$10 \lesssim \frac{x_e}{d} \lesssim 60 \ . \tag{3.216}$$

For calculation of the heat and mass transfer it is sufficient to consider turbulent flow to be hydrodynamically fully developed after an entry-length of $x_e/d \approx 10$. Then the small deviations from the end value of the velocity profile hardly affect the heat and mass transfer coefficients.

3.8.1.1 Hydrodynamic, fully developed, laminar flow

In the calculation of the velocity profile of a hydrodynamic, fully developed, laminar flow we will presume the flow to be incompressible and all properties to be constant. The velocity profile of a fully developed, tubular flow is only dependent on the radius r, $w_x = w_x(r)$ and $w_y = 0$. Therefore the acceleration term $\varrho d w_j/dt$ in the Navier-Stokes equation (3.59) disappears; body forces are not present, so $k_j = 0$. An equilibrium develops between the pressure and friction forces. Balancing the forces on an annular fluid element, Fig. 3.32, gives

$$-\tau_r \, 2r\pi \, dx + \tau_{r+dr} 2(r + dr)\pi \, dx + p \, 2r\pi \, dr - \left(p + \frac{dp}{dx}\right) 2r\pi \, dr = 0 \ .$$

This simplifies, with $\tau_{r+dr} = \tau_r + (d\tau_r/dr)\,dr$ to

$$\tau_r + r\frac{d\tau_r}{dr} - r\frac{dp}{dx} = 0$$

or

$$\frac{1}{r}\frac{d}{dr}(r\tau_r) = \frac{dp}{dx} \ .$$

With Newton's law

$$\tau = \eta\frac{dw_x}{dr}$$

and under the assumption of constant viscosity, we obtain

$$\frac{\eta}{r}\frac{d}{dr}\left(r\frac{dw_x}{dr}\right) = \frac{dp}{dx} \ . \tag{3.217}$$

The axial pressure drop dp/dx is independent of the radial coordinate r; it can also not be a function of the length x, as the left hand side of (3.217) only depends on the coordinate r. So that both sides agree, each must be constant. This means that the pressure can only change proportionally to the length x. Taking into account $(dw_x/dr)_{r=0} = 0$ integration yields

$$r\frac{dw_x}{dr} = \frac{1}{\eta}\frac{dp}{dx}\frac{r^2}{2} \ ,$$

from which, under consideration of $w_x(r = R) = 0$, and a further integration we obtain the velocity profile

$$w_x(r) = -\frac{1}{4\eta}\frac{dp}{dx}R^2\left[1 - \left(\frac{r}{R}\right)^2\right] \ . \tag{3.218}$$

The mean velocity over a cross section is found from this to be

$$w_m = \frac{1}{R^2\pi}\int_0^R w_x(r)2r\pi\,dr = -\frac{R^2}{8\eta}\frac{dp}{dx} \ . \tag{3.219}$$

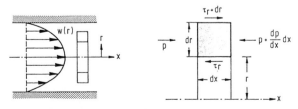

Fig. 3.32: Force balance on a ring element in hydrodynamic, fully developed, laminar flow

The velocity profile of a fully developed flow is therefore

$$\frac{w_x(r)}{w_\mathrm{m}} = 2\left[1 - \left(\frac{r}{R}\right)^2\right] \ . \tag{3.220}$$

Hydrodynamic, fully developed, tubular flows were first studied by Hagen and Poiseuille. The flow is therefore also called Hagen-Poiseuille flow. We can see from (3.220), that the velocity at the tube axis is twice the mean velocity, $w_x(r = 0) = 2w_\mathrm{m}$. From (3.219) we get the following for the pressure drop

$$-\frac{\mathrm{d}p}{\mathrm{d}x} = \frac{8\eta w_\mathrm{m}}{R^2} \ ,$$

from which, after introduction of the Reynolds number $Re = w_\mathrm{m}d/\nu$ and with $-\mathrm{d}p/\mathrm{d}x = \mathrm{const} = \Delta p/L$, we obtain the friction factor

$$\zeta = \frac{\Delta p}{(L/d)\varrho w_\mathrm{m}^2/2} = \frac{64}{Re} \ . \tag{3.221}$$

3.8.1.2 Thermal, fully developed, laminar flow

When a fluid at constant temperature ϑ_α enters, for example, a circular tube with an inner wall temperature $\vartheta_0 \neq \vartheta_\alpha$, it will be either heated up or cooled down. A thermal boundary layer develops, which increases downstream possibly up to the tube axis. The thickness of the thermal boundary layer grows to $d/2$. The heat transfer coefficient reaches it lowest value at this point $\alpha \sim \lambda/(d/2)$, or expressing it in a different manner $Nu = \alpha d/\lambda \sim 2$: The downstream Nusselt number approaches a final value, which according to our rough estimations has a magnitude of 2. A corresponding value exists for the Sherwood number in mass transfer. The following explanations for heat transfer can be applied accordingly to mass transfer. A flow is *thermally fully developed* when the Nusselt number has reached a value within a deviation of say 1 % of the end value.

The local heat transfer coefficient is defined by

$$\alpha := \frac{\dot{q}_0}{\vartheta_0 - \vartheta_\mathrm{F}} \ , \tag{3.222}$$

where ϑ_F is the adiabatic mixing temperature according to (1.30), in the case being considered

$$\vartheta_\mathrm{F} = \frac{1}{\dot{M}} \int\limits_{A_\mathrm{q}} \varrho \, w_x \, \vartheta \, \mathrm{d}A_\mathrm{q} \ .$$

In tubular flow, $\dot{M} = \varrho w_\mathrm{m} R^2 \pi$, $\mathrm{d}A_\mathrm{q} = 2r\pi \, \mathrm{d}r$ and therefore the adiabatic mixing temperature, under the assumption $\varrho = \mathrm{const}$, is

$$\vartheta_\mathrm{F} = \frac{2}{w_\mathrm{m}R^2} \int\limits_0^R w_x \vartheta \, \mathrm{d}r \ . \tag{3.223}$$

For the temperature profile for a thermal fully developed flow, from

$$Nu = \text{const} = \frac{\alpha d}{\lambda} = \frac{\dot{q}_0}{\vartheta_0 - \vartheta_F}\frac{d}{\lambda} = \frac{-\lambda(\partial\vartheta/\partial r)_R}{\vartheta_0 - \vartheta_F}\frac{d}{\lambda}$$

follows the relationship

$$\frac{(\partial\vartheta/\partial r)_R}{\vartheta_0 - \vartheta_F} = \text{const} . \tag{3.224}$$

This says that the temperature increase at the wall in a thermal fully developed flow changes with the length x in the same way as the difference between the wall and the adiabatic mixing temperature. The temperature profile that satisfies this condition is of the general form

$$\vartheta(x, r/R) = (\vartheta_0 - \vartheta_F)f_1(r/R) + f_2(x) ,$$

which can be ascertained by inserting it into (3.224). At the wall

$$\vartheta(x, 1) = \vartheta_0 = (\vartheta_0 - \vartheta_F)f_1(1) + f_2(x) .$$

Subtraction of both equations yields with the abbreviation $r^+ = r/R$:

$$\vartheta_0 - \vartheta(x, r^+) = (\vartheta_0 - \vartheta_F)[f_1(1) - f_1(r^+)]$$

or

$$\vartheta_0 - \vartheta(x, r^+) = (\vartheta_0 - \vartheta_F)f(r^+) . \tag{3.225}$$

This relationship serves in many publications as the definition of a thermally fully developed flow. It is, as we have already seen, a result of the fact that the heat transfer coefficient reaches its asymptotic, constant end value downstream.

We will now look at examples of two particular thermally, fully developed flows, namely that at constant heat flux $\dot{q}_0 = \text{const}$ at the wall, and that at constant wall temperature $\vartheta_0 = \text{const}$.

a) In the *case of constant heat flux*, it follows with

$$\vartheta_0 - \vartheta_F = \frac{\dot{q}_0}{\alpha}$$

from (3.225) that

$$\vartheta_0 - \vartheta(x, r^+) = \frac{\dot{q}_0}{\alpha}f(r^+) .$$

Here, the left hand side is independent of the length x, and so it holds that

$$\frac{d\vartheta_0}{dx} = \frac{\partial\vartheta(x, r^+)}{\partial x} .$$

The temperature at any arbitrary point r^+ changes in the same way with the length x as the wall temperature.

Furthermore, in (3.224) because of $\dot{q}_0 = \text{const}$, we also have $(\partial\vartheta/\partial r)_R = \text{const}$ and therefore $\vartheta_0 - \vartheta_F = \text{const}$. It follows from this that the adiabatic

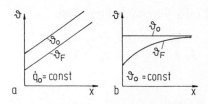

Fig. 3.33: Dependence of the wall temperature $\vartheta_0(x_1)$ and adiabatic mixing temperature $\vartheta_F(x_1)$ on the length x in laminar tubular flow, **a** $\dot{q}_0 = \text{const}$; **b** $\vartheta_0 = \text{const}$

mixing temperature changes in the same way as the wall temperature with the length

$$\frac{d\vartheta_0}{dx} = \frac{d\vartheta_F}{dx} .$$

Therefore we have

$$\frac{\partial\vartheta(x,r^+)}{\partial x} = \frac{d\vartheta_0}{dx} = \frac{d\vartheta_F}{dx} = \frac{\dot{q}_0\,2R\pi}{\dot{M}c_p} = \text{const} . \qquad (3.226)$$

At a sufficient distance downstream all the temperatures change linearly with the length, Fig. 3.33a.

b) In the *case of constant wall temperature*, the energy balance

$$\alpha 2R\pi\,dx[\vartheta_0 - \vartheta_F(x)] = \dot{M}c_p\,d\vartheta_F$$

yields the relationship

$$\frac{d\vartheta_F}{\vartheta_0 - \vartheta_F(x)} = \frac{\alpha d\pi}{\dot{M}c_p}\,dx$$

which can be integrated as the wall temperature is constant. With $\vartheta_F(x = 0) = \vartheta_\alpha$ we obtain

$$\frac{\vartheta_0 - \vartheta_F(x)}{\vartheta_0 - \vartheta_\alpha} = \exp\left(-\frac{\alpha_m d\pi}{\dot{M}c_p}x\right) , \qquad (3.227)$$

where α_m is the mean heat transfer coefficient over the length x. The difference between the wall temperature and the adiabatic mixing temperature decreases exponentially with the length x, Fig. 3.33b. According to (3.225)

$$\frac{\partial\vartheta(x,r^+)}{\partial x} = \frac{d\vartheta_F}{dx}f(r^+) \qquad (3.228)$$

with

$$f(r^+) = \frac{\vartheta_0 - \vartheta}{\vartheta_0 - \vartheta_F} .$$

The temperature change in the direction of flow now depends on the radial coordinate as well.

3.8.1.3 Heat transfer coefficients in thermally fully developed, laminar flow

In the following we will show how heat transfer coefficients are calculated for thermally fully developed, laminar flow. In a corresponding manner the mass transfer coefficients with regard to fully developed concentration profile can be obtained. In order to show this fundamentally we will consider tubular flow. The explanations can easily be transferred to cover other types of channel flow.

When we neglect the dissipation and the thermal conduction in the direction x of flow and under the assumption that the properties of the materials are temperature independent, the energy balance in cylindrical coordinates x, r is

$$w_x \frac{\partial \vartheta}{\partial x} + w_r \frac{\partial \vartheta}{\partial r} = a \frac{1}{r} \frac{\partial}{\partial r} \left(r \frac{\partial \vartheta}{\partial r} \right) . \qquad (3.229)$$

In fully developed flow we have $w_r = 0$, so that the second term on the left hand side drops out. The velocity profile $w_x(r)$ is given by the Hagen-Poiseuille parabola (3.220).

a) We will discuss the case of *constant heat flux* at the wall first. Then, according to (3.226)

$$\frac{\partial \vartheta}{\partial x} = \frac{\partial \vartheta_F}{\partial x} = \text{const} ,$$

and the energy equation is transformed into

$$2w_m \left[1 - \left(\frac{r}{R} \right)^2 \right] \frac{d\vartheta_F}{dx} = a \frac{1}{r} \frac{d}{dr} \left(r \frac{d\vartheta}{dr} \right) .$$

Separating the variables gives, after integration, the general solution of the differential equation

$$\vartheta(x, r) = \frac{2w_m}{a} \frac{d\vartheta_F}{dx} \left(\frac{r^2}{4} - \frac{r^4}{16 R^2} \right) + c_1 \ln r + c_2 .$$

This still has to satisfy the boundary conditions

$$\vartheta(x, r = R) = \vartheta_0(x) .$$

In addition, the temperature in the tube axis $\vartheta(x, r = 0)$ has to be finite, which is only possible if $c_1 = 0$. This gives a temperature profile

$$\vartheta(x, r) = \vartheta_0(x) - \frac{2w_m R^2}{a} \frac{d\vartheta_F}{dx} \left[\frac{3}{16} + \frac{1}{16} \left(\frac{r}{R} \right)^4 - \frac{1}{4} \left(\frac{r}{R} \right)^2 \right] . \qquad (3.230)$$

The adiabatic mixing temperature (3.223) is found by inserting the temperature profile into

$$\vartheta_F(x) = \frac{2}{w_m R^2} \int_{r=0}^{R} 2w_m \left[1 - \left(\frac{r}{R} \right)^2 \right] \vartheta(x, r) r \, dr$$

to give
$$\vartheta_F(x) = \vartheta_0(x) - \frac{11}{48}\frac{w_m R^2}{a}\frac{d\vartheta_F}{dx} .$$

So, with (3.226)
$$\vartheta_0(x) - \vartheta_F(x) = \frac{11}{48}\frac{\dot{q}_0 d}{\lambda} .$$

On the other hand the heat transfer coefficient is defined by $\dot{q}_0 = \alpha(\vartheta_0(x) - \vartheta_F(x))$ and therefore
$$\frac{\alpha d}{\lambda} = Nu = \frac{48}{11} = 4{,}36\overline{36} . \tag{3.231}$$

b) In the case of *constant wall temperature* the energy balance (3.229) is once again valid, into which $w_r = 0$ for fully developed flow has to be inserted. With the velocity profile (3.220) for Hagen-Poiseuille flow, and taking (3.228) into account, we obtain
$$2w_m\left[1 - \left(\frac{r}{R}\right)^2\right]\frac{d\vartheta_F}{dx}\frac{\vartheta_0 - \vartheta}{\vartheta_0 - \vartheta_F} = a\frac{1}{r}\frac{\partial}{\partial r}\left(r\frac{\partial\vartheta}{\partial r}\right) . \tag{3.232}$$

This equation can be further simplified with the dimensionless temperature
$$\vartheta^+ := \frac{\vartheta_0 - \vartheta}{\vartheta_0 - \vartheta_F} , \qquad \frac{d\vartheta^+}{dr} = -\frac{1}{\vartheta_0 - \vartheta_F}\frac{\partial\vartheta}{\partial r}$$

and with the energy balance
$$\alpha d\pi \, dx(\vartheta_0 - \vartheta_F) = \dot{M}c_p \, d\vartheta_F$$

which provides us with the following relationship
$$\frac{\alpha d\pi}{\dot{M}c_p} = \frac{4\alpha}{w_m d\varrho c_p} = \frac{1}{\vartheta_0 - \vartheta_F}\frac{d\vartheta_F}{dx} .$$

With this (3.232) is transformed into
$$-2Nu(1 - r^{+2})\vartheta^+ = \frac{1}{r^+}\frac{d}{dr^+}\left(r^+\frac{d\vartheta^+}{dr^+}\right) \tag{3.233}$$

with the Nusselt number $Nu = \alpha d/\lambda$ and $r^+ := r/R$. The solution has to satisfy the boundary conditions $\vartheta^+(r^+ = 1) = 0$. The equation (3.233) cannot be solved analytically. It is solved using a series expansion
$$\vartheta^+ = \sum_{n=0}^{\infty} C_{2n} r^{+2n} .$$

By inserting this into (3.233) and comparing terms of equal powers in r^+ we find that the coefficients are given by
$$\frac{C_2}{C_0} = -\frac{Nu}{2} \quad \text{and} \quad \frac{C_{2n}}{C_0} = \frac{Nu}{2n^2}\left(\frac{C_{2n-4}}{C_0} - \frac{C_{2n-2}}{C_0}\right) \quad n \geq 2 .$$

This means that all coefficients can be expressed in terms of the Nusselt number. In order to fulfill the boundary condition $\vartheta^+(r^+ = 1) = 0$, the following equation has to be fulfilled

$$\sum_{n=0}^{\infty} \frac{C_{2n}}{C_0} = 0 , \qquad (3.234)$$

and this is only possible for certain values of the Nusselt number. These values can be found, for example, by estimating the Nusselt number, then calculating the coefficients with the recursive formula given above and then checking whether the condition (3.234) has been satisfied. In this way we find that

$$Nu = 3.6568 .$$

Table 3.3: Nusselt numbers $Nu = \alpha d_h/\lambda$ in thermal, fully developed, laminar flow and resistance numbers $\zeta = \Delta p/(L/d_h \, \varrho/2 \, w_m^2)$ in channels with different cross sections. Nu_T is the Nusselt number at constant wall temperature, Nu_q that at constant heat flux at the wall and $Re = w_m d_h/\nu$ is the Reynolds number.

Channel	d_h	Nu_T	Nu_q	ζRe
Circular tube				
Diameter d	d	3.657	4.364	64
Parallel plates				
Plate distance $2b$	$4b$	7.541	8.235	96
Square				
Side length a	a	2.976	3.091	56.91
Rectangle				
Side ratio				
$b/a = 1/2$	$4b/3$	3.391	3.017	62.19
Equilateral triangle				
Side length a	$a/\sqrt{3}$	2.49	1.892	53.33
Ellipse				
Large semiaxis a				
Small semiaxis b				
$b/a = 1/2$	$\pi b/1.2111$	3.742	3.804	84.11

A corresponding calculation for channels with different cross sections leads to the values presented in Table 3.3. These represent the lower limits for the Nusselt number and it cannot fall below these values even in a flow that is not thermally fully developed. The resistance values are also exhibited in Table 3.3

$$\zeta = \Delta p/ \left(\frac{L}{d_h} \varrho \frac{w_m^2}{2} \right) , \qquad (3.235)$$

and these are formed with the hydraulic diameter

$$d_h = 4A/U , \qquad (3.236)$$

where A is the cross sectional area of the channel and U is its perimeter. Further end values of the Nusselt number can be found in Kakač [3.25] among others.

3.8.1.4   The thermal entry flow with fully developed velocity profile

In a laminar tubular flow the velocity profile shall be fully developed and can therefore be described by Hagen-Poiseuille's law (3.220). In contrast the temperature profile is not fully developed. This could be imagined as a heated tube that has an unheated length, long enough to allow the velocity profile to become fully developed before the fluid enters the heated section. The velocity profile forms much faster then the temperature profile if the Prandtl number of the fluid is very large, $Pr \to \infty$, as in the case of highly viscous oils. The high viscosity means that the friction in the fluid propagates fast, whilst the low thermal diffusivity only permits a slow change in the temperature, so the velocity profile rapidly reaches its end value in comparison with the temperature profile that only changes slowly along the flow path.

We presume negligible axial heat conduction, constant wall temperature ϑ_0 and a constant temperature ϑ_α of the fluid at the inlet of the tube. All material properties are temperature independent.

This problem was first dealt with by Graetz (1850–1891) in 1883 [3.26], later in 1910 by Nusselt (1882–1957) [3.27] and by many other authors. It is also known as a Graetz or Graetz-Nusselt problem. It is described by the energy equation (3.229), in which, according to the suppositions made, the radial velocity component disappears, $w_r = 0$, and the axial velocity is that of a Hagen-Poiseuille flow (3.220). With that the energy equation becomes

$$2w_{\mathrm{m}}\left[1 - \left(\frac{r}{R}\right)^2\right]\frac{\partial \vartheta}{\partial x} = a\frac{1}{r}\frac{\partial}{\partial r}\left(r\frac{\partial \vartheta}{\partial r}\right) \tag{3.237}$$

or after the introduction of dimensionless quantities

$$r^+ := \frac{r}{R} \quad , \qquad x^+ := \frac{x}{d\,Pe} \quad \text{with} \quad Pe = RePr = \frac{w_{\mathrm{m}}d}{a}$$

and

$$\vartheta^+ := \frac{\vartheta - \vartheta_0}{\vartheta_\alpha - \vartheta_0} \quad :$$

$$\frac{1}{2}\left(1 - r^{+2}\right)\frac{\partial \vartheta^+}{\partial x^+} = \frac{1}{r^+}\frac{\partial}{\partial r^+}\left(r^+\frac{\partial \vartheta^+}{\partial r^+}\right) \ . \tag{3.238}$$

The temperature profile should satisfy the following boundary conditions

$$\vartheta^+(x^+, r^+ = 1) = 0 \tag{3.239}$$

and

$$\vartheta^+(x^+ = 0, r^+) = 1 \ . \tag{3.240}$$

To allow for solution, the temperature profile $\vartheta^+(x^+, r^+)$ is written as the product of two functions

$$\vartheta^+(x^+, r^+) = \varphi(x^+)\psi(r^+) \ ,$$

of which one only depends on the length x^+ and the other only depends on the radial coordinate r^+. With this statement, the partial differential equation (3.238) can be converted into two ordinary differential equations

$$\varphi' + 2\beta^2\varphi = 0 \tag{3.241}$$

and

$$\psi'' + \frac{1}{r^+}\psi' + \beta^2(1 - r^{+2})\psi = 0 \ . \tag{3.242}$$

The general solution of (3.241) is

$$\varphi = c \exp\{-2\beta^2 x^+\} \ ,$$

where c and β^2 are arbitrary constants. Equation (3.242) cannot be solved explicitly. A particular solution can be found using the series

$$\psi = \sum_{n=0}^{\infty} C_{2n} r^{+2n} \ .$$

Inserting this into (3.242) and comparing the coefficients of equal powers of r^+ yields

$$\frac{C_2}{C_0} = -\frac{\beta^2}{2^2}$$

and for $n \geq 2$ the recursive formula for the coefficients

$$\frac{C_{2n}}{C_0} = \frac{\beta^2}{(2n)^2}\left(\frac{C_{2n-4}}{C_0} - \frac{C_{2n-2}}{C_0}\right) \ .$$

According to this the coefficients C_{2n}/C_0 are still dependent on the quantity β. This has to be determined such that the boundary condition

$$\psi(r^+ = 1) = 0$$

or

$$1 + \sum_{n=1}^{\infty} \frac{C_{2n}}{C_0} = 0$$

is satisfied. In this way we obtain a power series for β, the solution of which delivers an infinite number of values, the so-called eigenvalues β_i, $i = 1, 2, \ldots \infty$. There are therefore an infinite number of particular solutions

$$\psi_i(r^+) = f(\beta_i, r^+) = \sum_{n=0}^{\infty} C_{2n}(\beta_i)r^{+2n} \ ,$$

and the general solution is of the form

$$\vartheta^+ = \sum_{n=0}^{\infty} a_n \exp\{-2\beta_n^2 x^+\}\psi_n(r^+) \ . \tag{3.243}$$

It still has to satisfy the boundary condition $\vartheta^+(x^+ = 0, r^+) = 1$

$$1 = \sum_{n=0}^{\infty} a_n \psi_n(r^+) \ . \tag{3.244}$$

The constants a_n are obtained, see also Appendix A 6 for this, as

$$a_n = \frac{\int_0^1 \psi_n(r^+)(1 - r^{+2})r^+ \, dr^+}{\int_0^1 \psi_n^2(r^+)(1 - r^{+2})r^+ \, dr^+} \ . \tag{3.245}$$

The adiabatic mixing temperature follows from

$$\frac{\vartheta_F - \vartheta_0}{\vartheta_\alpha - \vartheta_0} = \vartheta_F^+ = 2 \int_0^1 \frac{w_x}{w_m}\vartheta^+ r^+ \, dr^+ = 4 \int_0^1 (1 - r^{+2})\vartheta^+ r^+ \, dr^+ \ ,$$

which produces, after inserting the temperature profile (3.243) and integrating,

$$\vartheta_F^+ = \sum_{n=0}^{\infty} B_n \exp\{-2\beta_n^2 x^+\} \ . \tag{3.246}$$

The constants B_n and the eigenvalues β_n of the first five terms of the series are presented in the following Table 3.4, according to calculations by [3.28].

Table 3.4: Eigenvalues β_n and constants B_n according to Brown [3.28]

n	β_n	B_n
0	2.70436	0.81905
1	6.67903	0.09753
2	10.67338	0.03250
3	14.67108	0.01544
4	18.66987	0.00879

The local Nusselt number is obtained from the adiabatic mixing temperature with the help of the energy balance

$$\alpha 2R\pi \, dx(\vartheta_0 - \vartheta_F) = \dot{M}c_p \, d\vartheta_F = w_m R^2 \pi \varrho c_p \, d\vartheta_F$$

to be

$$\alpha = \frac{w_m d c_p}{4} \frac{1}{\vartheta_0 - \vartheta_F} \frac{d\vartheta_F}{dx}$$

or

$$Nu = \frac{\alpha d}{\lambda} = -\frac{1}{4} \frac{d \ln \vartheta_F^+}{dx^+} . \tag{3.247}$$

The mean Nusselt number comes from the mean heat transfer coefficient

$$\alpha_m = \frac{1}{L} \int_0^L \alpha \, dx$$

as

$$Nu_m = \frac{\alpha_m d}{\lambda} = \frac{1}{X^+} \int_0^{X^+} Nu \, dx^+$$

and with $X^+ := L/(d\, Pe)$

$$Nu_m = -\frac{1}{4X^+} \ln \vartheta_F^+(X^+) . \tag{3.248}$$

A plot of the mean Nusselt number is shown in Fig. 3.35; it is identical to the curve for $Pr \to \infty$

For sufficiently large values of the length $x^+ \to \infty$ the adiabatic mixing temperature (3.246) may be calculated using only the first term of the series. The mean Nusselt number according to (3.248) is then transformed into

$$
\begin{aligned}
Nu_m &= -\frac{1}{4X^+} \ln\left(B_0 \exp\{-2\beta_0^2 X^+\}\right) \\
&= \frac{2\beta_0^2}{4} - \frac{1}{4X^+} \ln B_0 .
\end{aligned}
$$

With the values in Table 3.4 we obtain

$$Nu_m = 3.6568 + \frac{0.0499}{X^+} , \qquad \text{valid for} \quad X^+ \geq 0.05 . \tag{3.249}$$

For $X^+ \to \infty$, the known end value of 3.6568 for the Nusselt number for thermal fully developed flow is obtained. The calculation of the Nusselt number for small values of the length $X^+ = L/(d\, Pe)$ requires many terms of the series for the adiabatic mixing temperature (3.246). Therefore the exact solution has been approximated by empirical equations. According to Stephan [3.29] through

$$Nu_m = \frac{3.657}{\tanh(2.264\, X^{+1/3} + 1.7\, X^{+2/3})} + \frac{0.0499}{X^+} \tanh X^+ \tag{3.250}$$

the exact values in the total region $0 \leq X^+ \leq \infty$ will be reproduced with the largest error of 1 %. For large values $X^+ \geq 0.05$ the equation is transformed into the asymptotic solution (3.249), whilst for small values $X^+ \leq 5 \cdot 10^{-6}$ we obtain the so-called Lévêque solution [3.30] which is valid for short lengths

$$Nu_\mathrm{m} = 1.615 \, (X^+)^{-1/3} \, . \tag{3.251}$$

As we have already explained, in the calculation of the heat flow, the temperature change in the fluid between inlet and outlet has to taken into account. According to the previous explanations, (3.213), the heat flow is

$$\dot{Q} = \alpha_\mathrm{m} A_0 \Delta \vartheta_\mathrm{log} \, , \tag{3.252}$$

where A_0 is the surface of the tube over which heat is released and $\Delta\vartheta_\mathrm{log}$ is the logarithmic mean temperature difference, cf. (3.214),

$$\Delta\vartheta_\mathrm{log} = (\vartheta_\mathrm{F} - \vartheta_\alpha)/\ln\frac{\vartheta_0 - \vartheta_\alpha}{\vartheta_0 - \vartheta_\mathrm{F}} \, . \tag{3.253}$$

3.8.1.5 Thermally and hydrodynamically developing flow

In flow that is neither hydrodynamically nor thermally fully developed the velocity and temperature profiles change along the flow path. Fig. 3.34 shows qualitatively some velocity and temperature profiles, under the assumption that the fluid flows into the tube at constant velocity and temperature. The wall temperature of the tube is lower than the inlet temperature of the fluid.

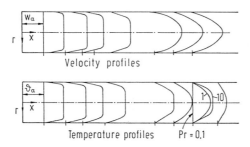

Velocity profiles

Temperature profiles   Pr = 0,1

Fig. 3.34: Velocity and temperature profiles of a hydrodynamically or thermally developing flow

The calculation of the velocities and temperatures requires the solution of the continuity, momentum and energy equations. An explicit solution of the system of equations is not possible. A numerical solution has been communicated by Stephan [3.23] and later by several other authors see [3.25]. As a result the pressure drop $\Delta p = p_\mathrm{i} - p$ between the pressure p_i at the inlet and the pressure p at any tube cross section is obtained. It can be approximated by an empirical correlation [3.29]

$$\frac{\Delta p}{\varrho w_\mathrm{m}^2/2} = (1.25 + 64X) \tanh(11.016X^{1/2}) \tag{3.254}$$

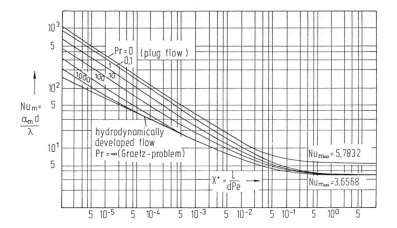

Fig. 3.35: Mean Nusselt numbers in laminar tubular flow

with $X := L/(d\,Re)$ and $Re := w_{\mathrm{m}}d/\nu$. The pressure drops calculated using this method deviate by less than 1.65 % from the values given by the numerical solution for the entire region $0 \leq X \leq \infty$.

In the same way as shown in the previous section, the heat transfer coefficient and from that the mean Nusselt number $Nu_{\mathrm{me}} = \alpha_{\mathrm{me}}d/\lambda$ (the index e stands for entry flow) can be obtained from the temperature profile. The Nusselt number can be calculated from an empirical equation of the form

$$\frac{Nu_{\mathrm{me}}}{Nu_{\mathrm{m}}} = \frac{1}{\tanh(2.432\,Pr^{1/6}X^{+1/6})} \ , \tag{3.255}$$

where Nu_{m} is the mean Nusselt number according to (3.250) for hydrodynamically fully developed flow. Equation (3.255) is valid for $0.1 \leq Pr \leq \infty$ and $0 \leq X^+ \leq \infty$. The deviation from the numerically calculated mean Nusselt numbers amounts to less than 5 % for $1 \leq Pr \leq \infty$, and increases for small Prandtl numbers $0.1 \leq Pr < 1$ to around 10 %.

Fig. 3.35 illustrates the pattern of the mean Nusselt number over the dimensionless length $X^+ = L/(d\,Pe)$. All the lines are arranged between those for $Pr = 0$ and $Pr \to \infty$. $Pr = 0$ in this case means that the viscosity vanishes, but the thermal conductivity is finite. As no friction forces act in the fluid, the velocity at the inlet remains constant. This type of flow is known as plug flow. In the limiting case $Pr \to \infty$, because the viscosity is large in comparison to the thermal diffusivity, the flow is hydrodynamically but not thermally fully developed. At the limit $Pr \to \infty$ equation (3.255) yields $Nu_{\mathrm{me}}/Nu_{\mathrm{m}} = 1$.

Vanishing Prandtl numbers $Pr = 0$ can also mean that the thermal diffusivity is approaching infinity, whilst the viscosity remains finite. Then the flow is already thermally fully developed at the inlet but not yet hydrodynamically fully developed. As the Péclet number disappears, $Pe = w_{\mathrm{m}}d/a = 0$,

$X^+ = L/(d\,Pe) = \infty$. The Nusselt number is equal to that for thermal fully developed flow $Nu_{\mathrm{m}} = 3.6568$.

3.8.2   Turbulent flow in circular tubes

An exclusively analytical treatment of heat and mass transfer in turbulent flow in pipes fails because to date the turbulent shear stress $\tau_{ij} = -\varrho\overline{w'_i w'_j}$, heat flux $\dot{q}_i = -\varrho c_p \overline{w'_i T'}$ and also the turbulent diffusional flux $j^*_{Ai} = -\varrho\overline{w'_i \xi'_A}$ cannot be investigated in a purely theoretical manner. Rather, we have to rely on experiments. In contrast to laminar flow, turbulent flow in pipes is both hydrodynamically and thermally fully developed after only a short distance $x/d \geq 10$ to 60, due to the intensive momentum exchange. This simplifies the representation of the heat and mass transfer coefficients by equations. Simple correlations, which are sufficiently accurate for the description of fully developed turbulent flow, can be found by using the analogy between momentum and heat or mass transfer, (3.199),

$$St\,Pr^{2/3} = \frac{c_{\mathrm{f}}}{2} \tag{3.256}$$

with $St = \alpha/(\varrho c_p w_{\mathrm{m}})$. The friction factor c_{f} is linked to the resistance factor ζ because

$$\tau_0 d\pi L = \Delta p\,(d^2\pi/4) \ ,$$

from which

$$\tau_0 = \Delta p\,(d/4L)$$

follows and

$$c_{\mathrm{f}} = \frac{\tau_0}{\varrho\,(w_{\mathrm{m}}^2/2)} = \frac{\Delta p}{(L/d)\varrho\,(w_{\mathrm{m}}^2/2)}\frac{1}{4} = \frac{\zeta}{4} \ . \tag{3.257}$$

Therefore

$$St\,Pr^{2/3} = \frac{\zeta}{8} \ . \tag{3.258}$$

With the resistance factor [3.32]

$$\zeta = 0.184\,Re^{-1/5} \quad \text{valid for} \quad Re \geq 10^4$$

for turbulent tubular flow, we obtain the following for the Nusselt number

$$Nu = 0.023\,Re^{4/5}\,Pr^{1/3} \ . \tag{3.259}$$

A relationship that approximately agrees with equation (3.259) was recommended by Kraussold [3.33] in 1933, based upon his own experiments and those of others. The equation was first communicated in the form presented here by McAdams [3.34] in 1942. It presumes small temperature differences between the wall and the fluid, and is valid in the region

$$0.7 \leq Pr \leq 160 \quad , \qquad Re = w_{\mathrm{m}}d/\nu \geq 10^4 \quad \text{and} \quad L/d \geq 100 \ .$$

All the material properties are taken at the adiabatic mixing temperature. With large differences in the temperature of the fluid and the wall, the influence of the viscosity that changes markedly with the temperature has to be taken into consideration. Hufschmidt and Burck [3.35] found, based on experiments by Sieder and Tate [3.36], that it is sufficient to multiply the right hand side of (3.259) with the factor

$$(Pr/Pr_0)^{0.11} \, ,$$

where the Prandtl number Pr is that at the mean temperature $\vartheta_m = (\vartheta_i + \vartheta_e)/2$ and Pr_0 is that at the wall temperature ϑ_0.

An equation that is also valid for large Reynolds numbers, was developed by Petukhov and Kirilov [3.37]. It has been modified by Gnielinski [3.38] so that in addition the region below $Re = 10^4$ is correctly described. It reads

$$Nu = \frac{(\zeta/8)(Re - 1\,000)Pr}{1 + 12.7\sqrt{\zeta/8}(Pr^{2/3} - 1)} \left[1 + \left(\frac{d}{L}\right)^{2/3} \right] \qquad (3.260)$$

with the resistance factor

$$\zeta = \frac{\Delta p}{(L/d)\varrho\,(w_m^2/2)} = \frac{1}{(0.79 \ln Re - 1.64)^2} \; . \qquad (3.261)$$

It is valid in the region

$$2300 \le Re \le 5 \cdot 10^6 \quad , \qquad 0.5 \le Pr \le 2000 \quad \text{and} \quad L/d > 1 \; .$$

All material properties are formed with the mean temperature $\vartheta_m = (\vartheta_i + \vartheta_e)/2$. The influence of a viscosity that changes starkly with temperature is once again accounted for by the factor $(Pr/Pr_0)^{0.11}$ on the right hand side of (3.260), where Pr is at the mean temperature ϑ_m, and Pr_0 is formed at the wall temperature.

The equations (3.259) and (3.260) can also be used for the calculation of mass transfer coefficients. As has already been explained, this merely requires the replacement of the Nusselt with the Sherwood number and the Prandtl with the Schmidt number.

3.8.3 Packed beds

A packed bed is understood to be the ordered or irregular arrangement of individual bodies of different shapes. As an example of this Fig. 3.36 shows a packed bed of particles of different sizes. A pipe register is also a packed bed in the sense of this definition.

In *fluidized* beds the particles are mixed up by a flowing fluid and kept in suspension. They then have properties similar to that of a fluid. Chemical reactions, drying or other mass transfer processes take place rapidly in fluidized beds as a result of the brisk movement of the particles.

Packed beds serve as regenerators in heat transfer. As so-called packed columns they are freqently implemented as mass transfer apparatus. This normally involves the introduction of a liquid mixture at the top of a column with a gas of different composition flowing in the opposite direction, as illustrated in Fig. 3.37. Through mass transfer one or more components of the gas are transferred into the liquid phase and the same occurs in reverse. In this way, toxic substances can be taken from the gas into a washing liquid falling down the column. The sizing of this type of apparatus is the object of thermal process engineering. This requires knowledge of the basics of heat and mass transfer which will be explained in the following.

In order to characterise the flow space within filling we will consider packing of equally sized spheres of diameter d_P. A suitable parameter for the description of packing is the *void fraction*

$$\varepsilon := V_G/V \ , \tag{3.262}$$

formed with the void space space filled with gas V_G and the total volume of the bed

$$V = V_G + V_S \ .$$

In this the volume V_S of the solids is found from the volume V_P of the individual particles and the number n of them, $V_S = nV_P$. Therefore

$$\frac{V_S}{V} = \frac{nV_P}{V} = 1 - \varepsilon$$

or for the number of particles per unit volume

$$n_V := \frac{n}{V} = \frac{1 - \varepsilon}{V_P} \ . \tag{3.263}$$

This is ascertained when the void fraction and the volume of the particles are known.

The two quantities, void fraction ε and particle diameter d_P, are not sufficient to clearly describe the flow and with that also the heat and mass transfer. Imagine a cubic packing of equally sized spheres placed one behind the other, as shown in Fig. 3.38a. This arrangement allows the fluid to flow along the gaps between the spheres. In an irregular packing, with the same void fraction, made up of spheres

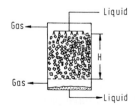

Fig. 3.36: Packed bed of par-   **Fig. 3.37**: Packed column in
ticles of different sizes   countercurrent operation

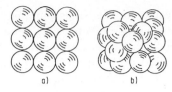

Fig. 3.38: Packing of spheres of the same size and void fraction **a** cubic packing **b** irregular packing

of the same diameter, the spaces between the spheres can become partially blocked and the throughflow will be more strongly hindered at some points than at others, as we can see in Fig. 3.38b. Despite having the same void fraction and particle diameter the flow pattern can still be different. Nevertheless the flow, heat and mass transfer can still be described by the two parameters ε and d_P, but only because in sufficiently large packings these differences are balanced out in the statistical average.

The specific surface area a_P of particles of any shape is defined by

$$a_P := n A_P / V \tag{3.264}$$

(SI units m^2/m^3), in which A_P is the surface area of an individual particle. The specific surface area a_P is a characteristic property of the packing. For packings of different forms and shapes $n A_P$ is the sum of the individual particle surface areas put into the volume V. Further, with (3.263)

$$a_P = \frac{n A_P}{V} = \frac{A_P}{V_P}(1 - \varepsilon) \ , \tag{3.265}$$

where A_P/V_P is the specific surface area of an individual particle. For spherical particles it is equal to $6/d$, so that for this

$$a_P = \frac{6}{d}(1 - \varepsilon) \tag{3.266}$$

is valid.

An example: A cubic packing of spheres, Fig. 3.38a, with z spheres per row and r rows behind and above each other has a volume $V = (zd)(rd)(rd) = zr^2d^3$. The volume of all the spheres is $V_S = nd^3\pi/6 = zr^2d^3\pi/6$, and the void fraction is

$$\varepsilon = 1 - \frac{V_S}{V} = 1 - \frac{\pi}{6} = 0.476 \ .$$

An irregular spherical packing has a void fraction of $\varepsilon \approx 0.4$. Its specific surface area is, according to (3.266), $a_P \approx 3.6/d$. Spheres of 1 cm diameter in an irregular arrangement already have a specific surface area $a_P \approx 360 \mathrm{m}^2/\mathrm{m}^3$. According to (3.263) the number of particles per unit volume is yielded to be $n_V \approx 1.15 \cdot 10^6/\mathrm{m}^3$. A powder of spherical particles of 100 μm yields $a_P \approx 3.6 \cdot 10^4 \mathrm{m}^2/\mathrm{m}^3$ and $n_V \approx 1.15 \cdot 10^{12}/\mathrm{m}^3$. These numbers clearly show that the surface area available for heat and mass transfer in packings can be extraordinarily large.

When a fluid flows through a packing the heat transfer coefficients increase rapidly in the first few rows and then reach fixed values, as Gillespie et al. [3.39] indicated for the example of heat transfer in an irregular, spherical packing. The end values for the Nusselt number are significantly larger in packing than for flow around a single sphere. This is due to the frequent eddies and changes in direction of the fluid. The corresponding findings would be expected for mass transfer. The end values of the mean Nusselt number Nu_m in spherical packing and of the mean Nusselt number Nu_{mS} in flow around a single sphere have a certain relationship with each other that is only dependent on the void fraction. This has been shown by experimentation. This relationship is

$$Nu_m = f_\varepsilon Nu_{mS} , \qquad (3.267)$$

where f_ε is an arrangement factor that is only dependent on the void fraction. The mean Nusselt number for flow over a sphere is given in 3.7.4, No.5. The Reynolds number used there was formed with the mean effective velocity $w_{eff} = w_m/\varepsilon$. This is found from the material balance $w_m A_0 = w_{eff} A_G$ to be $w_{eff} = w_m A_0/A_G = w_m/\varepsilon$, where A_0 is the cross sectional area of the empty column and A_G is the void cross sectional area through which the gas flows.

The factor f_ε can be calculated with sufficient accuracy for the range $0.26 < \varepsilon < 1$ according to Schlünder [3.40], from the simple formula

$$f_\varepsilon = 1 + 1.5(1 - \varepsilon) . \qquad (3.268)$$

Eq. (3.267) also holds for packings with non-spherical particles. Values giving some idea of the arrangement factor f_ε are contained in Table 3.5. The values hold over the range $10^2 < Re < 10^4$, if the Reynolds number is formed with the effective mean velocity $w_{eff} = w_m/\varepsilon$ and an equivalent sphere diameter d_P. This is calculated from the mean particle surface area

$$A_{mP} = a_P/n_V \qquad (3.269)$$

with the specific surface area a_P from (3.264) and the number n_V of particles per unit volume according to (3.263), by comparison with a sphere of the same surface area, to be

$$d_P = \sqrt{A_{mP}/\pi} . \qquad (3.270)$$

The heat flow is calculated using the method already known with

$$\dot{Q} = \alpha_m n A_P \Delta\vartheta_{\log}$$

Table 3.5: Arrangement factor f_ε for packings of non-spherical particles

Particle	f_ε	checked for
Cylinder, Length L, Diameter d	1.6	$0.24 < L/d < 1.2$
Cube	1.6	$0.6 \le Pr, Sc \le 1300$
Raschig rings	2.1	$Pr = 0.7, Sc = 0.6$
Berl saddle	2.3	$Sc = 2.25$

and the logarithmic mean temperature difference (3.214)

$$\Delta\vartheta_{\log} = (\vartheta_e - \vartheta_i)/\ln\frac{\vartheta_0 - \vartheta_i}{\vartheta_0 - \vartheta_e} ,$$

where ϑ_i is the inlet, ϑ_e the outlet (exit) temperature of the fluid and ϑ_0 is the surface temperature of the particle.

The equations also hold in a corresponding manner for mass transfer. This merely requires the Nusselt number to be replaced by the Sherwod number and the Prandtl by the Schmit number. Prerequisite for the validity of the equations is however, a sufficiently large value for the Péclet number $Pe = RePr \overset{\scriptscriptstyle >}{\sim} 500$ to 1000, because otherwise the flow and therefore the heat and mass transfer are distributed unequally over the cross section. It may also be possible for flow reversal to occur, such that the mean heat and mass transfer coefficients could be smaller than those for flow over a single body.

3.8.4   Fluidized beds

Fluidized beds consist of solid particles which are maintained in a suspension by a fluid flowing upwards through them. They were first used in the Winkler process (German patent DRP 437 970 from 28th Sept. 1922) for carbon gasification. Nowadays fluidized beds are used in a variety of ways; for chemical reactions and among others for the mixing, agglomeration or drying of solids. Fig. 3.39 shows schematically a gas-solid fluidized bed.

In order to generate a fluidized bed, the velocity of the fluid flowing upwards must be just great enough, so that a balance between the forces exerted by the fluid and the weight of the particles exists. Then

$$\Delta p = \varrho g H = [\varrho_S(1 - \varepsilon) + \varrho_F \varepsilon]g H , \qquad (3.271)$$

where H is the height of the solid particles in the column, ϱ_S the density of the solid material and ϱ_F that of the fluid.

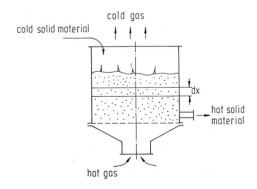

Fig. 3.39: Fluidized bed

As the mass of the solid remains constant, independent of how high the solid particles are forced upwards by the fluid, it holds that

$$M_S = \varrho_S(1 - \varepsilon)A_0 H = \text{const}$$

and in a column with a constant cross section

$$\varrho_S(1 - \varepsilon)H = \text{const} .$$

As we can see the void fraction increases with the height of the fluidized bed. On the other hand in gas-solid fluidized beds it is almost always the case that $\varrho_F \ll \varrho_S$ and therefore also

$$\Delta p \cong \varrho_S(1 - \varepsilon)gH = \text{const} .$$

The pressure drop in a fluidized bed is independent of its height in good approximation constant.

In order to bring the solid particles into suspension, the fluid has to exceed a certain minimum velocity, the so-called minimum fluidization velocity w_{mf}, so $w_G \geq w_{mf}$. This is yielded, with the help of the approximation equation from the VDI-Heat Atlas [3.41], from

$$Re_{mf} = \frac{w_{mf}d_P}{\nu} \cong 42.9\frac{(1 - \varepsilon_{mf})}{\varphi_S}\left[(1 + 3.11 \cdot 10^{-4}\frac{\varphi_S^3\varepsilon_{mf}^3}{(1 - \varepsilon_{mf})^2}Ar)^{1/2} - 1\right] \quad (3.272)$$

with the *Archimedes number Ar* as the ratio of the lift to the frictional forces

$$Ar := \frac{\varrho_S - \varrho_F}{\varrho_F}\frac{d_P^3 g}{\nu^2}$$

and the *sphericity* defined as

$$\varphi_S := \frac{\text{Surface area of a sphere of the same volume as a particle}}{\text{Particle surface area}} .$$

Numerical values for the sphericity are contained in the VDI-Heat Atlas [3.41]; for example, fire-blasted sand has $\varphi_S = 0.86$. ε_{mf} is the void fraction at the fluidization point.

At velocities above the minimum fluidization velocity the fluidized bed expands. Its height and void fraction increase. Solid-liquid systems expand continuously with increasing liquid velocity, and the solid particles are homogeneously distributed. In solid-gas systems gas bubbles form inside the fluidized bed, so that a heterogeneous system exists.

The fluid velocity or the Reynolds number of the fluid and the void fraction are not independent of each other in fluidized beds as they are in packed beds. Rather the void fraction increases with the Reynolds number, from the value of $\varepsilon \cong 0.4$ for a packed bed to a value of $\varepsilon \rightarrow 1$, which will then be approximately reached, if only one particle is located in a large volume.

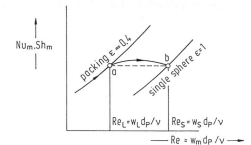

Fig. 3.40: Heat and mass transfer at the
transition from packed to fluidized bed

Heat and mass trasnfer coefficients in a fluidized bed lie between the values for a packed bed and those for a single particle. The fundamental pattern of the Nusselt or Sherwood number as functions of the Reynolds number is illustrated in Fig. 3.40. In this the Nusselt number $Nu = \alpha d_P/\lambda$ or the Sherwood number $Sh = \beta d_P/\eta$ and the Reynolds number $Re = w_m d_P/\nu$ are all formed with the particle diameter, which for non-spherical particles is the same as the equivalent sphere diameter according to (3.270). In the Reynolds number w_m is the mean velocity in the imaginary empty packing.

If the Reynolds number is increased in a packed bed, the Nusselt and Sherwood numbers increase corresponding to the left branch of the curve in Fig. 3.40.

Once the upward flowing fluid has reached the minimum fluidization velocity w_{mf} and with that the Reynolds number the value $Re_{mf} = w_{mf}d_P/\nu$, point a in Fig. 3.40, a fluidized bed is formed. The heat and mass transfer coefficients hardly change with increasing fluid velocity: The Nusselt and Sherwood numbers are only weakly dependent on the Reynolds number, corresponding to the slightly upwardly arched line a b in Fig. 3.40. After a certain fluid velocity has been reached, indicated here by point b in Fig. 3.40, the particles are carried upwards. At point b the heat and mass transfer coefficients are about the same as those for flow around a single sphere of diameter d_P.

In an homogeneous (bubble free) fluidized bed the velocity with which the particles are just carried at point b is equal to the fluid velocity. Points on the rising branch of the line for flow around a single sphere to the right of point b are attributed to a fluid velocity greater than the particle velocity. They lie in the region of pneumatic transport.

An observer moving with the fluid would, upon reaching point b, have the impression that the particle was about to start falling. The particle velocity in a homogeneous fluidized bed at point b is the same as the falling velocity w_S of a particle in a quiescent fluid. This can be calculated from the balance of the buoyancy F_B and resistances forces F_R. This is

$$F_B = (\varrho_S - \varrho_F)\frac{d_P^3\pi}{6}g = F_R = c_R\frac{d_P^2\pi}{4}\varrho_F\frac{w_S^2}{2} \ . \tag{3.273}$$

The resistance factor follows as

$$c_R = \frac{4}{3} \frac{(\varrho_S - \varrho_F) d_P g}{\varrho_F w_S^2} \quad . \tag{3.274}$$

With the Reynolds number $Re_S = w_S d_P / \nu$ and the previously defined Archimedes number, the resistance factor can also be written as

$$c_R = \frac{4}{3} \frac{Ar}{Re_S^2} \quad . \tag{3.275}$$

As the resistance factor c_R in flow around a sphere is dependent on the Reynolds number, according to (3.275), a relationship between the Reynolds number formed with the falling velocity w_S and the Archimedes number can be derived, $Re_S = f(Ar)$.

In process engineering the main interest lies in the region $Re_S \leq 10^4$. The area above $Re_S = 10^5$, in which the transition from laminar to turbulent boundary layer occurs, can generally be disregarded for chemical engineering tasks. By solving the Navier-Stokes equation up to $Re = 80$ and comparing this with experiments enabled Brauer and co-workers [3.42] to find a very accurate emprical law, for values up to $Re = 10^4$, for the resistance factor

$$c_R = \frac{24}{Re} + \frac{5.48}{Re^{0.573}} + 0.36 \quad . \tag{3.276}$$

Our case with $Re = w_S d_P / \nu = Re_S$ has to be put into this. It includes Stoke's law $c_R = 24/Re$ which is valid for low Reynolds numbers $0 \leq Re \leq 0.1$, and reaches, at the limit of validity $Re = 10^4$, the value $c_R = 0.39$. Now, putting (3.276) into (3.275), gives a transcendental equation of the form $Ar (Re_S)$, from which, at a given Archimedes number Ar, the Reynolds number Re_S can be developed by iteration. In order to reduce the computation time, without a significant loss in accuracy, (3.276) is replaced by the following approximation, which is valid for the range $0 \leq Re \leq 10^4$,

$$c_R \cong 0.36 \left[1 + \left(\frac{24}{0.36 Re} \right)^{1/1.8} \right]^{1.8} \quad . \tag{3.277}$$

Then, inserting this into (3.275), taking into account $Re = Re_S$, we obtain an explicit expression for the Reynolds number

$$Re_S \cong 19.14 \left[\left(1 + \frac{Ar^{1/1.8}}{12.84} \right)^{1/2} - 1 \right]^{1.8} \quad , \tag{3.278}$$

valid for homogeneous fluidized beds over the range $0 \leq Re_S \leq 10^4$ or $0 \leq Ar \leq 2.9 \cdot 10^7$.

In heterogeneous fluidized beds, like those where gas flows through the bed, (3.278) is likewise valid for $0 \leq Re_S \leq 1.2 \cdot 10^2$ or $0 \leq Ar \leq 10^4$. In regions beyond

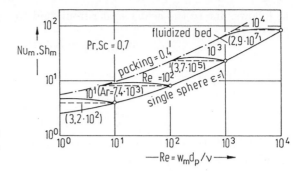

Fig. 3.41: Heat and mass transfer
fluid/particle in the fluidized bed,
according to Martin [3.43]

this, i.e. $Re_S \geq 1.2 \cdot 10^2$ or $Ar \geq 10^4$, according to Reh [3.44], the following holds
for heterogeneous fluidized beds

$$Re_S \cong \left(\frac{4}{3}Ar\right)^{1/2} . \tag{3.279}$$

The Reynolds numbers calculated according to (3.278) barely deviate from
those calculated using a different equation communicated by Martin [3.43], but
reproduce the relationship $Re_S(Ar)$, which is gained by inserting the accurate
resistance law (3.276) from Brauer into (3.275), somewhat better. The maximum
error in the Reynolds number from the approximation equation (3.278) is around
3.5 %. With assistance from (3.278) or (3.279), the Reynolds number Re_S at the
point b, for stipulated Archimedes numbers, can be calculated. Fig. 3.41 shows
the plot of the heat and mass transfer coefficients in fluidized beds for different
Archimedes numbers. After a suggestion by Martin [3.43] the slightly curved
lines for the heat and mass transfer in a fluidized bed are replaced by the dotted
horizontal lines, Fig. 3.40 and 3.41. The heat and mass transfer coefficients are
the same there as those for flow around a single sphere, point b in Fig. 3.40. The
equation valid for this is taken from section 3.7.4, No.5, in which Re_S is used for
the Reynolds number.

Finally, some observations on the *energy balance for a fluidized bed reactor*.
The temperature pattern in solid particles is, as explained for transient heat conduction in solids, determined decisively by the Biot number $Bi = \alpha d_P/\lambda_s$. It is
the ratio of the thermal resistance in the solid to that in the gas. Due to the
small dimensions of the particle and the good thermal conductivity of the solid in
comparison to that of the gas, the thermal resistance of the particle can usually
be neglected compared to that of the gas. This means that the Biot numbers in
fluidized beds are very small. Local temperature variations in the particle can
therefore be neglected. On the other hand, if the particles are well mixed in the
fluidized bed, the initial temperature changes will rapidly disappear. This therefore allows us to assign a unified temperature to the particles in a fluidized bed,
through which the energy balance is significantly simplified.

If we consider a volume element $dV = A_0\,dx$, of height dx, where A_0 is the
free cross section of a fluidized bed, see Fig. 3.39, the fluid temperature changes

as follows, due to the heat transfer from fluid to solid

$$- \dot{M}_F c_{pF}\, d\vartheta_F = \alpha n\, dA_p(\vartheta_F - \vartheta_S) \tag{3.280}$$

with $n\, dA_P = a_P A_0\, dx$. Then, because $\dot{M}_F = w_m A_0 \varrho_F$,

$$\frac{- d\vartheta_F}{\vartheta_F - \vartheta_S} = \frac{\alpha a_p}{\varrho_F w_m c_{pF}}\, dx \ .$$

Integration between the inlet cross section i and the exit cross section e of a
fluidized bed of height H yields

$$\ln \frac{\vartheta_{Fi} - \vartheta_S}{\vartheta_{Fe} - \vartheta_S} = \frac{\alpha_m a_p}{\varrho_F w_m c_{pF}} H$$

or with (3.266), when the equivalent sphere diameter d_P according to (3.270), and
the mean Nusselt number $Nu_m = \alpha_m d_P/\lambda$, the Reynolds number $Re = w_m d_P/\lambda$
and the Prandtl number $Pr = \nu/a$ of the fluid are used,

$$\ln \frac{\vartheta_{Fi} - \vartheta_S}{\vartheta_{Fe} - \vartheta_S} = \frac{Nu_m}{RePr} \frac{6(1 - \varepsilon)H}{d_P} \ . \tag{3.281}$$

Corresponding considerations also hold for mass transfer. The diffusion resistance of the particle can be neglected when compared to that of the gas, and
so a unified composition can be assigned to the particle. When only one component A is transferred between a fluid, consisting of components A and B, and the
solid S, the calculation can be usefully made using the mass content of the fluid
$X_F = (\dot{M}_A/\dot{M}_B)_F$. Then, under the assumption of a not too large mass content
of component A in the solid, see equation (1.195a)

$$- \dot{M}_B\, dX_F = \beta \frac{p}{R_B T}\, dA_p(X_F - X_S) = \beta \varrho_B a_p A_0\, dx(X_F - X_S) \ .$$

Here $X_S = (\dot{M}_A/\dot{M}_B)_S$ is the mass content of the fluid on the solid surface. After
integration between the inlet cross section i and the outlet cross section e of a
fluidized bed of height H we obtain, with $\dot{M}_B = w_m A_0 \varrho_B$, under consideration of
(3.266) and the definitions $Sh_m = \beta_m d_P/D$, $Re = w_m d_P/\nu$, $Sc = \nu/D$:

$$\ln \frac{X_{Fi} - X_S}{X_{Fe} - X_S} = \frac{Sh_m}{ReSc} \frac{6(1 - \varepsilon)H}{d_P} \ . \tag{3.282}$$

Example 3.12: In a fluidized bed, wet sand is to be dried by air which has a lower moisture
content but is at the same temperature as the sand. During this the water vapour content of
the air increases from the initial value X_{Fi} to the value X_{Fe}. At the surface of the sand the air
is saturated with water. Its water content there is X_S, so that the driving content difference
$X_S - X_{Fi}$ falls to the end value $X_S - X_{Fe}$. How high must the fluidized bed be, such that
the driving difference $X_S - X_{Fe}$ at the outlet cross section is only 5 % of the driving difference
$X_S - X_{Fi}$ at the inlet cross section?

The velocity w_m at the free cross section is 10 times the fluidization velocity w_L. Given are: particle diameter $d_P = 0.5\,\text{mm}$, density of the sand $\varrho_S = 1500\,\text{kg/m}^3$, density of the air $\varrho_F = 0.8\,\text{kg/m}^3$, kinematic viscosity of the air $\nu = 3\cdot 10^{-5}\,\text{m}^2/\text{s}$, Schmidt number $Sc = 0.7$, sphericity of sand $\varphi_S = 0.86$, void fraction $\varepsilon_S = 0.73$, void fraction at minimum fluidization point $\varepsilon_{mf} = 0.4$.

We have $X_S - X_{Fe} = 0.05\,(X_S - X_{Fi})$ and therefore in (3.282)

$$\ln\frac{X_{Fi} - X_S}{X_{Fe} - X_S} = \ln 20 = 2.996 = \frac{Sh_m}{Re\,Sc}\,\frac{6\,(1-\varepsilon)\,H}{d_P}\quad.$$

For all values

$$\frac{H}{d_P} \sim 0.5\,\frac{Re\,Sc}{Sh_m\,(1-\varepsilon)}$$

the driving difference at the outlet cross section is reduced to less than 5 % of the value at the inlet cross section.

The calculation of the Sherwood number necessitates investigation of the Archimedes number

$$Ar = \frac{(\varrho_S - \varrho_F)\,d_P^3\,g}{\varrho_F\,\nu^2} = \frac{(1500 - 0{,}8)\,\text{kg/m}^3 \cdot (0.5\cdot 10^{-3})^3\,\text{m} \cdot 9.81\,\text{m/s}^2}{0.8\,\text{kg/m}^3\,(3\cdot 10^{-5})^2\,\text{m}^4/\text{s}^2} = 2\,553\quad.$$

According to (3.278)

$$Re_S \approx 19.14\left[\left(\frac{1 + Ar^{1/1.8}}{12.84}\right)^{1/2} - 1\right]^{1.8} = 19.14\left[\left(\frac{1 + 2553^{1/1.8}}{12.84}\right)^{1/2} - 1\right]^{1.8} = 47.75\quad.$$

According to section 3.7.4, No. 5,

$$Sh_{m,\,\text{lam}} = 0.664\,Re_S^{1/2}\,Sc^{1/3} = 0.664\cdot 47.75^{1/2}\cdot 0.7^{1/3} = 4.07$$

$$Sh_{m,\,\text{turb}} = \frac{0.037\,Re_S^{0.8}\,Sc}{1 + 2.443\,Re_S^{-0.1}\,(Sc^{2/3} - 1)} = \frac{0.037\cdot 47.75^{0.8}\cdot 0.7}{1 + 2.443\cdot 47.75^{-0.1}\cdot (0.7^{2/3} - 1)} = 0.88$$

$$Sh_m = 2 + \left(Sh_{m,\,\text{lam}}^2 + Sh_{m,\,\text{turb}}^2\right)^{1/2} = 2 + (4.07^2 + 0.88^2)^{1/2} = 6.16\quad.$$

According to (3.272) the minimum fluidization velocity follows from

$$Re_{mf} = \frac{w_{mf}\,d_P}{\nu} = 42.9\,\frac{1 - \varepsilon_{mf}}{\varphi_S}\left[\left(1 + 3.11\cdot 10^{-4}\,\frac{\varphi_S^3\,\varepsilon_{mf}^3}{(1 - \varepsilon_{mf})^2}\,Ar\right)^{1/2} - 1\right]$$

$$Re_{mf} = 42.9\,\frac{1 - 0.4}{0.86}\left[\left(1 + 3.11\cdot 10^{-4}\,\frac{0.86^3\cdot 0.4^3}{(1 - 0.4)^2}\,2553\right)^{1/2} - 1\right] = 1.31\quad.$$

Due to $Re = 10\,Re_{mf} = 13.1$ we have

$$w_m = Re\,\frac{\nu}{d_P} = 13.1\,\frac{3\cdot 10^{-5}\,\text{m}^2/\text{s}}{0.5\cdot 10^{-3}\,\text{m}} = 0.786\,\text{m/s}\quad.$$

With that we get

$$\frac{H}{d_P} \approx 0.5\,\frac{13.1\cdot 0.7}{6.16\,(1 - 0.73)} = 2.73 \quad\text{and}\quad H \approx 1.38\cdot 10^{-3}\,\text{m}\quad.$$

It can be seen from this that the driving difference in content drop is starkly reduced. The minimum required height of the fluidized bed is accordingly very small.

Example 3.13: Calculate the actual height of the fluidized bed from example 3.12, when the fill height is 1 m and show how far the driving content difference between inlet and exit cross sections is really reduced.

The mass of the solid is the same at the fluidization point as the operating point. It follows from this that

$$A_0\, H_{\mathrm{mf}}\, \varrho_S\, (1 - \varepsilon_{\mathrm{mf}}) = A_0\, H\, \varrho_S\, (1 - \varepsilon_S)$$

or

$$\frac{H}{H_{\mathrm{mf}}} = \frac{1 - \varepsilon_{\mathrm{mf}}}{1 - \varepsilon_S} = \frac{1 - 0.4}{1 - 0.73} = 2.22 \ .$$

The fill height is the same as the height at the minimum fluidization point $H_{\mathrm{mf}} = 1\,\mathrm{m}$, from which $H = 2.22\,\mathrm{m}$. With (3.282) we then have

$$\ln \frac{X_{\mathrm{Fi}} - X_S}{X_{\mathrm{Fe}} - X_S} = \frac{6.16}{13.1 \cdot 0.7}\, \frac{6\,(1 - 0.73) \cdot 2.22\,\mathrm{m}}{0.5 \cdot 10^{-3}\,\mathrm{m}} = 4\,831$$

$$X_{\mathrm{Fe}} - X_S \approx 0 \quad \text{or} \quad X_{\mathrm{Fe}} \approx X_S \ .$$

By the exit cross section the driving mass content difference has completely vanished.

Example 3.14: Coffee beans that have been heated to 300 °C by roasting are to be cooled, in a fluidized bed with ambient air at a temperature of 20 °C, to a temperature of 30 °C. Calculate the time required for this to happen.

The following values are given: For the coffee beans: density $\varrho_S = 630\,\mathrm{kg/m^3}$, specific heat capacity $c_S = 1.70\,\mathrm{kJ/kgK}$, equivalent particle diameter $d_P = 6\,\mathrm{mm}$, thermal conductivity $\lambda_S = 0.6\,\mathrm{W/Km}$, mass of beans $M_S = 100\,\mathrm{kg}$, void fraction of the fluidized bed $\varepsilon = 0.7$. For the air the following values are given: mass flow rate $\dot{M}_G = 5\,\mathrm{kg/s}$, density $\varrho_G = 1.188\,\mathrm{kg/m^3}$, specific heat capacity $c_{pG} = 1.007\,\mathrm{kJ/kgK}$, thermal conductivity $\lambda_G = 0.0257\,\mathrm{W/Km}$, kinematic viscosity $\nu_G = 15.3 \cdot 10^{-6}\,\mathrm{m^2/s}$, Prandtl number $Pr = 0.715$, cross section of the column $1.4\,\mathrm{m^2}$.

Hints for the solution: The temperatures $\vartheta_{\mathrm{Fe}} = \vartheta_{\mathrm{Ge}}$ of the gas and ϑ_S of the surface of the particle in (3.281) are functions of time and the energy balance

$$M_S c_S \frac{d\vartheta_{\mathrm{mS}}}{dt} = \dot{M}_G c_{pG} (\vartheta_{\mathrm{Gi}} - \vartheta_G)$$

is available as a further equation for these two temperatures. The violent motion of the solid means that its mean temperature ϑ_{mS} is only a function of time and not position, so that both the left and right hand sides are independent of position. This means that the gas temperature ϑ_G on the right hand side can be replaced by the value ϑ_{Ge} at the exit cross section. M_S is the total amount of solid.

The Archimedes number has to be found before we can calculate the Nusselt number Nu_m according to (3.281):

$$Ar = \frac{\varrho_S - \varrho_F}{\varrho_F}\, \frac{d_P^3 g}{\nu^2} \ .$$

Here, $\varrho_F = \varrho_G$ and $\nu = \nu_G$. Which gives

$$Ar = \frac{(630 - 1.188)\,\mathrm{kg/m^3}}{1.188\,\mathrm{kg/m^3}}\, \frac{(6 \cdot 10^{-3})^3\,\mathrm{m^3} \cdot 9.81\,\mathrm{m/s^2}}{(15.3 \cdot 10^{-6})^2\,\mathrm{m^4/s^2}} = 4.79 \cdot 10^6 \ .$$

It then follows from (3.279) that

$$Re_S = (\frac{4}{3}\, Ar)^{1/2} = (\frac{4}{3} \cdot 4.79 \cdot 10^6)^{1/2} = 2.527 \cdot 10^3 \ .$$

According to section 3.7.4, No. 5,

$$Nu_{m,\,lam} = 0.664\,Re_S^{1/2}\,Pr^{1/3} = 0.664\,(2.527\cdot10^3)^{1/2}\,(0,715)^{1/3} = 29.85$$

$$Nu_{m,\,turb} = \frac{0.037\,Re_S^{0.8}\,Pr}{1 + 2.443\,Re_S^{-0.1}\,(Pr^{2/3} - 1)}$$

$$= \frac{0.037\cdot(2.527\cdot10^3)^{0.8}\cdot0.715}{1 + 2.443\cdot(2.527\cdot10^3)^{-0.1}\cdot(0.715^{2/3} - 1)} = 17.97$$

$$Nu_m = 2 + \left(Nu_{m,\,lam}^2 + Nu_{m,\,turb}^2\right)^{1/2} = 2 + (29.85^2 + 17.97^2)^{1/2} = 36.84\ .$$

The velocity in the empty column is

$$w_m = \frac{\dot{M}_G}{\varrho_G\,A_0} = \frac{5\,\text{kg/s}}{1.188\,\text{kg/m}^3\cdot1.4\,\text{m}^2} = 3\,\text{m/s}\ .$$

With that we have

$$Re = \frac{w_m\,d_P}{\nu_G} = \frac{3\,\text{m/s}\cdot6\cdot10^{-3}\,\text{m}}{15.3\cdot10^{-6}\,\text{m}^2/\text{s}} = 1.176\cdot10^3\ .$$

In (3.281), with $\vartheta_F = \vartheta_G$

$$\ln\frac{\vartheta_{Gi} - \vartheta_S}{\vartheta_{Ge} - \vartheta_S} = \frac{36.84}{1.176\cdot10^3\cdot0.715}\frac{6(1 - 0.7)H}{6\cdot10^{-3}\,\text{m}}\ .$$

The height H of the fluidized bed is yielded from the mass of the coffee beans $M_S = (1 - \varepsilon)\,A_0\,\varrho_S\,H$ to be

$$H = \frac{M_S}{(1 - \varepsilon)\,A_0\,\varrho_S} = \frac{100\,\text{kg}}{(1 - 0.7)\cdot1.4\,\text{m}^2\cdot630\,\text{kg/m}^3} = 0.378\,\text{m}\ .$$

With this we obtain

$$\frac{\vartheta_{Gi} - \vartheta_S}{\vartheta_{Ge} - \vartheta_S} = 143.8 = c_0 \tag{3.283}$$

with $\vartheta_{Gi} = 293.15\,\text{K}$. The equation still contains the unknown temperatures $\vartheta_{Ge}(t)$ at the exit cross section and the temperature $\vartheta_S(t)$ of the solid. As a further balance equation

$$M_S\,c_S\,\frac{d\vartheta_{mS}}{dt} = \dot{M}_G\,c_{pG}\,(\vartheta_{Gi} - \vartheta_{Ge}) \tag{3.284}$$

is still available. Here, $\vartheta_{mS}(t)$ is the mean temperature of the solid. It is related to the surface temperature $\vartheta_S(t)$ as follows

$$\alpha_{mS}\,(\vartheta_{mS} - \vartheta_S) = \alpha_{mG}\,(\vartheta_S - \vartheta_{Ge})$$

$$\vartheta_{mS} = \frac{\alpha_{mG}}{\alpha_{mS}}\,(\vartheta_S - \vartheta_{Ge}) + \vartheta_S\ .$$

With $\alpha_{mS} \approx \lambda_S/(d_P/2)$ and $Bi = \alpha_{mG}\,d_P/\lambda_S$ we have

$$\vartheta_{mS} \approx \frac{Bi}{2}\,(\vartheta_S - \vartheta_{Ge}) + \vartheta_S\ . \tag{3.285}$$

From (3.283) and (3.285), by eliminating ϑ_S, we obtain

$$\vartheta_{Gi} - \vartheta_{Ge} = \frac{(\vartheta_{Gi} - \vartheta_{mS})\,(c_0 - 1)}{c_0 + Bi/2}\ .$$

Eq. (3.285) is transformed into

$$\frac{1}{\vartheta_{Gi} - \vartheta_{mS}} \frac{d\vartheta_{mS}}{dt} = \frac{\dot{M}_G \, c_{pG}}{M_S \, c_S} \frac{c_0 - 1}{c_0 + Bi/2} \, .$$

Integration between $t = 0$ and $t = t_1$ yields

$$\ln \frac{\vartheta_{Gi} - \vartheta_{mS}(t = 0)}{\vartheta_{Ge} - \vartheta_{mS}(t = t_1)} = \frac{\dot{M}_G \, c_{pG}}{M_S \, c_S} \frac{c_0 - 1}{c_0 + Bi/2} t_1 \, .$$

Here

$$Bi = \frac{\alpha_{mG} d_P}{\lambda_S} = \frac{\alpha_{mG} \, d_P}{\lambda_G} \frac{\lambda_G}{\lambda_S} = Nu_m \frac{\lambda_G}{\lambda_S} = 36.84 \frac{0.0257 \, \text{W/Km}}{0.6 \, \text{W/Km}} = 1.58 \, .$$

With that we get

$$\ln \frac{293.15 - 673.15}{293.15 - 303.15} = \frac{5 \, \text{kg/s} \cdot 1.007 \, \text{kJ/kgK} \cdot (143.8 - 1)}{100 \, \text{kg} \cdot 1.70 \, \text{kJ/kgK} \cdot (143.8 + 1.58/2)} t_1 \, ,$$

and so

$$t_1 = 124.4 \, \text{s} \approx 2.1 \, \text{min} \, .$$

3.8.5 Some empirical equations for heat and mass transfer in flow through channels, packed and fluidized beds

What follows is a summary of the previous equations and some supplementary equations.

1. Flow inside a tube

Turbulent flow:

$$Nu_m = 0.023 Re^{0.8} Pr^{0.4} (\eta/\eta_0)^{0.14} \, ,$$

valid for $0.5 < Pr < 120$, $10^4 < Re < 10^5$, $L/d \sim 60$.

$$Nu_m = 0.037(Re^{0.75} - 180) Pr^{0.42} \left[1 + (d/L)^{2/3}\right] (\eta/\eta_0)^{0.14} \, ,$$

valid for $2300 < Re < 10^5$, $0.5 < Pr < 500$, $L/d \sim 10$.

Over a wide range of characteristic numbers it holds that

$$Nu_m = \frac{(\zeta/8)(Re - 1000)Pr}{1 + 12.7\sqrt{\zeta/8}(Pr^{2/3} - 1)} \left[1 + \left(\frac{d}{L}\right)^{2/3}\right]$$

with the resistance factor

$$\zeta = \frac{\Delta p}{(L/d)\varrho(w_m^2/2)} = \frac{1}{(0.79 \ln Re - 1.64)^2} \, ,$$

valid for $2300 \leq Re \leq 5 \cdot 10^6$, $0.5 \leq Pr \leq 2000$, $L/d > 1$.

All material properties are based on the arithmetic mean $(\vartheta_i + \vartheta_e)/2$ of the mean inlet temperature ϑ_i and the mean exit temperature ϑ_e, only η_0 is based on the wall temperature. The Nusselt and Reynolds numbers are formed with the tube diameter.

Laminar flow:
Hydrodynamically fully developed laminar flow, constant wall temperature:

$$Nu_m = \frac{3.657}{\tanh(2.264 X^{+1/3} + 1.7\ X^{+2/3})} + \frac{0.0499}{X+}\tanh X^+ \ ,$$

valid for $0 \leq X^+ = L/(dPe) \leq \infty$, $Re \leq 2300$.
Laminar flow that is developing hydrodynamically or thermally, (entry flow) constant wall temperature:

$$\frac{Nu_{me}}{Nu_m} = \frac{1}{\tanh(2.432 Pr^{1/6})X^{+1/6}} \ ,$$

valid for $0 < X^+ \leq \infty$, $0 < Pr \leq \infty$, $Re \leq 2300$.
Nu_{me} is the Nusselt number for developing flow, Nu_m is that for hydrodynamically fully developed laminar flow.

2. **Non-circular tubes**
The previous equations still hold; we merely have to replace the tube diameter d in the Nusselt and Reynolds numbers with the hydraulic diameter

$$d_h = 4A/C$$

with the cross sectional area A through which the fluid flows and the wetted perimeter C.

3. **Annular space between two concentric tubes**
With a concentric annular space there are three cases of heat transfer which have to be differentiated:
3.1: Heat is transferred at the inner tube. The outer tube is insulated.
3.2: Heat is transferred at the outer tube. The inner tube is insulated.
3.3: Heat is transferred at both the inner and outer tubes.
The fluid flows in the axial direction in the annular space between the two tubes. The external diameter of the inner tube is d_i, the internal diameter of the outer tube is d_o. The hydraulic diameter is $d_h = d_o - d_i$.
Turbulent flow
The mean Nusselt Nu_m is based on $(Nu_m)_{tube}$ of the circular tube through which the turbulent fluid flows according to No. 1a. Both Nusselt numbers are formed with the hydraulic diameter, $Nu_m = \alpha_m d_h/\lambda$. The Reynolds

number is $Re = w_m d_h/\nu$. It holds for the case above

3.1 : $\quad \dfrac{Nu_m}{(Nu_m)_{tube}} = 0.86 \left(\dfrac{d_i}{d_o}\right)^{-0.16}$

3.2 : $\quad \dfrac{Nu_m}{(Nu_m)_{tube}} = 1 - 0.14 \left(\dfrac{d_i}{d_o}\right)^{0.6}$

3.3 : $\quad \dfrac{Nu_m}{(Nu_m)_{tube}} = \dfrac{0.86(d_i/d_o) + [1 - 0.14(d_i/d_o)^{0.6}]}{1 + (d_i/d_o)}$.

The equations are valid over the same range as for $(Nu_m)_{tube}$ from No. 1a and for $0 \le d_i/d_o \le 1$.

Laminar, hydrodynamically fully developed flow

We have

$$\frac{Nu_m - Nu_\infty}{(Nu_m)_{tube} - 3.657} = f\left(\frac{d_i}{d_o}\right) .$$

In the equation for laminar pipe flow from Nr. 1b, the quantity X^+ is now formed with the hydraulic diameter, $X^+ = L/(d_h Pe)$ with $Pe = w_m d_h/a$. It holds for the case given above that

3.1 : $\quad Nu_\infty = 3.657 + 1.2(d_i/d_o)^{-0.8}$,

$\qquad f(d_i/d_o) = 1 + 0.14(d_i/d_o)^{-1/2}$.

3.2 : $\quad Nu_\infty = 3.657 + 1.2(d_i/d_o)^{1/2}$,

$\qquad f(d_i/d_o) = 1 + 0.14(d_i/d_o)^{1/3}$.

3.3 : $\quad Nu_\infty = 3.657 + \left(4 - \dfrac{0.102}{0.02 + (d_i/d_o)}\right)\left(\dfrac{d_i}{d_o}\right)^{0.04}$,

$\qquad f(d_i/d_o) = 1 + 0.14(d_i/d_o)^{0.1}$.

The equations are valid for $0 \le X^+ \le \infty$, $Re \le 2300$ and $0 \le d_i/d_o \le 1$.

4. Flo *ı* **through spherical or other packing**

$$Nu_m = f_\varepsilon Nu_{mS} .$$

The mean Nusselt number Nu_m can be traced back to the Nusselt number Nu_{mS} for flow around a single sphere (see 3.7.4). The factor f_ε is dependent on the void fraction $\varepsilon = V_G/V$, so

$$f_\varepsilon = 1 + 1.5(1 - \varepsilon) \quad \text{in the region} \quad 0.26 < \varepsilon < 1 .$$

For packed beds of spheres of different diameters or of non-spherical particles, and equivalent diameter has to be formed

$$d_P = \sqrt{A_{mP}/\pi}$$

with average particle surface area

$$A_{mP} = a_P/n_V , \quad \text{with} \quad a_P = 6(1 - \varepsilon)/d_P ,$$

where a_P is the specific surface area of the particle (SI units m^2/m^3) and n_V is the number of particles per unit volume (SI units $1/m^3$). With that the Nusselt number can be calculated as Nu_{mS}. Data on the factor f_ε can be found in Table 3.5.

5. Fluidized beds

The Nusselt number is the same as that for flow around a single sphere Nu_{mS}, section 3.7.4, No.5. The Reynolds number used there is yielded from (3.278)

$$Re_\mathrm{P} \cong 19.14 \left[\left(1 + \frac{Ar^{1/1.8}}{12.84} \right)^{1/2} - 1 \right]^{1.8}$$

with the Archimedes number

$$Ar = \frac{\varrho_\mathrm{S} - \varrho_\mathrm{F}}{\varrho_\mathrm{F}} \frac{d_\mathrm{P}^3 g}{\nu^2} \ ,$$

valid for homogeneous fluidized beds for

$$0 \le Re_\mathrm{S} \le 10^4 \quad \text{or} \quad 0 \le Ar \le 2.9 \cdot 10^7 \ .$$

The range over which this is valid for heterogeneous fluidized beds is

$$0 \le Re_\mathrm{S} \overset{<}{\sim} 1.2 \cdot 10^2 \quad \text{or} \quad 0 \le Ar \overset{<}{\sim} 10^4 \ .$$

Over the range $Re_\mathrm{S} > 1.2 \cdot 10^2$ or $Ar \ge 10^4$ we have

$$Re_\mathrm{S} \cong (4\,Ar/3)^{1/2} \ .$$

3.9 Free flow

Whilst forced flow is caused by external forces, for example the pressure from a pump or blower, free flows occur because of body forces in a fluid in which density gradients are present. As an example of this we will consider a vertical wall with a higher temperature than the fluid adjacent to it. The fluid which is heated at the wall will be specifically lighter and experiences, in comparison to the fluid surrounding it, a lift in the gravity field. A free flow originates. In this case the density gradients and the body force created by the gravity field are perpendicular to each other. Another example of free flow is shown in Fig. 3.42, a flow of hot air is discharged as a horizontal jet which enters colder infinitely extended quiescent fluid. The jet moves upwards due to the buoyancy forces.

The necessary prerequisites for the existence of free flow are body forces and density gradients that are not directed parallel to each other. This holds for a fluid flowing over a heated horizontal plate.

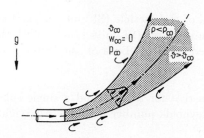

Fig. 3.42: Deflection of a hot air stream by lift forces

In order to see that a free flow can only exist if the density gradient and body forces are not parallel to each other, we will discuss a fluid that is initially at rest, $w_j = 0$. According to the momentum equation (3.48) it holds for this that

$$0 = \varrho \, k_j + \frac{\partial \tau_{ji}}{\partial x_i} \quad ,$$

where because $w_j = 0$, according to (3.46) and (3.47), $\tau_{ji} = -p \, \delta_{ji}$. For a quiescent fluid, the hydrostatic equilibrium equation follows from this as

$$\frac{\partial p}{\partial x_j} = \varrho \, k_j \quad . \tag{3.286}$$

The body forces k_j generate a pressure gradient. By differentiation it follows that

$$\frac{\partial p}{\partial x_i \, \partial x_j} = \frac{\partial \varrho}{\partial x_i} k_j + \varrho \frac{\partial k_j}{\partial x_i} \quad .$$

As the pressure in a quiescent fluid is a quantity of state, this expression can be equally represented by

$$\frac{\partial p}{\partial x_j \, \partial x_i} = \frac{\partial \varrho}{\partial x_j} k_i + \varrho \frac{\partial k_i}{\partial x_j} \quad .$$

Therefore

$$\varrho \left(\frac{\partial k_j}{\partial x_i} - \frac{\partial k_i}{\partial x_j} \right) + \frac{\partial \varrho}{\partial x_i} k_j - \frac{\partial \varrho}{\partial x_j} k_i = 0 \quad .$$

This equation is satisfied, if the body force possesses a potential ϕ, $k_j = -\partial\phi/\partial x_j$, as the expression in the brackets disappears, and so in addition it holds that

$$\frac{\partial \varrho}{\partial x_i} k_j - \frac{\partial \varrho}{\partial x_j} k_i = \nabla \varrho \times k = 0 \quad . \tag{3.287}$$

A necessary and sufficient condition for this is either the density is spatially constant or the body forces disappear or the density gradient and body forces are parallel to each other. If this is not the case, the density gradient and body forces are not in equlibrium according to (3.286). It then holds that

$$\frac{\partial p}{\partial x_j} \neq \varrho \, k_j \quad .$$

From the Navier-Stokes equation (3.59)

$$\varrho \frac{dw_j}{dt} = -\frac{\partial p}{\partial x_j} + \eta \frac{\partial^2 w_j}{\partial x_i{}^2} + \varrho k_j \quad ,$$

but also from Cauchy's equation of motion (3.48) together with (3.46) it follows that at least one of the expressions which contain the velocity does not vanish. A flow is inevitably initiated because the condition (3.286) of hydrostatic equilibrium has been violated.

It can be simultaneously recognised that free flow cannot exist in an incompressible fluid because density gradients do not exist.

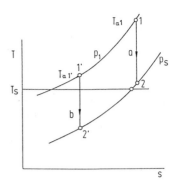

Fig. 3.43: Reversible adiabatic expansion of hot air from p_1 and initial temperature $T_{\alpha 1}$ or $T_{\alpha 1'}$, to the ambient pressure p_S. **a** Temperature difference $T_{\alpha 1} - T_S$ is large; $\varrho_2 < \varrho_S$, a ball of air rises further. **b** Temperature difference $T_{\alpha 1'} - T_S$ is small, $\varrho_{2'} > \varrho_S$, a ball of air sinks again

Density gradients are normally caused by temperature gradients, less often by concentration or pressure gradients. However in multicomponent mixtures the concentration differences can also create notable density gradients, so that under the preconditions named before, a free flow can also develop. The body force is frequently gravity, and less often centrifugal or electromagnetic forces.

The hydrostatic equilibrium in a fluid does not have to be stable. As an example of this we will now consider initially quiescent, cold air over a horizontal heated plate. The condition for hydrostatic equilibrium is satisfied as the density gradient is parallel to the gravity force. If, as a result of a disturbance, a ball of warm air rises from the plate, it will quickly assume the pressure of its surroundings. The density of the ball of air reduces approximately adiabatically and reversibly, in accordance with $p/\varrho^\kappa = p_0/\varrho_0^\kappa$.

If the density of the ball of air is lower than that of its surroundings, or, synonymous with that, the ball is hotter than its surroundings, then it will rise even further, see for this the plot $T_2 > T_S$ in Fig. 3.43. The arrangement is unstable, as a small disturbance, introduced by the rising ball of air, does not die away by itself.

If, in contrast, the density of the ball of air is greater than that of its surroundings after the expansion, corresponding to the plot $T_{2'} < T_S$ in Fig. 3.43, then it will once again sink. The stratification is stable.

This type of stable stratification can also occur when warmer and thereby lighter air masses move over cold ground air. The heavier and often exhause gas polluted air can no longer exchange with the lighter air above it. This type of layering is known as a *temperature inversion*. A stable layer settles over the area being considered and effectively restricts the removal and dispersion of the pollutants. There is a risk of smog.

3.9.1   The momentum equation

The momentum equation for a Newtonian fluid is yielded from Navier-Stokes equation (3.59)

$$\varrho \frac{dw_j}{dt} = \varrho\, k_j - \frac{\partial p}{\partial x_j} + \eta \frac{\partial^2 w_j}{\partial x_i^2} \ . \qquad (3.288)$$

As free flow occurs because of density gradients, we will assume a variable density whilst all other material properties shall be taken as constant. Gravity will act as the body force $k_j = g_j$. In areas where the density is constant, for example far away from a heated wall, hydrostatic equilibrium is reached. Here it holds that

$$\frac{\partial p_\infty}{\partial x_j} = \varrho_\infty\, g_j \ , \qquad (3.289)$$

where p_∞ is the hydrostatic pressure and ϱ_∞ is the associated density. Heat or mass transfer cause temperature or concentration gradients and therefore also density gradients to develop, so that the local values of pressure and density deviate from the values at hydrostatic equilibrium

$$p = p_\infty + \Delta p \ , \qquad \varrho = \varrho_\infty + \Delta \varrho \ . \qquad (3.290)$$

We assume that the deviations Δp and $\Delta \varrho$ are small in comparison to the values at hydrostatic equilibrium p_∞ and ϱ_∞.

As an example we will consider the vertical heated wall in Fig. 3.44, at a temperature ϑ_0, in front of which there is a fluid that before heating was in hydrostatic equilibrium and had a temperature ϑ_∞ throughout. After a short time steady temperature and velocity profiles develop. These are sketched in Fig. 3.44. With (3.290), the momentum equation (3.288) is transformed into

$$(\varrho_\infty + \Delta \varrho)\frac{dw_j}{dt} = (\varrho_\infty + \Delta \varrho)\, g_j - \frac{\partial (p_\infty + \Delta p)}{\partial x_j} + \eta \frac{\partial^2 w_j}{\partial x_i^2} \ . \qquad (3.291)$$

Considering that on the left hand side $\Delta \varrho \ll \varrho_\infty$ and on the right hand side according to (3.289) $\partial p_\infty / \partial x_j = \varrho_\infty g_j$ is valid, we obtain

$$\varrho_\infty \frac{dw_j}{dt} = \Delta \varrho\, g_j - \frac{\partial \Delta p}{\partial x_j} + \eta \frac{\partial^2 w_j}{\partial x_i^2} \ . \qquad (3.292)$$

In steady-state flow, which we want to presume, we can write the term

$$\frac{dw_j}{dt} = w_i \frac{\partial w_j}{\partial x_i} \ .$$

In (3.292) we estimate the magnitude of the individual terms. To this we consider the buoyancy flow on a flat plate which is sloped at an angle φ to vertical. The inertia forces are of the order

$$F_\mathrm{I} \sim \varrho_\infty \frac{w_\mathrm{ref}^2}{L} \ , \tag{3.293}$$

where w_ref is a characteristic velocity, for example the maximum velocity in Fig. 3.44. It develops purely due to the buoyancy and is not prescribed as in the case of longitudinal flow on a flat plate. Its magnitude still has to be estimated. With the reference velocity the magnitude of the friction forces is yielded as

$$F_\mathrm{F} \sim \eta \frac{w_\mathrm{ref}}{\delta^2} \ . \tag{3.294}$$

The buoyancy forces F_B have magnitude of

$$F_\mathrm{B} \sim -(\varrho_0 - \varrho_\infty)\, g \, \cos\varphi \ . \tag{3.295}$$

In the case of a heated, vertical plate according to Fig. 3.44, the density gradient and gravity are perpendicular to each other, $\cos\varphi = \cos 0 = 1$; in contrast, on a horizontal plate the two gradients are parallel to each other and we have $\cos\varphi = \cos\pi/2 = 0$. The negative sign in (3.295) stems from the fact that in heating $\varrho_0 - \varrho_\infty$ is negative, but the buoyancy force in Fig. 3.44 points in the direction of the x-coordinate.

In the layer adjacent to the wall, the inertia forces are of the same magnitude as the friction forces and therefore

$$\frac{w_\mathrm{ref}^2}{L} \sim \nu \frac{w_\mathrm{ref}}{\delta^2} \quad \text{oder} \quad \frac{\delta}{L} \sim \left(\frac{\nu}{w_\mathrm{ref} L}\right)^{1/2} \ . \tag{3.296}$$

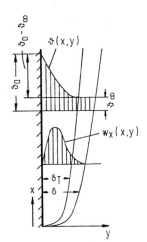

Fig. 3.44: Temperature and velocity profile in free flow on a heated, vertical wall

In addition the buoyancy forces are also of the same magnitude as the inertia forces

$$-(\varrho_0 - \varrho_\infty) g \cos \varphi \sim \varrho_\infty \frac{w_{\text{ref}}^2}{L} .$$

It follows from this that

$$w_{\text{ref}} \sim \left(\frac{-(\varrho_0 - \varrho_\infty) g \cos \varphi L}{\varrho_\infty} \right)^{1/2} . \tag{3.297}$$

This means that we can eliminate the reference velocity in (3.296), giving

$$\frac{\delta}{L} \sim \left(\frac{\nu^2}{(1 - \varrho_0/\varrho_\infty) g \cos \varphi L^3} \right)^{1/4} . \tag{3.298}$$

If the density change is only caused by temperature changes, under the assumption that the temperature change is sufficiently small, it holds that

$$\varrho_0 = \varrho_\infty + \left(\frac{\partial \varrho}{\partial \vartheta} \right)_p (\vartheta_0 - \vartheta_\infty) . \tag{3.299}$$

The quantity

$$\frac{1}{v} \left(\frac{\partial v}{\partial \vartheta} \right)_p = -\frac{1}{\varrho} \left(\frac{\partial \varrho}{\partial \vartheta} \right)_p = \beta$$

is the thermal expansion coefficient (SI units 1/K). It has to be put into (3.299) for the temperature ϑ_∞. We then obtain

$$\varrho_0 - \varrho_\infty = -\varrho_\infty \beta_\infty (\vartheta_0 - \vartheta_\infty) . \tag{3.300}$$

With ideal gases, because $v = RT/p$, the thermal expansion coefficient is

$$\beta = \frac{1}{v} \frac{R}{p} = \frac{1}{T}$$

and $\beta_\infty = 1/T_\infty$.

Then with (3.300), we obtain from (3.298)

$$\frac{\delta}{L} \sim \left(\frac{\nu^2}{\beta_\infty (\vartheta_0 - \vartheta_\infty) g L^3} \right)^{1/4} \frac{1}{(\cos \varphi)^{1/4}} . \tag{3.301}$$

Here, the dimensionless quantity

$$\frac{\beta_\infty (\vartheta_0 - \vartheta_\infty) g L^3}{\nu^2} := Gr \tag{3.302}$$

is the *Grashof number* Gr. For a vertical plate $\varphi = 0$, so

$$\frac{\delta}{L} \sim Gr^{-1/4} . \tag{3.303}$$

For large Grashof numbers $Gr \gg 1$, we get $\delta/L \ll 1$: A boundary layer flow exists. The boundary layer thickness grows according to

$$\delta \sim L^{1/4} \ .$$

A sloped plate requires the introduction of a modified Grashof number

$$Gr_\varphi = \frac{\beta_\infty \left(\vartheta_0 - \vartheta_\infty \right) g \, L^3 \, \cos \varphi}{\nu^2} \ .$$

This gives

$$\frac{\delta}{L} \sim Gr_\varphi^{-1/4} \ . \qquad (3.304)$$

If the plate is sloped at an angle φ to the vertical, a boundary layer forms once again, if $Gr_\varphi \gg 1$, whilst on a horizontal plate, $\varphi = \pi/2$, a boundary layer cannot exist.

As we can see from (3.293) and (3.295), the inertia and buoyancy forces are of equal magnitude, if

$$\frac{-(\varrho_0 - \varrho_\infty) \, g \, \cos \varphi}{\varrho_\infty \, w_{\text{ref}}^2/L} \sim 1$$

is valid. With (3.300) and the Reynolds number $w_{\text{ref}} L/\nu$, we can also write this as

$$\frac{\beta_\infty \left(\vartheta_0 - \vartheta_\infty \right) g \, L^3 \, \cos \varphi}{\left(w_{\text{ref}} L/\nu \right)^2 \nu^2} = \frac{Gr_\varphi}{Re^2} \sim 1 \ . \qquad (3.305)$$

Therefore the following approximation is valid: If $Gr_\varphi/Re^2 \ll 1$, then the buoyancy forces will be much smaller then the inertia forces. The flow will be determined by the inertia and friction forces. If, however, $Gr_\varphi/Re^2 \gg 1$, then the buoyancy forces will be much larger than the inertia force. The flow will be determined by the buoyancy and friction forces.

3.9.2 Heat transfer in laminar flow on a vertical wall

We will now deal with free flow on a vertical, flat wall whose temperature ϑ_0 is constant and larger than the temperature in the semi-infinite space. The coordinate origin lies, in accordance with Fig. 3.43, on the lower edge, the coordinate x runs along the wall, with y normal to it. Steady flow will be presumed. All material properties are constant. The density will only be assumed as temperature dependent in the buoyancy term, responsible for the free flow, in the momentum equation, in all other terms it is assumed to be constant. These assumptions from Oberbeck (1879) and Boussinesq (1903), [3.45], [3.46] are also known as the Boussinesq approximation; although it would be more correct to speak of the Oberbeck-Boussinesq approximation. It takes into account that the locally variable density is a prerequisite for free flow. The momentum equation (3.291) in the x-direction, taking into consideration that on the left hand side $\Delta\varrho \ll \varrho_\infty$, on the right hand side $p_\infty + \Delta p = p$, $\varrho_\infty + \Delta\varrho = \varrho$ and $g_j = -g$, is now

$$\varrho_\infty w_x \frac{\partial w_x}{\partial x} + \varrho_\infty w_y \frac{\partial w_x}{\partial y} = -\varrho\, g - \frac{\partial p}{\partial x} + \eta \frac{\partial^2 w_x}{\partial y^2} \ . \tag{3.306}$$

The momentum equation in the direction of the y-axis is reduced to

$$\frac{\partial p}{\partial y} = 0 \ , \tag{3.307}$$

as already shown in the boundary layer simplifications in section 3.5, because the pressure at a point x is constant perpendicular to the wall. As the core flow is quiescent, the condition (3.286) for hydrostatic equilibrium is valid there

$$-\frac{\partial p}{\partial x} = \varrho_\infty\, g \ , \tag{3.308}$$

and with that the momentum equation (3.306) simplifies to

$$w_x \frac{\partial w_x}{\partial x} + w_y \frac{\partial w_x}{\partial y} = \frac{\varrho_\infty - \varrho}{\varrho_\infty} g + \nu \frac{\partial^2 w_x}{\partial y^2} \ . \tag{3.309}$$

As we have presumed small density changes, we have

$$\varrho = \varrho_\infty + \left(\frac{\partial \varrho}{\partial \vartheta}\right)_p (\vartheta - \vartheta_\infty) = \varrho_\infty - \varrho_\infty \beta_\infty (\vartheta - \vartheta_\infty)$$

with the thermal expansion coefficient

$$\frac{1}{v}\left(\frac{\partial v}{\partial \vartheta}\right)_p = -\frac{1}{\varrho}\left(\frac{\partial \varrho}{\partial \vartheta}\right)_p = \beta \ ,$$

that has to be put in at the temperature ϑ_∞. With this we get the following term in the momentum equation (3.309)

$$\frac{\varrho_\infty - \varrho}{\varrho_\infty} = \beta_\infty (\vartheta - \vartheta_\infty) \ .$$

This leaves the following system of equations to solve:
The continuity equation

$$\frac{\partial w_x}{\partial x} + \frac{\partial w_y}{\partial y} = 0 \ , \tag{3.310}$$

the momentum equation

$$w_x \frac{\partial w_x}{\partial x} + w_y \frac{\partial w_x}{\partial y} = g\,\beta_\infty (\vartheta - \vartheta_\infty) + \nu \frac{\partial^2 w_x}{\partial y^2} \tag{3.311}$$

and the energy equation

$$w_x \frac{\partial \vartheta}{\partial x} + w_y \frac{\partial \vartheta}{\partial y} = a \frac{\partial^2 \vartheta}{\partial y^2} \ . \tag{3.312}$$

The dissipated energy is neglected here. The boundary conditions are

$$y = 0 \quad : \quad w_x = w_y = 0 \quad ; \quad \vartheta = \vartheta_0$$

$$y \to \infty \quad : \quad w_x = 0 \quad ; \quad \vartheta = \vartheta_\infty \ . \tag{3.313}$$

In order to solve this, we introduce a stream function $\psi(x, y)$, cf. (3.180):

$$w_x = \frac{\partial \psi}{\partial y} \quad \text{and} \quad w_y = -\frac{\partial \psi}{\partial x} \ ,$$

through which the continuity equation (3.310) is satisfied identically. The momentum equation (3.311) is transformed into

$$\frac{\partial \psi}{\partial y} \frac{\partial^2 \psi}{\partial y \partial x} - \frac{\partial \psi}{\partial x} \frac{\partial^2 \psi}{\partial y^2} = g \, \beta_\infty \, (\vartheta - \vartheta_\infty) + \nu \frac{\partial^3 \psi}{\partial y^3} \ . \tag{3.314}$$

As the boundary layer thickness according to (3.303) at a point x is given by

$$\delta \sim x/Gr_x^{1/4} \quad \text{with} \quad Gr_x = \frac{\beta_\infty \, (\vartheta_0 - \vartheta_\infty) \, g \, x^3}{\nu^2} \ ,$$

the y coordinate is normalised with the local boundary layer thickness, just as for flow along a flat plate, introducing a dimensionless coordinate normal to the wall

$$\eta^+ := \frac{y}{\delta} = \frac{y}{x} \left(\frac{Gr_x}{4} \right)^{1/4} = g(x, y) \ . \tag{3.315}$$

The factor of 4 is introduced in agreement with the similarity solution from Ostrach [3.47], because through this, no fractions will appear as factors in the following equations.

A characteristic velocity w_{ref} is found from (3.297) and (3.300) for the vertical wall ($\varphi = \pi/2$) to be

$$w_{\text{ref}} = 2 \left[\beta_\infty \, (\vartheta_0 - \vartheta_\infty) \, g \, x \right]^{1/2} = 2 \left(Gr_x \frac{\nu^2}{x^2} \right)^{1/2} \ . \tag{3.316}$$

The factor of 2 is once again chosen in agreement with Ostrach. With this we obtain for the stream function

$$\psi = \int_0^y w_x \, dy = w_{\text{ref}} \, \delta \int_0^y \frac{w_x}{w_{\text{ref}}} \, d \left(\frac{y}{\delta} \right) \tag{3.317}$$

or

$$\psi = 2 \left(Gr_x \frac{\nu^2}{x^2} \right)^{1/2} \frac{x}{(Gr_x/4)^{1/4}} \int_0^{\eta^+} w_x^+ \, d\eta^+ \ . \tag{3.318}$$

Therefore we now have

$$\psi = 4 \, \nu \left(\frac{Gr_x}{4} \right)^{1/4} \int_0^{\eta^+} w_x^+ \, d\eta^+ = 4 \, \nu \left(\frac{Gr_x}{4} \right)^{1/4} f(\eta^+) \ . \tag{3.319}$$

It is also assumed here that the velocity profile $w_x^+ = w_x/w_{\text{ref}}$ can be represented as a function of the boundary layer coordinate η^+. The velocities are yielded from this as

$$w_x = \frac{\partial \psi}{\partial y} = \frac{2\nu}{x} Gr_x^{1/4} f'(\eta^+)$$

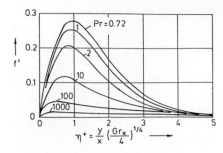
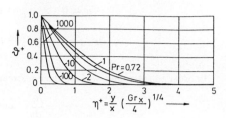

Fig. 3.45: Velocities in free flow on a vertical wall, according to [3.47]

Fig. 3.46: Temperatures in free flow on a vertical wall, according to [3.47]

and

$$w_y = -\frac{\partial \psi}{\partial x} = -\frac{\nu}{x}\left(\frac{Gr_x}{4}\right)^{1/4}\left[\eta^+ f'(\eta^+) - 3f(\eta^+)\right] \ . \tag{3.320}$$

In a correspondong way we introduce a normalised temperature

$$\vartheta^+ := \frac{\vartheta - \vartheta_\infty}{\vartheta_0 - \vartheta_\infty} = \vartheta^+(\eta^+) \ . \tag{3.321}$$

Using these equations the momentum equation (3.314) can be transformed into

$$f''' + 3ff'' - 2(f')^2 + \vartheta^+ = 0 \tag{3.322}$$

and the energy equation (3.312) into

$$\vartheta^{+''} + 3\,Pr\,f\,\vartheta^{+'} = 0 \ . \tag{3.323}$$

In place of the partial differential equations (3.310) to (3.312), two ordinary non-linear differential equations appear. The continuity equation is no longer required because it is fulfilled by the stream function. The solution has to satisfy the following boundary conditions:

$$\eta^+ = 0 \ : \ f = f' = 0 \ ; \quad \vartheta^+ = 1$$

$$\eta^+ \to \infty \ : \ f' = 0 \ ; \quad \vartheta^+ \to 1 \ .$$

These equations were first solved by Pohlhausen [3.48] for air with a Prandtl number $Pr = 0.733$ by series expansion and later numerically by Ostrach [3.47] for a wide range of Prandtl numbers, $0.008\,35 \le Pr \le 1\,000$. Fig. 3.45 illustrates the normalised velocities in the form

$$\frac{w_x\,x}{\nu}\,\frac{1}{2\,Gr_x^{1/2}} = f'(\eta^+)$$

and Fig. 3.46 the normalised temperatures $\vartheta^+(\eta^+)$ based on the calculations by Ostrach [3.47]. The heat transfer coefficients are yielded in the usual manner from the temperature profiles

$$\alpha\,(\vartheta_0 - \vartheta_\infty) = \dot{q} = -\lambda\left(\frac{\partial \vartheta}{\partial y}\right)_{y=0} \ .$$

In which

$$
\begin{aligned}
\left(\frac{\partial \vartheta}{\partial y}\right)_{y=0} &= (\vartheta_0 - \vartheta_\infty) \left(\frac{\mathrm{d}\vartheta^+}{\mathrm{d}\eta^+} \frac{\partial \eta^+}{\partial y}\right)_{y=0} \\
&= (\vartheta_0 - \vartheta_\infty) \frac{1}{x} \left(\frac{Gr_x}{4}\right)^{1/4} \left(\frac{\mathrm{d}\vartheta^+}{\mathrm{d}\eta^+}\right)_{\eta^+=0}
\end{aligned}
\tag{3.324}
$$

and therefore

$$
Nu_x = \frac{\alpha\, x}{\lambda} = \left(\frac{Gr_x}{4}\right)^{1/4} \left(\frac{\mathrm{d}\vartheta^+}{\mathrm{d}\eta^+}\right)_{\eta^+=0} = \left(\frac{Gr_x}{4}\right)^{1/4} \varphi(Pr) \ .
\tag{3.325}
$$

The temperature increase $(\mathrm{d}\vartheta^+/\mathrm{d}\eta^+)_{\eta^+=0}$ is, as Fig. 3.46 also shows, a function φ of the Prandtl number Pr. The numerical results from Ostrach have been reproduced by Le Fèvre [3.49] through an interpolation equation of the following form, which deviates from the exact numerical solution by no more than 0.5 %:

$$
\varphi(Pr) = \frac{0.849\, Pr^{1/2}}{(1 + 2.006\, Pr^{1/2} + 2.034\, Pr)^{1/4}} \ .
\tag{3.326}
$$

The solution includes the following special cases:
In the limiting case $Pr \to 0$ we get $\varphi(Pr) = 0.849\, Pr^{1/2}$ and with that

$$
Nu_x = 0.600 \left(Gr_x\, Pr^2\right)^{1/2} \ .
$$

In this case the friction forces are negligible compared to the inertia and buoyancy forces, so the heat transfer coefficient is independent of the viscosity of the fluid. In the limiting case of $Pr \to \infty$, we get $\varphi(Pr) = 0.7109\, Pr^{1/4}$ and with that

$$
Nu_x = 0.5027 \left(Gr_x\, Pr\right)^{1/4} \ .
$$

The inertia forces are negligible in comparison to the friction and buoyancy forces.
The mean heat transfer coefficient α_m is yielded by integration of (3.325) to be

$$
\alpha_\mathrm{m} = \frac{1}{L} \int_0^L \alpha \,\mathrm{d}x = \frac{\lambda}{L} \left[\frac{\beta_\infty\,(\vartheta_0 - \vartheta_\infty)\,g}{4\,\nu^2}\right]^{1/4} \varphi(Pr) \int_0^L \frac{\mathrm{d}x}{x^{1/4}} \ .
$$

The mean Nusselt number is

$$
Nu_\mathrm{m} = \frac{\alpha_\mathrm{m} L}{\lambda} = \frac{4}{3} \left(\frac{Gr}{4}\right)^{1/4} \varphi(Pr) = \frac{4}{3} Nu_x(x = L) \ .
\tag{3.327}
$$

Here the Grashof number is formed with the length L of the vertical wall. The equation holds, like (3.325), only for laminar flow. This develops, as experiments have shown, for Rayleigh numbers $Gr\, Pr \leq 10^9$.

An equation which reproduces measured and calculated values in *laminar and turbulent flow* on *vertical plates and cylinders* for all Rayleigh and Prandtl numbers has been communicated by Churchill and Chu [3.50]. In the region of laminar flow $Gr\,Pr \leq 10^9$ it is certainly less accurate than the equation from Le Fèvre we discussed earlier, and it reads

$$Nu_m = \left\{ 0.825 + \frac{0.387\,Ra^{1/6}}{[1+(0.492/Pr)^{9/16}]^{8/27}} \right\}^2 . \tag{3.328}$$

Here the mean Nusselt and Rayleigh numbers are formed with the height of the plate L:

$$Nu_m = \frac{\alpha_m\,L}{\lambda} \;, \quad Ra = Gr\,Pr = \frac{\beta_\infty\,(\vartheta_0 - \vartheta_\infty)\,g\,L^3}{\nu^2}\,Pr \;.$$

The equation also holds for vertical cylinders, if $d/L \geq 35\,Gr^{-1/4}$. It is also valid for constant heat flux, this merely requires that the factor of 0.492 in the denominator of (3.328) be replaced by 0.437.

3.9.3   Some empirical equations for heat transfer in free flow

In the following some further empirical correlations, besides the relationships (3.325) and (3.328) for a vertical plate which we have already discussed, for heat transfer in free flow will be communicated.

1. Horizontal plate

The characteristic length in the mean Nusselt and the Rayleigh number is an equivalent length $L = A/U$ made up of the heat transfer area A and the perimeter U of the external edges of the plate. We can differentiate between the following cases: *Upper side of the plate is heated or underside is cooled* in accordance with Figs. 3.47 and 3.48:

$$Nu_m = 0.54\,Ra^{1/4} \quad \text{valid for} \quad 10^4 \leq Ra \leq 10^7$$

$$Nu_m = 0.15\,Ra^{1/3} \quad \text{valid for} \quad 10^7 \leq Ra \leq 10^{11} \;.$$

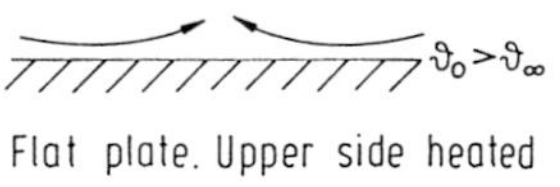

Flat plate. Upper side heated   Flat plate. Lower side cooled

Fig. 3.47: Flat plate. Upper side heated

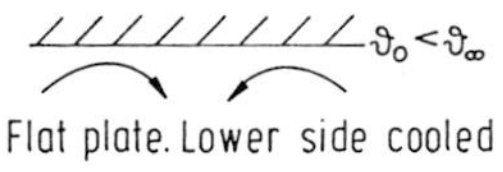

Fig. 3.48: Flat plate. Underside cooled

Upper side of the plate cooled or underside heated in accordance with Figs. 3.49 and 3.50:

$$Nu_m = 0.27\,Ra^{1/4} \quad \text{valid for} \quad 10^5 \leq Ra \leq 10^{10} \;.$$

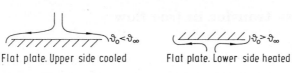

Flat plate. Upper side cooled   Flat plate. Lower side heated

Fig. 3.49: Flat plate. Upper   **Fig. 3.50**:   Flat plate.
side cooled   Underside heated

2. Inclined plate

Equation (3.328) holds for laminar and turbulent flow. The acceleration due to gravity g has to be replaced by its component parallel to the wall $g \cos \varphi$, in which φ is the angle of incline to the vertical. The equation is now valid for inclined plates, as long as $\varphi \leq \pi/3 = 60°$.

3. Horizontal cylinder

of diameter d and length $L \gg d$:

$$Nu_\mathrm{m} = \left\{ 0.60 + \frac{0.387 \, Ra^{1/6}}{[1 + (0.559/Pr)^{9/16}]^{8/27}} \right\}^2 \, ,$$

valid for $10^{-5} \leq Ra \leq 10^{12}$. It means that:

$$Nu_\mathrm{m} = \frac{\alpha_\mathrm{m} d}{\lambda} \; ; \quad Ra = Gr \, Pr = \frac{\beta_\infty (\vartheta_0 - \vartheta_\infty) \, g \, d^3}{\nu^2} \, Pr \, .$$

4. Sphere

of diameter d:

$$Nu_\mathrm{m} = 2 + \frac{0.589 \, Ra^{1/4}}{[1 + (0.469/Pr)^{9/16}]^{4/9}} \, ,$$

valid for $Pr \geq 0.7$ und $Ra \geq 10^{11}$.
This means that:

$$Nu_\mathrm{m} = \frac{\alpha_\mathrm{m} d}{\lambda} \; ; \quad Ra = Gr \, Pr = \frac{\beta_\infty (\vartheta_0 - \vartheta_\infty) \, g \, d^3}{\nu^2} \, Pr \, .$$

5. Vertical plate

being heated with constant heat flux:

$$Nu_x = 0.616 \, Ra_x^{1/5} \left(\frac{Pr}{0.8 + Pr} \right)^{1/5} \, ,$$

valid for $0.1 \leq Pr \leq \infty$; $Ra_x \leq 10^9$.
This means that:

$$Nu_x = \frac{\alpha x}{\lambda} \; ; \quad Ra_x = Gr_x \, Pr = \frac{\beta_\infty \dot{q} \, g \, x^4}{\nu^2 \lambda} \, Pr \, .$$

3.9.4   Mass transfer in free flow

In multicomponent mixtures density gradients can exist not only due to temperature and pressure gradients but also because of gradients in the composition. As long as the density gradient and body forces are not parallel to each other, a free flow develops which causes mass transfer. The mass transfer coefficients can be calculated from the equations given earlier. The only change is replacing the Nusselt number by the Sherwood number. The Grashof number is formed with the density difference

$$Gr = \frac{(\varrho_\infty - \varrho_0)\, g\, L^3}{\varrho_\infty \nu^2} \ . \tag{3.329}$$

In binary mixtures, when the density gradient is created by the concentration gradient alone, it follows from $\varrho = \varrho(p, T, \xi)$ that

$$\varrho_0 = \varrho_\infty + \left(\frac{\partial \varrho}{\partial \xi}\right)_{p,T} (\xi_0 - \xi_\infty) \ . \tag{3.330}$$

We call

$$\frac{1}{v}\left(\frac{\partial v}{\partial \xi}\right)_{T,p} = -\frac{1}{\varrho}\left(\frac{\partial \varrho}{\partial \xi}\right)_{T,p} := \gamma$$

the mass expansion coefficient. With this we get

$$\varrho_0 = \varrho_\infty - \varrho_\infty\, \gamma_\infty\, (\xi_0 - \xi_\infty) \ .$$

The Grashof number (3.329) is transformed into

$$Gr' = \frac{\gamma_\infty\, (\xi_0 - \xi_\infty)\, g\, L^3}{\nu^2} \ . \tag{3.331}$$

With mixtures of ideal gases the mass expansion coefficient can be found from the thermal equation of state

$$pv = RT$$

with the specific gas constant of a binary mixture

$$R = R_1\, \xi + R_2\, (1 - \xi)$$

to be

$$\gamma = (R_1 - R_2)/R \ ,$$

from which, with the universal gas constant $R_{\mathrm{m}} = \tilde{M}_i R_i$, the simple relationships

$$\gamma = \frac{\tilde{M}_2 - \tilde{M}_1}{\tilde{M}_2 \xi + \tilde{M}_1\, (1 - \xi)}$$

and

$$\gamma_\infty = \frac{\tilde{M}_2 - \tilde{M}_1}{\tilde{M}_2\, \xi_\infty + \tilde{M}_1 (1 - \xi_\infty)} \tag{3.332}$$

follow.

In *simultaneous heat and mass transfer* in binary mixtures, mean mass transfer coefficients can likewise be found using the equations from the previous sections. Once again this requires that the mean Nusselt number Nu_m is replaced by the mean Sherwood number Sh_m, and instead of the Grashof number a modified Grashof number is introduced, in which the density $\varrho(p, T, \xi)$ is developed into a Taylor series,

$$\varrho_0 = \varrho_\infty + \left(\frac{\partial \varrho}{\partial T}\right)_{p,\xi} (\vartheta_0 - \vartheta_\infty) + \left(\frac{\partial \varrho}{\partial \xi}\right)_{p,T} (\xi_0 - \xi_\infty) \ ,$$

so

$$\varrho_0 = \varrho_\infty - \varrho_\infty \beta_\infty (\vartheta_0 - \vartheta_\infty) - \varrho_\infty \gamma_\infty (\xi_0 - \xi_\infty)$$

is put into the Grashof number (3.329). The modified Grashof number obtained is

$$Gr = \frac{\beta_\infty (\vartheta_0 - \vartheta_\infty) g \, L^3}{\nu^2} + \frac{\gamma_\infty (\xi_0 - \xi_\infty) g \, L^3}{\nu^2} \ . \tag{3.333}$$

As Saville and Churchill [3.51] showed, the results obtained with this method are only sufficiently accurate if the Schmidt and Prandtl numbers of the mixtures are the same. In any other case the mutual influence of the mass transfer and the flow field will not be sufficiently taken into account.

3.10 Overlapping of free and forced flow

In the discussion of forced flow we neglected the influence of free flow and in reverse the effect of forced flow was neglected in our handling of free flow. However, frequently a free flow will overlap a forced flow as a result of density gradients. As we have already seen in 3.9.1, eq. (3.305), the decisive quantity for this is Gr/Re^2. If it is of the order 1, the buoyancy and inertia forces are equal, whilst for $Gr/Re^2 \ll 1$ the forced, and for $Gr/Re^2 \gg 1$, the free flow predominates.

Forced and free flow can, depending on the direction of the inertia and buoyancy forces, either mutually stimulate or dampen each other. In a forced flow overlapping a free flow, the heat and mass transfer can either be improved or inhibited. As an example of this we will look at a heated plate, Fig. 3.51. A free flow in the upwards direction develops, which can be strengthened Fig. 3.51a, or weakened, Fig. 3.51b, by a forced flow generated by a blower. Experiments have shown that the heat transfer coefficient can be calculated well by using equations of the form

$$Nu^n = |\, Nu_C^n \pm Nu_F^n\,| \ , \tag{3.334}$$

where Nu_C is the Nusselt number of the forced convection and Nu_F is that of the free flow. The positive sign is valid when the two flows are in the same direction. This is replaced by a negative sign if they are flowing countercurrent to each

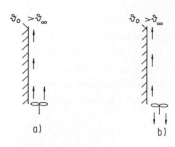

Fig. 3.51: Overlapping of free and forced flow. **a** Strengthening **b** Weakening of the free flow by the forced flow

other. The exponent n is generally $n = 3$. For longitudinal flow over horizontal plates, cylinders or spheres $n = 4$ is used. A corresponding relationship to (3.334) holds for masss transfer; this merely means replacing the Nusselt by the Sherwood number.

Example 3.15: A vertical metal plate of 0.5 m height and 1 m depth has a temperature of 170 °C. Calculate the heat flow by free flow to the surrounding air, which is at a temperature of 90 °C. In the solution the material properties should all be based on the mean boundary layer temperature $(170 \,°\mathrm{C} + 90 \,°\mathrm{C})/2 = 130 \,°\mathrm{C}$.

The following material properties are given: thermal conductivity $\lambda = 0.0336 \,\mathrm{W/Km}$, kinematic viscosity $\nu = 2.639 \cdot 10^{-5} \,\mathrm{m^2/s}$, Prandtl number $Pr = 0.697$. Furthermore $\beta_\infty = 1/T_\infty = 2.754 \cdot 10^{-3} \,\mathrm{K^{-1}}$.

We have

$$Gr = \frac{\beta_\infty \, (\vartheta_0 - \vartheta_\infty) \, g \, L^3}{\nu^2} = \frac{2.754 \cdot 10^{-3} \, 1/\mathrm{K} \cdot (170 - 90) \,\mathrm{K} \cdot 9.81 \,\mathrm{m/s^2} \cdot 0.5^3 \,\mathrm{m^3}}{(2.639 \cdot 10^{-5})^2 \,\mathrm{m^4/s^2}} \quad ,$$

and so

$$Gr = 3.879 \cdot 10^8 \quad \text{and} \quad Gr \, Pr = 2.704 \cdot 10^8 \quad .$$

The flow is laminar, so (3.327) is valid, with $\varphi(Pr)$ according to (3.326)

$$
\begin{aligned}
Nu_\mathrm{m} &= \frac{4}{3} Nu(x = L) = \frac{4}{3} \left(\frac{Gr}{4} \right)^{1/4} \varphi(Pr) \\[2mm]
&= \frac{4}{3} \left(\frac{3.879 \cdot 10^8}{4} \right)^{1/4} \frac{0.849 \cdot 0.679^{1/2}}{(1 + 2.006 \cdot 0.697^{1/2} + 2.034 \cdot 0.697)^{1/4}} = 65.08 \\[2mm]
\alpha_\mathrm{m} &= Nu_\mathrm{m} \frac{\lambda}{L} = 65.08 \frac{0.0336 \,\mathrm{W/Km}}{0.5} = 4.37 \,\mathrm{W/m^2K} \\[2mm]
\dot{Q} &= \alpha_\mathrm{m} \, A \, (\vartheta_0 - \vartheta_\infty) = 4.37 \,\mathrm{W/m^2K} \cdot 0.5 \,\mathrm{m} \cdot 1 \,\mathrm{m} \cdot (170 - 90) \,\mathrm{K} = 175 \,\mathrm{W} \quad .
\end{aligned}
$$

3.11 Compressible flows

In free flow the buoyancy is caused by a density change together with a body force, whilst forced laminar and turbulent flows have, up until now, been dealt with under the assumption of constant density. In liquids, this assumption of constant density is only in good approximation satisfied when the liquid is not close to its critical state. In the case of gases a moderate velocity with Mach number $Ma = w/w_S \ll 1$ has to be additionally presumed. If the velocity of the gas is large, or the fluid is close to its critical state, the density of a volume element of the material changes in the course of its motion. This is described by $d\varrho/dt \neq 0$. Therefore, in steady flow the density is locally variable.

As a result of heating through dissipation of kinetic energy, local temperature and therefore local density changes can exist in adiabatic flows. A flow, in which the density of a volume element of the material changes in the course of its motion is known as compressible. In the following we will discuss of heat transfer phenomena in these compressible flows. We will restrict ourselves to steady flows.

As the thermodynamic temperature T dependent velocity of sound w_S plays a role in this section, it is reasonable to use the thermodynamic temperature T rather than the temperature ϑ in calculations.

3.11.1 The temperature field in a compressible flow

In order to explain the basic properties of compressible flow, we will look at a two-dimensional, steady, boundary layer flow of a pure fluid. In a change from the previous discussion, in addition to the density, the viscosity and thermal conductivity will also be locally variable. The continuity equation is

$$\frac{\partial(\varrho\, w_x)}{\partial x} + \frac{\partial(\varrho\, w_y)}{\partial y} = 0 \ . \tag{3.335}$$

The momentum equation (3.58), with the boundary layer simplifications explained in section 3.5, is transformed into

$$\varrho\, \frac{dw_x}{dt} = -\frac{\partial p}{\partial x} + \frac{\partial \tau_{xy}}{\partial y} \ ,$$

$$\frac{\partial p}{\partial y} = 0 \tag{3.336}$$

with

$$\frac{dw_x}{dt} = w_x\, \frac{\partial w_x}{\partial x} + w_y\, \frac{\partial w_x}{\partial y} \quad \text{and} \quad \tau_{xy} = \eta\, \frac{\partial w_x}{\partial y} \ .$$

With the boundary layer simplifications from section 3.5, the enthalpy form of the energy equation (3.83) becomes

$$\varrho\, \frac{dh}{dt} = -\frac{\partial \dot{q}}{\partial y} + \frac{dp}{dt} + \phi \tag{3.337}$$

with $dp/dt = w_x\, \partial p/\partial x$, $\dot{q} = \dot{q}_y$ and the viscous dissipation $\phi = \eta\, (\partial w_x/\partial y)^2 = \tau_{xy}\, \partial w_x/\partial y$. Multiplication of the momentum equation (3.336) with w_x and addition to the energy equation (3.337) yields, after introduction of the so-called total enthalpy,

$$h_{tot} := h + \frac{w_x^2}{2} \, , \tag{3.338}$$

the relationship

$$\varrho \frac{dh_{tot}}{dt} = \frac{\partial}{\partial y} \left(w_x \tau_{xy} - \dot{q} \right) \, . \tag{3.339}$$

Here it is possible to replace the heat flux by the enthalpy $h = h(p, T)$, then taking $\partial p/\partial y = 0$ and $\dot{q} = -\lambda \, \partial T/\partial y$ into account, it holds that

$$\frac{\partial h}{\partial y} = c_p \frac{\partial T}{\partial y} = -c_p \frac{\dot{q}}{\lambda} \, . \tag{3.340}$$

Then, together with (3.338) and with $\lambda/c_p = \eta/Pr$ it follows that

$$\dot{q} = -\frac{\lambda}{c_p} \frac{\partial h}{\partial y} = -\frac{\eta}{Pr} \frac{\partial h_{tot}}{\partial y} + \frac{\eta}{Pr} w_x \frac{\partial w_x}{\partial y}$$

or

$$\dot{q} = -\frac{\eta}{Pr} \frac{\partial h_{tot}}{\partial y} + \frac{1}{Pr} w_x \tau_{xy} \, . \tag{3.341}$$

With that, the energy balance (3.339) becomes

$$\varrho \frac{dh_{tot}}{dt} = \frac{Pr - 1}{Pr} \frac{\partial (w_x \tau_{xy})}{\partial y} + \frac{1}{Pr} \frac{\partial}{\partial y} \left(\eta \frac{\partial h_{tot}}{\partial y} \right) \, , \tag{3.342}$$

when we presume constant laminar or turbulent Prandtl numbers. Equation (3.342) is suggested, for reasons of simplicity, for the investigation of laminar and turbulent flows of air or other gases, because they have Prandtl numbers $Pr \approx 1$. Then the first term on the right hand side, which describes the influence of the work done by the shear stress, disappears. With this equation (3.342) simplifies to

$$\varrho \frac{dh_{tot}}{dt} = \frac{\partial}{\partial y} \left(\eta \frac{\partial h_{tot}}{\partial y} \right) \, . \tag{3.343}$$

This equation has two important particular solutions:

a) **Adiabatic flow on a solid wall**

For this $\dot{q}(y = 0) = 0$, and as follows from (3.340) and (3.338) we also have

$$\left(\frac{\partial h_{tot}}{\partial y} \right)_{y=0} = 0 \, . \tag{3.344}$$

Further, at the edge of the boundary layer

$$h_{tot} = h_{tot\delta} = \text{const} \, . \tag{3.345}$$

At the leading edge of the plate

$$h_{tot}(x = 0, y) = \text{const} \, . \tag{3.346}$$

A solution of (3.343) which satisfies these boundary conditions is

$$h_{\text{tot}}(x, y) = \text{const} .$$
(3.347)

Therefore

$$h_{\text{tot}} = h + \frac{w_x^2}{2} = h_\delta + \frac{w_\delta^2}{2} .$$
(3.348)

From this, with $h - h_\delta = c_p(T - T_\delta)$ we obtain

$$\frac{T}{T_\delta} = 1 + \frac{1}{c_p T_\delta} \frac{w_\delta^2}{2} \left[1 - \left(\frac{w_x}{w_\delta} \right)^2 \right] .$$
(3.349)

The temperature profile in the boundary layer of an adiabatic flow, under the assumption that $Pr = 1$, is linked with the velocity profile. The relationship is valid independent of the pressure drop $\partial p / \partial x$. It can be rearranged for ideal gases, as the associated velocity of sound at the temperature T_δ for these gases is

$$w_{S\delta} = \sqrt{\kappa R T_\delta} = \sqrt{c_p (\kappa - 1) T_\delta}$$

and therefore

$$c_p T_\delta = \frac{w_{S\delta}^2}{\kappa - 1} .$$
(3.350)

This means that for ideal gases, (3.349) is transformed into

$$\frac{T}{T_\delta} = 1 + \frac{\kappa - 1}{2} M a_\delta^2 \left[1 - \left(\frac{w_x}{w_\delta} \right)^2 \right]$$
(3.351)

with the Mach number $Ma_\delta = w_\delta / w_{S\delta}$. Fig. 3.52 illustrates the dependence of the temperature profile T/T_δ on the velocities w_x/w_δ according to (3.351). If the velocity profile w_x/w_δ of incompressible flow in (3.351) is approximated by Blasius' solution of the plate boundary layer, section 3.7.1.1, we obtain the temperatures shown in Fig. 3.53. The diagram only represents the approximate temperature pattern. As we can see from (3.351), at the wall $y = 0$, because $w_x = 0$, the temperature is $T_0 = T_e$, which is also known as the *adiabatic wall temperature* or the *eigentemperature* T_e:

$$\frac{T_e}{T_\delta} = 1 + \frac{\kappa - 1}{2} M a_\delta^2 .$$
(3.352)

It is the one temperature attained at the surface of an adiabatic, insulated body in a flow. Through the relationship (3.352) the strong heating of the boundary layer at large Mach numbers becomes evident. Some eigentemperatures T_e are reproduced in Table 3.6.

Table 3.6: Eigentemperatures of an air flow of $T_\delta = 300\,\mathrm{K}$, $\kappa = 1{,}4$

Ma_δ	0	1	2	3	5
T_e/K	300	360	540	840	1800

The eigentemperature is identical to the temperature at the stagnation point in a flow at velocity w_δ on a body. It holds for this, that

$$h_{\mathrm{St}} = h_\delta + \frac{w_\delta^2}{2} \ ,$$

from which, with

$$h_{\mathrm{St}} - h_\delta = c_p \left(T_{\mathrm{St}} - T_\delta \right)$$

the relationship

$$\frac{T_{\mathrm{St}}}{T_\delta} = 1 + \frac{w_\delta^2}{2\,c_p T_\delta} = 1 + \frac{\kappa - 1}{2}\, Ma_\delta^2 \qquad (3.353)$$

which agrees with (3.352), follows.

The result (3.352) was only valid under the precondition that the Prandtl number of a gas which is presumed to be ideal is $Pr = 1$. In the general case of flow of an ideal gas with a Prandtl number $Pr \neq 1$, a different eigentemperature is present which is still dependent on the Prandtl number. In order to calculate this the so-called *"Recovery Factor"* r is introduced. It is defined by

$$\frac{T_e - T_\delta}{T_{\mathrm{St}} - T_\delta} = \frac{T_e - T_\delta}{w_\delta^2/2\,c_p} = r \ , \qquad (3.354)$$

which is still dependent on the Prandtl number. By definition, $r = 1$ for $Pr = 1$. A solution of the boundary layer equations by Eckert and Drake [3.52], for longitudinal flow along a plate, yielded, for laminar flow in the region

$$0.6 < Pr < 15 \ ,$$

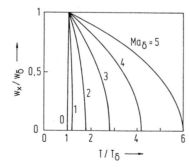

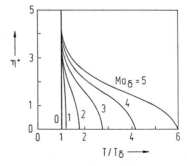

Fig. 3.52: Temperature and velocity profile in a compressible flow of ideal gases, according to (3.351)

Fig. 3.53: Temperature profiles in compressible flow of ideal gases. Adiabatic wall

the recovery factor can be approximated well by

$$r \cong \sqrt{Pr} \ , \tag{3.355}$$

and for turbulent flow in the region

$$0.25 < Pr < 10$$

by

$$r \cong \sqrt[3]{Pr} \ . \tag{3.356}$$

b) **Flow with vanishing pressure drop $\partial p / \partial x = 0$**

In a flow with a vanishing pressure drop $\partial p/\partial x$ the momentum equation
(3.336) simplifies to

$$\varrho \frac{dw_x}{dy} = \frac{\partial \tau_{xy}}{\partial y} = \frac{\partial}{\partial y} \left(\eta \frac{\partial w_x}{\partial y} \right) \ . \tag{3.357}$$

Its form agrees with the energy equation (3.342) for a fluid with $Pr = 1$

$$\varrho \frac{dh_{tot}}{dt} = \frac{\partial}{\partial y} \left(\eta \frac{\partial h_{tot}}{\partial y} \right) \ .$$

Therefore a particular solution

$$h_{tot} = a_0 \, w_x + a_1 \ , \tag{3.358}$$

exists, as $w_x(x, y)$ is a solution of the momentum equation, then $h_{tot}(x, y)$
is a solution of the energy equation. The constants a_0 and a_1 have to be
determined from the following boundary conditions:

$$\begin{aligned} h_{tot}(y = 0) &= h_0 \\ h_{tot}(y = \delta) &= h_{tot\delta} \ . \end{aligned} \tag{3.359}$$

With these the solution of (3.358) becomes

$$h_{tot} - h_0 = \frac{w_x}{w_\delta} \left(h_{tot\delta} - h_0 \right) \tag{3.360}$$

or with $h_{tot} = h + w_x^2/2$:

$$c_p \left(T - T_0 \right) + \frac{w_x^2}{2} = \frac{w_x}{w_\delta} \left[c_p \left(T_\delta - T_0 \right) + \frac{w_\delta^2}{2} \right] \ ,$$

where T_0 is the wall temperature. Together with (3.350) and the Mach
number $Ma_\delta = w_\delta/w_{S\delta}$, this yields the relationship valid for ideal gases
with a Prandtl number $Pr = 1$

$$\frac{T}{T_\delta} = \frac{T_0}{T_\delta} + \frac{T_\delta - T_0}{T_\delta} \frac{w_x}{w_\delta} + \frac{\kappa - 1}{2} Ma_\delta^2 \frac{w_x}{w_\delta} \left(1 - \frac{w_x}{w_\delta} \right) \ . \tag{3.361}$$

Again in this case the temperature profile is clearly linked to the velocity profile. In the event of very low Mach numbers $Ma \to 0$, i.e. incompressible flow of an ideal gas, this yields

$$T = T_0 + (T_\delta - T_0) \frac{w_x}{w_\delta} \ . \tag{3.362}$$

From that follows the heat flux

$$\dot{q} = \alpha \left(T_0 - T_\delta \right) = -\lambda \left(\frac{\partial T}{\partial y} \right)_{y=0} \ .$$

With this we obtain

$$\alpha = \frac{\lambda}{w_\delta} \left(\frac{\partial w_x}{\partial y} \right)_{y=0} = \frac{\lambda}{\eta \, w_\delta} \tau_0 \ .$$

Still taking into account the definition of the friction factor $c_f := \tau_0 / (\varrho \, w_\delta^2 / 2)$ and introducing the Reynolds number $Re = w_\delta \, x / \nu$, provides us with the already well known analogy between heat and momentum exchange

$$\frac{\alpha \, x}{\lambda} = Nu_x = Re \, \frac{c_f}{2} \ , \quad \text{valid for} \quad Pr = 1 \ . \tag{3.363}$$

As a special case, (3.361) also includes the solution for an adiabatic, insulated flat wall. This is

$$\left(\frac{\partial T}{\partial y} \right)_{y=0} = \left(\frac{\partial T}{\partial w_x} \right)_{y=0} \left(\frac{\partial w_x}{\partial y} \right)_{y=0} = 0 \ .$$

As the velocity gradient $(\partial w_x / \partial y)_{y=0}$ at the wall is not zero, $(\partial T / \partial w_x)_{y=0}$ has to disappear. By differentiating (3.361) we find

$$\left(\frac{\partial T}{\partial w_x} \right)_{y=0} = 0 = \frac{T_\delta - T_0}{T_\delta} \frac{1}{w_\delta} + \frac{\kappa - 1}{2} Ma_\delta^2 \frac{1}{w_\delta} \ . \tag{3.364}$$

From which we obtain the wall temperature

$$\frac{T_0}{T_\delta} = 1 + \frac{\kappa - 1}{2} Ma_\delta^2 \ .$$

As a comparison with (3.352) shows, it is equal to the eigentemperature T_e. Fig. 3.54 illustrates the dependence of the temperatures T/T_δ according to (3.361), on the velocities w_x / w_δ.
The temperature profile $\dot{q} = 0$, $T_0 = T_e$ is the bold line in Fig. 3.54. If heat flows from the wall to the fluid, the wall temperature T_0 has to lie above the eigentemperature T_e, so

$$\frac{T_0}{T_\delta} > 1 + \frac{\kappa - 1}{2} Ma_\delta^2 \quad \text{(heating the fluid)}$$

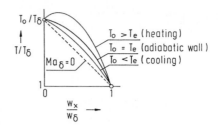

Fig. 3.54: Temperature and velocity profiles in compressible flow of an ideal gas; $\partial p/\partial x_1 = 0$; according to (3.361)

whilst in cooling the wall temperature T_0 has to be lower than the eigentemperature T_e, and so it holds that

$$\frac{T_0}{T_\delta} < 1 + \frac{\kappa - 1}{2} \, Ma_\delta^2 \quad \text{(cooling the fluid)} .$$

The curve $T_0 = T_e$ for adiabatic flow separates the line $T_0 > T_e$ for heating the fluid from that $T_0 < T_e$ for cooling. In the limiting case of $Ma_\delta = 0$, (3.361) is transformed into the straight dotted line.

In order to differentiate the three cases — heating, cooling or an adiabatic wall — a heat transfer parameter is introduced into (3.361)

$$\vartheta^+ := \frac{T_e - T_0}{T_e - T_\delta} . \tag{3.365}$$

Its sign tells us whether the wall is adiabatically insulated, or if the fluid is being heated or cooled. In heating we have $T_0 > T_e$ and $\vartheta^+ < 0$, for the adiabatic, insulated wall $T_0 = T_e$ and $\vartheta^+ = 0$ and in cooling $T_0 < T_e$ and $\vartheta^+ > 0$.

A simple relationship exists between the wall temperature T_0 and the heat transfer parameter, because it follows from (3.365) that

$$\frac{T_0}{T_\delta} = \frac{T_e}{T_\delta} - \vartheta^+ \frac{T_e}{T_\delta} + \vartheta^+ = 1 + \left(\frac{T_e}{T_\delta} - 1 \right) (1 - \vartheta^+) .$$

With the eigentemperature according to (3.352) we can write for this

$$\frac{T_0}{T_\delta} = 1 + \frac{\kappa - 1}{2} \, Ma_\delta^2 (1 - \vartheta^+) . \tag{3.366}$$

Inserting (3.366) into (3.361) yields an alternative form of (3.361), which now contains the heat transfer parameter:

$$\frac{T}{T_\delta} = 1 + \frac{\kappa - 1}{2} \, Ma_\delta^2 \left(1 - \frac{w_x}{w_\delta} \right) \left(1 - \vartheta^+ + \frac{w_x}{w_\delta} \right) . \tag{3.367}$$

According to these equations, which are valid for $Pr = 1$ and ideal gases, the temperature profile can be described by the velocity profile and the parameter ϑ^+. If we put in Blasius' solution from section 3.7.1.1, as an approximation for the velocity profile we obtain, for example, the temperature profile in Fig. 3.55 for $Ma_\delta = 2$, as a function of the distance from the wall η^+. This only reproduces an approximate temperature profile because of the simplifications made.

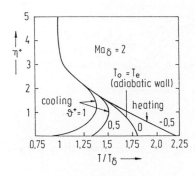

Fig. 3.55: Temperature profile in a compressible flow of ideal gas. Curve parameter $\vartheta^+ = (T_e - T_0)/(T_e - T_\delta)$. In heating $T_0 > T_e$ and $\vartheta^+ < 0$, in cooling $T_0 < T_e$ and $\vartheta^+ > 0$.

3.11.2   Calculation of heat transfer

The heat flux transferred from a surface at a given temperature $\vartheta_0 \neq \vartheta_e$ is obtained by solving the momentum and energy equations, taking into account the dissipation. This requires the introduction of boundary layer coordinates, as explained in section 3.1.1 about the solution of boundary layer equations. In addition to this the partial differential equations are transformed into ordinary differential equations, which can, however, be solved numerically. A solution of this type has been given by Eckert and Drake [3.52], for the incompressible flow of a fluid along a plate with a constant wall temperature. According to this the heat flux transferred locally from the plate can be described in good approximation by

$$\dot{q}(x) = 0.332\, Re_x^{1/2}\, Pr^{1/3}\, \frac{\lambda}{x}\, (T_0 - T_e) \quad \text{valid for} \quad 0.6 \le Pr \le 10 \ . \tag{3.368}$$

This result suggests the heat transfer coefficient α can be defined by

$$\dot{q} = \alpha\, (T_0 - T_e) \ . \tag{3.369}$$

The local Nusselt number formed with this heat transfer coefficient

$$Nu_x = \frac{\alpha\, x}{\lambda} = 0.332\, Re_x^{1/2}\, Pr^{1/3} \tag{3.370}$$

agrees with that found for incompresible flow, (3.196), for $0.6 < Pr < 10$.

To calculate the heat transferred, the heat transfer coefficient is found, according to a suggestion from Eckert and Drake [3.52], from the Nusselt relationships and then multiplied by the difference between the wall temperature and the eigen-temperature. As the Nusselt relationships were derived under the assumption of constant material properties, but in reality the density, viscosity and thermal conductivity of a compressible flow will change strongly due to the large temperature variation, the material properties should be those based on a reference temperature T_{ref}. Eckert [3.53] suggested the following empirical equation which is based on numerical calculations

$$T_{\text{ref}} = T_\infty + 0.5\, (T_0 - T_\infty) + 0.22\, (T_e - T_\infty) \ , \tag{3.371}$$

where T_∞ is the temperature at a great distance from the wall. In boundary layer flow it is equal to the temperature T_δ at the edge of the boundary layer. The eigentemperature is obtained from (3.354), with the recovery factor according to (3.355) or (3.356). In this way the heat transfer in laminar or turbulent flow can be calculated.

Example 3.16: Air at pressure $0.1\,\mathrm{MPa}$ and temperature $20\,^{\circ}\mathrm{C}$ flows at a velocity of $600\,\mathrm{m/s}$ over a $1\,\mathrm{m}$ long and $1\,\mathrm{m}$ wide, flat plate whose temperature is maintained at $60\,^{\circ}\mathrm{C}$. Calculate the heat trasnferred.
The following properties for air are given: kinematic viscosity $\nu(40\,^{\circ}\mathrm{C}) = 17.26 \cdot 10^{-6}\,\mathrm{m^2/s}$, $\nu(73\,^{\circ}\mathrm{C}) = 20.6 \cdot 10^{-6}\,\mathrm{m^2/s}$, $\nu(74.9\,^{\circ}\mathrm{C}) = 20.82 \cdot 10^{-6}\,\mathrm{m^2/s}$, gas constant $R = 0.2872\,\mathrm{kJ/kgK}$, thermal conductivity $\lambda(73\,^{\circ}\mathrm{C}) = 0.0295\,\mathrm{W/Km}$, $\lambda(74.9\,^{\circ}\mathrm{C}) = 0.0297\,\mathrm{W/Km}$, adiabatic exponent $\kappa = 1.4$, Prandtl number $Pr(74.9\,^{\circ}\mathrm{C}) = 0.709$.
First of all we calculate the Reynolds number at the end of the plate, to ascertain whether the flow there is laminar or turbulent. The reference temperature for the material properties is $(\vartheta_0 + \vartheta_\delta)/2 = (60 + 20)\,^{\circ}\mathrm{C}/2 = 40\,^{\circ}\mathrm{C}$. With that

$$Re = \frac{w_m\,L}{\nu} = \frac{600\,\mathrm{m/s} \cdot 1\,\mathrm{m}}{17.26 \cdot 10^{-6}\,\mathrm{m^2/s}} = 3.476 \cdot 10^7 \ .$$

The flow at the end of the plate is turbulent, at the start of the plate it is laminar. Furthermore

$$w_{S_\delta} = \sqrt{\kappa\,RT_\delta} = \sqrt{1.4 \cdot 0.2872 \cdot 10^3\,\mathrm{Nm/kgK} \cdot 293.15\,\mathrm{K}} = 343.3\,\mathrm{m/s}$$

and

$$Ma_\delta = \frac{w_\delta}{w_{S_\delta}} = \frac{600\,\mathrm{m/s}}{343.3\,\mathrm{m/s}} = 1.748 \ .$$

The stagnation point temperature, according to (3.353), is

$$T_{\mathrm{St}} = T_\delta\left[1 + \frac{\kappa - 1}{2}Ma_\delta^2\right] = 293.15\,\mathrm{K}\left[1 + \frac{1.4 - 1}{2}1.748^2\right] = 472.2\,\mathrm{K} \ .$$

The eigentemperature follows from (3.354) with the recovery factor according to (3.355) in the region of laminar flow with the estimated value $Pr \approx 0.7$:

$$T_e = \sqrt{Pr}\,(T_{\mathrm{St}} - T_\delta) + T_\delta = \sqrt{0.7} \cdot (472.2 - 293.15)\,\mathrm{K} + 293.15\,\mathrm{K} = 443\,\mathrm{K} \ .$$

For this the Prandtl number is $Pr = 0.705$; the assumption $Pr \approx 0.7$ does not have to be corrected. The reference temperature T_{ref} for the material properties, according to (3.371), is

$$T_{\mathrm{ref}} = 293.15\,\mathrm{K} + 0.5 \cdot (333.15 - 293.15)\,\mathrm{K} + 0.22 \cdot (443 - 293.15)\,\mathrm{K} = 346.12\,\mathrm{K} = 73\,^{\circ}\mathrm{C} \ .$$

The critical Reynolds number for the laminar-turbulent transition is

$$Re_{\mathrm{cr}} = \frac{w_\delta\,x_{\mathrm{cr}}}{\nu} = 5 \cdot 10^5 \ ,$$

from which

$$x_{\mathrm{cr}} = \frac{5 \cdot 10^5 \cdot 20.6 \cdot 10^{-6}\,\mathrm{m^2/s}}{600\,\mathrm{m/s}} = 0.0172\,\mathrm{m} \ .$$

The mean Nusselt number, the mean heat transfer coefficient and the mean heat flow in the laminar region according to section 3.7.4, No. 1, are

$$Nu_{m,\,\mathrm{lam}} = 0.664 \cdot Re_{\mathrm{cr}}^{1/2}\,Pr^{1/3} = 0.664 \cdot (5 \cdot 10^5)^{1/2} \cdot 0.709^{1/3} = 418.7$$

$$\alpha_{m,\,\mathrm{lam}} = Nu_{m,\,\mathrm{lam}}\frac{\lambda}{x_{\mathrm{cr}}} = 418.7 \cdot \frac{0.0295\,\mathrm{W/Km}}{0.0172\,\mathrm{m}} = 718\,\mathrm{W/m^2K} \ .$$

$$\dot{Q}_{\mathrm{lam}} = \alpha_{m,\,\mathrm{lam}}\,A\,(\vartheta_0 - \vartheta_e)$$

$$= 718\,\mathrm{W/m^2K} \cdot 0.0172\,\mathrm{m} \cdot 1\,\mathrm{m} \cdot (333.15 - 443)\,\mathrm{K} = -1.357\,\mathrm{kW} \ .$$

In the turbulent section we will once again estimate $Pr \approx 0.7$. The recovery factor is then $r = \sqrt[3]{Pr} = \sqrt[3]{0.7} = 0.8879$. The eigentemperature follows from (3.354) with (3.356) as

$$T_e = \sqrt[3]{Pr}\,(T_{St} - T_\delta) + T_\delta = 0.8879 \cdot (472.2 - 293.15)\,\text{K} + 293.15\,\text{K} = 452\,\text{K} = 179\,°\text{C} \ .$$

The Prandtl number for this is $Pr = 0.705$; the original guess does not have to be corrected. The reference temperature for the material properties, according to (3.371), is

$$T_{\text{ref}} = 293.15\,\text{K} + 0.5 \cdot (333.15 - 293.15)\,\text{K} + 0.22 \cdot (452 - 293.15)\,\text{K} = 348.1\,\text{K} = 74.9\,°\text{C} \ .$$

The Reynolds number in the turbulent region is

$$Re = \frac{600\,\text{m/s} \cdot (1 - 0.0172)\,\text{m}}{20.82 \cdot 10^{-6}\,\text{m}^2/\text{s}} = 2.833 \cdot 10^7 \ .$$

The quantities we are looking for in the turbulent region are, according to section 3.7.4, No. 1,

$$
\begin{aligned}
Nu_{m,\text{turb}} &= \frac{0.037\,Re^{0.8}\,Pr}{1 + 2.443\,Re^{-0.1}\,(Pr^{2/3} - 1)} \\[2mm]
&= \frac{0.037 \cdot (2.883 \cdot 10^7)^{0.8} \cdot 0.709}{1 + 2.443 \cdot (2.833 \cdot 10^7)^{-0.1} \cdot (0.709^{2/3} - 1)} = 2.68 \cdot 10^4 \\[2mm]
\alpha_{m,\text{turb}} &= Nu_{m,\text{turb}}\,\lambda/L = 2.68 \cdot 10^4 \cdot \frac{0.0297\,\text{W/Km}}{(1 - 0.0172)\,\text{m}} = 810\,\text{W/m}^2\text{K} \ . \\[2mm]
\dot{Q}_{\text{turb}} &= \alpha_{m,\text{turb}}\,A\,(T_0 - T_e) \\[2mm]
&= 810\,\text{W/m}^2\text{K} \cdot (1 - 0.0172)\,\text{m} \cdot 1\,\text{m} \cdot (333.15 - 452)\,\text{K} = -94.6\,\text{kW} \ .
\end{aligned}
$$

The total heat fed to the plate is $\dot{Q} = \dot{Q}_{\text{lam}} + \dot{Q}_{\text{turb}} = -95.96\,\text{kW}$. In order to keep the plate temperature constant this heat flow has to be removed.

3.12   Exercises

3.1: Plaster and sugar crystals conduct heat differently along the direction of the coordinate axes. Their thermal conductivity is given by the following tensor

$$\lambda_{ii} = \begin{pmatrix} \lambda_{11} & \lambda_{12} & 0 \\ \lambda_{21} & \lambda_{22} & 0 \\ 0 & 0 & \lambda_{33} \end{pmatrix} .$$

How large are the heat fluxes in the direction of the individual coordinate axes, and what is the differential equation for steady-state conduction through a thin, flat plate parallel to the x_2-direction?

3.2: The mean heat transfer coefficient on a small sphere of diameter d_0 with chloroform flowing over it is to be determined. This involves a number of experiments with water on a sphere with a diameter ten times larger.

a) At what temperature do the experiments have to be run, if the mean temperature of the chloroform is $T_0 = 293\,\mathrm{K}$?

b) The main interest is in heat transfer coefficients in the chloroform for flow velocities in the region $0.2\,\mathrm{m/s} \leq w_\alpha \leq 2\,\mathrm{m/s}$. Over what range of velocities do the experiments with water have to be carried out?

c) In the model experiments a mean heat transfer coefficient of $\alpha_M = 250\,\mathrm{W/m^2K}$ is found for a certain state. How big is the mean heat transfer coefficient on the sphere with the chloroform flowing over it for the same Reynolds and Prandtl numbers?

The following material data for chloroform at $T_0 = 293\,\mathrm{K}$ are known: $\nu = 0.383 \cdot 10^{-6}\,\mathrm{m^2/s}$, $\lambda = 0.121\,\mathrm{W/Km}$, $Pr = 4.5$. The properties for water can be read off the graph, Fig. 3.56.

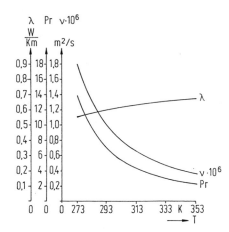

Fig. 3.56: Material properties of water

3.3: The heat loss from a valve due to laminar free flow to the surrounding air is to be determined from experiments using a reduced scale 1 : 2 model. On the model, with a temperature difference $\Delta\vartheta_M = 20\,\mathrm{K}$ between the model surface and the surrounding air, a heat loss of $\dot{Q}_M = 200\,\mathrm{W}$ is discovered. How large is the heat loss $\dot{Q}_O$ of the original if the temperature difference between the surface and the surrounding air is $\Delta\vartheta_O = 15\,\mathrm{K}$? The relevant thermophysical properties of the air can be assumed to be constant.

3.4: From measurements of the heat transfer in flow over a body it is known that the mean heat trasnfer coefficient α_m is dependent on the following quantities

$$\alpha_m = f(L, w_m, \varrho, \lambda, \nu, c) \ .$$

Using similarity theory show that the mean heat transfer coefficient can be represented as a function of three variables

$$Nu = f(Re, Pr)$$

with $Nu = \alpha_m L/\lambda$, $Re = w_m L/\nu$ and $Pr = \nu/a$.

3.5: The investigation of gas bubbles rising in vertical, cylindrical tubes filled with liquid is of technical interest, for example for the design of air-lift pumps and circulation evaporators among others. However the calculation of this process is difficult, even when it is highly idealised.

With the help of dimensional analysis, a number of far-reaching statements can be made about the form of the solution in this case. The problem will be idealised as follows:

 – It will be based on a very large air bubble in water, which virtually fills the entire cross section of the tube, see Fig. 3.57.

air bubble   **Fig. 3.57:** Air bubble in water

 – Air and water will be taken as frictionless and incompressible.

 – Capillary forces will be neglected.

 – The bubble moves at a steady velocity w up the tube.

Under these preconditions the physical process can be described by the following influencing quantities:

$$w, g, d, \varrho_W, \varrho_A \ .$$

g is the acceleration due to gravity, d the tube diameter, ϱ_A the density of air, ϱ_W the density of water.

a) Determine the dimensionless groups which describe the process $\pi_1, \pi_2, \ldots, \pi_n$.

b) What statements can be made from the relationship $\pi_1 = f(\pi_2, \pi_3, \ldots, \pi_n)$ between the dimensionless groups, about the form of the equation $w = f(g, d, \varrho_W, \varrho_A)$?

3.6: The velocity profile in the boundary layer can be approximately described by the statement

$$\frac{w_x}{w_\delta} = \sin\left(\frac{\pi}{2}\frac{y}{\delta}\right) \quad,$$

which likewise satisfies the boundary conditions

$$w_x(y = 0) = 0; \quad w_x(y = \delta) = w_\delta; \quad (\partial w_x/\partial y)_{y=0} = 0 \quad \text{und} \quad (\partial^2 w_x/\partial y^2)_{y=0} = 0 \quad.$$

Calculate the thickness of the boundary layer $\delta(x)$.

3.7: Hot air at 300 °C and a pressure of 0.01 MPa flows at a velocity of 10 m/s over a 1 m long, flat plate. What heat flow must be removed per m^2 of the plate surface area if we want to maintain the plate temperature at 25 °C?
Given are kinematic viscosity $\nu_1(p_1 = 0.1\,\text{MPa}, \vartheta_m = 162.5\,°\text{C}) = 30.84 \cdot 10^{-6}\,\text{m}^2/\text{s}$, the Prandtl number $Pr(\vartheta_m = 162.5\,°\text{C}) = 0.687$ and the thermal conductivity of air $\lambda(\vartheta = 162.5\,°\text{C}) = 3.64 \cdot 10^{-2}\,\text{W/Km}$.

3.8: Humid air at 20 °C and a relative humidity $\varphi = 0.5$ flows over a lake which is also at 20 °C. The lake is 200 m · 50 m big and the air flows over the longitudinal area with a velocity of 2 m/s. How much water evaporates per hour?
Given are the viscosity of air $\nu(20\,°\text{C}) = 1.535 \cdot 10^{-5}\,\text{m}^2/\text{s}$, the Schmidt number for water vapour-air $Sc = \nu/D = 0.6$, the saturation pressure of water $p_{WS}(20\,°\text{C}) = 2.337 \cdot 10^{-3}\,\text{MPa}$ and the saturation density $\varrho''(20\,°\text{C}) = 0.01729\,\text{kg/m}^3$.

3.9: A thin walled, steel pipe of 6 cm length and 50 mm internal diameter is heated by condensing steam from the outside, which maintains the internal wall temperature at 100 °C. The tube has $\dot{V} = 2.5 \cdot 10^{-4}\,\text{m}^3/\text{s}$ of water flowing through it, that is heated from 20 °C to 60 °C. How large is the mean heat transfer coefficient over the pipe length?
Properties of water at 20 °C: density $\varrho = 998.3\,\text{kg/m}^3$, specific heat capacity $c_p = 4.178\,\text{kJ/kgK}$.

3.10: A solar collector, see Fig. 3.58, consists of a parabolic reflector and an absorber pipe whose axis is located at the focusing point of the reflector. The reflector concentrates the solar radiation received onto the absorber pipe which has water running through it at a velocity of 0.05 m/s which is heated up. How long do the reflector and pipe have to be if the water is to be heated from 20 °C to 80 °C and the solar radiation amounts to $\dot{q}_S = 800\,\text{W/m}^2$? How high is the internal temperature of the pipe wall at the outlet?

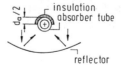

Fig. 3.58: Solar collector with reflector and absorber

The reflector has side length $s = 1\,\text{m}$, the internal diameter of the tube is $d_i = 60\,\text{mm}$, the outer diameter $d_o = 65\,\text{mm}$.
Further, the following properties for water are given: specific heat capacity $c_p = 4.181\,\text{kJ/kgK}$, density at 20 °C: $\varrho = 998.3\,\text{kg/m}^3$. The local heat transfer coefficient at the outlet will be $\alpha = 987\,\text{W/m}^2\text{K}$.

3.11: 0.2 kg/s of superheated steam at a pressure 0.1 MPa flow through a non-insulated steel pipe of inner diameter 25 mm and 5 m length. The steam enters the pipe at a temperature of $\vartheta_i = 150\,°C$ and is cooled to $\vartheta_e = 120\,°C$ by the time it reaches the outlet. The pipe is surrounded by cold air at a temperature of $\vartheta_0 = 0\,°C$; the mean heat transfer coefficient from the pipe surface to the air is $\alpha_e = 15\,W/m^2K$. Calculate the heat lost to the surroundings and the surface temperature of the pipe at the outlet. The thermal resistances are negligible. The following properties for steam are given: specific heat capacity $c_p(135\,°C) = 1.995\,kJ/kgK$, thermal conductivity $\lambda(135\,°C) = 0.0276\,W/Km$, Prandtl number $Pr(135\,°C) = 0.986$, dynamic viscosity $\eta(135\,°C) = 13.62 \cdot 10^{-6}\,kg/sm$

3.12: Air at $\vartheta_L = 40\,°C$ and relative humidity $\varphi = 0.2$ flows over an irregular, spherical packing whose surface temperature is kept constant at the wet bulb temperature $21.5\,°C$ by evaporating water. The air becomes loaded with water vapour. Calculate

a) the heat transferred and

b) the amount of water transferred.

The following values are given: sphere diameter $d = 0.02\,m$, channel cross section without the spheres $A_0 = 1\,m^2$, air velocity over A_0: $w_m = 2\,m/s$, void fraction $\varepsilon = 0.4$, packing height $H = 0.65\,m$, saturation pressure of water vapour $p_{WS}(40\,°C) = 73.75\,mbar$, thermal conductivity of air at $(\vartheta_A + \vartheta_0)/2 = 30.75\,°C$: $\lambda = 0.0262\,W/Km$, Prandtl number $Pr = 0.72$, kinematic viscosity $\nu = 1.64 \cdot 10^{-5}\,m^2/s$, Enthalpy of evaporation of water at $21.5\,°C$: $\Delta h_v = 2450.8\,kJ/kg$.

3.13: In a fluidized bed, hot sand at $850\,°C$ serves as the heating medium for atmospheric air at a pressure of 0.1 MPa, from its initial temperature $20\,°C$ to practically $850\,°C$. The hot air is then fed to a furnace. Calculate the blower power and the heat fed to the air. Given: inner diameter of the fluidizing chamber 3 m, kinematic viscosity of air $\nu(850\,°C) = 1.513 \cdot 10^{-4}\,m^2/s$, gas constant for air $0.2872\,kJ/kgK$, mean specific heat capacity between $30\,°C$ and $850\,°C$: $c_p = 1.163\,kJ/kgK$, specific heat capacity $c_p(20\,°C) = 1.007\,kJ/kgK$, density of sand $\varrho_S = 2500\,kg/m^3$, particle diameter $d_P = 0.5\,mm$, sphericity $\varphi_S = 0.86$, height H_0 of the quiescent sand layer $H_0 = 0.5\,m$, void fraction of the quiescent sand layer $\varepsilon_S = 0.36$, void fraction at the minimum fluidizing point $\varepsilon_{nf} = 0.44$, void fraction at 10 times the minimum fluidization velocity $\varepsilon = 0.55$, blower efficiency $\eta_v = 0.7$.

3.14: The temperature of a layer of air close to the ground is constant at $15\,°C$ up to a height of 100 m. Exhaust gas from a chimney containing mainly CO_2 has an initial temperature of $170\,°C$. Check if this exhaust gas can rise to a height greater than 100 m. Given are: gas constant for air $R_A = 0.2872\,kJ/kgK$, gas constant the exhaust gas $R_G = 0.1889\,kJ/kgK$, adiabatic exponent of the exhaust gas $\kappa = 1.3$, air pressure at the ground $p_1 = 0.1\,MPa$.

3.15: A $60\,°C$ hot, vertically standing, square plate of dimensions $L = b = 1\,m$ has air at a pressure of 0.1 MPa and a temperature of $20\,°C$ flowing over it from top to bottom at a velocity of 1 m/s. How much heat does the plate release? The following properties of air at the mean boundary layer temperature of $40\,°C$ are given: thermal conductivity $\lambda = 0.02716\,W/Km$, kinematic viscosity $\nu = 17.26 \cdot 10^{-6}\,m^2/s$, Prandtl number $Pr = 0.7122$. Further, the thermal expansion coefficient at $20\,°C$ is $\beta_\infty = 1/T_\infty = 3.411 \cdot 10^{-3}\,K^{-1}$.

3.16: A 0.4 m high, vertical plate being heated electrically with $\dot{q} = 15\,\text{W/m}^2$ has atmospheric air at a pressure of 0.1 MPa and a temperature of 10 °C flowing over it.
Calculate the plate temperature over its height.
As the mean boundary layer temperature is unknown, calculate approximately with the material properties of air at 10 °C. These are: thermal conductivity $\lambda = 0.02494\,\text{W/Km}$, kinematic viscosity $\nu = 14.42 \cdot 10^{-6}\,\text{m}^2/\text{s}$, Prandtl number $Pr = 0.716$, thermal expansion coefficient $\beta_\infty = 1/T_\infty = 3.532 \cdot 10^{-3}\,\text{K}^{-1}$.

3.17: Beer in cylindrical cans of 150 mm height and 60 mm diameter has a temperature of 27 °C and shall be cooled in a refrigerator which has an air temperature of 4 °C.
When does the beer cool more quickly: if the cans are laid down in the refrigerator, or if they are standing?
Calculate approximately with the properties of air at the initial mean temperature $(27 + 4)\,°\text{C}/2 = 15.5\,°\text{C}$. For this the thermal conductivity is $\lambda = 0.0235\,\text{W/Km}$, the kinematic viscosity is $\nu = 14.93 \cdot 10^{-6}\,\text{m}^2/\text{s}$, the Prandtl number is $Pr = 0.715$, further, the thermal expansion coefficient is $\beta_\infty = 1/T_\infty = 3.608 \cdot 10^{-3}\,\text{K}^{-1}$.

3.18: Calculate the temperature of the external skin of an aeroplane surface. The aeroplane is flying at a height of 10 000 m with a velocity of 700 km/h. The air temperature at 10 000 m height is $-50\,°\text{C}$.
What heat flow per m^2 aeroplane surface area is fed from the air conditioning to maintain the internal temperature T_i at 20 °C?
Given are the following properties for air: adiabatic exponent $\kappa = 1.4$, gas constant $R = 0.2872\,\text{kJ/kgK}$. Furthermore, the thermal resistance of the aeroplane skin is $\lambda/\delta = 5\,\text{m}^2\text{K/W}$ and the heat transfer coefficient inside is $\alpha_\text{i} = 10\,\text{W/m}^2\text{K}$.

3.19: Air flows over a wing at supersonic velocity. The Mach number Ma_δ is 2, the temperature of the external skin of the wing should not rise above 300 °C.
From what air temperature onwards does the wing have to be cooled?

4 Convective heat and mass transfer. Flows with phase change

Some of the convective heat and mass transfer processes with phase change that we will deal with in the following have already been explained in the previous chapters. This includes the evaporation of a liquid at the interface between a gas and a liquid or the sublimation at a gas-solid interface. They can be described using the methods for convective heat and mass transfer.

However in many heat and mass transfer processes in fluids, condensing or boiling at a solid surface play a decisive role. In thermal power plants water at high pressure is vaporised in the boiler and the steam produced is expanded in a turbine, and then liquified again in a condenser. In compression or absorption plants and heat pumps, boilers and condensers are important pieces of equipment in the plant. In the separation of mixtures, the different composition of vapours in equilibrium with their liquids is used. Boiling and condensing are, therefore, characteristic for many separation processes in chemical engineering. As examples of these types of processes, the evaporation, condensation, distillation, rectification and absorption of a fluid should all be mentioned.

In order to vaporise a liquid or condense a vapour, the enthalpy of vaporisation has to either be added to or removed from it. Under the precondition of thermo-dynamic equilibrium, the phase change demands that there is no difference in the temperatures of the two phases. However, in reality an imbalance is necessary for the phase change to occur, even if it is only a small temperature difference.

Heat transfer coefficients in condensation and boiling are, in general, much larger than those for convective heat transfer without a phase change. In addition the difference in density of the vapour and the liquid is large, as long as the phase change takes place far away from the critical region. This causes strong buoyancy forces $(\varrho_L - \varrho_G)g$ to appear, so that heat and mass transfer is supported by free flow. In the following sections we will study these processes.

4.1 Heat transfer in condensation

When vapour comes into contact with a wall, which is at a lower temperature than the saturation temperature of the vapour, the vapour will be liquified at the surface of the wall. A condensate is formed, which is subcooled as a result of its contact with the surface of the wall. This then induces further vapour to condense on the previously formed condensate. So, condensation of vapours is always linked to a mass transport, in which the vapour flows to the phase interface and is then transformed into the liquid phase.

The course of this process can be subdivided into several steps, in which a series of resistances have to be overcome. The fraction of these individual resistances in the total resistance can be very different. First, as a result of flow (convective transport) and molecular motion (diffusion transport), the vapour reaches the phase interface. In the next step the vapour condenses at the phase interface, and finally the enthalpy of condensation released at the interface is transported to the cooled wall by conduction and convection. Accordingly, three resistances in series have to be overcome: the thermal resistance in the vapour phase, the thermal resistance during the conversion of the vapour into the liquid phase, and finally the resistance to heat transport in the liquid phase.

Of these resistances, the thermal resistance in the liquid phase is generally decisive. The thermal resistance in the vapour is often low due to good mixing, but in the event of the vapour being superheated, this resistance has to be considered. This is also the case in the condensation of vapour mixtures or of mixtures containing inert gases because of the inhibitive effect of the diffusion. The conversion of vapour to liquid phase requires a temperature drop at the phase interface between vapour and liquid. However this is very small, normally only a few hundredths of a Kelvin, discounting very low pressures, which for water lie below 0.01 bar [4.1]. Correspondingly, the associated "molecular kinetic" resistance at the phase interface of vapour to liquid is almost always negligible.

4.1.1   The different types of condensation

Actual condensation can take place in many different ways. If the condensate forms a continuous film, Fig. 4.1, we speak of *film condensation*. The condensate film can be quiescent, or be in laminar or turbulent flow. The thermal resistance is decisive for the condensation rate, if the molecular kinetic resistance can be neglected. In calculations it is sufficient to just investigate this resistance, as Nusselt [4.2] first did for a flowing laminar film.

Instead of a film the condensate can also exist in the form of droplets, as shown in Fig. 4.2. This type of condensation is called *drop condensation*. Whether film or drop condensation prevails depends on whether the wall is completely or incompletely wetted. The decisive factor for this are the forces acting on a liquid droplet, which are illustrated in Fig. 4.3. σ_{LG} is the interfacial tension (SI units N/m) of the liquid (index L) against its own vapour (index G), σ_{SL} is the tension of the solid wall (index S) against the liquid and σ_{SG} is the interfacial tension of the wall with the vapour, so at equilibrium the contact angle β_0 is formed according

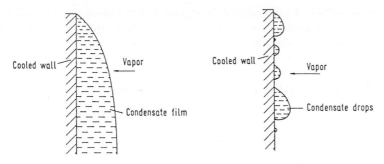

Fig. 4.1: Film condensation   **Fig. 4.2**: Drop condensation

to the equation

$$\sigma_{SG} - \sigma_{SL} = \sigma_{LG} \cos \beta_0 \ . \tag{4.1}$$

Finite values of this contact angle imply incomplete wetting and droplet formation. If, on the other hand, $\beta_0 = 0$, the droplets spread out over the entire wall. If the so-called wetting tension is $\sigma_{SG} - \sigma_{SL} > \sigma_{LG}$, an equilibrium according to that in (4.1), can no longer occur. Complete wetting takes place and a film develops.

Fig. 4.3: Interfacial tension at the droplet edge in incomplete wetting. The indices S, L, G represent the solid, liquid and gaseous phases, β_0 is the contact angle

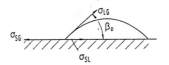

In practice, mixtures of vapours whose liquid phases are immiscible often have to be condensed. This results in the formation of a mix of drop and film condensation, Fig. 4.4. In an extended liquid film of one phase, large drops of the other phase develop, some of which reach the wall whilst others are enclosed by the liquid film or float on top of it.

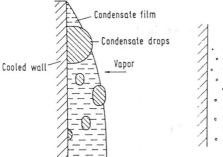

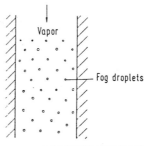

Fig. 4.4: Condensation of vapours of immiscible liquids   **Fig. 4.5**: Condensate formation as a droplet mist

Finally, by sufficient subcooling of a vapour and the presence of "condensation nuclei", that is tiny particles on which condensate can be deposited, a mist forms, as Fig. 4.5 shows schematically.

4.1.2   Nusselt's film condensation theory

When a vapour condenses on a vertical or inclined surface a liquid film develops, which flows downwards under the influence of gravity. When the vapour velocity is low and the liquid film is very thin, a laminar flow is created in the condensate. The heat will mainly be transferred by conduction from the surface of the condensate to the wall. The heat transferred by convection in the liquid film is negligibly small.

In 1916, Nusselt [4.2] had already put forward a simple theory for the calculation of heat transfer in laminar film condensation in tubes and on vertical or inclined walls. This theory is known in technical literature as *Nusselt's film condensation theory*. It shall be explained in the following, using the example of condensation on a vertical wall.

As Fig. 4.6 shows, saturated steam at a temperature ϑ_s is condensing on a vertical wall whose temperature ϑ_0 is constant and lower than the saturation temperature. A continuous condensate film develops which flows downwards under the influence of gravity, and has a thickness $\delta(x)$ that constantly increases. The velocity profile $w(y)$, with w for w_x, is obtained from a force balance. Under the assumption of steady flow, the force exerted by the shear stress are in equilibrium with the force of gravity, corresponding to the sketch on the right hand side of Fig. 4.6

$$\varrho_L g \, dV + \tau(y + dy) \, dx \, dy + p(x) \, dy \, dz = \tau(y) \, dx \, dz + p(x + dx) \, dy \, dz \; . \quad (4.2)$$

With

$$\tau(y + dy) - \tau(y) = (\partial\tau/\partial y) \, dy \; ,$$

$$p(x) \, dy \, dz - p(x + dx) \, dy \, dz = -(dp/dx) \, dx \, dy \, dz$$

and

$$dV = dx \, dy \, dz$$

we obtain

$$\frac{\partial\tau}{\partial y} = -\varrho_L g + \frac{dp}{dx} \; .$$

Considering only the vapour space, it is valid there that

$$\frac{dp}{dx} = \varrho_G g \; .$$

With that, the force balance is

$$\frac{\partial\tau}{\partial y} = -(\varrho_L - \varrho_G)g \; . \quad (4.3)$$

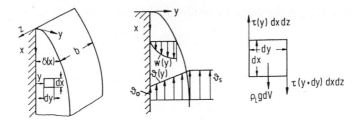

Fig. 4.6: Laminar condensate film on a vertical wall. Velocity and temperature profiles. Force balance

If the condensate is a Newtonian fluid, it holds that

$$\tau = \eta_{\mathrm{L}} \frac{\partial w}{\partial y} \ . \tag{4.4}$$

Under the assumption of temperature independent dynamic viscosity, (4.3) is transformed into

$$\eta_{\mathrm{L}} \frac{\partial^2 w}{\partial y^2} = -(\varrho_{\mathrm{L}} - \varrho_{\mathrm{G}})g \ , \tag{4.5}$$

from which, after integration, assuming constant density, the velocity profile parabolic in coordinate y

$$w = -\frac{(\varrho_{\mathrm{L}} - \varrho_{\mathrm{G}})g}{2\eta_{\mathrm{L}}} y^2 + c_1 y + c_0 \tag{4.6}$$

is obtained. The coefficients c_1 and c_0 can still depend on the coordinate x. Two boundary conditions are available for their determination:

At the wall, $y = 0$, the velocity is $w = 0$, and if we further assume that the vapour velocity is not very large, and as a result of this that the shear stress exerted by the vapour on the condensate film is low, then at the surface of the condensate $y = \delta$

$$\partial w / \partial y = 0 \ .$$

With these two boundary conditions we have

$$c_0 = 0 \quad \text{and} \quad c_1 = (\varrho_{\mathrm{L}} - \varrho_{\mathrm{G}})g\delta/\eta_{\mathrm{L}} \ ,$$

so that we obtain the following for the velocity profile

$$w = \frac{(\varrho_{\mathrm{L}} - \varrho_{\mathrm{G}})g}{\eta_{\mathrm{L}}} \delta^2 \left(\frac{y}{\delta} - \frac{y^2}{2\delta^2} \right) \ . \tag{4.7}$$

The mean velocity, $w_{\mathrm{m}}(x)$, over the thickness of the film, $\delta(x)$, is found by integration to be

$$w_{\mathrm{m}} = \frac{1}{\delta} \int_0^{\delta} w \, dy = \frac{(\varrho_{\mathrm{L}} - \varrho_{\mathrm{G}})g}{3\eta_{\mathrm{L}}} \delta^2 \ . \tag{4.8}$$

The mass flow rate of the condensate film follows as

$$\dot{M} = w_m \varrho_L b \delta = \frac{\varrho_L(\varrho_L - \varrho_G)gb}{3\eta_L}\delta^3 \ , \tag{4.9}$$

and the change in the mass flow rate with the film thickness is

$$\frac{\mathrm{d}\dot{M}}{\mathrm{d}\delta} = \frac{\varrho_L(\varrho_L - \varrho_G)gb}{\eta_L}\delta^2 \ . \tag{4.10}$$

The formation of the condensate flow $\mathrm{d}\dot{M}$, requires a heat flow $\mathrm{d}\dot{Q} = \Delta h_v \, \mathrm{d}\dot{M}$ to be removed, where Δh_v is the enthalpy of vaporisation. This heat will, by presumption, be transferred purely by heat conduction through the condensate film. The heat transferred through the condensate film by convection can be neglected. If, in addition to this, we presume constant thermal conductivity of the condensate and with that a linear temperature profile, like that in Fig. 4.6, then along surface segment $b \, \mathrm{d}x$ the heat flow will be

$$\mathrm{d}\dot{Q} = \lambda_L \frac{\vartheta_s - \vartheta_0}{\delta} b \, \mathrm{d}x \ .$$

On the other hand, because of $\mathrm{d}\dot{Q} = \Delta h_v \, \mathrm{d}\dot{M}$, with the condensate mass flow rate $\mathrm{d}\dot{M}$ given by (4.10), the last equation is transformed into

$$\lambda_L \frac{\vartheta_s - \vartheta_0}{\delta} b \, \mathrm{d}x = \Delta h_v \, \mathrm{d}\dot{M} = \Delta h_v \frac{\varrho_L(\varrho_L - \varrho_G)gb}{\eta_L}\delta^2 \, \mathrm{d}\delta \ ,$$

or

$$\delta^3 \frac{\mathrm{d}\delta}{\mathrm{d}x} = \frac{\lambda_L \eta_L}{\varrho_L(\varrho_L - \varrho_G)g\Delta h_v}(\vartheta_s - \vartheta_0) \ ,$$

from which, taking $\delta(x = 0) = 0$ into account, the film thickness

$$\delta = \left[\frac{4\lambda_L \eta_L (\vartheta_s - \vartheta_0)}{\varrho_L(\varrho_L - \varrho_G)g\Delta h_v} x\right]^{1/4} \tag{4.11}$$

is found by integration. The film thickness grows with the fourth root of the length x.

As the temperature profile has been assumed to be linear, the local heat transfer coefficient α is given by

$$\alpha = \frac{\lambda_L}{\delta} = \left[\frac{\varrho_L(\varrho_L - \varrho_G)g\Delta h_v \lambda_L^3}{4\eta_L(\vartheta_s - \vartheta_0)}\frac{1}{x}\right]^{1/4} \ , \tag{4.12}$$

and the mean heat transfer coefficient for a wall of height H is

$$\alpha_m = \frac{1}{H}\int_0^H \alpha \, \mathrm{d}x = \frac{4}{3}\alpha(x = H) = 0.943\left[\frac{\varrho_L(\varrho_L - \varrho_G)g\Delta h_v \lambda_L^3}{\eta_L(\vartheta_s - \vartheta_0)}\frac{1}{H}\right]^{1/4} \ . \tag{4.13}$$

All material properties are based on those for the condensate, and are best evaluated at the mean temperature $\vartheta_{\mathrm{m}} = (\vartheta_0 + \vartheta_{\mathrm{s}})/2$.

Using the energy balance for the condensate mass flow along the height H,

$$\dot{M}\Delta h_{\mathrm{v}} = \alpha_{\mathrm{m}}(\vartheta_{\mathrm{s}} - \vartheta_0)bH \ ,$$

allows the temperature difference $\vartheta_{\mathrm{s}} - \vartheta_0$ to be eliminated. A relationship equivalent ot that in equation (4.13) is obtained

$$\frac{\alpha_{\mathrm{m}}}{\lambda_{\mathrm{L}}} \left(\frac{\eta_{\mathrm{L}}^2}{\varrho_{\mathrm{L}}(\varrho_{\mathrm{L}} - \varrho_{\mathrm{G}})g} \right)^{1/3} = 0.925 \left(\frac{\dot{M}/b}{\eta_{\mathrm{L}}} \right)^{-1/3} . \tag{4.13a}$$

As we can see from the equations given above, large heat transfer coefficients are achieved when the temperature difference $\vartheta_{\mathrm{s}} - \vartheta_0$ and the height of the wall are small. In both cases the condensate film is thin and so the resistances to heat transfer are low. The results from above are also valid for condensation of vapours on the internal and external walls of vertical tubes, if the tube diameter is large in comparison to the film thickness. The width $b = \pi d$ has to be inserted into (4.13a).

The previous derivations apply to a vertical wall or tube with a large enough diameter. If the wall is inclined at an angle γ to the vertical, the acceleration due to gravity g, in the earlier equations, has to be replaced by its component $g \cos \gamma$ parallel to the wall. The heat transfer coefficient α_γ is then related to that for the vertical wall from (4.12), by the following

$$\alpha_\gamma = \alpha(\cos \gamma)^{1/4} \quad \text{and correspondingly} \quad \alpha_{\mathrm{m}_\gamma} = \alpha_{\mathrm{m}}(\cos \gamma)^{1/4} . \tag{4.14}$$

These results cannot be applied to inclined tubes because the liquid flow is not evenly distributed over the circumference of the tube.

Nusselt also calculated the heat transfer in laminar film condensation for *horizontal tubes*. A differential equation is then obtained for the film thickness. This can be solved numerically. If the outer diameter of the tube is indicated by d, the result found for the heat transfer coefficient averaged over the circumference $\alpha_{\mathrm{m,hor}}$ of a horizontal tube can be represented by

$$\alpha_{\mathrm{m,hor}} = 0.728 \left[\frac{\varrho_{\mathrm{L}}(\varrho_{\mathrm{L}} - \varrho_{\mathrm{G}})g\Delta h_{\mathrm{v}}\lambda_{\mathrm{L}}^3}{\eta_{\mathrm{L}}(\vartheta_{\mathrm{s}} - \vartheta_0)} \frac{1}{d} \right]^{1/4} = 0.864 \,\alpha_{\mathrm{m}}(H = \pi d/2) \ , \tag{4.15}$$

where α_{m} is the mean heat transfer coefficient on a vertical wall according to (4.13).[1] With the help of the energy balance for the condensate formed

$$\dot{M}\Delta h_{\mathrm{v}} = \alpha_{\mathrm{m,hor}}(\vartheta_{\mathrm{s}} - \vartheta_0)\pi \, d \, L$$

---

[1]Nusselt found, by graphical integration in eq. (4.15) instead of the value 0.728, the slightly less accurate value of 0.725, which was usd by all later authors.

the temperature difference can be eliminated from the previous equation, and can be replaced by the condensate mass flow rate. This produces a relationship equivalent ot the one in (4.15)

$$\frac{\alpha_{m,hor}}{\lambda_L}\left(\frac{\eta_L^2}{\varrho_L(\varrho_L - \varrho_G)g}\right)^{1/3} = 0.959\left(\frac{\dot{M}/L}{\eta_L}\right)^{-1/3} . \tag{4.15a}$$

A comparison of (4.15) and (4.13) shows that the mean heat transfer coefficient on a tube in a horizontal position is related to that for a vertical tube by

$$\alpha_{m,hor}/\alpha_m = 0.772\,(L/d)^{1/4} . \tag{4.16}$$

So, if a tube of length 3 m and diameter 0.029 m was chosen, then $\alpha_{m,hor} = 2.46\,\alpha_m$. Around 2.5 times more vapour condenses on a tube positioned vertically compared to on the same tube lying horizontally.

If n tubes lie one underneath the other in a horizontal tube bank, the condensate film in the lower tubes thickens because of the condensate coming from above, thereby reducing the heat transfer coefficient. On the other hand, the condensate falling from above improves the convection in the liquid film thereby improving the heat transfer. As the mean value for the heat transfer coefficients α_{mn} for n tubes lying on top of each other, we obtain, according to Nusselt, $\alpha_{mn}/\alpha_{m1} = n^{-1/4}$, where α_{m1} is the mean heat transfer coefficient for the uppermost tube according to (4.15). However, the improved heat transfer caused by the stronger convection due to the falling condensate is not considered in this hypothesis. The heat transfer coefficient α_{mn} calculated in this way is a little too small.

More exact values can be obtained, according to Chen [4.3], if we additionally take into account that the liquid film mixes with the condensate falling from above and is subcooled because of this, so that new vapour can condense. The result of this is

$$\alpha_{mn} = \left[1 + 0.2\,\frac{c_{pL}(\vartheta_s - \vartheta_0)}{\Delta h_v}(n-1)\right]n^{-1/4}\alpha_{m1} , \tag{4.17}$$

where α_{m1} is once again the mean heat transfer coefficient for the uppermost tube from (4.15). This relationship reproduces very well measured values on tubes arranged vertically in line, in the region $c_{pL}(\vartheta_s - \vartheta_0)(n-1)/\Delta h_v < 2$.

4.1.3 Deviations from Nusselt's film condensation theory

Experiments on film condensation of saturated vapour on vertical walls yield deviations by as much as +25% from the heat transfer coefficients according to Nusselt's film condensation theory. There are different reasons that are decisive for this.

a) Wave formation on the film surface

Nusselt's film condensation theory presumes an even increase in the thickness of the film due to further condensation. However experiments, among others [4.4] to [4.6], have shown that even in a flow that is clearly laminar, waves can develop at the film surface. These types of waves were not only observed on rough but also on polished surfaces. Obviously this means that the disturbances in the velocity that are always present in a stream are not damped under certain conditions, and so waves form. They lead to an improvement in the heat transfer of 10 to 25 % compared to the predictions from Nusselt's theory. According to Grimley [4.7], waves and ripples appear above a critical Reynolds number

$$Re = \frac{w_m \delta}{\nu_L} = \frac{\dot{M}/b}{\eta_L} = 0.392 \left[\left(\frac{\sigma}{\varrho_L g} \right)^{1/2} \left(\frac{g}{\nu_L^2} \right)^{1/3} \right]^{3/4} , \qquad (4.18)$$

where σ is the surface tension and $\nu_L = \eta_L/\varrho_L$ is the kinematic viscosity of the liquid. Based on a calculation of the disturbances van der Walt and Kröger [4.8] have proved that this relationship represents a good approximation of the point at which waves appear in water and some other fluids, such as the refrigerant R12. However in liquid sodium waves first develop at critical Reynolds numbers that are around five times larger than those calculated according to (4.18). If the formation of waves is ignored and the calculations are done according to Nusselt's film condensation theory, then the heat transfer coefficients obtained are too small, meaning that the condenser would be too big. Unfortunately at the moment there is no reliable theory for calculating the influence of wave formation on the heat transfer. In practical applications, the heat transfer coefficient α according to Nusselt's film condensation theory is multiplied by a correction factor f,

$$\alpha_{\text{waves}} = f\alpha , \qquad (4.19)$$

which accounts for the influence of wave formation. As the waves lead to a smaller film thickness, in terms of the statisical mean, the heat transfer is improved by wave formation. The correction factor above is therefore larger than one, and according to experiments by van der Walt und Kröger [4.8] is independent to a large degree of the amount of condensate. A good mean value is $f = 1.15$.

b) Temperature dependent properties

In (4.12) and (4.13) of Nusselt's film condensation theory the material properties of the condensate are presumed to be independent of temperature. This assumption is well met if the temperature drop $\vartheta_s - \vartheta_0$ in the condensate film is sufficiently small. In any other case the change in the dynamic viscosity, the thermal conductivity, and on a lower scale the density of the condensate film with the temperature has to be considered. In place of (4.5) the momentum balance appears

$$\frac{\partial}{\partial y} \left[\eta_L(\vartheta) \frac{\partial w}{\partial y} \right] = -[\varrho_L(\vartheta) - \varrho_G] g , \qquad (4.20)$$

and in the condensate film, under the assumption that heat is only transferred by conduction in the condensate film, the temperature profile is yielded from the energy balance

$$\frac{\partial}{\partial y} \left[\lambda_L(\vartheta) \frac{\partial \vartheta}{\partial y} \right] = 0 . \qquad (4.21)$$

These equations are to be solved under the boundary conditions

$$w(y = 0) = 0 , \qquad \frac{\partial w(y = \delta)}{\partial y} = 0 ,$$

$$\vartheta(y = 0) = \vartheta_0 \quad \text{and} \quad \vartheta(y = \delta) = \vartheta_s .$$

The heat transfer coefficient follows, by definition, from

$$\mathrm{d}\dot{Q} = \alpha b \, \mathrm{d}x \, (\vartheta_{\mathrm{s}} - \vartheta_0) = -\lambda_0 \left(\frac{\partial \vartheta}{\partial y}\right)_{y=0} b \, \mathrm{d}x = \Delta h_{\mathrm{v}} \, \mathrm{d}\dot{M} \ . \tag{4.22}$$

The detailed solution of (4.20) and (4.21) under the given boundary conditions shall not be presented here. Information can be found for this in a paper by Voskresenskij [4.9].

With the additional assumption that the density of the liquid film only slightly depends on the temperature, and that it is much larger than that of the vapour, $\varrho_{\mathrm{L}} \gg \varrho_{\mathrm{G}}$, we obtain the result that the relationship between the heat transfer coefficient α, and the heat transfer coefficient α_{Nu}, according to Nusselt's film condensation theory, can be represented by

$$\frac{\alpha}{\alpha_{Nu}} = f\left(\frac{\lambda_{\mathrm{s}}}{\lambda_0}; \frac{\eta_{\mathrm{s}}}{\eta_0}\right) \ . \tag{4.23}$$

The index s signifies the material properties of the condensate film at the saturation temperature, whilst index 0, indicates those properties that are formed at the wall temperature. The heat transfer coefficient α_{Nu} according to Nusselt's film condensation theory, (4.12), is calculated with the mean material properties

$$\eta_{\mathrm{L}} = \frac{1}{2}(\eta_{\mathrm{s}} + \eta_0) \quad \text{and} \quad \lambda_{\mathrm{L}} = \frac{1}{2}(\lambda_{\mathrm{s}} + \lambda_0) \ .$$

With the abbreviations

$$\eta^* = \eta_{\mathrm{s}}/\eta_0 \quad \text{and} \quad \lambda^* = \lambda_{\mathrm{s}}/\lambda_0$$

(4.23) in its complete form is

$$\frac{\alpha}{\alpha_{Nu}} = \left\{\frac{1+\eta^*}{10(1+\lambda^*)^3}\left[5 + \lambda^*(14 + 11\lambda^*) + \frac{\lambda^*}{\eta^*}(1 + 4\lambda^* + 5\lambda^{*2})\right]\right\}^{1/4} \ . \tag{4.24}$$

In the limitng case $\lambda^* = \eta^* = 1$ we obtain $\alpha = \alpha_{Nu}$.

Fig. 4.7 illustrates (4.24). As we can see from the graph, the temperature dependence of the dynamic viscosity and the thermal conductivity can have a marked influence on the heat transfer, as far as they change starkly with the temperature. In condensing steam, for a temperature difference between the saturation and wall temperatures of $\vartheta_{\mathrm{s}} - \vartheta_0 \leq 50\,\mathrm{K}$, the material properties vary for $\lambda_{\mathrm{s}}/\lambda_0$ between 0.6 and 1.2 and for η_{s}/η_0 between 1 and 1.3. This region is hatched in Fig. 4.7. It is clear that within this region the deviations from Nusselt's film condensation theory are less than 3%.

c) Subcooling the condensate and superheating the vapour

As the wall temperature is lower than the saturation temperature, not only the condensation enthalpy will be released at the wall, another heat flow from the subcooling of the condensate exists. In a cross section at point x the enthalpy flow of the falling condensate is

$$\dot{M}(h'' - h_{\mathrm{L}}) = \dot{M}\Delta h_{\mathrm{v}} + \int_0^\delta \varrho_{\mathrm{L}} w \, b \, c_{p\mathrm{L}}(\vartheta_{\mathrm{s}} - \vartheta) \, \mathrm{d}y \ , \tag{4.25}$$

where h_{L} is the mean specific enthalpy of the liquid. If the liquid over a cross section was at the saturation temperature $\vartheta - \vartheta_{\mathrm{s}}$, then we would have $h'' - h_{\mathrm{L}} = h'' - h'$. Presuming again a linear temperature profile

$$\vartheta_{\mathrm{s}} - \vartheta = (\vartheta_{\mathrm{s}} - \vartheta_0)\left(1 - \frac{y}{\delta}\right) \ , \tag{4.26}$$

and assuming the properties of the condensate are independent of temperature, after inserting the temperature from (4.26) and the velocity from (4.7) into (4.25), we obtain as a result of the integration

$$(h'' - h_{\mathrm{L}}) = \Delta h_{\mathrm{v}} + \frac{3}{8}c_{p\mathrm{L}}(\vartheta_{\mathrm{s}} - \vartheta_0) = \Delta h_{\mathrm{v}}^* \ . \tag{4.27}$$

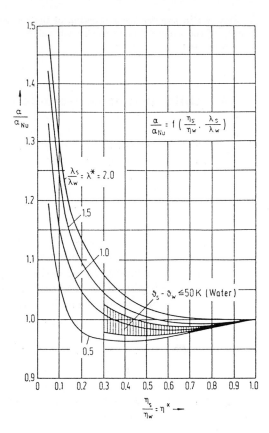

Fig. 4.7: Influence of temperature dependent material properties on heat transfer in film condensation [4.9]

It follows from this that, under the assumptions made, the specific enthalpy h_L of the flowing condensate is independent of the film thickness. The equation further shows that the enthalpy of vaporisation Δh_v in the equations for Nusselt's film condensation theory has to be replaced by the enthalpy difference Δh_v^*. If we additionally consider that the temperature profile in the condensate film is slightly curved, then according to Rohsenow [4.10] in place of (4.27), we obtain for Δh_v^* the more exact value

$$\Delta h_v^* = \Delta h_v + 0.68\,c_{pL}(\vartheta_s - \vartheta_0) \ . \tag{4.28}$$

This relationship holds for Prandtl numbers $Pr > 0.5$ and $c_{pL}(\vartheta_s - \vartheta_0)/\Delta h_v \leq 1$. The best agreement with experiments is yielded when the material properties, in particular the dynamic viscosity η_L, are eveluated at a temperature $\vartheta_L = \vartheta_0 + 1/4(\vartheta_s - \vartheta_0)$.

A further, and likewise small, deviation from Nusselt's film condensation theory is found for *superheated vapour*. In addition to the enthalpy of vaporisation, the superheat enthalpy $c_{pG}(\vartheta_G - \vartheta_s)$ has to be removed in order to cool the superheated vapour from a temperature ϑ_G to the saturation temperature ϑ_s at the phase interface. Instead of the enthalpy difference Δh_v according to (4.28), the enthalpy difference

$$\Delta h_{vs}^* = c_{pG}(\vartheta_G - \vartheta_s) + \Delta h_v + 0.68\,c_{pL}(\vartheta_s - \vartheta_0) \tag{4.29}$$

is used in (4.13). Furthermore, as the temperature difference in the condensate film is $\vartheta_s - \vartheta_0$, the heat flux is found from $\dot{q} = \alpha(\vartheta_s - \vartheta_0)$ and the mass flux of the condensate from $\dot{M}/A = \dot{q}/\Delta h_{vs}^*$.

However in practice the subcooling of the condensate, the buoyancy forces and the superheating of the vapour seldom reach such values that the improved equations (4.28) or (4.29) have to be used.

4.1.4   Influence of non-condensable gases

If a vapour condenses in the presence of a non-condensable gas it has to diffuse through this gas to the phase interface. This means that a drop in the partial pressure to the phase interface is required. As can be seen in Fig. 4.8, the partial pressure p_1 of the vapour drops from a constant value p_{1G} away from the phase interface to a lower value p_{1I} at the phase interface. Correspondingly, the associated saturation temperature $\vartheta_s(p_1)$ also falls to the value ϑ_I at the phase interface. The pressure p_0 of the inert gas rises towards the phase interface, so the sum $p_1 + p_0$ always yields a constant total pressure p.

The saturation temperature ϑ_I at the phase interface can, depending on the gas content, lie considerably below the saturation temperature $\vartheta_s(p)$ associated with the pressure p, which would occur if no inert gas was present. The temperature difference between the phase interface and the wall is lowered because of the inert gas and with that the heat transfer is also reduced. In order to avoid or prevent this, it should be possible to remove the inert gas through valves. Large condensers are fitted with steam-jet apparatus which suck the inert gas away. In other cases, for example the condensation of water out of a mixture of steam and air or in the condensation of ammonia from a mixture with air, it is inevitable that inert gases are always present. Therefore their influence on heat transfer has to be taken into account.

The influence of the inert gases on heat transfer can be determined with the aid of the energy balance

$$\dot{Q} = \alpha_L A(\vartheta_I - \vartheta_0) = \dot{M}\Delta h_v + \alpha_G^{\bullet} A(\vartheta_G - \vartheta_I) \ . \qquad (4.30)$$

The heat transfer coefficient $\alpha_G^{\bullet}$ signifies the heat flow transferred from the vapour-gas mixture to the phase interface. The superscript point indicates that the heat

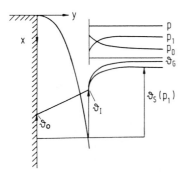

Fig. 4.8: Influence of the inert gas on the partial pressure and temperature profiles. ϑ_G vapour temperature, ϑ_s saturation temperature, p_1 partial pressure of the condensing vapour, p_0 partial pressure of the inert gas, $p = p_1 + p_0$ total pressure

is not only transferred by conduction but also by a material flow normal to the wall. In a flowing gas, the size of $\alpha_G^{\bullet}$ is determined from the material properties of the gas and the flow velocity.

The mass flow rate $\dot{M}$ of the vapour transported at the condensate surface increases with the difference in the partial pressure $p_{1G} = y_{1G}p$ in the vapour space and p_{1I} at the phase interface.

According to the laws of mass transfer, see (1.195),

$$\dot{M} = \varrho_G \beta_G A \ln \frac{p - p_{1I}}{p - p_{1G}} \; , \tag{4.31}$$

where ϱ_G is the density of the gas-vapour mixture at pressure p and temperature ϑ_G. Strictly speaking this relationship is only valid if the temperature up to the phase interface is constant. Disregarding low temperatures, this assumption of constant temperature rarely causes an error in the quantity $\dot{M}$. Further, p_{1I} is the partial pressure of the condensing gas at the phase interface and p_{1G} is that in the core of the gas-vapour flow.

The Lewis relationship (1.198), which we have already discussed, allows us to relate the mass transfer coefficient β_G approximately to the heat transfer coefficient

$$\beta_G = \frac{\alpha_G}{c_{pG} \varrho_G} Le^{-2/3} \; .$$

The values β_G and α_G are, as already explained, mass and heat transfer coefficients for a vapour flow, in which the condensate surface is seen as a quiescent, solid wall. Inserting the following into (4.30)

$$\alpha_G^{\bullet} = \alpha_G \zeta \; , \tag{4.32}$$

where the correction factor ζ, the so-called 'Ackermann correction' [4.11] takes into account that, in reality, vapour does not flow along a solid wall, rather part of the flow vanishes at the phase interface. The Ackermann correction is obtained in a similar manner to the Stefan correction (1.194) for mass transfer, in which the film theory from section 1.5.1 is applied to the processes of heat transfer in the condensation or sucking being considered here, to

$$\zeta = \frac{-\phi}{\exp(-\phi) - 1} \quad \text{with} \quad \phi = \frac{|\dot{M}|c_{pG}}{A\alpha_G} \; . \tag{4.33}$$

Using (4.31), we obtain from the energy balance (4.30)

$$\vartheta_I - \vartheta_0 = \frac{\alpha_G}{\alpha_L} \left[\frac{\Delta h_v}{c_{pG}} Le^{-2/3} \ln \frac{p - p_{1I}}{p - p_{1G}} + \zeta(\vartheta_G - \vartheta_I) \right] \tag{4.34}$$

as a determination equation for the unknown temperature ϑ_I at the phase interface. This equation can only be solved iteratively. For small inert gas contents

$p_0 \ll p$ or $p_1 \to p$, the quotient $(p - p_{1I})/(p - p_{1G})$ and with that the first sum in the brackets will be large. Equation (4.34) simplifies to

$$\vartheta_I - \vartheta_0 = \frac{\alpha_G \Delta h_v}{\alpha_L c_{pG}} Le^{-2/3} \ln \frac{p - p_{1I}}{p - p_{1G}} . \tag{4.35}$$

To be able to estimate the influence of the inert gas on the heat transfer we will consider a condensate film of thickness δ. In front of this there is vapour with an inert gas. If there was no inert gas present, the surface of the film would be at the saturation temperature $\vartheta_s(p)$ and the heat flux released would be

$$\dot{q} = \frac{\lambda_L}{\delta}(\vartheta_s - \vartheta_0) .$$

The presence of the inert gas means that with the same film thickness a smaller heat flux

$$\dot{q}_G = \frac{\lambda_L}{\delta}(\vartheta_I - \vartheta_0)$$

will be transferred. The ratio

$$\frac{\dot{q}_G}{\dot{q}} = \frac{\vartheta_I - \vartheta_0}{\vartheta_s - \vartheta_0} \leq 1 \tag{4.36}$$

shows by how much the heat flux is reduced by the existence of the inert gas. With (4.35) the approximate relationship below is yielded

$$\frac{\dot{q}_G}{\dot{q}} = \frac{\alpha_G \Delta h_v Le^{-2/3}}{(\vartheta_s - \vartheta_0)\alpha_L c_{pG}} \ln \frac{p - p_{1I}}{p - p_{1G}} . \tag{4.37}$$

As we can see from this, the flow velocity in the vapour phase has to be chosen to be sufficiently high, in particular with large temperature differences $\vartheta_s - \vartheta_0$, so that the heat transfer coefficient α_G of the vapour will be large and with that the heat flux $\dot{q}_G$ will not be too small.

The results of an accurate calculation are illustrated in Figs. 4.9 and 4.10. The ratio of the two heat fluxes $\dot{q}_G/\dot{q}$ are plotted over the temperature difference $\vartheta_s - \vartheta_0$ with the inert gas fraction as a parameter. Both diagrams are valid for the condensation of steam out of mixture with air. Fig. 4.9 is valid for condensation in forced flow, when the influence of gravity can be neglected compared to that of the inertia forces, as is the case of flow over a horizontal plate. Fig. 4.10 shows the reverse case, in which the inertia forces are negligible compared to the gravitational forces. This corresponds to free flow on a vertical plate. It is clearly visible in both pictures that the heat transfer reduces with increasing inert gas fractions, and that in free flow it is more significantly reduced due to the inert gas than in forced flow.

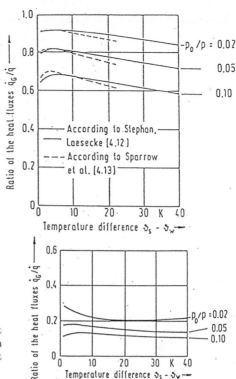

Fig. 4.9: Influence of inert gas on the heat transfer in condensation of a forced flow of steam and air; $\dot{q}_G$ heat flux with, $\dot{q}$ without inert gas

Fig. 4.10: Influence of inert gas on the heat transfer in condensation of a free flow of steam and air [4.12]; $\dot{q}_G$ heat flux with, $\dot{q}$ without inert gas

Example 4.1: Saturated steam at a pressure $9.8 \cdot 10^{-3}$ MPa, condenses on a vertical wall. The wall temperature is 5 K below the saturation temperature. Calculate the following quantities at a distance of $H = 0.08$ m from the upper edge of the wall: the film thickness $\delta(H)$, the mean velocity w_m of the downward flowing condensate, its mass flow rate $\dot{M}/b$ per m plate width, the local and the mean heat transfer coefficients.
At what point H_1 is the mass flow rate $\dot{M}/b$ double that at the point H? Why is the heat transfer coefficient over the length $H < x \leq H_1$ smaller than $\alpha(x \leq H)$?
Given are: saturation temperature at $9.58 \cdot 10^{-3}$ MPa: $\vartheta_s = 45.4$ °C, enthalpy of vaporisation $\Delta h_v = 2392$ kJ/kg, density of the liquid $\varrho_L = 991$ kg/m³, thermal conductivity $\lambda_L = 0.634$ W/Km, dynamic viscosity $\eta_L = 6.54 \cdot 10^{-4}$ kg/sm, density of the vapour $\varrho_G \ll \varrho_L$.
According to (4.11) the film thickness is

$$\delta(H) = \left[\frac{4 \cdot 0.634 \text{ W/K m} \cdot 6.54 \cdot 10^{-4} \text{ kg/s m} \cdot 5 \text{ K} \cdot 0.08 \text{ m}}{991^2 \text{ kg}^2/\text{m}^6 \cdot 9.81 \text{ m/s}^2 \cdot 2392 \cdot 10^3 \text{ J/kg}} \right]^{1/4}$$

$$\delta(H) = 7.325 \cdot 10^{-5} \text{ m} \approx 0.073 \text{ mm} \ .$$

The mean velocity, according to (4.8), is

$$w_m = \frac{991 \text{ kg/m}^3 \cdot 9.81 \text{ m/s}^2}{3 \cdot 6.54 \cdot 10^{-4} \text{kg/s m}} \cdot (7.325 \cdot 10^{-5})^2 \text{ m}^2 = 0.0266 \text{ m/s} \ .$$

From (4.9) we have

$$\dot{M}/b = w_m \, \varrho_L \, \delta \, 0.0266 \text{ m/s} \cdot 991 \text{ kg/m}^3 \cdot 7.325 \cdot 10^{-5} \text{ m} = 1.93 \cdot 10^{-3} \text{ kg/s m} \ .$$

The local heat transfer coefficient at point H is

$$\alpha(H) = \frac{\lambda_L}{\delta} = \frac{0.634 \,\text{W/K}\,\text{m}}{7.325 \cdot 10^{-5}\,\text{m}} = 8655 \,\text{W/m}^2\text{K} \quad.$$

The mean heat transfer coefficient follows from (4.13) as

$$\alpha_m(H) = \frac{4}{3}\,\alpha(H) = 11540 \,\text{W/m}^2\text{K} \quad.$$

Furthermore

$$\frac{\dot{M}(H_1)}{b} = 2\,\frac{\dot{M}(H)}{b} = 3.86 \cdot 10^{-3}\,\text{kg/sm} \quad.$$

According to (4.9) the associated film thickness is

$$\delta(H_1) = \left(\frac{\dot{M}(H_1) \cdot 3\eta_L}{\varrho_L^2\, g\, b}\right)^{1/3} = \left(\frac{3.86 \cdot 10^{-3}\,\text{kg/sm} \cdot 3 \cdot 6.54 \cdot 10^{-4}\,\text{kg/sm}}{991^2\,\text{kg}^2/\text{m}^6 \cdot 9.81\,\text{m/s}^2}\right)^{1/3}$$

$$\delta(H_1) = 9.23 \cdot 10^{-5}\,\text{m} \quad.$$

It follows from (4.11), with $x = H_1$, that

$$H_1 = \delta^4 \frac{\varrho_L^2\, g\, \Delta h_v}{4\,\lambda_L\, \eta_L\, (\vartheta_s - \vartheta_0)} = (9.23 \cdot 10^{-5})^4\,\text{m}^4 \cdot \frac{991^2\,\text{kg}^2/\text{m}^6 \cdot 9.81\,\text{m/s}^2 \cdot 2392 \cdot 10^3\,\text{J/kg}}{4 \cdot 0.634\,\text{W/K}\,\text{m} \cdot 6.54 \cdot 10^{-4}\,\text{kg/s}\,\text{m} \cdot 5\,\text{K}}$$

$$H_1 = 0.202\,\text{m} \quad.$$

Over the length $H < x \leq H_1$ the film is thicker than over the length $x \leq H$, and therefore the the heat transfer coefficient is smaller.

Example 4.2: In a tube bundle condenser, like that sketched in Fig. 4.11, $7 \cdot 10^3$ kg/h saturated vapour of refrigerant R22, at a pressure of 1.93 MPa is to be condensed. Cooling water is available at a temperature of 18 °C, and this can be heated by $10K$. Copper tubes of length 1.5 m, 16 mm outer diameter and 1 mm wall thickness are to be used in the construction of the condenser.
How many tubes are required? How large is the cooling water flow rate?
The following data are given for R22 at the saturation pressure 1.93 MPa: saturation temperature $\vartheta_s = 50$ °C, density of the liquid $\varrho_L = 1.0839 \cdot 10^3$ kg/m^3, enthalpy of vaporisation $\Delta h_v = 154.08$ kJ/kg, specific heat capacity of the liquid $c_{pL} = 1.38$ kJ/kg K, dynamic viscosity $\eta_L = 173 \cdot 10^{-6}$ kg/sm, thermal conductivity $\lambda_L = 75.4 \cdot 10^{-3}$ W/K m. The heat transfer coefficient on the water side is $\alpha_{mi} = 2000$ W/m^{2}K, the specific heat capacity of water is $c_{pW} = 4.1805$ kJ/kg K. The resistance to heat of the copper wall can be neglected.

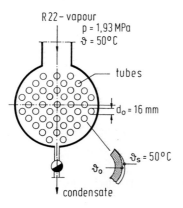

R 22 - vapour
p = 1,93 MPa
ϑ = 50°C

tubes

d_o = 16 mm

ϑ_s = 50°C

ϑ_0

condensate

Fig. 4.11: Tube bundle condenser

The required heat flow is

$$\dot{Q} = \dot{M}\,\Delta h_v = \frac{7 \cdot 10^3\,\text{kg/h}}{3600\,\text{s/h}} \cdot 154.08\,\text{kJ/kg} \approx 300\,\text{kW} \ .$$

The amount of cooling water follows from

$$\dot{M}_W = \frac{\dot{Q}}{c_{pW}\,\Delta\vartheta_W} = \frac{300\,\text{kW}}{4.1805\,\text{kJ/kg\,K} \cdot 10\,\text{K}} = 7.18\,\text{kg/s} \ .$$

The required area comes from $\dot{Q} = k_m\,A\,\Delta\vartheta_m$ with

$$\Delta\vartheta_m = \frac{(\vartheta_s - \vartheta_{\text{Win}}) - (\vartheta_s - \vartheta_{\text{Wout}})}{\ln[(\vartheta_s - \vartheta_{\text{Win}})/(\vartheta_s - \vartheta_{\text{Wout}})]} = \frac{(50-18)\,\text{K} - (50-28)\,\text{K}}{\ln[(50-18)\,\text{K}/(50-28)\,\text{K}]} = 26.69\,\text{K} \ ,$$

and with that

$$k_m\,A = \frac{\dot{Q}}{\Delta\vartheta_m} = \frac{300\,\text{kW}}{26.69\,\text{K}} = 11.24 \cdot 10^3\,\text{W/K} \ .$$

On the other hand

$$\frac{1}{k_m\,A} = \frac{1}{\alpha_{\text{mi}}\,A_i} + \frac{1}{\alpha_{\text{mo}}\,A_o}$$

and therefore with $A = A_o$

$$\frac{1}{k_m} = \frac{1}{\alpha_{\text{mi}}\,(d_i/d_o)} + \frac{1}{\alpha_{\text{mo}}} \ .$$

The heat transfer coefficient α_m on the condensate side is calculated from (4.15a) and (4.17). It still contains in (4.15a) the mass flow rate of the condensate $\dot{M}/L$ on a single tube, and can be determined if we knew the number of tubes or their surface area A. Therefore we will estimate the mean outside heat transfer coefficient α_{mo}, and use that to find the surface area A and then check the estimate at the end. Our estimate is $\alpha_{\text{mo}} = 1200\,\text{W/m}^2\text{K}$. With that we get

$$\frac{1}{k_m} = \frac{1}{2\,000\,\text{W/m}^2\text{K} \cdot (14\,\text{mm}/16\,\text{mm})} + \frac{1}{1\,200\,\text{W/m}^2\text{K}} = 1.4047 \cdot 10^{-3}\frac{\text{m}^2\,\text{K}}{\text{W}}$$

$$k_m = 711.9\,\text{W/m}^2\text{K} \ .$$

Giving an area of

$$A = k_m A/k_m = \frac{11.24 \cdot 10^3\,\text{W/K}}{711.9\,\text{W/m}^2\text{K}} = 15.79\,\text{m}^2 \ .$$

The number of tubes is

$$z = \frac{A}{d_o\,\pi\,L} = \frac{15.79\,\text{m}^2}{16 \cdot 10^{-3}\,\text{m} \cdot \pi \cdot 1.5\,\text{m}} = 209 \ .$$

We will choose 220 tubes. The mean wall temperature ϑ_0 follows from $\dot{Q} = \alpha_{\text{m}_o} A(\vartheta_s - \vartheta_0)$ as

$$\vartheta_0 = \vartheta_s - \frac{\dot{Q}}{\alpha_{\text{mo}}\,A}\,50\,°\text{C} - \frac{300 \cdot 10^3\,\text{W}}{1200\,\text{W/m}^2\text{K} \cdot 15.79\,\text{m}^2} = 34.2\,°\text{C} \ .$$

Examination of α_{mo}: According to (4.15a)

$$\alpha_{\text{m, hor}} = \alpha_{\text{m1}} = \lambda_L \left(\frac{\eta_L^2}{\varrho_L^2 g}\right)^{-1/3} 0.959 \left(\frac{\dot{M}/L}{\eta_L}\right)^{-1/3}$$

$$= 75.4 \cdot 10^{-3}\,\text{W/Km} \cdot \left[\frac{(173 \cdot 10^{-6})^2\,\text{kg}^2/\text{s}^2\text{m}^2}{(1.0839 \cdot 10^3)^2\,\text{kg}^2/\text{m}^6 \cdot 9.81\,\text{m/s}^2}\right]^{-1/3}$$

$$\cdot 0.959 \cdot \left(\frac{1.94\,\text{kg/s}}{220 \cdot 1.5\,\text{m} \cdot 173 \cdot 10^{-6}\,\text{kg/s\,m}}\right)^{-1/3}$$

$$\alpha_{\text{m, hor}} = 1624\,\text{W/m}^2\text{K} \ .$$

Around $\sqrt{220} \approx 15$ tubes lie one above each other. From (4.17) we get

$$\alpha_{mn} = \alpha_{mo} = \left[1 + 0.2 \cdot \frac{1.38\,\text{kJ/kg K} \cdot (50 - 34.2)\,\text{K}}{154.08\,\text{kJ/kg}} \cdot (15 - 1)\right] \cdot 15^{-1/4} \cdot 1624\,\text{W/m}^2\text{K}$$

$$\alpha_{mo} = 1152\,\text{W/m}^2\text{K} \ .$$

A correction of the estimate for α_{mo} is unnecessary.

Note: In the calculation of α_{mo} according to (4.15a) and (4.17) an error is made, because both equations presume a constant wall temperature. In this case this is variable, the cooling water temperature and with that the wall temperature increases in the direction of flow of the cooling water. However, because $\alpha_{m, hor}$ according to (4.15a) only changes with $\sim (\vartheta_s - \vartheta_0)^{-1/4}$, in this example the error hardly has any effect on the result.

4.1.5 Film condensation in a turbulent film

Nusselt's film condensation theory presumes a laminar film flow. As the amount of condensate increases downstream, the Reynolds number formed with the film thickness increases. The initially flat film becomes wavy and is eventually transformed from a laminar to a turbulent film; the heat transfer is significantly better than in the laminar film. The heat transfer in turbulent film condensation was first calculated approximately by Grigull [4.14], who applied the Prandtl analogy for pipe flow to the turbulent condensate film. In addition to the quantities for laminar film condensation the Prandtl number appears as a new parameter. The results can not be represented explicitly. In order to obtain a clear representation, we will now define the Reynolds number of the condensate film

$$Re := \frac{w_m \delta}{\nu_L} = \frac{w_m \delta \varrho_L b}{\nu_L \varrho_L b} = \frac{\dot{M}}{\eta_L b} \ . \tag{4.38}$$

In the framework of Nusselt's film condensation theory, using the mass flow rate for the condensate, eq. (4.9),

$$\dot{M} = w_m \varrho_L b \delta = \frac{\varrho_L (\varrho_L - \varrho_G) g b}{3 \eta_L^2} \delta^3$$

and taking $\varrho_L \gg \varrho_G$ into account, it can be rearranged into

$$Re = \frac{\dot{M}/b}{\eta_L} = \frac{\varrho_L^2 g}{3 \eta_L^2} \delta^3 \quad \text{or} \quad (3\,Re)^{1/3} = \left(\frac{g}{\nu_L^2}\right)^{1/3} \delta \ .$$

By eliminating the film thickness δ using $\alpha = \lambda_L/\delta$, the heat transfer equation for Nusselt's film condensation theory can also be written as

$$Nu = \frac{\alpha (\nu_L^2/g)^{1/3}}{\lambda_L} = (3\,Re)^{-1/3} \ , \tag{4.39}$$

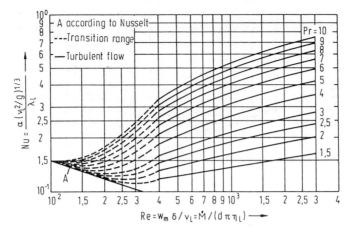

Fig. 4.12: Local Nusselt number as a function of the Reynolds number, according to [4.15], in condensation on a vertical tube or a vertical flat wall

where the Reynolds number is given by (4.38). In turbulent condensate films, as Grigull showed, we get

$$Nu = f(Re, Pr) \ .$$

The results of an analytical solution of the differential equations for a turbulent condensate film on a vertical tube are reproduced in Fig. 4.12, [4.15]. Line A represents Nusselt's film condensation theory according to (4.39).

The Reynolds number $Re = w_{\mathrm{m}} \delta / \nu_{\mathrm{L}}$ for the condensation on a vertical tube can, due to the mass balance $\dot{M} = w_{\mathrm{m}} d \pi \delta \varrho_{\mathrm{L}}$ and with $\eta_{\mathrm{L}} = \nu_{\mathrm{L}} \varrho_{\mathrm{L}}$, also be written as $Re = \dot{M}/(d \pi \eta_{\mathrm{L}})$. As the tube diameter d, on which the calculations from Fig. 4.12 are based, is much larger than the assumed thickness of the film, the curvature of the condensate film has no effect on the heat transfer. The results also hold for condensation on a vertical flat plate, with the Reynolds number defined in (4.38).

In the transition region between laminar and turbulent condensation, the condensate film is wavy. In Fig. 4.12 this transition region is represented by the dotted lines, according to results from Henstock and Hanratti [4.16].

The literature contains various comments concerning the size of the Reynolds number for this laminar-turbulent transition. As Fig. 4.12 shows, it is not possible to estabish a definite critical Reynolds number, although this has occasionally been attempted. Rather, a transition region with a wavy film, follows on directly from the laminar region. The Nusselt number in this region deviates from the values from Nusselt's film condensation theory. This transition region starts at very low Reynolds numbers Re_{trans}, if the Prandtl number is high enough, whilst at low Prandtl numbers, laminar flow is transformed into turbulent flow via a very short transition region. Therefore, we can say that the critical Reynolds number depends on the Prandtl number and can lie well below the critical Reynolds number of 400, at which the transition region roughly ends.

If a deviation of 1 % from the values for Nusselt's film condensation theory is

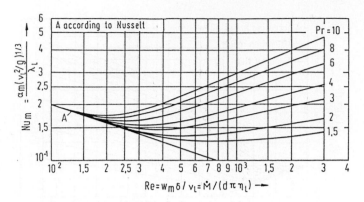

Fig. 4.13: Mean Nusselt number as a function of the Reynolds number, according to [4.15], in condensation on a vertical tube or a vertical flat wall

permitted, this leads to a Reynolds number of

$$Re_{\text{trans}} = 256 \, Pr^{-0.47} \ , \tag{4.40}$$

valid for $1 \leq Pr \leq 10$. In addition to this the film condensation theory can only be applied if the shear stress exerted by the vapour is small enough. In the turbulent region, at Reynolds numbers above $Re = 400$, the Nusselt numbers reach values many times greater than those predicted by the film condensation theory, in particular when the Prandtl numbers are large.

In the transition region, the Nusselt number initially decreases with Reynolds number and once it has gone through a minimum it increases again. At first the increased film thickness, which grows with the Reynolds number, causes a reduction in heat transfer, but this is overcome by the increasing influence of the turbulence which leads to improved heat transfer.

It should be noted that the local Nusselt numbers are plotted in Fig. 4.12. In a condenser, the Reynolds number formed with the film thickness changes from a value of zero, at the beginning of the condensation, to a final or end value. In between, within the transition region there may be states where the heat transfer is not particularly good, the mean heat transfer coefficient could be smaller than the local coefficient at the outlet cross section. Fig. 4.13 reproduces the mean heat transfer coefficients associated with Fig. 4.12, according to

$$\alpha_{\text{m}} \, (\vartheta_{\text{s}} - \vartheta_0) = \Delta h_{\text{v}} \, \dot{M}/A \ ,$$

where A is the surface area of the tube and $\dot{M}$ is the condensate mass flow rate at the bottom end of the tube.

As we can see from this, at small Prandtl numbers good heat transfer can be achieved by operating the condenser in the region of Nusselt's film condensation theory, that is at low Reynolds numbers. This can be attained by using short tubes. If, in contrast, the Prandtl number is large, then good heat transfer predominates in the turbulent region, and this can be achieved by using tubes of an adequate length.

Isashenko [4.17], has developed simple fomulae, originating from the solution of the momentum and energy equations for turbulent flow, for the turbulent film region at Reynolds numbers $Re \geq 400$. These formulae reproduce well known measured values given in the literature. According to this, the local Nusselt number is given by

$$Nu = \frac{\alpha \, (\nu_L^2/g)^{1/3}}{\lambda_L} = 0.0325 \, Re^{1/4} \, Pr^{1/2} \qquad (4.41)$$

with the Reynolds number $Re = (\dot{M}/b)/\eta_L = \Gamma/\eta_L$ and the Prandtl number Pr of the condensate. The equation is valid in the region $1 \leq Pr \leq 25$ and $400 \leq Re \leq 7 \cdot 10^5$. The Reynolds number of the condensate is obtained from

$$Re = \left[89 + 0.024 \, Pr^{1/2} \left(\frac{Pr}{Pr_0} \right)^{1/4} (Z - 2300) \right]^{4/3} \qquad (4.42)$$

with

$$Z = \frac{c_{pL} \, (\vartheta_s - \vartheta_0)}{\Delta h_v} \, \frac{1}{Pr} \, \frac{x}{(\nu_L^2/g)^{1/3}} \, . \qquad (4.42a)$$

In these equations all the fluid properties are evaluated at saturation temperature, except the Prandtl number Pr_0, which is formed with the wall temperature. From investigations by Labunzov [4.18] the measured heat transfer coefficients deviate at the most by $\pm 12\%$ from those calculated.

In the calculation of heat transfer in the transition region between laminar and turbulent film condensation, empirical interpolation formulae are well established. One of these types of formulae is

$$\alpha = \sqrt[4]{(f \, \alpha_{lam})^4 + \alpha_{turb}^4} \, . \qquad (4.43)$$

The factor f takes into account here the waviness of the laminar condensate film, $f \approx 1.15$; α_{lam} is the heat transfer coefficient for laminar film condensation from Nusselt's film condensation theory and α_{turb} is that for a turbulent condensate film, which is found, for example from (4.41) together with (4.42).

4.1.6 Condensation of flowing vapours

Nusselt extended his film condensation theory to take into account the influence of vapour flowing along the condensate film on the velocity of the condensate. The boundary conditions for (4.6) are no longer $\partial w/\partial y = 0$ for $y = \delta$, instead the velocity profile ends with a finite gradient at the free surface of the film corresponding to the shear stress exerted by the flowing vapour. In (4.6) for the velocity profile

$$w = -\frac{(\varrho_L - \varrho_G) \, g}{2 \, \eta_L} \, y^2 + c_1 \, y + c_0$$

the coefficients c_1 and c_0 now have to be determined, so that the boundary conditions

$$w(y = 0) = 0 \quad \text{and} \quad \eta_\mathrm{L}(\partial w/\partial y)_{y=\delta} = \pm \tau_\delta \tag{4.44}$$

are satisfied, where the plus sign holds for downstream flowing vapour, the negative sign is used when the vapour is flowing upstream. With that we obtain the following for the velocity

$$w = \frac{(\varrho_\mathrm{L} - \varrho_\mathrm{G})\, g}{\eta_\mathrm{L}} \delta^2 \left(\frac{y}{\delta} - \frac{y^2}{2\,\delta^2} \right) \pm \frac{\tau_\delta\, y}{\eta_\mathrm{L}} \; . \tag{4.45}$$

In the calculation of the shear stress which develops at the phase interface, it is assumed that the pressure and frictional forces in the vapour space are equal. The pressure drop along the flow direction $\mathrm{d}x$ is $\mathrm{d}p$. For pipe flow we now have

$$\tau_\delta\, d\,\pi = \frac{d^2\,\pi}{4} \frac{\mathrm{d}p}{\mathrm{d}x} \; . \tag{4.46}$$

On the other hand, for the pressure drop it holds that

$$\frac{\mathrm{d}p}{\mathrm{d}x} = \zeta\, \frac{\varrho_\mathrm{G}\, w_\mathrm{G}^2}{d} \; . \tag{4.47}$$

With that we have

$$\tau_\delta = \zeta\, \frac{\varrho_\mathrm{G}\, w_\mathrm{G}^2}{4} \; . \tag{4.48}$$

The mean velocity is found to be

$$w_\mathrm{m} = \frac{1}{\delta} \int_0^\delta w\, \mathrm{d}y = \frac{(\varrho_\mathrm{L} - \varrho_\mathrm{G})\, g}{3\,\eta_\mathrm{L}} \delta^2 \pm \frac{\tau_\delta\, \delta}{2\,\eta_\mathrm{L}} \tag{4.49}$$

and from that the mass flow rate of condensate, which flows through a plane perpendicular to the wall is

$$\dot M = w_\mathrm{m}\, \varrho_\mathrm{L}\, b\,\delta = \frac{\varrho_\mathrm{L}\,(\varrho_\mathrm{L} - \varrho_\mathrm{G})\, g\, b}{3\,\eta_\mathrm{L}} \delta^3 \pm \frac{\varrho_\mathrm{L}\, \tau_\delta\, \delta^2\, b}{2\,\eta_\mathrm{L}} \; . \tag{4.50}$$

The film thickness, as already shown in section 4.1.2, is found using the energy balance

$$\lambda_\mathrm{L} \frac{\vartheta_\mathrm{s} - \vartheta_0}{\delta} b\, \mathrm{d}x = \Delta h_\mathrm{v}\, \mathrm{d}\dot M$$

to be

$$\delta^4 \pm \frac{4}{3} \frac{\tau_\delta\, \delta^3}{g(\varrho_\mathrm{L} - \varrho_\mathrm{G})} = \frac{4\,\lambda\,\eta_\mathrm{L}\,(\vartheta_\mathrm{s} - \vartheta_0)}{\varrho_\mathrm{L}\,(\varrho_\mathrm{L} - \varrho_\mathrm{G})\, g\, \Delta h_\mathrm{v}}\, x \; , \tag{4.51}$$

and from the calculation of the film thickness δ, the heat flux $\dot q = \lambda_\mathrm{L}\,(\vartheta_\mathrm{s} - \vartheta_0)/\delta$ transferred is yielded. The shear stress τ_δ is given by (4.48).

If subcooling of the condensate or superheating of the vapour is considered, then in place of the enthalpy of vaporisation Δh_v the enthalpy difference Δh_vs^*

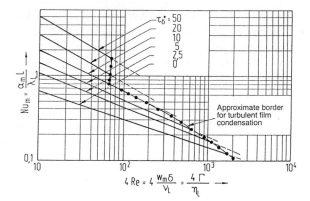

Fig. 4.14: Influence of the vapour shear stress on laminar film condensation, according to [4.19]

according to (4.29) is used. If the condensation surface is inclined at an angle γ to the vertical, the acceleration due to gravity g is replaced by its component $g \cos \gamma$ parallel to the wall.

Rohsenow et al. [4.19] rearranged (4.51) by introducing dimensionless quantities and then determined the heat transfer coefficient $\alpha = \lambda_L / \delta$ and also the mean heat transfer coefficient α_m from the calculated film thickness. Fig. 4.14 shows as a result of this the mean Nusselt number

$$Nu_m = \frac{\alpha_m L}{\lambda}$$

with the characteristic length

$$L := \left[\frac{\varrho_L \, \nu_L^2}{(\varrho_L - \varrho_G) \, g} \right]^{1/3}$$

plotted against the Reynolds number

$$Re = \frac{w_m \, \delta}{\nu_L} = \frac{\dot{M}}{b \, \eta_L} \ ,$$

where b is the width of the flowing condensate film, which for condensation on a vertical tube is $b = \pi \, d$. As a parameter, the dimensionless shear stress

$$\tau_\delta^* := \frac{\tau_\delta}{L \, (\varrho_L - \varrho_G) \, g}$$

is registered. The results are valid for vapour flowing upstream. The dotted lines indicate the approximate boundary to turbulent film condensation.

The laws of conversation of mass, momentum and energy apply to turbulent film condensation of flowing vapours as well as to laminar film condensation. However, additional information about the mechanisms of the turbulent exchange of mass, momentum and energy is required. The results of this type of calculation provide us, once again, with a temperature profile and from that the heat transfer coefficient. As an example of this Fig. 4.15 shows the results from Dukler [4.19], for the local Nusselt number as a function of the Reynolds number, at a Prandtl number $Pr = 5$ and a vapour flow in the downstream direction. The dimensionless quantities are defined in exactly the same manner as those in Fig. 4.14 for laminar flow, thereby enabling the comparison of heat transfer in laminar and turbulent flow. It is also not possible to obtain an analytical expression for turbulent flow, only numerical results are available for the calculated heat transfer coefficient. These can be represented by empirical equations. As a good example for practical usage the particularly simple equation from Shah [4.21] is presented. It contains the relationship (3.259) for convective heat transfer in turbulent single phase flow, in which the exponent for the Prandtl number has merely been increased from $1/3$ to 0.4, and an additional term taking the phase change and the effect of vapour flow into account. For the local heat transfer we have

$$Nu = 0.023\, Re^{0.8}\, Pr^{0.4} \left\{ (1 - x^*)^{0.8} + \frac{3.8\,(1 - x^*)^{0.04}\, x^{*0.76}}{p^{+0.38}} \right\} \qquad (4.52)$$

with

$$Nu = \frac{\alpha\, d}{\lambda_L}, \quad Re = \frac{w_m\, d}{\nu_L}, \quad w_m = \dot{M} / \left(\varrho_L \frac{\pi\, d^2}{4} \right), \quad Pr = \frac{\nu_L}{a_L}, \quad p^+ = \frac{p}{p_{cr}}.$$

The quantity $x^* = \dot{M}_G / \dot{M}$ is the mass quality or simply quality in the flow at a cross section of the tube. This equation reproduces the measurements for water, refrigerants R11, R12, R113 and methanol, ethanol, benzene, toluene and trichloroethylene in condensation in vertical, horizontal and inclined tubes with 7 to 40 mm internal diameter, at normalised pressures $0.002 \leq p^+ \leq 0.44$, saturation temperatures $21\,°C \leq \vartheta_s \leq 310\,°C$, vapour velocities $3\,m/s \leq w_G \leq 300\,m/s$, mass fluxes $10.8\,kg/m^2\,s \leq \dot{m} \leq 210.6\,kg/m^2\,s$, heat fluxes $158\,W/m^2 \leq \dot{q} \leq 1.893 \cdot 10^6\,W/m^2$, Reynolds numbers $100 \leq Re \leq 63\,000$, and Prandtl numbers $1 \leq Pr \leq 13$. The mean deviation from the experimental values has been shown to be $\pm 15.4\,\%$.

For *practical calculations* it is recommended that the condenser tube be subdivided into sections in which the change Δx^* in the quality of the flow is equal. If these sections are not chosen to be too large, a linear drop in the quality can be assumed in each section and then the local heat transfer coefficient in the middle of the section Δx^* can be calculated according to (4.52). Shah [4.21] states that a subdivision in sections $\Delta x^* < 0.4$ is sufficient. The heat transfer coefficient in the middle of the section is approximately equal to the mean heat transfer coefficient $\alpha = \alpha_m$ of the section. The area A of the section in question is obtained from

$$\dot{Q} = k_m\, A\, \Delta \vartheta_m \qquad (4.53)$$

with

$$\frac{1}{k_m\, A} = \frac{1}{\alpha_m\, A_C} + \frac{\delta}{\lambda\, A_m} + \frac{1}{\alpha_{mo}\, A_o},$$

or, if the heat transfer coefficient is based on the condensate surface $A_C = A$, with

$$\frac{1}{k_m} = \frac{1}{\alpha_m} + \frac{\delta\, d}{\lambda\, d_m} + \frac{1}{\alpha_{mo}} \frac{d}{d_o} = \frac{1}{\alpha_m} + \frac{1}{k'_m} \qquad (4.54)$$

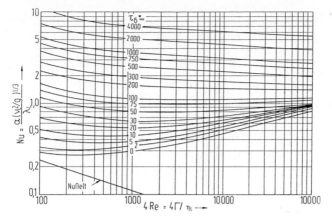

Fig. 4.15: Film condensation of downward flowing saturated vapour, according to [4.20], Prandtl number $Pr = 5.0$

and

$$\Delta\vartheta_{\mathrm{m}} = \frac{\vartheta_{\mathrm{e}} - \vartheta_{\mathrm{i}}}{\ln[(\vartheta_{\mathrm{G}} - \vartheta_{\mathrm{i}})/(\vartheta_{\mathrm{G}} - \vartheta_{\mathrm{e}})]} \quad , \tag{4.55}$$

where ϑ_{i} is the inlet temperature of the cooling medium in the relevant section, ϑ_{e} is the outlet (exit) temperature and ϑ_{G} is the mean vapour temperature.

Example 4.3: In a tube bundle condenser made up of 200 vertical tubes each with 25 mm inner diameter, $M_{\mathrm{G}} = 8\,\mathrm{kg/s}$ of saturated toluene vapour at a pressure of 0.1 MPa is to be condensed. The tubes will be cooled from outside by $\dot{M}_{\mathrm{W}} = 60\,\mathrm{kg/s}$ of water flowing counter-current, fed into the condenser at a temperature of 45 °C. The overall heat transfer coefficient of the cooling water up to the tube wall is $k'_{\mathrm{m}} = 2500\ \mathrm{W/m^2K}$. Calculate the required tube length.

The following properties are given: liquid toluene: density $\varrho_{\mathrm{L}} = 782\,\mathrm{kg/m^3}$, specific heat capacity $c_{p\mathrm{L}} = 2.015\,\mathrm{kJ/kg\ K}$, thermal conductivity $\lambda_{\mathrm{L}} = 0.126\,\mathrm{W/K\,m}$, dynamic viscosity $\eta_{\mathrm{L}} = 2.52 \cdot 10^{-4}\,\mathrm{kg/sm}$, Prandtl number $Pr = 4.03$. Further, for toluene $\varrho_{\mathrm{G}} \ll \varrho_{\mathrm{L}}$, the enthalpy of vaporisation $\Delta h_{\mathrm{v}} = 356\,\mathrm{kJ/kg}$, saturation temperature at 0.1 MPa: $\vartheta_{\mathrm{s}} = 383.75\,\mathrm{K} = 110.6\,°\mathrm{C}$, critical pressure $p_{\mathrm{cr}} = 4.11\,\mathrm{MPa}$. The specific heat capacity of the cooling water is $c_{p\mathrm{W}} = 4.10\,\mathrm{kJ/kg\ K}$.

The condenser power is $\dot{Q} = \dot{M}_{\mathrm{G}}\,\Delta h_{\mathrm{v}} = 8\,\mathrm{kg/s} \cdot 356\,\mathrm{kJ/kg} = 2848\,\mathrm{kW}$, the power for each tube $\dot{Q}/200 = 14.24\,\mathrm{kW}$. The outlet temperature of the cooling water follows out of $\dot{Q} = \dot{M}_{\mathrm{W}}\,c_{p\mathrm{W}}\,(\vartheta_{\mathrm{We}} - \vartheta_{\mathrm{Wi}})$ to be

$$\vartheta_{\mathrm{We}} = \vartheta_{\mathrm{Wi}} + \frac{\dot{Q}}{\dot{M}_{\mathrm{W}}\,c_{p\mathrm{W}}} = 45.0\ °\mathrm{C} + \frac{2\,848\,\mathrm{kW}}{60\,\mathrm{kg/s} \cdot 4.10\,\mathrm{kJ/kgK}} = 56.58\ °\mathrm{C} \quad .$$

In the calculation of the necessary tube length we will subdivide into four sections with $\Delta x^* = 0.25$. The condensate side heat transfer coefficient comes from (4.52). In this with a condensate amount $\dot{M} = 8(\mathrm{kg/s})/200 = 0.04\,\mathrm{kg/s}$ per tube, the Reynolds number is

$$Re = \frac{\dot{M}}{\eta_{\mathrm{L}}\,d\,\pi/4} = \frac{0.04\,\mathrm{kg/s}}{2.52 \cdot 10^{-4}\,\mathrm{kg/s\,m} \cdot (25 \cdot 10^{-3}) \cdot \pi/4} = 8\,084 \quad .$$

According to (4.52) we get

$$
\begin{aligned}
\alpha &= \frac{\lambda_L}{d}\, 0.023\, Re^{0.8}\, Pr^{0.4}\left\{(1-x^*)^{0.8} + \frac{3.8\,(1-x^*)^{0.04}\, x^{*0.76}}{p^{+0.38}}\right\} \\
&= \frac{0.126\,\text{W/K m}}{25\cdot 10^{-3}\,\text{m}}\cdot 0.023\cdot 8\,084^{0.8}\cdot 4.03^{0.4}\cdot\left\{(1-x^*)^{0.8} + \frac{3.8\cdot(1-x^*)^{0.04}\cdot x^{*0.76}}{(2.433\cdot 10^{-2})^{0.38}}\right\}\;, \\
\alpha &= 270.63\,(\text{W/m}^2\text{K})\left\{(1-x^*)^{0.8} + 15.579\,(1-x^*)^{0.04}\, x^{*0.76}\right\}\;.
\end{aligned}
$$

In the first section $1.0 \ge x^* \ge 0.75$ the mean quality is $(1.0+0.75)/2 = 0.875$. This yields from the previous equation a mean heat transfer coefficient of

$$
\begin{aligned}
\alpha = \alpha_m &= 270.63\,\text{W/m}^2\text{K}\left\{(1-0.875)^{0.8} + 15.597\cdot(1-0.875)^{0.04}\cdot 0.875^{0.76}\right\} \\
&= 3\,557\,\text{W/m}^2\text{K}\;.
\end{aligned}
$$

From (4.54), the mean overall heat transfer coefficient of this section is

$$
\frac{1}{k_m} = \frac{1}{\alpha_m} + \frac{1}{k'_m} = \frac{1}{3\,557\,\text{W/m}^2\text{K}} + \frac{1}{2\,500\,\text{W/m}^2\text{K}}
$$

$$
k_m = k_{m1} = 1\,468\,\text{W/m}^2\text{K}\;.
$$

The heat flow transferred in the first section is $14.24\,\text{kW}/4 = 3.56\,\text{kW} = \dot{Q}_1$. The cooling water enters the first section at a temperature of

$$
\vartheta_{Wi1} = \vartheta_{We} - \frac{\dot{Q}_1}{\dot{M}_{W1}\, c_{pW}} = 56.58\,°\text{C} - \frac{3.56\,\text{kW}}{0.3\,\text{kg/s}\cdot 4.10\,\text{kJ/kg K}} = 53.69\,°\text{C}\;.
$$

The logarithmic mean temperature according to (4.55) is

$$
\Delta\vartheta_{m1} = \frac{56.58 - 53.69}{\ln[(110.6 - 53.69)/(110.6 - 56.58)]}\,\text{K} = 55.4\,\text{K}\;.
$$

With that, the area of the first section follows from (4.53)

$$
A_1 = \frac{\dot{Q}_1}{k_{m1}\Delta\vartheta_{m1}} = \frac{3.56\cdot 10^3\,\text{W}}{1\,468\,\text{W/m}^2\text{K}\cdot 55.4\,\text{K}} = 4.377\cdot 10^{-2}\,\text{m}^2\;.
$$

x^*	x_m^*	α_m	k_m	ϑ_W	$\Delta\vartheta_m$	A	L
		W/m^{2}K	W/m^{2}K	°C	°C	m^2	m
1.0				56.58			
	0.875	3557	1468		55.4	$4.377\cdot 10^{-2}$	0.56
0.75				53.69			
	0.625	2960	1355		58.3	$4.507\cdot 10^{-2}$	0.57
0.5				50.80			
	0.375	2149	1156		61.0	$5.05\;\cdot 10^{-2}$	0.64
0.25				47.90			
	0.125	1106	768		63.9	$7.25\;\cdot 10^{-2}$	0.92
0.0				45			

The associated tube length is

$$
L_1 = \frac{A_1}{d\,\pi} = \frac{4.377\cdot 10^{-2}\,\text{m}^2}{25\cdot 10^{-3}\,\text{m}\cdot\pi} = 0.56\,\text{m}\;.
$$

A corresponding calculation for all the sections yields the values listed in the table. The total length of the tubes is $\sum L = L_{\text{tot}} = 2.69\,\text{m} \approx 2.7\,\text{m}$.

4.1.7   Dropwise condensation

As we have already explained in section 4.1.1., if the condensate does not completely wet the wall, individual liquid droplets form instead of a continuous condensate film. Heat transfer coefficients in dropwise condensation are significantly larger than in film condensation. In the condensation of steam, the heat transfer coefficients measured have been a factor of four to eight times larger. However, it has been shown that all investigated substances, in particular water, which condense on commonly used heating surfaces, will completely wet the surface. This is true as long as the material of the heated surface and the liquid have not been contaminated. This also corresponds to the experience that the formation of a water film is taken to be an indication that laboratory equipment is well cleaned.

The type of condensation is chiefly influenced by foreign substances absorbed on the solid surface. They lead to locally finite values of the contact angle and so cause incomplete wetting. Foreign substances that come into question are those added to water as antiwetting agents (injected substances, promoters). Occasionally such foreign substance may be unintentionally present in the vapour, for example, the lubricating oil from a boiler feed pump, even a small amount dissolved in the condensate is sufficient to generate dropwise concentration. The addition of waxy substances has also been suggested [4.22]. The maintenance of a stable dropwise condensation requires a constant or periodic feed of the injected substance, as over the course of time it is removed from the surface by washing or dissolution in the condensate. The formation of droplets is also influenced by the roughness of the wall and its height, as well as the heat flux and the temperature difference.

Thin gold layers applied by electrolysis and noble metal plating of gold, rhodium, palladium and platinum have all proved to be effective as promoters. However, as experiments have shown [4.23], these have to be of a certain minimum thickness, which for gold layers is around $0{,}2\,\mu m$, in order to maintain a lasting dropwise condensation. Although this layer thickness is only half the size of the wave length of visible light the necessary amount of gold is in no way small. It amounts to $3{,}8\,g$ per m^2 of heating surface. This means, at a gold price of around $15\,000$ US-\$ per kilo, this would cost about 57 US-\$ per m^2 of heating surface for the plating alone. This produces five to seven times better heat transfer for the condensation of steam, [4.23]. Nevertheless, the high expenditure means that the gold plating would only be made use of in very special cases, otherwise film condensation will be preferred.

Even when the type of condensation is not precisely known, film condensation is assumed for the calculation of condensers, so that the condenser area will be sufficient.

Experiments on dropwise condensation are difficult as they entail the measurement of temperature differences of 1 K or less for the determination of the heat transfer coefficient. At these

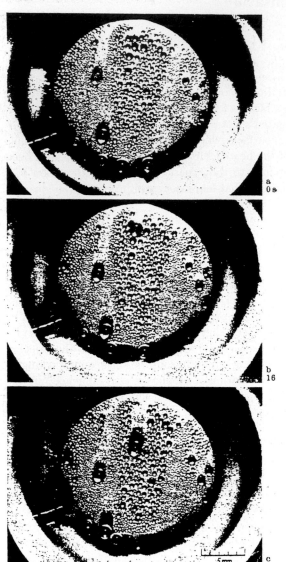

Fig. 4.16: Dropwise condensation on a vertical condensation surface, according to Krischer and Grigull [4.24]. **a** Start of condensation; **b** Condensation form after 16 s; **c** Condensation form after 32 s. The traces of the droplets rolling downwards are clearly recogniseable. Above, in picture **b** a drop is just leaving. $\dot{q} = 0.12\,\mathrm{W/cm^2}$, $\Delta\vartheta = 0.05\,\mathrm{K}$. Diameter of the condensation surface: 18 mm

small temperature differences the wall temperature fluctuates with time and also locally. The cost of the measuring techniques for the achievement of accurate results is, therefore, considerable.

The photographs in Fig. 4.16 are a good illustration of dropwise condensation on a vertical surface. The small droplets grow due to subsequent condensation from the vapour space and coalescing. As soon as a certain drop size has been reached, the drop rolls downwards and in the process of this takes the drops lying along its path with it. Behind the rolling drop new drops appear immediately. They form predominantly on the water left behind and on scratches on the condensation area. According to experiments by Krischer and Grigull [4.24], the nucleation site density is mainly dependent on the subcooling of the heating area. Experiments with condensing steam on copper and brass surfaces, with subcooling of the surface by 0.1 K, yield a nucleation site density of approximately $3 \cdot 10^3$ nuclei per mm^2 of the heating area, whilst with subcooling of 0.4 K the nucleation site density lay at $15 \cdot 10^3$ nuclei per mm^2.

Fig. 4.17 shows the results for heat transfer in dropwise condensation that have been produced over the last 20 years. Most condensation areas were made of copper. Different liquids served as promoters; they were spread over the cooling surface. Depending on the promoter and the material for the condensation surface very different results were found. As we can see, individual experimental results for the heat flux vary by as much as a factor of 30 from each other. Curves 2, 5, and 6 from Wenzel [4.26], Le Fèvre and Rose [4.29] and Tanner et al. [4.30], have weakly increasing gradients and only deviate slightly from one another. In Fig. 4.18, the results from [4.30] and [4.32] together with those from Krischer and Grigull [4.24] are drawn separately.

Several *theories on dropwise condensation* have been developed for the calculation of heat transfer. One of the oldest theories, that from Eucken [4.33], starts with the concept that the first droplets arise from an adsorbed monomolecular condensate layer, this layer is favoured by nuclei, and that the new condensate flows to the droplets by means of surface diffusion which primarily occurs at the edge of the drop. This theory was taken on later by other authors [4.34], [4.35] and developed further. Other theories presuppose that a thin, unstable water film exists

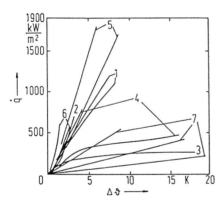

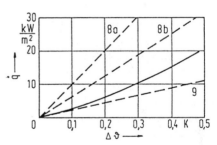

Fig. 4.17: Experimental results for dropwise condensation of water at a condensation pressure of about 1 bar according to [4.24]. 1 Hampson and Özisik 1952, graph for two different promoters [4.25]; 2 Wenzel 1957 [4.26]; 3 Welch and Westwater 1961 [4.27]; 4 Kast 1965, chrome-plated (upper line) and plain copper surfaces [4.28]; 5 Le Fèvre and Rose 1965, different promoters [4.29]; 6 Tanner et al. 1968, different promoters [4.30]; 7 Griffith and Lee 1967, gold-plated condensation surfaces: upper curve, copper, middle curve, zinc and lower curve, steel [4.31]

Fig. 4.18: Experimental results at around 0.03 bar condensation pressure, according to [4.24]. Dotted lines: interpolated values 8 from measurements by Tanner et al. [4.30] a) Promoter 'Montan wax'and b) Promoter 'Dioctadecyldisulfide'; 9 Measurements from Brown and Thomas [4.32]

between the drops, that, once a critical thickness of a few μm has been reached, collapses and vanishes into the droplets [4.36]. This concept has since been disproved by experiment [4.37], [4.38].

The idea that agrees best with experiments is that initially tiny drops form at nuclei sites, at depressions in the condensation surface or on the remnants of liquids. Their growth rate is determined by the resistance to thermal conduction in the drops and also partly by the thermal conduction resistances at the phase interface with the vapour. The growth rate is therefore only dependent on the specific drop radius and the driving temperature difference. This has also been confirmed by experiment [4.24]. The main reasons behind the lack of success in developing an explicit theory are that the nucleation site density is unknown, and it is difficult to predict the radius of the rolling drops, as this depends on the purity and smoothness of the condensation surface as well as the interfacial tension.

4.1.8 Condensation of vapour mixtures

In the process industry, vapour mixtures are frequently liquified with all the components being present in the liquid phase. Alternatively, the vapour mixture may contain inert gases that do not condense. The influence of this type of mixture on the heat transfer in condensation has already been dealt with in section 4.1.4. Therefore, the variation in the heat transfer for a condensate that contains all the components, in larger or smaller amounts, still has to be discussed.

Depending on the purpose for the application the vapour can either be completely or only partially condensed so that the vapour leaving the condenser generally has a different composition from that entering it. This type of apparatus is known as a *partial condenser* or in rectification columns it is called a *dephlegmator*. Its purpose is to separate the higher boiling point components from a vapour by condensation. These apparatus operate with smaller temperature differences between the vapour and the cooling medium than a condenser, in which the vapour should be totally condensed.

As the components with the higher boiling points normally condense first out of a vapour mixture, and the vapour left is lacking in these components, a concentration profile which varies along the flow direction forms. This decisively affects the heat transfer.

This suggests, that for the calculation of the heat and mass transfer, it would be sensible to subdivide the flow path into individual sections, and then solve the mass and energy balances for each section, taking the laws of heat and mass transfer into consideration. Calculations of this type for binary mixtures are only possible with the assistance of a computer. In the discussions presented here we want to limit ourselves to an explanation of the fundamental physical processes along with an illustration of the decisive balance equations. As the process of condensation in multicomponent mixtures with more than two components is similar to that in binary mixtures, we will limit ourselves to the consideration of binary mixtures.

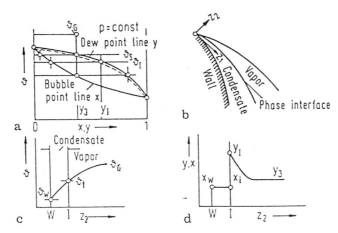

Fig. 4.19: Temperature-concentration diagram for a binary mixture as well as the temperature and concentration profiles in the vapour and the condensate. Indices: 0 cold wall, I interface, G core flow of vapour (G Gas). **a** boiling and dew point lines; **b** condensate and vapour boundary layer; **c** temperature profile; **d** concentration profile

When a binary mixture, whose boiling and dew point lines are shown in Fig. 4.19, condenses on a cooled wall of temperature ϑ_0, a condensate forms, Fig. 4.19b, which is bounded by the vapour. At the phase interface, a temperature ϑ_I develops, which lies between the temperature ϑ_G of the vapour far away from the wall and the wall temperature ϑ_0. If the vapour is saturated, its temperature is $\vartheta_G = \vartheta_s$ corresponding to Fig. 4.19a. The temperature profile in the vapour and condensate are illustrated in Fig. 4.19c.

In general, at the phase interface the components with the higher boiling points preferentially change from vapour to condensate. As a result of this the vapour mixture at the phase interface contains more of the more volatile component than at a large distance away from it. A concentration profile like that presented in Fig. 4.19d develops. The concentration of the more volatile component increases towards the phase interface. At steady-state, the molecules of the more volatile component that have not condensed at the condensate surface will be transported back into the vapour by diffusion. In the condensate the momentum, heat and mass transfer by convection can be neglected compared to that by conduction and diffusion, due to the low velocity of the condensate in comparison to that of the vapour. This applies as long as the Prandtl or Schmidt number is not too small. In the condensate film Nusselt's assumptions are once more valid, according to which the flow is only determined by the frictional and field (gravity) forces, whilst the temperature profile is mainly determined by heat conduction. The concentration over a cross section of the condensate film is constant because the wall does not let any of the material through and convective mass transport in the liquid film is negligible. This can easily be seen, as the diffusion equation for

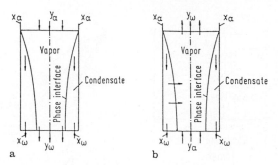

Fig. 4.20: Condensation in **a** co- and **b** countercurrent. In case **a** the falling liquid contains more of the more volatile components than in **b**. In **b** molecules of the more volatile components cross into the vapour, because the less volatile components rising up the column condense and drive out the more volatile molecules

these preconditions, known as Fick's second law, is

$$\frac{\partial}{\partial y}\left(D\,\frac{\partial c}{\partial y}\right) = 0 \;,$$

where y is the coordinate perpendicular to the wall, D is the diffusion coefficient and $c = N/V$ is the concentration of one of the components in a binary mixture. Integrating this equation, under the boundary conditions

$$\left(\frac{\partial c}{\partial y}\right)_{y=0} = 0 \quad \text{and} \quad c(y = \delta) = c_{\mathrm{I}} \;,$$

it immediately follows that $c = c_0 = c_{\mathrm{I}} = $ const. Therefore, we also have the case illustrated in Fig. 4.19d, $\tilde{x} = \tilde{x}_0 = \tilde{x}_{\mathrm{I}} = $ const.

The concentration profile is decisively influenced by the flow and the type of flow configuration. In the usual co-current flow of vapour and condensate, presented in Fig. 4.20a, the components with the higher boiling points go preferentially from vapour to condensate. The fraction of the components with lower boiling points increases downstream in the vapour from the initial composition $\tilde{y}_\alpha$ to the final composition $\tilde{y}_w$. The amount of condensate at the inlet is large, the result of which is that a great deal of the high boiling point components will be drawn out of the vapour.

The behaviour in countercurrent flow, Fig. 4.20b, like the flow which exists in falling film columns or in the reflux condensers in rectification columns, is different. Here, the vapour, of initial composition $\tilde{y}_\alpha$ meets a thick condensate film. The amount of vapour that condenses is smaller, and less of the high boiling point components are drawn out of the vapour than in the co-current case. The rising vapour comes into contact with an increasingly thinner condensate film of the falling liquid. The amount of condensate increases and the higher boiling point components of the vapour condense preferentially. The condensation enthalpy released causes evaporation of the low boiling point components. This is indicated

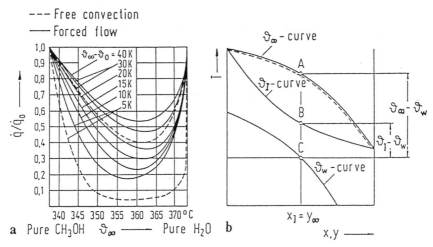

Fig. 4.21: Reduction of the heat flow rate in the condensation of methanol/water. **a** relative heat flux $\dot{q}/\dot{q}_0$ as a function of the temperature ϑ_∞; **b** boiling diagram

by the horizontal arrows in Fig. 4.20b. Mass exchange takes place between the liquid and vapour by "rectification", through which the vapour leaving contains more, and correspondingly the condensate leaving at the bottom fewer, of the more volatile components, than in the case of co-current flow according to Fig. 4.20b. So, if a condensate rich in the less volatile components is desired, a countercurrent flow configuration is preferable. This effect was first described by Claude [4.39] more than 70 years ago. The vapour flowing downstream contains more of the more volatile components in countercurrent flow than in co-current flow. However, in co-current flow the condensate is held up as a result of the shear stress exerted by the vapour, and with that the heat transfer worsens.

It is obvious that the flow configuration is decisive for mass transfer and therefore also for heat transfer. In general it holds that due to the concentration field at the phase interface, a resistance to mass transfer develops that hinders the mass flow towards the phase interface. As a result of this, the heat transfer in condensing vapour mixtures is lower than that in pure vapours, for the same temperature drop. This reduction can assume considerable values. As an example of this, Fig. 4.21a shows the relative heat flux $\dot{q}/\dot{q}_0$ plotted against the temperature ϑ_∞, which in this case shall be equal to the saturation temperature along the dew point line. As we can see, the heat flux $\dot{q}$ transferred is significantly smaller than the heat flux $\dot{q}_0$, that would be transferred if no resistance to mass transfer in the vapour had to be overcome, so that the condensate surface would be at the saturation temperature ϑ_∞. The continuous lines are valid for condensation of a vapour mixture of methanol (CH_3OH) and water (H_2O) with negligible gravitational effects, i.e. on a horizontal plate or when the vapour velocity is large enough, and so the *Froude number* $Fr = w_\infty^2/g\,x$ is very large. The dashed lines are valid for condensation at small Froude numbers, for example on a vertical wall with free convection of the vapour.

Similarly to Nusselt's film condensation theory, in the condensation of vapour mixtures, the heat flux transferred increases with the driving temperature difference $\vartheta_\infty - \vartheta_0$. According to Nusselt's film condensation theory, the heat transfer coefficient decreases with the driving temperature difference according to $\alpha \sim (\vartheta_\infty - \vartheta_0)^{-1/4}$, (4.12). The heat flux increases in accordance with $\dot{q} \sim (\vartheta_\infty - \vartheta_0)^{3/4}$. Fig. 4.21a shows clearly that a minimum for the transferred heat flux exists at a certain temperature ϑ_∞. This is because the temperature difference $\vartheta_I - \vartheta_0$ between the condensate surface and the wall, which is decisive for heat transfer, also assumes a minimum; this can be explained by the boiling diagram, Fig. 4.21. We will presume a sufficiently large temperature difference $\vartheta_\infty - \vartheta_0$, such that the vapour in its initial state A condenses, and a condensate accumulates. This is indicated by point B in Fig. 4.21b. The temperature ϑ_I at the phase interface is then equal to the boiling point of the liquid mixture and the composition of the accumulated condensate is identical to that of the vapour. This is known as *local total condensation*. The wall temperature, which is assumed to be constant, is characterised by point C. The line BC corresponds to the temperature difference $\vartheta_I - \vartheta_0$ that is decisive for the heat flux $\dot{q}$. If Nusselt's film condensation theory was also valid for vapour mixtures, then $\dot{q} \sim (\vartheta_I - \vartheta_0)^{3/4}$. If $\vartheta_\infty - \vartheta_0$ is kept constant and the temperature ϑ_∞, is increased, then by following the curve in Fig. 4.21a, $\vartheta_\infty - \vartheta_0 = \text{const}$ in the direction of increasing temperature ϑ_∞, then the distance BC in Fig. 4.21b can once again be measured, if a line parallel to the dewpoint line is drawn through point C. The distance AC $\hat{=} \vartheta_\infty - \vartheta_0$ remains, by presumption, unchanged, whilst BC will first be smaller and then larger. The relative heat flux $\dot{q}/\dot{q}_0$ changes in the same ratio. If, on the other hand, the temperature of the vapour ϑ_∞ is kept constant and the temperature difference $\vartheta_\infty - \vartheta_0$ is made smaller by raising the wall temperature, the ϑ_0-line running through point C in Fig. 4.21b has to be moved upwards. This means that the values of the heat flux $\dot{q}$ and the relative heat flux $\dot{q}/\dot{q}_0$ will be smaller.

4.1.8.1 The temperature at the phase interface

As the previous discussions have shown, the calculation of heat transfer in film condensation of vapour mixtures requires that the temperature ϑ_I at the phase interface is known. Thermodynamic equilibrium exists between liquid and vapour at the phase interface of a condensate film. The temperature ϑ_I is easy to calculate if the accumulated condensate has the same composition as the vapour at every position. In this case it agrees with respective temperature on the boiling line, and for example is determined by point B in Fig. 4.21b. In order to reach this temperature, the wall temperature $\vartheta_0 = \vartheta_C$ according to Fig. 4.21b, has to lie sufficiently below the boiling point ϑ_B. As a general rule, it should be $\vartheta_B - \vartheta_C > 2(\vartheta_A - \vartheta_B)$. The condensation rate has, according to this, to be adequately large, which is why we speak of local total condensation. In technical condensers this condition is satisfied, though there are cases, such as in partial condensation, where the wall temperature is deliberately chosen to be higher so that the more volatile component is barely or not present in the condensate. In order to show how the temperature at the phase interface is found, we will consider a binary mixture and establish the mass balance at the condensate surface. The mass flow $\dot{M}_G$ of the vapour perpendicular to the phase interface condenses there and $\dot{M}_L$ is removed as condensate. We have

$$\dot{M}_G = \dot{M}_L \ .$$

For the sake of simplicity we will assume the film theory to be valid, according to which the velocity and concentration profiles are only dependent on the coordinate y normal to the wall. With this assumption the continuity equation for the component 1 being considered is

$$\frac{\partial \dot{M}_1}{\partial y} = 0 \ ,$$

where y signifies the coordinate normal to the wall. This equation can, due to $\dot{M}_1 = \tilde{M}_1 \dot{N}_1$ with the molar mass $\tilde{M}_1$ and the molar flow rate $\dot{N}_1$ (SI units mol/s), also be written as

$$\frac{\partial \dot{N}_1}{\partial y} = 0 \ . \tag{4.56}$$

The mass flow in the gas is made up of the diffusional flow $j_1 A$ (SI units of j_1: mol/m²s) and the convective flow $\tilde{y}_1 \dot{N}$, where $\dot{N}$ is the total molar flow rate and $\tilde{y} = \tilde{y}_1$ is

$$\dot{N}_1 = j_1 A + \tilde{y} \dot{N} \ . \tag{4.57}$$

According to Fick's law, we obtain

$$j_1 = -D c \frac{\partial \tilde{y}}{\partial y} \tag{4.58}$$

with $c = N/V$. The coordinate y runs perpendicular to the condensate surface and points from it into the vapour space. With (4.58), equation (4.57), after division by $\dot{N}$, can also be written as

$$\frac{\dot{N}}{A} = \frac{j_1}{\dot{N}_1/\dot{N} - \tilde{y}} = -D c \frac{\partial \tilde{y}/\partial y}{\dot{N}_1/\dot{N} - \tilde{y}} \ . \tag{4.59}$$

As the molar flux of each of the two components is independent of the position coordinate y, the total molar flux $\dot{N}/A$ and likewise the quotient $\dot{N}_1/\dot{N}$ are also independent of y. Therefore (4.59) can be integrated easily. The integration extends from the condensate surface (index I) to the vapour space (index G). The thickness of the vapour boundary layer will be δ. We assume constant values for the pressure and temperature. Under the assumption that the gas phase exhibits ideal behaviour, the diffusion coefficient and the molar concentration $c = N/V = p/(R_m T)$ are likewise independent of the coordinate y. The integration yields

$$\frac{\dot{N}}{A}\delta = -D\,c\,\ln\frac{\dot{N}_1/\dot{N} - \tilde{y}_G}{\dot{N}_1/\dot{N} - \tilde{y}_I}$$

and after the introduction of the mass transfer coefficient $\beta_G = D/\delta$

$$\frac{\dot{N}}{A} = \dot{n} = \beta_G\,c\,\ln\frac{\dot{N}_1/\dot{N} - \tilde{y}_G}{\dot{N}_1/\dot{N} - \tilde{y}_I}\;. \tag{4.60}$$

This result can also be obtained from (1.192) of film theory, that was derived earlier, if we apply it to the vapour phase and solve for $\dot{n}$.

In (4.60) the molar flow rate at the phase interface I is given by $\dot{N}_1/A = (c_1 w_1)_G = (c_1 w_1)_L$ and $\dot{N}/A = (cw)_G = (cw)_L$. With that the ratio is $\dot{N}_1/\dot{N} = (c_1 w_1)_L/(cw)_L$. As no diffusion takes place in the liquid phase, $w_1 = w$ there and as a result of this it also holds that $\dot{N}_1/\dot{N} = c_1/c = n_1/n = \tilde{x}_1$. This value is to be taken at the phase interface, so that we can write

$$\frac{\dot{N}_1}{\dot{N}} = \tilde{x}_{1I} = \tilde{x}_I\;.$$

If a vapour condenses out of a mixture with an inert gas, then $\tilde{x}_I = 1$ and from (4.60) the known equation (4.31) is obtained.

As the quotient $(\tilde{x}_I - \tilde{y}_G)/(\tilde{x}_I - \tilde{y}_I)$ in (4.60) is smaller than one, the molar flux will be negative. The mass flow of the condensing vapour is in the opposite direction to the surface coordinate y chosen by us. As we are only interested in the absolute size of the mass flux, we can write

$$|\dot{n}| = \beta_G\,c\,\ln\frac{\tilde{y}_I - \tilde{x}_I}{\tilde{y}_G - \tilde{x}_I}\;. \tag{4.61}$$

(4.61) contains two limiting cases:

a) In local total condensation the composition of the liquid is identical to that of the vapour $\tilde{x}_I = \tilde{y}_G$. This gives $|\dot{n}| \to \infty$ for the molar flux of the vapour flowing towards the condensate surface. In order to achieve local total condensation, the wall temperature has to be far enough below the boiling point of all the components, so that they all condense and the condensate has the same composition as the vapour.

b) For vanishingly small molar flux of the existing condensate $|\dot{n}| \to 0$, according to (4.61) the vapour composition is $\tilde{y}_G = \tilde{y}_I$. Hardly any condensate forms, the result of which is that no concentration profile develops in the vapour space.

Both limiting cases are presented in Fig. 4.22. Actual condensation rates lie between the two extremes.

The temperature ϑ_I at the phase interface lies, as can be seen in Fig. 4.22, between the temperature on the dew point line (case b) and the temperature on

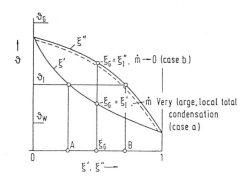

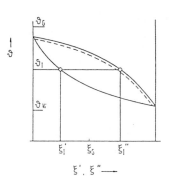

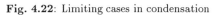

Fig. 4.22: Limiting cases in condensation

Fig. 4.23: Mole fractions and temperature at the condensate surface

the boiling line (case a). The associated vapour and liquid compositions can be read off the abcissa, points A and B in Fig. 4.22.

For the calculation of the temperature ϑ_I at the phase interface, the energy equation at the condensate surface is required as a further balance equation

$$\dot{q}_L = \dot{q}_G + |\dot{n}|\,\Delta\tilde{h}_v \ , \tag{4.62}$$

where $\dot{q}_L$ is the heat flux removed from the condensate by convection, $\dot{q}_G$ is the heat flux fed to the condensate from the vapour by convection and $\Delta\tilde{h}_v$ is the molar enthalpy of vaporisation of the mixture. We have

$$\dot{q}_L = \alpha_L\,(\vartheta_I - \vartheta_0) \ . \tag{4.63}$$

On the other hand, this heat flux is also transferred through the wall to the cooling medium, which is at a temperature of ϑ_c. The overall heat transfer coefficient between the wall and the cooling medium is k'. Then

$$\dot{q}_L = k'\,(\vartheta_0 - \vartheta_c) \ . \tag{4.64}$$

Using (4.63), ϑ_0 can be eliminated, giving

$$\dot{q}_L = \frac{\alpha_L}{(\alpha_L/k') + 1}\,(\vartheta_I - \vartheta_c) \ . \tag{4.65}$$

In the case of a very large overall heat transfer coefficient $k' \to \infty$ the wall temperature ϑ_0 is constant and equal to the temperature ϑ_c of the cooling medium.

The heat flux transferred from the vapour to the condensate surface is

$$\dot{q}_G = \alpha_G^{\bullet}\,(\vartheta_G - \vartheta_I) \ . \tag{4.66}$$

It contains the fraction for the heat transfer at the phase interface due to the temperature drop $\vartheta_G - \vartheta_I$ and the fraction for the energy transported through the vapour to the condensate surface.

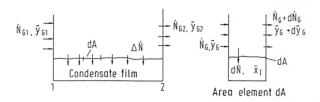

Fig. 4.24: Material balance in a section of a condenser

If the heat flux $\dot{q}_G$ from (4.66) together with the heat flux $\dot{q}_L$ from (4.65), are inserted into the energy balance (4.62), and, noting that the molar flux of the condensate is given by (4.61), then the following is obtained for the energy balance

$$\frac{\alpha_L}{(\alpha_L/k')+1}\left(\vartheta_I - \vartheta_K\right) = \alpha_G^\bullet \left(\vartheta_G - \vartheta_I\right) + \beta_G\, c \ln \frac{\tilde{y}_I - \tilde{x}_I}{\tilde{y}_G - \tilde{x}_I}\,\Delta \tilde{h}_v\ . \qquad (4.67)$$

The unknown temperature ϑ_I can be calculated from this equation. However, it still has to be considered that the mole fractions $\tilde{x}_I, \tilde{y}_I$ depend on the temperature ϑ_I, as Fig. 4.23 indicates for a binary mixture.

In addition to this the heat transfer coefficient α_L is dependent on the temperature ϑ_I. As free convection frequently occurs in the vapour, the term $\alpha_G^\bullet \left(\vartheta_G - \vartheta_I\right)$ can often not be neglected, but can be of the same magnitude as the other expressions in (4.67), in particular when the temperature at the phase interface ϑ_I is close to the wall temperature. This means that the driving temperature drop $\vartheta_I - \vartheta_0$ will be small in the condensate, whilst in contrast, that in the vapour $\vartheta_G - \vartheta_I$ will be large. The *practical calculation of the temperature ϑ_I at the condensate surface* is time consuming and, even in the case discussed here for a binary mixture, cannot be easily carried out without a computer.

4.1.8.2   The material and energy balance for the vapour

In the calculation of the temperature ϑ_I at the phase interface, the area A of the condenser is subdivided into sections ΔA. Each section is assigned unified (mean) values for the temperatures ϑ_I, ϑ_G and the composition $\tilde{y}_G$. With help from (4.67) the temperature ϑ_I can be obtained for given values of $\vartheta_G, \tilde{y}_G$. In order to calculate the values $\vartheta_G, \tilde{y}_G$ of any section from the values for the previous section, the material and energy balances have to be solved.

In setting up the material balance we will consider a section of area dA, as shown on the right hand side of Fig. 4.24. The material balance for the more volatile component is

$$\dot{N}_G\,\tilde{y}_G = \left(\dot{N}_G + d\dot{N}_G\right)\left(\tilde{y}_G + d\tilde{y}_G\right) + d\dot{N}_G\,\tilde{x}_I\ .$$

The mole fraction $\tilde{y}_G$ of the more volatile components here is, like in (4.67), an integral mean value over a cross section of the vapour space. From the last relationship with $\mathrm{d}\dot{N} = -\mathrm{d}\dot{N}_{\mathrm{G}}$ the so-called *Rayleigh equation* is obtained

$$\frac{\mathrm{d}\dot{N}_{\mathrm{G}}}{\dot{N}_{\mathrm{G}}} = \frac{-\mathrm{d}\tilde{y}_{\mathrm{G}}}{\tilde{y}_{\mathrm{G}} - \tilde{x}_{\mathrm{I}}} \ . \tag{4.68}$$

After integration between cross sections 1 and 2, Fig. 4.24 left, this delivers the expression

$$\ln\frac{\dot{N}_{\mathrm{G1}}}{\dot{N}_{\mathrm{G2}}} = \int_{\tilde{y}_{\mathrm{G1}}}^{\tilde{y}_{\mathrm{G2}}} \frac{\mathrm{d}\tilde{y}_{\mathrm{G}}}{\tilde{y}_{\mathrm{G}} - \tilde{x}_{\mathrm{I}}} \ ,$$

from which the accumulated condensate in the section of area ΔA is calculated

$$|\dot{n}| \, \Delta A = \dot{N}_{\mathrm{G1}} - \dot{N}_{\mathrm{G2}} = \dot{N}_{\mathrm{G1}} \left(1 - \exp\left(-\int_{\tilde{y}_{\mathrm{G1}}}^{\tilde{y}_{\mathrm{G2}}} \frac{\mathrm{d}\tilde{y}_{\mathrm{G}}}{\tilde{y}_{\mathrm{G}} - \tilde{x}_{\mathrm{I}}} \right) \right) \ . \tag{4.69}$$

Once the temperature ϑ_{I} at the phase interface has been found using (4.67), the molar condensate flow according to (4.61) is also known. Then, from (4.69), the mole fraction $\tilde{y}_{\mathrm{G2}}$ of the vapour in the next section ΔA can be calculated.

The energy balance for the area shown in Fig. 4.25 is

$$\dot{N}_{\mathrm{G}} \tilde{h}_{\mathrm{G}} = (\dot{N}_{\mathrm{G}} + \mathrm{d}\dot{N}_{\mathrm{G}})(\tilde{h}_{\mathrm{G}} + \mathrm{d}\tilde{h}_{\mathrm{G}}) + \mathrm{d}\dot{N} \, \tilde{h}_{\mathrm{GI}} + \dot{q}_{\mathrm{G}} \, \mathrm{d}A \ .$$

With $\mathrm{d}\dot{N} = -\mathrm{d}\dot{N}_{\mathrm{G}}$, it follows that

$$0 = -\mathrm{d}\dot{N} \, (\tilde{h}_{\mathrm{G}} - \tilde{h}_{\mathrm{GI}}) + \dot{N}_{\mathrm{G}} \, \mathrm{d}\tilde{h}_{\mathrm{G}} + \dot{q}_{\mathrm{G}} \, \mathrm{d}A \ .$$

Now, we have $\mathrm{d}|\dot{N}| = |\dot{n}| \, \mathrm{d}A$ with $|\dot{n}|$ according to (4.61). This yields

$$-\dot{N}_{\mathrm{G}} \, \mathrm{d}\tilde{h}_{\mathrm{G}} = |\dot{n}| \, (\tilde{h}_{\mathrm{G}} - \tilde{h}_{\mathrm{GI}}) \, \mathrm{d}A + \dot{q}_{\mathrm{G}} \, \mathrm{d}A \tag{4.70}$$

with $\mathrm{d}\tilde{h}_{\mathrm{G}} = \tilde{c}_{p\mathrm{G}} \, \mathrm{d}\vartheta_{\mathrm{G}}$, $\tilde{h}_{\mathrm{G}} - \tilde{h}_{\mathrm{GI}} = \tilde{c}_{p\mathrm{G}} (\vartheta_{\mathrm{G}} - \vartheta_{\mathrm{I}})$ and $\dot{q}_{\mathrm{G}} = \alpha_{\mathrm{G}}^{\bullet} (\vartheta_{\mathrm{G}} - \vartheta_{\mathrm{I}})$. This equation cannot be integrated analytically, as the enthalpies $\tilde{h}_{\mathrm{G}}$ und $\tilde{h}_{\mathrm{GI}}$ depend on the varying temperature and the mole fraction along the flow path. As long as the section of area ΔA is not chosen to be too large, a good approach is to work with forwards differences. Then

$$-\dot{N}_{\mathrm{G1}} \, (\tilde{h}_{\mathrm{G2}} - \tilde{h}_{\mathrm{G1}}) = |\dot{n}| \, (\tilde{h}_{\mathrm{G1}} - \tilde{h}_{\mathrm{GI}}) \, \Delta A + \dot{q}_{\mathrm{G}} \, \Delta A \ . \tag{4.71}$$

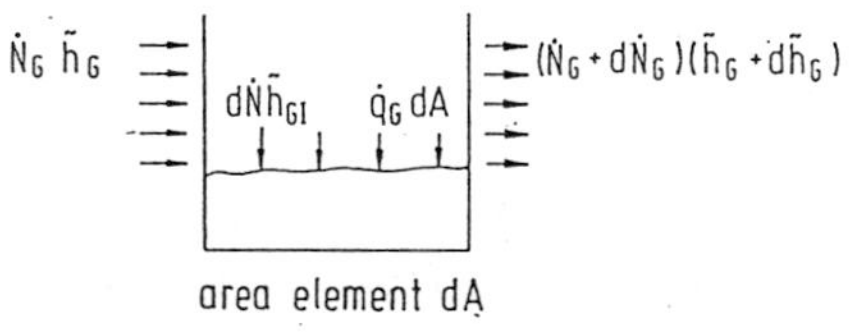

Fig. 4.25: Energy balance in a section of a condenser

The enthalpy $\tilde{h}_{G2}$ of the following section is obtained from this relationship, and with that also its temperature, because we have

$$\tilde{h}_G = \tilde{h}_{01}(\vartheta_G)\,\tilde{y}_G + \tilde{h}_{02}(\vartheta_G)\,(1 - \tilde{y}_G) \quad \text{and}$$

$$\tilde{h}_{GI} = \tilde{h}_{01}(\vartheta_I)\,\tilde{y}_I + \tilde{h}_{02}(\vartheta_I)\,(1 - \tilde{y}_I)\ , \tag{4.72}$$

where $\tilde{h}_{01}$ and $\tilde{h}_{02}$ are the temperature dependent molar enthalpies of the pure substances 1 and 2. As we have assumed the gas phase as ideal, the pressure dependence of the enthalpy is not considered. In (4.71) the heat flux released by the gas phase is $\dot{q}_G = \alpha_G^\bullet\,(\vartheta_G - \vartheta_I)$ the heat transfer coefficient $\alpha_G^\bullet$ of the gas at the phase interface is different from that for a vapour flow without condensation, α_G. According to (4.32) it is

$$\alpha_G^\bullet = \alpha_G\,\zeta_G\ ,$$

in which the correction factor ζ is given by the "Ackermann correction" (4.33).

4.1.8.3   Calculating the area of a condenser

The practical calculation for binary mixtures, based on the equations presented, is now possible in various ways. The following procedure is recommended

- Take a section of area ΔA and estimate the temperature ϑ_I at the surface of the condensate in this section. It is $\vartheta_G < \vartheta_I < \vartheta_c$, where ϑ_G is the temperature of the vapour and ϑ_c is that of the cooling medium. With this estimate of ϑ_I the mole fractions $\tilde{x}_I$ and $\tilde{y}_I$ at the condesate surface are fixed based on the phase equilibrium.
- Estimate the change $\Delta\tilde{y}_G$ in the vapour composition in the relevant section and approximate

$$\tilde{y}_G = (\tilde{y}_{G1} + \tilde{y}_{G2})/2 \quad \text{with} \quad \tilde{y}_{G2} = \tilde{y}_{G1} + \Delta\tilde{y}_G\ ,$$

where $\tilde{y}_{G1}$ is the vapour composition in the inlet cross section 1 and $\tilde{y}_{G2}$ is that in the outlet 2 of the section being considered.
- With these estimated values the vapour side heat transfer coefficient $\alpha_G^\bullet$ and the mass transfer coefficient β_G can be calculated. Just as the heat transfer coefficient α_L of the condensate film is also known, which in laminar film condensation is yielded from Nusselt's film condensation theory (4.39), and for turbulent film condensation from (4.41). From (4.67) the temperature ϑ_I at the condensate surface can be determined. If this does not agree with the estimate, then a new estimate of the temperature ϑ_I has to be made, until the calculated and estimated values agree with each other to a sufficient degree of accuracy.
- In the next step the accumulated flow of the condensate $|\dot{n}|$ is calculated using (4.69). If this agrees with the value found from (4.61), then the initial estimate with respect to the change $\tilde{y}_G$ in the composition of the vapour

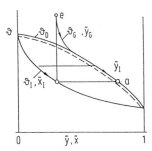

Fig. 4.26: Temperatures ϑ_G, ϑ_I and composition $\tilde{y}_G$, $\tilde{y}_I$, $\tilde{x}_I$ along a condenser. e = inlet, a = outlet state of the vapour

was correct. If this does not occur the calculation has to be repeated until the estimate and calculated value agree with each other.

- Once all these steps have been carried out the temperatures and the composition in the following segment can be investigated.
- As long as the vapour is superheated, $\vartheta_G > \vartheta_D$, ($\vartheta_D$ is the dew point), as shown in Fig. 4.26, it is advisable to calculate the vapour temperature in the following section using (4.71) and then go through the steps explained above to find the other quantities. If the vapour is cooled to the dew point ϑ_D, the energy equation (4.71) no longer has to be solved, because the vapour temperature $\vartheta_G = \vartheta_D$, as is apparent in Fig. 4.26, is only dependent on the mole fraction $\tilde{y}_G$.

How the vapour temperature ϑ_G the mole fractions $\tilde{y}_G$, $\tilde{y}_I$, and the temperature ϑ_I can vary along the condenser is illustrated schematically in Fig. 4.26.

4.1.9 Some empirical equations

In the following the equations introduced previously are summarised and supplemented by further equations.

1. **Laminar film condensation on vertical or inclined plates, and on the inside or outside of a vertical tube.**

    According to Nusselt's film condensation theory, from (4.13a), under the assumption $\varrho_G \ll \varrho_L$, the heat transfer coefficient follows as

$$Nu_m = \frac{\alpha_m}{\lambda_L}\left(\frac{\nu_L^2}{g}\right)^{1/3} = 0.925\left(\frac{\dot{M}/b}{\eta_L}\right)^{-1/3} = 0.925\,Re^{-1/3}\ .$$

    The local heat transfer coefficient is $\alpha = 3\,\alpha_m/4$.

    If the wall is inclined at an angle γ to the vertical then the acceleration due to gravity g has to be replaced by its component parallel to the wall $g\cos\gamma$ with $0 \le \gamma \le \pi/2$. The equation also holds for condensation of quiescent vapours on the inside or outside of a vertical tube, if the diameter of the tube is large in comparison to the film thickness. The width b has to be replaced by $b = \pi d$. If a deviation of 1 % from the values from Nusselt's film condensation theory is permitted, then the equation is valid up to a Reynolds number

$$Re_{\text{trans}} = 256\, Pr^{-0.47} \quad \text{with} \quad 1 \leq Pr \leq 10 \ .$$

2. **Laminar film condensation on horizontal tubes.**
According to (4.15a), with $\varrho_G \ll \varrho_L$, the mean heat transfer coefficient is

$$Nu_{\text{m}} = \frac{\alpha_{\text{m,hor}}\,(\nu_L^2/g)^{1/3}}{\lambda_L} = 0.959 \left(\frac{\dot{M}/L}{\eta_L}\right)^{-1/3} = 0.959\, Re^{-1/3} \ .$$

L is the tube length.
3. **Turbulent film condensation on vertical or inclined plates and on the inside or outside of vertical tubes.**
The local heat transfer coefficient, for $Re \geq 400$ is found from (4.41)

$$Nu = \frac{\alpha\,(\nu_L^2/g)^{1/3}}{\lambda_L} = 0.0325\, Re^{1/4}\, Pr^{1/2}$$

with

$$Re = \frac{\dot{M}/b}{\eta_L} = \left[89 + 0.024\, Pr^{1/2} \left(\frac{Pr}{Pr_0}\right)^{1/4} (Z - 2300)\right]^{4/3}$$

and

$$Z = \frac{c_{pL}\,(\vartheta_s - \vartheta_0)}{\Delta h_v}\, \frac{1}{Pr}\, \frac{x}{(\nu_L^2/g)^{1/3}} \ .$$

4. **Transition region between laminar and turbulent film condensation.**
In the transition region $256\, Pr^{-0.44} \leq Re_{\text{trans}} \leq 400$ between laminar and turbulent film condensation, the heat transfer coefficient according to (4.43) is obtained from

$$\alpha = \sqrt[4]{(f\,\alpha_{\text{lam}})^4 + \alpha_{\text{turb}}^4} \ .$$

The factor $f \approx 1.15$ accounts for the wave formation on the laminar condensate film.
5. **Turbulent film condensation of vapour flowing in tubes.**
According to Shah, (4.52), for the local heat transfer we have

$$Nu = 0.023\, Re^{0.8}\, Pr^{0.4} \left\{(1 - x^*)^{0.8} + \frac{3.8\,(1 - x^*)^{0.04}\, x^{*0.76}}{p^{+0.38}}\right\}$$

with

$$Nu = \frac{\alpha\, d}{\lambda_L}, \quad Re = \frac{w_m\, d}{\nu_L}, \quad w_m = \dot{M}/\left(\varrho_L\, \frac{\pi\, d^2}{4}\right), \quad Pr = \frac{\nu_L}{a_L} \quad \text{and} \quad p^+ = \frac{p}{p_{\text{cr}}} \ .$$

The region of validity is

$$100 \leq Re \leq 63\,000, \quad 1 \leq Pr \leq 13, \quad 0.002 \leq p^+ \leq 0.44 \ .$$

Further details regarding the region of validity can be found on pg. 424

7. Stratified flow in horizontal tubes.

In horizontal tubes, at low flow velocities, the liquid film that develops is not annular, but stratified. The condensate collects in the lower part of the tube, whilst the upper walls are wetted by the liquid. Stratified flow appears if the dimensionless vapour velocity is

$$w_G^* = \frac{x^* \dot{M}}{A \left[g \, d \, \varrho_G \left(\varrho_L - \varrho_G \right) \right]^{1/2}} \leq 1$$

and if the following holds for the liquid fraction

$$\frac{A_L}{A_G} = \frac{1 - \varepsilon}{\varepsilon} \quad \text{with} \quad \varepsilon = \frac{A_G}{A} \quad \text{and} \quad A = A_G + A_L \ .$$

$$\frac{1 - \varepsilon}{\varepsilon} \leq 0.5 \ .$$

The mean heat transfer coefficient is given by

$$\alpha_m = 0.728 \, \varepsilon \left[\frac{\varrho_L^2 \, g \, \Delta h_v \, \lambda_L^3}{\eta_L \left(\vartheta_s - \vartheta_0 \right)} \frac{1}{d} \right]^{1/4} \ .$$

4.2 Heat transfer in boiling

Whilst heat transfer in convection can be described by physical quantities such as viscosity, density, thermal conductivity, thermal expansion coefficients and by geometric quantities, in boiling processes additional important variables are those linked with the phase change. These include the enthalpy of vaporisation, the boiling point, the density of the vapour and the interfacial tension. In addition to these, the microstructure and the material of the heating surface also play a role. Due to the multiplicity of variables, it is much more difficult to find equations for the calculation of heat transfer coefficients than in other heat transfer problems. An explicit theory is still a long way off because the physical phenomena are too complex and have not been sufficiently researched.

The cause of this is not only that there are many influencing quantities that play a role in boiling processes, but also different types of heat transfer depending on the flow configuration and superheating. These different types of heat transfer will be considered first, followed by an explanation of the physical fundamentals of boiling phenomena. The final part of this section will consist of the calculation of the heat transfer.

4.2.1 The different types of heat transfer

Depending on the type of boiling, we differentiate between evaporation, nucleate boiling and convective boiling. We will consider *evaporation* first.

If a liquid on a heated wall is superheated to a temperature just above the saturation temperature, no or only a few vapour bubbles form. In a vessel filled with liquid, that is heated from below, a temperature profile develops, like that schematically illustrated in Fig. 4.27. A boundary layer forms over the heated wall which has a temperature of ϑ_0. The boundary layer has thickness of the order of 1 mm, with a stark temperature drop, whilst in the core of the liquid the temperature is almost constant (mean value ϑ_L) over the height z. At the free surface the temperature falls, in a thin layer, to the value ϑ_I, that lies slightly above the saturation temperature ϑ_s. The difference $\vartheta_I - \vartheta_s$ was first measured by Prüger [4.42] for water at 1.01 bar to be around 0.03 K, whereas for non-polar liquids like carbon tetrachloride this values is around 0.001 K. As important as this liquid superheating at the surface is for the kinetic considerations in evaporation, it can be ignored in technical calculations. In the following, therefore, the temperature at the vapour forming surface will always be given as the saturation temperature $\vartheta_I = \vartheta_s$.

In the thin layer adjacent to the wall the temperature drops steeply, as can be seen in Fig. 4.27. Conduction is the predominant heat transfer process in this layer. In the liquid, rising and falling convection streams provide for the transport of heat. They generate the uniform temperature field in the core of the liquid. The two boundary layers at the top and bottom are differentiated by the fact that the free surface can move because of the vapour formation, and so in contrast to the liquid at the wall, finite velocities parallel to the surface can also appear.

The vaporisation at the surface acts as a heat sink, which could be imaginarily replaced by another process, for example radiation. As the vaporisation takes place at the surface, we speak of "*stagnant boiling*". By its nature this process belongs to the phenomena of convection in closed spaces. Heat transfer coefficients from the heated surface to the liquid are formed with the driving temperature difference $\vartheta_0 - \vartheta_L$, where ϑ_0 is the wall temperature of the heated surface and ϑ_L is the liquid temperature. As the liquid temperature ϑ_L is not known in advance and, as explained above, only deviates slightly from the saturation temperature, it is sensible to form the heat transfer coefficients with the temperature difference $\Delta\vartheta = \vartheta_0 - \vartheta_s$. In stagnant boiling this law of heat transfer is valid in free flow. Thus we have $\alpha = c_1 \Delta\vartheta^{1/4}$ in laminar flow and $\alpha = c_2 \Delta\vartheta^{1/3}$ in turbulent flow over a horizontal plate. As the heat flux is given by $\dot{q} = \alpha\Delta\vartheta$, for laminar flow it

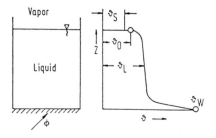

Fig. 4.27: Temperature pattern in the liquid during surface evaporation

holds that

$$\alpha = c_1 \Delta\vartheta^{1/4} \quad \text{or} \quad \alpha = c_1' \dot{q}^{1/5} \tag{4.73}$$

or in turbulent flow

$$\alpha = c_2 \Delta\vartheta^{1/3} \quad \text{or} \quad \alpha = c_2' \dot{q}^{1/4} . \tag{4.74}$$

If the wall temperature is raised by increasing the heat flow to the wall, above a certain wall temperature vapour bubbles begin to form. Observations have shown that these bubbles only appear at certain points on the heated surface. The number of bubbles increases with the heat being fed to the surface. This type of heat transfer is known as *nucleate boiling*.

Fig. 4.28 shows a typical temperature profile over a horizontal plate from measurements by Jakob et al. [4.43] to [4.49], to whom we owe the first fundamental investigations into this process. In comparison to Fig. 4.27 it is immediately apparent that the temperature difference $\vartheta_0 - \vartheta_L$ is larger, and $\vartheta_L - \vartheta_s$ is smaller now. The motion of the bubbles at the surface does not allow exact measurement of the boundary layer. The heat transfer coefficient is formed, in the same way as for stagnant boiling, with the temperature difference $\Delta\vartheta = \vartheta_0 - \vartheta_s$. Heat transfer is much better than in stagnant boiling and is approximately proportional to the cube of the temperature difference $\Delta\vartheta$. If we consider that the heat flux transferred is given by $\dot{q} = \alpha\Delta\vartheta$, then in the nucleate boiling region, it is approximately valid that

$$\alpha = c_3 \Delta\vartheta^3 \quad \text{or} \quad \alpha = c_3' \dot{q}^{3/4} . \tag{4.75}$$

If $\alpha(\dot{q})$ from (4.73) or (4.74) and (4.75) are graphically illustrated, two straight lines are produced, if the ordinate and abcissa have logarithmic scales. Two clearly separate areas are obtained, one for stagnant the other for nucleate boiling, as shown in Fig. 4.29, which reproduces the measured values from Jakob et al. [4.44], [4.45].

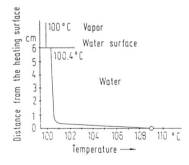

Fig. 4.28: Temperature profile over a horizontal heated surface, from Jakob und Linke [4.46]. Heat flux $\dot{q} = 22\,440 \text{ W/m}^2$, temperature of heated surface $\vartheta_0 = 109.1$ °C

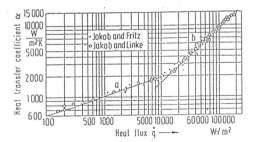

Fig. 4.29: Heat transfer in boiling water at 100 °C on a horizontal heated surface, according to Jakob et al. [4.43], [4.45]. Curve **a** stagnant boiling region, curve **b** nucleate boiling region

In industrial apparatus boiling normally takes place in *forced convection*. The flow behaviour is largely determined by the pressure difference along the heated surface. The vapour content increases constantly along the flow path until complete vaporisation sets in. Correspondingly, the decreasing amount of liquid yields various boiling phenomena, upon whose heat transfer properties the local boiling temperature is dependent. In general, a subcooled liquid enters a heating channel. The vapour bubbles formed at the wall condense back into the core of the liquid. If the core liquid is heated to the saturation temperature then nucleate boiling predominates. The heat transfer coefficient will be principally determined by the heat flux. In forced flow it only has a slight dependence on the mass flux, and in free flow it is virtually independent of the mass flux. The individual vapour bubbles coalesce into large bubbles, forming *slug flow*, as shown by Fig. 4.30.

With increasing vapour content, the large bubbles grow together such that a *semi-annular flow* develops, and subsequently a liquid film forms at the tube wall with vapour and liquid drops in the core. This is known as *annular flow*. As more heat is added to the tube the liquid film disappears downstream, leaving a vapour containing liquid droplets, so-called *spray flow*, flowing through the tube. Fig. 4.30 shows these flow patterns in a vertical tube. More complicated flow patterns exist in horizontal or inclined tubes. Bubble, slug, semi-annular and annular flows represent different forms of *convective boiling*.

In practice annular or at very low flow velocities, slug flow frequently occur. Bubble flow only appears at very low vapour contents and high flow velocities. Increasing pressure and with that a decreasing density difference between vapour and liquid extends the bubble flow region.

Whilst in nucleate boiling the heat transfer coefficient is chiefly dependent on the heat flux $\dot{q}$ and is barely dependent on the flow velocity, curve b in Fig. 4.29, in convective boiling the heat transfer is determined by the flow velocity or the mass flux $\dot{m}$, with the heat flux having vitually no influence. This is shown by Fig. 4.31, in which the nucleate boiling and convective boiling regions are distinctly separate from each other.

A further independent variable is the quality $x^* = \dot{M}_G/\dot{M}$. With increasing quality, the curves for convective boiling in Fig. 4.31 are shifted to larger heat transfer coefficients α.

The fundamental dependence of the heat transfer coefficients on the quality is illustrated in Fig. 4.32. In regions of low quality x^* nucleate boiling occurs and the heat transfer coefficient is principally dependent on the heat flux. The quality increases downstream, and with that the flow velocity also increases. The heat fed into the system is mainly transferred by convection from the tube wall to the vapour-liquid flow. A conversion from nucleate boiling to convective boiling occurs, as the arrow on the curves $\dot{q}_1$ and $\dot{m}_1$ in Fig. 4.32 indicates. In the convective boiling region the local heat transfer coefficient is practically independent of the heat flux $\dot{q}$, and depends strongly on the mass flow rate and the quality. At high quality the heat transfer coefficient decreases because of the low thermal conductivity of the vapour compared to that of the liquid.

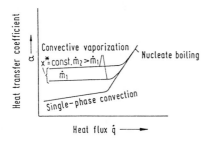

Fig. 4.31: Heat transfer coefficient in nucleate and convective boiling (qualitative)

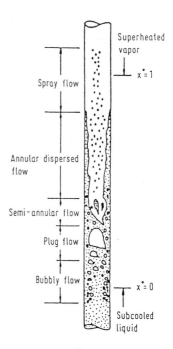

Fig. 4.30: Flow patterns in a vertical, heated tube

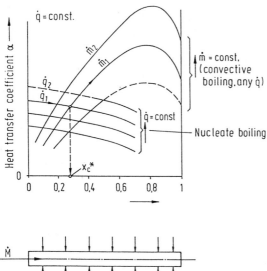

Fig. 4.32: Trends of the heat transfer coefficient for a horizontal evaporator tube

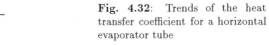

The calculation of the heat transfer coefficient can be carried out using equations of the form

$$\alpha = c \, \dot{q}^{\,n} \, \dot{m}^{\,s} \, f(x^*) \ ,$$

where c depends on the material properties. In the convective boiling region $n \approx 0$, with s lying between 0.6 und 0.8. In nucleate boiling n is approximately $3/4$ and s is around 0.1 to 0.3.

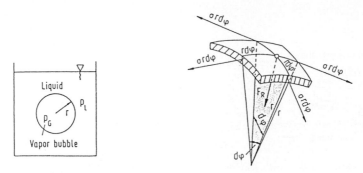

Fig. 4.33: Mechanical Equilibrium between a spherical vapour bubble and the liquid surrounding it

4.2.2   The formation of vapour bubbles

Heat transfer in boiling is more easily understood when we know how the vapour bubbles form on the hot surface. The following consists of a discussion of the formation and growth of vapour bubbles, with a subsequent explanation of the different types of heat transfer.

The following considerations hold for the equilibrium of a vapour bubble, assumed to be spherical, Fig. 4.33, with the liquid surrounding it. Between the gaseous bubble (gas = index G) and the surrounding liquid (liquid = index L), *thermal equilibrium* exists

$$\vartheta_G = \vartheta_L = \vartheta \ . \tag{4.76}$$

If a surface element of the spherical shell is cut out of the vapour bubble, as depicted in the right hand side of Fig. 4.33, with side lengths $r \, d\varphi$, the forces $\sigma r \, d\varphi$ exerted by the surface tension σ (σ is force per unit length) act upon the edges. The resultant F_R of these forces is given by

$$d^2 F_R = 2\sigma r \, d\varphi^2 \ .$$

The forces resulting from the gas and liquid pressure are also of influence

$$p_L(r \, d\varphi)^2 + d^2 F_R = p_G(r \, d\varphi)^2 \ .$$

From this, the condition of *mechanical equilibrium* follows

$$p_G = p_L + 2\sigma/r \ . \tag{4.77}$$

Finally the condition for *equilibrium with respect to mass exchange* between the gaseous and liquid phase also holds. This leads, as was illustrated, for instance, in [4.50], to

$$p_L = p_0 - \frac{\varrho'}{\varrho' - \varrho''} \frac{2\sigma}{r} \tag{4.78}$$

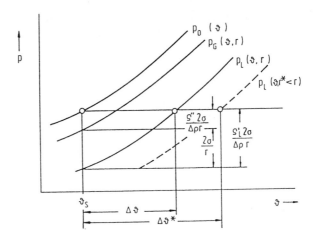

Fig. 4.34: Vapour and liquid pressure between a liquid and a spherical vapour bubble

or with (4.77) to

$$p_G = p_0 - \frac{\varrho''}{\varrho' - \varrho''}\frac{2\sigma}{r} \; . \tag{4.79}$$

Equation (4.78) or (4.79) is known as *Thomson's equation*. It produces a relationship between the vapour pressure $p_0(\vartheta)$ at a flat phase interface, the liquid pressure $p_L(\vartheta, r)$ and the vapour pressure $p_G(\vartheta, r)$ at the surface of a vapour bubble of radius r. These relationships are illustrated in Fig. 4.34. At a given temperature ϑ, the vapour pressure p_G, corresponding to (4.79), is smaller by

$$p_0 - p_G = \Delta p_G = \frac{\varrho''}{\varrho' - \varrho''}\frac{2\sigma}{r} \; ,$$

than the vapour pressure p_0 at the flat phase boundary. As the surface tension σ is temperature dependent, the curves for the vapour pressure p_G and the liquid pressure p_L do not run exactly but only approximately parallel to the vapour pressure curve p_0 at the phase interface.

If, instead of stipulating the boiling temperature ϑ, the pressure p_0 of a liquid-vapour bubble system is given, then the liquid has to be superheated by $\Delta\vartheta$ in comparison to the system with a flat phase boundary, so that a vapour bubble of radius r is in equilibrium with the liquid, as Fig. 4.34 shows. In addition it is clear that the required degree of superheating $\Delta\vartheta$ is larger, the smaller the radius r of the vapour bubble is, so for small radii $r^* < r$ the curves for the vapour pressure $p_G(\vartheta, r^*)$ and the liquid pressure $p_L(\vartheta, r^*)$ in Fig. 4.34 are shifted further to the right. Conversely, for a given degree of superheating $\Delta\vartheta$, a vapour bubble of definite radius r is in equilibrium with the superheated liquid. For the approximate calculation of the required superheating, we assume that the curves $p_0(\vartheta)$ and $p_L(\vartheta, r)$ in Fig. 4.34 run parallel. This gives

$$\frac{dp_L}{d\vartheta} = \frac{dp_0}{d\vartheta} \; .$$

The differential $dp_0/d\vartheta$ is the slope of the vapour pressure curve $p_0(\vartheta)$. It can be calculated from the Clausius-Clapeyron equation

$$\frac{dp_0}{d\vartheta} = \frac{\Delta h_v \varrho'' \varrho'}{T_s(\varrho' - \varrho'')} \;.$$ (4.80)

On the other hand with (4.78)

$$\frac{dp_L}{d\vartheta} \simeq \frac{p_0 - p_L}{\Delta\vartheta} \simeq \frac{1}{\Delta\vartheta}\frac{\varrho'}{\varrho' - \varrho''}\frac{2\sigma}{r} \;.$$ (4.81)

From these equations, the bubble radius r is calculated as a function of the superheating $\Delta\vartheta$ to be approximately

$$r \cong \frac{2\sigma T_S}{\varrho'' \Delta h_v \Delta\vartheta} \;.$$ (4.82)

According to this, for a particular degree of liquid superheating $\Delta\vartheta$ a definite bubble radius exists, at which the bubble is in equilibrium with the liquid. Bubbles whose radii are $r^* < r$ are in equilibrium with the liquid only if the superheating is $\Delta\vartheta^* > \Delta\vartheta$, as Fig. 4.34 shows. A liquid superheated by $\Delta\vartheta$ is too cold. Therefore bubbles that are too small will condense again. Bubbles of radius $r^* > r$ are in liquid that is superheated, and they can continue to grow. However, in reality the residence time of bubbles, in particular those close to the wall, is so small that equilibrium is never reached and the actual superheating of the fluid is many times greater than $\Delta\vartheta$. A particular critical bubble radius also belongs to this actual superheating. In boiling water at 1 bar, according to (4.82), the bubble diameter is $2r \cong 0.155\,\mathrm{mm}$, based on superheating in the core of the liquid by 0.4 K. A bubble of this size is then able to form and can then grow. A bubble of this type contains around $3 \cdot 10^{20}$ water molecules. However it is difficult for so many molecules with the high energy of vapour molecules to collect coincidentally at a certain position inside the liquid, form a bubble and then grow. This raises the question of how vapour bubbles actually form.

Observations have taught us that no bubbles form in a completely pure, carefully degassed liquid, unless the liquid is extremely superheated or, for example, ionising beams are sent through it. Furthermore, over a long time, the bubbles reappear at the same place on the heated surface, with a varying frequency that can be approximated to an error function. Obviously this has something to do with highly active centres that catalyse the transformation from unstable superheated liquid to stable vapour. These centres are the remains of gas or vapour in depressions in the surface that have not been driven out by the liquid, because even with good wetting ability it cannot completely fill the fine depressions on the surface. When heat is added the gas or vapour remnants expand until a critical size is reached that corresponds to the size of a viable bubble. Then a vapour bubble can grow further as a result of the superheating of the fluid, until finally the adhesion force becomes smaller than the buoyancy and dynamic forces and the bubble detaches itself from the heated surface. After the bubble breaks off, further gas or vapour remains enclosed in the depression. This will be cooled by the cold liquid flowing from the centre of the fluid to the wall, and then subsequently heated by the addition of heat from the wall. A new nucleus for the growth of a vapour bubble forms. These considerations explain why the surface structure is an influential quantity for heat transfer.

Vapour bubbles almost always develop at particularly favourable positions on solid surfaces or on suspended particles. Therefore, it is generally *heterogeneous nuclei formation* that prevails. The *homogeneous nuclei formation*, with bubbles formed by 'themselves' as a result of the natural fluctuations of the molecules, plays a very minor role.

4.2.3 Bubble frequency and departure diameter

The bubble formed at the wall grows by vaporisation until it reaches a limiting volume when it detaches itself from the wall and rises. In slow growth this limiting volume is determined by the equilibrium of buoyancy and surface forces and the adhesion conditions at the wall. The decisive differential equation for this was first solved numerically by Bashfort and Adams [4.51]. Based on their solution, Fritz [4.52] later showed that there is a greatest volume V_A for a vapour bubble, that can be represented in the form

$$\left(\frac{V_A}{b^3}\right)^{1/3} = f(\beta_0) \ , \tag{4.83}$$

where β_0 is the contact angle between the bubble and the heated wall. The so-called *Laplace constant* serves as a parameter

$$b = \sqrt{\frac{2\sigma}{g(\varrho_L - \varrho_G)}} \ . \tag{4.84}$$

If this is formed with the saturation values ϱ' and ϱ'' of the densities, then, for example, for water at 100 °C the value $b = 3.54$ mm is obtained, and for the refrigerant R134a at 25 °C $b = 1.18$ mm. Some values for air bubbles in a liquid are presented in Table 4.1. Eq. (4.83) has been confirmed by Fritz and Ende [4.53]

Table 4.1: Laplace constant b for air bubbles in liquids at 20 °C

	Water	Ethanol	Mercury
b in mm	3.82	2.26	2.69

and by Kabanow and Frumkin [4.54] from the evaluation of recordings of films, Fig. 4.35.

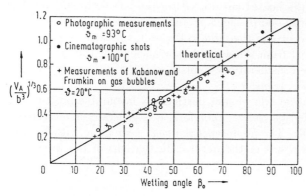

Fig. 4.35: Detachment volume V_A of vapour and gas bubbles over a horizontal heated surface, according to Fritz and Ende [4.53]. b Laplace constant

From (4.83), for a spherical bubble $d_A = (6V_A/\pi)^{1/3}$, the following is obtained for the departure diameter

$$d_A = 0.851 \beta_0 \sqrt{\frac{2\sigma}{g(\varrho_L - \varrho_G)}} \qquad (4.85)$$

with the contact angle β_0 in radians. According to Fritz [4.52] the contact angles of vapour bubbles in water at atmospheric pressure lie between 40° and 45°, and those in refrigerants of halogenated hydrocarbons at around 35°. Actual departure diameters deviate somewhat from those according to (4.85), and are also dependent on the shape of the depressions [4.55] and the superheating of the wall. The shape of the bubble before it breaks off is additionally influenced by the intensive evaporation at the heated surface. This produces a bubble shape like that in Fig. 4.36 a. A 'bubble neck' forms between the heated surface and the part of the bubble away from it. According to Mitrović [4.57], the bubble breaks off when the bubble neck is completely constricted, Fig. 4.36 b. The total vapour mass splits into two different sized amounts. A small amount of vapour remains attached to the heated surface. The larger amount of vapour is found in the detached bubble. Due to the small radius of curvature at the point at the bottom of the bubble a high capillary pressure forms which seeks to equalise itself and in the process of this, provided the bubble is not extremely small, causes the bubble to oscillate. In the same way a capillary pressure acts on the vapour remains on the surface. As Mitrović [4.57] showed with an example this can reach 4 bar. As a result of this overpressure the vapour remains will be partly or even completely liquified. In particular with little superheating of the wall the vapour remains will be condensed to a great extent. Therefore, it takes a long time until the next bubble forms, whilst with sufficient superheating the next bubble will develop out of the vapour remains without any waiting time. The breaking down of the capillary forces causes the peak in Fig. 4.36 b to disappear, and the bubble shape shown in Abb. 4.36 c forms with diverging streamlines underneath the bubble. In the liquid below the bubble an underpressure develops that keeps the detached bubbles close to the wall for a short time. This underpressure causes the liquid to be in a metastable state with high superheating. The superheated liquid can also serve as a nucleus for new bubbles, so that, in addition to the heterogeneous nucleation on the depressions in the wall, homogeneous nucleation close to the wall is also possible. This, along with the liquid flow, quickly removes the underpressure.

These thoughts show that convection is also of influence on the growth and detachment of the bubbles. Previous results for the influence of convection are still inconsistent, as emerges from the summaries from [4.58] among others. Tokuda [4.59] proved in a theoretical study that the growth of the bubble is only determined by heat conduction immediately following its formation, after that, with increasing bubble size the radial convection will be decisive. After the break off of the bubble, heat conduction and radial and axial convection all influence the bubble growth.

Equations for the calculation of the frequency f of vapour bubbles of departure dianeter d_A, originally started with the assumption that $f d_A = \text{const}$, where the constants for water and carbon tetrachloride were found to be $100 \, \text{mm/s}$ [4.60]. The constants were later expressed in terms of the physical properties of the boiling liquid,

$$ f d_A = 0.59 \left(\frac{g\sigma(\varrho' - \varrho'')}{\varrho'^2} \right)^{1/4} , $$

whilst other authors, [4.61] to [4.63], suggested equations of the form $f d_A^n = \text{const}$ with $n = 1/2$. More accurate investigations, [4.64], have, however, shown that the exponent n is not constant but assumes values between 0.5 and 2.

Under the assumption that at the moment of detachment the bubbles do not accelerate and their velocity is the same as their rising velocity in the fluid,

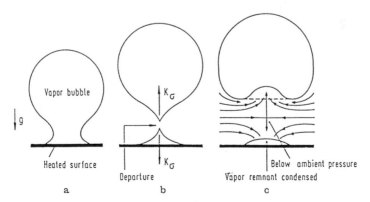

Fig. 4.36: Detachment process, according to Mitrović [4.57]. **a** Shape of the attached bubble; **b** constriction and break off of the bubble; **c** condensation of the remaining vapour and the liquid flow

Malenkov [4.65] found the following expression for the frequency

$$f = \frac{1}{d_A \pi} \left[\frac{d_A g(\varrho' - \varrho'')}{2(\varrho' + \varrho'')} + \frac{2\sigma}{d_A(\varrho' + \varrho'')} \right]^{1/2} , \qquad (4.86)$$

which can be simplified for regions well below the critical pressure, with $\varrho' \gg \varrho''$, to

$$f = \frac{g^{1/2}}{d_A^{1/2} \pi \sqrt{2}} \left(1 + \frac{4\sigma}{d_A^2 \varrho' g} \right)^{1/2} . \qquad (4.87)$$

These equations are only valid if the heat flux is so small that the bubbles do not noticeably influence each other. In any other case, the frequency increases by a heat flux dependent factor of

$$1 + \frac{\dot{q}}{\varrho'' \Delta h_v w} ,$$

where

$$w = \left[\frac{d_A g(\varrho' - \varrho'')}{2(\varrho' + \varrho'')} + \frac{2\sigma}{d_A(\varrho' + \varrho'')} \right]^{1/2}$$

is a velocity. The frequency is obtained to be

$$f = \frac{1}{d_A \pi} \left(1 + \frac{\dot{q}}{\varrho'' \Delta h_v w} \right) w , \qquad (4.88)$$

an equation in which the velocity w can be simplified, if $\varrho' \gg \varrho''$. In the limiting case of sufficiently small heat flux, this relationship transforms into the equation given above for single bubbles. These equations from Malenkov have been checked by numerous measurements by different authors and have shown good agreement.

As can be seen in (4.86), for large departure diameters, the second term in the square brackets is small, then we have $f d_A^{1/2}$ = const. If, however the departure diameter is sufficiently small the second sum in the square brackets will be large, giving $f d_A^{3/2}$ = const. In a middle region of the departure diameter the expression in the square brackets is independent of the departure diameter, and with that $f d_A$ = const. The equations contain the relationships, ascertained by different authors, between the frequency and the departure diameter.

According to (4.88) the bubble frequency increases with heat flux. The size of the nucleation site is also definitely of influence. Nucleation sites of small opening diameters release bubbles at a higher frequency than nucleation sites with larger opening diameters. The equations above do not take this effect into account, but reproduce mean values over nucleation sites of different sizes. Likewise, the influence of subcooling is also not considered. Increased subcooling produces bubbles less frequently because the bubble growth is hindered by condensation.

4.2.4   Boiling in free flow. The Nukijama curve

We will now consider boiling in free flow, and in this we will assume that the dimensions of the vessel in which the vapour is present are large in comparison to the vapour bubbles. The flow is a result of the buoyancy of the generated bubbles and of the density differences. For simplification we speak of *pool boiling*. The two phase flow in a narrow evaporator tube, even when it is caused by the buoyancy of the vapour bubbles in a horizontal tube, shall not be investigated. As was already explained in section 4.2.1, at small superheating of the wall 'stagnant boiling' develops. Heat is only transferred by free convection, curve a in Fig. 4.29. Nucleate boiling sets in only after a certain degree of superheating of the wall compared to the boiling temperature. We will look at the more frequent case where the liquid is at the saturation temperature. The driving temperature difference for heat transfer is $\vartheta_0 - \vartheta_s = \Delta\vartheta$ between the wall and the saturation temperature. Jakob and Linke [4.66] were the first to discover that the heat flux transferred in fully developed nucleate boiling, curve b in Fig. 4.29, can be described by simple empirical equations of the form

$$\dot{q} = c'\Delta\vartheta^m \ . \tag{4.89}$$

If the heat transfer coefficient is defined by

$$\alpha = \dot{q}/\Delta\vartheta \ , \tag{4.90}$$

then (4.89) can also be written as

$$\alpha = c\dot{q}^n \ , \tag{4.91}$$

with $c = (c')^{1/m}$ and $n = (m-1)/m$. In this equation, as many measurements show, the quantity n and therefore also m principally depend on the type of boiling liquid, but also on the material, structure and shape of the heated surface and the pressure. In general

$$0.6 < n < 0.8 \ .$$

Lower values $n \approx 0.5$ have only been found for substances with low boiling points, such as helium. The quanitity c is strongly dependent on the material properties of the boiling liquid and the structure of the heating surface. If, for example, c is known for a particular liquid that boils at a given pressure on a flat, polished, steel tube, then the heat transfer coefficient calculated with this quantity may not be used if the same liquid was boiling on a rough steel tube or a copper tube.

Equation (4.89) or (4.91) is only valid in regions of intensive nucleate boiling. However, as Nukijama [4.67] was the first to show there are further regions of vaporisation. Fig. 4.37 illustrates the individual regions that appear in boiling water at atmospheric pressure and in free flow. The heat flux $\dot{q}$ is plotted against the temperature difference $\Delta\vartheta = \vartheta_0 - \vartheta_s$. The rising left hand branch of the curve indicates the region of fully developed nucleate boiling. The heat flux increases with rising temperature. Once a maximum has been reached the transferred heat flux decreases again despite the rising temperature. This falling region of the 'boiling line' is known as partial film boiling, because the heated surface is partly covered by vapour. A minimum heat flux is passed and once again the heat flux starts to increase with the wall temperature. This region on the right hand end of the boiling line is designated as *film boiling*, because there the heated surface is completely covered by a vapour film.

The N-shaped curve, also called the *Nukijama* curve, which at first glance seems rather strange, is physically plausible: With increasing wall temperature more vapour bubbles form at the heated wall, which set the liquid close to the wall into motion. The result of this is that bubble formation promotes heat transfer from the wall to the liquid.

Because the vapour insulates the wall from the liquid, the heat transfer will be hindered the greater the amount of vapour present. It is easy to imagine that with increasing wall temperature the insulating effect of the vapour dominates and so despite the rising temperature the heat flux decreases. Once a sufficiently high wall temperature has been attained the heat flux will begin to increase again, as the thickness of the vapour film only increases slightly with the wall temperature. The maximum on the boiling line is frequently called the critical or *maximum heat flux*. Useful, but less approporiate designations are burnout or DNB (departure from nucleate boiling).

4.2.5 Stability during boiling in free flow

If the wall temperature of the heated surface is raised above the value associated with the maximum heat flux, the wall is suddenly covered by a vapour film. Within fractions of a second the temperature of the wall sharply increases and can even reach the melting temperature of the wall. An example of this is illustrated in Figs. 4.38 and 4.39. Fig. 4.38 shows nucleate boiling of the refrigerant R11 on a

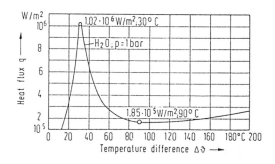

Fig. 4.37: $(\dot{q}, \Delta\vartheta)$ diagram for boiling water

horizontal copper tube that has been electrically heated with a heat flux of around $2 \cdot 10^5$ W/m². This heat flux is only slightly less than the maximum. After a minor increase in the heat flux, the temperature rises so far that the tube melts through. This is clearly recognisable in parts of the tube shown in Fig. 4.39.

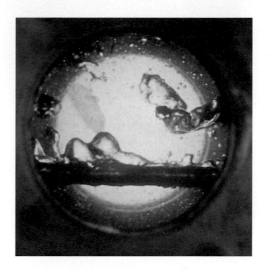

Fig. 4.38: Electrially heated copper tube immediately before the maximum heat flux is reached. $\dot{q}_{max} = 2.2 \cdot 10^5$ W/m². Boiling R11, $p = 0.1$ MPa

In order to prevent this phenomenon, to be avoided above all in nuclear reactors, a sufficiently large 'safety margin' should be maintained from the maximum heat flux. Obviously not all of the states along the boiling line illustrated in Fig. 4.37 are stable, rather under certain conditions instabilities can occur.

In the following we will explain these instability phenomena. The boiling line, like that in Fig. 4.37, represents all pairs of values $(\dot{q}, \Delta\vartheta)$ that are possible in the steady-state operation of a boiler. Of these pairs of values only quite specific ones come into question as operating points of the evaporator. Similarly to the way the operating point of a compressor is determined by the compressor and tube operating lines, the operating point of a boiler is determined by the boiling curve and the operating line of the boiler. When the material properties of the heated surface are presumed to be independent of temperature, the operating line of the boiler is a straight line in a $(\dot{q}, \Delta\vartheta)$-diagram. This can be explained with the example of a flat wall, Fig. 4.40.

Using the notation in Fig. 4.40, the heat flux transferred from the wall is

$$\dot{q} = \frac{1}{R_W}(\vartheta_L - \vartheta_0) \tag{4.92}$$

with the thermal resistance

$$R_W = \frac{1}{\alpha_i} + \frac{s}{\lambda} \, , \tag{4.93}$$

where α_i is the heat transfer coefficient between the heating fluid and the wall, s is the wall thickness and λ is the thermal conductivity of the wall. (4.92) can also

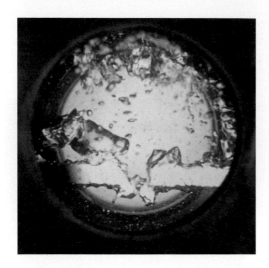

Fig. 4.39: Electrically heated tube from Fig. 4.38 at the moment of melting (burnout)

be written as

$$\dot{q} = \frac{-1}{R_{\mathrm{W}}}(\vartheta_0 - \vartheta_{\mathrm{s}}) + \frac{1}{R_{\mathrm{W}}}(\vartheta_{\mathrm{L}} - \vartheta_{\mathrm{s}}) \ . \tag{4.94}$$

If additional internal energy is generated in the wall, from nuclear fission or

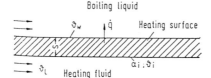

Fig. 4.40: Heat transfer in a boiling liquid

electrical dissipation, then the heat flux released increases by $\dot{q}_0$, and we obtain

$$\dot{q} = \frac{-1}{R_{\mathrm{W}}}(\vartheta_0 - \vartheta_{\mathrm{s}}) + \frac{1}{R_{\mathrm{W}}}(\vartheta_{\mathrm{L}} - \vartheta_{\mathrm{s}}) + \dot{q}_0 \ . \tag{4.95}$$

The operating line being sought is given by this equation. Its graphical presentation is a straight line in a $(\dot{q}, \Delta\vartheta)$-diagram. All pairs of values $(\dot{q}, \Delta\vartheta)$ that are possible in the steady-state operation of the evaporator have to lie along this line. As (4.95) corresponds, in terms of its structure, to Ohm's law, the operating line is called the 'resistance line'. Due to its negative slope $-1/R_{\mathrm{W}}$, the heat flux released by the heated surface decreases with rising wall temperature ϑ_0. This surprising result is understandable if we consider that with increasing wall temperature ϑ_0, the difference between wall temperature ϑ_0 and fluid temperature ϑ_{L} becomes smaller, and so the heat flux decreases.

The operating point of the boiler is the point at which the boiling and resistance lines meet. As Fig. 4.41 shows, the two curves can have one intersection and one contact point or three intersection points in common. These are the possible

operating points of the evaporator. The stability consideration [4.68], which will
not derived here, shows that the operating point is only stable if the gradient of
the boiling line at that point is greater than the gradient of the resistance line.
The following statement for the N-shaped curve has the same meaning: When
three intersection points exist only the two outer ones are stable whilst the middle
point characterises an unstable state. If only one intersection point occurs this is
always stable.

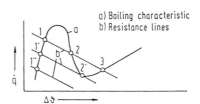

Fig. 4.41: Operating point of
boiler

These explanations make the so-called burn-out phenomenon easier to understand. If the resistance line for nuclear energy generation or electrical resistance
heating is drawn in a $(\dot{q}, \vartheta)$-diagram the boiling and resistance lines have three
intersection points close to the maximum in common, Fig. 4.42, of which points
1 and 2 are very close to each other. A small disturbance is sufficient to transform the system from the stable state 1 into the unstable state 2. Then because
of the inertia of the system it will move further to point 3 where a new stable
state is attained. In most systems, this new steady-state is associated with a wall
temperature that is larger than the melting temperature of the heated surface, so
it will melt before a stable state is reached. In order to prevent this somewhat
undesirable event, the resistance line has to be chosen so steep such that it does
not intersect the boiling line a second time near the maximum. This procedure
is clearly illustrated by curve a in Fig. 4.42. A steep resistance line is yielded,
in accordance with (4.93) and (4.94), if α_i is chosen to be large and the thermal
resistance of the wall is small. This also allows a stable operating point at the
maximum heat flux to be realised.

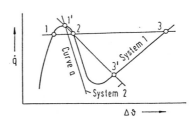

Fig. 4.42: Stability at maximum heat
flux

4.2.6   Calculation of heat transfer coefficients for boiling in free flow

Current known models for heat transfer in boiling only partially describe the process and so do not lead to a complete theory. Obviously, heat transfer in nucleate boiling is so complex that several exchange mechanisms are active simultaneously. Depending on the conditions present, the heat flux, pressure, wetting, forced convection, subcooling, etc., one or more of the mechanisms will dominate. This is probably a good explanation for the fact that the theories presented up until now agree with measured values in certain regions for particular substances, but fail for other regions and materials.

Basically, it has been ascertained that the current models contain one or more of the following mechanisms for heat transfer:

1. Microconvection in the boundary layer adjacent to the wall as a result of the rapid growth of the vapour bubble and its collapse in the subcooled liquid.
2. Displacement of hot liquid from the wall by the growing and detaching vapour bubbles, and the back-flow of cold liquid from the core to the wall.
3. Transient heat conduction to the liquid transported in the wake of a bubble.
4. Transient heat conduction in the liquid adjacent to the wall during the growth of a bubble in its vicinity.
5. Vapour formation from the thin, superheated liquid layer under the growing bubble. If the core liquid is subcooled then vapour will simulataneously condense at the top of the bubble.

It is not uncommon for the heat transfer coefficients calculated according to the different equations, to deviate by up to a factor of 2 from the experimental values. However these deviations cannot be purely traced back to shortcomings in the theory, but can also be partly explained by experimental errors and inaccurate material property data. Moreover, the departure diameter of the bubbles, the contact angle and the bubble frequency at fixed external conditions are not, as is often assumed, constant but are subject to statistical fluctuations. It is not yet clear how these fluctuations have to be taken into consideration in the theory. It is also disadvantageous that, with a few exceptions, in the equations the structure of the heating surface is disregarded, although it has been known for some time that heat transfer in nucleate boiling is noticeably better on rough than on smooth surfaces.

In order to find an equation with a wide application range, it is expedient to combine the decisive properties for heat transfer into dimensionless groups. For the general law of heat transfer it is advisable to choose an exponential equation in these quantities, because equations of this nature have proven themselves for the representation of heat transfer.

Regression analysis represents an effective tool for establishing a relationship between the Nusselt number and the other dimensionless variables. This analysis allows us to decide which of the characteristic quantities are important in heat transfer and which are less meaningful. Stephan and Abdelsalam [4.69] have critically examined around 5000 known data for heat transfer in nucleate boiling and established empirical correlations using this method. Regression analysis showed that some dimensionless quantities are important for some substances, but are meaningless for others. The data could best be reproduced when the substances were split into four groups (hydrocarbons, cryogenic fluids, refrigerants and water) and a different set of dimensionless quantities were used for each of these groups. In addition, an equation valid for all substances has been derived, although it is less accurate than the equations for the particular substance groups named above.

As most experiments have been carried out near to atmospheric pressure, which for organic liquids corresponds to an approximate, normalised boiling pressure of $p/p_{cr} = 0.03$, the accuracy of these correlations can be increased if their validity is restricted to pressures close to ambient pressure. Such a relationship has been established by Stephan and Preußer [4.70]. It reads:

$$
Nu = 0.0871 \left(\frac{\dot{q}d_A}{\lambda'T_s}\right)^{0.674} \left(\frac{\varrho''}{\varrho'}\right)^{0.156} \left(\frac{\Delta h_v d_A^2}{a'^2}\right)^{0.371} \left(\frac{a'^2 \varrho'}{\sigma d_A}\right)^{0.350} (Pr')^{-0.162} \ .
$$

(4.96)

The quantities with ' relate to the saturated boiling liquid, and those with " relate to the saturated vapour. The Nusselt number is defined by $Nu = \alpha d_A/\lambda'$. The departure diameter d_A is given by (4.85), in which $\varrho_L = \varrho'$ and $\varrho_G = \varrho''$ have to be substituted. For the contact angle β_0 in water $\pi/4$ rad $= 45°$, in cryogenic liquids 0.01745 rad $= 1°$ and for other liquids 0.611rad $= 35°$ have to be used.

Measurements for *water* in the region of fully developed nucleate boiling with heat fluxes between

$$
10^4 \, \text{W/m}^2 < \dot{q} < 10^6 \, \text{W/m}^2
$$

and boiling pressures between

$$
0.5 \, \text{bar} < p < 20 \, \text{bar}
$$

can, according to investigations by Fritz [4.71], be reproduced well by the simple empirical equation

$$
\alpha = 1.95 \, \dot{q}^{0.72} p^{0.24} \ ,
$$

(4.97)

where α is in W/m²K, $\dot{q}$ in W/m² and p in bar.

From (4.96), for a given heat flux $\dot{q}_0$, the heat transfer coefficients α_0 can be obtained, in accordance with the validity of the equation only in the vicinity of atmospheric pressure. If a pair of values $\alpha_0, \dot{q}_0$ have been found for a moderate reference pressure p_0, then, according to a suggestion by Danilowa [4.72], from

$$
\frac{\alpha}{\alpha_0} = F(p^+) \left(\frac{\dot{q}}{\dot{q}_0}\right)^n \ ,
$$

(4.98)

where $F(p^+ = p/p_{cr})$ is a pressure function, the heat transfer coefficients at other pressures can be calculated. The pressure function has to fulfill the condition $F(p_0^+ = p_0/p_{cr}) = 1$. According to Gorenflo [4.73], for organic liquids, sulphur hexafluoride and ammonia, it reads

$$F(p^+) = 2.1\, p^{+0.27} + \left(4.4 + \frac{1.8}{1 - p^+}\right) p^+ \tag{4.99}$$

and for water and cryogenic liquids

$$F(p^+) = 2.55\, p^{+0.27} + \left(9 + \frac{1}{1 - p^{+2}}\right) p^{+2} . \tag{4.100}$$

Both equations were established for a heat flux $\dot{q}_0 = 20\,000\ \mathrm{W/m^2}$. The reference pressure p_0 is chosen such that, for the normalised pressure $p_0^+ = 0.03$, the function will be $F(p_0^+) = 1$. Correspondingly $\alpha_0, \dot{q}_0$ in (4.98) are values belonging to a pair of heat transfer coefficient and heat flux at the normalised pressure $p_0^+ = 0.03$.

Numerous experiments have yielded that the exponent n of the heat flux in (4.98) is not constant, but decreases with rising boiling pressure. According to Gorenflo [4.73] for organic liquids, sulphur hexafluoride and ammonia, it is

$$n = 0.9 - 0.3\, p^{+0.3} . \tag{4.101}$$

According to that $n = 0.8$ for $p_0^+ = 0.03$ and $n = 0.62$ for $p_1^+ = 0.8$. The pressure dependence for water and cryogenic liquids is somewhat weaker

$$n = 0.9 - 0.3\, p^{+0.15} . \tag{4.102}$$

For water at a boiling pressure of 1 bar corresponding to $p_1^+ = 0.004532$, it follows that $n = 0.77$, for $p_0^+ = 0.03$ we have $n = 0.72$, and for $p_2^+ = 0.8$, $n = 0.61$.

With this, the *data for water*, with the values $\alpha_0 = 3\,800\ \mathrm{W/m^2K}$, $\dot{q}_0 = 20\,000\ \mathrm{W/m^2}$ at $p_0^+ = 0.03$, can be reproduced from (4.98) by the following empirical equation

$$\frac{\alpha}{3\,800\ \mathrm{W/m^2K}} = F(p^+) \left(\frac{\dot{q}}{20\,000\ \mathrm{W/m^2}}\right)^{0.9 - 0.3\, p^{+0.15}} , \tag{4.103}$$

where $F(p^+)$ is given by (4.100). This equation is valid in the fully developed nucleate boiling region and at normalised boiling pressures of

$$10^{-4} \le p^+ \le 0.9 ,$$

which correspond to boiling pressures of

$$0.0221\ \mathrm{bar} \le p \le 199\ \mathrm{bar} .$$

It extends over a wider range of pressures than the simple equation (4.97) by Fritz and agrees well with this equation within its range of validity.

4.2.7 Some empirical equations for heat transfer during nucleate boiling in free flow

In the following the equations introduced previously for heat transfer in nucleate boiling in free flow are summarised and supplemented by further equations.

1. Nucleate boiling of water in free flow.
According to (4.97) it holds for $0.5\,\text{bar} < p < 20\,\text{bar}$ that

$$\alpha = 1.95\,\dot{q}^{0.72}p^{0.24}$$

with α in $\text{W/m}^2\text{K}$, $\dot{q}$ in W/m^2 and p in bar, and from (4.103), for $0.0221\,\text{bar} \leq p \leq 199\,\text{bar}$, it is also valid that

$$\frac{\alpha}{3\,800\,\text{W/m}^2} = F(p^+)\left(\frac{\dot{q}}{20\,000\,\text{W/m}^2}\right)^{0.9-0.3\,p^{+0.15}}$$

with

$$F\left(p^+ = \frac{p}{p_{\text{cr}}}\right) = 2.55\,p^{+0.27}\left(9 + \frac{1}{1-p^{+2}}\right)p^{+2}\ .$$

2. Nucleate boiling, general equation.
For pressures close to atmospheric, according to (4.96), it holds that

$$Nu = \frac{\alpha d_{\text{A}}}{\lambda}$$
$$= 0.0871\left(\frac{\dot{q}d_{\text{A}}}{\lambda' T_{\text{s}}}\right)^{0.674}\left(\frac{\varrho''}{\varrho'}\right)^{0.156}\left(\frac{\Delta h_{\text{v}}d_{\text{A}}^2}{a'^2}\right)^{0.371}\left(\frac{a'^2\varrho'}{\sigma d_{\text{A}}}\right)^{0.350}(Pr')^{-0.162}$$

with the departure diameter from (4.85):

$$d_{\text{A}} = 0.851\beta_0\sqrt{\frac{2\sigma}{g(\varrho_{\text{L}} - \varrho_{\text{G}})}}\ .$$

The contact angle β_0, in radians, for water lies between 0.22π ($=40°$) and 0.25π ($=45°$), and for the halogenated hydrocarbons used as refrigerants it is around 0.194π ($=35°$). For other pressures, α_0 for the reference heat flux $\dot{q}_0 = 20\,000\,\text{W/m}^2$ at a reference pressure p_0 close to the ambient pressure, is first calculated from the equation above. The desired heat transfer coefficient α at pressure p and heat flux $\dot{q}$ is obtained from (4.98):

$$\frac{\alpha}{\alpha_0} = F(p^+)\left(\frac{\dot{q}}{\dot{q}_0}\right)^n\ .$$

For organic liquids, sulphur hexafluoride and ammonia

$$F(p^+) = 2.1\,p^{+0.27} + \left(4.4 + \frac{1.8}{1-p^{+2}}\right)p^+$$

and

$$n = 0.9 - 0.3\,p^{+0.3} \ .$$

For water and cryogenic liquids

$$F(p^+) = 2.55\,p^{+0.27} + \left(9 + \frac{1}{1 - p^{+2}}\right)p^{+2}$$

and

$$n = 0.9 - 0.3\,p^{+0.15} \ .$$

3. Nucleate boiling on horizontal copper tubes.
According to Gorenflo [4.74]

$$\frac{\alpha}{\alpha_0} = \frac{F(p^+/\sqrt{\varphi})}{F(p_0^+/\sqrt{\varphi})} \cdot 5^{0.1H/t_1} \left(\frac{\dot{q}}{\dot{q}_0}\right)^{n_f} \ .$$

In which, α_0 is the heat transfer coefficient at pressure p_0 and reference heat flux $\dot{q}_0 = 20\,000$ W/m² calculated from the equation given in 2 for a flat tube. $F(p^+/\sqrt{\varphi})$ is obtained from the function $F(p^+)$ in 2, when in place of $p^+ = p/p_{\text{cr}}$ the quantity $p^+/\sqrt{\varphi}$ is used with the area ratio $\varphi = A_f/A$. A_f is the surface area of the finned tube, A that of a flat tube of the same core diameter as the finned tube. It is

$$n_f = n - 0.1\,H/t_1 \ .$$

n is the exponent for the heat flux given in 2. H is the height of the fin, t_1 is the internal fin spacing, see Fig. 4.43.

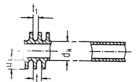

                                    Fig. 4.43: Dimensions of a finned tube

4. Nucleate boiling in a horizontal bank of smooth or finned tubes.
Heat transfer coefficients in boiling on the outside of smooth or finned tubes in a bundle are greater than those for an individual tube. The heat transfer is improved by the oncoming flow at the lowest tube row, in the tubes themselves the ascending vapour bubbles produce better convection. The influence of convection on the total heat transfer is clearly noticeable, especially if the heat flux is not too large. As a result of this the mean heat transfer coefficient α_m of the bundle is significantly larger than that of a single tube. According to [4.73] it holds that

$$\frac{\alpha_m}{\alpha_1} = 1 + \frac{1}{2 + \dot{q}\varphi/(1\,000\,\text{W}/\text{m}^2)} \ ,$$

where α_1 is the heat transfer coefficient of the lowest tube row. The equation is equally valid for smooth and finned tubes in the range

$$1\,000\mathrm{W/m^2} \leq \dot{q}\varphi \leq 20\,000\mathrm{W/m^2}$$

and for pressures close to ambient. The heat flux $\dot{q}$ is related to the total external surface area of the tube. The quantity φ is the area ratio for a finned tube; for a smooth tube, we have $\varphi = 1$. The heat transfer coefficients α_1 at the bottom row of tubes are obtained, according to a suggestion by Slipčević [4.75], from the heat transfer coefficients α_B for nucleate boiling on a single tube, see No. 2, and the heat transfer coefficients α_C for free flow without bubble formation, see 3.9.3 No. 3, also for a single tube, according to

$$\alpha_1 = \alpha_B + f\alpha_K \ ,$$

where f is a factor, which depends on the magnitude of the inlet flow velocity. For small bundles f is around 0.5, whilst for large bundles $f = 1$.

5. **Maximum heat flux.**
According to [4.76] and [4.77], it holds that

$$\dot{q}_{\mathrm{max}} = K\Delta h_v\sqrt{\varrho''}[\sigma(\varrho' - \varrho'')g]^{1/4} \ ,$$

where K is a constant to be found by experimentation, that lies between 0.13 and 0.16. A mean value is $K = 0.145$.

6. **Film boiling.**
In film boiling, the influence of radiation on the heat transfer cannot be neglected because of the high wall temperature. According to [4.78], the mean heat transfer coefficient α_m in film boiling is given by

$$\frac{\alpha_m}{\alpha_{mG}} = 1 + \frac{1}{5}\frac{\alpha_R}{\alpha_{mG}}\left(4 + \frac{1}{1 + 3\alpha_{mG}/\alpha_R}\right)$$

with

$$\alpha_{mG} = \frac{2}{3}\left(\frac{\varrho_G(\varrho_L - \varrho_G)g\Delta h_v\lambda_G^3}{\eta_G\Delta\vartheta}\frac{1}{H}\right)^{1/4}$$

and

$$\alpha_R = \frac{\sigma_S}{\dfrac{1}{\varepsilon_0} + \dfrac{1}{\varepsilon_L} - 1}\frac{T_0^4 - T_s^4}{\vartheta_0 - \vartheta_s} \ .$$

H is the height of the plate or $H = d\pi/2$ for horizontal tubes, $\sigma_S = 5.67 \cdot 10^{-8}\,\mathrm{W/m^2K^4}$ is the Stefan-Boltzmann constant, cf. section 1.1.5, page 25, and 5.2.3, page 548, ε_0 is the emissivity of the wall, ε_L is the emissivity of the liquid surface.

Example 4.4: In a saucepan 1 l of water at atmospheric pressure 0.1013 MPa is to be boiled on an electric cooker. The power of the hotplate of the electric cooker is $\dot{Q} = 3\,\text{kW}$, its diameter is the same as that of the saucepan, namely 0.3 m.

a) How long does it take until the water starts to boil, if the initial temperature is 20 °C? The heat losses to the surroundings amount to 30 % of the heat input.

b) What is the temperature at the bottom of the saucepan when the water begins to boil?

c) How long does it take for all the water to be vaporised?

d) Calculate the maximum heat flux.

The following data are given: saturation temperature $\vartheta_s = 100$ °C at a pressure of 0.1013 MPa, density of the liquid $\varrho_L = 958.1\,\text{kg/m}^3$, specific heat capacity of the liquid $c_{pL} = 4.216$ kJ/kgK, vapour density $\varrho'' = 0.5974\,\text{kg/m}^3$, surface tension $\sigma = 58.92 \cdot 10^{-3}\,\text{N/m}$, enthalpy of vaporisation $\Delta h_v = 2257.3\,\text{kJ/kg}$.

a) Until the boiling point is reached the following amount of heat has to be fed to the water

$$Q = M\,c_{pL}\,\Delta\vartheta = \varrho_L\,V\,c_{pL}\,\Delta\vartheta = 958.1\,\text{kg/m}^3 \cdot 10^{-3}\,\text{m}^3 \cdot 4.216\,\text{kJ/kgK} \cdot 80\,\text{K} = 323\,\text{kJ} \ .$$

We have $\dot{Q}\,\Delta t_0 = Q$, i.e. the time taken to reach the boiling temperature is

$$\Delta t_0 = \frac{Q}{\dot{Q}}\frac{323\,\text{kJ}}{3\,\text{kW} \cdot 0.7} = 154\,\text{s} = 2.56\,\text{min} \ .$$

b) According to (4.97) the heat transfer coefficient in boiling in obtained as

$$\alpha = 1.95\,\dot{q}^{0.72}\,p^{0.24} \quad \text{with} \quad \dot{q} = \frac{\dot{Q}}{A} = \frac{0.7 \cdot 3\,\text{kW}}{0.3^2\,\text{m}^2 \cdot \pi/4} = 2.971 \cdot 10^4\,\text{W/m}^2 \ .$$

So

$$\alpha = 1.95 \cdot (2.971 \cdot 10^4)^{0.72} \cdot (1.013)^{0.24}\,\text{W/m}^2\text{K} = 3250\,\text{W/m}^2\text{K}$$

and

$$\Delta\vartheta = \frac{\dot{q}}{\alpha} = \frac{2.971 \cdot 10^4\,\text{W/m}^2}{3250\,\text{W/m}^2\text{K}} = 9.1\,\text{K} \ .$$

The temperature at the bottom of the pan is $\vartheta_0 = \vartheta_s + \Delta\vartheta = 100\,°\text{C} + 9.1\,°\text{C} = 109.1\,°\text{C}$.

c) 70 % of the heat released is fed to the water. In order to completely vaporise the mass $M = \varrho_L\,V$, the following amount of heat is required

$$Q = \varrho_L\,V\,\Delta h_v = 958.1\,\text{kg/m}^3 \cdot 10^{-3}\,\text{m}^3\,2257.3\,\text{kJ/kg} = 2163\,\text{kJ} \ .$$

The time required for this is

$$\Delta t = Q/\dot{Q} = \frac{2163\,\text{kJ}}{0.7 \cdot 3\,\text{kW}} = 1\,030\,\text{s} = 17.2\,\text{min} \ .$$

d) The maximum heat flux is found, using 4.2.7, No. 5, to be

$$\begin{aligned}
\dot{q}_{max} &= 0.145\,\Delta h_v\,\sqrt{\varrho''}\,[\sigma\,(\varrho' - \varrho'')\,g]^{1/4} \\
&= 0.145 \cdot 2.2573 \cdot 10^6\,\text{J/kg} \cdot \sqrt{0.5974\,\text{kg/m}^3} \\
&\quad \cdot [58.92 \cdot 10^{-3}\,\text{N/m} \cdot (958.1 - 0.5974)\,\text{kg/m}^3 \cdot 9.81\,\text{m/s}^2]^{1/4} \\
\dot{q}_{max} &= 1.23 \cdot 10^6\,\text{W/m}^2 \ .
\end{aligned}$$

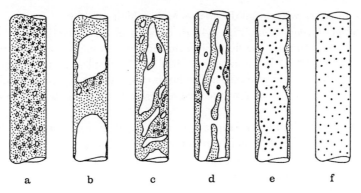

Fig. 4.44: Flow types in a vertical, unheated tube with upward flow. **a** Bubble flow; **b** Plug flow; **c** Churn flow; **d** Wispy-annular flow; **e** Annular flow; **f** Spray or drop flow

4.2.8   Two-phase flow

In boiling in free flow the flow pattern and heat transfer are determined by the difference in the hot surface and boiling temperatures, as well as by the properties of the fluid and the heating wall. Meanwhile in boiling in forced flow the velocity of the vapour and liquid phases and the distribution of the phases are additional factors that influence the flow patterns and heat transfer. As has already been discussed in section 4.2.1, the heat transfer coefficient can no longer be represented by a simple empirical correlation of the form $\alpha = c\dot{q}^n$, but the mass flux $\dot{m}$ and the quality x^* come into play as influencing quantities, such that empirical heat transfer relationships of the form $\alpha = c\dot{q}^n\dot{m}^s f(x^*)$ are used. The form of these relationships is significantly determined by the flow pattern, which will be discussed in the following section.

4.2.8.1   The different flow patterns

The numerous forms of two-phase flow can be divided into certain basic types, between which transition and mixed states are possible. The basic types of flow pattern in *upward two-phase flow in a vertical, unheated tube* are shown in Fig. 4.44.

In *bubble flow*, Fig. 4.44 a, the gas or vapour phase is uniformly dispersed in the continuous liquid phase. Only very small bubbles are spherical the larger ones are oblate. This type of flow pattern occurs when the gas fraction is small. In *plug flow*, Fig. 4.44 b, large bubbles (plugs) almost fill the entire tube cross section. Between the plugs, the liquid is interspersed with small bubbles.

At high mass fluxes the bubble structure increasingly disintegrates. This produces *churn flow*, Fig. 4.44 c. It consists more or less of large irregular gas or vapour fragments and has a very unstable character. This type of flow develops especially in tubes with large diameters and at high pressure.

The *wispy-annular flow*, Fig. 4.44 d, consists of a relatively thick liquid film at the wall, even though the liquid fraction in the vapour or gas core of the flow is still large. The film is interspersed with small bubbles, and the liquid phase in the core flow is mainly made up of large drops that sometimes coalesce into liquid strands. This type of flow pattern is normally observed when the mass flux is large.

A pattern that frequently appears is *annular flow*, Fig. 4.44 e. It is characterised by the fact that the main portion of the liquid mass is at the wall and the gas or vapour phase, that is interspersed with drops, flows in the core of the tube at a significantly higher velocity.

As a result of evaporation and especially at high velocites of vapour or gas, the liquid film at the wall disintegrates and a *spray or drop flow* is formed, Fig. 4.44 f. This occurs particularly in evaporation at high pressure.

All these two-phase flow patterns are not only observed in cylindrical tubes, but also in channels with other cross sections, as long as they do not deviate completely from the circular shape, for example a channel with a very flat, oblate cross section.

All these flow types appear more or less in a series one after the other during the evaporation of a liquid in a vertical tube, as Fig. 4.30 illustrates. The structure of a non-adiabatic vapour-liquid flow normally differs from that of an adiabatic two-phase flow, even when the local flow parameters, like the mass flux, quality, etc. agree with each other. The cause of this are the deviations from thermodynamic equilibrium created by the radial temperature differences, as well as the deviations from hydrodynamic equilibrium. Processes that lead to a change in the flow pattern, such as bubbles coalescing, the dragging of liquid drops in fast flowing vapour, the collapse of drops, and the like, all take time. Therefore, the quicker the evaporation takes place, the further the flow is away from hydrodynamic equilibrium. This means that certain flow patterns are more pronounced in heated than in unheated tubes, and in contrast to this some may possibly not appear at all.

The force of gravity causes the liquid in a *horizontal, unheated tube* or channel, to flow mainly in the lower section and the vapour in the upper part of the tube. The smaller the inertia forces are in comparison to the gravitational forces, the greater the difference between the flow in horizontal and vertical tubes. Thus, at low velocities flow patterns develop in horizontal tubes that are never seen in vertical tubes. Fig. 4.45 shows the characteristic patterns of two-phase flow in a horizontal tube.

The vapour bubbles collect in the upper part of the tube due to buoyancy forces, this is known as *bubble flow*, Fig. 4.45 a. They can then grow into plugs, producing *plug flow*, Fig. 4.45 b.

At low velocities the two phases are completely separate. This is called *stratified flow*, Fig. 4.45 c. If the gas or vapour velocity increases then waves will form at the surface of the liquid, *wavy flow*, Fig. 4.45 d. The wave peaks will get larger

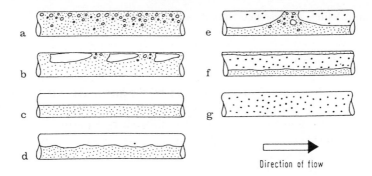

Fig. 4.45: Flow patterns in a horizontal, unheated tube **a** bubble flow; **b** plug flow; **c** stratified flow; **d** wavy flow; **e** slug flow; **f** annular flow; **g** spray or drop flow

and eventually wet the upper channel wall in an irregular sequence, *slug flow*, Fig. 4.45 e. At even higher velocities an *annular flow* develops, Fig. 4.45 f, where the thickness of the film in the top half of the tube is generally smaller than that in the lower part of the tube. With a further increase in the vapour velocity this is transformed into *spray flow*, Fig. 4.45 g.

In a *horizontal, heated tube* these flow patterns appear one after the other. For the same reasons as in a vertical evaporator tube, thermodynamic equilibrium is not achieved because of the radial temperature profile. Fig. 4.46 shows the different flow patterns in a horizontal evaporator tube, under the assumption that the liquid enters the tube at a sufficiently low velocity, below 1 m/s. It is clear

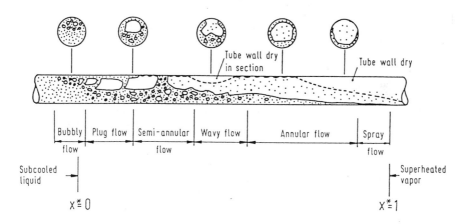

Fig. 4.46: Flow patterns in horizontal evaporator tube according to Collier [4.79]

here, that in slug flow the upper wall of the tube alternates between being dry and wetted. In the region of annular flow, the upper wall of the tube dries out further downstream. The higher the mass flux, the more symmetrical the phase distribution, which approaches that in a vertical evaporator tube.

4.2.8.2   Flow maps

As the different flow regimes are determined by the forces between the two phases, above all by the inertia and gravitational forces, it is appropriate to mark the dependence of the boundaries between the different flow patterns on these forces in diagrams, so-called flow maps. This type of flow map was first presented by Baker [4.80]. Therefore, we also speak of *Baker-diagrams*. These diagrams only provide a rough orientation because the decisive forces in the different flow regions are not known with sufficient certainty. In particular, for a two-phase flow with heat input, large uncertainty has to be reckoned with, as most flow maps have been developed for adiabatic flows. In addition to this, the transition regions between the different flow regimes are not very clear, a definite boundary with a marked transition does not exist.

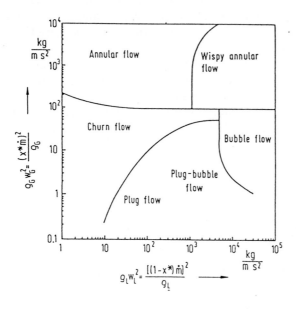

Fig. 4.47: Flow pattern diagram for vertical, two-phase flow according to Hewitt and Roberts [4.81]

Fig. 4.47 is an example of one of these flow maps. It was developed by Hewitt and Roberts [4.81] from observations of the flow in an air-water mixture at ambient pressure and in water-steam mixtures at high pressure in vertical tubes of diameters between 10 and 30 mm. The momentum fluxes $\varrho_G w_G^2$ of the vapour and $\varrho_L w_L^2$ of the liquid are plotted on the ordinates. These have the same dimensions as pressure and are formed with the apparent gas and liquid velocities, that would appear if each of the two phases filled the entire cross section,

$$w_G = x^* \dot{m}/\varrho_G \quad \text{and} \quad w_L = (1 - x^*)\dot{m}/\varrho_L \; .$$

A corresponding flow map for horizontal and inclined tubes has been developed by Taitel and Dukler [4.82].

4.2.8.3 Some basic terms and definitions

Some basic terms and defintions from two-phase flow theory are required for the description of heat transfer in boiling. For this we will consider the section of a channel shown in Fig. 4.48, in which the gas and liquid are flowing. An annular

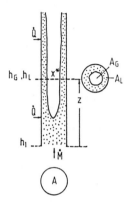

Fig. 4.48: Two-phase flow in a heated channel

flow is shown in the picture for the sake of simplicity, but the following terms are also valid for the other flow patterns. In any cross section A, the gas phase fills the fraction

$$\varepsilon := A_G/A \; , \tag{4.104}$$

and the liquid phase the corresponding fraction

$$1 - \varepsilon = A_L/A \; , \tag{4.105}$$

because $A_G + A_L = A$. These fractions do not change within a sufficiently small tube section Δz. Therefore

$$\varepsilon = A_G\Delta z/A\Delta z \; .$$

The volume fraction of the vapour in the tube section under consideration is then

$$\varepsilon := V_G/V \tag{4.106}$$

and the volume fraction of the liquid is

$$1 - \varepsilon = V_L/V \; . \tag{4.107}$$

The quantity ε is called the *volumetric vapour content*. This is to be distinguished from the *volumetric quality*

$$\varepsilon^* := \dot{V}_G/\dot{V} \; , \tag{4.108}$$

which is the volumetric flow of the vapour over the total flow of vapour and liquid. A further term is the *quality*, that represents the ratio of the mass flow rate $\dot{M}_G$ of the vapour to the total mass flow rate $\dot{M} = \dot{M}_G + \dot{M}_L$:

$$x^* := \dot{M}_G/\dot{M} \ . \tag{4.109}$$

Thus

$$1 - x^* = \dot{M}_L/\dot{M} \ . \tag{4.110}$$

The quality has to distinguished from the specific vapour content in thermodynamics, which is defined by $x = M_G/M$. With the mass flow rates $\dot{M}_G$ of the gas and $\dot{M}_L$ of the liquid the mean velocity of both phases in any cross section can be obtained

$$w_G = \frac{\dot{M}_G}{\varrho_G A_G} = \frac{x^* \dot{M}}{\varrho_G \varepsilon A} = \frac{x^* \dot{m}}{\varrho_G \varepsilon} \ , \tag{4.111}$$

$$w_L = \frac{\dot{M}_L}{\varrho_L A_L} = \frac{(1 - x^*)\dot{M}}{\varrho_L (1 - \varepsilon)A} = \frac{(1 - x^*)\dot{m}}{\varrho_L (1 - \varepsilon)} \ . \tag{4.112}$$

The ratio of the two velocities is known as *slip* or the *slip factor*

$$s := \frac{w_G}{w_L} = \frac{x^*}{1 - x^*} \frac{1 - \varepsilon}{\varepsilon} \frac{\varrho_L}{\varrho_G} \ . \tag{4.113}$$

The quality x^* and the volumetric quality ε^* are linked to each other by

$$x^* = \frac{\dot{M}_G}{\dot{M}_G + \dot{M}_L} = \frac{\dot{V}_G \varrho_G}{\dot{V}_G \varrho_G + \dot{V}_L \varrho_L} = \frac{\varepsilon^* \dot{V} \varrho_G}{\varepsilon^* \dot{V} \varrho_G + (1 - \varepsilon^*)\dot{V} \varrho_L}$$

or

$$x^* = \frac{\varepsilon^*}{\varepsilon^* + (1 - \varepsilon^*)\varrho_L/\varrho_G} \ . \tag{4.114}$$

The volumetric quality ε^* and the vapour content ε coincide when the slip factor is equal to one, i.e. when $w_G = w_L$, which then gives

$$\varepsilon^* = \frac{\dot{V}_G}{\dot{V}} = \frac{w_G A_G}{w_G A_G + w_L A_L} = \frac{A_G}{A} = \varepsilon \ .$$

The quality x^* is normally known or is easy to find out.

In general, for flow in unheated channels, the mass flow rates of the individual phases are stipulated. In a heated channel the quality is yielded from an energy balance, Fig. 4.48. We assume here that the liquid enters the evaporator tube in a subcooled state. Its specific enthalpy at the inlet is h_1. The input heat flow rate $\dot{Q}$ heats the liquid to the saturation temperature where it begins to evaporate. The energy balance between the inlet and another cross section at the point z, disregarding the kinetic and potential energy, yields

$$\dot{M} h_1 + \dot{Q} = \dot{M}[x^* h_G + (1 - x^*)h_L] \ .$$

If we further assume that the vapour and liquid are in thermal equilibrium in the cross section at point z, then their specific enthalpies in the saturated state have to be calculated at pressure $p(z)$. This gives $h_G = h''$ and $h_L = h'$ as well as $h'' - h' = \Delta h_v$ the enthalpy of vaporisation at pressure p in the cross section being considered. As we have assumed thermodynamic equilibrium, the quality is indicated by x^*_{th} and is called the *thermodynamic quality*. This yields

$$\dot{M}h_1 + \dot{Q} = \dot{M}(x^*_{th}\Delta h_v + h')$$

and

$$x^*_{th} = \frac{1}{\Delta h_v}\left(\frac{\dot{Q}}{\dot{M}} + h_1 - h'\right) \ . \tag{4.115}$$

The assumptions made mean that this equation is only valid if the vapour and liquid in a cross section are at the same temperature. It fails close to the inlet, where the quality x^* is still low, and also with a high vapour content. In the vicinity of the inlet, vapour bubbles can already form on the hot wall, even when the core flow is still subcooled. So the vapour and liquid are at different temperatures. The quality is thus positive, whilst according to (4.115) a negative value

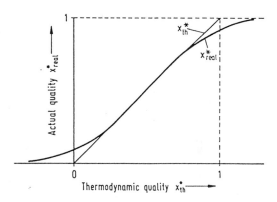

Fig. 4.49: Qualitative plot of the real quality x^*_{real} against the thermodynamic quality x^*_{th}

for the thermodynamic quality is obtained, because the liquid is still subcooled, and therefore $\dot{Q} + \dot{M}h_1 < \dot{M}h'$. At high vapour content spray flow develops. Heat is principally transferred to the vapour which is superheated, although fluid drops are still present in the core flow, which only evaporate slowly. The quality is therefore lower than one, even though the thermodynamic quality according to (4.115) has already reached a value of one.

However, in regions of intermediate quality, where neither subcooled boiling or spray flow occur, (4.115) delivers exact values. Fig. 4.49 shows a qualitative plot of the real quality x^*_{real} against the thermodynamic quality x^*_{th} based on (4.115).

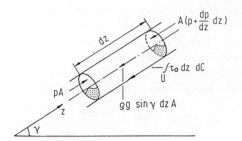

Fig. 4.50: Force balance on a two-phase fluid element

4.2.8.4 Pressure drop in two-phase flow

In two-phase flow, the boiling temperature falls in the direction of flow as a result of the pressure drop. This results in a change in the driving temperature drop decisive for heat transfer along the flow path. Calculation of the heat transfer without simultaneous investigation of the pressure drop is therefore impossible. The fundamentals steps for this shall be explained in the following.

In the calculation of the pressure drop, we will consider a channel inclined at an angle γ to the horizontal, through which a two-phase fluid flows, Fig. 4.50.

The forces exerted by the pressure, the friction forces at the channel wall and gravity all act on a volume element of length $\mathrm{d}z$, as illustrated in the picture. We are presuming the flow to be steady and one-dimensional and that the cross section of the channel is constant. The pressure, density and momentum flux over a cross section are only dependent on the flow path z. The momentum balance implies that the sum of the pressure, friction and gravitational forces are equal to the change in momentum

$$\left[p - \left(p + \frac{\mathrm{d}p}{\mathrm{d}z}\,\mathrm{d}z\right)\right] - \tau_0 C\,\mathrm{d}z - \varrho\,g A \sin\gamma\,\mathrm{d}z = \frac{\mathrm{d}}{\mathrm{d}z}(\dot{M}w)\,\mathrm{d}z$$

or

$$-\frac{\mathrm{d}p}{\mathrm{d}z} = \tau_0 \frac{C}{A} + \frac{1}{A}\frac{\mathrm{d}}{\mathrm{d}z}(\dot{M}w) + \varrho g \sin\gamma \ , \tag{4.116}$$

where τ_0 is the shear stress at the channel wall and C is the circumference of the channel. The density of a two-phase mixture is calculated from

$$\varrho = \varepsilon\varrho_\mathrm{G} + (1-\varepsilon)\varrho_\mathrm{L} \ . \tag{4.117}$$

The flow momentum is made up of that of the gas and the liquid, according to

$$\dot{M}w = \dot{M}_\mathrm{G}w_\mathrm{G} + \dot{M}_\mathrm{L}w_\mathrm{L} \ . \tag{4.118}$$

Together with (4.111) and (4.112), this can also be written as

$$\dot{M}w = \frac{\dot{M}^2}{A}\left[\frac{x^{*2}}{\varepsilon\varrho_\mathrm{G}} + \frac{(1-x^*)^2}{(1-\varepsilon)\varrho_\mathrm{L}}\right] \ . \tag{4.119}$$

Putting (4.117) and (4.119) into (4.116) yields

$$-\frac{dp}{dz} = \tau_0 \frac{C}{A} + \dot{m}^2 \frac{d}{dz} \left[\frac{x^{*2}}{\varepsilon \varrho_G} + \frac{(1-x^*)^2}{(1-\varepsilon)\varrho_L} \right] + \left[\varepsilon \varrho_G + (1-\varepsilon)\varrho_L \right] g \sin\gamma \ . \quad (4.120)$$

The total pressure drop is composed of three parts, the pressure drop due to friction

$$-\left(\frac{dp}{dz} \right)_f = \tau_0 \frac{C}{A} \ , \quad (4.121)$$

the acceleration pressure drop

$$-\left(\frac{dp}{dz} \right)_a = \dot{m}^2 \frac{d}{dz} \left[\frac{x^{*2}}{\varepsilon \varrho_G} + \frac{(1-x^*)^2}{(1-\varepsilon)\varrho_L} \right] \quad (4.122)$$

and the pressure drop as a result of gravity, which is also called the geodetic pressure drop

$$-\left(\frac{dp}{dz} \right)_g = \left[\varepsilon \varrho_G + (1-\varepsilon)\varrho_L \right] g \sin\gamma \ . \quad (4.123)$$

In channels with bends or constrictions additional pressure drops occur, which will not be dealt with here. The relevant literature is suggested for further reading on this topic [4.83].

The pressure drop due to friction exists because of the shear stress between the fluid flow and the channel wall. The acceleration pressure drop is yielded from the change in momentum in both phases, as in vapour-liquid flows evaporation due to flashing occurs because of the loss in pressure, and in heated channels, evaporation occurs in addition due to the heat input. This causes a change in the mass and velocity and therefore in the momentum flux of both phases. The geodetic pressure drop is caused by the gravitational force acting on the fluid. It disappears in horizontal flows. Geodetic and acceleration pressure drops are often negligible in comparison to the frictional pressure drop.

However, in heated channels with large heat and mass fluxes the *acceleration pressure drop* can be considerable and no longer assumes negligible values. In an adiabatic, two-phase flow, the acceleration pressure drop only exists as a result of further evaporation or expansion of the vapour or gas phase, and is small. A simple rule for adiabatic two-phase flow in refrigeration plants say that the acceleration pressure drop does not play a role as long as $\Delta p_f/p_s < 0.2$ holds for the frictional pressure drop. If this is not the case, then the acceleration pressure drop has to be determined, according to (4.122), over the channel length between inlet 1 and outlet 2:

$$(p_1 - p_2)_a = \dot{m}^2 \left[\frac{x_2^{*2}}{\varepsilon_2 \varrho_{G2}} - \frac{x_1^{*2}}{\varepsilon_1 \varrho_{G1}} + \frac{(1-x_2^*)^2}{(1-\varepsilon_2)\varrho_{L2}} - \frac{(1-x_1^*)^2}{(1-\varepsilon_1)\varrho_{L1}} \right] \ . \quad (4.124)$$

Accurate calculations are only possible here, if a reliable relationship for the volumetric vapour content is available.

If the slip factor can be set to $s = 1$, (homogeneous flow), then, according to (4.113), the volumetric vapour content will be

$$\varepsilon = \left(\frac{\varrho_G}{\varrho_L} \frac{1 - x^*}{x^*} + 1 \right)^{-1} . \tag{4.125}$$

In complete vaporisation, with a change in the quality from $x_1^* = 0$ to $x_2^* = 1$, $\varepsilon_1 = 0$ and $\varepsilon_2 = 1$ and the acceleration pressure drop will be

$$(p_1 - p_2)_a = \dot{m}^2 \left[\frac{1}{\varrho_{G2}} - \frac{1}{\varrho_{L1}} \right] .$$

The *frictional pressure drop* normally constitutes the largest fraction of the total pressure drop. However, only empirical methods are available for its calculation. It includes not only the momentum transfer between the fluid and the wall, but also the momentum transfer between the individual phases. These two processes cannot be measured separately and can only be estimated for simple flows. Thus, only imprecise ideas of the influence of momentum transfer between the phases exist.

At constant mass flux, the frictional pressure drop does not show linear dependence on the quality between the limits $x^* = 0$ and $x^* = 1$, but rises with increasing x^* until it reaches a maximum between $x^* = 0.7$ and $x^* = 0.9$, and then falls back to the pressure drop of a pure vapour stream, Fig. 4.51. This maximum is more pronounced the larger the difference in the densities of the two phases.

Methods for the determination of the frictional pressure drop usually start with simple models. For the most part, either *homogeneous flow* (homogeneous distribution of the phases $s \to 1$) or *heterogeneous flow* (heterogeneous distribution of the phases, $s > 1$) are presumed. Less common are methods based upon specific flow patterns and are only applicable if this flow pattern is present.

In the calculation of frictional pressure drop it is advantageous to define a few parameters that are suitable for the representation of two-phase frictional pressure drop and the volumetric quality. The frictional pressure drop is often reduced to the pressure drop for single phase flow, using the definitions from Lockhart and Martinelli [4.84]

$$\left(\frac{dp}{dz} \right)_f := \Phi_L^2 \left(\frac{dp}{dz} \right)_L \tag{4.126}$$

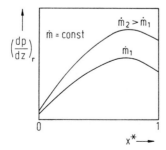

Fig. **4.51**: Profile of the frictional pressure drop

or

$$\left(\frac{dp}{dz}\right)_f := \Phi_G^2 \left(\frac{dp}{dz}\right)_G \ . \tag{4.127}$$

Here $(dp/dz)_L$ is the frictional pressure drop of the liquid, $(dp/dz)_G$ is that of the vapour, under the assumption that each of the two phases is flowing by itself through the tube. The factors Φ_L^2 and Φ_G^2 are defined by th equations above. If these factors are known, only the pressure drop of the individual phases has to be determined, to allow the frictional pressure drop for two-phase flow to be calculated.

The pressure drop in the two phases can be determined using well known methods. For the frictional pressure drop of a liquid stream, it holds that

$$\left(\frac{dp}{dz}\right)_L = -\zeta_L \frac{1}{d}\frac{\varrho_L w_L^2}{2} = -\zeta_L \frac{1}{d}\frac{\dot{m}_L^2}{2\varrho_L} \tag{4.128}$$

with $\dot{m}_L = \dot{m}(1 - x^*)$, where d is the diameter of the channel und ζ is the friction factor dependent on the Reynolds number. According to Blasius, for example,

$$\zeta_L = c_1/Re_L^n \ , \tag{4.129}$$

where c_1 and n are functions of the flow, i.e. whether it is laminar or turbulent, and the roughness of the tube. The Reynolds numbers is calculated under the assumption that only the liquid is flowing through the tube

$$Re_L = \frac{w_L \varrho_L d}{\eta_L} = \frac{\dot{m}_L d}{\eta_L} = \frac{\dot{m}(1 - x^*)d}{\eta_L} \ . \tag{4.130}$$

Another possibility for the investigation of the pressure drop in two-phase flows exists in the determination of the frictional pressure drop of the gas and the liquid, under the assumption that the total mass of the fluid would be flowing through the tube either as liquid or as gas. In this case the factors Φ_{L0} and Φ_{G0} are defined by

$$\left(\frac{dp}{dz}\right)_f := \Phi_{L0}^2 \left(\frac{dp}{dz}\right)_{L0} \tag{4.131}$$

and

$$\left(\frac{dp}{dz}\right)_f := \Phi_{G0}^2 \left(\frac{dp}{dz}\right)_{G0} \ . \tag{4.132}$$

This gives

$$\left(\frac{dp}{dz}\right)_{L0} = -\zeta_{L0} \frac{1}{d}\frac{\dot{m}^2}{2\varrho_L} \ , \tag{4.133}$$

and the resistance factor ζ_{L0} has to be determined with the Reynolds number

$$Re_{L0} = \frac{\dot{m}d}{\eta_L} \ . \tag{4.134}$$

The homogeneous model

In flows with a high proportion of small bubbles, $x^* \to 0$, in spray or drop flow, $x^* \to 1$, or in flows with small density differences, like those near the critical state, the frictional pressure drop can be well represented by the homogeneous model. The heterogeneous model delivers inaccurate results for these conditions.

Calculations according to the homogeneous model are similar to those for single-phase flow, although they involve suitably defined mean property values. The following holds for the frictional pressure drop in a homogeneous two-phase flow

$$\left(\frac{dp}{dz}\right)_f := -\zeta \frac{1}{d} \frac{\varrho w^2}{2} = -\zeta \frac{1}{d} \frac{\dot{m}^2}{2\varrho} \ . \tag{4.135}$$

With the assumption of equal liquid and vapour velocities, the volumetric vapour content is obtained from (4.125) and with that the density of the homogeneous flow is

$$\varrho = \varepsilon \varrho_G + (1 - \varepsilon)\varrho_L = \left(\frac{x^*}{\varrho_G} + \frac{1 - x^*}{\varrho_L}\right)^{-1} \ . \tag{4.136}$$

The friction factor is calculated in the same way as for single-phase flow, for example, according to Blasius

$$\zeta = c_1/Re^n \ . \tag{4.137}$$

The dynamic viscosity of the homogeneous two-phase flow now has to be put into the Reynolds number $Re = \dot{m}d/\eta$. Empirical equations for this have been suggested in the literature [4.85] to [4.87]. These include the limits $\eta(x^* = 0) = \eta_L$ and $\eta(x^* = 1) = \eta_G$. According to McAdams et al. [4.85] it holds that

$$\frac{1}{\eta} = \frac{x^*}{\eta_G} + \frac{1 - x^*}{\eta_L} \ , \tag{4.138}$$

according to Cicchitti et al. [4.86]

$$\eta = x^* \eta_G + (1 - x^*)\eta_L \tag{4.139}$$

and from Dukler et al. [4.87]

$$\eta = \varrho \left[x^* \frac{\eta_G}{\varrho_G} + (1 - x^*)\frac{\eta_L}{\varrho_L}\right] \ . \tag{4.140}$$

With (4.136), the frictional pressure drop in homogeneous two-phase flow is obtained from (4.135) to be

$$\left(\frac{dp}{dz}\right)_f = -\zeta \frac{1}{d} \frac{\dot{m}^2}{2} \left(\frac{x^*}{\varrho_G} + \frac{1 - x^*}{\varrho_L}\right) \ . \tag{4.141}$$

On the other hand, the pressure drop, under the assumption that only liquid is flowing through the tube, is

$$\left(\frac{dp}{dz}\right)_L = -\zeta_L \frac{1}{d} \frac{\dot{m}^2(1 - x^*)^2}{2\varrho_L} \ . \tag{4.142}$$

The friction factor ζ_L of the liquid stream, according to Blasius' law, is

$$\zeta_L = c/Re_L^n \quad \text{with} \quad Re_L = \dot{m}(1-x^*)d/\eta_L \ . \tag{4.143}$$

With (4.142), (4.135) and (4.126), the factor Φ_L^2 for homogeneous two-phase flow is found to be

$$\Phi_L^2 = \frac{\zeta}{\zeta_L}\left[1 + x^*\left(\frac{\varrho_L}{\varrho_G} - 1\right)\right](1+x^*)^{-2} \ . \tag{4.144}$$

The friction factor, still using Blasius' law (4.137), and (4.143), can be eliminated, giving

$$\Phi_L^2 = (1+x^*)^{n-2}\left[1 + x^*\left(\frac{\varrho_L}{\varrho_G} - 1\right)\right]\bigg/\left[1 + x^*\left(\frac{\eta_L}{\eta_G} - 1\right)\right]^n \ , \tag{4.145}$$

where n is the exponent of Blasius' law (4.137).

The heterogeneous model

In contrast to the homogeneous model, the two phases flow separately to each other and have different velocities, so that slip exists between the phases.

A particularly simple and frequently used method comes from Lockhart and Martinelli [4.84]. It is based on measurements of air-water and air-oil mixtures in horizontal tubes at low pressure. However the procedure has also proved itself in upward, vertical flow of two-phase single and multicomponent mixtures. The basic idea of the Lockhart-Martinelli method is that the frictional pressure drop in a two-phase flow can be determined, with use of a correction factor, from the frictional pressure drop in the individual phases. This means that the two-phase mulitipliers Φ_L^2 and Φ_G^2 are defined according to (4.126) and (4.127).

We put

$$X^2 := \frac{\Phi_G^2}{\Phi_L^2} = \frac{(dp/dz)_L}{(dp/dz)_G} \tag{4.146}$$

with $(dp/dz)_L$ according to (4.142) and

$$\left(\frac{dp}{dz}\right)_G = -\zeta_G \frac{1}{d}\frac{\dot{m}^2 x^{*2}}{2\varrho_G} \ . \tag{4.147}$$

It follows from (4.142) and (4.147) that

$$X^2 = \frac{(dp/dz)_L}{(dp/dz)_G} = \frac{\zeta_L}{\zeta_G}\left(\frac{1-x^*}{x^*}\right)^2\frac{\varrho_G}{\varrho_L} \ . \tag{4.148}$$

In general, the quantity X defined by (4.146) is known as the *Lockhart-Martinelli parameter*. It assumes different values depending on the type of flow for the two phases, whether laminar or turbulent. The following combinations, indicated by indices on X, are possible:

Flow state of the phases		Indices of X
Gas	Liquid	
laminar	laminar	ll
laminar	turbulent	lt
turbulent	laminar	tl
turbulent	turbulent	tt

As an example we will calculate X_{tt} for the case where both phases are turbulent, a state that occurs frequently. Presuming the validity of Blasius' law for the friction factor

$$\zeta_L = c/Re_L^n = c/\left[\frac{\dot{m}(1 - x^*)d}{\eta_L}\right]^n \tag{4.149}$$

and

$$\zeta_G = c/Re_G^n = c/\left(\frac{\dot{m}x^*d}{\eta_G}\right)^n \ , \tag{4.150}$$

then, according to (4.148)

$$X_{tt}^2 = \left(\frac{1 - x^*}{x^*}\right)^{2-n} \left(\frac{\eta_L}{\eta_G}\right)^n \frac{\varrho_G}{\varrho_L} \ . \tag{4.151}$$

For turbulent flow in technically smooth tubes, the exponent n lies between 0.2 and 0.25. With $n = 0.2$ we have

$$X_{tt} = \left(\frac{1 - x^*}{x^*}\right)^{09} \left(\frac{\eta_L}{\eta_G}\right)^{01} \left(\frac{\varrho_G}{\varrho_L}\right)^{05} \ . \tag{4.152}$$

Lockhart and Martinelli assumed that each of the two factors Φ_G and Φ_L can be represented as a function of the parameter X. Fig. 4.52 shows the quantities Φ_L and Φ_G, determined by Lockhart and Martinelli for the different types of flow, plotted against the parameter X.

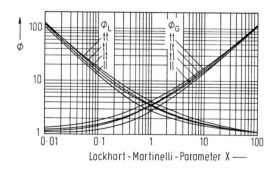

Fig. 4.52: Investigation of the frictional pressure drop according to Lockhart and Martinelli [4.84]

The curves are reproduced well by the following equations

$$\Phi_L^2 = 1 + \frac{C}{X} + \frac{1}{X^2} \ , \tag{4.153}$$

and
$$\Phi_G^2 = 1 + CX + X^2 \ . \tag{4.154}$$

The values from Table 4.2 are to be used for the constant C.

Table 4.2: Constants C in the equations (4.153) and (4.154)

Gas/Vapour	Liquid	Index	C
laminar	laminar	ll	5
laminar	turbulent	lt	10
turbulent	laminar	tl	12
turbulent	turbulent	tt	20

The flow of a phase can be assumed to be laminar when $Re < 1000$, and turbulent when $Re > 2000$. The transition region $1000 < Re < 2000$ is more difficult to predict but to be on the safe side the values for turbulent flow should be taken.

The Lockhart-Martinelli method is especially simple and clear. In certain parameter regions its accuracy is surpassed by other methods, but it delivers satisfactory values for the pressure drop, irrespective of the application, within a range of uncertainty of around ±50%. Larger deviations are to be expected for tube diameters $d > 0.1$ m. Furthermore, it should also be taken into account, that the two-phase mulitipliers Φ_L and Φ_G were determined from measurements at low pressures. Other equations have been developed for higher pressure and greater demands on the accuracy of the result. They are normally only valid for certain substances and in a narrow range of parameters. The extensive literature on two-phase flow, in particular the summary in [4.83], is suggested for further information on this subject.

4.2.8.5   The different heat transfer regions in two-phase flow

We shall now consider subcooled liquid fed into the bottom of a vertical evaporator tube, that is uniformly heated along its entire length. The heat flux $\dot{q}$ is assumed to be low and the tube should be long enough such that the liquid can be completely evaporated. Fig. 4.53 shows, on the left, alongside the various heat exchange regions that have already been explained, the profiles of the liquid and wall temperatures.

As long as the wall temperature stays below that required for the formation of vapour bubbles, heat will be transferred by single-phase, forced flow. If the wall is adequately superheated, vapour bubbles can form even though the core liquid is still subcooled. This is a region of subcooled boiling. In this area, the wall temperature is virtually constant and lies a few Kelvin above the saturation

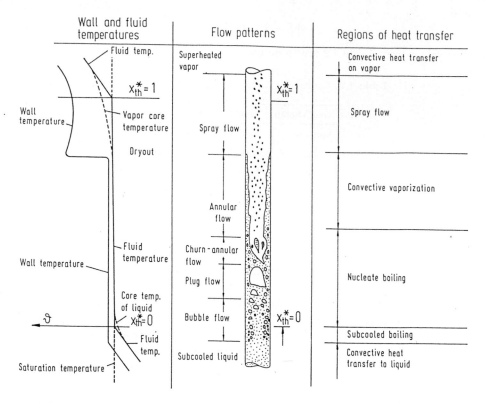

Fig. 4.53: Wall and liquid temperatures, flow pattern and the associated heat transfer regions, heated tube

temperature. The transition to *nucleate boiling*, is, by definition, at the point where the liquid reaches the saturation temperature at its centre, and with that the thermodynamic quality is $x_{th}^* = 0$. In reality, as Fig. 4.53 indicates, the liquid at the core is still subcooled due to the radial temperature profile, whilst at the same time vapour bubbles form at the wall, so that the mean enthalpy is the same as that of the saturated liquid. As explained in the previous section, the saturation temperature in the core is reached further downstream from the point $x_{th}^* = 0$.

In the nucleate boiling region heat transfer is chiefly determined by the formation of vapour bubbles and only to a small extent by convection. This region encompasses the bubble, plug, churn and a part of the annular flow regimes. The vapour content constantly increases downstream, and at a sufficiently high vapour content the churn flow converts into an annular flow, with a liquid film at the wall and vapour, with liquid droplets, in the core. The entire nucleate boiling region is characterised by the formation of vapour bubbles at the wall. However in annular flow, the liquid film downstream is so thin and its resistance to heat transfer is so low, that the liquid close to the wall is no longer sufficiently superheated, and the formation of bubbles at the wall is suppressed. Heat is conducted principally by the liquid that is evaporating at its surface. Heat is transferred by 'convective evaporation'.

As soon as the liquid film at the wall is completely evaporated, the temperature of a wall being heated with constant heat flux rises. This transition is known as dryout. The spray flow region is entered, followed by a region where all the liquid droplets being carried along by the vapour are completely evaporated, in which heat is transferred by convection to the vapour.

The transition regions shown in Fig. 4.53, can extend over very different areas depending on the size of the heat and mass flux.

4.2.8.6 Heat transfer in nucleate boiling and convective evaporation

Nucleate boiling in saturated liquids, so-called *saturated boiling*, and convective evaporation occur most frequently in evaporators, and shall therefore be explained in some depth here. The extensive publications concerning the other types of heat transfer in two-phase flows are suggested for further information, including [4.50]. By definition, saturated boiling begins when the quality, calculated assuming thermodynamic equilibrium from (4.115), is $x_{th}^* = 0$. The mean liquid temperature is then equal to the saturation temperature. However, in reality at the start of saturated boiling, the liquid core is still subcooled and the liquid layer adjacent to the wall is superheated. If bubbles have already formed in the region of subcooled boiling, bubble formation is suppressed again even when the mean temperature approaches the saturation temperature. This is the case when, as a result of improved heat transfer, the heat removed from a two-phase flow is equal to that supplied by the wall. Saturated boiling begins, when the heat supplied by the wall is greater than the heat that can be removed from the two-phase flow without additional bubble formation. A sufficiently large heat flux is necessary for this to happen. Its calculation is explained in [4.50], pg. 192.

In the calculation of the heat transfer in saturated boiling it is presumed that the heat transfer is predominantly determined by the bubble formation. In addition, convection also has some influence on the heat transfer, although this may only be small. Models and empirical equations start from the heat transfer in boiling in free flow. The influence of forced flow is taken into account with an extra term. This is based on the concept that the growth of the bubbles is only slightly influenced by the flow, as long as the bubbles are smaller than the superheated boundary layer. The heat transfer coefficient is a combination of two parts

$$\alpha_{2Ph} = \alpha_B + \alpha_C \ , \tag{4.155}$$

where α_B is the heat transfer coefficient in boiling in free flow, section 4.2.6, page 457 to 459, and α_C is the heat transfer coefficient for forced flow, if only the liquid had been flowing through the tube. This is calculated from the Colburn relationship for heat transfer in a single-phase forced flow

$$Nu = \frac{\alpha_C d}{\lambda_L} = 0.023 \, Re^{0.7} Pr^{1/3} \qquad (4.156)$$

with the Reynolds number $Re = \dot{m}_L d/\eta_L = \dot{m}(1-x^*)d/\eta_L$ and the Prandtl number $Pr = \nu_L/a_L$ of the liquid. In (4.155), the heat transfer by forced flow α_C only amounts to a few percent of the total heat transfer. This calculation method does not take into account the weak effect of the quality on the heat transfer in the saturated boiling region, but reproduces measured values with sufficient accuracy. More exact equations that take this effect into account have been communicated by Chawla [4.88], as well as by Stephan and Auracher [4.89].

When vapour and liquid flow downwards in a vertical tube, the slip between the two phases is reduced because of the buoyancy. This leads to a deterioration in the heat transfer, and according to measurements made by Pujol [4.90] using the refrigerant R113, the heat transfer coefficient α_{down} of the downward flow is smaller by a factor 0.75 than that for upward flow α_{up}

$$\alpha_{down} = 0.75 \, \alpha_{up} \ . \qquad (4.157)$$

The value α_{up} can, for example, be calulated from (4.155).

Although in (4.155) no differentiation was made between vertical and *horizontal tubes*, more recent measurements, by Steiner and colleagues [4.91], have confirmed the expectation that in horizontal tubes the vapour and liquid are always non-symmetrically distributed, disregarding very high flow velocities. This means that heat transfer in saturated boiling in vertical and horizontal tubes cannot be calculated from unified equations. The upper side of a horizontal tube is wetted less by the liquid than the lower side. This effect is especially noticeable when the tube wall is thin and a poor conductor of heat, because the temperatures over the perimeter are not as uniform as in thick walled tubes that conduct heat well. The heat transfer coefficients are, therefore, variable in the flow direction and over the perimeter. Local heat transfer measurements have not yet been carried out. Only heat transfer coefficients averaged over the perimeter have been measured by several authors. Steiner [4.91] has evaluated these and his own data, and developed equations for the calculation of the heat transfer in horizontal tubes with good and also with poor thermal conductivity. Further reference should be made to the representation [4.91] in the VDI-Heat Atlas. Information about heat transfer in saturated boiling in inclined tubes and tube bends can also be found in this publication.

The region of saturated boiling is follwed by that of *convective evaporation*. With the increasing vapour content the heat transfer from the wall to the fluid improves. The thermal resistance of the boundary layer decreases in comparison to the thermal resistances in nucleate boiling. Likewise, the wall temperature drops, cf. Fig. 4.53, so that only a few or no bubbles are formed at the wall. The heat transfer is predominantly or exclusively determined by evaporation at the phase boundary between the liquid at the wall and the vapour in the core flow.

Of the many different methods for calculating heat transfer, that from Chen [4.92] will be discussed here, as it was established with a model that is plausible in physical terms. It also has the advantage that it is not only valid for convective evaporation but also for saturated boiling. Similarly to saturated boiling, it is assumed that the heat transfer coefficient is a combination of two parts which are independent of each other. These are the part associated with bubble formation α'_B and that for convection α'_C:

$$\alpha_{2Ph} = \alpha'_B + \alpha'_C \ . \tag{4.158}$$

The part α'_B emanating from the bubble formation is based on the heat transfer coefficient α_B in nucleate boiling in free flow. However because the temperature rise in the boundary layer of a forced flow is steeper than in free flow nucleate boiling, more heat will be released from the wall by conduction and the bubble formation will be partially suppressed in comparison to that in free flow. Chen accounted for this effect with a suppression factor $S \leq 1$, which the heat transfer coefficient α_B in nucleate boiling in free flow is multiplied by, $\alpha'_B = S\alpha_B$.

The factor S approaches one for vanishingly small mass flux, the heat transfer coefficient α'_B is then the same as that in boiling in free flow. It approaches zero at large mass flux because then the heat transfer is exclusively determined by the convective part α'_C in (4.158).

The convective part α'_C of the heat transfer coefficient includes a contribution for the heat transfer to the single-phase liquid. The rapidly flowing vapour and the vapour bubbles that are still present mean that the heat transfer coefficient α_C of the single-phase forced liquid flow will be greater and has to be improved by an *enhancement factor* $F \geq 1$:

$$\alpha'_C = F\alpha_C \ .$$

The factor F is principally determined by the shear stress exerted by the vapour on the liquid and, as Chen showed, may be expressed by the Lockhart-Martinelli parameter (4.152). With that equation (4.158) becomes

$$\alpha_{2Ph} = S\alpha_B + F\alpha_C \ . \tag{4.159}$$

In the limiting case of $S = F = 1$, this equation converts into that for saturated boiling, (4.155). The heat transfer coefficients α_B for nucleate boiling in free flow were taken by Chen from an equation from Forster and Zuber [4.93]. More recent investigations [4.94], etc. showed however, that this gives somewhat inaccurate results. It therefore seems more sensible to calculate α_B using one of the formulae from section 4.2.6, pages 457 to 459.

The factor S in (4.159) depends on the mass flux of the liquid, and can be expressed in terms of the fluid's Reynolds number

$$Re = \dot{m}(1 - x^*)d/\eta_L \ .$$

Good agreement with known experimental data for vertical tubes is obtained [4.95], if the factors S and F in (4.159) are calculated with the following correlations

$$S = \left(1 + 1.15 \cdot 10^{-6} F^2 Re^{1.17}\right)^{-1} , \tag{4.160}$$

$$F = 1 + 2.4 \cdot 10^4 Bo^{1.16} + 1.37 X_{tt}^{-0.86} \tag{4.161}$$

with $Bo = \dot{q}/\dot{m}\Delta h_v$, $Re = \dot{m}(1 - x^*)d/\eta_L$ and the Martinelli parameter X_{tt}. The mean deviation in the heat transfer coefficient α_{2Ph}, from around 4300 measured values which were used in setting up (4.160) and (4.161), amounts to $\pm 21.4\%$.

For *convective evaporation in the annular space between two tubes heated from only one side*, an equivalent diameter, d_e, has to be introduced. It is given by

$$d_e = 4A/C_w$$

for annular gaps over 4 mm and by

$$d_e = 4A/C_h$$

for annular gaps less than 4 mm. A is the flow cross section, C_w the wetted and C_h the heated circumference.

Calculation procedure

In the calculation of the heat transfer coefficient according to (4.159) from Chen it is sensible to take the following steps:

- The Martinelli parameter should be determined first

$$X_{tt} = \left(\frac{1 - x^*}{x^*}\right)^{0.9} \left(\frac{\eta_L}{\eta_G}\right)^{0.1} \left(\frac{\varrho_G}{\varrho_L}\right)^{0.5} .$$

- The heat transfer coefficient α_B in nucleate boiling, for the given heat flux $\dot{q}$ is then found using the formulae presented in section 4.2.6, pages 457 to 459. Likewise the factors S and F are then calculated from (4.160) and (4.161).
- With these values the heat transfer coefficient α_{2Ph} according to (4.159) is fixed and then the wall temperature $\vartheta_0 = \vartheta_s + \dot{q}/\alpha_{2Ph}$ can be determined.

4.2.8.7 Critical boiling states

As has already been explained in connection with boiling in free flow, the heat transfer coefficient falls quickly once a maximum heat flux has been exceeded. The liquid at the wall is displaced by vapour. If the surface is heated electrically, by nuclear heating or thermal radiation, then the wall temperature rises rapidly after the maximum heat flux has been reached. If in contrast, the wall temperature is determined by another fluid, such as in a heat exchanger or condenser, a small increase in the temperature leads to a drastic fall in the heat flux. These

occurrences come together under the term of *critical boiling states*. This is understood to be a reduction in the heat transfer coefficient after a critical heat flux
is exceeded. In general for the case of boiling in forced convection this is not
identical to the maximum heat flux in free flow and can, as the following explains,
be caused by different mechanisms. Therefore we speak of critical heat flux to
distinguish it from the maximum heat flux in free flow boiling.

The phenomena that are observed in free flow also appear in forced flow, but
are more complex. There are two fundamental types of boiling crisis:

a) With small volumetric vapour content *film boiling* occurs. The liquid is
   the continuous phase, and once a critical heat flux has been reached, a
   vapour film forms at the wall that separates the liquid from the hot wall.
   The higher thermal resistance of the vapour film leads to a drop in the
   heat flux if the wall temperature is imposed, or to an increase in the wall
   temperature if the heat flux is stipulated. The critical heat flux is larger,
   the smaller the volumetric vapour content.

b) With a larger volumetric vapour content annular flow develops. At the
   wall there is chiefly liquid present, and in the core the vapour forms a
   continuous phase. Once the critical heat flux has been reached the liquid
   film disappears at the wall and this becomes covered by vapour. This is
   known as *dry-out*.

If the vapour content is sufficiently large the heated surface drys out at very
low heat flux. Downstream, small liquid drops can reach the heated surface from
the core flow, and because of the low heat flux they are only partially vaporised.
This type of dry-out with subsequent spray cooling of the wall with liquid droplets
is called "deposition controlled burnout".

Fig. 4.54 illustrates both types of critical boiling, namely film boiling and dry-
out. A qualitative diagram of the change in the critical heat flux with the quality
is presented in Fig. 4.55.

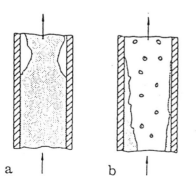

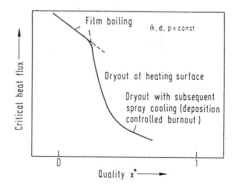

Fig. 4.54: Types of boiling crisis. **a** film   **Fig. 4.55**: Critical heat flux as a function of
boiling; **b** dry-out   quality

In the film boiling region, the critical heat flux decreases approximately linearly with the quality. In dry-out the critical heat flux drops markedly with the quality. It is only at high quality in a dry-out region with subsequent spray cooling that it falls weakly with the quality.

A lower limit for the critical heat flux can be found from the energy balance

$$\dot{q}d\pi z = \dot{M}c_{pL}\left[\vartheta_L(z) - \vartheta_1\right] = \alpha d\pi z\left[\vartheta_0 - \vartheta_L(z)\right] \quad,$$

where ϑ_1 is the inlet temperature. From this follows

$$\vartheta_0 = \vartheta_L(z) + \frac{\dot{q}}{\alpha} = \vartheta_1 + \frac{\dot{q}d\pi z}{\dot{M}c_{pL}} + \frac{\dot{q}}{\alpha}$$

or

$$\dot{q} = \frac{\vartheta_0 - \vartheta_1}{\dfrac{d\pi z}{\dot{M}c_{pL}} + \dfrac{1}{\alpha}} \quad.$$

As the wall temperature is significantly greater than the saturation temperature once the critical heat flux has been reached, then the lower limit for the critical heat flux is

$$\dot{q}_{cr} > \frac{\vartheta_s - \vartheta_1}{\dfrac{d\pi z}{\dot{M}c_{pL}} + \dfrac{1}{\alpha}} = \frac{\Delta\vartheta_u}{\dfrac{d\pi z}{\dot{M}c_{pL}} + \dfrac{1}{\alpha}} \quad, \tag{4.162}$$

where $\Delta\vartheta_u$ is the subcooling of the liquid at the inlet.

Using this in the example of water with an initial subcooling of $\Delta\vartheta_u = 50\,\mathrm{K}$, mass flux $\dot{m} = 1000\,\mathrm{kg/m^2s}$ and specific heat capacity $c_{pL} = 4.186\,\mathrm{kJ/kgK}$, which boils in a 1 m long tube of $d = 25\,\mathrm{mm}$ inner diameter, assuming the heat transfer coefficient is $\alpha = 10\,000\,\mathrm{W/m^2K}$, the critical heat flux has to be

$$\dot{q}_{cr} > 3.62 \cdot 10^5\,\mathrm{W/m^2} \quad.$$

Even when the critical heat flux is reached not all the liquid is vaporised. Therefore an upper limit for the critical heat flux is obtained, which is necessary for complete evaporation of the liquid. Once again this is found from an energy balance

$$\dot{q}d\pi z = \dot{M}(h'' - h_1) = \dot{M}(\Delta h_v + h' - h_1) \quad.$$

Substituting in

$$\dot{M}(h' - h_1) = \dot{M}c_{pL}\Delta\vartheta_u \quad,$$

where c_{pL} is the mean specific heat capacity between the inlet temperature ϑ_1 and the boiling temperature ϑ_s, gives, because $\dot{q}_{cr} < \dot{q}$:

$$\dot{q}_{cr} < \frac{\dot{M}\Delta h_v}{d\pi z}\left(1 + \frac{c_{pL}\Delta\vartheta_u}{\Delta h_v}\right) \quad. \tag{4.163}$$

With the data from the example above and an enthalpy of vaporisation of 2100 kJ/kg we have

$$\dot{q}_{cr} < 1.4 \cdot 10^7\,\mathrm{W/m^2} \quad,$$

so that, for the example, the critical heat flux has to lie between $3.62 \cdot 10^5$ and $1.4 \cdot 10^7\,\mathrm{W/m^2}$. In view of the large number of empirical methods for the calculation of the critical heat flux and the many influential quantities upon which this depends, it is recommended that the interval in which the critical heat flux must lie is estimated. Application formulae for the calculation of the critical heat flux can be found in the relevant literature, e.g. [4.50] pg. 215.

4.2.8.8 Some empirical equations for heat transfer in two-phase flow

In the following the equations presented previously for heat transfer in two-phase flow shall be summarised and supplemented by further equations.

1. **Subcooled boiling**
   According to [4.96], bubble growth begins when the wall is superheated to a certain critical value with respect to the boiling point:

$$(\vartheta_0 - \vartheta_s)_{cr} = 2 \left(\frac{2 \dot{q} \sigma T_s}{\varrho'' \Delta h_v \lambda_L} \right)^{1/2} .$$

For water, this gives the dimension equation [4.97]

$$(\vartheta_0 - \vartheta_s)_{cr} = \frac{5}{9} \left[\left(\frac{\dot{q}}{1120 \text{ W/m}^2} \right)^{0.436} p^{0.535} \right]^{p^{0.0234}}$$

valid for 1.03 bar $\leq p \leq 138$ bar, with $\dot{q}$ in W/m^2, p in bar und $(\vartheta_0 - \vartheta_s)_{cr}$ in K. Both equations presume a minimum roughness of the heated surface. The radius of the opening of the smallest pore required for bubble formation is

$$(r_0)_{cr} = \left(\frac{2 \sigma T_s \lambda_L}{\varrho'' \Delta h_v \dot{q}} \right)^{1/2} .$$

The transferred heat flux in the subcooled boiling region follows from

$$\dot{q} = \dot{q}_B + \dot{q}_C = \alpha_B (\vartheta_0 - \vartheta_s) + \alpha_C (\vartheta_0 - \vartheta_L) .$$

Here α_B is the heat transfer coefficient for nucleate boiling from section 4.2.6, pages 457 to 459, α_C is that by forced, single phase flow, sections 3.7.4, page 335, and 3.9, page 380, ϑ_L is the adiabatic mixing temperature of the liquid.

2. **Saturated boiling**
   Subcooled boiling changes to saturated boiling when the quality, according to (4.115), is $x_{th}^* = 0$. Heat transfer is strongly dependent on the heat flux, but weakly dependent on the mass flux and the quality; it holds that

$$\alpha_{2Ph} = \alpha_B + \alpha_C .$$

Here α_B is the heat transfer coefficient for nucleate boiling from section 4.2.6, pages 457 to 459, and α_C is that for forced, single phase flow, sections 3.7.4, page 335, and 3.9, page 380.

3. **Convective boiling in vertical tubes** Heat transfer is highly dependent on the mass flux and the quality, but only shows weak dependence on the heat flux. It holds that

$$\alpha = S \alpha_B + F \alpha_C .$$

Here α_B is the heat transfer coefficient for nucleate boiling from section 4.2.6, pages 457 to 459, and α_C is that for forced, single phase flow, sections

3.7.4 and 3.9. The special case of $S = F = 1$ in saturated boiling heat transfer is included in the equations above. The factors S (suppression factor) and F (enhancement factor) are yielded from

$$S = (1 + 1.15 \cdot 10^{-6} F^2 Re^{1.17})^{-1}$$

$$F = 1 + 2.4 \cdot 10^4 Bo^{1.16} + 1.37 X_{tt}^{-0.86}$$

with

$$Bo = \dot{q}/\dot{m}\Delta h_v \quad , \qquad Re = \dot{m}(1 - x^*)d/\eta_L$$

and

$$X_{tt} = \left(\frac{1 - x^*}{x^*}\right)^{0.9} \left(\frac{\eta_L}{\eta_G}\right)^{0.1} \left(\frac{\varrho_G}{\varrho_L}\right)^{0.5} .$$

4. **Convective evaporation in horizontal tubes**
Under the influence of gravity the liquid predominantly collects at the bottom of the tube, assuming the axial velocity is not too great, whilst the top of the tube is barely or not wetted. Stratified flow exists if the Froude number is

$$Fr = \dot{m}^2/(\varrho_L^2 gd) < 0.04 \quad ;$$

at large Froude numbers, the influence of gravity is negligible. Annular flow occurs, and the equation in 3 for a vertical tube is valid. In stratified flow, [4.98], [4.99] are valid:

$$\frac{\alpha_{2Ph}}{\alpha_C} = 3.9 \, Fr^{0.24} \left(\frac{x^*}{1 - x^*}\right)^{0.64} \left(\frac{\varrho_L}{\varrho_G}\right)^{0.4}$$

with

$$\alpha_C = \frac{\lambda_L}{d} 0.023 \left(\frac{\dot{m}(1 - x^*)}{\eta_L}\right)^{0.8} Pr^{0.4} .$$

4.2.9   Heat transfer in boiling mixtures

In industrial processes liquid mixtures of two or more components often have to be vaporised in order to separate the components from each other. Examples of this include concentrating solutions, the recovery of solvents, distillation of sea water to gain drinking water or the separation of substances by boiling in distillation. Heat and mass transfer are closely linked with each other in evaporation, and the amount of vapour generated will, in contrast to the evaporation of pure substances, be determined by the mass transfer. It is known from earlier experiments that the heat transfer coefficients in the evaporation of mixtures can be considerably smaller than those for the pure components of the mixture. On the other hand, significant improvements in heat transfer have been noticed when one of the components in the mixture was very surface active. This leads to a reduction

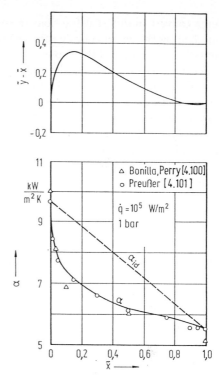

Fig. 4.56: Heat transfer coefficients in boiling ethanol-water. $\tilde{y}$ mole fraction of ethanol in the vapour; $\tilde{x}$ mole fraction of ethanol in the liquid

in the surface tension, and with that a larger bubble density, as was explained in section 4.2.2. At the same time the bubble frequency and therefore the heat transfer increase. However, mixtures of organic and inorganic liquids only contain surface active components in certain cases, (soap, addition of wetting media), so, in general, a reduction in the heat transfer coefficient in comparison to those for the pure components should be reckoned with. In order to explain the fundamental phenomena we will only consider here the heat transfer in boiling of a binary mixture in free flow.

A large number of binary mixtures of organic liquids and of water with organic liquids were first investigated by Bonilla and Perry [4.100]. In Fig. 4.56 the heat transfer coefficients found from their measurements with an ethanol-water mixture are presented along with the results from Preußer [4.101]. Heat flux and pressure are indicated in the diagram. As we can see, the heat transfer coefficients α of the mixture are noticeably smaller than the values α_{id}, which would be obtained, if we had interpolated linearly between the heat transfer coefficients of the pure components. A significant reduction of the heat transfer coefficient in the region where a large difference exists between the vapour and fluid composition $\tilde{y} - \tilde{x}$ can also be seen from a comparison with the curve in Fig. 4.56.

This decrease in the heat transfer in a region of large difference between vapour and liquid composition can be observed in many mixtures. It can be explained by the fact that during the formation of vapour bubbles the individual components convert in different fractions from liquid to vapour. The more the vapour is enriched by the more volatile components, the lower the fraction of these components in the vicinity of the vapour bubble. This means that we have to distinguish between processes close to the wall and those in the core of the liquid. At the heated surface the vapour bubbles form at preferred places, which are suitable for the nucleation. The bubbles grow as a result of the heat input, until the buoyancy forces are large enough so that the bubble detaches from the surface. As the departure diameter is relatively small, only a part of the heat input serves for the bubble formation at the wall. Another part of the heat flow is released by the heated surface into the liquid and serves at the surface of bubbles inside the liquid for further vapour formation. The hot wall thereby delivers the necessary nuclei for nucleate boiling and also releases heat into the liquid boundary layer, which is transmitted by convection or conduction via the liquid columns, to the vapour bubbles. This process is no different for mixtures than for pure substances. However there are basic differences with respect to the heat transport to the rising bubbles, because the heat transport in mixtures is also determined by mass transfer.

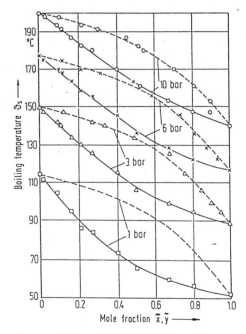

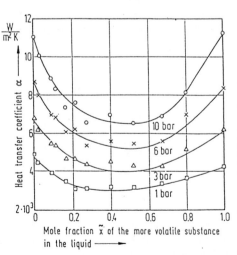

Fig. 4.57: Boiling diagram acetone/n-butanol. Solid line: mole fraction $\tilde{x}$ of the more volatile component in the liquid; dotted line: mole fraction $\tilde{y}$ of the most volatile component in the vapour

Fig. 4.58: Heat transfer coefficient α for an acetone/n-butanol mixture. Heat flux $\dot{q} = 10^5 \, \mathrm{W/m^2}$

The influence of the composition of the mixture on the heat transfer is shown in the following diagrams taken from a paper by Stephan and Körner [4.102]. In a mixture that in a certain region shows large differences between vapour and liquid composition, as indicated by the boiling diagram Fig. 4.57 for an aceton/n-butanol mixture ($(CH_3)_2CO/C_4H_9OH$), the heat transfer is considerably reduced. This can be seen in Fig. 4.58, in which the heat transfer coefficient is plotted against the composition. Where the difference $\tilde{y} - \tilde{x}$ is largest, the greatest fall in the heat transfer coefficient exists.

This is particularly conspicuous in boiling mixtures that contain an azeotropic point. With regard to this we will consider a methanol/benzene mixture (CH_3OH/C_6H_6), Fig. 4.59 and Fig. 4.60. The mixture shows a sharp reduction in heat transfer in the region of large differences $\tilde{y} - \tilde{x}$. Towards the azeotropic point the heat transfer increases again. In the vicinity of an azeotropic point the heat transfer coefficient decreases. Obviously it is not the sign of the difference $\tilde{y} - \tilde{x}$, but its absolute value that is decisive.

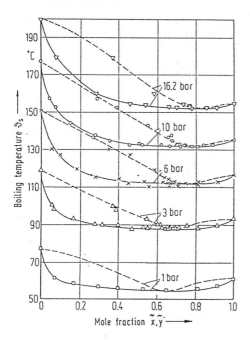

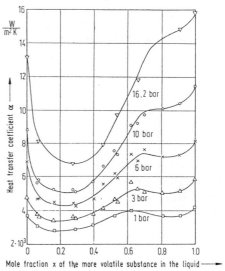

Fig. 4.59: Boiling diagram methanol/benzene. Solid line: mole fraction $\tilde{x}$ of the most volatile component in the liquid; dashed line: mole fraction $\tilde{y}$ of the most volatile component in the vapour

Fig. 4.60: Heat transfer coefficient α for the methanol/benzene mixture. Heat flux $\dot{q} = 10^5 \, \text{W/m}^2$

Two methods have proved themselves in the reproduction of heat transfer measurements. One starts from empirical correlations for the pure substances.

These correlations normally contain dimensionless numbers, that now have to be formed with the properties of the binary mixture. The reduction in heat transfer because of inhibited bubble growth caused by diffusion is taken into account by the introduction of an extra term. This type of equation has been presented by Preußer [4.101]

$$\alpha = \alpha_0 \left[1 + \left| (\tilde{y} - \tilde{x}) \left(\frac{\partial \tilde{y}}{\partial \tilde{x}} \right)_{\mathrm{P}} \right| \right]^{-0.0733} , \qquad (4.164)$$

in which the heat transfer coefficient α_0 is calculated from (4.96) established for pure substances. The property data for the mixture should be used for this case. Equation (4.164) is based on measurements at ambient pressure, and so is only valid for that.

As Preußer showed a considerable fraction of the reduction in heat transfer compared to pure substances is caused by the change in the thermal properties, whilst the additional term in the square brackets provides a comparatively small contribution. For most hydrocarbon mixtures and those of hydrocarbons with water it lies between 0.8 and almost 1. All the other methods and models presented in the following are used to avoid the extensive calculations of the property data for the mixture.

One of these correlations, which avoids the difficult and time consuming calculation of the properties of the mixture, has been suggested by Stephan and Körner [4.102]. It is based on the fact that the transfer of a given heat flux to the mixture, requires a larger superheating of the wall $\Delta\vartheta = \vartheta_0 - \vartheta_{\mathrm{s}}$, than in the vaporisation of the pure substances. The saturation temperature ϑ_{s} here is the boiling temperature of the mixture at the mean composition $\tilde{x}$ of the liquid. In order to calculate the superheat, an 'ideal' wall superheat $\Delta\vartheta_{\mathrm{id}}$ is defined by

$$\Delta\vartheta_{\mathrm{id}} := \tilde{x}_1 \Delta\vartheta_1 + \tilde{x}_2 \Delta\vartheta_2 , \qquad (4.165)$$

wherein the temperature differences $\Delta\vartheta_1$ and $\Delta\vartheta_2$ between the wall and saturation temperature are yielded from the heat transfer coefficients α_1 and α_2 of the pure substances 1 and 2 at the heat flux $\dot{q}$ of the mixture, in accordance with

$$\Delta\vartheta_1 = \dot{q}/\alpha_1 \quad \text{and} \quad \Delta\vartheta_2 = \dot{q}/\alpha_2$$

and can also be calculated with, for example, the aid of (4.96). According to (4.165) the driving temperature differences that would be yielded from the vaporisation of the pure substances, are added together corresponding to the mole fractions of the two components. The actual driving temperature difference $\Delta\vartheta$ is different from the ideal value $\Delta\vartheta_{\mathrm{id}}$ by an extra temperature difference $\Delta\vartheta_{\mathrm{E}}$. We put

$$\Delta\vartheta = \Delta\vartheta_{\mathrm{id}} + \Delta\vartheta_{\mathrm{E}} \quad \text{or} \quad \Delta\vartheta = \Delta\vartheta_{\mathrm{id}}(1 + \theta) \qquad (4.166)$$

with

$$\theta = \Delta\vartheta_{\mathrm{E}}/\Delta\vartheta_{\mathrm{id}} .$$

The dimensionless extra term θ principally depends on the difference between the vapour and liquid compositions and is always positive, due to the reduction in heat transfer in the mixture. Experiments with numerous mixtures yielded a linear relationship

$$\theta = K_{12} |\tilde{y} - \tilde{x}| \ , \tag{4.167}$$

where K_{12} is a positive number, virtually independent of the composition. K_{12} can also be interpreted as a binary interaction parameter, that has to be found for each mixture at every pressure. In the pressure range between 1 and 10 bar, the pressure dependence of K_{12} can be represented approximately by the empirical equation

$$K_{12} = K_{12}^0 \left(0.88 + 0.12\, p/p_0\right) \tag{4.168}$$

with $p_0 = 1$ bar. The value K_{12}^0 is different for each mixture, but is independent of pressure. A good mean value for binary mixtures of organic liquids and of water with organic liquids is $K_{12}^0 = 1.4$.

4.3 Exercises

4.1: Saturated steam at a pressure of 0.101325 MPa condenses on a horizontal tube of 25 mm outer diameter and a wall temperature of 60 °C. Estimate by what factor the accumulated condensate is reduced, if the steam is mixed with 10 mass-% = 6.47 mol % air at the same total pressure of 0.101325 MPa.
The following property data are given for water at the saturation temperature of 100 °C: liquid density $\varrho_L = 958.1\,\text{kg/m}^3$, vapour density $\varrho_G = 0.5974\,\text{kg/m}^3$, enthalpy of vaporisation $\Delta h_v = 2257.3\,\text{kJ/kg}$, liquid thermal conductivity $\lambda_L = 0.677\,\text{W/K m}$, dynamic viscosity of the liquid $\eta_L = 0.2822 \cdot 10^{-3}\,\text{kg/s m}$, specific heat capacity of the steam-air mixture $c_{pG} = 1.93\,\text{kJ/kgK}$. Further, the heat transfer coefficient between the steam-air mixture and the condensate film is $\alpha_G = 30\,\text{W/m}^2\text{K}$.

4.2: A horizontal, electrically conductive, cylindrical metal rod of 5 mm diameter, 0.5 m length and a surface temperature of 300 °C is placed in a boiling water bath at 100 °C. Calculate the thermal power of the rod.
The following material data are given: density of the boiling water at 100 °C: $\varrho_L = 958.1\ \text{kg/m}^3$, enthalpy of vaporisation $\Delta h_v = 2257.3\ \text{kJ/kgK}$, density of steam at 300 °C: $\varrho_G = 46.255\ \text{kg/m}^3$, specific heat capacity $c_{pG} = 6.144\ \text{kJ/kgK}$, thermal conductivity $\lambda_G = 0.0718\ \text{W/K m}$, kinematic viscosity $\nu_G = 0.427 \cdot 10^{-6}\ \text{m}^2/\text{s}$, emissivity $\varepsilon_0 = \varepsilon_L = 1$.

4.3: In a process at a chemical plant, steam at 0.25 MPa is to be generated in a steam generator. This generator is a horizontal tube bundle, 25 kg/s of a heat carrying oil is flowing through the tubes. The oil is to be cooled from 200 °C to 150 °C. The water will evaporate on the outside of the tubes. The heat transfer coefficient between the oil and the wall is $\alpha_i=700\,\text{W/m}^2\text{K}$. The thermal resistance of the tube wall may be neglected. The water enters the generator at approximately its boiling temperature. How much steam can be produced per hour? What is the area of the steam generator? The following values are given: specific heat capacity of the heat carrying oil $c_{p\text{oil}}=2.4\,\text{kJ/kgK}$, boiling temperature of water at 0.25 MPa: $\vartheta_s=127.4$ °C, enthalpy of vaporisation $\Delta h_v=2160\,\text{kJ/kg}$.

4.4: A vertical, 3.5 m long evaporator tube of 12 mm internal diameter has water flowing through it from bottom to top. The water enters the tube at a pressure of 5.5 MPa, subcooled with a temperature $\vartheta_i = 210\,°C$. It is initially heated to the saturation temperature and then partly vaporised. The tube is heated with a constant heat flux $\dot{q} = 7.58 \cdot 10^5\,W/m^2$. Calculate

a) the tube length at which the water reaches the saturation temperature at its centre and

b) the vapour content at the oulet.

The following data are given: saturation temperature $\vartheta_s(5.5\,MPa) = 269.9\,°C \approx 270\,°C$, enthalpies: $h_1(210\,°C;\ 5.5\,MPa) = 940\,kJ/kg$, $h'(270\,°C) = 1184.5\,kJ/kg$, enthalpy of vaporisation $\Delta h_v = 1605\,kJ/kg$; density of the saturated liquid $\varrho'(270\,°C) = 768.0\,kg/m^3$, density of the sturated vapour $\varrho''(270\,°C) = 28.07\,kg/m^3$. The calculations may be done with constant densities. Mass flux of water $\dot{m} = 10^3\,kg/m^2s$.

4.5: Calculate the frictional pressure drop in the evaporation zone of the evaporator tube in exercise 4.4.
The friction factor is $\zeta_L = 0.038$, the dynamic viscosities are $\eta_L = 97.4 \cdot 10^{-6}\,kg/sm$ and $\eta_G = 18.38 \cdot 10^{-6}\,kg/sm$. Both phases are flowing turbulently. The saturation temperature is reached in the centre at a length $\Delta z = 0.97\,m$.

4.6: Water at a temperature of $\vartheta_1 = 250\,°C$ and a pressure of 6 MPa (associated saturation temperature $\vartheta_s = 275.56\,°C$), flows into a vertical tube of internal diameter 20 mm. The mass flux is $\dot{m} = 1000\,kg/m^2s$. The heat flux, due to the condensing steam on the outside of the tube, is approximately constant and amounts to $\dot{q} = 8 \cdot 10^5\,W/m^2$.

a) How long does the tube have to be, if the quality at the outlet is to be $x^* = 0.25$? The pressure drop can be disregarded in this case.

b) What is the temperature of the wall at the outlet, under the assumption that the pressure drop is 500 hPa?

The following data are given: liquid density $\varrho_L = 758.3\,kg/m^3$, vapour density $\varrho_G = 30.85\,kg/m^3$, dynamic viscosity of the liquid $\eta_L = 97.7 \cdot 10^{-6}\,kg/s\,m$, dynamic viscosity of the vapour $\eta_G = 18.45 \cdot 10^{-6}\,kg/sm$, Prandtl number of the liquid $Pr_L = 0.875$, thermal conductivity of the liquid $\lambda_L = 0.581\,W/K\,m$, specific enthalpy at the inlet $h_1 = 1085.8\,kJ/kg$, of the saturated liquid $h'(6\,MPa) = 1213.7\,kJ/kg$, of the saturated vapour $h''(6\,MPa) = 2785.0\,kJ/kg$, $p_{cr} = 22.064\,MP.$

5 Thermal radiation

Thermal radiation differs from heat conduction and convective heat transfer in its fundamental laws. Heat transfer by radiation does not require the presence of matter; electromagnetic waves also transfer energy in empty space. Temperature gradients or differences are not decisive for the transferred flow of heat, rather the difference in the fourth power of the thermodynamic (absolute) temperatures of the bodies between which heat is to be transferred by radiation is definitive. In addition, the energy radiated by a body is distributed differently over the single regions of the spectrum. This wavelength dependence of the radiation must be taken as much into account as the distribution over the different directions in space.

In the first section the physical quantities, which are necessary for the formulation of the laws of thermal radiation, will be introduced. These laws also have to cover the directional and wavelength dependence of the radiant energy. The second section is devoted to the ideal radiator, the black body. The discovery of the laws governing black body radiation by M. Planck (1900) stands at the origin of modern physics, namely quantum theory. The third section is a discussion of the properties and material laws for real radiators. In section 5.4 we will deal with solar radiation and how it is weakened on passing through the Earth's atmosphere. The following section concerns heat transfer between radiating bodies, so-called radiative exchange. The final section in this chapter offers an introduction to gas radiation which plays an important role in heat transfer in furnaces and combustion chambers.

5.1 Fundamentals. Physical quantities

As the laws for thermal radiation are different to those valid for heat conduction and convective heat transfer, the essential terms and physical quantities, from which the thermal radiation laws are formulated, are introduced in the following sections.

5.1.1   Thermal radiation

All the considerations that follow are only valid for radiation that is stimulated thermally. Radiation is released from all bodies and is dependent on their material properties and temperature. This is known as heat or thermal radiation. Two theories are available for the description of the emission, transfer and absorption of radiative energy: the classical theory of electromagnetic waves and the quantum theory of photons. These theories are not exclusive of each other but instead supplement each other by the fact that each describes individual aspects of thermal radiation very well.

According to quantum theory, radiation consists of photons (= light particles), that move at the velocity of light and have no rest mass. They transfer energy, whereby each photon tranports the energy quantum

$$e_{\mathrm{Ph}} = h\nu \ .$$

Here, $h = (6.626\,075\,5 \pm 0.000\,004\,0) \cdot 10^{-34}$ J s is the Planck constant, also known as Planck's action quantum; ν is the frequency of the photons. Quantum theory is required to calculate the spectral distribution of the energy emitted by a body. Other aspects of heat transfer can, in contrast, be covered by classical theory, according to which the radiation is described as the emission and propagation of electromagnetic waves.

Electromagnetic waves are transverse waves that oscillate perpendicular to the direction of propagation. They spread out in a straight line and in a vaccuum at the velocity of light $c_0 = 299\,792\,458$ m/s. Their velocity c in a medium is lower than c_0, whilst their frequency ν remains unchanged; the ratio $n := c_0/c > 1$ is the refractive index of the medium. The wavelength λ is linked to the frequency ν by

$$\lambda \cdot \nu = c \ .$$

The energy transported by the electromagnetic waves depends on λ. This also has to be considered for heat transfer.

Fig. 5.1 shows the electromagnetic spectrum that extends from $\lambda = 0$ to very large wave lengths ($\lambda \to \infty$). At small wave lengths ($\lambda < 0.01$ μm) we have gamma-rays and x-rays, neither of which are thermally stimulated and so therefore do not belong to thermal radiation. The same is true for the region of large wavelengths, ($\lambda > 10^3$ μm), that is determind by the oscillations of electronic switching networks (radar, television and radio waves). Neither region has any meaning for thermal radiation. The thermal radiation region is the middle of the range of wavelengths between around 0.1 μm and 1000 μm. Within this region bodies, whose temperatures lie between a few Kelvin and $2 \cdot 10^4$ K, radiate. This includes the visible light region between 0.38 μm (violet) and 0.78 μm (red). The designation of this radiation as light has no physical reason, but instead is based on the peculiarity that the human eye can "see" in this wavelength range. The

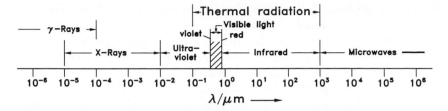

Fig. 5.1: Electromagnetic wave spectrum

wavelength interval 0.01 μm $\leq \lambda \leq$ 0.38 μm is the range of ultraviolet (UV) radiation. Between 0.78 μm and 1000 μm lies infrared (IR) radiation. This is the wavelength range in which most earthly bodies radiate.

The limits for thermal radiation at $\lambda = 0.1$ μm and $\lambda = 1000$ μm are somewhat arbitrary. Very hot bodies, e.g. stars, also radiate in the region $\lambda < 0.1$ μm. However they only release more than a few percent of their energy in this region if their temperature lies above 15000 K. Thermal radiation can also be emitted in the range $\lambda > 1000$ μm. If this is to make up more than a few percent of the total radiation, then the radiator has to be colder than around 12 K. An ideal radiator, the so-called black body, cf. section 5.2, at the temperature of boiling helium (4.22 K) has its maximum emission at $\lambda = 686$ μm, and in the wavelength region $\lambda > 1000$ μm more than 30% of the total radiation will be emitted.

Thermal radiation is not only dependent on the wavelength; in numerous problems, particularly in radiative exchange between different bodies, its distribution in space must also be considered. This holds for the emission of radiative energy in the same way as for reflection and absorption of radiation incident on a body. This double dependency — on the wavelength and the direction in space — makes the quantitative description of heat radiation quite complicated. It requires four different types of physical radiation quantities:

- *Directional spectral quantities.* These describe the directional and wavelength distribution of the radiative energy in a detailed manner. They are of fundamental meaning, but are very difficult to determine experimentally or theoretically. This is why we frequently employ radiation quantities that only include one effect, either the dependence on the wavelength or the direction.
- *Hemispherical spectral quantities* average the radiation into all directions of the hemisphere over a surface element and so are only dependent on the wavelength.
- *Directional total quantities* average the radiation over all wavelengths and describe the dependence on the directions in the hemisphere.
- *Hemispherical total quantites* combine the radiation over all wavelengths and from all directions. They do not provide information on the spectral distribution and the directional dependence of the radiation; but are frequently sufficient to provide the solution to radiative heat transfer problems.

This variety of physical quantities of radiation with their different significations causes some difficulties for the beginner, even though exact relationships exist between the four groups. The fundamental directional spectral quantities are used to calculate the other three by integration over the wavelength or over all the solid angles in the hemisphere, or finally over the two independent variables. In the following sections these four groups of quantities and their relationships for the specific cases of emission of radiation, the irradiation of an area, as well as the absorption and reflection of radiative energy, will be dealt with. All these cases are based on the same train of thought, only the expressions and the symbols are different. It is, therefore, sufficient to discuss the emission of radiation in depth, whilst limiting ourselves to the exact definitions of the quantities and the establishment of the associated equations for the other cases.

5.1.2 Emission of radiation

By the emission of thermal radiation, internal energy of the emitting body is converted into energy of the electromagnetic waves or, in the language of quantum theory, the energy of photons, which leave the surface of the radiating body. In this emission process the atoms or molecules of the body change from a state of higher energy to one of lower energy. However we do not need to go into these intramolecular processes for the formulation of the important phenomenological laws of heat transfer.

Matter emits radiation in all its aggregate states. In gases or solids which allow radiation go through (e.g. glass), the radiation emitted from a finite volume is the combination of the local emissions within the volume being considered. We will come back to a discussion of these volumetric emissions in section 5.6. In most solids and liquids the radiation released from molecules within the body is strongly absorbed by neighbouring molecules, so that it cannot reach the surface. Therefore, the radiation emitted by solids and liquids comes normally from molecules in a layer immediately below the surface. As the thickness of this layer only amounts to around 1 μm the emission can be associated with the surface, and we speak of radiating surfaces rather than radiating bodies.

5.1.2.1 Emissive power

We consider an element of the surface of a radiating body, that has a size of dA. The energy flow (heat flow) dΦ, emitted into the hemisphere above the surface element, is called *radiative power* or *radiative flow*, Fig. 5.2. Its SI-unit is the Watt. The radiative power devided by the size of the surface element

$$M := \mathrm{d}\Phi/\,\mathrm{d}A \qquad\qquad (5.1)$$

Fig. 5.2: Radiation flow $d\Phi$, emitted from a surface element

is called the (hemispherical total) *emissive power*. This is the heat flux released by radiation; the SI units for M are the same as the SI units for $\dot{q}$, i.e. $\mathrm{W/m^2}$. The emissive power M belongs to the group of hemispherical total quantities, as it combines the radiation energy emitted over the total range of wavelengths and into all the directions of the hemisphere. M is a property of the radiator; it changes above all with the thermodynamic temperature of the radiator, $M = M(T)$, and depends on the nature of its surface.

5.1.2.2 Spectral intensity

We will now investigate how the emitted radiation $d\Phi$ is distributed over the spectrum of wavelengths and the directions in the hemisphere. This requires the introduction of a special distribution function, the spectral intensity L_λ. It is a directional spectral quantity, with which the wavelength and direction distribution of the radiant energy is described in detail.

A certain direction in space is determined by two angular coordinates β and φ, Fig. 5.3. β is the polar angle measured outwards from the surface normal ($\beta = 0$) and φ is the circumferential angle with an arbitrarily assumed position for $\varphi = 0$. The radiative flux, that falls on a small area dA_n at a distance r from the surface element dA, *perpendicular* to the radiation direction, Fig. 5.4, is proportional to the *solid angle element*

$$d\omega = dA_\mathrm{n}/r^2 \ . \tag{5.2}$$

The small area dA_n in Fig. 5.4 and with that the solid angle element $d\omega$ result

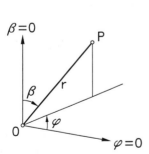

Fig. 5.3: Spherical coordinates of the point P: distance from origin r, polar angle β, circumferential angle φ

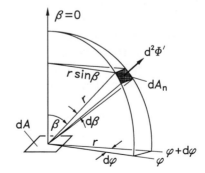

Fig. 5.4: Radiative flux $d^2\Phi'$ into a solid angle element $d\omega$ in the direction of the polar angle β and the circumferential angle φ

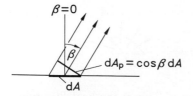

Fig. 5.5: Projection dA_p of the surface element
dA perpendicular to the radiation direction

from the fact that the polar angle β changes by $d\beta$ and the circumferential angle
φ by $d\varphi$. So $dA_n = r\,d\beta \cdot r\sin\beta\,d\varphi = r^2\sin\beta\,d\beta\,d\varphi$, and we obtain for the solid
angle element

$$d\omega = \sin\beta\,d\beta\,d\varphi \ . \tag{5.3}$$

The *solid angle* indicates the contents of a cone shaped section, whereby the apex of the
cone coincides with the vertex of the solid angle. If a sphere of any radius R is placed around
the vertex, the surface of the cone cuts an area of A_K out of the surface of the sphere. The size
of the solid angle is defined by

$$\omega = A_K/R^2 \ ,$$

cf. DIN 1315 [5.1]. The solid angle units are m^2/m^2 and are called stere radians (unit symbol
sr). They may also be replaced by the number 1.

Now $d^3\Phi$ signifies the radiative flow that the surface element dA emits into a
solid angle element $d\omega$, that lies in the direction indicated by β and φ; additionally
$d^3\Phi$ contains only a part of this radiative flux that is emitted at a certain wave
length λ, in an infinitesimal wave length interval $d\lambda$. This restriction of the
radiative flux to a solid angle element and a wavelength interval serves to describe
the directional and wavelength dependence of the radiant energy. The following
formulation can be made for $d^3\Phi$

$$d^3\Phi = L_\lambda\left(\lambda,\beta,\varphi,T\right)\cos\beta\,dA\,d\omega\,d\lambda \ . \tag{5.4}$$

This is the defining equation for the fundamental material function L_λ, the *spectral
intensity*; it describes the directional and wavelength dependence of the energy
radiated by a body and has the character of a distribution function. The (thermodynamic) temperature T in the argument of L_λ points out that the spectral
intensity depends on the temperature of the radiating body and its material properties, in particular on the nature of its surface. The adjective spectral and the
index λ show that the spectral intensity depends on the wavelength λ and is a
quantity per wavelength interval. The SI-units of L_λ are $W/(m^2\mu m\,sr)$. The units
μm and sr refer to the relationship with $d\lambda$ and $d\omega$.

The factor $\cos\beta$ that appears in (5.4) is a particularity of the definition of
L_λ: the spectral intensity is not relative to the size dA of the surface element
like in $M(T)$, but instead to its projection $dA_p = \cos\beta\,dA$ perpendicular to the
radiation direction, Fig. 5.5. It complies with the geometric fact that the emission
of radiation for $\beta = \pi/2$ will be zero and will normally be largest in the direction
of the normal to the surface $\beta = 0$. An area that appears equally "bright" from
all directions is characterised by the simple condition that L_λ does not depend on

β and φ. This type of surface with $L_\lambda = L_\lambda(\lambda, T)$ is known as a *diffuse radiating surface*, cf. 5.1.2.4.

5.1.2.3 Hemispherical spectral emissive power and total intensity

The spectral intensity $L_\lambda(\lambda, \beta, \varphi, T)$ characterises in a detailed way the dependence of the energy emitted on the wavelength and direction. An important task of both theoretical and experimental investigations is to determine this distribution function for as many materials as possible. This is a difficult task to carry out, and it is normally satisfactory to just determine the radiation quantities that either combine the emissions into all directions of the hemisphere or the radiation over all wavelengths. The quantities, the hemispherical spectral emissive power M_λ and the total intensity L, characterise the distribution of the radiative flux over the wavelengths or the directions in the hemisphere.

The *hemispherical spectral emissive power* $M_\lambda(\lambda, T)$ is obtained by integrating (5.4) over all the solid angles in the hemisphere. This yields

$$d^2\Phi = M_\lambda(\lambda, T)\, d\lambda\, dA \tag{5.5}$$

with

$$M_\lambda(\lambda, T) = \int_\cap L_\lambda(\lambda, \beta, \varphi, T) \cos\beta\, d\omega \ . \tag{5.6}$$

Here $d^2\Phi$ is the radiation flow emitted from the surface element dA in the wavelength interval $d\lambda$ into the hemisphere. The symbol $\cap$ in (5.6) signifies that the integration should be carried out over all the solid angles in the hemisphere. The hemispherical spectral emissive power $M_\lambda(\lambda, T)$ with the SI-units $W/m^2\mu m$ belongs to the hemispherical spectral quantities; it represents the wavelength distribution of the emissive power, Fig. 5.6. The area under the isotherm of $M_\lambda(\lambda, T)$ in Fig. 5.6 corresponds to the emissive power, because integration of (5.5) over all the wavelengths leads to

$$d\Phi = \int_0^\infty M_\lambda(\lambda, T)\, d\lambda\, dA \ , \tag{5.7}$$

from which, due to (5.1)

$$M(T) = \int_0^\infty M_\lambda(\lambda, T)\, d\lambda \tag{5.8}$$

follows.

The integration in (5.6) over all the solid angles of the hemisphere corresponds to a double integration over the angular coordinates β and φ. With $d\omega$ according

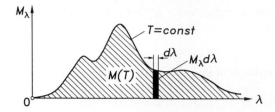

Fig. 5.6: Hemispherical spectral emissive power $M_\lambda(\lambda, T)$ as a function of wavelength λ at constant temperature T (schematic). The hatched area under the curve represents the emissive power $M(T)$

to (5.3) we obtain

$$M_\lambda(\lambda, T) = \int\limits_{\varphi=0}^{2\pi} \int\limits_{\beta=0}^{\pi/2} L_\lambda(\lambda, \beta, \varphi, T) \cos\beta \sin\beta \, \mathrm{d}\beta \, \mathrm{d}\varphi \ . \tag{5.9}$$

In most cases, the spectral intensity L_λ only depends on the polar angle β and not on the circumferential angle φ. We then obtain the more simple relationship

$$M_\lambda(\lambda, T) = 2\pi \int\limits_{0}^{\pi/2} L_\lambda(\lambda, \beta, T) \cos\beta \sin\beta \, \mathrm{d}\beta \ . \tag{5.10}$$

The directional distribution of the emission integrated over all the wave lengths is described by the *total intensity* $L(\beta, \varphi, T)$. This is found by integrating (5.4) over λ, yielding

$$\mathrm{d}^2\Phi' = L(\beta, \varphi, T) \cos\beta \, \mathrm{d}\omega \, \mathrm{d}A \tag{5.11}$$

with the total intensity

$$L(\beta, \varphi, T) = \int\limits_{0}^{\infty} L_\lambda(\lambda, \beta, \varphi, T) \, \mathrm{d}\lambda \ . \tag{5.12}$$

Here $\mathrm{d}^2\Phi'$ is the radiation flow emitted by the surface element into the solid angle element $\mathrm{d}\omega$ in the direction of the angle β and φ. The total intensity L has units $\mathrm{W/m^2\,sr}$; it belongs to the directional total quantities and represents the part of the emissive power falling into a certain solid angle element.

If we integrate (5.11) over all the solid angles in the hemisphere then we obtain the radiation flow $\mathrm{d}\Phi$, emitted by the surface element in the entire hemisphere:

$$\mathrm{d}\Phi = \int\limits_{\triangle} L(\beta, \varphi, T) \cos\beta \, \mathrm{d}\omega \, \mathrm{d}A \ . \tag{5.13}$$

A comparison with (5.1) shows that the emissive power $M(T)$ according to

$$M(T) = \int_{\triangle} L(\beta, \varphi, T) \cos \beta \, \mathrm{d}\omega \tag{5.14}$$

can be calculated from the total intensity $L(\beta, \varphi, T)$.

To summarise, there are in total four radiation quantities for the characterisation of the emission of radiation from an area:

1. The spectral intensity $L_\lambda(\lambda, \beta, \varphi, T)$ describes the distribution of the emitted radiation flow over the wavelength spectrum and the solid angles of the hemisphere (directional spectral quantity).
2. The hemispherical spectral emissive power $M_\lambda(\lambda, T)$ covers the wavelength dependency of the radiated energy in the entire hemisphere (hemispherical spectral quantity).
3. The total intensity $L(\beta, \varphi, T)$ describes the directional dependency (distribution over the solid angles of the hemisphere) of the radiated energy at all wavelengths (directional total quantity).
4. The emissive power $M(T)$ combines the radiation flow emitted at all wavelengths and in the entire hemisphere (hemispherical total quantity).

The relationships between the four quantities are schematically represented and illustrated in Fig. 5.7. The spectral intensity $L_\lambda(\lambda, \beta, \varphi, T)$ contains all the information for the determination of the other three radiation quantities. Each arrow in Fig. 5.7 corresponds to an integration; on the left first over the solid

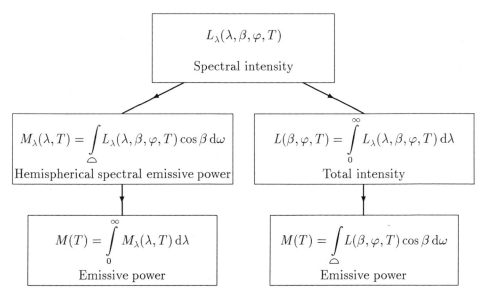

Fig. 5.7: Relationships between the four radiation quantities. Each arrow represents an integration

angles in the hemisphere and then over the wavelengths, on the right first over
the wavelengths and then over the solid angles. The result of the two successive
integrations each time is the emissive power $M(T)$.

Example 5.1: The spectral intensity L_λ of radiation emitted by a body shall not depend
on the circumferential angle φ and can be approximated by the function

$$L_\lambda(\lambda,\beta,T) = L_{\lambda,0}(\lambda,T)\cos\beta \; , \tag{5.15}$$

where $L_{\lambda,0}(\lambda,T)$ has the pattern shown in Fig. 5.8 for a particular temperature. Calculate the
intensity $L(\beta,T)$, the spectral emissive power $M_\lambda(\lambda,T)$ and the emissive power $M(T)$.
   The intensity is obtained from (5.12) and (5.15) by integrating over all the wavelengths

$$
\begin{aligned}
L(\beta,T) &= \int_0^\infty L_\lambda(\lambda,\beta,T)\,d\lambda = \cos\beta \int_0^\infty L_{\lambda,0}(\lambda,T)\,d\lambda\\
&= \cos\beta\, L_{\lambda,0}^{\max}[(4.0-3.0)\,\mu m + \tfrac{1}{2}(8.0-4.0)\,\mu m]\\
&= \cos\beta\, L_{\lambda,0}^{\max}3.0\,\mu m = L^{\max}\cdot\cos\beta = 1650\,(\text{W/m}^2\text{sr})\cos\beta \; ,
\end{aligned}
$$

cf. Fig. 5.8. For the spectral emissive power, it follows from (5.10) that

$$M_\lambda(\lambda,T) = 2\pi\int_0^{\pi/2} L_\lambda(\lambda,\beta,T)\cos\beta\sin\beta\,d\beta = 2\pi L_{\lambda,0}(\lambda,T)\int_0^{\pi/2}\cos^2\beta\sin\beta\,d\beta \; .$$

The integral that appears here has, because of

$$\int_0^\beta \cos^2\beta\sin\beta\,d\beta = -\left[\frac{\cos^3\beta}{3}\right]_0^\beta = \frac{1}{3}\left(1-\cos^3\beta\right) \; , \tag{5.16}$$

the value 1/3. With that we get

$$M_\lambda(\lambda,T) = \frac{2\pi}{3}L_{\lambda,0}(\lambda,T) = 2.094\,\text{sr}\,L_{\lambda,0}(\lambda,T) \; .$$

The spectral emissive power agrees with the function $L_{\lambda,0}(\lambda,T)$ from Fig. 5.8 except for the
factor 2.094 sr.
   The emissive power $M(T)$ is calculated according to (5.14) by integrating the intensity
$L(\beta,T)$ over the solid angles of the hemisphere. This yields

$$
\begin{aligned}
M(T) &= \int_{\varphi=0}^{2\pi}\int_{\beta=0}^{\pi/2} L(\beta,T)\cos\beta\sin\beta\,d\beta\,d\varphi = 2\pi L^{\max}\int_0^{\pi/2}\cos^2\beta\sin\beta\,d\beta\\
&= (2\pi/3)L^{\max} = 2.094\,\text{sr}\cdot 1650\,\text{W/m}^2\text{sr} = 3456\,\text{W/m}^2 \; .
\end{aligned}
$$

   The radiation emitted by a small surface element dA of the body being considered here is
absorbed by a sheet with a circular opening cf. Fig. 5.9. What proportion of the radiation flow
emitted by dA succeeds in passing through the opening?
   The radiation flow emitted from the surface element dA in Fig. 5.9 that goes through the
circular hole is indicated by $d\Phi'$. It holds for this, that

$$d\Phi' = \int_{\varphi=0}^{2\pi}\int_{\beta=0}^{\beta^\bullet} L(\beta,T)\cos\beta\sin\beta\,d\beta\,d\varphi\,dA = 2\pi L^{\max}\int_0^{\beta^\bullet}\cos^2\beta\sin\beta\,d\beta\,dA$$

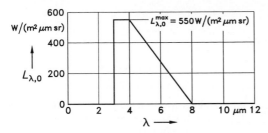
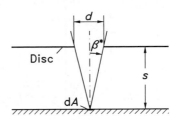

Fig. 5.8: Dependence of the spectral intensity $L_{\lambda,0}(\lambda, T)$ in the direction normal to the surface ($\beta = 0$) on the wave length λ for a constant temperature

Fig. 5.9: Radiation of a surface element through a circular opening in a sheet ($s = 50\,\text{mm}$, $d = 25\,\text{mm}$)

and with (5.16)

$$\mathrm{d}\Phi' = \frac{2\pi}{3} L^{\max} \left(1 - \cos^3 \beta^*\right) \mathrm{d}A \ .$$

The total radiation flow emitted from $\mathrm{d}A$ is $\mathrm{d}\Phi = M(T)\,\mathrm{d}A$. For the desired ratio $\mathrm{d}\Phi'/\mathrm{d}\Phi$ we obtain

$$\mathrm{d}\Phi'/\mathrm{d}\Phi = 1 - \cos^3 \beta^* \ .$$

According to Fig. 5.9, it holds for the angle β^* that

$$\cos \beta^* = \frac{s}{\sqrt{s^2 + (d/2)^2}} = \frac{50}{\sqrt{50^2 + 12.5^2}} = 0.9701 \ .$$

With that we have $\mathrm{d}\Phi'/\mathrm{d}\Phi = 0.0869$; only a small portion of the radiation succeeds in passing through the opening, even though it lies vertically above the surface element.

5.1.2.4 Diffuse radiators. Lambert's cosine law

No radiator exists that has a spectral intensity L_λ independent of the wave length. However, the assumption that L_λ does not depend on β and φ applies in many cases as a useful approximation. Bodies with spectral intensities independent of direction, $L_\lambda = L_\lambda(\lambda, T)$, are known as diffuse radiators or as bodies with diffuse radiating surfaces. According to (5.9), for their hemispherical spectral emissive power it follows that

$$M_\lambda(\lambda, T) = L_\lambda(\lambda, T) \int\limits_{\varphi=0}^{2\pi} \int\limits_{\beta=0}^{\pi/2} \cos\beta \sin\beta \,\mathrm{d}\beta \,\mathrm{d}\varphi \ . \tag{5.17}$$

The double integral here has the value π, so that for diffuse radiating surfaces

$$M_\lambda(\lambda, T) = \pi \, L_\lambda(\lambda, T) \tag{5.18}$$

is yielded as a simple relationship between spectral emissive power and spectral intensity.

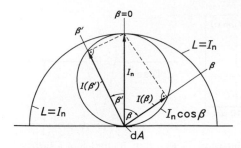

Fig. 5.10: Intensity $L = I_n(T)$ and directional emissive power $I = I_n(T)\cos\beta$ of a diffuse radiating surface

As the intensity L is also independent of β and φ, from (5.11) we obtain

$$d^2\Phi' = L(T)\cos\beta\,d\omega\,dA \tag{5.19}$$

for the radiative power of a diffuse radiating surface element into the solid angle element $d\omega$. The radiative power per area dA and solid angle $d\omega$ is known as the *directional emissive power*

$$I := \frac{d^2\Phi'}{dA\,d\omega}\;. \tag{5.20}$$

For a diffuse radiating surface, it follows from (5.19) that

$$I(\beta,T) = L(T)\cos\beta = I_n(T)\cos\beta\;, \tag{5.21}$$

where $I_n(T)$ is the directional emissive power in the direction normal to the surface ($\beta = 0$), Fig. 5.10. This relationship is called *Lambert's cosine law*[1] [5.2]; diffuse radiators are also called Lambert radiators. The emissive power of a Lambert radiator is found from (5.14) to be

$$M(T) = L(T)\int_\cap \cos\beta\,d\omega = \pi\,L(T) \tag{5.22}$$

in analogy to (5.18) for the corresponding spectral quantities.

5.1.3 Irradiation

When a radiation flow $d\Phi_{in}$ hits an element on the surface of a body, Fig. 5.11, the quotient

$$E := d\Phi_{in}/\,dA \tag{5.23}$$

---

[1]Johann Heinrich Lambert (1728–1777), mathematician, physicist and philosopher, was a tutor for the Earl P. v. Salis in Chur from 1748–1759, where he wrote his famous work on photometry [5.2]. In 1759 he became a member of the Bavarian Academy of Science and upon proposal by L. Euler became a member of the Berlin Academy of Science in 1765. Lambert wrote several philosophical works and dealt with subjects from all areas of physics and astronomy in his numerous publications. He presented the absolute zero point as a limit in the expansion of gases and constructed several air thermometers. In 1761 he proved that π and e are not rational numbers. His works on trigonometry were particularly important for the theory of map construction.

Fig. 5.11: Radiation flow $\mathrm{d}\Phi_{\mathrm{in}}$ of radiation incident on a surface element

is known as the *irradiance* of the surface element of size $\mathrm{d}A$. The irradiance E records the total heat flux incident by radiation as an integral value over all wavelengths and solid angles in the hemisphere. It belongs to the hemispherical total quantities; its SI units are W/m².

The description of the direction and wavelength distribution of the radiation flow is provided by radiation quantities that are defined analogous to those for the emission of radiation. For the radiation flow $\mathrm{d}^3\Phi_{\mathrm{in}}$, from a solid angle element $\mathrm{d}\omega$ in the direction of the angles β and φ incident on the surface element $\mathrm{d}A$, and which only contains the radiation in a wavelength interval $\mathrm{d}\lambda$, we can make a statement analogous to (5.4)

$$\mathrm{d}^3\Phi_{\mathrm{in}} = K_\lambda(\lambda, \beta, \varphi) \cos\beta \, \mathrm{d}A \, \mathrm{d}\omega \, \mathrm{d}\lambda \ . \tag{5.24}$$

The distribution function $K_\lambda(\lambda, \beta, \varphi)$, the *incident spectral intensity*, is defined by this. It describes the wavelength and directional distribution of the radiation flow falling onto the irradiated surface element. Like the corresponding quantity L_λ for the emission of radiation, K_λ is defined with the projection $\mathrm{d}A_\mathrm{p} = \cos\beta \, \mathrm{d}A$ of the irradiated surface element perpendicular to the direction of the incident radiation, Fig. 5.12. The SI units of K_λ are W/(m²µm sr); the relationship to the wavelength interval $\mathrm{d}\lambda$ and the solid angle element $\mathrm{d}\omega$ is also clear from this.

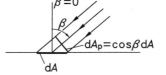

Fig. 5.12: Projection $\mathrm{d}A_\mathrm{p}$ of the surface element $\mathrm{d}A$ perpendicular to the direction of the radiation incident under the polar angle β

In contrast to L_λ, K_λ is not a material property of the irradiated body, but a characteristic function of λ, β and φ for the incident radiative energy: It is the spectral intensity of the incident radiation. The spectral intensity remains *constant* along the radiation path from source to receiver, as long as the medium between the two neither absorbs nor scatters radiation and also does not emit any radiation itself[2]. If this applies, and the radiation comes from a source with a temperature T^*, it holds that

$$K_\lambda(\lambda, \beta, \varphi) = L_\lambda(\lambda, \beta^*, \varphi^*, T^*) \ . \tag{5.25}$$

---

[2]The proof for the constancy of the spectral intensity along a path through a medium that does not influence the radiation can be found in R. Siegel u. J.R. Howell [5.37], pg. 518–520.

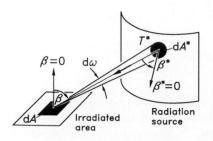

Fig. 5.13: Irradiated surface with surface element dA, the radiation is obtained from the surface element dA^* of a radiation source at a temperature T^*

Here $L_\lambda(\lambda, \beta^*, \varphi^*, T^*)$ is the spectral intensity of a surface element dA^* of the radiation source, from which the solid angle $d\omega$ starting at dA stretches out in the direction (β, φ), Fig. 5.13. The angles β^* and φ^* indicate the direction at which the *irradiated* surface element dA appears to the radiation source. The incident spectral intensity, K_λ of dA, therefore depends *indirectly* on the temperature T^* and the other properties of the radiation source. The directional and wavelength distributions of the radiation energy incident on the irradiated surface element dA are, however, completely described by the function $K_\lambda(\lambda, \beta, \varphi)$, without further knowledge of the properties of the radiation source being required. The statement of its temperature T^* in K_λ is, therefore, unnecessary. With known properties of the radiation source, $K_\lambda(\lambda, \beta, \varphi)$ can be found from the spectral intensity of the radiation source. However in many cases this is very difficult or even impossible, for instance when the radiation hitting dA comes from several sources or when the source of radiation is unknown. The incident spectral intensity $K_\lambda(\lambda, \beta, \varphi)$ has then to be measured in situ, i.e. on the surface element dA.

By integrating (5.24) over all the solid angles in the hemisphere, the radiation flows that come from all the directions in the wavelength interval $d\lambda$ are combined, giving

$$d^2\Phi_{\text{in}} = E_\lambda(\lambda)\, d\lambda\, dA \tag{5.26}$$

with the *spectral irradiance*

$$E_\lambda(\lambda) = \int_\cap K_\lambda(\lambda, \beta, \varphi) \cos\beta\, d\omega \ . \tag{5.27}$$

It belongs to the hemispherical spectral quantities. Integration of (5.26) over all wavelengths leads to

$$d\Phi_{\text{in}} = E\, dA \ , \tag{5.28}$$

giving the irradiance already introduced by (5.23):

$$E = \int_0^\infty E_\lambda(\lambda)\, d\lambda \ . \tag{5.29}$$

The spectral irradiance E_λ describes the distribution of the incident energy over the spectrum, whereby the radiation from all directions in the hemisphere is combined. If, on the contrary, the directional distribution of the radiation falling on the body is to be described, without considering the dependence on the wavelength, then (5.24) should be integrated over all λ. This gives the following for the radiation flow which falls on the surface element $\mathrm{d}A$ from a particular solid angle element $\mathrm{d}\omega$,

$$\mathrm{d}^2\Phi'_{\mathrm{in}} = K(\beta,\varphi)\cos\beta\,\mathrm{d}\omega\,\mathrm{d}A \tag{5.30}$$

with the *incident intensity*

$$K(\beta,\varphi) = \int\limits_0^\infty K_\lambda(\lambda,\beta,\varphi)\,\mathrm{d}\lambda \ . \tag{5.31}$$

Through integration of the incident intensity over all solid angles the irradiance is finally obtained as

$$E = \int\limits_\cap K(\beta,\varphi)\cos\beta\,\mathrm{d}\omega \ . \tag{5.32}$$

Just as for the emission of radiation, cf. 5.1.2, four radiation quantities are used for the characterisation of the incident radiation flow on a surface:

- The incident spectral intensity $K_\lambda(\lambda,\beta,\varphi)$ describes the distribution of the incident radiation flow over the solid angles of the hemisphere and the spectrum (directional spectral quantity).
- The spectral irradiance $E_\lambda(\lambda)$ describes the wavelength distribution of the radiation flow incident from the entire hemisphere (hemispherical spectral quantity).
- The incident intensity $K(\beta,\varphi)$ describes the directional distribution of the incident radiation flow (directional total quantity).
- The irradiance E combines the incident radiative power of all directions and wavelengths (hemispherical total quantity).

Fig. 5.14 shows the relationships between these four quantities. It is assembled in an analogous manner to Fig. 5.7, which contains the four quantities for emission of radiation.

5.1.4 Absorption of radiation

The radiation falling on a body can be partially reflected at its surface, whilst the portion that is not reflected penetrates the body. Here the radiative energy is absorbed and then converted into internal energy or part of it may be allowed to pass through the body. The absorbed portion is very important in terms of heat transfer. It is covered by the four absorptivities described in the following. These four belong to the four groups of physical radiation quantities introduced in 5.1.1.

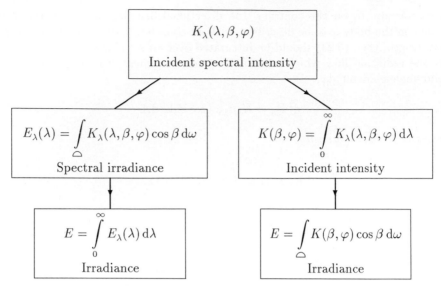

Fig. 5.14: Relationships between the four radiation quantities of irradiation

Just as in 5.1.3 we will consider a radiation flow $\mathrm{d}^3\Phi_\mathrm{in}$ according to (5.24), coming from a solid angle element $\mathrm{d}\omega$, that hits a surface element $\mathrm{d}A$ and only contains the radiation power within a wavelength interval $\mathrm{d}\lambda$. For the absorbed portion of the radiation flow we put

$$\mathrm{d}^3\Phi_\mathrm{in,abs} = a'_\lambda(\lambda,\beta,\varphi,T)\mathrm{d}^3\Phi_\mathrm{in} \tag{5.33}$$

and with that define the *directional spectral absorptivity* a'_λ. This dimensionless quantity, which has a value below one, is a material property of the absorbing body; it depends on the variables given in (5.33) namely wavelength λ, polar angle β, circumferential angle φ and the temperature T of the absorbing surface element. In addition to this the directional spectral absorptivity is also strongly influenced by the surface properties, e.g. the roughness of the surface.

Using a'_λ allows the absorbed portions of the integrated radiation flows introduced in 5.1.3 to be calculated. By integrating (5.33) over all solid angles in the hemisphere, the absorbed part of the hemispherical irradiation of the surface element $\mathrm{d}A$ in the wavelength interval $\mathrm{d}\lambda$ is obtained

$$\mathrm{d}^2\Phi_\mathrm{in,abs} = \int_\cap a'_\lambda(\lambda,\beta,\varphi,T)K_\lambda(\lambda,\beta,\varphi)\cos\beta\,\mathrm{d}\omega \quad \mathrm{d}\lambda\,\mathrm{d}A \ . \tag{5.34}$$

If, however, (5.33) is integrated over all wavelengths then the absorbed portion of the total radiative power from a solid angle element $\mathrm{d}\omega$ is obtained. This gives

$$\mathrm{d}^2\Phi'_\mathrm{in,abs} = \int_0^\infty a'_\lambda(\lambda,\beta,\varphi,T)K_\lambda(\lambda,\beta,\varphi)\,\mathrm{d}\lambda \quad \cos\beta\,\mathrm{d}\omega\,\mathrm{d}A \ . \tag{5.35}$$

Table 5.1: Definitions of the absorptivities and the relationships that exist between them

Directional spectral absorptivity

$$a'_\lambda(\lambda, \beta, \varphi, T) := \frac{d^3\Phi_{\text{in,abs}}}{d^3\Phi_{\text{in}}}$$

a'_λ is a material property of the absorbing body and gives, for every wavelength λ and for each direction (β, φ), the absorbed part of the incident radiation flow $d^3\Phi_{\text{in}}$, which, in a wavelength interval $d\lambda$, comes from a solid angle element $d\omega$.

Hemispherical spectral absorptivity

$$a_\lambda(\lambda, T) := \frac{d^2\Phi_{\text{in,abs}}}{d^2\Phi_{\text{in}}} = \frac{1}{E_\lambda(\lambda)} \int_{\cap} a'_\lambda(\lambda, \beta, \varphi, T) \cdot K_\lambda(\lambda, \beta, \varphi) \cos\beta\, d\omega$$

a_λ covers the radiation flow $d^2\Phi_{\text{in}}$, which comes from the entire hemisphere within a certain wavelength interval, and provides, for each wavelength λ, the absorbed portion of the spectral irradiance $E_\lambda(\lambda)$.

Directional total absorptivity

$$a'(\beta, \varphi, T) := \frac{d^2\Phi'_{\text{in,abs}}}{d^2\Phi'_{\text{in}}} = \frac{1}{K(\beta, \varphi)} \int_0^\infty a'_\lambda(\lambda, \beta, \varphi, T) K_\lambda(\lambda, \beta, \varphi)\, d\lambda$$

a' describes the radiation flow $d^2\Phi'_{\text{in}}$ of all wavelengths that comes from a certain solid angle element, and gives, for each direction (β, φ), the absorbed portion of the incident intensity $K(\beta, \varphi)$.

Hemispherical total absorptivity

$$a(T) := \frac{d\Phi_{\text{in,abs}}}{d\Phi_{\text{in}}} = \frac{1}{E} \int_0^\infty a_\lambda(\lambda, T) E_\lambda(\lambda)\, d\lambda$$

$$= \frac{1}{E} \int_0^\infty \left[\int_\cap a'_\lambda(\lambda, \beta, \varphi, T) K_\lambda(\lambda, \beta, \varphi) \cos\beta\, d\omega \right] d\lambda$$

a covers the radiation flow $d\Phi_{\text{in}}$ of all wavelengths that comes from the entire hemisphere and gives the absorbed part of the irradiance E.

Finally, (5.34) can be integrated over all wavelengths or (5.35) over all solid angles in the hemisphere. This gives the absorbed part of the total radiation flow hitting the surface element dA:

$$d\Phi_{\text{in,abs}} = \int_0^\infty \left[\int_\cap a'_\lambda(\lambda, \beta, \varphi, T) K_\lambda(\lambda, \beta, \varphi) \cos\beta\, d\omega \right] d\lambda\, dA \ . \tag{5.36}$$

Now, by putting these absorbed energy flows in relation to the associated incident radiation flows $d^2\Phi_{\text{in}}$ from (5.26), $d^2\Phi'_{\text{in}}$ from (5.30) and $d\Phi_{\text{in}}$ from (5.28), the absorptivities presented in Table 5.1 are obtained. These describe either the absorption of radiation coming from all directions in the hemisphere or over all wavelengths or finally the absorption of the total radiation on the surface element.

All absorptivities are less than one, but in contrast to the directional spectral absorptivity a'_λ, $a_\lambda(\lambda, T)$, $a'(\beta, \varphi, T)$ and $a(T)$ are not material properties of the absorbing body. They also depend on the direction and wavelength distribution of the radiation falling on the body, which are given by the incident spectral intensity $K_\lambda(\lambda, \beta, \varphi)$. The four absorptivities from Table 5.1 correspond to the four radiation quantities in 5.1.3 used to describe the irradiation of a surface element quantitatively. This multitude of absorptivities is required for applications where the selectivity of the absorbing surface with respect to the direction and the wavelength region as well as the directional and wavelength dependency of the incident radiation have to be considered.

Example 5.2: The hemispherical spectral absorptivity of a surface which appears bright to the eye is strongly simplified given by

$$a_\lambda(\lambda, T) = \begin{cases} \alpha_{\lambda 1} = 0.10 & \text{for} \quad 0 \leq \lambda \leq \lambda_1 \\ \\ \alpha_{\lambda 2} = 0.80 & \text{for} \quad \lambda_1 < \lambda < \infty \end{cases}$$

with $\lambda_1 = 1.50 \, \mu\text{m}$; Fig. 5.15. So, at small wavelengths, only a low amount of radiative energy will be absorbed, whilst at large wavelengths significantly more radiation energy will be absorbed. Radiation hits the surface, from a source at a temperature T^* with spectral irradiance

$$E_\lambda(\lambda, T^*) = \frac{c_1}{\lambda^5} \exp(-c_2/\lambda T^*) \ . \tag{5.37}$$

Here $c_2 = 14.5 \cdot 10^3 \, \mu\text{m K}$ and c_1 is a constant of proportionality. For $T^* = 1000 \, \text{K}$ and $T^* = 5777 \, \text{K}$ (temperature of the sun's surface) calculate the hemispherical total absorptivity $a(T)$.

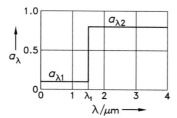

Fig. 5.15: Approximation of the hemispherical spectral absorptivity a_λ of a surface

The wavelength and temperature dependency given by (5.37) correspond to a relationship found by W. Wien [5.3] in 1896 to be approximately valid for the hemispherical spectral emissive power $M_{\lambda s}(\lambda, T^*)$ of an ideal radiator, a black body, with a temperature T^*. We will come back to the properties of black bodies in section 5.1.6 and more extensively in 5.2.2. In our example a spectral irradiance $E_\lambda \sim M_{\lambda s}$ has been assumed, so that its indirect dependence on T^* appears explicitly in (5.37).

Fig. 5.16 shows the spectral irradiance E_λ according to (5.37) for $T^* = 1000 \, \text{K}$ and $T^* = 5777 \, \text{K}$. Here, the proportionality constants c_1 were each chosen so that for both temperatures the maximum of E_λ, which appears at $\lambda_{\max} = c_2/5T^*$, was the same. The hatched areas in Fig. 5.16 are proportional to the absorbed fraction of the incident radiation flow, whilst the areas under the E_λ curves correspond to the irradiance E. The desired hemispherical total absorptivity $a(T)$ is, according to (5.28) and (5.36) the ratio of these areas. For $T^* = 1000 \, \text{K}$ an absorptivity close to $a_{\lambda 2}$ will be expected. In contrast the largest portion of the solar radiation ($T^* = 5777 \, \text{K}$) falls in the region of small wavelengths ($\lambda < \lambda_1$); therefore for solar radiation a total absorptivity only slightly larger than $a_{\lambda 1}$ is expected.

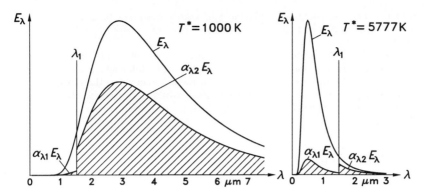

Fig. 5.16: Spectral irradiance $E_\lambda(\lambda, T^*)$ according to (5.37) for $T^* = 1000$ K and $T^* = 5777$ K. The ordinates of the maxima were arbitrarily chosen to be the same. The hatched areas represent the absorbed part of the irradiance $E(T^*)$.

The results found from Fig. 5.16 can be confirmed by the following calculations. According to Table 5.1 the hemispherical total absorptivity is obtained to be

$$a(T) = \frac{\int\limits_0^\infty a_\lambda(\lambda, T) E_\lambda(\lambda)\, d\lambda}{E} = \frac{a_{\lambda 1} \int\limits_0^{\lambda 1} E_\lambda(\lambda)\, d\lambda + a_{\lambda 2} \int\limits_{\lambda 1}^\infty E_\lambda(\lambda)\, d\lambda}{\int\limits_0^\infty E_\lambda(\lambda)\, d\lambda} \quad . \tag{5.38}$$

The evaluation of the integrals is facilitated by the introduction of the dimensionless variable

$$\zeta := c_2/\lambda T^*$$

which with

$$E_\lambda(\zeta, T^*) = \left(c_1/c_2^5\right) T^{*5} \zeta^5 e^{-\zeta}$$

and

$$d\lambda = -\frac{c_2}{T^*} \frac{d\zeta}{\zeta^2}$$

gives for the absorptivity

$$a(\zeta_1) = \frac{a_{\lambda 1} \int\limits_{\zeta_1}^\infty \zeta^3 e^{-\zeta}\, d\zeta + a_{\lambda 2} \int\limits_0^{\zeta_1} \zeta^3 e^{-\zeta}\, d\zeta}{\int\limits_0^\infty \zeta^3 e^{-\zeta}\, d\zeta} \quad .$$

It depends on $\zeta_1 = c_2/\lambda_1 T^*$ and therefore indirectly on the temperature T^* of the radiation source. The integrals that appear here can be calculated from

$$\int \zeta^3 e^{-\zeta}\, d\zeta = C - e^{-\zeta} \left(\zeta^3 + 3\zeta^2 + 6\zeta + 6\right) = C - F(\zeta) \quad .$$

This then gives

$$a(\zeta_1) = \frac{1}{6} \left\{ a_{\lambda 1} F(\zeta_1) + a_{\lambda 2} \left[6 - F(\zeta_1)\right] \right\} \quad . \tag{5.39}$$

For the radiation from the source at $T^* = 1000$ K, $\zeta_1 = 9.667$, from which, with $F(\zeta_1) = 0.0791$, the total absorptivity $a = 0.791$ is yielded. The result expected from Fig. 5.16 $a \approx$

$a_{\lambda 2} = 0.80$ is therefore confirmed. The incident solar radiation with $T^* = 5777\,\mathrm{K}$ yields the value $\zeta_1 = 1.6733$ and from that $F(\zeta_1) = 5.4646$. This yields, from (5.39), the much smaller absorptivity $a = 0.162$.

The hemispherical total absorptivity is not only a property of the absorbing surface. Rather, it depends on the spectral distribution of the incident radiation energy. This is shown by the different values of a for the mainly short-wave solar radiation, in which the absorption properties at small wavelengths are decisive, and for the incident radiation from an earthly source, for which the long-wave portion of the absorption spectrum $a_\lambda(\lambda, T)$ is of importance.

5.1.5 Reflection of radiation

The radiation flow reflected from the surface of a body can be described using dimensionless reflectivities, in the same manner as for the absorbed power with the absorptivities dealt with in the last section. However, this involves further complications if we do not only want to find out what proportion of the radiation from a certain direction is reflected but also in which direction the reflected energy is sent back. The possible reflective behaviour of a surface can be idealised by two limiting cases: mirrorlike (or specular) reflection and diffuse reflection.

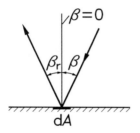

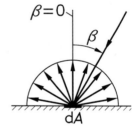

Fig. 5.17: Mirrorlike reflection of radiation incident from the polar angle β

Fig. 5.18: Diffuse reflection of the radiation incident from the polar angle β

In mirrorlike reflection the ray incident from the angles β and φ will be reflected at an equal polar angle $\beta_r = \beta$, but at a circumferential angle $\varphi_r = \varphi + \pi$, Fig. 5.17. On a diffuse reflecting surface the radiation falling from a certain direction (β, φ) generates a reflected radiation whose total intensity is equal over all radiating directions away from the surface (β_r, φ_r), cf. Fig. 5.18. The reflective behaviour of real surfaces lies somewhere between these two limits. Shiny, polished metal surfaces reflect in approximately mirror fashion. Rough and matt surfaces reflect in good approximation in a diffuse manner. Here the scale of the roughness must also be seen in relation to the wavelength of the radiation. A surface may be rough for short wave radiation, whilst for long wave radiation it may be considered to be smooth. The complex relationships for the consideration of the directional distribution of the reflected radiation have to be described by bidirectional reflectivities, which are dependent on two pairs of angles (β, φ) and

Table 5.2: Definitions of the reflectivities and the relationships existing between them

Directional spectral reflectivity

$$r'_\lambda(\lambda, \beta, \varphi, T) := \frac{d^3\Phi_{in,ref}}{d^3\Phi_{in}}$$

r'_λ is a material property of the reflecting body and gives, for each wavelength λ and for each direction (β, φ), the reflected portion of the incident radiation flow $d^3\Phi_{in}$, which, in the wavelength interval $d\lambda$, comes from a solid angle element.

Hemispherical spectral reflectivity

$$r_\lambda(\lambda, T) := \frac{d^2\Phi_{in,ref}}{d^2\Phi_{in}} = \frac{1}{E_\lambda(\lambda)} \int_\Omega r'_\lambda(\lambda, \beta, \varphi, T) K_\lambda(\lambda, \beta, \varphi) \cos\beta \, d\omega$$

r_λ covers the radiation flow $d^2\Phi_{in}$, which comes from the entire hemisphere, within a certain wavelength interval, and provides, for each wavelength λ the reflected part of the spectral irradiance $E_\lambda(\lambda)$.

Directional total reflectivity

$$r'(\beta, \varphi, T) := \frac{d^2\Phi'_{in,ref}}{d^2\Phi'_{in}} = \frac{1}{K(\beta,\varphi)} \int_0^\infty r'_\lambda(\lambda, \beta, \varphi, T) K_\lambda(\lambda, \beta, \varphi) \, d\lambda$$

r' covers the radiation flow $d^2\Phi'_{in}$ of all wavelengths which comes from a certain solid angle element, and gives, for each direction (β, φ), the reflected portion of the incident intensity $K(\beta, \varphi)$.

Hemispherical total reflectivity

$$r(T) := \frac{d\Phi_{in,ref}}{d\Phi_{in}} = \frac{1}{E} \int_0^\infty r_\lambda(\lambda, T) E_\lambda(\lambda) \, d\lambda$$

$$= \frac{1}{E} \int_0^\infty \left[\int_\Omega r'_\lambda(\lambda, \beta, \varphi, T) K_\lambda(\lambda, \beta, \varphi) \cos\beta \, d\omega \right] d\lambda$$

r covers the radiation flow $d\Phi_{in}$, of all wavelengths, that comes from the entire hemisphere and gives the reflected fraction of the irradiance.

(β_r, φ_r). We will not look at this here, but suggest the extensive discussion of this subject by R. Siegel et al [5.4] for further reading. In the following we will restrict ourselves to the introduction of reflectivities which only provide us with the portion of the incident radiation that is reflected, without specifying what proportion of the reflected energy is sent back in which direction. However, in the two limiting cases of mirrorlike and diffuse reflection, this question has already been answered, see Fig. 5.17 and 5.18. As the reflectivities are defined in a completely analogous way to the absorptivities discussed in 5.1.4, it is sufficient to consider Table 5.2, which is the analogue of Table 5.1, without further explanations. In addition if the body does not allow radiation to pass through it, then the balance is

$$d^3\Phi_{in,ref} + d^3\Phi_{in,abs} = d^3\Phi_{in} \ , \tag{5.40}$$

according to which, the incident radiation will either be reflected or absorbed Dividing (5.40) by $d^3\Phi_{in}$ gives the relationship

$$r'_\lambda(\lambda,\beta,\varphi,T) + a'_\lambda(\lambda,\beta,\varphi,T) = 1 \ . \tag{5.41}$$

The two material functions r'_λ and a'_λ of an opaque body are not independent of each other. The directional spectral reflectivity r'_λ is determined by the directional spectral absorptivity a'_λ. The similar relationship between the different absorptivities and reflectivities from Tables 5.1 and 5.2, respectively, mean that equations analogous to (5.41) are valid, with which the three other reflectivities can be found from the corresponding absorptivities.

5.1.6   Radiation in an enclosure. Kirchhoff's law

Thermodynamic relationships exist between the emission and absorption capabilities of a body. These were first discovered in 1860 by G.R. Kirchhoff[3] [5.5]. These relationships also show that an upper limit exists for the emitted radiation flow.

In order to derive these we will consider an adiabatic evacuated enclosure, like that shown in Fig. 5.19, with walls of any material. In this enclosure a state of thermodynamic equilibrium will be reached: The walls assume the same temperature T overall and the enclosure is filled with radiation, which is known as *hollow enclosure radiation*. In the sense of quantum mechanics this can also be interpreted as a photon gas in equilibrium. This equilibrium radiation is fully homogeneous, isotropic and non-polarised. It is of equal strength at every point in the hollow enclosure and is independent of direction; it is determined purely by the temperature T of the walls. Due to its isotropic nature, the spectral intensity L_λ^* of the hollow enclosure radiation does not depend on β and φ, but is, as Kirchhoff was the first to recognise, a universal function of wavelength and temperature: $L_\lambda^* = L_\lambda^*(\lambda,T)$, which is also called Kirchhoff's function. As the enclosure is filled with the same diffuse radiation, the incident spectral intensity K_λ for every element of any area that is oriented in any position, will, according to (5.25), be the same as the spectral intensity of the hollow enclosure radiation:

$$K_\lambda = L_\lambda^*(\lambda,T) \ . \tag{5.42}$$

---

[3]Gustav Robert Kirchhoff (1824–1887) first formulated and published the laws named after him for electrical networks when he was still a student at university in Königsberg. In 1850 he was made a professor in Breslau and in 1854 he became a professor in Heidelberg. It was here that he worked with R. Bunsen for over 10 years and carried out investigations into the emission and absorption of radiation. Their results became known as Kirchhoff's radiation laws and as Bunsen-Kirchhoff spectral analysis. In 1875 he was called to the University of Berlin to be Professor of Theoretical Physics. Alongside his teacher F. Neumann, Kirchhoff was a founder of mathematical (theoretical) physics in Germany.

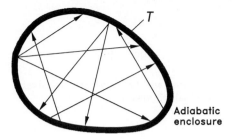

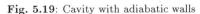

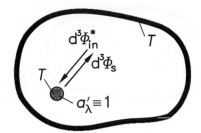

Fig. 5.19: Cavity with adiabatic walls

Fig. 5.20: Black body ($a'_\lambda \equiv 1$) in an adiabatic and isothermal cavity

A small body is located in the enclosure that, once thermodynamic equilibrium has been reached, assumes the same temperature T as the walls, Fig. 5.20. This body shall have the special property that it completely absorbs all incident radiation from every direction at every wavelength. Its directional spectral absorptivity is (independent of λ, β, φ and T) always one:

$$a'_\lambda(\lambda, \beta, \varphi, T) \equiv 1 . \tag{5.43}$$

Then, according to 5.1.4, its three other absorptivities a_λ, a' and a are also equal to one. Kirchhoff named this ideal absorber a *black body*. According to the 2nd law, the state of this equilibrium system consisting of radiation in an enclosure and the black body cannot change despite the absorption of the radiation in the enclosure. The black body therefore has to replace the radiation it absorbs by its own emission of radiation. This holds for every wavelength interval and every solid angle element; otherwise the distribution of the hollow enclosure radiation, given by $L^*_\lambda(\lambda, T)$, would change, and the thermodynamic equilibrium in the enclosure would be disturbed. So the radiation flow $\mathrm{d}^3\Phi_\mathrm{s}$, emitted by the black body in a particular solid angle element $\mathrm{d}\omega$ and within a certain wavelength interval $\mathrm{d}\lambda$ has to be the same as the radiation flow $\mathrm{d}^3\Phi^*_\mathrm{in}$ of the hollow enclosure radiation that hits the black body.

According to (5.4), the radiation flow emitted from the black body is

$$\mathrm{d}^3\Phi_\mathrm{s} = L_{\lambda\mathrm{s}}(\lambda, \beta, \varphi, T) \cos\beta \, \mathrm{d}\omega \, \mathrm{d}\lambda \, \mathrm{d}A ,$$

where $L_{\lambda\mathrm{s}}$ is its spectral intensity. The incident radiation flow of the hollow enclosure radiation is, from (5.24) and (5.42),

$$\mathrm{d}^3\Phi^*_\mathrm{in} = L^*_\lambda(\lambda, T) \cos\beta \, \mathrm{d}\omega \, \mathrm{d}\lambda \, \mathrm{d}A . \tag{5.44}$$

From the condition of thermodynamic equilibrium, $\mathrm{d}^3\Phi_\mathrm{s} = \mathrm{d}^3\Phi^*_\mathrm{in}$, it follows that

$$L_{\lambda\mathrm{s}}(\lambda, \beta, \varphi, T) = L^*_\lambda(\lambda, T) .$$

According to this, the spectral intensity of the black body is independent of direction and is the same as the spectral intensity of hollow enclosure radiation at

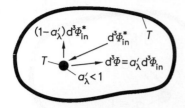

Fig. 5.21: Body with any directional spectral absorptivity a'_λ in an adiabatic enclosure

the same temperature:

$$L_{\lambda s}(\lambda, T) = L^*_\lambda(\lambda, T) \ .$$ (5.45)

Hollow enclosure radiation and radiation of a black body ($a'_\lambda \equiv 1$) have identical properties. The black body radiates diffusely; from (5.18) it holds for its hemispherical spectral emissive power that

$$M_{\lambda s}(\lambda, T) = \pi L_{\lambda s}(\lambda, T) \ .$$ (5.46)

We will now consider an enclosure with a body that has any radiation properties, Fig. 5.21. Thermodynamic equilibrium means that this body must also emit exactly the same amount of energy in every solid angle element and in every wavelength interval as it absorbs from the hollow enclosure radiation. It therefore holds for the emitted radiative power that

$$d^3\Phi = a'_\lambda(\lambda, \beta, \varphi, T)d^3\Phi^*_{in}$$ (5.47)

with a'_λ as its directional spectral absorptivity. For $d^3\Phi$ we use the expression from (5.4); the radiation flow $d^3\Phi^*_{in}$ of the hollow enclosure radiation is given by (5.44). With (5.45), for the spectral intensity of any radiator, it follows that

$$L_\lambda(\lambda, \beta, \varphi, T) = a'_\lambda(\lambda, \beta, \varphi, T)L_{\lambda s}(\lambda, T) \ .$$ (5.48)

This is the law from G.R. Kirchhoff [5.5]: *Any body at a given temperature T emits, in every solid angle element and in every wavelength interval, the same radiative power as it absorbs there from the radiation of a black body (= hollow enclosure radiation) having the same temperature.* Therefore, a close relationship exists between the emission and absorption capabilities. This can be more simply expressed using this sentence: A good absorber of thermal radiation is also a good emitter.

As the black body is the best absorber, $a'_\lambda \equiv 1$, it also emits the most. From (5.48) with $a'_\lambda \leq 1$ the inequality follows

$$L_\lambda(\lambda, \beta, \varphi, T) \leq L_{\lambda s}(\lambda, T) \ .$$

At a given temperature the black body emits the maximum radiative power at each wavelength and in every direction. It is not possible for any other body at the same temperature to emit more radiative power in any wavelength interval or any solid angle than that emitted by a black body at that temperature. The black body is also the ideal emitter. The determination of the spectral intensity $L_{\lambda s}(\lambda, T)$ of a black body as the thermodynamic upper limit of emission is therefore an important task of the theory of thermal radiation; we will consider this in the following sections.

5.2 Radiation from a black body

As shown in 5.1.6, the laws of thermodynamics demand that there must be an upper limit for the spectral intensity $L_\lambda(\lambda, \beta, \varphi, T)$ for all bodies. This maximum emission is associated with an ideal radiator, the black body. Its radiation properties shall be dealt with in the following.

5.2.1 Definition and realisation of a black body

A black body is defined as a body where all the incident radiation penetrates it and is completely absorbed within it. No radiation is reflected or allowed to pass through it. This holds for radiation of all wavelengths falling onto the body from all angles. In addition to this the black body is a diffuse radiator. Its spectral intensity $L_{\lambda s}$ does not depend on direction, but is a universal function $L_{\lambda s}(\lambda, T)$ of the wavelength and the thermodynamic temperature. The hemispherical spectral emissive power $M_{\lambda s}(\lambda, T)$ is linked to Kirchhoff's function $L_{\lambda s}(\lambda, T)$ by the simple relationship

$$M_{\lambda s}(\lambda, T) = \pi L_{\lambda s}(\lambda, T) \ , \tag{5.49}$$

see (5.46).

The determination of the universal functions $L_{\lambda s}(\lambda, T)$ or $M_{\lambda s}(\lambda, T)$ is a fundamental task of physics. This was solved experimentally towards the end of the 19th century, but a theoretical basis for the measured data was first found by M. Planck[4] in 1900. Through assumptions, which later formed one of the foundations of quantum theory, he was able to formulate the law for the spectral intensity named after him, which has been confirmed experimentally and is considered to be correct until today.

---

[4]Max Karl Ernst Ludwig Planck (1858–1947) became Professor of Theoretical Physics in Kiel in 1885; from 1888 to 1920 he taught at the University of Berlin, where he was successor to G.R. Kirchhoff. From 1894 onwards he was a member of the Prussian Academy of Science. Planck dealt with thermodynamic problems in his PhD thesis (1879), in particular with the 2nd law and the concept of entropy. He found his famous radiation law in 1900, by linking thermodynamic laws for the energy and entropy of hollow enclosure radiation with the electromagnetic radiation theory, statistical methods and the assumption that energy was made up of a large number of discrete, small energy elements (quanta). Planck also dealt with the theory of relativity and the philosophical basis of science. In 1918 he was awarded the Nobel prize for Physics.

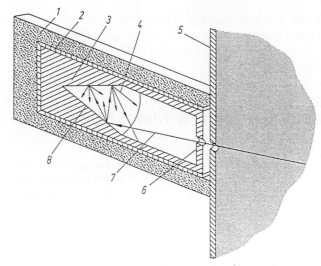

Fig. 5.22: Isothermal hollow enclosure for the realisation of a black body. 1 insulation; 2 heating; 3 copper cylinder; 4 reflected radiation; 5 polished surface; 6 black surface; 7 incident beam; 8 strongly absorbing surface

The black body got its name because of the property of a good absorber to appear black to the eye when visible light falls on its surface. However, the eye is only an indicator of absorption capability of a body within the very small wavelength region of visible light, see Fig. 5.1. Therefore, surfaces which appear to be black, such as soot or black platinum only approximately attain the complete absorption of radiation at all wavelengths, which is the prerequisite for a black body.

In order to realise the black body as a reference standard for radiation measurements a cavity with a small opening, like that shown in Fig. 5.22, is used. A beam coming through this opening hits the enclosure wall, and will be partly absorbed there. The portion that is reflected hits another point on the wall where it will once again be absorbed with a small part being reflected, and so on. With a sufficiently small opening to the enclosure only a tiny part of the beam entering the enclosure will be able to leave it. The condition of complete absorption of the entering radiation is fulfilled to a high degree. The opening of the enclosure therefore absorbs (almost) like a black body. So, according to Kirchhoff's law it radiates black body radiation.

5.2.2 The spectral intensity and the spectral emissive power

We refrain from deriving the equations for the spectral intensity and the hemispherical spectral emissive power of a black body, found by M. Planck [5.6], for this see [5.7]. They are

$$M_{\lambda s}(\lambda, T) = \pi L_{\lambda s}(\lambda, T) = \frac{c_1}{\lambda^5 \left[\exp\left(c_2/\lambda T\right) - 1\right]} \, . \tag{5.50}$$

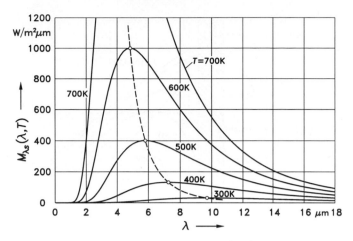

Fig. 5.23: Hemispherical spectral emissive power $M_{\lambda s}(\lambda, T)$ of a black body according to Planck's radiation law (5.50)

The two radiation constants c_1 and c_2 are made up of fundamental natural constants, from the velocity of light c_0 in a vacuum, the Planck constant h and the Boltzmann constant k. With the best values for these quantities, according to [5.8],

$$c_1 = 2\pi hc_0^2 = (3.741\,774\,9 \pm 0.000\,002\,2) \cdot 10^{-16}\mathrm{Wm^2} \qquad (5.51\mathrm{a})$$

and

$$c_2 = hc_0/k = (14\,387.69 \pm 0.12)\,\mu\mathrm{m\,K} \qquad (5.51\mathrm{b})$$

are obtained.

Planck's law (5.50) is illustrated in Fig. 5.23 for several isotherms. These have a horizontal tangent at $\lambda = 0$. The emission of radiation at small wavelengths is initially very low, but increases steeply with increasing λ, runs through a maximum and then falls back to a lower value. For $\lambda \to \infty$ the limit value $M_{\lambda s} = 0$ is yielded. A characteristic is the displacement of the maximum to small wavelengths with increasing temperature. Here, the maximum value $M_{\lambda s}(\lambda_{\mathrm{max}}, T)$ increases rapidly with T. The different isotherms do not cross each other (except at $\lambda = 0$); at every wavelength the emission of radiation increases with rising temperature

In Fig. 5.24, $M_{\lambda s}$ is reproduced with a logarithmic scale. The region of visible light is also indicated. It is only at sufficiently high temperatures that a significant portion of the emissive power is emitted within this wavelength interval. First, at the so-called Draper point at 798 K (525 °C), [5.9], will a heated body in dark surroundings appear as a dark red object to the human eye. The sun emits radiation with a hemispherical spectral emissive power approximately corresponding to a black body at 5777 K. As Fig. 5.24 shows, at this temperature the maximum of $M_{\lambda s}$ lies in the visible spectral region. The human eye has adapted itself to this and is at its most sensitive at these wavelengths.

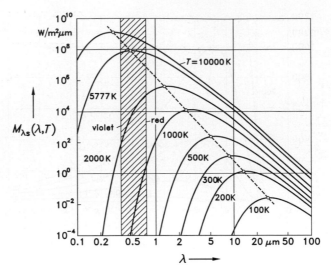

Fig. 5.24: Hemispherical spectral emissive power $M_{\lambda s}(\lambda, T)$ of a black body, according to (5.50), with a logarithmic scale. The hatched wavelength region corresponds to the region of visible light.

The position of the maximum of $M_{\lambda s}$ on an isotherm is found from the condition

$$\frac{\partial M_{\lambda s}}{\partial \lambda} = 0, \quad T = \text{const} .$$

This leads to the transcendental equation

$$\left(1 - \frac{c_2}{5\lambda T}\right) \exp\left(c_2/\lambda T\right) = 1$$

with the solution $(c_2/\lambda_{\max} T) = 4.965\,114\,23$ or

$$\lambda_{\max} T = (2897.756 \pm 0.024)\mu\text{m K} . \tag{5.52}$$

This is a form of Wien's[5] [5.10] displacement law. It is frequently used to calculate the temperature of a radiator from its measured wavelength $\lambda_{\max}$, under the assumption that it behaves like a black body.

The isotherms of Planck's radiation law (5.50) may be reproduced by a single curve, if instead of $M_{\lambda s}$, $M_{\lambda s}/T^5$ over the product λT or the dimensionless quantity

$$x := \lambda T/c_2 \tag{5.53}$$

---

[5]Wilhelm Carl Werner Otto Fritz Franz Wien (1864–1928) became an assistant to Hermann v. Helmholtz at the Physikalisch-Technische Reichsanstalt in Berlin in 1890. It was there that he discovered the displacement law in 1893, and also published an equation for $M_{\lambda s}$ in 1896, that only slightly differed from Planck's law. Wien became Professor of Physics at the TH in Aachen in 1896, moved in 1899 to become a professor in Würzburg, and once again changed to the University of Munich in 1920. In 1911 he was awarded the Nobel prize for Physics as an acknowledgement of his work on thermal radiation.

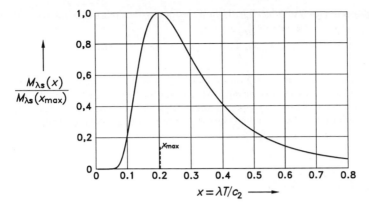

Fig. 5.25: Dimensionless representation of Planck's radiation law from (5.55) and (5.57)

is plotted. This gives

$$\frac{M_{\lambda s}(\lambda T)}{T^5} = \frac{c_1}{(\lambda T)^5 \left[\exp\left(c_2/\lambda T\right) - 1\right]} \tag{5.54}$$

or

$$\frac{M_{\lambda s}(x)}{T^5} = \frac{c_1/c_2^5}{x^5 \left(e^{1/x} - 1\right)} \tag{5.55}$$

with $c_1/c_2^5 = (0.606\,907 \pm 0.000\,025) \cdot 10^{-12}\,\mathrm{W}/(\mu\mathrm{m\,m^2\,K^5})$. The maximum for this function appears at

$$x_{\max} = \frac{\lambda_{\max} T}{c_2} = 0.201\,405\,2 \tag{5.56}$$

and has the value

$$\frac{M_{\lambda s}(x_{\max})}{T^5} = 21.201\,44\frac{c_1}{c_2^5} = (0.128\,673 \pm 0.000\,005) \cdot 10^{-10}\,\frac{\mathrm{W}}{\mu\mathrm{m\,m^2\,K^5}} \ . \tag{5.57}$$

So, the maximum value of $M_{\lambda s}$ increases with the fifth power of T. Fig. 5.25 shows the ratio $M_{\lambda s}(x)/M_{\lambda s}(x_{\max})$. In Table 5.3 the values of $M_{\lambda s}(\lambda T)/T^5$ calculated using (5.54) as a function of λT are presented.

The relationships communicated for $M_{\lambda s}(\lambda, T)$ are exactly valid for the radiation of a black body in a *vacuum*. With radiation in a medium with refractive index $n > 1$ the propagation velocity c and the wavelength λ_M ar smaller than in a vacuum; although the frequency ν remains the same. In Planck's law, some change has to be made. The velocity of light in a vacuum c_0 that appears in the constants c_1 and c_2 has to be replaced by c, and in (5.50) λ is replaced by λ_M. Taking into account the relationship $c = c_0/n$, instead of (5.54), we obtain

$$\frac{M_{\lambda_M,s}(\lambda_M T)}{n^3 T^5} = \frac{c_1}{(n\lambda_M T)^5 \left[\exp\left(c_2/n\lambda_M T\right) - 1\right]} \ ,$$

where c_1 and c_2 have their original meanings according to (5.50). The values in Table 5.3 can be used, if λT is interpreted as $n\lambda_M T$ and $M_{\lambda s}/T^5$ as $M_{\lambda_M,s}/n^3 T^5$.

The emissive power of a black body in a medium with a refractive index n will be

$$M_{s,n}(T) = \int\limits_0^\infty M_{\lambda_M,s}(\lambda_M, T)\, d\lambda_M = n^2 M_s(T) \ ,$$

where $M_s(T)$ is the emissive power in a vacuum. This will be calculated in the next section.

As the refractive index n of air and other gases lies very close to one, the increase in the emission compared to emission into a vacuum ($n = 1$) is of very little importance. Exceptions to this are the radiation effects in molten glass with $n \approx 1.5$ and similar semi-transparent materials, see [5.11].

5.2.3   The emissive power and the emission of radiation in a wavelength interval

According to (5.8) the emissive power $M_s(T)$ of a black body is obtained by integration of $M_{\lambda s}(\lambda, T)$ over all wavelengths. A surprisingly simple result comes out of the fairly complex equation (5.50). With

$$d\lambda = \frac{c_2}{T}\, dx$$

according to (5.53), the following is obtained from (5.55)

$$M_s(T) = \int\limits_0^\infty M_{\lambda s}(\lambda, T)\, d\lambda = c_2 T^4 \int\limits_0^\infty \frac{M_{\lambda s}(x)}{T^5}\, dx = \frac{c_1}{c_2^4}\, T^4 \int\limits_0^\infty \frac{dx}{x^5\left(e^{1/x} - 1\right)} \ .$$

The definite integral has the value $\pi^4/15$, so

$$M_s(T) = \sigma T^4 \tag{5.58}$$

with

$$\sigma = \frac{c_1}{c_2^4}\frac{\pi^4}{15} = \frac{2\,\pi^5 k^4}{15\,c_0^2 h^3} = (5.670\,51 \pm 0.000\,19)\cdot 10^{-8}\frac{W}{m^2 K^4} \tag{5.59}$$

is obtained, [5.8]. The best directly measured, but not so accurate value is $\sigma = (5.669\,59 \pm 0.000\,76)\cdot 10^{-8}\,W/m^2 K^4$ from [5.12].

Equation (5.58) is the famous law from Stefan and Boltzmann, cf. also section 1.1.5. J. Stefan [5.13] presented this law in 1879 based on experimental results, whilst L. Boltzmann [5.14] derived this relationship between emissive power and temperature in 1884, by thermodynamic reasoning and using results from the classic electromagnetic theory of radiation. Of course neither researcher could reduce the constant σ, also called the Stefan-Boltzmann constant, to fundamental constants of nature, as the appearance of the Planck constant h in (5.59) shows that this is only possible using quantum theory.

The emissive power $M_s(T)$ of a black body according to (5.34) is illustrated in Fig. 5.26 by the total area under the isotherm of $M_{\lambda s}(\lambda, T)$. In many calculations

Table 5.3: Hemispherical spectral emissive power of a black body divided by the fifth power of the temperature, according to (5.54) and fraction function $F(0, \lambda T)$ according to (5.60) as a function of the product λT

λT μm K	$M_{\lambda s}/T^5$ $\dfrac{10^{-10}\text{W}}{\text{m}^2\mu\text{mK}^5}$	$F(0, \lambda T)$	λT μm K	$M_{\lambda s}/T^5$ $\dfrac{10^{-10}\text{W}}{\text{m}^2\mu\text{mK}^5}$	$F(0, \lambda T)$	λT μm K	$M_{\lambda s}/T^5$ $\dfrac{10^{-10}\text{W}}{\text{m}^2\mu\text{mK}^5}$	$F(0, \lambda T)$
1000	0.00211	0.00032	2750	0.12781	0.21660	4500	0.08641	0.56431
1050	0.00328	0.00056	2800	0.12830	0.22789	4550	0.08482	0.57186
1100	0.00485	0.00091	2850	0.12859	0.23922	4600	0.08325	0.57927
1150	0.00686	0.00142	2900	0.12867	0.25056	4650	0.08169	0.58654
1200	0.00933	0.00213	2950	0.12857	0.26191	4700	0.08016	0.59367
1250	0.01230	0.00308	3000	0.12831	0.27323	4750	0.07864	0.60067
1300	0.01573	0.00432	3050	0.12788	0.28453	4800	0.07715	0.60754
1350	0.01963	0.00587	3100	0.12730	0.29578	4850	0.07568	0.61428
1400	0.02395	0.00779	3150	0.12659	0.30697	4900	0.07423	0.62089
1450	0.02864	0.01011	3200	0.12576	0.31810	4950	0.07280	0.62737
1500	0.03365	0.01285	3250	0.12482	0.32915	5000	0.07140	0.63373
1550	0.03892	0.01605	3300	0.12377	0.34011	5050	0.07002	0.63996
1600	0.04438	0.01972	3350	0.12263	0.35097	5100	0.06866	0.64608
1650	0.04998	0.02388	3400	0.12141	0.36173	5150	0.06732	0.65207
1700	0.05563	0.02854	3450	0.12011	0.37238	5200	0.06601	0.65795
1750	0.06130	0.03369	3500	0.11875	0.38291	5250	0.06472	0.66371
1800	0.06691	0.03934	3550	0.11733	0.39332	5300	0.06346	0.66937
1850	0.07242	0.04549	3600	0.11586	0.40360	5350	0.06222	0.67491
1900	0.07778	0.05211	3650	0.11434	0.41375	5400	0.06100	0.68034
1950	0.08295	0.05920	3700	0.11279	0.42377	5450	0.05981	0.68566
2000	0.08790	0.06673	3750	0.11120	0.43364	5500	0.05863	0.69089
2050	0.09261	0.07469	3800	0.10959	0.44338	5550	0.05748	0.69600
2100	0.09704	0.08306	3850	0.10796	0.45297	5600	0.05636	0.70102
2150	0.10119	0.09180	3900	0.10631	0.46241	5650	0.05525	0.70594
2200	0.10504	0.10089	3950	0.10465	0.47172	5700	0.05417	0.71077
2250	0.10859	0.11031	4000	0.10297	0.48087	5750	0.05311	0.71550
2300	0.11182	0.12003	4050	0.10130	0.48987	5800	0.05207	0.72013
2350	0.11475	0.13002	4100	0.09962	0.49873	5850	0.05105	0.72468
2400	0.11737	0.14026	4150	0.09794	0.50744	5900	0.05005	0.72914
2450	0.11969	0.15071	4200	0.09626	0.51600	5950	0.04907	0.73351
2500	0.12172	0.16136	4250	0.09459	0.52442	6000	0.04812	0.73779
2550	0.12346	0.17217	4300	0.09293	0.53269	6050	0.04718	0.74199
2600	0.12493	0.18312	4350	0.09128	0.54081	6100	0.04626	0.74611
2650	0.12613	0.19419	4400	0.08965	0.54878	6150	0.04536	0.75015
2700	0.12709	0.20536	4450	0.08802	0.55662	6200	0.04448	0.75411

Table 5.3: (continued)

λT μm K	$M_{\lambda s}/T^5$ $\dfrac{10^{-10}\text{W}}{\text{m}^2\mu\text{mK}^5}$	$F(0,\lambda T)$	λT μm K	$M_{\lambda s}/T^5$ $\dfrac{10^{-10}\text{W}}{\text{m}^2\mu\text{mK}^5}$	$F(0,\lambda T)$	λT μm K	$M_{\lambda s}/T^5$ $\dfrac{10^{-10}\text{W}}{\text{m}^2\mu\text{mK}^5}$	$F(0,\lambda T)$
6250	0.04362	0.75800	10000	0.01164	0.91416	17000	0.00198	0.97765
6300	0.04278	0.76181	10100	0.01128	0.91618	17200	0.00190	0.97834
6350	0.04195	0.76554	10200	0.01094	0.91814	17400	0.00182	0.97899
6400	0.04115	0.76921	10300	0.01061	0.92004	17600	0.00175	0.97962
6450	0.04035	0.77280	10400	0.01029	0.92188	17800	0.00168	0.98023
6500	0.03958	0.77632	10500	0.00998	0.92367	18000	0.00162	0.98081
6600	0.03808	0.78317	10600	0.00969	0.92540	18200	0.00156	0.98137
6700	0.03665	0.78976	10700	0.00940	0.92709	18400	0.00150	0.98191
6800	0.03527	0.79610	10800	0.00913	0.92872	18600	0.00144	0.98243
6900	0.03395	0.80220	10900	0.00886	0.93031	18800	0.00139	0.98293
7000	0.03269	0.80808	11000	0.00861	0.93185	19000	0.00133	0.98341
7100	0.03148	0.81374	11200	0.00812	0.93480	19200	0.00129	0.98387
7200	0.03033	0.81918	11400	0.00767	0.93758	19400	0.00124	0.98431
7300	0.02922	0.82443	11600	0.00725	0.94021	19600	0.00119	0.98474
7400	0.02816	0.82949	11800	0.00686	0.94270	19800	0.00115	0.98516
7500	0.02714	0.83437	12000	0.00649	0.94505	20000	0.00111	0.98555
7600	0.02617	0.83907	12200	0.00615	0.94728	20500	0.00102	0.98649
7700	0.02523	0.84360	12400	0.00583	0.94939	21000	0.00093	0.98735
7800	0.02434	0.84797	12600	0.00552	0.95139	21500	0.00085	0.98814
7900	0.02348	0.85219	12800	0.00524	0.95329	22000	0.00079	0.98886
8000	0.02266	0.85625	13000	0.00498	0.95509	22500	0.00072	0.98952
8100	0.02187	0.86018	13200	0.00473	0.95681	23000	0.00067	0.99014
8200	0.02111	0.86397	13400	0.00450	0.95843	23500	0.00062	0.99070
8300	0.02038	0.86763	13600	0.00428	0.95998	24000	0.00057	0.99123
8400	0.01969	0.87116	13800	0.00407	0.96145	24500	0.00053	0.99172
8500	0.01902	0.87457	14000	0.00388	0.96285	25000	0.00049	0.99217
8600	0.01838	0.87787	14200	0.00369	0.96419	26000	0.00043	0.99297
8700	0.01776	0.88105	14400	0.00352	0.96546	27000	0.00037	0.99368
8800	0.01717	0.88413	14600	0.00336	0.96667	28000	0.00032	0.99429
8900	0.01660	0.88711	14800	0.00321	0.96783	29000	0.00028	0.99482
9000	0.01606	0.88999	15000	0.00306	0.96893	30000	0.00025	0.99529
9100	0.01553	0.89278	15200	0.00292	0.96999	32000	0.00020	0.99607
9200	0.01503	0.89547	15400	0.00280	0.97100	34000	0.00016	0.99669
9300	0.01455	0.89808	15600	0.00267	0.97196	36000	0.00013	0.99719
9400	0.01408	0.90060	15800	0.00256	0.97289	38000	0.00010	0.99759
9500	0.01363	0.90305	16000	0.00245	0.97377	40000	0.00008	0.99792
9600	0.01320	0.90541	16200	0.00234	0.97461	45000	0.00005	0.99851
9700	0.01279	0.90770	16400	0.00225	0.97542	50000	0.00004	0.99890
9800	0.01239	0.90992	16600	0.00215	0.97620	55000	0.00002	0.99917
9900	0.01201	0.91207	16800	0.00206	0.97694	60000	0.00002	0.99935

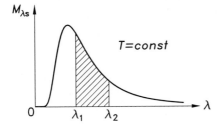

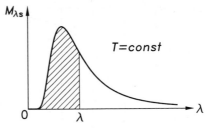

Fig. 5.26: Isotherm of $M_{\lambda s}(\lambda, T)$. The emissive power in the wavelength interval (λ_1, λ_2) corresponds to the hatched area

Fig. 5.27: Function $F(0, \lambda)$ according to (5.60), is represented by the ratio of the hatched area to the total area under the isotherm of $M_{\lambda s}(\lambda, T)$

for radiative transfer, the portion of the emissive power from a certain wavelength interval (λ_1, λ_2) is required. This part corresponds to the hatched area in Fig. 5.26. Its calculation requires the definition of the fraction function

$$F(\lambda_1, \lambda_2) := \frac{\int\limits_{\lambda_1}^{\lambda_2} M_{\lambda s}(\lambda, T)\, d\lambda}{\int\limits_0^\infty M_{\lambda s}(\lambda, T)\, d\lambda} = \frac{1}{\sigma T^4} \int\limits_{\lambda_1}^{\lambda_2} M_{\lambda s}(\lambda, T)\, d\lambda \ .$$

Using the function

$$F(0, \lambda) := \frac{1}{\sigma T^4} \int\limits_0^\lambda M_{\lambda s}(\lambda, T)\, d\lambda \ , \tag{5.60}$$

which is illustrated in Fig. 5.27, $F(\lambda_1, \lambda_2)$ can be expressed as

$$F(\lambda_1, \lambda_2) = F(0, \lambda_2) - F(0, \lambda_1) \ . \tag{5.61}$$

If, instead of λ, the dimensionless variable x from (5.53) is introduced, then $F(0, \lambda)$ can be represented as a function only of x:

$$F(0, x) = F(0, \frac{\lambda T}{c_2}) = \frac{15}{\pi^4} \int\limits_0^x \frac{dx}{x^5 \left(e^{1/x} - 1\right)} \ , \tag{5.62}$$

Fig. 5.28. As x and λT only differ by the radiation constant c_2, the function $F(0, \lambda T)$ is used for applications. It is presented in Table 5.3. For $\lambda_{\max} T$ according to (5.52) this gives $F(0, \lambda_{\max} T) = 0.250\,05$. Therefore, just over a quarter of the emissive power of a black body falls in the wavelength region $\lambda \leq \lambda_{\max}$, at all temperatures.

The analytical representation of the integral (5.62) is possible by integrating a series expansion of the integrand. With $y = 1/x = c_2/\lambda T$ we obtain, for all y, the convergent series

$$F(0, y) = \frac{15}{\pi^4} \sum_{m=1}^\infty \frac{e^{-my}}{m^4} \left\{ [(my + 3)my + 6]\, my + 6 \right\} \ .$$

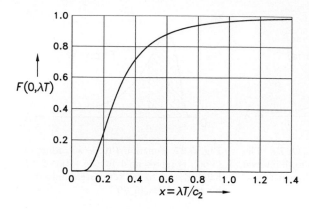

For $y < 2$ it is more convenient to use the series

$$F(0, y) = 1 - \frac{15}{\pi^4} y^3 \left(\frac{1}{3} - \frac{y}{8} + \frac{y^2}{60} - \frac{y^4}{5040} + \frac{y^6}{272160} - \frac{y^8}{13305600} + \cdots \right) ,$$

cf. [5.15], [5.16].

Example 5.3: The sun can be approximated as a radiating sphere of radius $R_S = 6.96 \cdot 10^8$ m. The distance between the earth and the sun is $D_{ES} = 1.496 \cdot 10^{11}$ m. At this distance, the irradiance on an area normal to the sun and at the outer edge of the earth's atmosphere has a value of $E_0 = 1367$ W/m^2, that will be known as the *solar constant*. The extraterrestrial solar radiation shall be considered to be black body radiation. Under these assumptions determine the emissive power M_S of the sun and its surface temperature T_S. What proportion of the radiation leaving the sun falls within the region of visible light ($0.38 \,\mu m \leq \lambda \leq 0.78 \,\mu m$)?

The radiative power emitted by the sun is

$$\Phi_S = 4\pi R_S^2 M_S(T_S) .$$

This radiative power also penetrates an (imaginary) spherical surface lying at a distance D_{ES} concentric around the sun, Fig. 5.29. So, it holds that

$$\Phi_S = 4\pi D_{ES}^2 E_0$$

with E_0 as the solar constant. This gives the emissive power of the sun as

$$M_S(T_S) = \left(\frac{D_{ES}}{R_S} \right)^2 E_0 = 63.16 \frac{MW}{m^2} .$$

A black body with this emissive power has, according to the Stefan-Boltzmann law, a temperature of

$$T_S = \left(\frac{M_S}{\sigma} \right)^{1/4} = \left(\frac{D_{ES}}{R_S} \right)^{1/2} \left(\frac{E_0}{\sigma} \right)^{1/4} = 5777 \text{ K} .$$

As a first approach, extraterrestrial solar radiation can be taken to be radiation from a black body at this temperature, see also section 5.3.5.

The proportion of the radiation emitted by this black body within the wavelength interval $\lambda_1 = 0.38 \,\mu m$ to $\lambda_2 = 0.78 \,\mu m$, according to (5.61) and (5.62), is

$$F(\lambda_1, \lambda_2) = F(0, \lambda_2 T_S) - F(0, \lambda_1 T_S) .$$

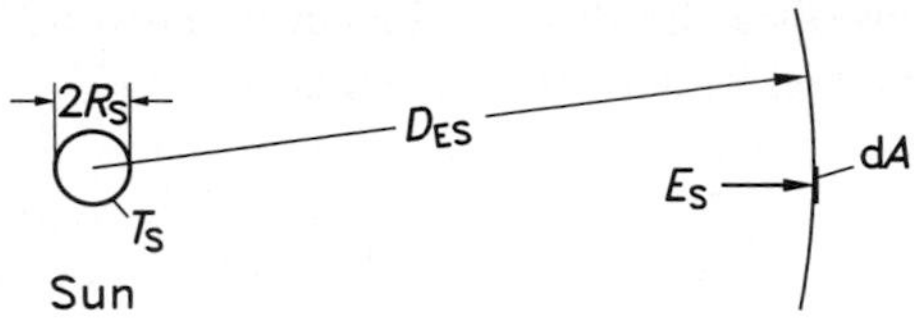

Fig. 5.29: Geometry of the sun-earth
system (not to scale) and schematic representation of the irradiation of a surface
element dA at a distance D_{ES} from the
centre of the sun, R_{S} radius of sun

For $\lambda_2 T_{\mathrm{S}} = 4506\,\mu\mathrm{m}\,\mathrm{K}$ and $\lambda_1 T_{\mathrm{S}} = 2195\,\mu\mathrm{m}\,\mathrm{K}$

$$F(\lambda_1, \lambda_2) = 0.5652 - 0.1000 = 0.4652$$

is found from Table 5.3 as an approximate value for the proportion of the solar radiation in the
visible wavelength region.

5.3 Radiation properties of real bodies

In the following sections we will look at the radiation properties of real bodies,
which, with respect to the directional dependence and the spectral distribution of
the radiated energy, are vastly different from the properties of the black body. In
order to record these deviations the emissivity of a real radiator is defined. Kirchhoff's law links the emissivity with the absorptivity and suggests the introduction
of a "semi-ideal" radiator, the diffuse radiating grey body, that is frequently used
as an approximation in radiative transfer calculations. In the treatment of the
emissivities of real radiators we will use the results from the classical electromagnetic theory of radiation. In the last section the properties of transparent bodies,
(e.g. glass) will be dealt with.

5.3.1 Emissivities

According to Kirchhoff's law a black body emits the maximum radiation energy at
every wavelength in every direction in the hemisphere. It therefore suggests itself
to relate the four radiation quantities, used to characterise the emission of any
radiator in 5.1.2, to their maximum values, namely the corresponding quantities
for a black body at the same temperature. This leads to the definition of four
dimensionless quantities that are smaller than one and are called *emissivities*.
These four emissivities are material properties of the radiation emitting body.

The definitions of the four emissivities are brought together in Table 5.4. It
additionally contains the relationships which are used in the calculation of the
other three emissivities from the directional spectral emissivity $\varepsilon'_\lambda(\lambda, \beta, \varphi, T)$. This
emissivity describes the directional and wavelength distributions of the emitted
radiation flow, whilst the hemispherical spectral emissivity $\varepsilon_\lambda(\lambda, T)$ only gives
the spectral energy distribution. The directional total emissivity $\varepsilon'(\beta, \varphi, T)$ only

describes the distribution over the solid angles in the hemisphere. In contrast, the hemispherical total emissivity $\varepsilon(T)$ — normally just called the emissivity — does not have a distribution function character. It merely gives the temperature dependent factor by which the emissive power $M(T)$ of a real body is smaller than the emissive power $M_{\mathrm{s}}(T)$ of a black body at the same temperature.

The following relationships for the four radiation quantities for emission from a real body are obtained from the defining equations for the emissivities. The body's spectral intensity L_λ is

$$L_\lambda(\lambda, \beta, \varphi, T) = \varepsilon'_\lambda(\lambda, \beta, \varphi, T) L_{\lambda\mathrm{s}}(\lambda, T) \ , \qquad (5.63)$$

Table 5.4: Definitions of the emissivities and the relationships existing between them

---

Directional spectral emissivity

$$\varepsilon'_\lambda(\lambda, \beta, \varphi, T) := \frac{L_\lambda(\lambda, \beta, \varphi, T)}{L_{\lambda\mathrm{s}}(\lambda, T)}$$

ε'_λ describes the directional and wavelength distribution of the emitted radiation flow by comparing the spectral intensity L_λ with that of a black body.

Hemispherical spectral emissivity

$$\varepsilon_\lambda(\lambda, T) := \frac{M_\lambda(\lambda, T)}{M_{\lambda\mathrm{s}}(\lambda, T)}$$

$$\varepsilon_\lambda(\lambda, T) = \frac{1}{\pi} \int_{\cap} \varepsilon'_\lambda(\lambda, \beta, \varphi, T) \cos\beta \, \mathrm{d}\omega$$

ε_λ describes the wavelength distribution of the radiative power emitted in the hemisphere by comparison of the hemispherical spectral emissive power M_λ with that of a black body.

Directional total emissivity

$$\varepsilon'(\beta, \varphi, T) := \frac{L(\beta, \varphi, T)}{L_{\mathrm{s}}(T)} = \frac{\pi}{\sigma T^4} L(\beta, \varphi, T)$$

$$\varepsilon'(\beta, \varphi, T) = \frac{1}{L_{\mathrm{s}}(T)} \int_0^\infty \varepsilon'_\lambda(\lambda, \beta, \varphi, T) L_{\lambda\mathrm{s}}(\lambda, T) \, \mathrm{d}\lambda$$

ε' describes the directional distribution of the emitted radiation flow of all wavelengths by comparison of the total intensity L with that of a black body.

Hemispherical total emissivity

$$\varepsilon(T) := \frac{M(T)}{M_{\mathrm{s}}(T)} = \frac{M(T)}{\sigma T^4}$$

$$\varepsilon(T) = \frac{1}{M_{\mathrm{s}}(T)} \int_0^\infty \varepsilon_\lambda(\lambda, T) M_{\lambda\mathrm{s}}(\lambda, T) \, \mathrm{d}\lambda = \frac{1}{\pi} \int_{\cap} \varepsilon'(\beta, \varphi, T) \cos\beta \, \mathrm{d}\omega$$

ε compares the emissive power M with that of a black body.

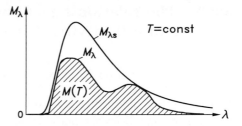

Fig. 5.30: Hemispherical spectral emissive power $M_\lambda(\lambda, T)$ of a real body compared to the hemispherical spectral emissive power $M_{\lambda s}(\lambda, T)$ of a black body at the same temperature. The hatched area represents the emissive power $M(T)$ of the real body

where the spectral intensity $L_{\lambda s}$ of the black body is given by Planck's function according to (5.50). As the black body is a Lambert radiator, ε_λ' alone covers the deviations of the real emission from Lambert's cosine law. This also applies to the directional total emissivity $\varepsilon'(\beta, \varphi, T)$, with which the intensity L of the real body is expressed by

$$L(\beta, \varphi, T) = \varepsilon'(\beta, \varphi, T) L_s(T) = \varepsilon'(\beta, \varphi, T)\frac{\sigma}{\pi}T^4 \ , \tag{5.64}$$

In general, the directional emissivities $\varepsilon_\lambda'(\lambda, \beta, \varphi, T)$ and $\varepsilon'(\beta, \varphi, T)$ do not depend on the circumferential angle φ. Integration over all solid angles in the hemisphere, which, according to Table 5.4, leads from ε_λ' to ε_λ and from ε' to ε, is then simplified. This produces

$$\varepsilon_\lambda(\lambda, T) = \frac{1}{\pi} \int_{\varphi=0}^{2\pi} \int_{\beta=0}^{\pi/2} \varepsilon_\lambda'(\lambda, \beta, T) \cos\beta \sin\beta \, d\beta \, d\varphi = 2 \int_{\beta=0}^{\pi/2} \varepsilon_\lambda'(\lambda, \beta, T) \cos\beta \sin\beta \, d\beta \ . \tag{5.65}$$

and likewise

$$\varepsilon(T) = 2 \int_{\beta=0}^{\pi/2} \varepsilon'(\beta, T) \cos\beta \sin\beta \, d\beta \ . \tag{5.66}$$

It holds for the hemispherical spectral emissive power of a real body that

$$M_\lambda(\lambda, T) = \varepsilon_\lambda(\lambda, T) M_{\lambda s}(\lambda, T) \ , \tag{5.67}$$

where the hemispherical spectral emissive power $M_{\lambda s}$ of a black body is given by (5.50). Fig. 5.30 illustrates (5.67) for a given temperature. Every ordinate of $M_{\lambda s}(\lambda, T)$ is reduced by a factor of $\varepsilon_\lambda(\lambda, T)$. The hatched area corresponds to the emissive power of the real body, since it holds that

$$M(T) = \int_0^\infty M_\lambda(\lambda, T) \, d\lambda = \int_0^\infty \varepsilon_\lambda(\lambda, T) M_{\lambda s}(\lambda, T) \, d\lambda = \varepsilon(T) M_s(T) = \varepsilon(T)\sigma T^4 \ . \tag{5.68}$$

534 5 Thermal radiation

5.3.2 The relationships between emissivity, absorptivity and reflectivity. The grey Lambert radiator

5.3.2.1 Conclusions from Kirchhoff's law

Kirchhoff's law shows that a close relationship exists between the emission and absorption capabilities of a body, see 5.1.6. From (5.48) we obtained the following equation for the spectral intensity of any body

$$L_\lambda(\lambda, \beta, \varphi, T) = a'_\lambda(\lambda, \beta, \varphi, T) L_{\lambda s}(\lambda, T)$$

as a quantitative expression of Kirchhoff's law. Here, the directional spectral absorptivity a'_λ is a material property that characterises the absorption capability of the body. The comparison of this relationship with (5.63) yields

$$a'_\lambda(\lambda, \beta, \varphi, T) = \varepsilon'_\lambda(\lambda, \beta, \varphi, T) \ . \tag{5.69}$$

The directional spectral absorptivity of a any radiator agrees with its directional spectral emissivity.

The directional spectral reflectivity r'_λ of an opaque body can also be traced back to the directional spectral emissivity ε'_λ. According to (5.41) and (5.69), it holds that

$$r'_\lambda(\lambda, \beta, \varphi, T) = 1 - a'_\lambda(\lambda, \beta, \varphi, T) = 1 - \varepsilon'_\lambda(\lambda, \beta, \varphi, T) \ . \tag{5.70}$$

This says that one single material function is sufficient for the description of the emission, absorption and reflective capabilities of an opaque body. Table 5.4 shows that it is possible to calculate the emissivities ε_λ, ε' and ε from ε'_λ. Correspondingly, with known incident spectral intensity K_λ of the incident radiation, this also holds for the calculation of a_λ, a' and a from a'_λ as well as of r_λ, r' and r from r'_λ, cf. Tables 5.1 and 5.2. So, only one single material function, e.g. $\varepsilon'_\lambda = \varepsilon'_\lambda(\lambda, \beta, \varphi, T)$, is actually necessary to record all the radiation properties of a real body[6]. This is an example of how the laws of thermodynamics limit the number of possible material functions (equations of state) of a system.

5.3.2.2 Calculation of absorptivities from emissivities

The equality resulting from Kirchhoff's law between the directional spectral absorptivity and the emissivity, $a'_\lambda = \varepsilon'_\lambda$, suggests that investigation of whether the other three (integrated) absorptivities a_λ, a' and a can be calculated from the

---

[6]However, this is valid with the restriction that the bidirectional reflectivities are not covered, which describe the directional distribution of the reflected radiation; see for this the comments in 5.1.5 as well as [5.4], pg. 71 ff.

corresponding emissivities ε_λ, ε' and ε should be carried out. This will be impossible without additional assumptions, as the absorptivities a_λ, a' and a are not alone material properties of the absorbing body, they also depend on the incident spectral intensity K_λ of the incident radiation, see Table 5.1. The emissivities ε_λ, ε' and ε are, in contrast, purely material properties. An accurate test is therefore required to see whether, and under what conditions, the equations analogous to (5.69), $a_\lambda = \varepsilon_\lambda$, $a' = \varepsilon'$ and $a = \varepsilon$ are valid.

If the hemispherical spectral absorptivity and emissivity shall agree, $a_\lambda(\lambda, T) = \varepsilon_\lambda(\lambda, T)$, then according to Table 5.1 and 5.4 the equation

$$\frac{1}{E_\lambda(\lambda)} \int_\triangle \varepsilon_\lambda'(\lambda, \beta, \varphi, T) K_\lambda(\lambda, \beta, \varphi) \cos\beta \, d\omega = \frac{1}{\pi} \int_\triangle \varepsilon_\lambda'(\lambda, \beta, \varphi, T) \cos\beta \, d\omega \qquad (5.71)$$

has to be satisfied. On the left hand side, according to Kirchhoff's law, a_λ' was replaced by ε_λ'. Equation (5.71) is only satisfied in two cases:

1. The body has a diffuse radiating surface (Lambert radiator); then $\varepsilon_\lambda' = \varepsilon_\lambda'(\lambda, T)$ is valid.

2. Diffuse irradiation with $K_\lambda = K_\lambda(\lambda)$ is present, see section 5.1.3.

If the directional total absorptivity a' equals the directional total emissivity, $a'(\beta, \varphi, T) = \varepsilon'(\beta, \varphi, T)$, then according to Tables 5.1 and 5.4

$$\frac{1}{K(\beta, \varphi)} \int_0^\infty \varepsilon_\lambda'(\lambda, \beta, \varphi, T) K_\lambda(\lambda, \beta, \varphi) \, d\lambda = \frac{1}{L_s(T)} \int_0^\infty \varepsilon_\lambda'(\lambda, \beta, \varphi, T) L_{\lambda s}(\lambda, T) \, d\lambda \qquad (5.72)$$

must hold. Once again, this is only possible in two cases:

1. The directional spectral emissivity is independent of the wavelength λ: $\varepsilon_\lambda' = \varepsilon_\lambda'(\beta, \varphi, T)$. A body with this property is called a *grey body* or a *grey radiator*.

2. The incident radiation satisfies the condition $K_\lambda(\lambda, \beta, \varphi) = C(\beta, \varphi) \cdot L_{\lambda s}(\lambda, T)$, where the factor C does not depend on λ.

Finally, it remains to investigate under which conditions the hemispherical total quantities a and ε are the same, such that $a(T) = \varepsilon(T)$ can be stated. According to Tables 5.1 and 5.4, the equation

$$\frac{\int_\triangle \left[\int_0^\infty \varepsilon_\lambda'(\lambda, \beta, \varphi, T) K_\lambda(\lambda, \beta, \varphi) \, d\lambda \right] \cos\beta \, d\omega}{\int_\triangle \left[\int_0^\infty K_\lambda(\lambda, \beta, \varphi) \, d\lambda \right] \cos\beta \, d\omega} = \frac{1}{M_s(T)} \int_\triangle \left[\int_0^\infty \varepsilon_\lambda'(\lambda, \beta, \varphi, T) L_{\lambda s}(\lambda, T) \, d\lambda \right] \cos\beta \, d\omega$$

$$(5.73)$$

has to be satisfied. This is possible in four cases:

1. The body is both a diffuse and grey radiator: $\varepsilon_\lambda' = \varepsilon_\lambda'(T)$.

2. The body is a diffuse radiator, $\varepsilon_\lambda' = \varepsilon_\lambda'(\lambda, T)$, and the spectral irradiance K_λ of the radiation has a wavelength dependence like black body radiation at the temperature of the absorbing body: $K_\lambda(\lambda, \beta, \varphi) = C(\beta, \varphi) \cdot L_{\lambda s}(\lambda, T)$.

3. The body is a grey radiator, $\varepsilon_\lambda' = \varepsilon_\lambda'(\beta, \varphi, T)$, and K_λ is independent of the direction (diffuse irradiation), $K_\lambda = K_\lambda(\lambda)$.

4. The body has any radiation properties. The spectral irradiance K_λ is independent of direction and is proportional to the spectral intensity of a black body at the temperature of the absorbing surface: $K_\lambda = C \cdot L_{\lambda s}(\lambda, T)$.

The proof of the correctness of these conditions, under which the equations (5.71) to (5.73) hold, is left to the reader.

The equality of the three pairs of absorptivities and emissivities, namely $a_\lambda(\lambda, T) = \varepsilon_\lambda(\lambda, T)$, $a'(\beta, \varphi, T) = \varepsilon'(\beta, \varphi, T)$ and $a(T) = \varepsilon(T)$, is only given if the absorbing and emitting surfaces have particular properties, or if the incident spectral intensity K_λ of the radiation satisfies certain conditions in terms of its directional and wavelength dependency. These conditions are satisfied by incident black body radiation, when the black body is at the same temperature as the absorbing body, which does not apply for heat transfer. In practice, the more important cases are those in which the directional spectral emissivity ε'_λ of the absorbing body at least approximately satisfies special conditions. We will once again summarise these conditions:

1. When $\varepsilon'_\lambda = \varepsilon'_\lambda(\lambda, T)$, (diffuse radiating surface), it holds that

$$a_\lambda(\lambda, T) = \varepsilon_\lambda(\lambda, T) = \varepsilon'_\lambda(\lambda, T) \ . \tag{5.74}$$

2. When $\varepsilon'_\lambda = \varepsilon'_\lambda(\beta, \varphi, T)$, (grey radiating surface), it holds that

$$a'(\beta, \varphi, T) = \varepsilon'(\beta, \varphi, T) = \varepsilon'_\lambda(\beta, \varphi, T) \ . \tag{5.75}$$

3. When $\varepsilon'_\lambda = \varepsilon'_\lambda(T)$, (diffuse and grey emitting surface), it holds that

$$a(T) = \varepsilon(T) = \varepsilon'_\lambda(T) \ . \tag{5.76}$$

If the conditions mentioned above for ε'_λ are satisfied, then (5.74) to (5.76) are valid for incident radiation with *any* incident spectral intensity K_λ.

5.3.2.3 The grey Lambert radiator

In radiative exchange calculations, it is preferable to use the model, described in the previous section, of a grey, diffuse radiating body as a simple approximation for the radiative behaviour of real bodies. As Lambert's cosine law is valid for this model, we denote these bodies as grey Lambert radiators. The energy radiated from them is distributed like that from a black body over the directions in the hemisphere and the spectrum; yet the spcetral emissive power $M_\lambda(\lambda, T)$ and the spectral intensity $L_\lambda(\lambda, T)$ are smaller by the purely temperature dependent emissivity $\varepsilon(T)$ than the corresponding functions for a black body, Fig. 5.31:

$$M_\lambda(\lambda, T) = \pi L_\lambda(\lambda, T) = \varepsilon(T) M_{\lambda s}(\lambda, T) \ . \tag{5.77}$$

Here, $M_{\lambda s}(\lambda, T)$ is the Planck function according to (5.50). The emissivity $\varepsilon(T)$ is the only material function of a grey Lambert radiator; all four emissivities are equal and the same as the four absorptivities:

$$\varepsilon(T) = \varepsilon'(T) = \varepsilon_\lambda(T) = \varepsilon'_\lambda(T) = a'_\lambda(T) = a_\lambda(T) = a'(T) = a(T) \ . \tag{5.78}$$

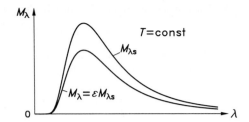

Fig. 5.31: Hemispherical spectral emissive power $M_\lambda(\lambda, T) = \varepsilon(T)M_{\lambda s}(\lambda, T)$ of a grey Lambert radiator at a certain temperature

Real bodies are often approximate Lambert radiators. In contrast, ε'_λ and ε_λ are, in general, highly dependent on the wavelength. The assumption of a *grey* radiating surface no longer applies. However it is a good approximation, if the wavelength region with significant spectral emissive power M_λ and spectral irradiance E_λ lies in a wavelength interval in which ε_λ is roughly constant. This is illustrated in Fig. 5.32, where is is assumed that at small wavelengths radiation is neither emitted, incident nor absorbed. This applies to radiation sources that are at not too high temperatures. However if solar radiation, with a maximum lying at small wavelengths $\lambda < \lambda_1$, falls on the body under consideration in Fig. 5.32, then $a \neq \varepsilon$; the assumption of a grey body is now incorrect, so special absorptivities a_S for solar radiation have to be used, cf. section 5.4.5.

Example 5.4: A material has a directional spectral emissivity that only depends on the polar angle β: $\varepsilon'_\lambda(\lambda, \beta, \varphi, T) = \varepsilon'(\beta)$. This directional dependence is given by

$$\varepsilon'(\beta) = \begin{cases} \varepsilon'_1 = 0.20 & \text{für} \quad 0 \leq \beta < \pi/4 \\ \varepsilon'_2 = 0.50 & \text{für} \quad \pi/4 \leq \beta \leq \pi/2 \end{cases}.$$

The external skin of a satellite is made of this material and is exposed to solar radiation with a flux density $\dot{q}_{sol} = 1500\,\text{W/m}^2$. What temperature does the surface of the satellite assume, when it is hit perpendicularly by the solar radiation? What temperature develops when the surface forms an angle of $\pi/6 = 30°$ with the direction of the solar radiation? A heat flow between the surface and the inside of the satellite can be neglected; the temperature of space may be assumed to be $T_W = 0\,\text{K}$.

Under the assumptions given, the emissive power $M(T) = \varepsilon\sigma T^4$ of the satellite surface has

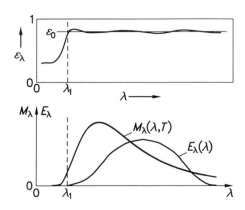

Fig. 5.32: Approximately constant spectral emissivity ε_λ for $\lambda > \lambda_1$ as well as the pattern of the hemispherical spectral emissive power M_λ and the spectral irradiance E_λ, such that a grey radiator can be assumed

to be equal to the energy flux $\dot{q}_{\mathrm{abs}}$ that it absorbs from the incident solar radiation:

$$\varepsilon \sigma T^4 = \dot{q}_{\mathrm{abs}} = a'(\beta)E = \varepsilon'(\beta)\dot{q}_{\mathrm{sol}} \cos \beta \ .$$

Here, (5.75) has been used because a grey radiator (ε'_λ independent of λ) is present. The temperature T of the satellite surface will then be

$$T = \sqrt[4]{\frac{\varepsilon'(\beta)\dot{q}_{\mathrm{sol}} \cos \beta}{\varepsilon \sigma}} \ . \tag{5.79}$$

In this equation, ε is the hemispherical total emissivity, calculated according to (5.66). This gives

$$\varepsilon = 2\,\varepsilon'_1 \int\limits_0^{\pi/4} \cos \beta \sin \beta \, \mathrm{d}\beta + 2\,\varepsilon'_2 \int\limits_{\pi/4}^{\pi/2} \cos \beta \sin \beta \, \mathrm{d}\beta \ .$$

The two integrals each have a value of $1/4$, yielding $\varepsilon = \frac{1}{2}(\varepsilon'_1 + \varepsilon'_2) = 0.35$.

For perpendicularly falling radiation, $\beta = 0$, the temperature obtained from (5.79) is $T = 351\,\mathrm{K}$. When the surface and the direction of the solar radiation form an angle of $\pi/6$, the polar angle is $\beta = \pi/2 - \pi/6 = \pi/3$. The directional emissivity $\varepsilon'(\beta)$, decisive for the absorption, now assumes the large value of ε'_2, whilst the irradiance is reduced by a factor of $\cos \beta = \cos(\pi/3) = 0.5$. The two effects partly compensate each other, such that (5.79) delivers the only slightly higher surface temperature of $T = 371\,\mathrm{K}$.

5.3.3   Emissivities of real bodies

According to 5.3.2.1, the radiation properties of an opaque body are determined by its directional spectral emissivity $\varepsilon'_\lambda = \varepsilon'_\lambda(\lambda, \beta, \varphi, T)$. In order to determine this material function experimentally numerous measurements are required, as the dependence on the wavelength, direction and temperature all have to be investigated. These extensive measurements have, so far, not been carried out for any substance. Measurements are frequently limited to the determination of the emissivity $\varepsilon'_{\lambda,\mathrm{n}}$ normal to the surface ($\beta = 0$), the emissivities for a few chosen wavelengths or only the hemispherical total emissivity ε is measured.

In addition to incomplete radiation measurements, the strong dependence of the results on the condition of the surface is a further difficulty. Impurities also play a role, alongside roughness. Even a very thin film of water or an oxide layer can completely change the radiation behaviour compared to the base material alone. It is therefore no surprise that the emissivities measured by various researchers often differ significantly. Unfortunately, in the description of the experiments the surface properties were inexactly or incompletely characterised, which frequently occurs due to the lack of quantitative measures for surface properties. The emissivities presented in Tables B12 and B13 of the Appendix must therefore be taken to be relatively uncertain.

In view of the experimental difficulties a theory for radiation properties is desirable. The classical theory of electromagnetic waves from J.C. Maxwell (1864), links the emissivity ε'_λ with the so-called optical constants of the material, the refractive index n and the extinction coefficient k, that can be combined into a complex refractive index $\bar{n} = n - ik$. The optical "constants" depend on the temperature, the wavelength and electrical properties, in particular the electrical resistivity r_e of the material. In addition, the theory delivers, in the form of Fresnels' equations, an explicit dependence of the emissivity on the polar angle β, whilst no dependence on the circumferential angle φ appears, as isotropy has been assumed.

Unfortunately the electromagnetic theory is only valid under a series of limiting suppositions, so that the emissivities calculated from it frequently differ from reality. Despite this, it provides important, qualititative statements that can be used for the extrapolation from measurements or to estimate for missing data. We will not discuss the electromagnetic theory, see for this [5.4], but will use some of its results in the treatment of emissivities of electrical insulators and electrical conductors (metals). These two material groups differ significantly in their radiation behaviour.

5.3.3.1 Electrical insulators

Materials that do not conduct electricity include construction materials, paints, oxide layers on metals and most liquids. For the application of electromagnetic theory they are idealised by the assumption that their specific electrical resistance is $r_e \rightarrow \infty$. These substances are called *dielectrics*. Their extinction coefficient k is zero; their refractive index is yielded, according to

$$n = \sqrt{\mu_r \gamma_r} \tag{5.80}$$

from their magnetic permeability μ_r and their electrical permittivity γ_r. In a vacuum both quantities are equal to one, so $n = 1$. The refractive indices of most dielectrics lie between $n = 1$ and $n = 3$.

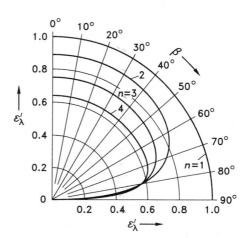

Fig. 5.33: Directional spectral emissivity $\varepsilon'_\lambda(\beta, n)$ of electrical insulators according to (5.82) in a polar diagram

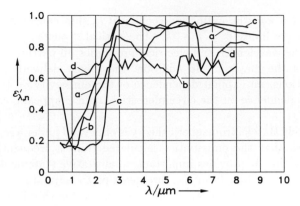

Fig. 5.34: Directional spectral emissivity $\varepsilon'_\lambda(\lambda, \beta \approx 0)$ as a function of wavelength for various insulators, according to measurements by W. Sieber [5.19]. **a** white paint on wood, **b** oak wood (smoothed), **c** white tiles (glazed), **d** concrete

According to theory, the spectral emissivity $\varepsilon'_{\lambda,n}$ normal to the surface simply depends on the refractive index as follows

$$\varepsilon'_{\lambda,n} = \frac{4n}{(n+1)^2} \; . \tag{5.81}$$

The complicated dependence $\varepsilon'_\lambda = \varepsilon'_\lambda(\beta, n)$ on the polar angle β is illustrated in Fig. 5.33, a polar diagram. According to that and (5.81) high emissivities, over 0.8, would be expected for insulators, which is confirmed by the data presented in Table B12. In Fig. 5.33 lines $n = \text{const}$ run almost circular up to large polar angles β. Therefore Lambert's cosine law holds in good approximation; dielectrics can be treated like diffuse radiators. Measurements from E. Schmidt [5.17] as well as from E. Schmidt and E. Eckert [5.18] essentially confirm the dependence of the emissivity on the polar angle β, as illustrated in Fig. 5.33.

The electromagnetic theory makes no explicit statements about the wavelength dependence of n. The high emissivities expected from (5.81) and Fig. 5.33 were only observed in radiation measurements at large wavelengths, around $\lambda > 2\,\mu\text{m}$. This is shown in Fig. 5.34, in which $\varepsilon'_\lambda(\lambda, \beta \approx 0)$ according to measurements by W. Sieber [5.19] is illustrated. In the regions of large wavelengths important in practice, as a first approximation, ε'_λ does not depend on λ; we can put $\varepsilon' \approx \varepsilon'_\lambda$ and $\varepsilon \approx \varepsilon_\lambda$, if the temperature is low enough such that the emission at wavelengths below around $2\,\mu\text{m}$ does not deliver any meaningful contribution. Dieletrics are approximate grey radiators.

Since frequently only the emissivity ε'_n or $\varepsilon'_{\lambda,n}$ normal to the surface are determined in radiation experiments, and because the hemispherical total emissivity ε is required for radiative exchange calculations, the ratio $\varepsilon/\varepsilon'_n$ is of interest. For dielectrics, we can put

$$\varepsilon/\varepsilon'_n \approx \varepsilon_\lambda(n) / \varepsilon'_{\lambda,n}(n) \; ,$$

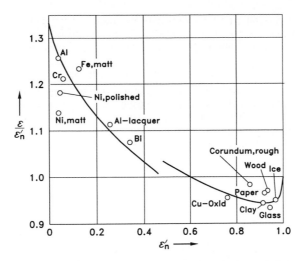

Fig. 5.35: Ratio $\varepsilon/\varepsilon'_n$ of the hemispherical total emissivity ε to the emissivity ε'_n normal to the surface as a function of ε'_n. Right hand line: Dielectrics from Table 5.5, left hand line: Metals from Table 5.6. Circles: data from [5.18]

so that the ratio can be calculated from the electromagnetic theory equations. It is presented in Tab. 5.5, and illustrated in Fig. 5.35. As Lambert's cosine law is roughly satisfied, $\varepsilon/\varepsilon'_n$ only deviates slightly from one

The equation of electromagnetic theory, upon which Fig. 5.33 is based, is

$$\varepsilon'_\lambda (\beta, n) = \frac{2a \cos \beta}{(a + \cos \beta)^2} \left[1 + \frac{n^2}{\left(a \cos \beta + \sin^2 \beta \right)^2} \right] \tag{5.82}$$

with $a = (n^2 - \sin^2 \beta)^{1/2}$, cf. [5.20]. By integration of ε'_λ from (5.66), the hemispherical emissivity ε_λ can be found, which, according to theory only depends on the refractive index n from (5.80). R.V. Dunkle [5.21] carried out this integration with the result

$$
\begin{aligned}
\varepsilon_\lambda (n) =\ & \frac{1}{2} - \frac{(3n + 1)(n - 1)}{6(n + 1)^2} + \frac{n^2 (n^2 - 1)^2}{(n^2 + 1)^3} \ln \frac{n + 1}{n - 1} \\
& + \frac{2n^3 (n^2 + 2n - 1)}{(n^2 + 1)(n^4 - 1)} - \frac{8n^4 (n^4 + 1)}{(n^2 + 1)(n^4 - 1)^2} \ln n \ .
\end{aligned}
\tag{5.83}
$$

Through division by $\varepsilon'_{\lambda,n}$ from (5.81) the values in Tab. 5.5 and Fig. 5.33 were obtained, which can be interpreted as $\varepsilon/\varepsilon'_n$.

5.3.3.2 Electrical conductors (metals)

In contrast to dielectrics, the electromagnetic waves that penetrate metals are damped; the extinction coefficient k in the complex refractive index $\bar{n} = n - ik$ is not equal to zero, but generally greater than n. For the spectral emissivity $\varepsilon'_{\lambda,n}$

Table 5.5: Hemispherical (spectral) emissivity ε_λ, ratio $\varepsilon_\lambda/\varepsilon'_{\lambda,n}$ and refractive index n as functions of the directional (spectral) emissivity $\varepsilon'_{\lambda,n}$ normal to the surface for electrical non-conductors according to electromagnetic theory. The ratio $\varepsilon_\lambda/\varepsilon'_{\lambda,n}$ can be set equal to $\varepsilon/\varepsilon'_n$.

$\varepsilon'_{\lambda,n}$	ε_λ	$\varepsilon_\lambda/\varepsilon'_{\lambda,n}$	n	$\varepsilon'_{\lambda,n}$	ε_λ	$\varepsilon_\lambda/\varepsilon'_{\lambda,n}$	n	$\varepsilon'_{\lambda,n}$	ε_λ	$\varepsilon_\lambda/\varepsilon'_{\lambda,n}$	n
0.50	0.5142	1.0284	5.8284	0.80	0.7647	0.9559	2.6180	0.92	0.8675	0.9429	1.7888
0.55	0.5575	1.0136	5.0757	0.82	0.7812	0.9526	2.4738	0.94	0.8868	0.9434	1.6488
0.60	0.5999	0.9999	4.4415	0.84	0.7978	0.9497	2.3333	0.96	0.9082	0.9461	1.5000
0.65	0.6417	0.9872	3.8973	0.86	0.8146	0.9472	2.1957	0.98	0.9342	0.9532	1.3294
0.70	0.6829	0.9756	3.4221	0.88	0.8317	0.9451	2.0600	0.99	0.9518	0.9615	1.2222
0.75	0.7238	0.9651	3.0000	0.90	0.8492	0.9436	1.9249	1.00	1.0000	1.0000	1.0000

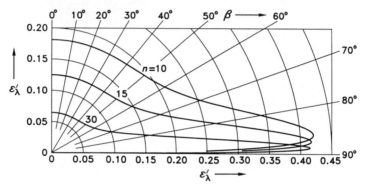

Fig. 5.36: Directional spectral emissivity $\varepsilon'_\lambda(\beta, n)$ of metals according to the simplified electromagnetic theory, eq. (5.87)

normal to the surface, the electromagnetic theory delivers the relationship

$$\varepsilon'_{\lambda,n}(n, k) = \frac{4n}{(n+1)^2 + k^2}\;. \tag{5.84}$$

According to this, with metals, significantly smaller emissivities than those for electrical insulators are to be expected.

At large wavelengths, around $\lambda > 5\,\mu$m, and small electrical resistivity r_e, the radiation properties of metals can be described by a simplified version of electromagnetic wave theory, which goes back to P. Drude [5.22]. It is also linked with the names E. Hagen and H. Rubens, who checked its applicability by experiment [5.23]. According to this theory n and k assume large values, and it holds that

$$n = k = \sqrt{\frac{c_0\mu_0}{4\pi}\frac{\lambda}{r_e}} = \sqrt{\frac{R_0\lambda}{r_e}}\;. \tag{5.85}$$

Here c_0 is the velocity of light in a vacuum and $\mu_0 = 4\pi\cdot10^{-7}\mathrm{N/A}^2$ is the magnetic field constant. These universal constants yield an electrical resistance of $R_0 = c_0\mu_0/4\pi = 29.979\,\Omega$.

Table 5.6: Hemispherical total emissivity ε, ratio $\varepsilon/\varepsilon_n'$ and product $r_e T$ as functions of the (directional) total emissivity ε_n' normal to the surface, for metals, calculated according to the simplified electromagnetic theory ($n = k$)

ε_n'	ε	$\varepsilon/\varepsilon_n'$	$r_e T$	ε_n'	ε	$\varepsilon/\varepsilon_n'$	$r_e T$	ε_n'	ε	$\varepsilon/\varepsilon_n'$	$r_e T$
			ΩcmK				ΩcmK				ΩcmK
0.00	0.0000	1.333	0.00000	0.12	0.1424	1.187	0.04992	0.30	0.3231	1.077	0.3999
0.02	0.0258	1.292	0.00123	0.14	0.1640	1.172	0.06967	0.35	0.3688	1.054	0.5904
0.04	0.0506	1.264	0.00505	0.16	0.1852	1.157	0.0934	0.40	0.4129	1.032	0.8417
0.06	0.0745	1.242	0.01163	0.18	0.2059	1.144	0.1213	0.45	0.4555	1.012	1.1722
0.08	0.0977	1.221	0.02114	0.20	0.2263	1.131	0.1537				
0.10	0.1203	1.203	0.03384	0.25	0.2757	1.103	0.2577				

According to Drude's theory, ε_λ' only depends on the polar angle β and n from (5.85), cf. Fig. 5.36. At polar angles larger than $80°$, ε_λ' assumes a distinct maximum; Lambert's cosine law is not fulfilled; metals cannot be treated as diffuse radiators. The ratio $\varepsilon/\varepsilon_n'$ illustrated in Fig. 5.35, is therefore considerably larger than one. Measurements by E. Schmidt and E. Eckert [5.18] confirm qualitatively the wavelength dependence of ε_λ' shown in Fig. 5.36.

As the refractive index n from (5.85) depends on the wavelength λ, this also applies to ε_λ'. At large wavelengths, the spectral emissivity of metals falls with increasing λ. Metals can therefore, not be viewed as grey radiators, so $a(T) = \varepsilon(T)$ cannot be used. According to E. Eckert [5.24], the following holds for the absorptivity a of a metal surface at temperature T for incident black or grey radiation from a source at temperature T^*

$$a = \varepsilon(\sqrt{T \cdot T^*}) \ . \tag{5.86}$$

However, this relationship is only correct when T^* is so low that the portion of the incident radiation in the region $\lambda < 5\,\mu$m can be neglected.

According to Drude's theory, the directional spectral emissivity ε_λ' of metals, see [5.4], is found to be

$$\varepsilon_\lambda'(\beta, n) = 2n \cos\beta \left[\frac{1}{(n\cos\beta)^2 + (1 + n\cos\beta)^2} + \frac{1}{n^2 + (n + \cos\beta)^2} \right] \ . \tag{5.87}$$

Fig. 5.36 is based on this equation. By integrating (5.87) corresponding to (5.66), gives the spectral emissivity that is only dependent on n

$$\varepsilon_\lambda(n) = 4n + \frac{2}{n} - \left(4n^2 + \frac{1}{n^2}\right) \ln\left(1 + \frac{1}{n} + \frac{1}{2n^2}\right) - \frac{2}{n^2} \ln(2n) \ . \tag{5.88}$$

As Drude's theory is only applicable for large n, $\varepsilon_\lambda(n)$ is developed into a power series of $(1/n)$. With (5.85), the numerical value equations valid for r_e in Ωcm and λ in μm follow as

$$\begin{aligned} \varepsilon_\lambda = \ & 48.70\sqrt{\frac{r_e}{\lambda}} \left\{ 1 + \left[31.62 + 6.849 \ln\frac{r_e}{\lambda}\right] \sqrt{\frac{r_e}{\lambda}} - 166.78\frac{r_e}{\lambda} \right. \\ & \left. + 3973.8\left(\frac{r_e}{\lambda}\right)^2 - 47628\left(\frac{r_e}{\lambda}\right)^{5/2} + \cdots \right\} \ , \end{aligned} \tag{5.89}$$

valid for $r_e/\lambda < 5 \cdot 10^{-4}\,\Omega\mathrm{cm}/\mu\mathrm{m}$. According to that, at large wavelengths, ε_λ decreases proportionally to $\lambda^{-1/2}$. The electrical resistivity r_e increases with rising temperature, such that ε_λ goes up with T. However, this only holds at large wavelengths; at small wavelengths Drude's theory fails, and a fall in ε_λ with rising temperature is observed.

In order to obtain the hemispherical total emissivity $\varepsilon(T)$ from ε_λ, according to Table 5.4 ε_λ has to be multiplied by $M_{\lambda\mathrm{s}}$ and integrated over all wavelengths. This yields the numerical value equation (for r_e in $\Omega\mathrm{cm}$ and T in K)

$$\begin{aligned} \varepsilon(T) \;=\;& 0{,}7671\sqrt{r_e T} - (0.3091 - 0.08884\ln r_e T)\,(r_e T) - 0.02334\,(r_e T)^{3/2} \\ &+ 3.50\cdot 10^{-4}\,(r_e T)^{5/2} - 8.7\cdot 10^{-5}\,(r_e T)^3 + \cdots \;, \end{aligned} \tag{5.90}$$

valid for $r_e T < 1.2\,\Omega\mathrm{cmK}$. As Drude's theory is invalid at small wavelengths, the integration of $\varepsilon_\lambda \cdot M_{\lambda\mathrm{s}}$ still begins at $\lambda = 0$, then (5.90) is only applicable at such low temperatures that the contribution of small wavelengths to ε is neglibgible. In general, this is the case for $T < 550\,\mathrm{K}$.

Finally $\varepsilon'_{\lambda,\mathrm{n}}$ from (5.84) with $k = n$ can be developed into a power series of $(1/n)$ and by integrating over all wavelengths, the total emissivity ε'_n normal to the surface can be calculated to be

$$\begin{aligned} \varepsilon'_\mathrm{n}(T) \;=\;& 0.5753\sqrt{r_e T} - 0.1777\,(r_e T) + 0.0292\,(r_e T)^{3/2} \\ &- 0.00184\,(r_e T)^{5/2} + 0.000712\,(r_e T)^3 - \cdots \end{aligned} \tag{5.91}$$

The values given in Table 5.6 for $\varepsilon/\varepsilon'_\mathrm{n}$, which are also illustrated in Fig. 5.35 are obtained from (5.90) and (5.91).

5.3.4 Transparent bodies

Metals and most electrical insulators completely absorb incident radiation in a layer of only a few micrometres thickness close to the surface, so that they are opaque. Exceptions to this are formed by liquids and solids like glass and some minerals (rock salt, sylvite and fluorite); these selectively let radiation through even at large thicknesses, namely in a limited wavelength band, mostly at small wavelengths in the visible part of the spectrum.

In order to describe radiation transmitted by a body, the *spectral transmissivity* is defined as

$$\tau_\lambda(\lambda, T) := \mathrm{d}^2\Phi_{\mathrm{in,tr}}/\mathrm{d}^2\Phi_{\mathrm{in}} \;. \tag{5.92}$$

Here, just as in section 5.1.3

$$\mathrm{d}^2\Phi_{\mathrm{in}} = E_\lambda(\lambda)\,\mathrm{d}\lambda\,\mathrm{d}A$$

is the radiation flow, in the wavelength interval $\mathrm{d}\lambda$, incident on an element $\mathrm{d}A$ of the surface of the body; $\mathrm{d}^2\Phi_{\mathrm{in,tr}}$ is the radiation flow transmitted through the body. The spectral transmissivity τ_λ for every wavelength indicates the portion of the spectral irradiance E_λ that is allowed to pass through. With the spectral absorptivity $a_\lambda(\lambda, T)$ from Table 5.1 and the spectral reflectivity $r_\lambda(\lambda, T)$ from Table 5.2 the following balance is obtained

$$r_\lambda(\lambda, T) + a_\lambda(\lambda, T) + \tau_\lambda(\lambda, T) = 1 \;. \tag{5.93}$$

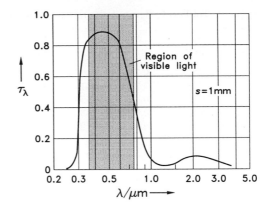

Fig. 5.37: Transmissivity τ_λ of a sheet of Jena-glass (colourless) of $s = 1\,\mathrm{mm}$ thickness, taking into account the multireflections on the two surfaces at $T \approx 300\,\mathrm{K}$

Fig. 5.37 illustrates the transmissivity τ_λ of Jena-glass. Glass possesses a high transparency within a very limited interval at small wavelengths. This selective transparency provides the possibility of collecting solar energy through glass windows. A significant proportion of the mostly short wave sunlight will be transmitted. Bodies located behind the window, which are at around ambient temperature, emit radiation at large wavelengths, where τ_λ assumes small values. This longwave radiation is not allowed to pass through the glass, but is mostly absorbed. This is what has become known as the *greenhouse effect*. It is not only used in plant nurseries, but also in collectors for the thermal use of solar energy, see [5.25]. A glass covering over the absorber surface of the collector allows solar radiation almost uninhibited through and at the same time reduces the undesirable release of heat by the collector to the surroundings.

The absorptivity a_λ and transmissivity τ_λ of a transparent body do not only depend on its capability for absorbing penetrating radiation, but also on its thickness s. In order to clarify these relationships, further material characteristics are used, which we will look at in the following, cf. also DIN 1349 [5.26].

A body of thickness s is struck by radiation with a spectral irradiance $E_\lambda(\lambda)$. From this, the spectral radiation flow

$$\Phi_\lambda\,(0) = (1 - r_\lambda)\,E_\lambda \qquad (5.94)$$

penetrates the body. Through absorption inside the body, the penetrated radiation flow is reduced and reaches the smaller value $\Phi_\lambda(s)$ at the other surface. This weakening is covered by the *spectral pure transmissivity*, defined by

$$\tau_i\,(\lambda, s) := \Phi_\lambda\,(s)\,/\Phi_\lambda\,(0) \quad . \qquad (5.95)$$

The index i indicates the weakening taking place *inside* the body. The portion of the penetrating radiation flow absorbed in the body is described by the *spectral pure absorptivity*

$$a_i\,(\lambda, s) := \frac{\Phi_\lambda\,(0) - \Phi_\lambda\,(s)}{\Phi_\lambda\,(0)} = 1 - \tau_i\,(\lambda, s) \quad . \qquad (5.96)$$

In order to separate the effect of the body thickness s from its specific absorption capability, the *spectral absorption coefficient* is defined as

$$\kappa\,(\lambda, s) := \frac{1}{s}\ln\frac{1}{\tau_i\,(\lambda, s)} \quad . \qquad (5.97)$$

Table 5.7: Spectral absorption coefficient κ_λ of water according to [5.27]

λ μm	κ_λ cm^{-1}	λ μm	κ_λ cm^{-1}	λ μm	κ_λ cm^{-1}	λ μm	κ_λ cm^{-1}	λ μm	κ_λ cm^{-1}
0.20	0.0691	0.70	0.0060	1.8	8.03	3.8	112.0	7.0	5740
0.25	0.0168	0.75	0.0261	2.0	69.1	4.0	145.0	7.5	5460
0.30	0.0067	0.80	0.0196	2.2	16.5	4.2	206.0	8.0	5390
0.35	0.0023	0.85	0.0433	2.4	50.1	4.4	294.0	8.5	5430
0.40	0.00058	0.90	0.0679	2.6	153.0	4.6	402.0	9.0	5570
0.45	0.00029	0.95	0.388	2.8	5160.0	4.8	393.0	9.5	5870
0.50	0.00025	1.0	0.363	3.0	11400.0	5.0	312.0	10.0	6380
0.55	0.000045	1.2	1.04	3.2	3630.0	5.5	265.0		
0.60	0.0023	1.4	12.4	3.4	721.0	6.0	2240.0		
0.65	0.0032	1.6	6.72	3.6	180.0	6.5	758.0		

For most materials κ does not depend on the layer thickness s, but is one of the material properties dependent on the wavelength: $\kappa = \kappa(\lambda)$. We will presume this, getting

$$\tau_i(\lambda, s) = \exp\left[-\kappa(\lambda)s\right] = 1 - a_i(\lambda, s) \quad . \tag{5.98}$$

The spectral pure transmissivity decreases exponentially with the thickness of the body. An opaque body of thickness s has such a large spectral absorption coefficient that the product $\kappa(\lambda)s$ reaches values over 7, then $\tau_i \approx 0$ and $a_i \approx 1$.

In Table 5.7 the spectral absorption coefficient of water is exhibited. At wavelengths in the visible part of the spectrum, $\kappa(\lambda)$ is very small. This gives, for $\lambda = 0.5\,\mu$m and $s = 1.0$ m

$$\tau_i = \exp\left(-0.00025\,\text{cm}^{-1} \cdot 1.0\,\text{m}\right) = 0.975 \quad .$$

The water layer transmits short wave radiation almost completely unweakened. This does not hold for wavelengths $\lambda \geq 0.95\,\mu$m. Here τ_i will be immeasureably small for $s = 1.0$ m. As the sun, according to Example 5.3, mainly radiates in a region of short wavelengths, 36% of the total solar radiation penetrating the water will still pass through it to a depth of 1.0 m. In contrast, the mainly long wave radiation from earthly sources will already be completely absorbed in a water layer of around 2 mm thickness [5.17].

The radiation allowed to pass through solid bodies, such as glass, is not only determined by the absorption internally, the *multiple reflection* at both surfaces also plays a role. The radiation flows that appear in multiple reflections are schematically illustrated in Fig. 5.38. At each of the two edges the radiation flow is divided by reflection into a reflected and a continuing part. The decisive reflectivity is known as the spectral Fresnel reflectivity $\bar{r}(\lambda)$ according to [5.26]. It is calculated from $\bar{r}(\lambda) = 1 - \varepsilon'_\lambda$ with the directional spectral emissivity ε'_λ from (5.82), if the body is bounded by a medium with the refractive index $n = 1$, that is, in a vacuum or in air. By each transmission from one edge to the other, the radiation flow is weakened by the factor τ_i. This gives the proportions drawn in Fig. 5.38 for the continuing and reflected fractions of an incident radiation flow set equal to one.

Addition of all transmitted radiation flows gives

$$\tau_\lambda = (1 - \bar{r})^2\,\tau_i\left[1 + \bar{r}^2\tau_i^2 + \bar{r}^4\tau_i^4 + \cdots\right]$$

or

$$\tau_\lambda = \frac{(1 - \bar{r})^2\,\tau_i}{1 - \bar{r}^2\tau_i^2} \quad . \tag{5.99}$$

Addition of the reflected radiation flows produces the spectral reflectivity

$$r_\lambda = \bar{r} + \bar{r}(1 - \bar{r})^2\,\tau_i^2\left[1 + \bar{r}^2\tau_i^2 + \bar{r}^4\tau_i^4 + \cdots\right]$$

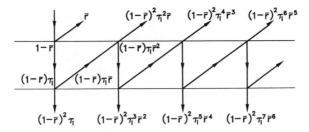

Fig. 5.38: Schematic of the radiation flow in multiple reflection at the edges of a plate that allows radiation to pass through it

or

$$r_\lambda = \bar{r} \left[1 + \frac{(1 - \bar{r})^2 \, \tau_i^2}{1 - \bar{r}^2 \tau_i^2} \right] \quad . \tag{5.100}$$

The multiple reflections increase r_λ compared to $\bar{r}$ and reduce τ_λ compared to τ_i. The fraction of radiation absorbed in the body can be calculated from (5.93). The spectral absorptivity is obtained as

$$a_\lambda = (1 - \bar{r}) \left[1 - \frac{(1 - \bar{r}) \, \tau_i}{1 - \bar{r}\tau_i} \right] \quad ; \tag{5.101}$$

it is smaller than the pure absorptivity $a_i = 1 - \tau_i$. For many electrical insulators, in particular for glass with $n = 1.5$, for polar angles $\beta < 60°$, Fresnels' reflectivity is $\bar{r} < 0.1$. Then $1 - \bar{r}^2\tau_i^2 = 1 - \bar{r}^2$ may be used, which has an error of less than 1%, giving from (5.99)

$$\tau_\lambda \approx \frac{1 - \bar{r}}{1 + \bar{r}} \tau_i \quad . \tag{5.102}$$

The transmissivity τ_λ of glass, illustrated in Fig. 5.37, reaches a maximum value of 0.88. With a refractive index $n = 1.55$, this yields, from (5.81), for perpendicular incident radiation, $\bar{r} = (n - 1)^2/(n + 1)^2 = 0.0465$. This then gives, from (5.101) a pure transmissivity $\tau_i = 0.97$. In the region of the maximum of τ_λ, glass is almost completely transparent; the reduction of τ_λ compared to the highest value $\tau_\lambda = 1$ can mainly be put down to the reflection on the two edges of the sheet of glass.

Example 5.5: The spectral transmissivity of a 2.0 mm thick glass sheet is approximated by

$$\tau_\lambda = \begin{cases} 0 & \text{for} \quad \lambda < \lambda_1 \\ 0.86 & \text{for} \quad \lambda_1 \leq \lambda \leq \lambda_2 \\ 0.03 & \text{for} \quad \lambda > \lambda_2 \end{cases}$$

with $\lambda_1 = 0.35\,\mu\text{m}$ and $\lambda_2 = 1.5\,\mu\text{m}$. Calculate the total transmissivity for perpendicularly incident solar radiation. The spectral irradiance $E_\lambda(\lambda)$ shall be assumed to be proportional to the hemispherical spectral emissive power $M_{\lambda s}$ of a black body with $T = T_S = 5777\,\text{K}$, see Example 5.3. How large will τ be for a glass sheet with $s = 4.0\,\text{mm}$ and refractive index $n = 1.52$?

The total transmissivity, in analogy to the total absorptivity a from Table 5.1, is given by

$$\tau = \frac{\int\limits_0^\infty \tau_\lambda(\lambda, T) E_\lambda(\lambda) \, d\lambda}{\int\limits_0^\infty E_\lambda(\lambda) \, d\lambda} = \frac{\int\limits_0^\infty \tau_\lambda(\lambda) M_{\lambda s}(\lambda, T_S) \, d\lambda}{\sigma T_S^4} \quad .$$

With the function $F(0, \lambda T)$ from (5.62) and from Table 5.3 this gives

$$\tau = 0{,}86\left[F(0, \lambda_2 T_S) - F(0, \lambda_1 T_S)\right] + 0{,}03\left[1 - F(0, \lambda_2 T_S)\right] \quad .$$

For $\lambda_1 T_S = 2022\,\mu\text{mK}$ and $\lambda_2 T_S = 8665.5\,\mu\text{mK}$, we find, from Table 5.3, $F(0, \lambda_1 T_S) = 0.0702$ and $F(0, \lambda_2 T_S) = 0.8800$. With that $\tau = 0.700$. The glass sheet allows 70% of the incident solar radiation to pass through it.

A doubling of the sheet thickness to $s = 4.0\,\text{mm}$ changes the spectral transmissivity τ_λ. Firstly, we will calculate the pure transmissivity τ_i for the glass sheet with 2.0 mm thickness. With $n = 1.52$ we obtain from (5.81) and $\bar{r} = 1 - \varepsilon'_{\lambda,n}$ Fresnel's reflectivity $\bar{r} = (n-1)^2/(n+1)^2 = 0.0426$. As $\bar{r} < 0.1$, then τ_i can be calculated from (5.102):

$$\tau_i = \frac{1+\bar{r}}{1-\bar{r}}\tau_\lambda = 1{,}089\,\tau_\lambda \quad .$$

This yields

$$\tau_i = \begin{cases} 0 & \text{for } \lambda < \lambda_1 \\ 0.937 & \text{for } \lambda_1 \le \lambda \le \lambda_2 \\ 0.033 & \text{for } \lambda > \lambda_2 \end{cases} .$$

With $\tau_i = \exp[-\kappa(\lambda)s]$ from this we get, for $s = 2.0\,\text{mm}$ the spectral absorption coefficients

$$\kappa = \begin{cases} \infty & \text{for } \lambda < \lambda_1 \\ 0.328\,\text{cm}^{-1} & \text{for } \lambda_1 \le \lambda \le \lambda_2 \\ 17.1\,\text{cm}^{-1} & \text{for } \lambda > \lambda_2 \end{cases} .$$

This then gives the pure transmissivity of the thicker glass sheet $s = 4.0\,\text{mm}$ as

$$\tau_i = \begin{cases} 0 & \text{for } \lambda < \lambda_1 \\ 0.877 & \text{for } \lambda_1 \le \lambda \le \lambda_2 \\ 0.001 & \text{for } \lambda > \lambda_2 \end{cases}$$

and finally the desired spectral transmissivity as

$$\tau_\lambda = \begin{cases} 0 & \text{for } \lambda < \lambda_1 \\ 0.805 & \text{for } \lambda_1 \le \lambda \le \lambda_2 \\ 0.001 & \text{for } \lambda > \lambda_2 \end{cases} .$$

The total transmissivity is then

$$\tau = 0.805\left[F(0, \lambda_2 T_S) - F(0, \lambda_1 T_S)\right] + 0.001\left[1 - F(0, \lambda_2 T_S)\right] = 0.652 \quad .$$

The glass sheet with double the thickness absorbs a somewhat larger part of the solar radiation, but still allows 65% of the incident radiation to pass through it compared to 70% with the thin sheet. The spectral reflectivity r_λ of the thick sheet from (5.100) is

$$r_\lambda = \begin{cases} 0.0426 & \text{for } \lambda < \lambda_1 \\ 0.0727 & \text{for } \lambda_1 \le \lambda \le \lambda_2 \\ 0.0426 & \text{for } \lambda > \lambda_2 \end{cases} .$$

It is increased between λ_1 and λ_2 by the multiple reflection at the edges, so that in total 6.7% instead of 4.26% of the solar energy is reflected.

5.4 Solar radiation

Without the radiation from the sun life on earth would be impossible; therefore the sun belongs to the most important radiation sources. The energetic use of the sun's radiation is still very low; however in the future it will have a far greater importance, in order to gain energy for heating and endothermic reactions and to generate electrical energy in photovoltaic and solar-thermal power stations. The following section does not contain a discussion on solar energy technology, further information can be found in [5.25] and [5.28] to [5.30]. We will, however, deal with the quantity and spectral distribution of the radiant energy provided by the sun and how it is weakened during its transmission through the earth's atmosphere. This allows us to calculate the solar irradiance at the surface of the earth. We will conclude this section with consideration of the absorptivities for solar radiation. They differ from the absorptivities for radiation from earthly sources because solar radiation is only incident at small wavelengths, namely below $4\,\mu$m and mainly in the region of visible light.

5.4.1 Extraterrestrial solar radiation

The sun is an almost spherical radiation source with a diameter of $1.392 \cdot 10^6$ km. It lies in one of the foci of the elliptical orbit of the earth. The solar radiation flow, which reaches the earth is inversely proportional to the square of the distance r between the sun and the earth. The mean distance is $r_0 = 149.6 \cdot 10^6$ km; this distance is called one astronomical unit (AU)[7]. The smallest distance lies at 0.983 AU and occurs on 3rd January, the largest separation between the sun and earth is 1.017 AU, this is reached on 4th July. The ratio required for radiation calculations, $(r_0/r)^2$ is known as the eccentricity factor, and according to J.A. Duffie and W.A. Beckmann [5.29], can be calculated approximately from

$$f_{\text{ex}} = (r_0/r)^2 = 1 + 0.033 \cos\left(2\pi d_{\text{n}}/365\right) \ , \tag{5.103}$$

where d_{n} is the day number of the year, starting with $d_{\text{n}} = 1$ on 1st January and ending with $d_{\text{n}} = 365$ on 31st December. A more accurate relationship from J.W. Spencer [5.31] can be found in M. Iqbal [5.34].

The large distance between sun and earth means that solar radiation forms a quasi-parallel bundle of rays. The radiation that is not yet weakened by scattering and absorption in the earth's atmosphere is called extraterrestrial radiation. If it is *perpendicularly* incident on a surface just outside the the earth's atmosphere, at a distance $r_0 = 1$ AU from the centre of the sun then the irradiance of the extraterrestrial solar radiation is called the *solar constant* E_0. By evaluating more recent measurements, C. Fröhlich and R.W. Brusa [5.33] determined the value

$$E_0 = (1367 \pm 1.6)\text{W/m}^2 \ ,$$

that was also accepted by the World Meteorological Organisation (WMO) in 1981 as the best value. With this value of E_0, the temperature $T_{\text{S}} = 5777$ K in Example

---

[7]The exact value is 1AU $= 149.597\,870 \cdot 10^6$ km. For the strict definition of the AU, see [5.32].

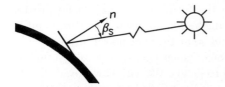

Fig. 5.39: Extraterrestrial solar radiation on a
surface, whose normal forms the polar angle β_S
with the direction of the solar rays

5.3 was calculated. This would be the temperature at the surface of the sun if it
radiated like a black body.

The irradiance of extraterrestrial radiation, that falls perpendicularly onto a
surface that is at the same distance r as the earth is from the sun, is given by

$$E_n^{sol} = E_0(r_0/r)^2 = E_0 f_{ex} \tag{5.104}$$

with f_{ex} from (5.103). If the direction of the sun's rays forms the polar angle β_S
with the surface normal, Fig. 5.39, then the irradiance will be

$$E^{sol} = E_0 f_{ex} \cos \beta_S \ . \tag{5.105}$$

The sun's polar angle β_S depends on the position and the orientation of the irradiated surface. This dependence is reproduced by trigonometric equations that
can be found in books about solar radiation and its uses, e.g. in [5.30] and [5.34].

The *spectral irradiance* $E_{\lambda,n}^{sol}$ of extraterrestrial solar radiation, that falls perpendicularly on a surface at a distance $r_0 = 1\,\mathrm{AU}$ from the sun, has been determined by several series of experiments using stratospheric aircraft. Their evaluation by C. Fröhlich and C. Werli at the World Radiation Centre in Davos,
Switzerland, yielded the spectrum reproduced in Fig. 5.40. The numerical values
upon which this diagram is based can be found in M. Iqbal [5.34]. The maximum
of $E_{\lambda,n}^{sol}$ lies in the visible light region at $\lambda \approx 0.45\,\mu\mathrm{m}$. 99 % of the irradiance falls
in the wavelength band $\lambda \le 3.8\,\mu\mathrm{m}$. Fig. 5.40 also shows the spectral irradiance
$E_{\lambda,s}$ of the radiation emitted by a 'black' sun at $T_S = 5777\,\mathrm{K}$. The areas under
the two curves (up to $\lambda \to \infty$) are equal — they each yield the solar constant E_0
—, but the spectrum of the extraterrestrial solar radiation deviates significantly
at some points, in particular at $\lambda < 0.6\,\mu\mathrm{m}$, from the spectrum of radiation from
a black body.

Example 5.6: Determine the irradiance of extraterrestrial solar radiation on a horizontal
area in Berlin (latitude $\psi = 52.52°$ North, longitude $\varphi = 13.35°$ East) on 1st September, 12.00
central European summertime.

The irradiance E^{sol} is given by (5.105), where the sun's polar angle β_S is still unknown. It
holds for a horizontal surface, cf. [5.34], that

$$\cos \beta_S = \sin \delta \sin \psi + \cos \delta \cos \psi \cos \omega \ . \tag{5.106}$$

The declination δ of the sun, for 1st September, that appears in this equation, can be taken
from tables presented by M. Iqbal [5.34]: $\delta = 8{,}51°$. The hour angle ω is calculated from the
local solar time t_S, according to

$$\omega = 15° \, (t_S/\mathrm{h}) \ .$$

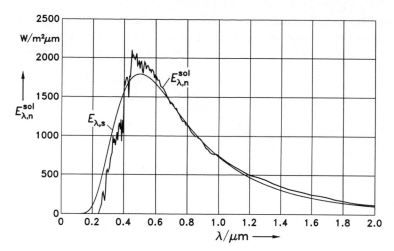

Fig. 5.40: Spectral irradiance $E_{\lambda,n}^{\rm sol}$ of extraterrestrial solar radiation falling perpendicularly on an area at a distance $r_0 = 1\,\mathrm{AU}$ from the sun

In the validity region of central European time $t_{\rm CET}$ it holds that[8]

$$t_{\rm S} = t_{\rm CET} - 12\mathrm{h} + (\varphi/15° - 1)\,\mathrm{h}\ .$$

12.00 hours central European summertime corresponds to $t_{\rm CET} = 11.0\,\mathrm{h}$. From this we get with $\varphi = 13.35°$ as the degree of longitude of Berlin, $t_{\rm S} = -1.11\,\mathrm{h}$ and $\omega = -16.65°$. With these values, it follows from (5.106) that $\cos\beta_{\rm S} = 0.6940$, i.e. $\beta_{\rm S} = 46.05°$.

For 1st September, $d_{\rm n} = 244$. This then gives the eccentricity factor from (5.103) of $f_{\rm ex} = 0.9838$. The irradiance of the extraterrestrial radiation is found to be

$$E^{\rm sol} = E_0 f_{\rm ex} \cos\beta_{\rm S} = 1367\,\frac{\mathrm{W}}{\mathrm{m}^2}\,0.9838 \cdot 0.6940 = 933\,\frac{\mathrm{W}}{\mathrm{m}^2}\ .$$

5.4.2 The attenuation of solar radiation in the earth's atmosphere

The extraterrestrial solar radiation incident on the outer edge of the earth's atmosphere is weakened as it travels through the atmosphere, so the solar irradiance at the surface of the earth is considerably lower than the extraterrestrial value from (5.105). Part of the incident radiant solar energy is removed from the bundle of rays by scattering on air molecules and aerosols; a further part is absorbed by the constituents of air. Around half of this scattered radiation reaches the ground in the form of so-called diffuse radiation, whilst the other half is radiated back into

---

[8]This equation yields the *mean* local solar time. In order to obtain the actual (date dependent) local solar time, a small correction, the so-called equation of time, has to be used. This correction amounts to only a few minutes.

space, cf. Fig. 5.41. The radiation absorbed by molecules in the atmosphere raises
the energy of the atmosphere which emits radiation itself. Part of this mostly long
wave radiation reaches the ground and is known as atmospheric counter-radiation.

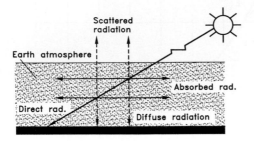

Fig. 5.41: Direct and diffuse solar radiation that passes through the atmosphere to the earth's surface (schematic)

5.4.2.1   Spectral transmissivity

We are now going to calculate the attenuation of direct (directional) solar radiation
through scattering and absorption in the atmosphere. The atmosphere is assumed
to be cloudless; for details on the complicated effect of clouds, [5.34] is suggested.
We will consider a bundle of rays that goes through an optically turbid, namely
absorbent and scattering medium, Fig. 5.42. The reduction $\mathrm{d}L_\lambda$ of its spectral
intensity L_λ according to the law from P. Bouguer[9] , is proportional to the distance
through which the radiation passes $\mathrm{d}l$ and the density ϱ of the scattering and
absorbing particles:

$$\frac{\mathrm{d}L_\lambda}{L_\lambda} = -\mu(\lambda, l)\varrho(l)\,\mathrm{d}l \ . \tag{5.107}$$

It is assumed, according to the so-called Beer's[10] law, that the attenuation coefficient μ is independent of the pathway. Integration between l and l_0 gives

$$L_\lambda(l) = L_\lambda(l_0)\exp[-\mu(\lambda)\textstyle\int_{l_0}^{l}\varrho(l)\,\mathrm{d}l] \ . \tag{5.108}$$

We will now apply (5.108) to solar radiation, Fig. 5.43. The path s of the
bundle of rays is slightly curved because of refraction in the atmosphere. It begins,

---

[9]Pierre Bouguer (1698–1758) was nominated at the age of 15 to be Professor of Hydrology as
a successor to his father who had passed away. In 1735, Bouguer became a member of the Paris
Académie Royale des Sciences. He wrote several books about ship building and navigation. In
his "Essai d'optique sur la gradation de la lumière"(Essay in optics on the gradation of light),
published in 1729, he was the first to develop methods of photometry and layed down the law
named after him, according to which the strength of a light ray in a homogeneous medium falls
according to an exponential law as it passes through.

[10]August Beer (1825–1863) became a Professor of Physics in Bonn in 1855. In his 'Einleitung
in die höhere Optik' (Introduction to higher optics), published in 1854, he summarised the
theory of light known at that time.

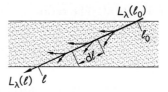

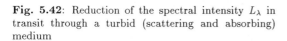

Fig. 5.42: Reduction of the spectral intensity L_λ in transit through a turbid (scattering and absorbing) medium

at large distance away, with $L_\lambda(l_0) = L_\lambda(s \to \infty) = L_\lambda^{\text{sol}}$, the spectral intensity of extraterrestrial solar radiation, and ends at the earth's surface ($s = 0$) with the intensity $L_\lambda(s = 0)$. It therefore follows from (5.108), taking into account $dl = -ds$, that

$$L_\lambda(s = 0) = L_\lambda^{\text{sol}} \exp[-\mu(\lambda) \int_0^\infty \varrho(s)\,ds] \ . \tag{5.109}$$

The ratio

$$\tau_\lambda = \frac{L_\lambda(s = 0)}{L_\lambda^{\text{sol}}} = \exp[-\mu(\lambda) \int_0^\infty \varrho(s)\,ds] \tag{5.110}$$

is the *spectral transmissivity* of the atmosphere ($\tau_\lambda \leq 1$). This is the factor by which the direct solar radiation reaching the ground is reduced compared to extraterrestrial solar radiation.

The integral which appears in (5.109) and (5.110)

$$m = \int_0^\infty \varrho(s)\,ds \tag{5.111}$$

is known as the *optical mass* of the atmosphere. Its units are kg/m², and it is proportional to the mass and therefore the number of atmospheric particles along the pathway of the solar ray bundle. If the sun is at its zenith, the pathway through the atmosphere is at its shortest and the optical mass attains its smallest value

$$m_{\text{n}} = \int_0^\infty \varrho(z)\,dz \ ,$$

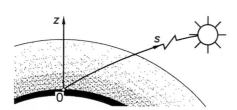

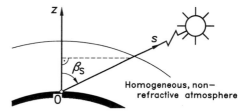

Fig. 5.43: Passage of directional solar radiation through the earth's atmosphere

Fig. 5.44: Passage of directional solar radiation through a homogeneous, non-refractive atmosphere

cf. Fig. 5.43. The ratio

$$m_{\mathrm{r}} = m/m_{\mathrm{n}} = \int\limits_0^\infty \varrho(s)\,\mathrm{d}s \Big/ \int\limits_0^\infty \varrho(z)\,\mathrm{d}z \qquad (5.112)$$

is called the *relative optical mass*, where $m_{\mathrm{r}} \geq 1$ is valid. According to Fig. 5.44, for a homogeneous and non-refractive atmosphere, the simple result below is obtained

$$m_{\mathrm{r}} = \frac{1}{\cos\beta_{\mathrm{S}}} \ . \qquad (5.113)$$

This equation shows errors because of refraction and the height dependent density ϱ. However, these errors only become significant at large polar angles $\beta_{\mathrm{S}} > 70°$. According to F. Kasten and A.T. Young [5.35], for an atmosphere of dry air

$$m_{\mathrm{r,L}} = \frac{1}{\cos\beta_{\mathrm{S}} + 0.5057\,(96.080 - \beta_{\mathrm{S}})^{-1.6364}} \qquad (5.114)$$

is obtained, where the sun's polar angle β_{S} is in degrees.

With the relative optical mass m_{r}, it follows for the argument of the exponential function in (5.110) that

$$\mu(\lambda) \int\limits_0^\infty \varrho(s)\,\mathrm{d}s = \mu(\lambda)m_{\mathrm{n}}m_{\mathrm{r}} \ .$$

If we combine the product of $\mu(\lambda)$ and m_{n} into the dimensionless attenuation coefficient

$$\kappa(\lambda) := \mu(\lambda)m_{\mathrm{n}} \ , \qquad (5.115)$$

then the spectral transmissivity of the atmosphere is found to be

$$\tau_\lambda = \exp\left[-\kappa(\lambda)m_{\mathrm{r}}(\beta_{\mathrm{S}})\right] \ . \qquad (5.116)$$

In the calculation of τ_λ it must be taken into account that several independent scattering and absorption processes act in the attenuation of direct solar radiation. It therefore holds that

$$\kappa(\lambda)m_{\mathrm{r}} = \sum_{i=1}^{j} \kappa_i(\lambda)m_{\mathrm{r},i} \ , \qquad (5.117)$$

where j is the number of attenuation processes. In general, five different types of process are considered: Rayleigh scattering on the molecules of the atmosphere, scattering and absorption on aerosols, the absorption by ozone, water vapour and other gases in the atmosphere. It follows from (5.116) and (5.117) that the spectral transmissivity is

$$\tau_\lambda = \tau_{\lambda,\mathrm{R}}\,\tau_{\lambda,\mathrm{A}}\,\tau_{\lambda,\mathrm{O}_3}\,\tau_{\lambda,\mathrm{W}}\,\tau_{\lambda,\mathrm{G}} \ , \qquad (5.118)$$

where the indices are R for Rayleigh scattering, A for aerosols, O_3 for ozone, W for water vapour and G for other gases in the atmosphere.

5.4.2.2   Molecular and aerosol scattering

As the molecular diameter ($\sim 10^{-4}\,\mu$m) is considerably smaller than the wavelength of the radiation, the scattering by the molecules of the atmosphere can be described by the theory of light scattering by small particles, first presented by Lord Rayleigh[11] [5.36], cf. also [5.37]. According to this, the attenuation coefficient is $\kappa_R(\lambda) \sim \lambda^{-4}$. This also explains the blue colour of a cloudless sky. The blue fraction of sunlight lies at the small wavelength end of the visible spectrum and therefore experiences strong Rayleigh scattering in all directions. Without this molecular scattering the sky would appear to be black with the exception of the bright disk of the sun.

The scatter coefficient $\kappa_R(\lambda)$ of Rayleigh scattering in dry air is reproduced well by an equation presented by R.E. Bird and C. Riordan [5.38]. This gives the following for the spectral transmissivity with λ in μm

$$\tau_{\lambda,R} = \exp\left[-m_{r,L}\lambda^{-4}\left(115.6406 - 1.335/\lambda^2\right)^{-1}\right] \ . \tag{5.119}$$

The relative optical mass $m_{r,L}$ is given by (5.114). As can be seen in Fig. 5.45, at small wavelengths Rayleigh scattering considerably weakens direct solar radiation. In contrast to this, for $\lambda > 1.2\,\mu$m no noticeable attenuation occurs. At large sun polar angles β_S, that is with large optical masses $m_{r,L} > 5$, the atmosphere is almost opaque for short wavelength light. This explains the reddy-yellow colour of the sun disc at sunrise and sunset.

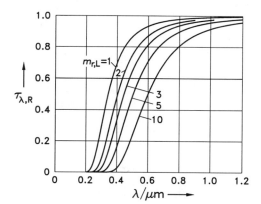

Fig. 5.45: Spectral transmissivity $\tau_{\lambda,R}$ of the atmosphere as a result of Rayleigh scattering from (5.119) for various relative optical masses $m_{r,L}$

Dust and small suspended water droplets form *aerosols*. They scatter and absorb solar radiant energy, whereby the scattered proportion predominates. The scattering and absorption by aerosols are difficult and inexact to model. The turbidity formula from A. Ångström [5.39] is frequently used

$$\tau_{\lambda,A} = \exp\left(-\beta^*\lambda^{-\alpha^*}m_{r,A}\right) \ . \tag{5.120}$$

Values between 0.8 and 1.8 are used for the exponent α^*; β^* varies from $\beta^* = 0$ (pure atmosphere) via 0.1 (clear) and 0.2 (cloudy) to 0.3 (very murky atmosphere), see for this [5.34]. The

[11] John William Strutt, Third Baron of Rayleigh (1842–1919) set up his own physical laboratory at his family seat, Terling Place in Essex, England. In 430 scientific publications he dealt with problems from all areas of classical physics, in particular acoustics, for which he wrote his famous work, 'The Theory of Sound' (1877/78). Together with W. Ramsey he discovered the element Argon (1892–95), for which he was awarded the Nobel prize for Physics in 1904. The chemist W. Ramsey was awarded the Nobel prize for Chemistry in the same year.

556 5 Thermal radiation

relative optical aerosol mass $m_{r,A}$ is generally unknown because of the large fluctuations in the size, distribution and composition of aerosol particles. This is why $m_{r,L}$ from (5.114) is often used in place of $m_{r,A}$.

5.4.2.3 Absorption

In contrast to scattering, absorption only takes place within certain narrow wavelength intervals, the so-called absorption bands. The main constituents of the atmosphere are N_2 and O_2, which dissociate to atomic N and O at heights above 100 km. These four gases only absorb, although very strongly, at small wavelengths. N and O absorb all radiation below $0.085\,\mu m$; O_2 and N_2 absorb solar radiation in several overlapping bands up to $\lambda = 0.20\,\mu m$, so no radiation of wavelength below $0.20\,\mu m$ reaches the earth's surface. O_2 has three additional distinctive but weak bands at 0.63, 0.69 and $0.76\,\mu m$.

Ozone (O_3) absorbs strongly between $0.2\,\mu m$ and $0.35\,\mu m$, through which the energy rich UV-B-radiation is kept away from the earth. Further absorption bands lie in the visible region between 0.47 and $0.76\,\mu m$. Ozone is generated in the stratosphere principally by solar UV radiation; close to the ground it exists due to the photo-chemical decomposition of nitrous oxides. Chlorine, which reaches the stratosphere mainly in the form of long living chlorofluorocarbons (CFCs), attacks and destroys the protective stratospheric ozone layer. This is why the production and use of CFC's, for example as refrigerants or as propellants for insulation foams, should be discontinued in the next few years, cf. [5.40].

The amount of ozone in the atmosphere is frequently indicated by the height h_{O_3} of a vertical column of gaseous ozone under standard conditions ($t_n = 0\,°C$, $p_n = 1.01325\,bar$). This quantity varies seasonally and with latitude; it has an average value of around 2.5 mm at the equator, 3.5 mm at medium latitudes and up to 4.5 mm at the poles. The amount of ozone has fallen over several years as a result of the discharge of CFCs. At the beginning of spring a reduction up to 20 % of the average value occurs over northern Europe. The "hole in the ozone layer" over the south pole, which appears in October, leads to a reduction at times of up to 75 %. Fig. 5.46 shows the spectral transmissivity

$$\tau_{\lambda,O_3} = \exp\left[-k_{O_3}(\lambda)\,h_{O_3}m_{r,O_3}\right] \quad . \tag{5.121}$$

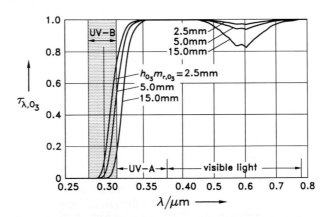

Fig. 5.46: Spectral transmissivity τ_{λ,O_3} of the absorption by ozone from (5.121). The region of energy rich UV-B-radiation is highlighted

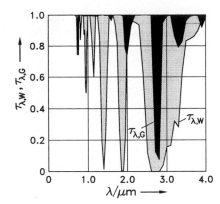

Fig. 5.47: Spectral transmissivities $\tau_{\lambda,\text{W}}$ and $\tau_{\lambda,\text{G}}$ as a result of absorption by water vapour, O_2 and CO_2 respectively, for $m_{\text{r,L}} = 1$

Tabulated values of absorption coefficients $k_{O_3}(\lambda)$ can be found in [5.34] and [5.38]. The relative optical mass of ozone is indicated by $m_{\text{r,O}_3}$. It only deviates significantly from $m_{\text{r,L}}$ for $\beta_{\text{S}} > 70°$ and can be calculated using a relationship given by N. Robinson [5.41].

The most important *absorbers in the infrared region* of the spectrum are water vapour and CO_2. Here the humidity of the atmosphere undergoes considerable fluctuations. The amount of water vapour is often described by the thickness w of a layer of water on the ground formed by the condensation of water vapour perpendicular to the ground. A typical value is $w = 20\,\text{mm}$. The law from Bouguer, (5.107), is not exactly valid for water vapour and CO_2. Therefore, relationships used for the transmission coefficients $\tau_{\lambda,\text{W}}$ and $\tau_{\lambda,\text{G}}$ have a form different from (5.116). These equations can be found in M. Iqbal [5.34], where the associated absorption coefficients are given as functions of the wavelength. Fig. 5.47 shows the pattern of $\tau_{\lambda,\text{W}}$ and $\tau_{\lambda,\text{G}}$. These spectral transmissivities have values close to one at wavelengths around 1.2 and $1.6\,\mu\text{m}$ as well as at 2.2 and $3.9\,\mu\text{m}$. These narrow wavebands are called atmospheric windows, as here the atmosphere allows solar radiation and also radiation from the earth's surface to pass through with virtually no attenuation.

5.4.3 Direct solar radiation on the ground

The spectral transmissivity τ_λ from (5.110) also gives the ratio of the spectral irradiance $E_{\lambda,\text{n}}$ of an area located on the ground and oriented perpendicular to the direction of radiation, to the spectral irradiance $E_{\lambda,\text{n}}^{\text{sol}}$ by extraterrestrial solar radiation. It therefore holds that

$$E_{\lambda,\text{n}} = \tau_\lambda E_{\lambda,\text{n}}^{\text{sol}} = \tau_{\lambda,\text{R}}\,\tau_{\lambda,\text{A}}\,\tau_{\lambda,\text{O}_3}\,\tau_{\lambda,\text{W}}\,\tau_{\lambda,\text{G}} E_{\lambda,\text{n}}^{\text{sol}}\ , \qquad (5.122)$$

and for an area whose normal forms an angle β_{S} with the suns rays, it follows that

$$E_\lambda = \tau_\lambda \cos\beta_{\text{S}} E_{\lambda,\text{n}}^{\text{sol}}\ . \qquad (5.123)$$

Fig. 5.48 shows the extraterrestrial spectrum $E_{\lambda,\text{n}}^{\text{sol}}$ and the associated pattern of $E_{\lambda,\text{n}}$. The upper edge of this curve represents the irradiance reduced purely by Rayleigh scattering. The deteriorations marked in black are caused by the

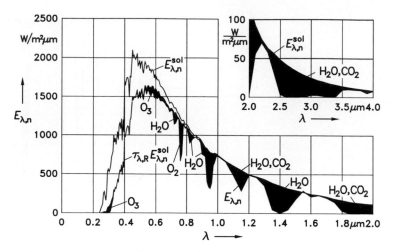

Fig. 5.48: Spectral irradiance $E_{\lambda,n}^{\text{sol}}$ of extraterrestrial solar radiation and $E_{\lambda,n}$ of direct solar radiation at the ground for a pure, cloudless atmosphere with $m_{r,L} = 1.5$. The curve indicated by $\tau_{\lambda,R} E_{\lambda,n}^{\text{sol}}$ represents the attenuation caused by Rayleigh scattering alone. The dark areas indicate the absorption by each of the gases written on the graph ($h_{O_3} = 0.30\,\text{cm}$, $w = 2.0\,\text{cm}$)

absorption by the gases O_3, O_2, H_2O and CO_2. Further diagrams of this type, which show the variation in the influencing quantities (water vapour and ozone content, turbidness due to aerosols, different optical masses), are available in M. Iqbal [5.34].

In most practical applications, it is normally sufficient to know just the *irradiance* E at an area on the ground. The most exact method for obtaining E is by integrating the spectral irradiance E_λ over all wavelengths, practically from $\lambda = 0.3\,\mu\text{m}$ to $\lambda = 4.0\,\mu\text{m}$. In order to avoid this somewhat difficult numerical integration, approximation formulae are used. Various authors have presented these formulae and they have been collected and compared by M. Iqbal [5.34].

A rather accurate calculation of E is permitted by the relationships given by R.E. Bird and R.L. Hulstrom [5.42] for the irradiance of direct (directional) solar radiation. The irradiance of a surface, whose normal forms an angle β_S with the sun's rays, is

$$E = E_n \cos \beta_S \quad . \tag{5.124}$$

The irradiance E_n of a surface normal to the sun's rays is, according to [5.42] and [5.34],

$$E_n = 0{,}975\, E_0\, \tau_R\, \tau_A\, \tau_{O_3}\, \tau_W\, \tau_G \quad , \tag{5.125}$$

where the transmissivities are calculated according to the following equations. The relative optical mass $m_{r,L}$ from (5.114) is uniformly used, simplified to here m_r.
Rayleigh scattering:

$$\tau_R = \exp\left[-0.0903\, m_r^{0.84}\left(1 + m_r - m_r^{1.01}\right)\right] \quad .$$

Aerosol scattering:

$$\tau_A = 0.1245\,\alpha^* - 0.0162 + (1.003 - 0.125\,\alpha^*)\exp\left[-\beta^* m_r\left(1.089\,\alpha^* + 0.5123\right)\right] \quad ,$$

where α^* and β^* are the parameters from (5.120). τ_A can also be given as a function of the (horizontal) visibility s_h in km:

$$\tau_A = \exp\left[m_r^{0.9}\ln\left(0.97 - \frac{1.265}{s_h^{0.66}}\right)\right] \quad , \quad 5\,\text{km} < s_h < 180\,\text{km} \ .$$

The utilisation of s_h replaces the estimation of the parameters α^* and β^*.
Absorption by Ozone[12]:

$$\tau_{O_3} = 1 - \frac{0.153\,h_{O_3}\,m_r}{(1 + 139.5\,h_{O_3}\,m_r)^{0.3035}} \quad ,$$

where h_{O_3} is entered in cm, cf. section 5.4.2.3.
Absorption by water vapour:

$$\tau_W = 1 - \frac{2.496\,w\,m_r}{6.385\,w\,m_r + (1 + 79.03\,w\,m_R)^{0.6828}}$$

with w in cm, cf. section 5.4.2.3.
Absorption by CO_2 and other gases:

$$\tau_G = \exp\left(-0.0127\,m_r^{0.26}\right) \ .$$

Example 5.7: Determine the irradiance of direct solar radiation on a horizontal area in Berlin on 1st September, 12.00 central European summertime, cf. Example 5.6. The sky is cloudless, $h_{O_3} = 0.30\,\text{cm}$, $w = 2.6\,\text{cm}$ and a horizontal visibility of $s_h = 40\,\text{km}$ may be assumed.

According to Example 5.6 the sun's polar angle is $\beta_S = 46.05°$. This gives, from (5.114), a relative optical mass $m_{r,L} = m_r = 1.439$. The transmissivities that appear in (5.125) have the following values:

$$\tau_R = 0.8852, \quad \tau_{O_3} = 0.9811, \quad \tau_W = 0.8715, \quad \tau_G = 0.9861, \quad \tau_A = 0.8100 \ .$$

It follows from (5.125) that $E_n = 805.8\,\text{W/m}^2$; with $\cos\beta_S = 0.694$, according to (5.124) $E = 559\,\text{W/m}^2$ is obtained. This value reaches only $59.9\,\%$ of the irradiance E^{sol} of extraterrestrial solar radiation calculated in Example 5.6. Scattering and absorption by aerosols are of great influence. If a less turbid atmosphere is assumed with $s_h = 100\,\text{km}$, then τ_A increases to $\tau_A = 0.8766$, and the irradiance reaches a value of $E = 605\,\text{W/m}^2$.

5.4.4 Diffuse solar radiation and global radiation

In addition to the direct solar radiation dealt with in the previous sections, part of the radiation scattered in the atmosphere also reaches the ground. This is known as diffuse solar radiation. Diffuse and direct solar radiation are partially reflected by the ground; the reflected radiation is also sent back to the ground by the atmosphere to a small extent. The reflection between the atmosphere and the earth's surface continues with increasingly smaller fractions of radiation, and leads in total, to an additional radiation flow towards the ground. This radiation and the diffuse solar radiation are together called sky-radiation. This should not be confused with the atmospheric counter-radiation mentioned in section 5.4.2;

---

[12]The equation for τ_{O_3} has been simplified, compared with [5.42], without any loss of accuracy.

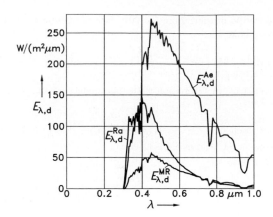

Fig. 5.49: Spectral irradiance of the three fractions of diffuse sky-radiation, calculated for $m_{r,L} = 1.5$, $h_{O_3} = 0.3\,\mathrm{cm}$, $w = 2.0\,\mathrm{cm}$ as well as $\alpha^* = 1.3$ and $\beta^* = 0.10$ in (5.120): $E_{\lambda,d}^{Ra}$ spectral irradiance of Rayleigh scattering $E_{\lambda,d}^{Ae}$ of scattering by aerosols, $E_{\lambda,d}^{MR}$ spectral irradiance due to multiple reflection.

this long wave radiation is emitted by molecules in the atmosphere, chiefly water vapour and CO_2. Sky-radiation, on the other hand, occurs at small wavelengths due to its formation in the scattering of directional solar radiation. This is shown by Fig. 5.49, which exemplarily illustrates the three fractions of the spectral irradiance $E_{\lambda,d}$ of diffuse sky-radiation, namely $E_{\lambda,d}^{R}$ of Rayleigh scattering, $E_{\lambda,d}^{Ae}$ of aerosol scattering and $E_{\lambda,d}^{MR}$ as a result of multiple reflections.

The direct solar radiation and the diffuse sky-radiation are combined under the term *global radiation*. The global irradiance E_G of a horizontal area on the ground is made up of the following parts:

$$E_G = E_n \cos \beta_S + E_d^{Ra} + E_d^{Ae} + E_d^{MR} \ . \tag{5.126}$$

Here, the first term on the right hand side is the irradiance due to direct solar radiation, from (5.124) and (5.125); E_d^{Ra} indicates the irradiance of diffuse radiation from Rayleigh scattering by air molecules, E_d^{Ae} the irradiance of diffuse radiation caused by aerosol scattering, and E_d^{MR} the irradiance due to multiple reflection. For areas inclined to the horizontal, more complicated relationships are yielded because the radiative exchange with the surroundings of the area being considered also have to be taken into account. M. Iqbal [5.34] is suggested for an extensive illustration of this.

The model from R.E. Bird und R.L. Hulstrom [5.42], see also [5.34], is once again mentioned for the calculation of the diffuse fraction of radiation in (5.126). The source of the scattered radiation is the non-absorbed direct solar radiation with the irradiance

$$E_n^{na} = 0.786 \, E_0 \cos \beta_S \, \tau_{O_3} \, \tau_W \, \tau_G \, \tau_A^{abs} \ .$$

Here, τ_A^{abs} is the transmissivity resulting from the absorption by aerosols alone:

$$\tau_A^{abs} = 1 - (1 - \omega_0)\,(1 - \tau_A)\,(1 - m_r + m_r^{1.06}) \ .$$

The quantity ω_0 is the ratio of the energy scattered from the aerosols to the energy scattered and absorbed by them. This fraction can only be estimated; values of around 0.9 are generally used.

Under the assumption that half of the energy scattered by molecules in the atmosphere reaches the ground, according to [5.42], we obtain

$$E_d^{Ra} = \frac{1}{2} E_n^{na} \frac{1 - \tau_R}{1 - m_r + m_r^{1.02}}$$

for the irradiance as a result of Rayleigh scattering. For the irradiance from aerosol scattering we get

$$E_d^{Ae} = F_A E_n^{na} \frac{1 - \tau_A/\tau_A^{abs}}{1 - m_r + m_r^{1.02}} \quad .$$

The factor F_A indicates what proportion of the energy scattered by the aerosols is scattered "forwards", i.e. the part which reaches the ground. F_A also has to be estimated. R.E. Bird and R.L. Hulstrom recommend $F_A = 0.84$. G.D. Robinson [5.43] determined F_A for aerosols over the British Isles; these values are well reproduced by

$$F_A = \begin{cases} 0.91 & \text{for} \quad 0 \leq \beta_S \leq 45° \\ 0.45 + 0.65 \cos \beta_S & \text{for} \quad 45° < \beta_S \leq 85° \end{cases} \quad .$$

Through considerations, analogous to the explanation of multiple reflection in section 5.3.4, we obtain

$$E_d^{MR} = \left(E_n \cos \beta_S + E_d^{Ra} + E_d^{Ae} \right) \frac{r_E r_{At}}{1 - r_E r_{At}} \quad . \tag{5.127}$$

Here, r_E is the reflectivity of the earth's surface for short wave radiation originating from the sun; it is also known as the Albedo in meteorology. r_E can be calculated from information about the absorptivity of solar radiation, given in the following section 5.4.5. The reflectivity of the atmosphere is indicated by r_{At}; it is small and, according to [5.42], can be calculated from

$$r_{At} = 0.0685 + (1 - F_A) \left(1 - \tau_A/\tau_A^{abs} \right) \quad .$$

With E_d^{MR} from (5.127), we obtain the following for the irradiance of global radiation on a horizontal area from (5.126)

$$E_G = \left(E_n \cos \beta_S + E_d^{Ra} + E_d^{Ae} \right) \frac{1}{1 - r_E r_{At}} \quad .$$

This equation is also valid for models in which E_n, E_d^{Ra}, E_d^{Ae} and r_{At} are determined using different relationships than those in [5.42].

Example 5.8: Calculate the irradiances of the diffuse solar radiation and the global radiation for the case dealt with in Example 5.7. Additional assumptions are: $\omega_0 = 0.90$, F_A according to G.D. Robinson [5.43] and $r_E = 0.25$.

With $m_r = 1.439$ and $\tau_A = 0.810$ from Example 5.7, we obtain $\tau_A^{abs} = 0.980$. With that and the other transmissivities calculated in Example 5.7, the irradiance of the non-absorbed direct solar radiation is found to be

$$E_n^{na} = 0.786 \cdot 1367 \frac{W}{m^2} 0.694 \cdot 0.9811 \cdot 0.8715 \cdot 0.9861 \cdot 0.980 = 616 \frac{W}{m^2} \quad .$$

For the Rayleigh fraction, the irradiance then follows as $E_d^{Ra} = 35.0 \, W/m^2$. With

$$F_A = 0.45 + 0.65 \cos 46.05° = 0.90$$

we obtain the irradiance for the aerosol fraction as $E_d^{Ae} = 95.3 \, W/m^2$.

The reflectivity of the atmosphere is $r_{At} = 0.086$, and with that the fraction due to multiple reflection is calculated to be

$$E_d^{MR} = (559 + 35.0 + 95.3) \frac{W}{m^2} \frac{0.25 \cdot 0.086}{1 - 0.25 \cdot 0.086} = 15.1 \frac{W}{m^2} \quad .$$

The irradiance of the diffuse sky-radiation is therefore

$$E_d^{Ra} + E_d^{Ae} + E_d^{MR} = 145 \, W/m^2 \quad ,$$

which is 26 % of the irradiance of the direct solar radiation. The global irradiance is $E_G = 704 \, W/m^2$.

If we assume a less turbid atmosphere, like in Example 5.7, with $\tau_A = 0.8766$, then $\tau_A^{abs} = 0.987$ and $r_{At} = 0.080$. This yields $E_d^{Ra} = 35.2 \, W/m^2$, $E_d^{Ae} = 61.8 \, W/m^2$ and $E_d^{MR} = 14.3 \, W/m^2$. The irradiance of the diffuse radiation, with $111 \, W/m^2$, is smaller, whilst the irradiance of the direct radiation, calculated as in Example 5.7, has increased to $605 \, W/m^2$. The global radiation is, with $E_G = 716 \, W/m^2$, only 1.7 % larger than in the more turbid atmosphere.

5.4.5 Absorptivities for solar radiation

The spectral emissivity $\varepsilon_\lambda(\lambda, T)$ of almost all substances assumes considerably different values at small wavelengths below $2 \, \mu m$, than at larger wavelengths. The spectral emissivities of electrical insulators are generally significantly smaller, and the spectral emissivities of metals a little larger than at large wavelengths. Assuming a diffuse radiating surface (Lambert radiator), this behaviour is also true of the spectral absorptivity $a_\lambda(\lambda, T)$, because $a_\lambda(\lambda, T) = \varepsilon_\lambda(\lambda, T)$. It is therefore to be expected that most substances behave differently in the absorption of solar radiation than in the absorption of predominantly long-wave radiation from earthly radiation sources.

The model of a *grey* Lambert radiator, with $a(T) = \varepsilon(T)$ can therefore not be applied to the absorption of solar radiation. Rather, it is to be expected that the absorptivities deviate vastly from the tabulated emissivities. These absorptivities a_S for the absorption of predominantly short-wave solar radiation, generally have to be determined by special measurements. Some results from such measurements are put together in Table 5.8. Further data for 'natural' surfaces like corn fields, different soil types, forests, snow and ice can be found in K.Y. Kondratyew [5.44], in the form of reflectivity $r_S = 1 - a_S$, that is also known as the albedo in meteorological circles, see also [5.34].

For solar technology applications, the ratio a_S/ε of a surface is of great importance. It should be large for solar collectors, so that the radiation to the surroundings proportional to ε is small compared to the absorbed solar radiation. If, in contrast, the surface under solar radiation is to assume a low temperature, a_S/ε should be as small as possible, which, for example, can be achieved by painting the surface white, ($a_S = 0.22$; $\varepsilon = 0.92$) giving $a_S/\varepsilon = 0.24$.

Table 5.8: Absorptivity a_S for solar radiation and (total) emissivity $\varepsilon = \varepsilon(300\,\mathrm{K})$ of different materials

Material	a_S	ε	a_S/ε	Material	a_S	ε	a_S/ε
Aluminium, polished	0.20	0.08	2.5	Asphalt, Road covering	0,93		
Chrome, polished	0.40	0.07	5.7	Leaves, green	0,71...0,79	0,86	0,83...0,92
Iron, galvanised	0.38			Tar paper, black	0,82	0,91	0,90
rough	0.75	0.82	0.91	Earth, ploughed	0,75		
Gold, polished	0.29	0.026	11.1	Paint			
Copper, polished	0.18	0.03	6.0	Zinc white	0,22	0,92	0,24
oxidised	0.70	0.45	1.56	Black oil paint	0,90	0,92	0,98
Magnesium, polished	0.19	0.12	1.6	Marble, white	0,46	0,90	0,51
Nickel, polished	0.36	0.09	4.0	Slate	0,88	0,91	0,97
Platinum, shiny	0.31	0.07	4.4	Snow, clean	0,20...0,35	0,95	0,21...0,37
Silver, polished	0.13	0.018	7.2	Brick, red	0,75	0,93	0,81

5.5 Radiative exchange

In heat transfer by radiation, energy is not only transported from hot to cold bodies; the colder body also emits radiation that strikes the warmer body and can be absorbed there. An *exchange* of energy takes place, in contrast to the transfer that occurs in heat conduction and convection. This radiative exchange depends on the mutual position and orientation of the radiating surfaces, their temperatures and there radiative properties. In the following sections it is assumed that the radiating surfaces are separated by a medium that has no effect on the radiative exchange, that neither absorbs, emits nor scatters radiation. This condition is exactly satisfied by a vacuum, although most gases also have little effect on radiative exchange. We will discuss gas radiation in section 5.6.

Even when the medium between the surfaces has no influence on the radiative exchange, the calculations are still difficult if the directional and wavelength dependence of the absorbed, emitted and reflected radiation is to be considered exactly. We will therefore use the grey Lambert radiator from 5.3.2.3 as a model in the following discussions, and, (with the exception of 5.5.4), assume that reflection is diffuse. The complex calculation for radiative exchange between two surfaces that are neither diffuse nor grey radiators is dealt with extensively by R. Siegel et al., [5.37], [5.45].

The geometric relationships for radiative exchange between grey Lambert radiators are described by the view factor, which we will look at in section 5.5.1. The section after that deals with radiative exchange between black bodies, a simple case due to the fact that no reflection takes place. Radiative exchange between any two isothermal areas that behave like grey Lambert radiators is explained in 5.5.3. Finally, the last section covers the insulatory effect of radiation shields where we will also consider specularly reflecting surfaces.

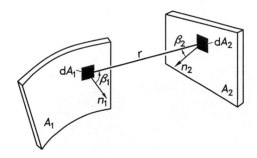

Fig. 5.50: Geometric quantities for
the calculation of the view factor

5.5.1 View factors

The calculation of radiative exchange between two surfaces requires a quantity
that describes the influence of their position and orientation. This is the *view
factor*, which is also known by the terms configuration factor or angle factor. The
view factor indicates to what extent one surface can be "seen" by another, or
more exactly, what proportion of the radiation from surface 1 falls on surface 2.

The first step in the calculation of the view factor is to determine the radiation
flow $d^2\Phi_{12}$, emitted from surface element dA_1 that strikes surface element dA_2,
Fig. 5.50. With L_1 as the intensity of the radiation emitted from dA_1, from (5.11),
we get

$$d^2\Phi_{12} = L_1 \cos\beta_1 \, dA_1 \, d\omega_2 \ .$$

Here, $d\omega_2$ is the solid angle at which the surface element dA_2 appears to dA_1:

$$d\omega_2 = \frac{dA_{2n}}{r^2} = \frac{\cos\beta_2 \, dA_2}{r^2} \ .$$

This produces

$$d^2\Phi_{12} = L_1 \frac{\cos\beta_1 \cos\beta_2}{r^2} dA_1 \, dA_2 \ . \tag{5.128}$$

This relationship is also known as the *photometric fundamental law*. According
to this, the radiation that reaches dA_2 decreases with the square of the distance
r between radiation source and receiver. In addition to this, the orientation of
the surface elements to the straight line between them is of importance. This is
expressed in terms of a cosine function of the two polar angles β_1 and β_2.

We will now calculate the radiation that is emitted by the finite surface 1 that
strikes surface 2, Fig. 5.50. This involves the assumption that the intensity L_1 is
constant over the entire surface 1. Integration of (5.128) over both surfaces yields

$$\Phi_{12} = L_1 \int\limits_{A_1} \int\limits_{A_2} \frac{\cos\beta_1 \cos\beta_2}{r^2} \, dA_1 \, dA_2 \ . \tag{5.129}$$

This is the radiation flow emitted by 1 that falls on 2. With

$$\Phi_1 = \pi L_1 A_1$$

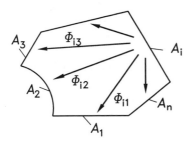

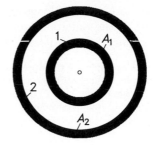

Fig. 5.51: Enclosure and radiation flows Φ_{ij}, emitted from the area A_i

Fig. 5.52: Enclosure formed by the concentric spherical areas 1 and 2

as the radiation flow emitted by surface 1 into the hemisphere, the *view factor* is obtained as

$$F_{12} := \frac{\Phi_{12}}{\Phi_1} = \frac{1}{\pi A_1} \int\limits_{A_1} \int\limits_{A_2} \frac{\cos\beta_1 \cos\beta_2}{r^2}\, \mathrm{d}A_1\, \mathrm{d}A_2 \ . \tag{5.130}$$

This quantity gives the proportion of the radiation emitted by surface 1 that falls on surface 2. The view factor is only dependent on the geometry. This is the result of the limiting asssumption of constant intensity L_1: Equation (5.130) is only valid if surface 1 radiates diffusely, has a constant temperature and the same radiation properties over the entire area.

If the indices 1 and 2 are exchanged in (5.130), then

$$F_{21} = \frac{\Phi_{21}}{\Phi_2} = \frac{1}{\pi A_2} \int\limits_{A_2} \int\limits_{A_1} \frac{\cos\beta_1 \cos\beta_2}{r^2}\, \mathrm{d}A_1\, \mathrm{d}A_2 \tag{5.131}$$

is obtained as the proportion of the radiation flow emitted by surface 2 (at constant intensity L_2!), that strikes surface 1. The equations (5.130) and (5.131) provide the important *reciprocity rule for view factors*,

$$A_1 F_{12} = A_2 F_{21} \ . \tag{5.132}$$

This means that only one of the two view factors has to be determined by the generally very complicated integration of (5.130) or (5.131).

A further relationship between view factors can be found when n areas, for each of which $L_i = \text{const}$ holds, form an enclosure such as that illustrated schematically in Fig. 5.51. From the radiation balance for area i,

$$\Phi_{i1} + \Phi_{i2} + \cdots + \Phi_{in} = \Phi_i \ ,$$

and by dividing by Φ_i, the summation rule

$$\sum_{j=1}^{n} F_{ij} = 1 \ , \qquad i = 1, 2, \ldots n \tag{5.133}$$

is obtained. F_{ii} also belongs to the view factors in the sum. It tells us what proportion of the radiation emitted by i strikes i. $F_{ii} \neq 0$ is only possible for a concave surface; it "sees itself". For flat and convex surfaces, we have $F_{ii} = 0$.

A simple example for the application of the relationships (5.132) and (5.133) is provided by radiation in an enclosure formed by two spherical surfaces 1 and 2, Fig. 5.52. There are four view factors in this case, F_{11}, F_{12}, F_{21} and F_{22}. The summation rule is applied to the inner sphere in order to calculate them:

$$F_{11} + F_{12} = 1 \ .$$

As surface 1 is convex, we have $F_{11} = 0$, and it follows that $F_{12} = 1$: All the radiation emitted by 1 strikes the outer sphere surface 2. The factor F_{21} is obtained using the reciprocity rule (5.132) with A_1 and A_2 as the surfaces of the two spheres:

$$F_{21} = \frac{A_1}{A_2} F_{12} = \frac{A_1}{A_2} = \left(\frac{r_1}{r_2}\right)^2 < 1 \ .$$

The fourth view factor is found by applying the summation rule $F_{21} + F_{22} = 1$ to the outer sphere

$$F_{22} = 1 - F_{21} = 1 - (A_1/A_2) = 1 - (r_1/r_2)^2 \ .$$

It is not equal to zero because part of the radiation emitted by the outer spherical surface also strikes it again.

View factors are not always as easy to find as for the simple geometry present in the example we have just looked at. Then the multiple integral in (5.130) has to be evaluated. However, not all the view factors have to be calculated in this manner. In an enclosure bounded by n surfaces there are n^2 view factors in total. From these, n view factors can be found by the application of the summation rule (5.133) on each of the n surfaces. In addition to this, $n(n-1)/2$ view factors can be determined using the reciprocity rule (5.132). Therefore the number of view factors that have to be calculated from (5.130) is only

$$n^2 - n - n(n-1)/2 = n(n-1)/2 \ .$$

This number is reduced even further by the number of flat or convex surfaces for which $F_{ii} = 0$ holds.

The multiple integral in (5.130) has been calculated for a large number of geometrical arrangements, most of which yield equations that are difficult to evaluate. The methods that should be applied to obtain these equations are reported by R. Siegel [5.45], and other authors. Some examples of calculated view factors are presented in Table 5.9. A larger collection of view factors can be found in R. Siegel and others [5.45] with numerous information on sources, in the VDI-Heat Atlas [5.46] and in J.R. Howell [5.47].

Example 5.9: Calculate the view factors for the inside of a cylinder, according to Fig. 5.53, with $r = 0.10\,\text{m}$ and $h = 0.25\,\text{m}$.

In an enclosure bounded by three surfaces there are 9 view factors. Of these only 3 need to be calculated according to (5.130). As the end areas 1 and 2 are flat, we have $F_{11} = 0$ and $F_{22} = 0$, so that only one view factor has to be determined by evaluating the double integral from (5.130). This is the view factor F_{12}; it is found from Table 5.9 (two equally sized, parallel concentric circular discs) with $z = 2 + (h/r)^2 = 8.25$ to be

$$F_{12} = \frac{1}{2}\left(z - \sqrt{z^2 - 4}\right) = 0.123 \ .$$

This means that $F_{13} = 1 - F_{12} = 0.877$.

From symmetry (or by applying the reciprocity rule), we find that $F_{21} = F_{12} = 0.123$ and

Table 5.9: View factors F_{12} for selected geometric arrangements

---

Two infinitely long, parallel strips, with centre lines that lie vertically above one another

$x = b_1/h \ ; \ y = b_2/h$

$$F_{12} = \frac{1}{2x}\left[\sqrt{(x+y)^2 + 4} - \sqrt{(y-x)^2 + 4}\right]$$

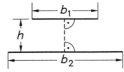

Two infinitely long strips, perpendicular to each other with a common edge

$$F_{12} = \frac{1}{2}\left[1 + \frac{b_1}{b_2} - \sqrt{1 + (b_2/b_1)^2}\right]$$

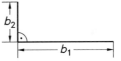

Two identical, parallel rectangles lying opposite each other

$x = a/h \ ; \ y = b/h$

$$F_{12} = \frac{2}{\pi xy}\left[\frac{1}{2}\ln\frac{(1+x^2)(1+y^2)}{1+x^2+y^2}\right.$$

$$+ x\sqrt{1+y^2}\arctan\frac{x}{\sqrt{1+y^2}}$$

$$\left. + y\sqrt{1+x^2}\arctan\frac{y}{\sqrt{1+x^2}} - x\arctan x - y\arctan y\right]$$

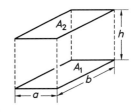

Two rectangles perpendicular to each other with a common edge

$x = b_1/a \ ; \ y = b_2/a$

$$F_{12} = \frac{1}{\pi x}\left[x\arctan\frac{1}{x} + y\arctan\frac{1}{y}\right.$$

$$- \sqrt{x^2+y^2}\arctan\frac{1}{\sqrt{x^2+y^2}}$$

$$+ \frac{1}{4}\ln\frac{(1+x^2)(1+y^2)}{1+x^2+y^2}$$

$$\left. + \frac{x^2}{4}\ln\frac{x^2(1+x^2+y^2)}{(1+x^2)(1+y^2)} + \frac{y^2}{4}\ln\frac{y^2(1+x^2+y^2)}{(1+x^2)(1+y^2)}\right]$$

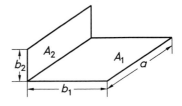

Table 5.9: (continued)

Two parallel circular disks with common central vertical

$$x = r_1/h \;\; ; \;\; y = r_2/h$$
$$z = 1 + \left(1 + y^2\right)/x^2$$

$$F_{12} = \frac{1}{2} \left[z - \sqrt{z^2 - 4\left(y/x\right)^2} \right]$$

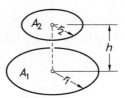

An infinitely long strip and an infinitely long cylinder parallel to it

$$F_{12} = \frac{r}{b-a} \left(\arctan \frac{b}{h} - \arctan \frac{a}{h} \right)$$

Two infinitely long, parallel cylinders with equal diameters

$$x = h/2r$$

$$F_{12} = F_{21} = \frac{1}{\pi} \left(\sqrt{x^2 - 1} + \arcsin \frac{1}{x} - x \right)$$

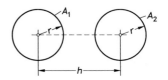

A sphere and a circular disk, whose central vertical goes through the centre of the sphere

$$F_{12} = \frac{1}{2} \left(1 - \frac{1}{\sqrt{1 + \left(r_2/h\right)^2}} \right)$$

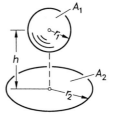

Two areas on the inner side of a hollow sphere

$$F_{12} = \frac{A_2}{4\pi r^2}$$

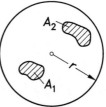

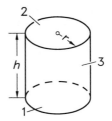

Fig. 5.53: Hollow cylinder with end areas 1 and 2 and body surface 3

$F_{23} = 1 - F_{21} = 0.877 = F_{13}$. From the reciprocity rule

$$A_3 F_{31} = A_1 F_{13}$$

follows

$$F_{31} = \frac{A_1}{A_3} F_{13} = \frac{\pi r^2}{2\pi rh} F_{13} = 0.1754 \ .$$

With $F_{32} = F_{31}$ (symmetry!), from

$$F_{31} + F_{32} + F_{33} = 1$$

we finally get

$$F_{33} = 1 - 2F_{31} = 0.649 \ .$$

5.5.2 Radiative exchange between black bodies

The calculation of radiative exchange is simplified if black bodies are considered, because no reflection occurs and the entire incident radiation flow is absorbed. Besides that, the intensity L_s of a black body is only dependent on its temperature. Therefore the intensity is constant on an isothermal surface of a black body. This was a prerequisite for the calculation of view factors in the previous section.

Fig. 5.54: Radiation flows Φ_{12} and Φ_{21} in direct radiative exchange between black bodies 1 and 2

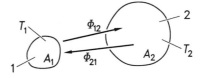

We will calculate first the *direct radiative interchange* between two black bodies of arbitrary shape with surfaces A_1 and A_2 and uniform temperatures T_1 and T_2, Fig. 5.54. In this case, all radiation flows emitted by one body that do not strike the other body will be ignored. The proportion of the radiation flow emitted by 1 and incident on 2 is given by

$$\Phi_{12} = A_1 F_{12} \sigma T_1^4 \ .$$

This energy flow is absorbed by black body 2. The radiation flow received and absorbed by black body 1, from body 2, is

$$\Phi_{21} = A_2 F_{21} \sigma T_2^4 \; .$$

The net radiation flow transferred by direct radiative interchange from 1 to 2 is therefore

$$\Phi_{12}^* = \Phi_{12} - \Phi_{21} = A_1 F_{12} \sigma T_1^4 - A_2 F_{21} \sigma T_2^4 \; .$$

If both bodies are at the same temperature $T_1 = T_2$, then no (net) energy flow will be transferred betweeen them: $\Phi_{12}^* = 0$. This then yields

$$A_1 F_{12} = A_2 F_{21} \; ,$$

i.e. the reciprocity rule, (5.132), for view factors, previously derived in another manner. This gives

$$\Phi_{12}^* = A_1 F_{12} \sigma \left(T_1^4 - T_2^4 \right) = A_2 F_{21} \sigma \left(T_1^4 - T_2^4 \right) \; . \tag{5.134}$$

The net radiation flow transferred by direct radiative exchange between two black bodies is proportional to the difference of the fourth powers of their thermodynamic temperatures.

We will now consider a hollow enclosure surrounded by walls consisting of several parts each with an *isothermal* surface, Fig. 5.55. According to H.C. Hottel and A.F. Sarofim [5.48], these isothermal sections of the surrounding walls are called *zones*. Non-isothermal walls with continuously changing temperatures can be approximated by a series of sufficiently small zones each at a different temperature. In the following all zones are assumed to be black surfaces. An opening in the enclosure is viewed as a zone with such a temperature that its radiation corresponds to the radiation, being assumed to be black, coming from outside through the opening into the enclosure.

In order to maintain a steady state, each zone has to have a heat flow supplied (or removed) from outside, to make up the difference between the emitted radiation flows and the sum of all the incident (and absorbed) radiation flows. The energy balance for zone i with surface A_i and temperature T_i is

$$\dot{Q}_i = A_i \sigma T_i^4 - \sum_{j=1}^{n} A_j F_{ji} \sigma T_j^4 \; . \tag{5.135}$$

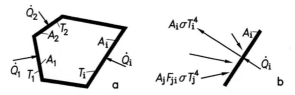

Fig. 5.55: a Hollow enclosure bounded by black radiating edges. b Illustration of the energy balance for the zone i

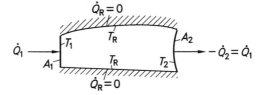

Fig. 5.56: Enclosure formed from radiation source 1, radiation receiver 2 and reradiating walls with $\dot{Q}_R = 0$

By applying the reciprocity rule (5.132) the following is obtained

$$\dot{Q}_i = A_i \sigma T_i^4 - \sum_{j=1}^{n} A_i F_{ij} \sigma T_j^4 \ .$$

Since, according to (5.133), $\sum_{j=1}^{n} F_{ij} = 1$, it also holds that

$$\dot{Q}_i = A_i \sigma \sum_{j=1}^{n} F_{ij} \left(T_i^4 - T_j^4 \right) \quad ; \qquad i = 1, 2, \ldots n \ . \tag{5.136}$$

The heat flow supplied to zone i from outside (or, with $\dot{Q}_i < 0$, released to the outside), is the sum of the net radiation flows Φ_{ij}^* from (5.134) between zone i and the other zones that bound the enclosure.

A zone with $\dot{Q}_i > 0$ is called a (net) *radiation source*, as it emits more radiation than it absorbs. A zone with $\dot{Q}_i < 0$ is a (net) *radiation receiver*, that absorbs more radiation than it emits. An adiabatic zone ($\dot{Q}_i = 0$) with respect to the outside is known as a *reradiating wall*. Its temperature is such that it emits just as much radiation as it absorbs from radiation incident upon it (radiative equilibrium).

The heat flows for all the zones can be found from (5.136) for given temperatures. If, on the contrary, some of the heat flows are known, the n balance equations (5.136) represent a linear system of equations, from which all the unknown temperatures and heat flows can be determined.

An enclosure with only three zones is often a good approximation for the case of a radiation source of area A_1 and temperature T_1 in radiative exchange with a radiation receiver of area A_2 and temperature $T_2 < T_1$, cf. Fig. 5.56. In addition to this walls that are adiabatic with respect to the outside also participate in the radiative exchange. These can be roughly assigned a unified temperature, T_R. The reradiating walls that enclose the space are combined here into a single zone with T_R and $\dot{Q}_R = 0$.

The following balance equations are valid for this hollow enclosure with three black radiating zones:

$$\dot{Q}_1 = A_1 \sigma \left[F_{12} \left(T_1^4 - T_2^4 \right) + F_{1R} \left(T_1^4 - T_R^4 \right) \right] \ , \tag{5.137a}$$
$$\dot{Q}_2 = A_2 \sigma \left[F_{21} \left(T_2^4 - T_1^4 \right) + F_{2R} \left(T_2^4 - T_R^4 \right) \right] \ , \tag{5.137b}$$
$$0 = A_R \sigma \left[F_{R1} \left(T_R^4 - T_1^4 \right) + F_{R2} \left(T_R^4 - T_2^4 \right) \right] \ . \tag{5.137c}$$

With the reciprocity rule (5.132) it follows from here that $\dot{Q}_2 = -\dot{Q}_1$, which is also yielded from the balance for the entire enclosure. The temperature of the reradiating zone is obtained from (5.137c) as

$$T_R^4 = \frac{A_1 F_{1R} T_1^4 + A_2 F_{2R} T_2^4}{A_1 F_{1R} + A_2 F_{2R}} \tag{5.138}$$

and by elimination of T_R^4 from (5.137a)

$$\dot{Q}_1 = -\dot{Q}_2 = A_1 \overline{F}_{12} \sigma \left(T_1^4 - T_2^4\right) \tag{5.139}$$

with the modified view factor

$$\overline{F}_{12} = F_{12} + \frac{F_{1R} F_{2R}}{(A_1/A_2)F_{1R} + F_{2R}} . \tag{5.140}$$

As comparison with (5.134) shows, the heat flow $\dot{Q}_1$ transferred from 1 to 2 is increased compared to the net radiation flow Φ_{12}^* due to the reradiating walls, because $\overline{F}_{12} > F_{12}$. If the radiation source and receiver have flat or convex surfaces ($F_{11} = 0$, $F_{22} = 0$), then the view factors F_{1R} and F_{2R} can lead back to F_{12} and instead of (5.140)

$$\overline{F}_{12} = \frac{1 - (A_1/A_2)F_{12}^2}{1 - 2(A_1/A_2)F_{12} + A_1/A_2} \tag{5.141}$$

is obtained. Only one view factor, namely F_{12} is required to calculate $\dot{Q}_1$.

Example 5.10: The hollow cylinder from Example 5.9 has black radiating walls. The two ends are kept at temperatures $T_1 = 550\,\mathrm{K}$ and $T_2 = 300\,\mathrm{K}$. The body area 3 is adiabatic, $\dot{Q}_3 = \dot{Q}_R = 0$. Calculate the heat flow $\dot{Q}_1$ and the temperature $T_3 = T_R$ of the reradiating body area, if this is taken to be an approximately isothermal area (zone).

In order to determine the heat flow $\dot{Q}_1$ from (5.139), the modified view factor $\overline{F}_{12}$ is required. This can be calculated according to (5.141), because the two ends are flat. With $F_{12} = 0.123$ from Example 5.9 and $A_1/A_2 = 1$, $\overline{F}_{12} = 0.5615$ is obtained. This yields the following from (5.139)

$$\dot{Q}_1 = -\dot{Q}_2 = \pi\, 0.10^2\,\mathrm{m}^2 \cdot 0.5615 \cdot 5.67 \cdot 10^{-8}\frac{\mathrm{W}}{\mathrm{m^2K^4}} \left(550^4 - 300^4\right)\mathrm{K}^4 = 83.4\,\mathrm{W} .$$

The temperature of the reradiating shell area can be found from (5.138). With $F_{1R} = F_{13} = F_{23} = F_{2R}$ (symmetry!) and $A_1 = A_2$ we get

$$T_R^4 = \frac{1}{2}\left(T_1^4 + T_2^4\right)$$

and from that $T_R = 472\,\mathrm{K}$.

This is the temperature at which the assumed isothermal body area emits as much energy as it absorbs. In reality the temperature of the body area varies continuously between T_1 and T_2, radiation with different irradiance strikes each annular strip of infinitesimal width. So each strip has a different emissive power and assumes a temperature accordingly. The correct treatment of this type of radiative exchange process with continuously varying temperature (corresponding to an infinite number of infinitesimal zones) is mathematically very involved; see [5.45], p. 107–132 for more information.

5.5.3 Radiative exchange between grey Lambert radiators

If the bodies participating in radiative exchange cannot be assumed to be black bodies, then the reflected radiation flows also have to be considered. In hollow enclosures, multiple reflection combined with partial absorption of the incident radiation takes place. A general solution for radiative exchange problems without simplifying assumptions is only possible in exceptional cases. If the boundary walls of the hollow enclosure are divided into isothermal zones, like in 5.5.2, then a relatively simple solution is obtained, if these zones behave like grey Lambert radiators. Each zone is characterised purely by its hemispherical total emissivity $\varepsilon_i = \varepsilon_i(T_i)$, whilst $a_i = \varepsilon_i$ is valid for its absorptivity, and for the reflectivity we have $r_i = 1 - \varepsilon_i$. In addition to this the intensity is constant for each zone. The reflected radiation also has constant intensity, if diffuse reflection is assumed, cf. section 5.1.5. The sum of the radiation emitted and reflected from one zone therefore obeys the cosine law, so just like for black bodies, view factors can be used to describe the radiative exchange between the zones.

We will now investigate radiative exchange between the isothermal walls (zones) of the enclosure illustrated in Fig. 5.57. The temperature of some of the zones is known, for others the heat flow supplied from or released to the outside is given. The heat flows of the zones with known temperatures and the temperature of each zone with stipulated heat flow are what we are seeking. There are as many unknown quantities (temperatures or heat flows) as there are zones.

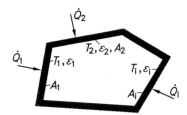

Fig. 5.57: Hollow enclosure bounded by isothermal surfaces (zones) each of which is a grey Lambert radiator

The energy balance equations for all the zones need to be established to solve this radiative exchange problem. This is done using the net-radiation method introduced by G. Poljak [5.49]. This yields a system of linear equations that, when solved, deliver the unknown temperatures and heat flows. With simple problems of only two or three zones, an electrical circuit analogy presented by A.K. Oppenheim [5.50] leads, in a simple manner, to the relationship between the temperatures and heat flows of the zones.

5.5.3.1 The balance equations according to the net-radiation method

According to G. Poljak, a new quantity has to be introduced when setting up the energy balance equation for a zone. It combines the radiation emitted and reflected by an isothermal surface i. It is made up by adding the emissive power M_i of the surface i and the reflected portion of its irradiance E_i:

$$H_i := M_i + r_i E_i = M_i + (1 - \varepsilon_i)E_i \ . \tag{5.142}$$

The quantity H_i is called the *radiosity* of the surface i, cf. E.R.G. Eckert [5.51].

Fig. 5.58: Illustration of the energy balance for
the zone i with area A_i

We will now set up the energy balance illustrated in Fig. 5.58 for a zone i. The
heat flow $\dot{Q}_i$ supplied from outside has to cover the difference between the emitted
and reflected radiation flow and the incident radiation flow. It holds, therefore,
that

$$\dot{Q}_i = A_i\,(H_i - E_i) \quad ; \tag{5.143}$$

the heat flow agrees with the net radiation flow. We now calculate the irradiance
E_i from (5.142) and put it into (5.143) with the result

$$\dot{Q}_i = \frac{A_i}{1-\varepsilon_i}\,(M_i - \varepsilon_i H_i) = \frac{A_i\varepsilon_i}{1-\varepsilon_i}\,\left(\sigma T_i^4 - H_i\right) \quad . \tag{5.144}$$

The emissivity ε_i appearing here is normally dependent on the temperature; it
has to be calculated at the temperature T_i of the zone i: $\varepsilon_i = \varepsilon_i(T_i)$.

A second relationship between $\dot{Q}_i$ and H_i is obtained, when the radiation flow
$A_i E_i$ incident on zone i is linked with the radiation flows emitted by the other
zones. The radiation flow $A_j H_j$ is sent out by zone j, but only the radiation flow
$A_j F_{ji} H_j$, multiplied by the view factor F_{ji}, strikes zone i. Therefore, the total
radiation striking zone i is

$$A_i E_i = \sum_{j=1}^{n} A_j F_{ji} H_j = A_i \sum_{j=1}^{n} F_{ij} H_j \quad ,$$

where the reciprocity rule (5.132) has been applied for the view factors. Putting
this expression into (5.143), and taking (5.133) into account, it follows that

$$\dot{Q}_i = A_i\left(H_i - \sum_{j=1}^{n} F_{ij} H_j\right) = A_i \sum_{j=1}^{n} F_{ij}\left(H_i - H_j\right) \quad . \tag{5.145}$$

The two balance equations (5.144) and (5.145) can be established for each
zone $(i = 1, 2, \ldots n)$. Therefore, $2n$ equations are available for the n unknown
radiosities H_i and the n required values for $\dot{Q}_i$ and T_i respectively. Before we go
into this equation system in 5.5.3.4, the next section offers solutions for the more
simple case of enclosures bounded by only two or three zones.

5.5.3.2 Radiative exchange between a radiation source, a radiation receiver and a reradiating wall

An enclosure surrounded by three isothermal surfaces (zones), like that shown
schematically in Fig. 5.59, serves as a good approximation for complicated cases

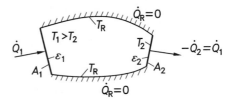

Fig. 5.59: Enclosure formed by a radiation
source 1, radiation receiver 2 and (adiabatic)
reradiating walls R

of radiative exchange. Zone 1 at temperature T_1 and with emissivity ε_1 is the
(net-) radiation source, it is supplied with a heat flow $\dot{Q}_1$ from outside. Zone 2
with temperature $T_2 < T_1$ and emissivity ε_2 is the radiation receiver, whilst the
third zone at temperature T_R, assumed to be spatially constant, is a reradiating
wall, ($\dot{Q}_R = 0$). The heat flow $\dot{Q}_1 = -\dot{Q}_2$ transferred by radiative exchange in the
enclosure is to be determined.

The solution of this problem starts with the writing of the two fundamental
balance equations (5.144) and (5.145) in the form

$$\dot{Q}_i = \frac{\sigma T_i^4 - H_i}{\dfrac{1 - \varepsilon_i}{A_i \varepsilon_i}} \tag{5.146}$$

and

$$\dot{Q}_i = \sum_{j=1}^{n} \frac{H_i - H_j}{\dfrac{1}{A_i F_{ij}}} \ . \tag{5.147}$$

We put these relationships in analogy to an electrical circuit. According to (5.146),
the "current" $\dot{Q}_i$, caused by the "potential difference" between σT_i^4 and H_i, flows
through a "conductor" with "resistance" $(1 - \varepsilon_i)/A_i \varepsilon_i$. This is illustrated in the
equivalent electrical circuit diagram in Fig. 5.60. Eq. (5.146) can be interpreted as
the current with a „potential" H_i splitting at a node into wires with the "geometric
resistances" $(1/A_i F_{ij})$ to the "potentials" H_j, see Fig. 5.61. The wire possible for
$F_{ii} \neq 0$ is missing, as due to $H_j = H_i$ no "current" flows.

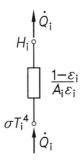

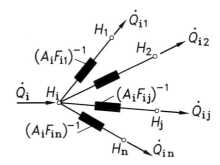

Fig. 5.60: Equivalent electrical circuit diagram for (5.146) with "reflec-
tion resistance" $(1 - \varepsilon_i)/A_i \varepsilon_i$

Fig. 5.61: Equivalent electrical circuit diagram
for (5.147): Current branching with the "geometric resistances" $(A_i F_{ij})^{-1}$

The radiative exchange in the enclosure from Fig. 5.59 can be replaced by the circuit diagram from Fig. 5.62. As the reradiating wall has no current, $(\dot{Q}_R = 0)$, the current $\dot{Q}_1$ of potential σT_1^4 flows to the node H_1, where it branches off, it flows directly and via H_R to the node H_2 and finally reaches the potential σT_2^4. Three resistances, placed one behind the other, lie between the end points with the potentials σT_1^4 and σT_2^4, whereby, because of the branching of the current, the central resistance $(A_1\overline{F}_{12})^{-1}$ between H_1 and H_2 is made up of the three individual resistances $(A_1 F_{12})^{-1}$, $(A_1 F_{1R})^{-1}$ and $(A_2 F_{2R})^{-1}$. As three series resistances are added, it holds that

$$\dot{Q}_1 = \frac{\sigma\left(T_1^4 - T_2^4\right)}{\dfrac{1-\varepsilon_1}{\varepsilon_1 A_1} + \dfrac{1}{A_1\overline{F}_{12}} + \dfrac{1-\varepsilon_2}{\varepsilon_2 A_2}} \ . \tag{5.148}$$

With parallel resistances the conductances are added together; it then follows that

$$A_1\overline{F}_{12} = A_1 F_{12} + \frac{1}{(A_1 F_{1R})^{-1} + (A_2 F_{2R})^{-1}} \ . \tag{5.149}$$

This relationship for $\overline{F}_{12}$ agrees with (5.140), which was derived in a different manner. If both surfaces 1 and 2 are flat or convex ($F_{11} = 0$ and $F_{22} = 0$), then $A_1\overline{F}_{12}$ can, according to (5.141), be calculated using only F_{12}, A_1 and A_2.

Equation (5.148) is often written in the form

$$\dot{Q}_1 = \varepsilon_{12} A_1 \sigma \left(T_1^4 - T_2^4\right) \ , \tag{5.150}$$

through which the *radiative exchange factor* ε_{12} is defined. From (5.148) we get

$$\frac{1}{\varepsilon_{12} A_1} = \frac{1-\varepsilon_1}{\varepsilon_1 A_1} + \frac{1}{A_1\overline{F}_{12}} + \frac{1-\varepsilon_2}{\varepsilon_2 A_2} \ . \tag{5.151}$$

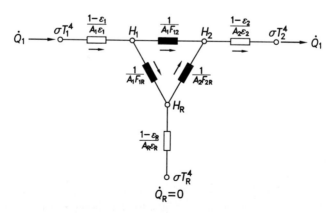

Fig. 5.62: Equivalent electrical circuit diagram for the radiative exchange in a hollow enclosure according to Fig. 5.59

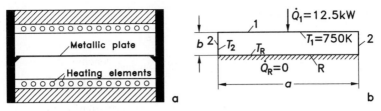

Fig. 5.63: **a** Electrically heated oven for the surface treatment of metal plates, **b** Hollow enclosure for the calculation of the radiative exchange of the top half of the oven

For emissivities dependent on the temperature, $\varepsilon_1 = \varepsilon_1(T_1)$ and $\varepsilon_2 = \varepsilon_2(T_2)$ should be used.

No current, $\dot{Q}_R = 0$, flows between the nodes with the potentials σT_R^4 and H_R, cf. Fig. 5.62. The resistance $(1 - \varepsilon_R)/\varepsilon_R A_R$ therefore has no effect and $\sigma T_R^4 = H_R$ is valid. The temperature T_R of a reradiating wall presents itself independent of its emissivity ε_R. The radiosity H_R required for its determination is found from the balance

$$A_1 F_{1R} (H_1 - H_R) = A_2 F_{2R} (H_R - H_2) \ ,$$

read off Fig. 5.62 as

$$H_R = \sigma T_R^4 = \frac{A_1 F_{1R} H_1 + A_2 F_{2R} H_2}{A_1 F_{1R} + A_2 F_{2R}} \ . \tag{5.152}$$

The radiosities H_1 and H_2 are obtained from (5.146) for $i = 1$ and $i = 2$, with $\dot{Q}_1$ from (5.150) and $\dot{Q}_2 = -\dot{Q}_1$.

Example 5.11: The electrically heated oven in Fig. 5.63a is used for the surface treatment of thin, square metal plates that are covered on both sides. The oven has a square base of side length $a = 1.50$ m. The radiation emitting surface of the heating elements has emissivity $\varepsilon = 0.85$; their distance from the metal plate is $b = 0.25$ m. 12.5 kW power is supplied to each of the two rows of heating elements being well insulated against the exterior. The non-insulated side walls of the oven have emissivity $\varepsilon = 0.70$. At steady-state the surface temperature of the heating elements reaches 750 K. Determine the temperature of the side walls and the temperature of the covered metal plate.

The symmetry of the construction means that it is sufficient to just consider the top half of the oven. It forms the schematically illustrated enclosure in Fig. 5.63b. It is bounded at the top by the heated square 1 with $\varepsilon_1 = 0.85$, at the side by the rectangular areas 2 with $\varepsilon_2 = 0.70$, which release heat to the outside, and below by the metal plate R. It is adiabatic as a result of symmetry, and represents a reradiating wall. We will assign the approximately uniform temperatures T_1, T_2 and T_R to these surfaces, such that the radiative exchange in a hollow enclosure bounded by three zones is to be calculated according to (5.148) or (5.151).

The first step is the determination of the view factors required for (5.149), F_{12}, F_{1R} and F_{2R}. The easiest to calculate is the view factor F_{1R} between two parallel squares lying one above the other (side length a) from Table 5.9. With $x = y = a/b = 6.0$ we obtain, from

$$F_{1R} = \frac{2}{\pi x^2} \left[\frac{1}{2} \ln \frac{(1 + x^2)^2}{1 + 2x^2} + 2x\sqrt{1 + x^2} \arctan \frac{x}{\sqrt{1 + x^2}} - 2x \arctan x \right] \ ,$$

the value $F_{1R} = 0.7326$. As $F_{11} = 0$, then $F_{12} = 1 - F_{1R} = 0.2674$. In order to determine F_{2R}, we consider that due to symmetry $F_{R2} = F_{12}$ is valid. Using the reciprocity rule (5.132), it then follows, with $A_R = A_1 = a^2$ and $A_2 = 4ab$, that

$$F_{2R} = \frac{A_R}{A_2} F_{R2} = \frac{A_1}{A_2} F_{12} = \frac{a}{4b} F_{12} = 0.4011 \ .$$

This yields from (5.149) the modified view factor $\overline{F}_{12} = 0.4633$.

We now calculate the radiative exchange factor ε_{12} from (5.151) and obtain

$$\frac{1}{\varepsilon_{12} A_1} = 1.3234 \ \mathrm{m}^{-2} \quad \text{and} \quad \varepsilon_{12} = 0.3358 \ .$$

The temperature T_2 of the four side walls is calculated from (5.150) with $\dot{Q}_1 = 12{,}5 \ \mathrm{kW}$:

$$T_2^4 = T_1^4 - \frac{\dot{Q}_1}{\varepsilon_{12} A_1 \sigma} \ .$$

This gives $T_2 = 396.3 \ \mathrm{K}$. The temperature T_R of the covered metal plates is obtained from their radiosity H_R as

$$T_R = (H_R/\sigma)^{1/4} \ . \tag{5.153}$$

The radiosities H_1 and H_2 are required for the calculation of H_R from (5.152). They are found from (5.146) to be

$$H_1 = \sigma T_1^4 - \frac{1 - \varepsilon_1}{\varepsilon_1 A_1} \dot{Q}_1 = 16.96 \ \frac{\mathrm{kW}}{\mathrm{m}^2}$$

and because $\dot{Q}_2 = -\dot{Q}_1$

$$H_2 = \sigma T_2^4 + \frac{1 - \varepsilon_2}{\varepsilon_2 A_2} \dot{Q}_1 = 4.970 \ \frac{\mathrm{kW}}{\mathrm{m}^2} \ .$$

It then follows from (5.152) that $H_R = 13.75 \ \mathrm{kW/m}^2$ and finally from (5.153), $T_R = 702 \ \mathrm{K}$.

5.5.3.3 Radiative exchange in a hollow enclosure with two zones

The relationships derived in the last section for the heat flow, $\dot{Q}_1$, transferred from a radiation emitter 1 to a receiver 2, are also valid for an enclosure that is only bounded by these two zones. As no reradiating zone is present, with $F_{1R} = 0$ and $F_{2R} = 0$ from (5.149), $\overline{F}_{12} = F_{12}$ is obtained. The heat flow transferred from 1 to 2 is

$$\dot{Q}_1 = \varepsilon_{12} A_1 \sigma \left(T_1^4 - T_2^4 \right) \ . \tag{5.154}$$

The radiative exchange factor ε_{12} is yielded from (5.151) to be

$$\frac{1}{\varepsilon_{12}} = \frac{1}{F_{12}} + \frac{1}{\varepsilon_1} - 1 + \frac{A_1}{A_2} \left(\frac{1}{\varepsilon_2} - 1 \right) . \tag{5.155}$$

These equations hold for several important, practical cases:

1. The area 1 is completely enclosed by the area 2, so that $F_{12} = 1$. For the radiative exchange factor we have now

$$\frac{1}{\varepsilon_{12}} = \frac{1}{\varepsilon_1} + \frac{A_1}{A_2} \left(\frac{1}{\varepsilon_2} - 1 \right) \ . \tag{5.156}$$

This result holds in particular for concentric spheres and very long concentric cylinders as here the assumption of *isothermal* surfaces applies more easily. If, however, body 1 lies eccentric in the enclosure surrounded by body 2, Fig. 5.64, then the two surfaces will generally not be isothermal, as the radiation flow is much higher in the regions where the two surfaces are close to each other than where a large distance exists between them.

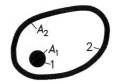

Fig. 5.64: Enclosure surrounded by body 2 with an eccentrically placed body 1

2. Surface 2 completely surrounds surface 1, ($F_{12} = 1$), and is black: $\varepsilon_2 = 1$. It now follows from (5.156) that $\varepsilon_{12} = \varepsilon_1$, the simple equation

$$\dot{Q}_1 = \varepsilon_1 A_1 \sigma \left(T_1^4 - T_2^4\right) \qquad (5.157)$$

for the transferred heat flow. As the surrounding shell does not reflect any radiation, the size A_2 of its surface has no effect on the radiative exchange.

3. Area 2 completely encloses area 1, ($F_{12} = 1$), and its surface is much larger than that of area 1: $A_2 \gg A_1$. With $A_1/A_2 \to 0$ it follows from (5.156) that $\varepsilon_{12} = \varepsilon_1$, with (5.157) for the heat flow. The radiative properties of a very large shell do not have any influence. The shell 2 appears like a black body to the small body 1.

4. For two very large, parallel plates, $F_{12} = 1$ is likewise valid, and in addition $A_1 = A_2$. From (5.156) the radiative exchange factor is found to be

$$\frac{1}{\varepsilon_{12}} = \frac{1}{\varepsilon_1} + \frac{1}{\varepsilon_2} - 1 \ . \qquad (5.158)$$

The second of these special cases of the body 1 completely enclosed by a black radiating shell 2 can also be generalised for an enclosed body with *any radiative properties*. The radiation flow emitted by it is

$$\Phi_1 = A_1 \varepsilon_1(T_1) \sigma T_1^4 \ .$$

The black shell absorbs this radiation flow completely. It emits the radiation flow $A_2 \sigma T_2^4$ itself, of which the proportion F_{21} strikes body 1. Body 1 absorbs the following radiation flow from the black radiation striking it

$$\Phi_{21} = a_1 \left(T_1, T_2\right) A_2 F_{21} \sigma T_2^4 \ .$$

Here, $a_1(T_1, T_2)$ is the hemispherical total absorptivity of body 1 for black radiation at temperature T_2. This gives

$$\dot{Q}_1 = \Phi_1 - \Phi_{21} = \varepsilon_1 \left(T_1\right) A_1 \sigma T_1^4 - a_1 \left(T_1, T_2\right) A_2 F_{21} \sigma T_2^4 \qquad (5.159)$$

for the heat flow transferred from body 1 to the black shell 2.

We will now consider the limiting case $T_2 = T_1$, for which $\dot{Q}_1 = 0$. Eq. (5.159) yields

$$a_1 \left(T_1, T_2\right) A_2 F_{21} = \varepsilon_1 \left(T_1\right) A_1 \ . \qquad (5.160)$$

According to section 5.3.2.2, the hemispherical total absorptivity of a body with any radiative properties is equal to its hemispherical total emissivity, if radiation from a black body at the same temperature strikes the body. This is the case here. It therefore follows from (5.160) that $A_2 F_{21} = A_1$. This corresponds to the reciprocity rule (5.132) with $F_{12} = 1$. Its application to this case was however not assured from the start as the intensity of body 1 is not constant.

We obtain now, as a generalisation of (5.157)

$$\dot{Q}_1 = A_1 \sigma \left[\varepsilon_1 \left(T_1 \right) T_1^4 - a_1 \left(T_1, T_2 \right) T_2^4 \right] \ . \tag{5.161}$$

For a grey Lambert radiator the absorptivity a_1 for every incident radiation is the same as the emissivity $\varepsilon_1(T_1)$, so that (5.157) is once again yielded.

5.5.3.4   The equation system for the radiative exchange between any number of zones

In complicated geometries the boundary walls of an enclosure must be divided into several zones. Non-isothermal walls also have to be split into a number of isothermal surfaces (= zones) in order to increase the accuracy of the results[13]. The equivalent electrical circuit diagram introduced in 5.5.3.2 would be confusing for this case. It is more sensible to set up and then solve a system of linear equation for the n radiosities of the n zones. The difficulty here is not the solving of the large number of equations in the system, but is the determination of the n^2 view factors that appear.

For n zones, $2n$ equations (5.144) and (5.145) with $i = 1, 2, \ldots n$ are valid. They are rearranged so that a system of n equations is yielded for the n radiosities H_i. The solution of this system of equations gives, from (5.144), the heat flow $\dot{Q}_i$ for each zone where the temperature T_i is given. The temperature of a zone with stipulated heat flow $\dot{Q}_i$ is found by solving (5.144) for T_i^4:

$$T_i^4 = \frac{1}{\sigma} \left[H_i + \frac{1 - \varepsilon_i \left(T_i \right)}{\varepsilon_i \left(T_i \right)} \frac{\dot{Q}_i}{A_i} \right] \ . \tag{5.162}$$

If, as this equation indicates, the emissivity ε_i depends on the temperature, iteration is required.

In order to obtain the equation system for the radiosities, (5.145) is rearranged:

$$H_i - \sum_{j=1}^{n} F_{ij} H_j = \dot{Q}_i / A_i \ . \tag{5.163}$$

With the Kronecker symbol

$$\delta_{ij} = \left\{ \begin{array}{ll} 0 & \text{for} \quad i \neq j \\ 1 & \text{for} \quad i = j \end{array} \right. \tag{5.164}$$

the linear equation follows

$$\sum_{j=1}^{n} \left(\delta_{ij} - F_{ij} \right) H_j = \dot{Q}_i / A_i \ . \tag{5.165}$$

---

[13]For demands of high accuracy, even this procedure of splitting the area with continuously varying temperature into a finite number of zones has to be avoided. However, the correct method for non-isothermal areas leads to complicated mathematical relationships (integral equations), which we will not go into here; [5.45], p. 107–132 is suggested for further reading.

It is used for the zones of the enclosure with given heat flow $\dot{Q}_i$ and for which the temperature T_i has to be determined.

For the zones where the temperature T_i is given, the unknown $\dot{Q}_i$ is eliminated by setting (5.144) equal to (5.145). After cancelling A_i this yields

$$H_i - \sum_{j=1}^{n} F_{ij} H_j = \frac{\varepsilon_i}{1 - \varepsilon_i} \left(\sigma T_i^4 - H_i \right) \quad ;$$

from which follows

$$H_i - (1 - \varepsilon_i) \sum_{j=1}^{n} F_{ij} H_j = \varepsilon_i \sigma T_i^4$$

or with the Kronecker symbol from (5.164)

$$\sum_{j=1}^{n} [\delta_{ij} - (1 - \varepsilon_i) F_{ij}] H_j = \varepsilon_i \sigma T_i^4 \quad . \tag{5.166}$$

This equation, linear in the radiosity, is used for the zones where the temperature T_i is given, with which $\varepsilon_i(T_i)$ can also be found, so all the coefficients and the right hand side are known.

The zones $1, 2, \ldots m$ shall have given temperatures and the zones $m+1, m+2, \ldots n$ shall have stipulated heat flows. The linear equation system for the radiosities becomes

$$\sum_{j=1}^{n} [\delta_{ij} - (1 - \varepsilon_i) F_{ij}] H_j = \varepsilon_i \sigma T_i^4 \quad , \qquad i = 1, 2, \ldots m \tag{5.167}$$

and

$$\sum_{j=1}^{n} (\delta_{ij} - F_{ij}) H_j = \dot{Q}_i / A_i \quad , \qquad i = m+1, m+2, \ldots n \quad . \tag{5.168}$$

This can be solved using known methods of linear algebra. With the radiosities H_i, the heat flows $\dot{Q}_i$ for $i = 1, 2, \ldots m$ are found using (5.144) and the temperatures T_i for $i = m+1, m+2, \ldots n$ from (5.162).

If some of the m areas with given temperatures are black, then the associated equations (5.167) with $\varepsilon_i = 1$ are reduced to $H_i = \sigma T_i^4$. These radiosities are known from the start so the number of unknown radiosities is accordingly reduced. Equ. (5.144) is not suitable for the calculation of heat flows $\dot{Q}_i$ for black bodies; $\dot{Q}_i$ is determined from the less convenient relationship (5.163).

Example 5.12: The radiative exchange in the hollow cylinder ($r = 0.10\,\mathrm{m}; h = 0.25\,\mathrm{m}$) from Examples 5.9 and 5.10 is to be investigated further. End 1 is black (as in Example 5.10), ($\varepsilon_1 = 1$) and has a temperature $T_1 = 550\,\mathrm{K}$. The other end has emissivity $\varepsilon_2 = 0.75$ and temperature $T_2 = 300\,\mathrm{K}$. The body area is adiabatic. In order to obtain higher accuracy the reradiating body area is split into equally sized zones 3 and 4, the position of which is indicated in Fig. 5.65. The heat flows $\dot{Q}_1$ and $\dot{Q}_2$ and the temperatures T_3 and T_4 of the two reradiating zones are to be determined.

Before we set up the equation system for the radiosities of the four zones we will calculate the 16 view factors F_{ij}. Obviously $F_{11} = F_{22} = 0$; the view factor $F_{12} = 0.1230$ is carried over from Example 5.9. In order to determine the other view factors the auxiliary area 5 in Fig. 5.65 with the dashed line is introduced. With $z = 2 + (h/2r)^2 = 3.5625$, F_{15} is calculated from the equation given in Example 5.9 as $F_{15} = 0.3072$. For the enclosure formed by 1, 3 and 5 we have $F_{13} + F_{15} = 1$, from which $F_{13} = 0.6928$. The summation for the entire cylinder is

$$F_{11} + F_{12} + F_{13} + F_{14} = 1 \quad ,$$

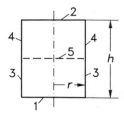

Fig. 5.65: Hollow cylinder with ends 1 and 2. The reradiating body area is divided into two equally sized zones 3 and 4. 5 imaginary auxiliary area for the calculation of the view factors

from which $F_{14} = 0.1841$ is obtained. Due to symmetry with respect to the auxiliary area 5, the view factors for area 2 are obtained from those calculated for 1 as $F_{21} = F_{12}$; $F_{22} = F_{11} = 0$; $F_{23} = F_{14}$; $F_{24} = F_{13}$.

The reciprocity rule is used for the determination of the view factors for area 3. This gives $F_{31} = (A_1/A_3)F_{13} = (r/h)F_{13} = 0.2771$ as well as $F_{32} = (A_2/A_3)F_{23} = (r/h)F_{14} = 0.0736$. Coming back to the enclosure formed by 1, 3 and 5, the summation rule

$$F_{31} + F_{33} + F_{35} = 1$$

gives, because of $F_{35} = F_{31}$ (symmetry!), the view factor $F_{33} = 1 - 2F_{31} = 0.4458$. The corresponding summation for the entire cylinder yields

$$F_{34} = 1 - F_{31} - F_{32} - F_{33} = 0.2035 .$$

Symmetry is also the reason for the view factors for area 4 being $F_{41} = F_{32}$; $F_{42} = F_{31}$; $F_{43} = F_{34}$ and $F_{44} = F_{33}$. Combining these results into a matrix makes things a lot clearer:

$$(F_{ij}) = \begin{bmatrix} 0 & 0.1230 & 0.6928 & 0.1841 \\ 0.1230 & 0 & 0.1841 & 0.6928 \\ 0.2771 & 0.0736 & 0.4458 & 0.2035 \\ 0.0736 & 0.2771 & 0.2035 & 0.4458 \end{bmatrix} .$$

As area 1 is black, the first of the two equations (5.167) is reduced to

$$H_1 = \sigma T_1^4 = 5188.9 \, \text{W/m}^2 .$$

The second equation from (5.167) with $i = m = 2$ and the two equations (5.168), with $\dot{Q}_3 = \dot{Q}_4 = 0$, take the form

$$
\begin{aligned}
H_2 &- (1 - \varepsilon_2) F_{23} H_3 &- (1 - \varepsilon_2) F_{24} H_4 &= \varepsilon_2 \sigma T_2^4 + (1 - \varepsilon_2) F_{21} H_1 \\
-F_{32} H_2 &+ (1 - F_{33}) H_3 &- &= F_{31} H_1 \\
-F_{42} H_2 &- F_{43} H_3 &+ (1 - F_{44}) H_4 &= F_{41} H_1 .
\end{aligned}
$$

With the values given and those already calculated we get

$$
\begin{aligned}
H_2 &- 0.0460 \, H_3 &- 0.1732 \, H_4 &= 504.0 \quad \text{W/m}^2 \\
-0.0736 \, H_2 &+ 0.5542 \, H_3 &- 0.2035 \, H_4 &= 1437.8 \quad \text{W/m}^2 \\
-0.2771 \, H_2 &- 0.2035 \, H_3 &+ 0.5542 \, H_4 &= 381.9 \quad \text{W/m}^2 .
\end{aligned}
$$

with the solution $H_2 = 1126.8 \, \text{W/m}^2$, $H_3 = 3703.3 \, \text{W/m}^2$ and $H_4 = 2612.3 \, \text{W/m}^2$.

The heat flow from the black area 1 is calculated from (5.163) to be

$$\dot{Q}_1 = A_1 (H_1 - F_{12}H_2 - F_{13}H_3 - F_{14}H_4) = 62.9 \, \text{W} .$$

The heat flow of the other end is yielded from (5.144) as

$$\dot{Q}_2 = \frac{A_2 \varepsilon_2}{1 - \varepsilon_2} \left(\sigma T_2^4 - H_2 \right) = -62.9 \, \text{W} .$$

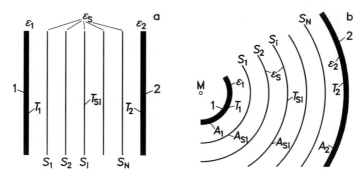

Fig. 5.66: a Flat radiation shields between two flat parallel walls 1 and 2. **b** Concentric radiation shields between concentric spheres or very long cylinders 1 and 2

It is the same, as it has to be, as $-\dot{Q}_1$. The temperatures of the reradiating walls 3 and 4 are found from $H_i = \sigma T_i^4$ as $T_3 = 506\,\mathrm{K}$ and $T_4 = 463\,\mathrm{K}$. As expected area 3, that is closer to the hot end 1, has a higher temperature than area 4.

We will now compare these results with the rough approximation in which the body area is treated as a single, reradiating zone of temperature T_R. From (5.150) and (5.151) we get $\dot{Q}_1 = -\dot{Q}_2 = 70.3\,\mathrm{W}$, a value greater by almost 12 %. The temperature of the body area of $T_R = 474\,\mathrm{K}$ lies between T_3 and T_4.

5.5.4 Protective radiation shields

Protective radiation shields are used to reduce the radiative exchange between walls at different temperatures: thin foils or sheets made of good reflecting materials are placed between the walls, Fig. 5.66. The spaces between the protective shields are normally evacuated so that heat transfer by convection is prevented. This multi-layer arrangement is used predominantly in cryogenic applications for the insulation of containers for very cold liquified gases.

The heat flux transferred between two very large, parallel, flat walls, according to (5.154) and (5.158), is given by

$$\dot{q} = \frac{\dot{Q}}{A} = \frac{\sigma\left(T_1^4 - T_2^4\right)}{\dfrac{1}{\varepsilon_1} + \dfrac{1}{\varepsilon_2} - 1} \ . \tag{5.169}$$

We will now consider the case of N radiation shields present between the walls 1 and 2. The emissivity ε_S shall have the same value on both sides of the shield and for all shields. As the shields are very thin, each shield can have a uniform temperature assigned to it. The following equations are obtained with T_{S_i} as the temperature of the i-th shield:

$$\dot{q}\left(\frac{1}{\varepsilon_1} + \frac{1}{\varepsilon_S} - 1\right) = \sigma\left(T_1^4 - T_{S1}^4\right) \ ,$$

$$\dot{q}\left(\frac{1}{\varepsilon_S} + \frac{1}{\varepsilon_S} - 1\right) = \sigma\left(T_{S1}^4 - T_{S2}^4\right) \ ,$$

$$\cdots\cdots\cdots\cdots\cdots\cdots\cdots\cdots , $$

$$\dot{q}\left(\frac{1}{\varepsilon_S} + \frac{1}{\varepsilon_2} - 1\right) = \sigma\left(T_{SN}^4 - T_2^4\right) \ .$$

The temperatures of the shields drop out of the right hand side when all the equations are added together, giving

$$\dot{q}\left[\frac{1}{\varepsilon_1} + \frac{1}{\varepsilon_2} - 1 + N\left(\frac{2}{\varepsilon_S} - 1\right)\right] = \sigma\left(T_1^4 - T_2^4\right)$$

or

$$\dot{q}(N) = \frac{\sigma\left(T_1^4 - T_2^4\right)}{\dfrac{1}{\varepsilon_1} + \dfrac{1}{\varepsilon_2} - 1 + N\left(\dfrac{2}{\varepsilon_S} - 1\right)} \ . \qquad (5.170)$$

As can immediately be seen, the heat flux is significantly reduced compared to the case without protective shields ($N = 0$). Table 5.10 shows examples for $\varepsilon_S = 0.05$, of how the ratio $\dot{q}(N)/\dot{q}(N = 0)$ decreases with N for different emissivities $\varepsilon_1 = \varepsilon_2$ of the outer walls. According to this, the effect of shielding is greater, the higher the emissivity $\varepsilon_1 = \varepsilon_2$.

Table 5.10: Ratio of the heat flux $\dot{q}(N)$ for N protective shields with $\varepsilon_S = 0.05$ to the heat flux $\dot{q}(N = 0)$ without protective shields between two flat walls with $\varepsilon_1 = \varepsilon_2 = \varepsilon$

ε	$N = 1$	2	5	10	20	50
0.1	0.3276	0.1959	0.0888	0.04645	0.02378	0.00965
0.2	0.1875	0.1034	0.0441	0.02256	0.01141	0.00459
0.4	0.0930	0.0488	0.0201	0.01015	0.00510	0.00205
0.6	0.0565	0.0290	0.0118	0.00595	0.00298	0.00120
0.8	0.0370	0.0189	0.0076	0.00383	0.00192	0.00077
1.0	0.0250	0.0127	0.0051	0.00256	0.00128	0.00052

We will now look at the radiative exchange between concentric cylinders or spheres. The heat flow transferred calculated from (5.154) and (5.156) is

$$\dot{Q} = \frac{A_1\sigma\left(T_1^4 - T_2^4\right)}{\dfrac{1}{\varepsilon_1} + \dfrac{A_1}{A_2}\left(\dfrac{1}{\varepsilon_2} - 1\right)} \ . \qquad (5.171)$$

N very thin protective shields with the same emissivity ε_S are placed concentrically between the cylinders or spheres. Following the same procedure as for flat radiation protection shields gives the reduced heat flow

$$\dot{Q} = \frac{A_1 \sigma \left(T_1^4 - T_2^4 \right)}{\dfrac{1}{\varepsilon_1} + \dfrac{A_1}{A_2} \left(\dfrac{1}{\varepsilon_2} - 1 \right) + \left(\dfrac{2}{\varepsilon_S} - 1 \right) \displaystyle\sum_{i=1}^{N} \dfrac{A_1}{A_{Si}}} \ . \tag{5.172}$$

Here, A_1 is the surface area of the *inner* wall and the index i of the shields rises from inside to outside.

Surfaces with low emissivities often exhibit approximately *mirrorlike or specular reflection* rather than diffuse reflection. We want to investigate how the assumption of mirrorlike reflection affects the heat transfer. The assumptions regarding the emission of diffuse and grey radiation remain unaltered. Grey Lambert radiators with mirrorlike reflection are therefore assumed.

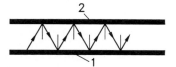

Fig. 5.67: Radiation pathways for mirrorlike reflection between two large, flat, parallel walls 1 and 2

As the pathways between two large, flat plates drawn schematically in Fig. 5.67 show, the radiation emitted by plate 1 always strikes plate 2 and continues to be reflected between the plates until it has been completely absorbed. The same is true for the radiation emitted by plate 2. The heat transfer is not different from that by diffuse reflection and equation (5.169) for the heat flux holds, regardless of whether one or both walls reflect diffusely or mirrorlike. The relevant relationship (5.170) for the heat flux with N radiation shields is applied without any changes for mirrorlike reflection. It is also valid when only the shields reflect mirrorlike and the plates diffusely.

Fig. 5.68 illustrates the radiation pathways between two concentric cylinders or spheres that reflect like mirrors. The radiation emitted from the inner area 1, (pathway a), always strikes the outer area 2 and is reflected such that it strikes area 1 again. $F_{12} = 1$ is valid. As in diffuse reflection, the radiation emitted by the inner area is continuously reflected between the two surfaces until it is completely absorbed.

This is different for the radiation emitted by the outer area 2; it either strikes the inner surface or bypasses it and falls back on the outer surface 2. This part, pathway b in Fig. 5.68, is specularly reflected in such a way that it never leaves area 2. It does not participate in the radiative exchange between the surfaces. The other part, given by the view factor $F_{21} = A_1 F_{12}/A_2 = A_1/A_2$, (pathway c), strikes area 1 and is mirrorlike reflected between the surfaces until it is completely absorbed. Therefore, the outer surface only contributes to the radiative exchange by mirrorlike reflection, on the scale as if its surface A_2 was reduced by the factor F_{21}: it has the effective surface area $A_2 F_{21} = A_1$.

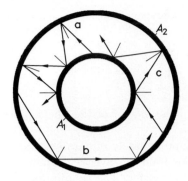

Fig. 5.68: Ray pathways in mirrorlike reflection between concentric spheres or very long cylinders 1 and 2

Therefore, if the outer surface 2 reflects mirrorlike, in the following equation for the heat flow, from (5.148) with $\bar{F}_{12} = F_{12} = 1$,

$$\dot{Q} = \frac{\sigma\left(T_1^4 - T_2^4\right)}{\dfrac{1 - \varepsilon_1}{\varepsilon_1 A_1} + \dfrac{1}{A_1} + \dfrac{1 - \varepsilon_2}{\varepsilon_2 A_2}} \ ,$$

in the third "resistance" of the denominator the area A_2 should be replaced by the effective area A_1, yielding

$$\dot{Q} = \frac{A_1 \sigma\left(T_1^4 - T_2^4\right)}{\dfrac{1}{\varepsilon_1} + \dfrac{1}{\varepsilon_2} - 1} \ . \tag{5.173}$$

The size of the outer area 2 is immaterial for radiative exchange. Eq. (5.173) also holds if the inner surface 1 reflects diffusely.

N concentric, thin radiation shields, with the same emissivity ε_{S}, are placed between the (diffuse or mirrorlike reflecting) inner area 1 and the diffuse reflecting outer area 2. The shields reflect mirrorlike. According to (5.173), the following balance equations hold

$$\frac{\dot{Q}}{A_1}\left(\frac{1}{\varepsilon_1} + \frac{1}{\varepsilon_{\mathrm{S}}} - 1\right) \qquad = \sigma\left(T_1^4 - T_{\mathrm{S}1}^4\right) \ ,$$

$$\frac{\dot{Q}}{A_{\mathrm{S}1}}\left(\frac{2}{\varepsilon_{\mathrm{S}}} - 1\right) \qquad = \sigma\left(T_{\mathrm{S}1}^4 - T_{\mathrm{S}2}^4\right) \ ,$$

$$\cdots\cdots\cdots\cdots\cdots\cdots\cdots\cdots\cdots\cdots$$

$$\frac{\dot{Q}}{A_{\mathrm{S}N-1}}\left(\frac{2}{\varepsilon_{\mathrm{S}}} - 1\right) \qquad = \sigma\left(T_{\mathrm{S}N-1}^4 - T_{\mathrm{S}N}^4\right) \ ,$$

$$\frac{\dot{Q}}{A_{\mathrm{S}N}}\left[\frac{1}{\varepsilon_{\mathrm{S}}} + \frac{A_{\mathrm{S}N}}{A_2}\left(\frac{1}{\varepsilon_2} - 1\right)\right] = \sigma\left(T_{\mathrm{S}N}^4 - T_2^4\right) \ .$$

The temperatures of the shields drop out when the equations are added together giving

$$\dot{Q}(N) = \frac{A_1 \sigma\left(T_1^4 - T_2^4\right)}{\dfrac{1}{\varepsilon_1} + \dfrac{A_1}{A_2}\left(\dfrac{1}{\varepsilon_2} - 1\right) + \left(\dfrac{2}{\varepsilon_{\mathrm{S}}} - 1\right)\displaystyle\sum_{i=1}^{N-1}\dfrac{A_1}{A_{\mathrm{S}i}} + \dfrac{1}{\varepsilon_{\mathrm{S}}}\left(1 + \dfrac{A_1}{A_{\mathrm{S}N}}\right) - 1} \ . \tag{5.174}$$

If the outer area 2 also reflects mirrorlike then A_2 in the second term of the denominator should be replaced by $A_{\mathrm{S}N}$ because this is the size of the effective surface area of 2 for mirrorlike reflection.

Example 5.13: A tube, with liquid nitrogen flowing through it, has an external diameter $d_1 = 30\,\text{mm}$. It emissivity is $\varepsilon_1 = 0.075$ and its temperature $T_1 = 80\,\text{K}$. The tube is surrounded by a second concentric tube with internal diameter $d_2 = 60\,\text{mm}$ with $\varepsilon_2 = 0.12$ and $T_2 = 295\,\text{K}$. The space between them has been evacuated. Determine the heat flow per tube length L that is transferred by radiation. The limiting cases of diffuse and mirrorlike reflection of the outer tube should be investigated.

With diffuse reflection, the desired heat flow is obtained from (5.171) to be

$$\frac{\dot{Q}}{L} = \frac{\pi d_1 \sigma \left(T_1^4 - T_2^4\right)}{\dfrac{1}{\varepsilon_1} + \dfrac{d_1}{d_2}\left(\dfrac{1}{\varepsilon_2} - 1\right)} = -2.368\,\frac{\text{W}}{\text{m}} \quad .$$

The minus sign indicates that the heat is transferred from outside to inside. The inner tube is cooled by the liquid nitrogen. In the case of mirrorlike reflection, from (5.173), the smaller value $\dot{Q}/L = -1.948\,\text{W/m}$ is obtained. As only a part of the outer area contributes to radiative exchange the insulation effect is greater.

A thin radiation protective shield of diameter $d_S = 45\,\text{mm}$ is introduced between the tubes to improve the insulation of the nitrogen pipe. This shield reflects like a mirror; its emissivity is $\varepsilon_S = 0.025$. Find the decrease in the incident heat flow.

Eqn. (5.174) with $N = 1$ is applied here, so that in the denominator the sum drops out:

$$\frac{\dot{Q}\,(N=1)}{L} = \frac{\pi d_1 \sigma \left(T_1^4 - T_2^4\right)}{\dfrac{1}{\varepsilon_1} + \dfrac{d_1}{d_2}\left(\dfrac{1}{\varepsilon_2} - 1\right) + \dfrac{1}{\varepsilon_S}\left(1 + \dfrac{d_1}{d_s}\right) - 1} = -0.487\,\frac{\text{W}}{\text{m}} \quad .$$

This holds for the diffuse reflecting outer cylinder. The heat flow has decreased significantly. The ratio $\dot{Q}(N=1)/\dot{Q}(N=0)$ has the value $0.487/2.368 = 0.206$. With a mirrorlike reflecting outer cylinder d_2 is replaced by d_S. This then gives $\dot{Q}(N=1)/L = -0.480\,\text{W/m}$ and $\dot{Q}(N=1)/\dot{Q}(N=0) = 0.480/1.948 = 0.246$. In specular reflection of the outer cylinder, the *relative* decrease of the heat flow caused by the protective shield is somewhat lower than that for a diffuse reflecting outer cylinder. However, the smallest absolute value of $\dot{Q}/L$ is yielded when both the shield and the outer cylinder reflect mirrorlike.

5.6 Gas radiation

A. Schack [5.52] was the first to recognise the technical importance of radiation from gases in 1924. He suggested that the radiation of the CO_2 and H_2O in combustion gases would contribute significantly to heat transfer in industrial ovens and in furnaces of steam generators. This was confirmed experimentally and between 1932 and 1942 the radiation of these gases was systematically investigated. The work by E. Schmidt [5.53] and E. Eckert [5.54], [5.55], in Germany and that by H.C. Hottel and his coworkers, [5.56] to [5.58], in the USA were of particular importance.

In technical applications, the gas radiation in the infrared, i.e. at wavelengths above $1\,\mu\text{m}$ is of interest. In this region, the main emitters are CO_2 and H_2O, although other gases like CO, SO_2, NH_3, CH_4 and further hydrocarbons also emit here. In contrast, N_2 and O_2, the main constituents of air allow radiation in the infrared region to pass through with virtually no attenuation; they do not absorb, and therefore according to Kirchhoff's law do not emit either.

Gases only absorb and emit radiation in narrow wavelength regions, so-called bands. Their spectral emissivities show a complex dependency on the wavelength, in complete contrast to solid bodies. This means that gases cannot be idealised as grey radiators without a loss of accuracy.

In the following we will deal with the absorption of radiation by gases and then look at the definitions of the absorptivities and emissivities of radiating gas spaces. These quantities are dependent on the size and shape of the gas space; their calculation entails complicated integrations. This is avoided by introducing the mean beam length of the gas space, the determination of which will also be explained here. This allows radiative exchange between isothermal gas volumes and their boundary walls to be calculated. Finally we will give some hints how radiative exchange in complicated cases, for example in combustion chambers and furnaces, can be determined.

5.6.1 Absorption coefficient and optical thickness

If radiation passes through an optically turbid gas or mixture of gases its energy is reduced due to absorption and scattering by the gas molecules. We will now assume that the radiation is only absorbed but not scattered. This applies to the infrared wavelengths because Rayleigh scattering by the molecules only takes place at very small wavelengths and is practically meaningless for $\lambda > 1\,\mu$m, cf. 5.4.2.2. In the gas mixture being considered, only one component may absorb radiant energy, for example CO_2 in the non-absorbing components N_2 and O_2.

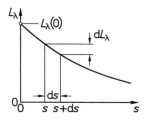

Fig. 5.69: Reduction in the spectral intensity L_λ with increasing beam length s as a result of absorption by a gas

Absorption causes a reduction in the spectral intensity L_λ with increasing beam length s of the gas, Fig. 5.69. This fall in the spectral intensity as radiation passes through the distance $\mathrm{d}s$ is described by

$$-\frac{\mathrm{d}L_\lambda}{L_\lambda} = k_G(\lambda, T, p, p_G)\,\mathrm{d}s \ , \tag{5.175}$$

through which the *spectral absorption coefficient* k_G of the absorbing gas is defined. It is dependent on the wavelength λ and the state of the gas, namely on its temperature T, the pressure p and the partial pressure p_G, which is a measure of the concentration of the absorbing gas in the mixture. In order to simplify the expression, we generally do without writing the dependency in an explicit way. We simply write k_G instead of $k_G(\lambda, T, p, p_G)$.

In a *non-homogeneous gas mixture* the properties vary along the path followed by the radiation. The absorption coefficient therefore has an indirect dependence on s. Integration of (5.175) between $s = 0$ and s yields

$$\ln \frac{L_\lambda(s)}{L_\lambda(s = 0)} = - \int_0^s k_G \, ds$$

or

$$L_\lambda(s) = L_\lambda(s = 0) \exp(- \textstyle\int_0^s k_G \, ds) \ . \tag{5.176}$$

The integral

$$\kappa_G := \int_0^s k_G(\lambda, T, p, p_G) \, ds \tag{5.177}$$

is known as the *optical thickness* of the gas layer with the (geometric) thickness given by s. In contrast to s, κ_G is a dimensionless quantity; it is a measure of the strength of the absorption of a gas layer of certain thickness s. A gas with $\kappa_G \to 0$ is called *optically thin*. With $\kappa_G = 7$, $L_\lambda(s)$ falls to less than 1 pro mille of the initial value $L_\lambda(s = 0)$; the gas almost completely absorbs the penetrating radiation.

A *homogeneous gas mixture* has constant intensive properties T, p and p_G over its entire volume. Its spectral absorption coefficient k_G is, therefore, independent of s. So, from (5.177), the optical thickness is found to be

$$\kappa_G = k_G(\lambda, T, p, p_G)s = k_G s \ , \tag{5.178}$$

and the decrease in the spectral intensity as

$$L_\lambda(s) = L_\lambda(s = 0) \exp\left(-k_G s\right) \ . \tag{5.179}$$

This equation corresponds to the law from P. Bouguer (1729), according to which the spectral intensity falls exponentially along the path radiation is passing through.

As the absorption of radiation takes place on the individual molecules of the gas, the assumption is suggested, that the spectral absorption coefficient κ_G is proportional to the molar concentration $c_G = N_G/V$ of the absorbing molecules or to the partial pressure $p_G = R_m T c_G$. The optical thickness of a homogeneous gas is then given by

$$\kappa_G = k_G s = k_G^*(\lambda, T, p)(p_G s) \ , \tag{5.180}$$

that is a proportionality to the product $(p_G s)$. This is known as the law from A. Beer (1854). It applies to some gases, e.g. CO_2, very well, but from other gases, in particular H_2O, it is not satisfied.

The spectral absorption coefficient k_G is also decisive for the radiation energy emitted by a volume element of the gas. $d^2\Phi_{\lambda,V}$ indicates the radiation flow emitted by a volume element dV of a homogenous gas in the wavelength interval $d\lambda$ into all directions. It holds for this that

$$d^2\Phi_{\lambda,V} = 4\pi k_G L_{\lambda s}(\lambda, T)\, dV\, d\lambda = 4\, k_G M_{\lambda s}(\lambda, T)\, dV\, d\lambda \ . \qquad (5.181)$$

The spectral intensity $L_{\lambda s}$ and the hemispherical spectral emissive power $M_{\lambda s}$ of a black body are given here by (5.50). A derivation of (5.181) can be found in R. Siegel and J.R. Howell [5.37], p. 531.

5.6.2   Absorptivity and emissivity

The enclosure schematically illustrated in Fig. 5.70 contains a *homogeneous* gas mixture with an absorbent component. The element dA on the surface of the gas volume, shown in Fig. 5.70, will be used for the definition and calculation of its directional spectral absorptivity $a'_{\lambda,G}$. The radiation emitted from dA, with the spectral intensity L_λ, is weakened by absorption. Depending on the direction, the path through the gas is of different lengths, which according to (5.179) leads to varying reductions in L_λ.

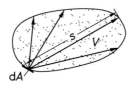

Fig. 5.70: For the calculation of the spectral absorptivity in a gas space

The spectral absorptivity $a'_{\lambda,G}$ belonging to a certain direction is the ratio of the energy absorbed in the distance s, to the energy emitted:

$$a'_{\lambda,G} := \frac{L_\lambda(s=0) - L_\lambda(s)}{L_\lambda(s=0)} = 1 - \frac{L_\lambda(s)}{L_\lambda(s=0)} = 1 - \tau'_{\lambda,G} \ . \qquad (5.182)$$

The ratio $L_\lambda(s)/L_\lambda(s=0)$ is the directional spectral transmissivity $\tau'_{\lambda,G}$. For a homogeneous gas or gas mixture, we obtain from (5.179)

$$a'_{\lambda,G}(\lambda, s, T, p, p_G) = a'_{\lambda,G}(k_G s) = 1 - \exp\left(-k_G s\right) \ . \qquad (5.183)$$

The directional spectral absorptivity is a property (variable of state) of the absorbing gas. Its direction dependence manifests itself in the dependence on the beam length s, through which the radiation passes in the gas.

According to Kirchhoff's law, cf. section 5.3.2.1., the *directional spectral emissivity* $\varepsilon'_{\lambda,G}$ of a gas is equal to its directional spectral absorptivity:

$$\varepsilon'_{\lambda,G}(\lambda, T, p, p_G, s) = \varepsilon'_{\lambda,G}(k_G s) = a'_{\lambda,G}(k_G s) = 1 - \exp\left(-k_G s\right) \ . \qquad (5.184)$$

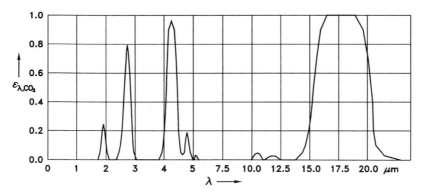

Fig. 5.71: Directional spectral emissivity $\varepsilon'_{\lambda,CO_2}$ of carbon dioxide at $T = 294$ K and $p = 10.13$ bar for a beam length $s = 0.38$ m according to D.K. Edwards [5.65]

$\varepsilon'_{\lambda,G}$ is a property of the gas that additionally depends on the (direction dependent) beam length s. As an example, Fig. 5.71 shows the directional emissivity of CO_2 at $T = 294$ K and $p = 10.13$ bar. The emission bands are clearly recognisable.

With $\varepsilon'_{\lambda,G}$, the radiation flow $d^3\Phi_{\lambda,G}$ received by an element dA on the surface of the gas volume, in the wavelength interval $d\lambda$, from the solid angle element $d\omega$ highlighted in Fig. 5.72, can be calculated. According to (5.4), the defining equation for the spectral intensity, it holds for this that

$$d^3\Phi_{\lambda,G} = \varepsilon'_{\lambda,G}\,(k_G s)\,L_{\lambda s}(\lambda, T)\,d\lambda\,d\omega\cos\beta\,dA \ , \qquad (5.185)$$

where $L_{\lambda s}$ is the spectral intensity of a black body.

In order to determine the radiation flow $d^2\Phi_{\lambda,G}$ that strikes dA from the entire gas space, $d^3\Phi_{\lambda,G}$ has to be integrated over all the solid angles coming from dA with their associated beam lengths s, that is over the whole gas space:

$$d^2\Phi_{\lambda,G} = \int_{\cap} \varepsilon'_{\lambda,G}(k_G s)\cos\beta\,d\omega\,L_{\lambda s}(\lambda, T)\,d\lambda\,dA \ . \qquad (5.186)$$

Analogous to the hemispherical spectral emissivity of a solid, cf. Table 5.4, through

$$\varepsilon^{\vee}_{\lambda,G}(k_G L_0) := \frac{1}{\pi}\int_{\cap} \varepsilon'_{\lambda,G}(k_G s)\cos\beta\,d\omega \qquad (5.187)$$

the *spectral emissivity of the gas volume* for the radiation on an element dA of its surface is defined. It depends on the shape of the gas space, the position of

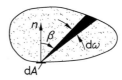

Fig. 5.72: Gas space with surface element dA and associated solid angle element $d\omega$

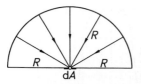

Fig. 5.73: Radiation of a gas hemisphere with the radius R on a surface element $\mathrm{d}A$ at the centre of the sphere

the surface element $\mathrm{d}A$ and on the optical thickness $\kappa_\mathrm{G} = k_\mathrm{G} L_0$. Here L_0 is a characteristic length for the gas space. The spectral irradiance of the surface element as a result of the gas radiation, with $M_{\lambda\mathrm{s}} = \pi L_{\lambda\mathrm{s}}$, is

$$E_{\lambda,\mathrm{G}} = \frac{\mathrm{d}^2\Phi_{\lambda,\mathrm{G}}}{\mathrm{d}A\,\mathrm{d}\lambda} = \varepsilon_{\lambda,\mathrm{G}}^\mathrm{V}(k_\mathrm{G}L_0)M_{\lambda\mathrm{s}}(\lambda,T) \ . \qquad (5.188)$$

Integration of $E_{\lambda,\mathrm{G}}$ over all wavelengths, taking $k_\mathrm{G} = k_\mathrm{G}(\lambda,T,p,p_\mathrm{G})$ into account, yields the (total) irradiance of $\mathrm{d}A$:

$$E_\mathrm{G} = \frac{\mathrm{d}\Phi_\mathrm{G}}{\mathrm{d}A} = \varepsilon_\mathrm{G}^\mathrm{V}(T,p,p_\mathrm{G},L_0)\,M_\mathrm{s}(T) = \varepsilon_\mathrm{G}^\mathrm{V}(T,p,p_\mathrm{G},L_0)\,\sigma T^4 \ . \qquad (5.189)$$

In which

$$\varepsilon_\mathrm{G}^\mathrm{V}(T,p,p_\mathrm{G},L_0) := \frac{1}{M_\mathrm{s}(T)}\int_0^\infty \varepsilon_{\lambda,\mathrm{G}}^\mathrm{V}(k_\mathrm{G}L_0)\,M_{\lambda\mathrm{s}}(\lambda,T)\,\mathrm{d}\lambda \qquad (5.190)$$

is the *total emissivity of the gas volume* for its radiation on an element $\mathrm{d}A$ of its surface. $\varepsilon_\mathrm{G}^\mathrm{V}$ also depends on the form of the gas space, which is expressed in the formula by the dependence on the charateristic length L_0.

As the emissivity of the gas radiation depends on the shape of the gas space, it is not purely a material property like the emissivity of solid surfaces. The dependence on the shape of the gas space is especially easy to consider for radiation from a *hemisphere of gas* on the surface element at the centre of the sphere, see Fig. 5.73. The directional spectral emissivity $\varepsilon'_{\lambda,\mathrm{G}}$ is independent of direction here, because the beam length s is equal to the radius R for all directions. It follows from (5.187) that simply $\varepsilon_{\lambda,\mathrm{G}}^\mathrm{V} = \varepsilon'_{\lambda,\mathrm{G}}(k_\mathrm{G}R)$. In this case $\varepsilon_{\lambda,\mathrm{G}}^\mathrm{V}$ is termed the *spectral emissivity* $\varepsilon_{\lambda,\mathrm{G}}$ of the gas, for which, according to (5.184)

$$\varepsilon_{\lambda,\mathrm{G}}(k_\mathrm{G}R) = 1 - \exp\left(-k_\mathrm{G}R\right) \qquad (5.191)$$

holds, treating $\varepsilon_{\lambda,\mathrm{G}}$ as a gas property, even though $\varepsilon_{\lambda,\mathrm{G}}$ is the spectral emissivity of a gas hemisphere that radiates on the surface element at its centre.

The total emissivity of the radiating hemisphere is yielded from (5.190), with $\varepsilon_{\lambda,\mathrm{G}}^\mathrm{V} = \varepsilon_{\lambda,\mathrm{G}}(k_\mathrm{G}R)$ from (5.191), as

$$\varepsilon_\mathrm{G}(T,p,p_\mathrm{G},R) = \frac{1}{\sigma T^4}\int_0^\infty [1 - \exp\left(-k_\mathrm{G}R\right)]\,M_{\lambda\mathrm{s}}(\lambda,T)\,\mathrm{d}\lambda \ . \qquad (5.192)$$

This emissivity is also viewed as a material property. The next section contains graphs from which ε_G for CO_2 and H_2O can be taken. The validity of Beer's law means that ε_G depends on the product $p_G R$, such that $\varepsilon_G = \varepsilon_G(T, p, p_G R)$ holds.

As we will show in section 5.6.4, the complicated determination of the emissivities $\varepsilon_{\lambda,G}^V$ and ε_G^V of any shape of gas space can be traced back to the "standard case" of the gas hemisphere we have just dealt with. A *mean beam length* s_m is determined for the gas space under consideration from the following condition: a gas hemisphere with the radius $R = s_m$ should give rise to the same spectral irradiance on a surface element at its centre as that for the radiation from any shaped gas volume on a certain element of its surface. As follows from (5.188) and (5.191)

$$\varepsilon_{\lambda,G}^V(k_G L_0) = \varepsilon_{\lambda,G}(k_G s_m) = 1 - \exp\left(-k_G s_m\right) \tag{5.193}$$

can then be set, and from (5.189) and (5.192) it follows that

$$\varepsilon_G^V(T, p, p_G, L_0) = \varepsilon_G\left(T, p, p_G, s_m\right) \ . \tag{5.194}$$

This avoids the integration required in (5.187) and (5.190) over all solid angles and wavelengths respectively. The graphs from the next section can be used to determine ε_G, thereby making it easy to find the irradiance E_G from (5.189) and (5.194), once the mean beam length s_m has been calculated for the problem.

5.6.3 Results for the emissivity

H.C. Hottel and R.B. Egbert [5.57], [5.58] have critically compared the results available for CO_2 and H_2O from radiation measurements, and offered best values of the total emissivity in graphs, in which ε_{CO_2} and ε_{H_2O} are plotted against the gas temperature T with the product $(p_{CO_2} s_m)$ or $(p_{H_2O} s_m)$ as curve parameters. These diagrams have formed the basis of the calculations for gas radiation from CO_2 and H_2O for the last 50 years. Uncertainties of at least 5 % have to be reckoned with in their application. Information concerning correction factors, based on more recent data can be found in [5.37], p. 636. The emissivities ε_{CO_2} and ε_{H_2O} are valid for the calculation of the radiation of a gas hemisphere of radius $R = s_m$ on a surface element in the centre of the sphere, cf. Fig. 5.73. How these results can be transferred to other shapes of gas space by using the mean beam length, s_m, is explained in the next section.

The hemispherical total emissivity $\varepsilon_{CO_2}(T, p_{CO_2} s_m)$ of CO_2 at $p = 100\,\mathrm{kPa}$ is illustrated in Fig. 5.74. ε_{CO_2} increases slightly with rising pressure. D. Vortmeyer [5.59] presents a particularly complex pressure correction factor, which can be neglected for pressures below around 200 kPa.

As H_2O does not follow Beer's law, ε_{H_2O} has to be determined from

$$\varepsilon_{H_2O}(T, p, p_{H_2O}, s_m) = C_{H_2O}(p + p_{H_2O}, p_{H_2O} s_m)\, \varepsilon_{H_2O}^*(T, p_{H_2O} s_m) \tag{5.195}$$

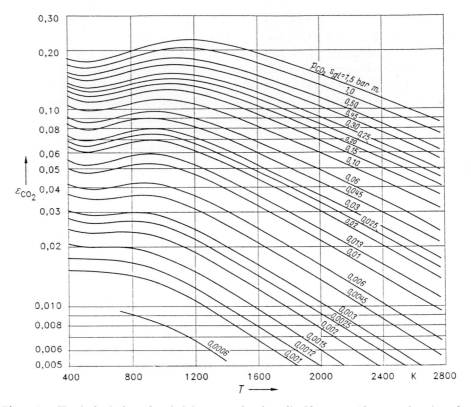

Fig. 5.74: Hemispherical total emissivity ε_{CO_2} of carbon dioxide at $p = 1$ bar as a function of temperature T with the product of the partial pressure p_{CO_2} and the mean beam length s_m as a parameter. 1 bar $= 100\,\text{kPa} = 0.1\,\text{MPa}$

with the help of a second graph. Fig. 5.75 shows the emissivity $\varepsilon_{H_2O}^*$ for $p = 100\,\text{kPa}$, extrapolated to $p_{H_2O} \to 0$. The partial pressure correction C_{H_2O} is taken from Fig. 5.76. For pressures p greater than 100 kPa, $\varepsilon_{H_2O}^*$ should be multiplied by the pressure correction, an equation for this is given in [5.59].

If, as in combustion gases, CO_2 and H_2O appear at the same time, then the emissivity of this sort of mixture is slightly less than the sum $\varepsilon_{CO_2} + \varepsilon_{H_2O}$, calculated at the respective partial pressures. This can be traced back to the fact that some of the absorption and emission bands of CO_2 and H_2O overlap. H.C. Hottel and R.B. Egbert [5.58] determined the correction factor $\Delta\varepsilon$ that has to be introduced into

$$\varepsilon_G = \varepsilon_{CO_2} + \varepsilon_{H_2O} - \Delta\varepsilon \tag{5.196}$$

and plotted it in graphs. These can also be found in [5.59].

Equations for the dependency of the emissivities ε_{CO_2} and ε_{H_2O} on T, p and $(p_G s_m)$ are required for model design and process simulations. Various authors have developed these equations, cf. [5.60] to [5.63] as well as [5.37], p. 639–641. Although they contain numerous terms, they can only reproduce nearly accurate results for limited ranges of the variables, and this is why they cannot be recommended without limitations.

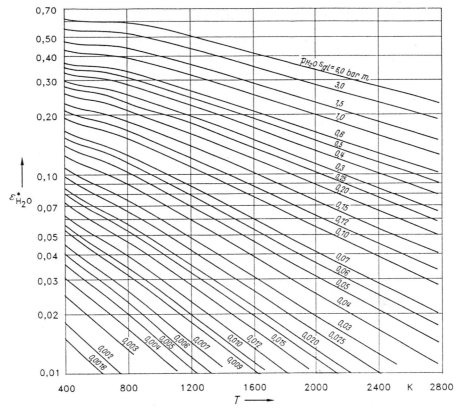

Fig. 5.75: Total emissivity $\varepsilon^*_{H_2O}$ for water vapour at $p = 1\,\mathrm{bar}$, extrapolated to $p_{H_2O} \to 0$, as a function of temperature T, with the product of the partial pressure p_{H_2O} and the mean beam length s_m as parameter. $1\,\mathrm{bar} = 100\,\mathrm{kPa} = 0.1\,\mathrm{MPa}$

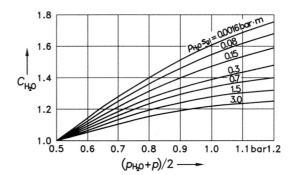

Fig. 5.76: Partial pressure correction factor C_{H_2O} for water vapour for use in (5.195)

The emissivities of other gases, namely SO_2, NH_3 and CH_4, are presented graphically in [5.59]. Similar diagrams for CO, HCl and NO_2 can be found in [5.48].

5.6.4   Emissivities and mean beam lengths of gas spaces

The emission of gas radiation depends on the size and shape of the gas space; it is described quantitatively by the irradiance, which the gas radiation generates at the surface of the gas space. The decisive equations (5.188) and (5.189) include the spectral emissivity $\varepsilon_{\lambda,G}^V$ and the emissivity integrated over all wavelengths ε_G^V, which, according to (5.193) and (5.194), can be replaced by the emissivity of a gas hemisphere with a radius the same as the mean beam length s_m of any shaped gas space.

In order to explain the determination of the mean beam length, we first have to look at the calculation of the spectral emissivity $\varepsilon_{\lambda,G}^V$ of a gas volume. We will then show how s_m is determined and put together in Table 5.11 values of s_m that have been calculated for different geometries. In addition a simple approximation formula will be derived with which s_m can be determined for gas spaces that are not covered in this collection.

The spectral emissivity $\varepsilon_{\lambda,G}^V$ from (5.187) covers the radiation coming from the entire gas space, which is incident on a surface element $dA = dA_2$ in Fig. 5.77. The solid angle element $d\omega_1$ which appears in

$$\varepsilon_{\lambda,G}^V = \frac{1}{\pi} \int_{\cap} \varepsilon_{\lambda,G}'(k_G s) \cos\beta_2 \, d\omega_1 \ , \tag{5.197}$$

and to which the beam length s belongs, is bounded by the surface element dA_1, such that

$$d\omega_1 = \cos\beta_1 \, dA_1/s^2$$

holds. With that we obtain the following from (5.197)

$$\varepsilon_{\lambda,G}^V = \frac{1}{\pi} \int_{A_1} \varepsilon_{\lambda,G}'(k_G s) \cos\beta_1 \cos\beta_2 \frac{dA_1}{s^2} \ . \tag{5.198}$$

The integration over all solid angles is replaced by the integration over all the surface elements dA_1 visible from dA_2. In general, in (5.198) the integration is carried out over the entire surface area A_1 of the gas space. We will show how the integration is carried out for the example of a gas sphere of diameter $D = 2R$, that radiates on an element dA_2 of its surface.

According to Fig. 5.78, $\beta_1 = \beta_2 = \beta$ has to be put into (5.198). We choose an annular surface element

$$dA_1 = 2\pi s \sin\beta \frac{s \, d\beta}{\cos\beta} = 2\pi s^2 \frac{\sin\beta}{\cos\beta} \, d\beta \ .$$

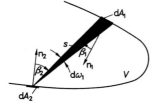

Fig. 5.77: Gas space with surface element dA_2, which receives radiation from the solid angle element $d\omega_1$ which is bounded by the surface element dA_1

According to Fig. 5.78, we have $\cos\beta = s/2R = s/D$, from which $\sin\beta\,d\beta = -\,ds/D$ follows. With that $dA_1 = -2\pi s\,ds$. With $\varepsilon'_{\lambda,\mathrm{G}}(k_\mathrm{G}s)$ from (5.184), we obtain out of (5.198)

$$\varepsilon^{\mathrm{V}}_{\lambda,\mathrm{G}} = \frac{2}{D^2}\int\limits_0^D [1 - \exp\left(-k_\mathrm{G}s\right)]\,s\,ds \ .$$

Carrying out the integration gives

$$\varepsilon^{\mathrm{V}}_{\lambda,\mathrm{G}}\left(k_\mathrm{G}D\right) = 1 - \frac{2}{\left(k_\mathrm{G}D\right)^2}\left[1 - (1 + k_\mathrm{G}D)\exp\left(-k_\mathrm{G}D\right)\right] \ . \tag{5.199}$$

The spectral emissivity of the gas sphere depends on its optical thickness $\kappa_\mathrm{G} = k_\mathrm{G}D$. The characteristic length L_0 of the gas space introduced in (5.187), is, as could be expected, the diameter D of the sphere.

The symmetry of the sphere means that $\varepsilon^{\mathrm{V}}_{\lambda,\mathrm{G}}$ is independent of the position of the irradiated surface element dA_2. The spectral irradiance $E_{\lambda,\mathrm{G}}$ from (5.188) is constant over the entire surface of the sphere. So, with $\varepsilon^{\mathrm{V}}_{\lambda,\mathrm{G}}(k_\mathrm{G}D)$ the mean irradiance $\overline{E}_{\lambda,\mathrm{G}}$ of an arbitrary sized piece A_2 of the sphere surface can be calculated. In general, the mean spectral irradiance $\overline{E}_{\lambda,\mathrm{G}}$ of a finitely large surface is obtained by integration of $E_{\lambda,\mathrm{G}}$ over all surface elements dA_2 that make up A_2. A corresponding emissivity $\overline{\varepsilon}^{\mathrm{V}}_{\lambda,\mathrm{G}}$ is yielded from the additional integration of (5.198) over all surface elements dA_2 followed by division by A_2.

The *mean beam length* s_m of a gas space of any shape that radiates on an element $dA = dA_2$ of its surface, is defined by the fact that the spectral irradiance $E_{\lambda,\mathrm{G}}$ of dA_2 has exactly the same magnitude as the spectral irradiance of a surface element in the centre of a gas hemisphere of radius $R = s_\mathrm{m}$. According to section 5.6.2, the spectral irradiance of this surface element is

$$E_{\lambda,\mathrm{G}} = \varepsilon_{\lambda,\mathrm{G}}(k_\mathrm{G}s_\mathrm{m})M_{\lambda\mathrm{s}}(\lambda,T) = [1 - \exp\left(-k_\mathrm{G}s_\mathrm{m}\right)]\,M_{\lambda\mathrm{s}}(\lambda,T) \ . \tag{5.200}$$

By setting this expression equal to $E_{\lambda,\mathrm{G}}$ from (5.188), the relationship (5.193) from section 5.6.2 is obtained, from which

$$\frac{s_\mathrm{m}(k_\mathrm{G}L_0)}{L_0} = -\frac{1}{k_\mathrm{G}L_0}\ln\left[1 - \varepsilon^{\mathrm{V}}_{\lambda,\mathrm{G}}\left(k_\mathrm{G}L_0\right)\right] \ . \tag{5.201}$$

follows. According to this, the mean beam length of a particular gas space is not constant, but depends on its optical thickness $k_\mathrm{G}L_0$. For a gas sphere $s_\mathrm{m}/D = f(k_\mathrm{G}D)$ can be calculated exactly from (5.199).

The dependence of the mean beam length s_m on the optical thickness of the gas space makes the use of this quantity more difficult. Therefore, a *constant* mean beam length is used. It is determined in such a way that (5.193) is satisfied in the best approximation for all technically

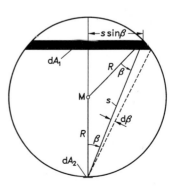

Fig. 5.78: Gas sphere of radius $R = D/2$ and the surface elements dA_1 and dA_2

important optical thicknesses. For constant mean beam length s_m, the integration carried out in (5.190) over all wavelengths can be replaced by the integration from (5.192) with $R = s_m$, so that (5.192) is approximately satisfied and the irradiance E_G from (5.190) with ε_G, instead of ε_G^V, can be calculated.

In the determination of a constant mean beam length s_m we first will consider the limiting case of an optically thin gas with $\kappa_G = k_G L_0 \to 0$. The spectral radiation flow emitted from a volume element of the gas in all directions is, according to (5.181),

$$d^2 \Phi_{\lambda,V} = 4 k_G M_{\lambda s}(\lambda, T) \, dV \, d\lambda \ .$$

It is not weakened as it passes through the optically thin gas. Therefore, the radiation flow emitted from the entire gas volume is

$$d\Phi_\lambda = 4 k_G V M_{\lambda s}(\lambda, T) \, d\lambda \ .$$

This generates, over the entire surface A of the gas volume, the mean spectral irradiance

$$\overline{E}_{\lambda,G} = \frac{1}{A} \frac{d\Phi_\lambda}{d\lambda} = 4 k_G \frac{V}{A} M_{\lambda s}(\lambda, T) \ .$$

An *optically thin gas hemisphere* of radius $R = s_m$ causes, according to (5.200), the spectral irradiance

$$\lim_{k_G s_m \to 0} E_\lambda (k_G s_m) = \lim_{k_G s_m \to 0} [1 - \exp(1 - k_G s_m)] M_{\lambda s}(\lambda, T) = (k_G s_m + \cdots) M_{\lambda s}(\lambda, T)$$

on the surface element in the centre of the sphere. Setting this irradiance equal to $\overline{E}_{\lambda,G}$ delivers the simple result

$$s_m^* = 4V/A \ , \qquad k_G L_0 \to 0 \ , \tag{5.202}$$

where s_m^* indicates the limit of s_m for an optically thin gas.

Values of s_m^* can be found easily for different gas spaces. For example a sphere has

$$s_m^* = 4 \frac{\pi D^3/6}{\pi D^2} = \frac{2}{3} D$$

and for the gas layer between infinitely large, flat plates with separation d we get

$$s_m^* = 4 \frac{d}{2} = 2d \ .$$

In the case of finite optical thickness s_m^* is corrected by a constant factor C, so that

$$s_m = C s_m^* = C 4V/A \tag{5.203}$$

holds. The correction factor C is determined in such a way that the spectral emissivity $\varepsilon_{\lambda,G}(k_G s_m)$ of the gas sphere agrees with $\varepsilon_{\lambda,G}^V(k_G L_0)$ within a few percent for a wide range of optical thicknesses, and equation (5.193) is satisfied in the average. This is shown by Fig. 5.79 for the example of a gas sphere of diameter D. The ratio $\varepsilon_{\lambda,G}^V(k_G D)/\varepsilon_{\lambda,G}(k_G C s_m^*)$ is plotted versus the optical thickness $k_G D$. For $C = 1$ only negative deviations from the ideal value occur. Correction factors $C < 1$ lead to smaller deviations and an optimal fit is obtained with $C = 0.96$, that is for $s_m = 0.64 D$ in place of $s_m^* = (2/3)D$.

As can be seen from Fig. 5.79, the choice of the correction factor C is arbitrary within certain limits. It also depends on the range of optical thickness in which a particularly good agreement between $\varepsilon_{\lambda,G}^V(k_G L_0)$ and $\varepsilon_{\lambda,G}(k_G s_m)$ is to be reached. This is why the values of s_m for some cases in the literature may differ slightly from one another.

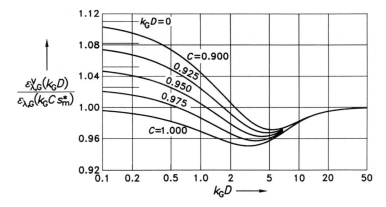

Fig. 5.79: Ratio of the spectral emissivity $\varepsilon^{V}_{\lambda,G}(k_G D)$ from (5.199) for a gas sphere with diameter D to the spectral emissivity $\varepsilon_{\lambda,G}(k_G C s^*_m)$ from (5.200), with $s^*_m = (2/3)D$ as a function of the optical thickness $k_G D$

$\varepsilon^{V}_{\lambda,G}$ has been calculated exactly for a series of gas spaces and the associated mean beam lengths s_m have also been determined. These values of s_m are put together in Table 5.11. This shows that the correction factors C, in (5.203), only a little deviate from 0.9. It can therefore be said that for those gas space geometries not included in Table 5.11

$$s_m \approx 0.9\, s^*_m = 3.6\ V/A \ . \tag{5.204}$$

With this mean beam length, the mean irradiance $\overline{E}_G$ of the total surface A of the radiating gas space of volume V is found to be

$$\overline{E}_G = \varepsilon_G(T, p, p_G, s_m)\, \sigma T^4 \ . \tag{5.205}$$

The emissivities ε_G for CO_2 and H_2O can be taken from Fig. 5.74 to 5.76 in section 5.6.3.

Example 5.14: A hemisphere of radius $R = 0.50\,\text{m}$ contains CO_2 at $p = 1\,\text{bar}$ and $T = 1200\,\text{K}$. Determine the mean irradiance $\overline{E}_{CO_2}$ of its surface and compare this value with the irradiance E_{CO_2} of a surface element at the centre of the sphere.

We obtain $\overline{E}_{CO_2}$ from (5.205), whereby ε_{CO_2} is taken from Fig. 5.74. As no value is given for s_m for a hemisphere in Table 5.11, s_m is calculated approximately from (5.204). With $V = (2\pi/3)R^3$ and $A = 2\pi R^2 + \pi R^2 = 3\pi R^2$, this gives $s^*_m = (8/9)R$ and $s_m = 0.80R = 0.40\,\text{m}$. With that we have $p_{CO_2} s_m = p s_m = 0.40\,\text{bar·m}$. $\varepsilon_{CO_2} = 0.16$ is read off Fig. 5.74. This yields, according to (5.205), the approximate value

$$\overline{E}_{CO_2} = 0.16 \cdot 5.67 \cdot 10^{-8}\frac{\text{W}}{\text{m}^2\text{K}^4}\, 1200^4\text{K}^4 = 18.8\ \frac{\text{kW}}{\text{m}^2} \ .$$

For the radiation of the gas hemisphere on a surface element at its centre, we have exactly $s_m = R = 0.50\,\text{m}$. Then from Fig. 5.74, the somewhat larger emissivity $\varepsilon_{CO_2} = 0.18$ and the correspondingly larger irradiance $E_{CO_2} = 21.2\,\text{kW/m}^2$ are obtained. This result with $E_{CO_2} > \overline{E}_{CO_2}$ is expected because the irradiance is not constant over the surface of the hemisphere.

Table 5.11: Mean beam lengths s_m^* for vanishingly small optical thicknesses and s_m for finite optical thickness. Further information in [5.37], [5.48] and [5.59]

Gas space	Characteristic length L_0	$\dfrac{s_m^*}{L_0}$	$\dfrac{s_m}{L_0}$	$\dfrac{s_m}{s_m^*}$
Sphere	Diameter	2/3	0.64	0.96
Infinitely long cylinder	Diameter	1	0.94	0.94
Cylinder, Length $L = D$,	Diameter			
Radiation on the whole surface		2/3	0.60	0.90
Radiation on the centre of the base		0.764	0.71	0.93
Cylinder, Length $L = D/2$	Diameter			
Radiation on the whole surface		1/2	0.45	0.90
Radiation on the body		0.525	0.46	0.88
Radiation on an end		0.475	0.43	0.91
Cylinder, Length $L = 2D$	Diameter			
Radiation on the whole surface		4/5	0.73	0.91
Radiation on the body		0.817	0.76	0.93
Radiation on both ends		0.730	0.60	0.82
Cylinder, Length $L \to \infty$	Diameter			
Radiation on the base		0.814	0.65	0.80
Cube	Side length	2/3	0.60	0.90
Cuboid, Side lengths 1:1:4	shortest side			
Radiation on the total area		0.89	0.81	0.91
Radiation on 1×4 area		0.90	0.82	0.91
Radiation on 1×1 area		0.86	0.71	0.83
Flat layer between two infinitely extended parallel walls	Distance	2	1.76	0.88

5.6.5   Radiative exchange in a gas filled enclosure

The radiative exchange in a gas filled enclosure is more difficult to calculate than the exchange dealt with in 5.5.3, without an absorbing and therefore self radiating gas. In the following we will consider two simple cases, in which an isothermal gas is involved in radiative interchange with its boundary walls that are likewise at a uniform temperature. At the end of this section we will point to more complex methods with which more difficult radiative exchange problems may be solved.

5.6.5.1   Black, isothermal boundary walls

An isothermal gas at temperature T_G is enclosed by isothermal walls at a temperature $T_W < T_G$. As a simplification the walls will be idealised as black bodies;

reflection does not need to be considered. A heat flow $\dot{Q}_{GW}$ from the gas to the colder walls is transferred by radiative exchange. If the gas is to maintain its temperature T_G, then this energy has to be supplied to the gas in another way, for example from a combustion process taking place in the gas space.

The radiation flow emitted by the gas generates a mean irradiance

$$E_G = \varepsilon_G(T_G, p, p_G, s_m)\sigma T_G^4$$

at the walls, cf. (5.205). The black walls with an area A_W absorb the radiation flow $A_W E_G$ completely. The radiation flow emitted by them

$$A_W M_s(T_W) = A_W \sigma T_W^4$$

is partially absorbed as it travels through the gas. The portion that is not absorbed strikes the walls once again and is absorbed there. As the walls have the same temperature T_W overall this part of the radiation flow does not contribute to the heat flow $\dot{Q}_{GW}$. This gives

$$\dot{Q}_{GW} = A_W\left[E_G(T_G) - a_G M_s(T_W)\right] = A_W \sigma \left(\varepsilon_G T_G^4 - a_G T_W^4\right) . \tag{5.206}$$

Here, a_G is the absorptivity of the gas for the radiation emitted by a black body at temperature T_W. As the gas is not a grey radiator, a_G is not the same as ε_G, except for the limiting case where $T_W = T_G$. H.C. Hottel and R.B. Egbert [5.58], see also [5.48], determined a_G for CO_2 and H_2O from absorption measurements, and related them to the emissivity from 5.6.3, as shown in the following equations:

$$a_{CO_2} = \left(\frac{T_G}{T_W}\right)^{0,65} \varepsilon_{CO_2}\left(T_W, p_{CO_2}\frac{T_W}{T_G}s_m\right) , \tag{5.207}$$

$$a_{H_2O} = \left(\frac{T_G}{T_W}\right)^{0,45} \varepsilon_{H_2O}^*\left(T_W, p_{H_2O}\frac{T_W}{T_G}s_m\right) \cdot C_{H_2O}\left(\frac{p_{H_2O}+p}{2}, p_{H_2O}\frac{T_W}{T_G}s_m\right) . \tag{5.208}$$

The emissivities are taken from Fig. 5.74 to 5.76, for a wall temperature T_W and the product $(p_G s_m)$ reduced by a factor of (T_W/T_G). For pressures p above 1 bar, the pressure corrections mentioned in section 5.6.3 have to be brought into play at ε_{CO_2} and $\varepsilon_{H_2O}^*$.

5.6.5.2 Grey isothermal boundary walls

If the gas filled enclosure is surrounded by grey walls with emissivity $\varepsilon_W = \varepsilon_W(T_W)$, then the energy reflected by the walls also has to be considered. Rough, oxidised and dirty walls have emissivites that are only slightly smaller than one. In this case, which often applies in furnaces, the reflected radiation proportion is of little importance. The heat flow $\dot{Q}_{GW}$ transferred from the gas to the walls is smaller, by a factor that lies between ε_W and 1, than the heat flow calculated according to (5.206) for black walls. At sufficiently large values of ε_W, around $\varepsilon_W > 0.8$, the approximation recommended by H.C. Hottel and A.F. Sarofim [5.48] of

$$\dot{Q}_{GW} = \frac{\varepsilon_W + 1}{2} A_W \sigma \left(\varepsilon_G T_G^4 - a_G T_W^4 \right) \tag{5.209}$$

is accurate enough.

In the consideration of the radiation reflected by the walls, the exchange of radiation can be calculated just like in section 5.5.3. According to (5.143), the heat released by the walls to the outside

$$\dot{Q}_{GW} = A_W \left(E_W - H_W \right) \tag{5.210}$$

is equal to the net flow of radiation. Here this is the difference between the mean irradiance E_W and the mean radiosity H_W of the area A_W. The irradiance E_W is made up of the fraction generated by the gas radiation, $E_G = \varepsilon_G \sigma T_G^4$, and the fraction emitted by the walls that is not absorbed by the gas, $(1 - a_G) H_W$:

$$E_W = \varepsilon_G \sigma T_G^4 + (1 - a_G) H_W \ . \tag{5.211}$$

The radiosity H_W includes, according to (5.142), the emissive power $M_W = \varepsilon_W \sigma T_W^4$ of the wall area and the reflected fraction of the irradiance E_W:

$$H_W = \varepsilon_W \sigma T_W^4 + (1 - \varepsilon_W) E_W \ . \tag{5.212}$$

Solving the two equations, (5.211) and (5.212), for E_W and H_W, and by putting the result into (5.210), gives

$$\dot{Q}_{GW} = \frac{\varepsilon_W A_W \sigma}{1 - (1 - a_G)(1 - \varepsilon_W)} \left(\varepsilon_G T_G^4 - a_G T_W^4 \right) \ . \tag{5.213}$$

This equation is also not exact, because of the restrictive assumptions that were made. Only symmetrically formed gas spaces (spheres, very long cylinders) have approximately constant values of E_W and H_W on their entire boundary walls, which is one of the presumptions made here. The radiation not absorbed by the gas, given by the term $(1 - a_G) H_W$ in (5.211), does not take into account in H_W that the radiation passing through the gas consists of radiation made up of fractions with different spectral distributions: This includes the grey radiation emitted by the walls and the radiation reflected by the walls after one, two or even multiple passes through the absorbing gas. The application of a uniform absorptivity a_G is therefore incorrect. As long as the reflected fractions are small (large ε_W), the use of the absorptivity calculated according to (5.207) and (5.208) is a good enough approximation. It is certainly better than the assumption of a *grey gas*, where $a_G = \varepsilon_G$ would be valid.

A more exact calculation of the radiative exchange, in which the absorption of the reflected radiation in the gas space has been extensively modelled, has been presented by K. Elgeti [5.64]. Another method for the consideration of the spectral absorption bands can be found in [5.37], section 17-7.

Example 5.15: A cylindrical combustion chamber with diameter $D = 0.40\,\mathrm{m}$ and length $L = 0.95\,\mathrm{m}$ contains a combustion gas at temperature $T_G = 2000\,\mathrm{K}$ and pressure $p = 1.1\,\mathrm{bar}$. The partial pressures of CO_2 and H_2O are $p_{CO_2} = 0.10\,\mathrm{bar}$ and $p_{H_2O} = 0.20\,\mathrm{bar}$ respectively. The walls of the chamber are at $T_W = 900\,\mathrm{K}$ and their emissivity is $\varepsilon_W = 0.75$. Determine the heat flow transferred from the gas to the chamber casing.

The relationships derived in this section for the radiative exchange only allow the calculation of the mean heat flux of the entire surface of the gas space. It follows from (5.213) as

$$\dot{q}_{GW} = \frac{\dot{Q}_{GW}}{A_W} = \frac{\varepsilon_W \sigma}{1 - (1 - a_G)(1 - \varepsilon_W)}\left(\varepsilon_G T_G^4 - a_G T_G^4\right) \ . \tag{5.214}$$

This provides us with an approximate value for the heat flow transferred from the gas to the chamber casing

$$\dot{Q}_{GW} = \dot{q}_{GW} A_M = \pi D L\, \dot{q}_{GW} \ . \tag{5.215}$$

The mean beam length has to be determined for the calculation of ε_G and a_G. Its limit for an optically thin gas is

$$s_m^* = 4\frac{V}{A_W} = \frac{D}{1 + D/2L} = 0.330\,\mathrm{m} \ .$$

According to Table 5.11, the ratio $C = s_m/s_m^*$ for a cylinder with $L/D = 2$ has a value of 0.91. This value applies with good accuracy for a combustion chamber with $L/D = 2.375$. We therefore have $s_m = 0.91 \cdot 0.330\,\mathrm{m} = 0.30\,\mathrm{m}$.

For $p_{CO_2} s_m = 0.030\,\mathrm{bar\,m}$, we read off from Fig. 5.74, an emissivity $\varepsilon_{CO_2} = 0.0362$. Then from Figs. 5.75 and 5.76, we obtain for $p_{H_2O} s_m = 0.060\,\mathrm{bar\,m}$,

$$\varepsilon_{H_2O} = \varepsilon_{H_2O}^* C_{H_2O} = 0.0360 \cdot 1.20 = 0.0432 \ .$$

As the gas pressure $p = 1.1\,\mathrm{bar}$ is only slightly greater than 1 bar, a pressure correction of ε_{CO_2} and ε_{H_2O} is not necessary. In order to obtain ε_G from (5.196), the correction $\Delta\varepsilon$ is required. This is found, using the graph reproduced in [5.59], to be $\Delta\varepsilon = 0.0025$, which gives $\varepsilon_G = 0.0769$.

The determination of the absorptivity a_G requires the emissivity at the reduced partial pressures, i.e. at

$$p_{CO_2}\frac{T_W}{T_G}s_m = 0.0135\,\mathrm{bar\,m} \quad \text{and} \quad p_{H_2O}\frac{T_W}{T_G}s_m = 0.027\,\mathrm{bar\,m} \ .$$

For $T = T_W = 900\,\mathrm{K}$, Fig. 5.74 gives $\varepsilon_{CO_2} = 0.060$, which, with (5.207) gives $a_{CO_2} = 0.101$. For H_2O, we find $\varepsilon_{H_2O}^* = 0{,}226$ and $C_{H_2O} = 1.20$, so $a_{H_2O} = 0.388$ is yielded from (5.208). With $\Delta\varepsilon = 0.003$ from [5.59] we get

$$a_G = a_{CO_2} + a_{H_2O} - \Delta\varepsilon = 0.486 \ .$$

The gas mixture absorbs the radiation coming from the walls far more strongly than it emits itself radiation in comparision to "black" gas radiation.

The mean heat flux transferred by radiation to the walls of the combustion chamber is found, with the values of ε_G and a_G, from (5.214), to be

$$\dot{q}_{GW} = \frac{0.75 \cdot 5.67 \cdot 10^{-8}\,\mathrm{W/(m^2K^4)}}{1 - (1 - 0.486)(1 - 0.75)}\left[0.0769\,(2000\,\mathrm{K})^4 - 0.468\,(900\,\mathrm{K})^4\right] = 45.1\,\frac{\mathrm{kW}}{\mathrm{m^2}} \ .$$

The heat flow we want to determine follows from (5.215) as

$$\dot{Q}_{GM} = \pi(0.40\,\mathrm{m})(0.95\,\mathrm{m})45.1(\mathrm{kW/m^2}) = 53.8\,\mathrm{kW} \ .$$

According to (1.64) in section 1.1.6, the heat flux $\dot{q}_{GW}$, corresponds to a heat transfer coefficient of radiation of

$$\alpha_{\mathrm{rad}} = \frac{\dot{q}_{GW}}{T_G - T_W} = 41.0\,\frac{\mathrm{W}}{\mathrm{m^2K}} \ .$$

It lies in the same order of magnitude as the heat transfer coefficient expected here for convection. The gas radiation may not be neglected in heat transfer calculations for combustion chambers and furnaces.

5.6.5.3   Calculation of the radiative exchange in complicated cases

If the gas filled enclosure is surrounded by *non-isothermal walls*, then these should
be split into several isothermal areas, the so-called zones. The radiative exchange
between several zones can then be calculated as in section 5.5.3. However, the
absorption of the radiation as it passes through the gas must also be taken into
account along with the radiation exchange between the gas and the zones. This
calculation requires the view factors introduced in section 5.5.1, as well as the
transmissivities of the gas volumes participating in radiative interchange between
the zones. As in section 5.5.3.1, a system of equations is obtained, from which the
radiosities of the zones, their temperatures and the heat flows supplied from or
released to the outside can be calculated. R. Siegel and J.R. Howell [5.37], Chap.
13, show how this equation system should be set up and how the transmissivities of
the gas volumes which the radiation passes through are calculated. A somewhat
different formulation of the radiative exchange equations can be found in H.C.
Hottel and A.F. Sarofim [5.48].

A further problem in the calculation of radiative exchange is the consideration
of the fact that gases only absorb and emit within certain wavelength intervals
or bands. Here, sometimes the highly simplified assumption will be made that
the gas behaves like a grey radiator. A better model of real radiation behaviour
is band approximation. This is extensively discussed in [5.37], p. 549–567 and
607–609.

In the more realistic calculations of radiative exchange in furnaces and combustion chambers, a *non-isothermal gas space* has to be considered. H.C. Hottel
and A.F. Sarofim [5.48] developed the so-called zone method for this case, cf.
also [5.37], p. 647–652. Other procedures for the consideration of the temperature
fields in the gas space have been extensively dealt with by R. Siegel and J.R. Howell [5.37], Chapter 15. The application of the Monte-Carlo method is suggested
in particular, cf. [5.37] and [5.66], which despite being mathematically complex,
produces results without making highly simplified assumptions.

In technical furnaces the radiation from soot, coal and ash particles has to
be considered as well as the gas radiation. Then the scattering of radiation by
the suspended particles becomes important, alongside absorption and emission.
P. Biermann and D. Vortmeyer [5.67], as well as H.-G. Brummel and E. Kakaras
[5.68] have developed models for this. A summary can be found in [5.69] and in
[5.37], p. 652–673. The calculations of heat transport in furnaces has been dealt
with by W. Richter and S. Michelfelder [5.70] as well as H.C. Hottel and A.F.
Sarofim [5.48].

5.7   Exercises

5.1: Radiation with the vacuum wavelength $\lambda = 3.0\,\mu m$ goes through glass (refractive index $n = 1.52$). What are the propagation velocity c and the wavelength λ_M in the glass? Calculate the energy of a photon in a vacuum and in glass.

5.2: A Lambert radiator emits radiation at a certain temperature only in the wavelength interval (λ_1, λ_2), where its spectral intensity

$$L_\lambda(\lambda) = L_\lambda(\lambda_m) - a\,(\lambda - \lambda_m)^2$$

with $L_\lambda(\lambda_m) = 72\,W/(m^2\mu m\,sr)$, $a = 50\,W/(m^2\mu m^2\,sr)$ and $\lambda_m = 3.5\,\mu m$, has positive values. For $\lambda \leq \lambda_1$ and $\lambda \geq \lambda_2$ we have $L_\lambda = 0$. Find λ_1 and λ_2. Calculate the intensity L and the emissive power M. What fraction of the emissive power falls into the solid angle bounded by $(\pi/3) \leq \beta \leq (\pi/2)$ and $0 \leq \varphi \leq (\pi/4)$?

5.3: What proportion ΔM of the emissive power M of a Lambert radiator falls in the portion of the hemisphere for which the polar angle is $\beta \leq 30°$?

5.4: An opaque body, with the hemispherical total reflectivity $r = 0.15$, reflects diffusely. Determine the intensity L_{ref} of the reflected radiation and the absorbed radiative power per area for an irradiance $E = 800\,W/m^2$.

5.5: An enclosure is kept at a constant temperature. Radiation enters from outside through a small, circular opening of diameter $d = 5.60\,mm$; the radiation flow is $\Phi = 2.35\,W$. What is the temperature of the enclosure walls?

5.6: A radiator emits its maximum hemispherical spectral emissive power at $\lambda_{max} = 2.07\,\mu m$. Estimate its temperature T and its emissive power $M(T)$, under the assumption that it radiates like a black body.

5.7: What temperature does a black body need to be at, so that a third of its emissive power lies in the visible light region ($0.38\,\mu m \leq \lambda \leq 0.78\,\mu m$)?

5.8: A diffuse radiating oven wall has a temperature $T = 500\,K$ and a spectral emissivity approximated by the following function:

$$\varepsilon_\lambda = \begin{cases} 0.12\,, & 0 < \lambda \leq 1.6\,\mu m\,, \\ 0.48\,, & 1.6\,\mu m < \lambda \leq 8\,\mu m\,, \\ 0.86\,, & \lambda > 8\,\mu m\,. \end{cases}$$

The oven wall is exposed to radiation from glowing coal; the spectral irradiance E_λ can be assumed to be proportional to the hemispherical spectral emissive power $M_{\lambda s}(T_K)$ of a black body at $T_K = 2000\,K$.

a) Calculate the total emissivity ε of the oven wall.

b) Calculate the total absorptivity a of the oven wall for the radiation emitted by the coal.

5.9: A very long cylinder is struck by radiation that comes from a single direction, perpendicular to its axis (parallel directed radiation). The surface of the cylinder behaves like a grey radiator with the directional total emissivity $\varepsilon'(\beta) = 0.85 \cos\beta$. Calculate the reflected fraction of the incident radiative power.

5.10: A plate which allows radiation to pass through with a hemispherical total absorptivity $a = 0.36$, is irradiated equally from both sides, whilst air with $\vartheta_L = 30\ ^\circ$C flows over both surfaces. They assume a temperature $\vartheta_W = 75\ ^\circ$C at steady-state. The heat transfer coefficient between the plate and the air is $\alpha = 35\ \mathrm{W/m^2 K}$. Using a radiation detector it is ascertained that the plate releases a heat flux $\dot{q}_{Str} = 4800\ \mathrm{W/m^2}$ from both sides. Calculate the irradiance E and the hemispherical total emissivity ε of the plate.

5.11: A smooth, polished platinum surface emits radiation with an emissive power of $M = 1.64\ \mathrm{kW/m^2}$. Using the simplified electromagnetic theory determine its temperature T, the hemispherical total emissivity ε and the total emissivity ε'_n in the direction of the surface normal. The specific electrical resistance of platinum may be calculated according to

$$r_e = (0.384\,6\,(T/K) - 6.94)\,10^{-7}\,\Omega\mathrm{cm}\ .$$

5.12: A long channel has a hemispherical cross section (circle diameter $d = 0.40$ m). Determine the view factors F_{11}, F_{12}, F_{21} and F_{22}, where index 1 indicates the flat surface and index 2 the curved surface.

5.13: A sphere 1 lies on an infinitely large plane 2. How large is the view factor F_{12}?

5.14: An enclosure is formed from three flat surfaces of finite width and infinite length. The three widths are $b_1 = 1.0$ m, $b_2 = 2.5$ m and $b_3 = 1.8$ m. Calculate the nine view factors F_{ij} $(i, j = 1, 2, 3)$.

5.15: A very long, cylindrical heating element of diameter $d = 25$ mm is $h = 50$ mm away from a reflective (adiabatic) wall, Fig. 5.80. The heating element has a temperature $T = 700$ K, the

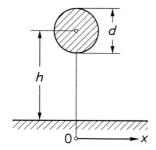

Fig. 5.80: Cylindrical heating element above a reflective wall

surroundings are at $T_S = 300$ K. The heating element, wall and surroundings are all assumed to be black bodies. Only the heat transfer by radiation is to be considered. Determine the temperature of the wall surface as a function of the coordinate x and calculate $T(x)$ for $x = 0$, $x = h$, $x = 2h$, $x = 10h$ and $x \to \infty$. Note: The view factor between a surface strip $\mathrm{d}A_1$

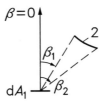

Fig. 5.81: Surface strip dA_1 of infinitesimal width and a ruled surface 2 generated by parallel, infinitely long straight lines perpendicular to the drawing plane

of infinitessimal width and any length (perpendicular to the drawing plane in Fig. 5.81) and a ruled or cylindrical surface 2 is given by

$$F_{12} = \frac{1}{2} \left(\sin \beta_2 - \sin \beta_1 \right) \ .$$

Here, the ruled surface 2 is produced by parallel straight lines of infinite length, perepndicular to the drawing plane in Fig. 5.81. The derivation of this equation is available in [5.37], p. 197–199.

5.16: A long, cylindrical nickel rod ($d_1 = 10$ mm) is heated electrically and releases a heat flow per length L of $\dot{Q}_1/L = 210$ W/m. The emissivity of nickel can be calculated from

$$\varepsilon(T) = 0.050 + 0.000\,10\,(T/K) \ .$$

The rod is surrounded by a concentrically arranged hollow cylinder, $d_2 = 25$ mm, $\varepsilon_2 = 0.88$. The hollow cylinder is cooled from outside to a temperature of $T_2 = 290$ K. What surface temperature does the nickel rod reach, if heat is only transferred by radiation?

5.17: A living room has a rectangular floor (width 3.5 m, depth 4.8 m) and a height of 2.8 m. Underfloor heating keeps the base area 1 at a constant temperature $\vartheta_1 = 29$ °C. One of the narrow side walls (3.5 m by 2.8 m) is the external wall and has been built into a window, the temperature of which is $\vartheta_2 = 17$ °C. The ceiling and the other walls can be considered as (adiabatic) reradiating walls. All surfaces are grey Lambert radiators with $\varepsilon = 0.92$. Determine the heat flow transferred by radiation from the floor to the external wall.

5.18: In order to reduce the radiative exchange between two large, parallel plates, a thin, flat radiation protection shield is introduced between the plates. However, the emissivities of the two surfaces of the shield are different; one surface has emissivity $\varepsilon_S < 0.4$, the other an emissivity of $2.5\,\varepsilon_S$.

a) The protection shield is to be orientated so that the heat flow transferred is as small as possible. Should the side with the emissivity ε_S be directed to the plate at the temperature $T_1 > T_2$, or is the opposite arrangement better?

b) Which of the two orientations leads to the higher temperature T_S of the protection shield?

5.19: Two very thin, radiation protection shields, A and B, are positioned parallel to and between two very large, parallel plates at temperatures $T_1 = 750$ K and $T_2 = 290$ K. All surfaces are grey radiators with the same emissivity $\varepsilon = 0.82$.

a) Calculate the temperatures T_A and T_B of the protection shields.

b) Calculate the heat flux $\dot{q}$ transferred by radiation between the two plates.

c) The two protection shields are removed, plate 1 is heated with the heat flux $\dot{q}$ from b), and plate 2 is kept at the temperature $T_2 = 290$ K. What is the temperature T_1 of plate 1?

5.20: The walls of older houses were often constructed as two shells, see Fig. 5.82. A column of air exists between two brick walls ($\lambda = 0.95\,\text{W/K m}$, $\varepsilon = 0.88$) of thicknesses $\delta_1 = 0.24\,\text{m}$ and $\delta_2 = 0.115\,\text{m}$. This column is $\delta = 0.060\,\text{m}$ wide. The heat transfer coefficients have the values $\alpha_i = 7.5\,\text{W/m}^2\text{K}$ and $\alpha_a = 18.0\,\text{W/m}^2\text{K}$. The temperatures are $\vartheta_i = 22.0\,°\text{C}$ and $\vartheta_a = -5.0\,°\text{C}$. The thermal conductivity of air is $\lambda_{\text{air}} = 0.024\,5\,\text{W/K m}$.

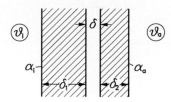

Fig. 5.82: Double shell brick walls with air gap of width δ

a) An effective thermal conductivity λ_{eff} of the air gap that also takes into account the effect of radiation is introduced. The free convection in the air gap can be neglected. Calculate λ_{eff} and the heat flux $\dot{q}$ transferred by the wall.

b) The air gap is filled with insulating foam ($\lambda_{\text{is}} = 0.040\,\text{W/K m}$). How does this change $\dot{q}$?

5.21: CO_2 at $p = 1$ bar flows through a long, cooled pipe with internal diameter $d = 0.10\,\text{m}$. Its velocity is $w = 20\,\text{m/s}$, and its mean temperature is $\vartheta_G = 1000\,°\text{C}$. The pipe wall temperature is $\vartheta_W = 500\,°\text{C}$, the emissivity of the pipe wall is $\varepsilon_W = 0.86$. In order to determine the contribution of the gas radiation to heat transfer, calculate the heat transfer cosfficients α for convection and α_{rad} for radiation. — Property data for CO_2 at $1000\,°\text{C}$: $\lambda = 0.0855\,\text{W/K m}$, $\nu = 117 \cdot 10^{-6}\,\text{m}^2/\text{s}$, $Pr = 0.736$.

Appendix A: Supplements

A.1 Introduction to tensor notation

In the derivation of balance equations the tensor notation is used because it allows the equations to be written in a clearer and simpler fashion. We have restricted ourselves in this to cartesian coordinates. In the following, the essential features of cartesian tensor notation are only illustrated to an extent required for the derivation of the balance equations; extensive publications are available for further reading. We will start with an example. The velocity $\boldsymbol{w}$ of a point of mass is known as a vector which can be set in a cartesian coordinate system using its components w_x, w_y, w_z:

$$\boldsymbol{w}(w_x, w_y, w_z) \ .$$

If the unit vector in a cartesian coordinate system is indicated by $\boldsymbol{e}_x$, $\boldsymbol{e}_y$, $\boldsymbol{e}_z$, it holds that

$$\boldsymbol{w} = w_x \, \boldsymbol{e}_x + w_y \, \boldsymbol{e}_y + w_z \, \boldsymbol{e}_z \ .$$

In tensor notation, the indices x, y, z are replaced by the indices 1, 2, 3 and we write instead

$$\boldsymbol{w} = w_1 \, \boldsymbol{e}_1 + w_2 \, \boldsymbol{e}_2 + w_3 \, \boldsymbol{e}_3 = \sum_{i=1}^{3} w_i \, \boldsymbol{e}_i \ .$$

The velocity vector $\boldsymbol{w}$ is characterised completely by its components w_i, $i = 1, 2, 3$. In tensor notation, the velocity vector is indicated by the abbreviation w_i with $i = 1, 2, 3$. Correspondingly, the position vector $\boldsymbol{x}(x, y, z)$ is determined by its components $x_1 = x$, $x_2 = y$, $x_3 = z$, and in tensor notation by the abbreviated x_i with $i = 1, 2, 3$. According to this, a vector is indicated by a single index.

It is possible to differentiate between tensors of different levels. Zero level tensors are scalars. They do not change by transferring to another coordinate system. Examples of scalars are temperature ϑ, pressure p and density ϱ. No index is necessary for their characterisation.

First level tensors are vectors. As explained above, they are indicated by one index. Second level tensors are characterised by two indices. The stress tensor is such a quantity. It has nine components τ_{11}, τ_{12}, τ_{13}, $\tau_{21} \ldots \tau_{33}$. The abbreviation normally written is τ_{ji}, where j and i each assume the values 1, 2, 3. In calculations with tensors, the following rule is used. If an index only appears once in a term of an equation it is called a *free index*. It can be replaced by any other index. All terms in an equation must agree in their free indices. The relationship

$$a_i = c \cdot b_i \; , \tag{A.1}$$

whereby c is a constant (scalar), means that the vectors a_i and b_i only differ in their amount, they have the same direction. So

$$a_1 = c\, b_1 \; ; \quad a_2 = c\, b_2 \; ; \quad a_3 = c\, b_3 \; .$$

The *internal product* (scalar product) of two vectors

$$\boldsymbol{a} \cdot \boldsymbol{b} = a_1\, b_1 + a_2\, b_2 + a_3\, b_3$$

looks like the folowing for the index notation

$$\boldsymbol{a} \cdot \boldsymbol{b} = \sum_{i=1}^{3} a_i\, b_i \; .$$

As internal products appear frequently the following "*summation convention*" was settled: *if an index appears twice in a term, then it should be summed over this index*. This index is called a *bound index*. It can not be replaced by any other index. The summation symbol is left out. With that we have

$$a_i\, b_i = a_1\, b_1 + a_2\, b_2 + a_3\, b_3 \; . \tag{A.2}$$

Differentiation leads to a tensor that is one order higher. So the gradient of a scalar p

$$\mathrm{grad} p = \nabla p = \boldsymbol{e}_1 \frac{\partial p}{\partial x_1} + \boldsymbol{e}_2 \frac{\partial p}{\partial x_2} + \boldsymbol{e}_3 \frac{\partial p}{\partial x_3} \tag{A.3}$$

is a vector with the three components $\partial p/\partial x_i$, $i = 1, 2, 3$, which is abbreviated to just $\partial p/\partial x_i$ in tensor notation. If we differentiate a vector w_j, each of the three components w_1, w_2, w_3 can be differentiated with respect to each position coordinate x_1, x_2, x_3. This gives a second level tensor

$$\frac{\partial w_j}{\partial x_i} \quad (i = 1, 2, 3 \; ; \; j = 1, 2, 3) \; , \tag{A.4}$$

that consists of 9 components. On the other hand, the divergence of a vector is a scalar,

$$\mathrm{div} \boldsymbol{w} = \nabla \cdot \boldsymbol{w} = \frac{\partial w_i}{\partial x_i} = \frac{\partial w_1}{\partial x_1} + \frac{\partial w_2}{\partial x_2} + \frac{\partial w_3}{\partial x_3} \; . \tag{A.5}$$

The formulation of a divergence produces a tensor one order lower than the original tensor. A sensible 'operator' is the *Kronecker delta* δ_{ij}, defined by

$$\delta_{ij} = 1 \text{ for } i = j \; , \quad \delta_{ij} = 0 \text{ for } i \neq j \; . \tag{A.6}$$

δ_{ij} is also called the *unit tensor*. It further holds that

$$\delta_{ij}\,b_j = b_i \ , \tag{A.7}$$

which can be confirmed by writing the whole equation out, because it is

$$\delta_{ij}\,b_j = \delta_{i1}\,b_1 + \delta_{i2}\,b_2 + \delta_{i3}\,b_3$$

that is, for $i = 1$: $\delta_{1j}\,b_j = \delta_{11}\,b_1 = b_1$, for $i = 2$: $\delta_{2j}\,b_j = \delta_{22}\,b_2 = b_2$ and for $i = 3$: $\delta_{3j}\,b_j = \delta_{33}\,b_3 = b_3$, and with that $\delta_{ij}\,b_j = b_i$.

A.2 Relationship between mean and thermodynamic pressure

By definition, the mean pressure $\bar{p} = -1/3\,\delta_{ji}\,\tau_{kk}$ only includes normal stresses. In order to create a link between mean and thermodynamic pressure we will consider a cubic fluid element at a temperature T and of specific volume v, Fig. A1. We will now assume that the cube is at rest at time $t = 0$, so that the thermodynamic pressure p prevails inside the element. Now let us assume the mean pressure $\bar{p}$ is being exerted on the element from outside. When $\bar{p} > p$ the cube is compressed, should $\bar{p} < p$ then it expands. So, work $-\bar{p}\,dV$ is carried out by the external pressure $\bar{p}$. This is equal to the work done during the volume change in the gas $-p\,dV$ and the dissipated work. It therefore holds that $dW = -\bar{p}\,dV = -p\,dV + dW_{\text{diss}}$ with the the dissipated work as

$$dW_{\text{diss}} = -(\bar{p} - p)\,dV \ .$$

This is always positive according to the second law, because for $\bar{p} > p$ we have $dV < 0$ and for $\bar{p} < p$, $dV > 0$. On the other hand, the increase in the volume dV is yielded from the transport theory (3.18) with $Z = V$ and $z = Z/M = V/M = 1/\varrho$ to be

$$\frac{dV}{dt} = \int_{V(t)} \frac{\partial w_i}{\partial x_i}\,dV \ .$$

The dissipated work can also be written as

$$dW_{\text{diss}} = -(\bar{p} - p)\left(\int_{V(t)} \frac{\partial w_i}{\partial x_i}\,dV \right) dt \ .$$

It is clearly reasonable that the speed dV/dt of the volume change or $\partial w_i/\partial x_i$ is a monotonically decaying function of $\bar{p} - p$, Fig. A2, as the larger the overpressure $\bar{p} - p$, the faster the volume of the cube reduces. It is therefore sugested, that

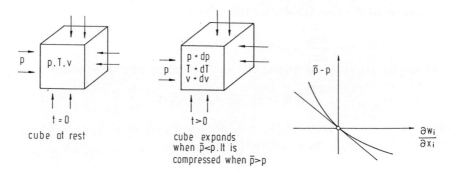

Fig. A.1: For the link between mean and thermodynamic pressure

Fig. A.2: Expansion as a
function of the over pressure

where the speed of the volume change is not that fast, the curve in Fig. A2 may
be replaced by a straight line:

$$\bar{p} - p = -\zeta \frac{\partial w_i}{\partial x_i} \; .$$

The factor defined by this, $\zeta > 0$, is the *volume viscosity* (SI units kg/s m).
It has to be determined either experimentally or using methods of statistical
thermodynamics, which is only possible for substances with simple molecules. It
can be seen that the mean and thermodynamic pressures only strictly agree when
$\zeta = 0$ or the fluid is incompressible, $\partial w_i/\partial x_i = 0$.

A.3   Navier-Stokes equations for an incompress-
ible fluid of constant viscosity in cartesian
coordinates

The mass force is the acceleration due to gravity $k_j = g_j$.
$x_1 = x$-direction:

$$\varrho \left(\frac{\partial w_1}{\partial t} + w_1 \frac{\partial w_1}{\partial x_1} + w_2 \frac{\partial w_1}{\partial x_2} + w_3 \frac{\partial w_1}{\partial x_3} \right) =$$
$$\varrho \, g_1 - \frac{\partial p}{\partial x_1} + \eta \left(\frac{\partial^2 w_1}{\partial x_1{}^2} + \frac{\partial^2 w_1}{\partial x_2{}^2} + \frac{\partial^2 w_1}{\partial x_3{}^2} \right) \; . \tag{A.8}$$

$x_2 = y$-direction:

$$\varrho \left(\frac{\partial w_2}{\partial t} + w_1 \frac{\partial w_2}{\partial x_1} + w_2 \frac{\partial w_2}{\partial x_2} + w_3 \frac{\partial w_2}{\partial x_3} \right) =$$
$$\varrho \, g_2 - \frac{\partial p}{\partial x_2} + \eta \left(\frac{\partial^2 w_2}{\partial x_1{}^2} + \frac{\partial^2 w_2}{\partial x_2{}^2} + \frac{\partial^2 w_3}{\partial x_3{}^2} \right) \; . \tag{A.9}$$

$x_3 = z$-direction:

$$\varrho \left(\frac{\partial w_3}{\partial t} + w_1 \frac{\partial w_3}{\partial x_1} + w_2 \frac{\partial w_3}{\partial x_2} + w_3 \frac{\partial w_3}{\partial x_3} \right) =$$

$$\varrho\, g_3 - \frac{\partial p}{\partial x_3} + \eta \left(\frac{\partial^2 w_3}{\partial x_1{}^2} + \frac{\partial^2 w_3}{\partial x_2{}^2} + \frac{\partial^2 w_3}{\partial x_3{}^2} \right) \,. \tag{A.10}$$

A.4 Navier-Stokes equations for an incompressible fluid of constant viscosity in cylindrical coordinates

The mass force is the acceleration due to gravity $k_j = g_j$.
r-direction:

$$\varrho \left(\frac{\partial w_r}{\partial t} + w_r \frac{\partial w_r}{\partial r} + \frac{w_\theta}{r} \frac{\partial w_r}{\partial \theta} - \frac{w_\theta^2}{r} + w_z \frac{\partial w_r}{\partial z} \right) =$$

$$\varrho\, g_r - \frac{\partial p}{\partial r} + \eta \left[\frac{\partial}{\partial r} \left(\frac{1}{r} \frac{\partial}{\partial r}(r\, w_r) \right) + \frac{1}{r^2} \frac{\partial^2 w_r}{\partial \theta^2} - \frac{2}{r^2} \frac{\partial w_\theta}{\partial \theta} + \frac{\partial^2 w_r}{\partial z^2} \right] \,. \tag{A.11}$$

θ-direction:

$$\varrho \left(\frac{\partial w_\theta}{\partial t} + w_r \frac{\partial w_\theta}{\partial r} + \frac{w_\theta}{r} \frac{\partial w_\theta}{\partial \theta} + \frac{w_r\, w_\theta}{r} + w_z \frac{\partial w_\theta}{\partial z} \right) =$$

$$\varrho\, g_\theta - \frac{1}{r} \frac{\partial p}{\partial \theta} + \eta \left[\frac{\partial}{\partial r} \left(\frac{1}{r} \frac{\partial}{\partial r}(r\, w_\theta) \right) + \frac{1}{r^2} \frac{\partial^2 w_\theta}{\partial \theta^2} + \frac{2}{r^2} \frac{\partial w_r}{\partial \theta} + \frac{\partial^2 w_\theta}{\partial z^2} \right] \,. \tag{A.12}$$

z-direction:

$$\varrho \left(\frac{\partial w_z}{\partial t} + w_r \frac{\partial w_z}{\partial r} + \frac{w_\theta}{r} \frac{\partial w_z}{\partial \theta} + w_z \frac{\partial w_z}{\partial z} \right) =$$

$$\varrho\, g_z - \frac{\partial p}{\partial z} + \eta \left[\frac{1}{r} \frac{\partial}{\partial r} \left(r \frac{\partial^2 w_z}{\partial r^2} \right) + \frac{1}{r^2} \frac{\partial^2 w_z}{\partial \theta^2} + \frac{\partial^2 w_z}{\partial \theta^2} \right] \,. \tag{A.13}$$

A.5 Entropy balance for mixtures

The Gibb's fundamental equation for mixtures,

$$\mathrm{d}u = T\, \mathrm{d}s - p\, \mathrm{d}v + \sum_K \frac{\mu_K}{\tilde{M}_K}\, \mathrm{d}\xi_K$$

or

$$\varrho \frac{du}{dt} = \varrho\, T \frac{ds}{dt} - \varrho\, p \frac{dv}{dt} + \varrho \sum_K \frac{\mu_K}{\tilde{M}_K} \frac{d\xi_K}{dt} \;, \qquad (A.14)$$

taking into account

$$\frac{dv}{dt} = -\frac{1}{\varrho^2} \frac{d\varrho}{dt} = \frac{1}{\varrho} \frac{\partial w_i}{\partial x_i}$$

and (3.25)

$$\varrho \frac{d\xi_K}{dt} = -\frac{\partial j^*_{K,i}}{\partial x_i} + \dot{\Gamma}_K \qquad (A.15)$$

can be rearranged into

$$\varrho \frac{du}{dt} = \varrho\, T \frac{ds}{dt} - p \frac{\partial w_i}{\partial x_i} - \sum_K \frac{\mu_K}{\tilde{M}_K} \frac{\partial j^*_{K,i}}{\partial x_i} + \sum_K \frac{\mu_K}{\tilde{M}_K} \dot{\Gamma}_K \;.$$

Due to $\mu_K/\tilde{M}_K = h_K - T\, s_K$ this delivers

$$\varrho \frac{du}{dt} = \varrho\, T \frac{ds}{dt} - p \frac{\partial w_i}{\partial x_i} - \sum_K h_K \frac{\partial j^*_{K,i}}{\partial x_i} + T \sum_K s_K \frac{\partial j^*_{K,i}}{\partial x_i} + \sum_K \frac{\mu_K}{\tilde{M}_K} \dot{\Gamma}_K \;.$$

Putting this into the energy equation (3.81), taking into account the following, yields

$$\dot{q}' = \dot{q}_i + \sum_K h_K\, j^*_{K,i}$$

$$-\frac{\partial \dot{q}_i}{\partial x_i} + \Phi + \sum_K j^*_{K,i}\, k_{K,i} = \varrho\, T \frac{ds}{dt} + \sum_K j^*_{K,i} \frac{\partial h_K}{\partial x_i} + T \sum_K s_K \frac{\partial j^*_{K,i}}{\partial x_i} + \sum_K \frac{\mu_K}{\tilde{M}_K} \dot{\Gamma}_K \;.$$
$$(A.16)$$

We write the following for

$$T \sum_K s_K \frac{\partial j^*_{K,i}}{\partial x_i} = T \sum_K \frac{\partial(j^*_{K,i}\, s_K)}{\partial x_i} - T \sum_K j^*_{K,i} \frac{\partial s_K}{\partial x_i}$$

and combine to give

$$\sum_K j^*_{K,i} \left(\frac{\partial h_K}{\partial x_i} - T \frac{\partial s_K}{\partial x_i} \right) = \sum_K j^*_{K,i} \left(\frac{\partial \mu_K}{\partial x_i} \right)_T \frac{1}{\tilde{M}_K} \;.$$

Then (A.16) becomes

$$-\frac{\partial \dot{q}_i}{\partial x_i} + \Phi + \sum_K j^*_{K,i}\, k_{K,i} = \varrho\, T \frac{ds}{dt} + T \sum_K \frac{\partial(j^*_{K,i}\, s_K)}{\partial x_i}$$
$$+ \sum_K j^*_{K,i} \left(\frac{\partial \mu_K}{\partial x_i} \right)_T \frac{1}{\tilde{M}_K} + \sum_K \frac{\mu_K}{\tilde{M}_K} \dot{\Gamma}_K \;.$$

And so

$$\varrho \frac{ds}{dt} = -\frac{\partial(\dot{q}_i/T)}{\partial x_i} - \sum_K \frac{\partial(j^*_{K,i} s_K)}{\partial x_i} - \frac{\dot{q}_i}{T^2} \frac{\partial T}{\partial x_i}$$
$$+ \frac{1}{T} \sum_K j^*_{K,i} \left[k_{K,i} - \frac{1}{\tilde{M}_K} \left(\frac{\partial \mu_K}{\partial x_i} \right)_T \right] + \frac{\Phi}{T} + \frac{1}{T} \sum_K \frac{\mu_K}{\tilde{M}_K} \dot{\Gamma}_K .$$

or

$$\varrho \frac{ds}{dt} = -\frac{\partial J_S}{\partial x_i} + \dot{\sigma} \qquad (A.17)$$

with the entropy flow J_S (SI units W/m²K)

$$J_S = \frac{\dot{q}_i}{T} + \sum_K j^*_{K,i} s_K \qquad (A.18)$$

and the entropy generation $\dot{\sigma}$ (SI units W/m³K)

$$\dot{\sigma} = -\frac{\dot{q}_i}{T^2} \frac{\partial T}{\partial x_i} + \frac{1}{T} \sum_K j^*_{K,i} \left[k_{K,i} - \frac{1}{\tilde{M}_K} \left(\frac{\partial \mu_K}{\partial x_i} \right)_T \right] + \frac{\Phi}{T} + \frac{1}{T} \sum_K \frac{\mu_K}{\tilde{M}_K} \dot{\Gamma}_K . \quad (A.19)$$

The entropy flow is based on a heat or material flow, the entropy generation on a heat flow in a temperature field, diffusion by mass forces and differences in the chemical potential, mechanical dissipation and chemical reactions.

A.6 Relationship between partial and specific enthalpy

When the following summation is written out with the abbreviation $\delta = \partial/\partial x_i$ it looks like

$$\sum_{K=1}^N j^*_{K,i} \frac{\partial h_K}{\partial x_i} = j^*_{1,i} \delta h_1 + j^*_{2,i} \delta h_2 + \ldots j^*_{N,i} \delta h_N .$$

Now $\sum_{K=1}^N j^*_{K,i} = 0$, also $j^*_{N,i} = -j^*_{1,i} - j^*_{2,i} - \ldots - j^*_{N-1,i}$. The sum can also be written as

$$j^*_{1,i} \delta(h_1 - h_N) + j^*_{2,i} \delta(h_2 - h_N) + \ldots j^*_{N-1,i} \delta(h_{N-1} - h_N) .$$

In the thermodynamics of mixtures [3.1], page 114, it can be shown, that for specific partial quantities of state the following relationship is valid

$$h_A - h_N = \left(\frac{\partial h}{\partial \xi_A} \right)_{T,p,K \neq A} , \qquad A = 1,2,\ldots N-1 .$$

It holds therefore that

$$\sum_{K=1}^{N} j_{K,i}^* \frac{\partial h_K}{\partial x_i} = \sum_{K=1}^{N-1} j_{K,i}^* \delta \left(\frac{\partial h}{\partial \xi_K} \right)_{T,p,A \neq K} = \sum_{K=1}^{N-1} j_{K,i}^* \frac{\partial}{\partial x_i} \left(\frac{\partial h}{\partial \xi_K} \right)_{T,p,A \neq K} .$$

On the other hand, the enthalpy of a mixture is given by

$$h = h(T, p, \xi_1, \xi_2, \ldots \xi_{N-1})$$

and therefore

$$\left(\frac{\partial h}{\partial \xi_N} \right)_{T,p,\xi_K} = 0 .$$

So the sum can also run from $K = 1$ to $K = N$, instead of from $K = 1$ to $K = N - 1$, as the last term with $K = N$ is zero.

A.7 Calculation of the constants a_n of a Graetz-Nusselt problem (3.243)

Multiplication of (3.244) with $\psi_m(r^+)(1 - r^{+2})r^+$ and integration between the limits $r^+ = 0$ and $r^+ = 1$ yields

$$\int_0^1 \psi_m (1 - r^{+2}) r^+ \, \mathrm{d}r^+ = \sum_{n=0}^{\infty} a_n \int_0^1 \psi_m \psi_n (1 - r^{+2}) r^+ \, \mathrm{d}r^+ .$$

All the integrals on the right hand side disappear when $m \neq n$, yielding (3.245). In order to show that this applies, we write (3.242) for the eigenvalues β_m and β_n:

$$\frac{\mathrm{d}}{\mathrm{d}r^+} (r^+ \psi_m') = -\beta_m^2 (1 - r^{+2}) r^+ \psi_m$$

$$\frac{\mathrm{d}}{\mathrm{d}r^+} (r^+ \psi_n') = -\beta_n^2 (1 - r^{+2}) r^+ \psi_n .$$

Multiplication of the first equation with ψ_n and of the second with ψ_m gives

$$\psi_n \frac{\mathrm{d}}{\mathrm{d}r^+} (r^+ \psi_m') = -\beta_m^2 (1 - r^{+2}) r^+ \psi_m \psi_n$$

$$\psi_m \frac{\mathrm{d}}{\mathrm{d}r^+} (r^+ \psi_n') = -\beta_n^2 (1 - r^{+2}) r^+ \psi_n \psi_m ,$$

which can also be written as

$$\frac{\mathrm{d}}{\mathrm{d}r^+} (r^+ \psi_n \psi_m') - r^+ \psi_n' \psi_m' = -\beta_m^2 (1 - r^{+2}) r^+ \psi_m \psi_n$$

$$\frac{\mathrm{d}}{\mathrm{d}r^+} (r^+ \psi_n' \psi_m) - r^+ \psi_n' \psi_m' = -\beta_n^2 (1 - r^{+2}) r^+ \psi_n \psi_m .$$

Subtraction of both equations and integration between the limits $r^+ = 0$ and $r^+ = 1$ yields

$$r^+ (\psi_n \psi_m' - \psi_n' \psi_m) \Big|_0^1 = (\beta_n^2 - \beta_m^2) \int_0^1 \psi_m \psi_n (1 - r^{+2}) r^+ \, \mathrm{d}r^+ \ .$$

The left hand side disappears because of $\psi_m(r^+ = 1) = \psi_n(r^+ = 1) = 0$ and $\psi_m'(r^+ = 0) = \psi_n'(r^+ = 0) = 0$. This means that the integral on the right hand side must also vanish when $n \neq m$.

Appendix B: Property data

Table B 1: Properties of air at pressure $p = 1$ bar

ϑ °C	ϱ kg/m^3	c_p kJ/kg K	β 10^{-3}/K	λ 10^{-3}W/K m	ν 10^{-7}m^2/s	a 10^{-7}m^2/s	Pr —
−200	5.106	1.186	17.24	6.886	9.786	11.37	0.8606
−180	3.851	1.071	11.83	8.775	17.20	21.27	0.8086
−160	3.126	1.036	9.293	10.64	25.58	32.86	0.7784
−140	2.639	1.010	7.726	12.47	35.22	46.77	0.7530
−120	2.287	1.014	6.657	14.26	46.14	61.50	0.7502
−100	2.019	1.011	5.852	16.02	58.29	78.51	0.7423
−80	1.807	1.009	5.227	17.74	71.59	97.30	0.7357
−60	1.636	1.007	4.725	19.41	85.98	117.8	0.7301
−40	1.495	1.007	4.313	21.04	101.4	139.7	0.7258
−20	1.377	1.007	3.968	22.63	117.8	163.3	0.7215
0	1.275	1.006	3.674	24.18	135.2	188.3	0.7179
20	1.188	1.007	3.421	25.69	153.5	214.7	0.7148
40	1.112	1.007	3.200	27.16	172.6	242.4	0.7122
80	0.9859	1.010	2.836	30.01	213.5	301.4	0.7083
100	0.9329	1.012	2.683	31.39	235.1	332.6	0.7070
120	0.8854	1.014	2.546	32.75	257.5	364.8	0.7060
140	0.8425	1.016	2.422	34.08	280.7	398.0	0.7054
160	0.8036	1.019	2.310	35.39	304.6	432.1	0.7050
180	0.7681	1.022	2.208	36.68	329.3	467.1	0.7049
200	0.7356	1.026	2.115	37.95	354.7	503.0	0.7051
300	0.6072	1.046	1.745	44.09	491.8	694.3	0.7083
400	0.5170	1.069	1.486	49.96	645.1	903.8	0.7137
500	0.4502	1.093	1.293	55.64	813.5	1131	0.7194
600	0.3986	1.116	1.145	61.14	996.3	1375	0.7247
700	0.3576	1.137	1.027	66.46	1193	1635	0.7295
800	0.3243	1.155	0.9317	71.54	1402	1910	0.7342
900	0.2967	1.171	0.8523	76.33	1624	2197	0.7395
1000	0.2734	1.185	0.7853	80.77	1859	2492	0.7458

Table B 2: Properties of water at pressure $p = 1$ bar

Liquid water

ϑ °C	ϱ kg/m^3	c_p kJ/kg K	β 10^{-3}/K	λ 10^{-3}W/K m	ν 10^{-6}m^2/s	a 10^{-6}m^2/s	Pr —
0	999.84	4.218	−0.0672	561.0	1.793	0.1330	13.48
5	999.97	4.203	0.0162	570.5	1.519	0.1358	11.19
10	999.70	4.192	0.0879	580.0	1.307	0.1384	9.443
15	999.10	4.185	0.1507	589.3	1.139	0.1409	8.082
20	998.21	4.181	0.2067	598.4	1.004	0.1434	7.001
25	997.05	4.179	0.2572	607.2	0.893	0.1457	6.128
30	995.65	4.177	0.3034	615.5	0.801	0.1480	5.414
35	994.03	4.177	0.3459	623.3	0.724	0.1501	4.823
40	992.22	4.177	0.3855	630.6	0.658	0.1521	4.328
45	990.21	4.178	0.4226	637.3	0.602	0.1540	3.909
50	988.04	4.180	0.4578	643.6	0.554	0.1558	3.553
55	985.69	4.182	0.4912	649.2	0.512	0.1575	3.248
60	983.20	4.184	0.5232	654.4	0.475	0.1591	2.983
65	980.55	4.187	0.5541	659.0	0.442	0.1605	2.754
70	977.77	4.190	0.5840	663.1	0.413	0.1619	2.553
75	974.84	4.193	0.6130	666.8	0.388	0.1631	2.376
80	971.79	4.197	0.6414	670.0	0.365	0.1643	2.221
85	968.61	4.201	0.6693	672.8	0.344	0.1653	2.082
90	965.31	4.206	0.6967	675.2	0.326	0.1663	1.959
95	961.89	4.211	0.7238	677.3	0.309	0.1672	1.849
99.63[a]	958.61	4.216	0.7487	678.9	0.295	0.1680	1.757

[a] Saturated state

Water vapour

ϑ °C	ϱ kg/m^3	c_p kJ/kg K	β 10^{-3}/K	λ 10^{-3}W/K m	ν 10^{-6}m^2/s	a 10^{-6}m^2/s	Pr —
100	0.5896	2.042	2.881	25.08	20.81	20.83	0.9990
150	0.5164	1.980	2.452	28.85	27.46	28.22	0.9733
200	0.4604	1.975	2.160	33.28	35.14	36.60	0.9600
250	0.4156	1.990	1.938	38.17	43.83	46.15	0.9497
300	0.3790	2.013	1.761	43.42	53.54	56.92	0.9406
350	0.3483	2.040	1.616	48.96	64.22	68.90	0.9321
400	0.3223	2.070	1.493	54.76	75.86	82.07	0.9243
450	0.2999	2.102	1.388	60.77	88.42	96.40	0.9172
500	0.2805	2.135	1.297	66.97	101.9	111.9	0.9107
600	0.2483	2.203	1.147	79.89	131.4	146.1	0.8993
700	0.2227	2.273	1.029	93.37	164.1	184.2	0.8899
800	0.2019	2.343	0.9327	107.3	199.9	226.8	0.8816
900	0.1847	2.412	0.8530	121.7	238.6	273.0	0.8739
1000	0.1702	2.478	0.7859	163.3	280.0	323.2	0.8665

Table B 3: Properties of water in the saturated state from the triple point to the critical point

ϑ °C	p bar	ϱ'	ϱ'' kg/m^3	c_p'	c_p'' kJ/kg K	β'	β'' 10^{-3}/K	Δh_v kJ/kg
0.01	0.006117	999.78	0.004855	4.229	1.868	−0.08044	3.672	2500.5
10	0.012281	999.69	0,009404	4.188	1.882	0.08720	3.548	2476.9
20	0.023388	998.19	0.01731	4.183	1.882	0.2089	3.435	2453.3
30	0.042455	995.61	0.03040	4.183	1.892	0.3050	3.332	2429.7
40	0.073814	992.17	0.05121	4.182	1.904	0.3859	3.240	2405.9
50	0.12344	987.99	0.08308	4.182	1.919	0.4572	3.156	2381.9
60	0.19932	983.16	0.13030	4.183	1.937	0.5222	3.083	2357.6
70	0.31176	977.75	0.19823	4.187	1.958	0.5827	3.018	2333.1
80	0.47373	971.79	0.29336	4.194	1.983	0.6403	2.964	2308.1
90	0.70117	965.33	0.42343	4.204	2.011	0.6958	2.919	2282.7
100	1.0132	958.39	0.59750	4.217	2.044	0.7501	2.884	2256.7
110	1.4324	951,00	0.82601	4.232	2.082	0.8038	2.860	2229.9
120	1.9848	943.16	1.1208	4.249	2.126	0.8576	2.846	2202.4
130	2.7002	934.88	1.4954	4.267	2.176	0.9123	2.844	2174.0
140	3.6119	926.18	1.9647	4.288	2.233	0.9683	2.855	2144.6
150	4.7572	917.06	2.5454	4.312	2.299	1.026	2.878	2114.1
160	6.1766	907.50	3.2564	4.339	2.374	1.087	2.916	2082.3
170	7.9147	897.51	4.1181	4.369	2.460	1.152	2.969	2049.2
180	10.019	887.06	5.1539	4.403	2.558	1.221	3.039	2014.5
190	12.542	876.15	6.3896	4.443	2.670	1.296	3.128	1978.2
200	15.536	864.74	7.8542	4.489	2.797	1.377	3.238	1940.1
250	39.736	799.07	19.956	4.857	3.772	1.955	4.245	1715.4
300	85.838	712.41	46.154	5.746	5.981	3.273	7.010	1404.7
350	165.21	574.69	113.48	10.13	16.11	10.37	22.12	893.03
373.976	220.55	322,00	322,00	∞	∞	∞	∞	0

Table B 3: (Continued)

ϑ °C	λ' 10^{-3}W/K m	λ''	ν' 10^{-6} m^2/s	ν''	a' 10^{-6} m^2/s	a''	Pr' —	Pr''	σ 10^{-3}N/m
0.01	561.0	17.07	1.792	1898.0	0.1327	1883.0	13.51	1,008	75.65
10	580.0	17.62	1.307	1006.0	0.1385	999.8	9.434	1,006	74.22
20	598.4	18.23	1,004	562.0	0.1433	559.6	7,005	1,004	72.74
30	615.4	18.89	0.8012	329.3	0.1478	328.3	5.422	1,003	71.20
40	630.5	19.60	0.6584	201.3	0.1519	200.9	4.333	1,002	69.60
50	643.5	20.36	0.5537	127.8	0.1558	127.7	3.555	1,001	67.95
60	654.3	21.18	0.4746	83.91	0.1591	83.92	2.983	1,000	66.24
70	663.1	22.07	0.4132	56.80	0.1620	56.85	2.551	0.9992	64.49
80	670.0	23.01	0.3648	39.51	0.1644	39.56	2.219	0.9989	62.68
90	675.3	24.02	0.3258	28.17	0.1664	28.20	1.958	0.9989	60.82
100	679.1	25.09	0.2941	20.53	0.1680	20.55	1.750	0.9994	58.92
110	681.7	26.24	0.2680	15.27	0.1694	15.26	1.582	1,001	56.97
120	683.2	27.46	0.2462	11.56	0.1705	11.53	1.444	1,003	54.97
130	683.7	28.76	0.2278	8.894	0.1714	8.840	1.329	1,006	52.94
140	683.3	30.14	0.2123	6.946	0.1720	6.869	1.234	1.011	50.86
150	682.1	31.59	0.1991	5.496	0.1725	5.399	1.154	1.018	48.75
160	680.0	33.12	0.1877	4.402	0.1727	4.285	1.087	1.027	46.60
170	677.1	34.74	0.1779	3.565	0.1727	3.430	1.030	1.039	44.41
180	673.4	36.44	0.1693	2.915	0.1724	2.764	0.9822	1.055	42.20
190	668.8	38.23	0.1619	2.405	0.1718	2.241	0.9423	1.073	39.95
200	663.4	40.10	0.1554	2,001	0.1709	0.825	0.9093	1.096	37.68
250	621.4	51.23	0.1329	0.8766	0.1601	0.6804	0.8299	1.288	26.05
300	547.7	69.49	0.1207	0.4257	0.1338	0.2517	0.9018	1.691	14.37
350	447.6	134.6	0.1146	0.2098	0.07692	0.07365	1.490	2.849	3.675
373.976	141.9	141.9	0.1341	0.1341	0	0	∞	∞	0

Table B 4: Properties of ammonia at pressure $p = 1$ bar

ϑ °C	ϱ kg/m^3	c_p kJ/kg K	β 10^{-3}/K	λ 10^{-3}W/K m	ν 10^{-7}m^2/s	a 10^{-7}m^2/s	Pr —
−50	702.1	4.434	1.685	—	—	—	—
−40	690.1	4.441	1.763	601.9	4.093	1.964	2.084
−30	0.8645	2.309	4.638	17.84	93.35	89.37	1.045
−20	0.8266	2.242	4.342	18.94	101.9	102.2	0.9973
−10	0.7925	2.202	4.100	20.06	110.8	115.0	0.9637
0	0.7615	2.178	3.896	21.22	120.1	127.9	0.9390
10	0.7330	2.165	3.719	22.39	129.8	141.1	0.9203
20	0.7068	2.160	3.563	23.59	139.9	154.4	0.9060
30	0.6826	2.161	3.423	24.80	150.4	168.1	0.8947
40	0.6600	2.166	3.297	26.04	161.3	182.1	0.8858
50	0.6390	2.174	3.181	27.29	172.6	196.4	0.8786
60	0.6193	2.185	3.075	28.56	184.2	211.1	0.8729
70	0.6009	2.197	2.977	29.84	196.3	226.0	0.8683
80	0.5835	2.210	2.886	31.13	208.7	241.4	0.8647
90	0.5672	2.225	2.801	32.44	221.5	257.1	0.8618
100	0.5517	2.241	2.721	33.76	234.7	273.1	0.8595
110	0.5371	2.257	2.646	35.09	248.3	289.5	0.8578
120	0.5233	2.274	2.576	36.43	262.2	306.2	0.8565
130	0.5101	2.291	2.509	37.79	276.6	323.3	0.8555
140	0.4976	2.309	2.446	39.15	291.3	340.7	0.8549
150	0.4858	2.328	2.386	40.53	306.3	358.4	0.8546
200	0.4340	2.422	2.127	47.59	387.1	452.6	0.8553
250	0.3923	2.518	1.921	54.93	477.0	556.0	0.8579
300	0.3580	2.612	1.751	62.54	576.0	668.9	0.8611

Table B 5: Properties of carbon dioxide at pressure $p = 1$ bar

ϑ °C	ϱ kg/m^3	c_p kJ/kg K	β 10^{-3}/K	λ 10^{-3}W/K m	ν 10^{-7}m^2/s	a 10^{-7}m^2/s	Pr —
−50	2.403	0.7825	4.682	11.10	46.69	59.05	0.7907
−40	2.296	0.7903	4.453	11.77	51.05	64.85	0.7873
−30	2.198	0.7988	4.248	12.45	55.60	70.92	0.7839
−20	2.109	0.8078	4.063	13.17	60.32	77.28	0.7805
−10	2.027	0.8172	3.896	13.90	65.21	83.94	0.7769
0	1.951	0.8267	3.742	14.66	70.28	90.89	0.7732
10	1.880	0.8363	3.601	15.43	75.51	98.13	0.7695
20	1.815	0.8459	3.471	16.22	80.92	105.7	0.7659
30	1.754	0.8555	3.351	17.03	86.49	113.5	0.7623
40	1.697	0.8650	3.239	17.84	92.22	121.5	0.7589
50	1.644	0.8744	3.134	18.67	98.12	129.8	0.7557
60	1.594	0.8837	3.037	19.50	104.2	138.4	0.7526
70	1.547	0.8929	2.945	20.34	110.4	147.2	0.7498
80	1.503	0.9018	2.859	21.18	116.8	156.3	0.7471
90	1.461	0.9107	2.778	22.03	123.3	165.5	0.7447
100	1.422	0.9193	2.702	22.87	129.9	175.0	0.7425
120	1.349	0.9361	2.561	24.57	143.7	194.6	0.7386
140	1.283	0.9523	2.435	26.27	158.1	215.0	0.7353
160	1.224	0.9678	2.321	27.96	173.0	236.1	0.7327
180	1.169	0.9827	2.217	29.64	188.4	257.9	0.7306
200	1.120	0.9971	2.122	31.31	204.4	280.5	0.7289
300	0.9238	1.061	1.749	39.47	291.8	402.6	0.7248
400	0.7864	1.114	1.488	47.26	390.6	539.4	0.7242
500	0.6846	1.159	1.294	54.70	499.5	689.7	0.7242
600	0.6061	1.196	1.146	61.84	617.7	853.2	0.7239
700	0.5438	1.227	1.028	68.69	744.3	1030	0.7229
800	0.4931	1.253	0.9320	75.30	878.9	1219	0.7212
900	0.4511	1.275	0.8525	81.69	1021	1420	0.7189

Table B 6: Properties of nitrogen at pressure $p = 1$ bar

ϑ °C	ϱ kg/m³	c_p kJ/kg K	β 10^{-3}/K	λ 10^{-3}W/K m	ν 10^{-7}m²/s	a 10^{-7}m²/s	Pr —
−210	867.9	1.951	4.287	176.4	2.497	1.04	2.386
−200	827.4	2.053	5.270	156.9	1.957	0.9238	2.118
−190	4.195	1.102	13.31	8.061	13.04	17.44	0.7477
−180	3.707	1.081	11.53	9.108	16.73	22.74	0.7357
−160	3.019	1.061	9.199	11.13	25.19	34.75	0.7248
−120	2.212	1.048	6.643	14.86	46.06	64.11	0.7184
−100	1.953	1.045	5.847	16.59	58.30	81.27	0.7173
−90	1.845	1.044	5.518	17.43	64.83	90.42	0.7170
−80	1.749	1.044	5.224	18.24	71.63	99.95	0.7167
−70	1.662	1.043	4.961	19.04	78.69	109.8	0.7165
−60	1.583	1.043	4.724	19.83	86.01	120.1	0.7163
−50	1.512	1.042	4.508	20.59	93.58	130.7	0.7162
−40	1.447	1.042	4.312	21.35	101.4	141.6	0.7160
−30	1.387	1.042	4.132	22.09	109.4	152.8	0.7159
−20	1.332	1.042	3.967	22.81	117.7	164.4	0.7159
−10	1.281	1.041	3.814	23.53	126.2	176.3	0.7158
0	1.234	1.041	3.673	24.23	134.9	188.5	0.7158
10	1.190	1.041	3.542	24.92	143.9	201.0	0.7157
20	1.150	1.041	3.420	25.60	153.1	213.8	0.7157
30	1.112	1.041	3.307	26.27	162.4	227.0	0.7157
40	1.076	1.041	3.200	26.93	172.0	240.3	0.7157
50	1.043	1.042	3.101	27.59	181.8	254.0	0.7157
60	1.011	1.042	3.007	28.23	191.8	268.0	0.7158
70	0.9818	1.042	2.918	28.87	202.0	282.2	0.7158
80	0.9539	1.042	2.836	29.50	212.4	296.7	0.7159
90	0.9276	1.043	2.757	30.13	223.0	311.5	0.7159
100	0.9027	1.043	2.683	30.75	233.8	326.5	0.7160
120	0.8568	1.044	2.546	31.97	255.9	357.3	0.7162
140	0.8153	1.046	2.422	33.18	278.8	389.1	0.7165
160	0.7776	1.048	2.310	34.37	302.4	421.9	0.7168
180	0.7433	1.050	2.208	35.55	326.8	455.6	0.7172
200	0.7118	1.053	2.114	36.72	351.8	490.2	0.7177
300	0.5876	1.070	1.745	42.47	487.1	675.8	0.7208
400	0.5003	1.092	1.485	48.12	638.4	880.9	0.7247
500	0.4356	1.116	1.293	53.68	804.8	1104	0.7288
600	0.3857	1.140	1.145	59.13	985.6	1345	0.7327
700	0.3461	1.162	1.027	64.45	1180	1603	0.7363
800	0.3139	1.182	0.9316	69.63	1388	1887	0.7394

Table B 7: Properties of oxygen at pressure $p = 1$ bar

ϑ °C	ϱ kg/m^3	c_p kJ/kg K	β 10^{-3}/K	λ W/K m	ν 10^{-7}m^2/s	a 10^{-7}m^2/s	Pr —
−210	1268	1,676	3.58	0.192	3.34	0.903	3.70
−200	1223	1.678	3.84	0.177	2.57	0.862	2.98
−190	1176	1.685	4.06	0.162	1.90	0.818	2.32
−180	4.254	0.9474	11.7	0.0860	16.0	21.3	0.75
−160	3.458	0.9304	9.24	0.0106	24.4	32.9	0.74
−140	2.921	0.9237	7.77	0.0126	34.3	46.7	0.73
−120	2.930	0.9192	6.67	0.0144	39.3	53.5	0.73
−100	2.233	0.9164	5.86	0.0162	57.9	79.2	0.73
−90	2.110	0.9154	5.53	0.0171	64.5	88.5	0.73
−80	1.999	0.9149	5.23	0.0179	71.5	97.9	0.73
−70	1.900	0.9146	4.97	0.0188	78.8	107	0.73
−60	1.810	0.9143	4.73	0.0196	86.2	118	0.73
−50	1.728	0.9141	4.50	0.0204	94.0	129	0.73
−40	1.653	0.9143	4.31	0.0211	102	140	0.73
−30	1.585	0.9147	4.13	0.0219	110	151	0.73
−20	1.522	0.9152	3.98	0.0227	119	163	0.73
−10	1.464	0.9159	3.82	0.0234	127	175	0.73
0	1.410	0.9167	3.67	0.0242	136	187	0.73
10	1.360	0.9177	3.54	0.0249	146	200	0.73
20	1.314	0.9189	3.43	0.0257	155	213	0.73
25	1.292	0.9195	3.38	0.0260	160	219	0.73

Table B 8: Properties of helium at pressure $p = 1.01325$ bar $= 1$ atm

T K	ϱ kg/m^3	c_p kJ/kgK	β 10^{-3}/K	λ 10^{-3}W/Km	ν 10^{-7}m^2/s	a 10^{-7}m^2/s	Pr —
4.222	16.84	9.144	628.8	9.038	0.7375	0.5869	1.26
5	11.98	6.770	329.2	9.537	1.022	1.179	0.867
6	9.164	6.025	223.2	11.17	1.564	2.023	0.773
8	6.433	5.581	143.9	13.97	2.779	3.891	0.714
10	5.016	5.429	108.5	16.38	4.181	6.015	0.695
20	2.440	5.251	50.65	26.15	13.70	20.41	0.671
40	1.216	5.206	25.00	41.07	43.19	64.88	0.666
60	0.8112	5.198	16.64	53.43	84.22	126.7	0.665
80	0.5952	5.196	12.20	64.45	135.2	203.7	0.665
100	0.4871	5.196	9.986	74.53	195.6	294.5	0.665
120	0.4060	5.194	8.323	84.09	265.0	398.8	0.665
140	0.3481	5.194	7.135	93.47	342.4	517.0	0.665
160	0.3046	5.193	6.244	101.8	428.1	643.6	0.665
180	0.2708	5.193	5.506	110.7	521.0	783.6	0.665
200	0.2437	5.193	4.996	118.2	621.3	934.0	0.665
220	0.2216	5.193	4.543	126.1	728.3	1096	0.665
240	0.2031	5.193	4.164	133.8	842.9	1269	0.665
260	0.1875	5.193	3.844	141.2	964.3	1450	0.665
280	0.1741	5.193	3.569	148.5	1092	1643	0.665
300	0.1625	5.193	3.332	155.7	1226	1845	0.655
350	0.1393	5.193	2.856	173.0	1589	2392	0.655
400	0.1219	5.193	2.499	189.6	1991	2995	0.655
500	0.09753	5.193	1.999	221.3	2904	4365	0.655
600	0.08128	5.193	1.666	251.3	3958	5954	0.655
700	0.06967	5.193	1.428	280.1	5147	7742	0.655
800	0.06096	5.193	1.250	307.9	6467	9726	0.665
900	0.05419	5.193	1.111	334.9	7911	11900	0.665
1000	0.04877	5.193	1.000	361.1	9479	14260	0.665
1100	0.04434	5.193	0.9090	386.7	11170	16790	0.665
1200	0.04065	5.193	0.8333	411.9	12910	19510	0.665
1300	0.03752	5.193	0.7692	436.5	14890	22400	0.665
1400	0.03484	5.193	0.7142	460.8	16940	25470	0.665
1500	0.03252	5.193	0.6666	484.7	19090	28700	0.665

Table B 9: Diffusion coefficients at pressure $p = 1.01325$ bar $= 1$ atm

a) Gases
The pressure and temperature dependency in the ideal gas state can be
estimated from $D \sim T^{1.75}/p$.

Substances	$N_2 - CO_2$	$N_2 - CH_4$	$N_2 - C_2H_6$	$N_2 - SF_6$
ϑ	D	D	D	D
°C	$10^{-4} m^2/s$	$10^{-4} m^2/s$	$10^{-4} m^2/s$	$10^{-4} m^2/s$
0	0.1391	0.1955	0.1302	0.0869
20	0.1583	0.2219	0.1481	0.0987
40	0.1785	0.2495	0.1669	0.1110
60	0.1997	0.2784	0.1865	0.1239
80	0.2217	0.3084	0.2069	0.1373
100	0.2446	0.3396	0.2281	0.1512
200	0.3714	0.5122	0.3455	0.2281
300	0.5175	0.7106	0.4806	0.3166
400	0.6814	0.9331	0.6322	0.4159
500	0.8621	1.1785	0.7993	0.5254

Substance	T	D
	K	$10^{-4} m^2/s$
$Air - CO_2$	276	0.144
	317	0.179
$Air - C_2H_5OH$	313	0.147
$Air - He$	276	0.632
$Air - H_2O$	313	0.292
$CO_2 - H_2O$	307	0.201
$He - H_2O$	352	1.136
$H_2 - H_2O$	307	0.927
$CH_4 - H_2O$	352	0.361

b) Diluted aqueous solutions

Substances	T	D
	K	$10^{-9} m^2/s$
$CH_4 - H_2O$	275	0.85
	333	3.55
$CO_2 - H_2O$	298	2,00
$CH_3OH - H_2O$	288	1.26
$C_2H_5OH - H_2O$	288	1.0
$O_2 - H_2O$	298	2.4
$N_2 - H_2O$	298	2.6
$H_2 - H_2O$	298	6.3

Table B 10: Thermophysical properties of non-metallic solids at 20 °C.

Substance	ϱ $10^3\,\mathrm{kg/m^3}$	c kJ/kgK	λ W/K m	a $10^{-6}\,\mathrm{m^2/s}$
Acrylic glass (plexiglass)	1.18	1.44	0.184	0.108
Asphalt	2.12	0.92	0.70	0.36
Bakelite	1.27	1.59	0.233	0.115
Concrete	2.1	0.88	1.0	0.54
Ice (0 °C)	0.917	2.04	2.25	1.203
Ground, coarse gravel	2.04	1.84	0.52	0.14
Sandy ground, dry	1.65	0.80	0.27	0.20
Sand ground, damp	1.75	1.00	0.58	0.33
Clay soil	1.45	0.88	1.28	1.00
Fat	0.91	1.93	0.16	0.091
Glass, window-	2.48	0.70	0.87	0.50
mirror-	2.70	0.80	0.76	0.35
Quartz-	2.21	0.73	1.40	0.87
Thermometer-	2.58	0.78	0.97	0.48
Plaster	1.00	1.09	0.51	0.47
Granite	2.75	0.89	2.9	1.18
Cork sheets	0.19	1.88	0.041	0.115
Marble	2.6	0.80	2.8	1.35
Mortar	1.9	0.80	0.93	0.61
Paper	0.7	1.20	0.12	0.14
Polyethylene	0.92	2.30	0.35	0.17
Polyamide	1.13	2.30	0.29	0.11
Polytetrafluoroethylene (Teflon)	2.20	1.04	0.23	0.10
PVC	1.38	0.96	0.15	0.11
Porcelain (95 °C)	2.40	1.08	1.03	0.40
Hard coal	1.35	1.26	0.26	0.15
Pine wood (radial)	0.415	2.72	0.14	0.12
Plasterwork	1.69	0.80	0.79	0.58
Celluloid	1.38	1.67	0.23	0.10
Brick	1.6...1.8	0.84	0.38...0.52	0.28...0.34

Table B 11: Thermophysical properties of metals and alloys at 20 °C.

Substance	ϱ $10^3\,\mathrm{kg/m^3}$	c kJ/kgK	λ W/Km	a $10^{-6}\,\mathrm{m^2/s}$
Metals				
Aluminium	2.70	0.888	237	98.8
Lead	1.34	0.129	35	23.9
Chromium	6.92	0.440	91	29.9
Iron	7.86	0.452	81	22.8
Gold	19.26	0.129	316	127.2
Iridium	22.42	0.130	147	50.4
Copper	8.93	0.382	399	117.0
Magnesium	1.74	1.020	156	87.9
Manganese	7.42	0.473	21	6.0
Molybdenum	10.2	0.251	138	53.9
Sodium	9.71	1.220	133	11.2
Nickel	8.85	0.448	91	23.0
Platinum	21.37	0.133	71	25.0
Rhodium	12.44	0.248	150	48.6
Silver	10.5	0.235	427	173.0
Titanium	4.5	0.522	22	9.4
Uranium	18.7	0.175	28	8.6
Tungsten	19.0	0.134	173	67.9
Zinc	7.10	0.387	121	44.0
Tin, white	7.29	0.225	67	40.8
Zirconium	6.45	0.290	23	12.3
Alloys				
Bronze (84 Cu, 9 Zn, 6 Sn, 1 Pb)	8.8	0.377	62	18.7
Duraluminium	2.7	0.912	165	67.0
Cast iron	7.8	0.54	42...50	10...12
Carbon steel (< 0.4 % C)	7.85	0.465	45...55	12...15
Cr-Ni-Steel (X12 CrNi 18,8)	7.8	0.50	15	3.8
Cr-Steel (X8 Cr17)	7.7	0.46	25	7.1

Substance	$\vartheta/\ ^\circ\mathrm{C}$	ε_n	ε
Acrylic glass (plexiglass)	20...60	0.97	
Concrete, rough	0...93		0.94
Beechwood	70	0.94	0.91
Tar roofing	20	0.91	
Oakwood, planed	0...93		0.90
Ice, smooth, thickness > 4 mm	−9.6	0.965	0.918
Enamel paint, white	20	0.91	
Tiles, light grey	25	0.92	
Floor covering (Pegulan)	20...60	0.94	
Rubber	20	0.92	
Glazed tiles, white	25	0.93	
Coal	150	0.81	
Cork	25	0.80	
Paint, black, gloss	25	0.88	
Marble, polished	0...93		0.90
Oil paints, 16 different colours	100	0.92...0.96	
Oil, thick layer	21	0.82	
Paper, white, matt	95	0.92	0.89
Plaster	0...200		0.91
Polytetrafluoroethylene (Teflon)	20...100	0.97	
Pyrex glass	−170...430	0.85	
White frost coating, rough	0	0.985	
Sand	20	0.76	
Fire clay	1000	0.75	
Emery (corundum), rough	84	0.855	0.842
Table glass, 6 mm thick	−60...0	0.910	
	60	0.913	
	120	0.919	
Water, thickness > 0.1 mm	10...50	0.965	0.91
Brick, red	0...93		0.93

Table B 13: Emissivities of metal surfaces. ε_n total emissivity in the direction of the surface normal, ε hemispherical total emissivity. Where the temperature interval is stated, the emissivity may be linearly interpolated between the given values.

Substance	$\vartheta / \,°C$	ε_n	ε
Aluminium, polished	20	0.045	
rough	75	0.055...0.07	
rolled smooth	170	0.039	0.049
commercial foil	100	0.09	
oxidised at 600 °C	200...600	0.11...0.19	
strongly oxidised	100...500	0,32...0.31	
Lead, not oxidised	127...227	0.06...0.08	
grey oxidised	20	0.28	
Chromium, polished	150	0.058	
oxidised by red heat	400...800	0.11...0.32	
Iron, polished	−73...727	0.04...0.19	0.06...0.25
oxidised	−73...727	0.32...0.60	
polished with emery	25	0.24	
electrolytically polished	150	0.128	0.158
casting skin	100	0.80	
rusted	25	0.61	
very rusted	20	0.85	
Gold, polished	227...627	0.020...0.035	
oxidised	−173...827		0.013...0.070
Copper, polished	327...727	0.012...0.019	
oxidised	130	0.76	0.725
highly oxidised	25	0.78	
	327	0.83	
	427	0.89	
	527...727	0.91...0.92	
Magnesium, rolled shiny	118	0.048	0.053
Brass, polished	25	0.038...0.043	
matt	50...350	0.22	
oxidised	200...600	0.60	
Nickel, polished	100	0.045	0.053
	127...1127	0.07...0.19	
	127...727		0.09...0.15
oxidised	227...627	0.37...0.47	
Platinum, polished	127...1127	0.05...0.16	
	127...1527		0.07...0.21
Platinum wire	227...1377	0.07...0.18	
Mercury, not oxidised	25...100		0.10...0.12
Silver, polished	127...527	0.020...0.030	
	127...927		0.020...0.047
Steel, material No. DIN 1.4301=AISI 304			
polished	50...200	0.111...0.132	
sand blasted, $R_a = 2.1\,\mu m$	−50...200	0.446...0.488	
Bismuth, smooth	80	0.340	0.366
Tungsten, aged	1327...2427		0.20...0.31
Zinc, polished	227...327	0.04...0.05	
grey oxidised	25	0.23...0.25	
Tin, shiny	25	0.064	

Appendix C: Solutions to the exercises

Chapter 1: Introduction. Technical Applications

1.1: $\dot{Q} = 484$ W; for the concrete blocks $\dot{Q} = 270$ W

1.3: By differentiation of $\dot{q} = -\lambda(\vartheta)\,\mathrm{d}\vartheta/\,\mathrm{d}x$ follows

$$\frac{\mathrm{d}\lambda}{\mathrm{d}\vartheta} = -\lambda\frac{\mathrm{d}^2\vartheta/\,\mathrm{d}x^2}{(\mathrm{d}\vartheta/\,\mathrm{d}x)^2} < 0 \ .$$

λ decreases with rising temperature.

1.4: $I_{\max} = 28.6$ A

1.5: $\vartheta_{\mathrm{W}1} = 38.9$ °C

1.6: a) $\vartheta_{\max} = \vartheta_0$ along the hypotenuse $y = x$; $\vartheta_{\min} = \vartheta_0 - \vartheta_1$ at $x = l, y = 0$. $\vartheta_1 = \vartheta_{\max} - \vartheta_{\min}$.

b) $\mathrm{grad}\vartheta = \frac{2\vartheta_1}{l}\left(-\frac{x}{l}e_x + \frac{y}{l}e_y\right)$, $\dot{q} = \frac{2\lambda\vartheta_1}{l}\left(\frac{x}{l}e_x - \frac{y}{l}e_y\right)$, $|\dot{q}| = \frac{2\lambda\vartheta_1}{l}\sqrt{\left(\frac{x}{l}\right)^2 + \left(\frac{y}{l}\right)^2}$ is largest at $x = y = l$.

c) $y = 0 : \dot{Q}_1 = 0$; $x = l : \dot{Q}_2 = -2\lambda h\vartheta_1$; $y = x : \dot{Q}_3 = 2\lambda h\vartheta_1 = -\dot{Q}_2$.

1.7: $\alpha = 6241$ W/m²K

1.8: $\alpha = 5.0$ W/m²K

1.9: $\alpha_{\mathrm{W}}/\alpha_{\mathrm{A}} = 578$

1.10: $\vartheta_{\mathrm{i}} = 552$ °C, $\vartheta_{\mathrm{o}} = 548$ °C, $\alpha_{\mathrm{rad}} = 5.33$ W/m²K

1.11: $\dot{q} = 72.7$ W/m², $\vartheta_{\mathrm{W}1} = 12.6$ °C, $\vartheta_{\mathrm{W}2} = -9.1$ °C

1.12: $\dot{q} = 15.35$ W/m², $\vartheta_{\mathrm{W}1} = 20.0$ °C

1.13: $\dot{Q}/L = 514$ W/m; $\vartheta_3 = 193$ °C lies below the permitted value of 250 °C.

1.14: $\vartheta_{\mathrm{F}} = 17.0$ °C; $t^* = 3.16$ h $= 3$ h 10 min. These values are valid under the assumption that one of the circular ends is adiabatic.

1.15: $\vartheta_1^+(z) := [\vartheta_1(z) - \vartheta_2']/(\vartheta_1' - \vartheta_2')$; $\vartheta_2^+(z) := [\vartheta_2(z) - \vartheta_2']/(\vartheta_1' - \vartheta_2')$;

$$C = 1 : \vartheta_1^+(z) = 1 - \frac{N}{1+N}\frac{z}{L} = 1 - \varepsilon\frac{z}{L} \ ; \ \vartheta_2^+(z) = \frac{N}{1+N}\left(1 - \frac{z}{L}\right) = \varepsilon\left(1 - \frac{z}{L}\right)$$

$$C \neq 1 : \vartheta_1^+(z) = 1 - \frac{1 - C_1\varepsilon_1}{1 - C_1}\left\{1 - \exp\left[N_1(C_1 - 1)\frac{z}{L}\right]\right\} \ ; \ \vartheta_2^+(z) = C_1\left[\vartheta_1^+(z) + \varepsilon_1 - 1\right]$$

$\varepsilon_1 = \varepsilon_1(N_1, C_1)$ is to be calculated according to Table 1.4.

1.16: Countercurrent: $kA = 423$ W/K; Cross-flow with one tube row: $kA = 461$ W/K, Cross countercurrent flow with two tube rows as in Fig. 1.59: $kA = 433$ W/K.

1.17: Under the assumptions mentioned $\dot{N}_A = {}_u j_A 2\pi r L = -D \partial c_A / \partial r 2\pi r L = $ const and with that

$$\frac{d\dot{N}_A}{dr} = 0 = \frac{\partial}{\partial r}\left(Dr\frac{\partial c_A}{\partial r}\right) \quad .$$

Integration between the limits $c_A(r_0) = c_{AW}$ and $c_A(r_0+\delta) = c_{A\delta}$ yields the concentration profile

$$\frac{c_A - c_{AW}}{c_{A\delta} - c_{AW}} = \frac{\ln r/r_0}{\ln[(r_0+\delta)/r_0]} \quad .$$

From this, by differentiation and introduction into the equation for $\dot{N}_A$ the expression given is found.

1.18: a) $\dot{n}_A = 3.848 \cdot 10^{-6}$ kmol/m^{2}s; $\dot{m}_A = \dot{n}_A \tilde{M}_A = 1.773 \cdot 10^{-4}$ kg/(m^2 s).
b) During the time dt an enthalpy amount of $\dot{m}_A A \, dt = \dot{n}_A \tilde{M}_A A \, dt$ evaporates. According to the question, the diffusion flow is calculated from that of the steady-state solution (1.174), in which we now have $y_1 = y(t)$. This is then

$$dt = -\frac{\varrho_L}{\tilde{M}_A}\frac{y_2 - y(t)}{pD/R_m T}\frac{1}{\ln(p_{B2}/p_{B1})}\,dy = B\left[y_2 - y(t)\right]dy \quad .$$

Integration and putting in the numerical values gives $t = 20.57$ h.

1.19: Equimolar counterdiffusion prevails in the tubes. It therefore follows from (1.176): $\dot{N}_A = -cDA\partial\tilde{x}_A/\partial y$. As $\dot{N}_A$ is constant, a linear concentration drop over the length L of the tubes develops. It then holds that

$$\dot{N}_A = -cDA\frac{\tilde{x}_{Aa} - \tilde{x}_{Ae}}{L} \quad .$$

$\tilde{x}_{Aa} = 1$ is the mole fraction of the ammonia in the pipes, $\tilde{x}_{Ae} = 0$ that of the ammonia in the air. The ammonia loss is found to be $\dot{N}_A = 3.99 \cdot 10^{-13}$ kmol/s, $\dot{M}_A = \tilde{M}_A \dot{N}_A = 6.78 \cdot 10^{-12}$ kg/s. The amount of ammonia flowing through the pipes is $\dot{M} = 1.91 \cdot 10^{-3}$ kg/s, which is $\dot{N} = 1.123 \cdot 10^{-4}$ kmol/s. The amount of air which gets into the ammonia is $\dot{N}_B = -\dot{N}_A = -3.99 \cdot 10^{-13}$ kmol/s, $\dot{M}_B = \tilde{M}_B \dot{N}_B = 1.16 \cdot 10^{-11}$ kg/s. The mole fraction of air in the pipes is exteremly small, namely $\tilde{x}_B = |\dot{N}_B|/\dot{N} = 3.55 \cdot 10^{-9}$.

1.20: Unidirectional mass transfer prevails. From (1.195), we have

$$\dot{m}_A = \tilde{M}_A \dot{n}_A = \tilde{M}_A \frac{p}{R_m T}\beta \ln\frac{1 - \tilde{x}_{A\delta}}{1 - \tilde{x}_{A0}} = 3.59 \cdot 10^{-5}\frac{\text{kg}}{\text{m}^2\text{s}} \quad .$$

In order to use (1.195a), the moisture content is required

$$X_A = 1.530 \cdot 10^{-2}, \quad \dot{m}_A = 3.635 \cdot 10^{-5}\,\text{kg/m}^2\text{s} \quad .$$

1.21: From (1.210) and (1.203), using steam tables, the solution is found by trial and error to be $\vartheta_I = 2.56\,°C$. It is $p_s(2.56\,°C) = 7.346$ mbar.

1.22: We will assume a small blowing rate. The lowest temperature is the wet bulb temperature ϑ_I. In (1.211), the factor is

$$\frac{\tilde{M}_B c\beta_m}{\alpha_m} = \frac{\tilde{M}_A c\beta_m c_{pA}}{\alpha_m}\frac{\tilde{M}_B}{\tilde{M}_A c_{pA}} = 1.097 \cdot 10^{-3}\,\text{mol K/J}$$

This allows (1.211) to be written as follows, as $X_{A\delta} = 0$,

$$(600 - \vartheta_I)\,°C = 1.097 \cdot 10^{-3} \cdot 2346 \cdot 10^3 \cdot X_{AI}\,°C \quad .$$

With (1.203), using a steam table, the wet bulb temperature is found through trial and error to be $\vartheta_I = 65.1\,°C$. This value also satisfies (1.206) very well. The amount of water fed to the chamber follows from (1.209) as $\dot{m}_A = 2.27 \cdot 10^{-2}$ kg/m^{2}s.

1.23: The amount of benzene transferred is $\Delta \dot{N}_B = 11.88$ kmol/h. The molar ratio of the gas in cross section e is $\tilde{Y}_e = 2.5 \cdot 10^{-3}$. According to (1.224), $\tilde{X}_o = 0.216$.

Chapter 2: Heat conduction and mass diffusion

2.2: At $x = 0$, $\vartheta(x)$ has a horizontal tangent. The tangent at $x = \delta$ intersects the horizontal $\vartheta = \vartheta_F$ at point R with the abcissa $x_R = \delta + \lambda/\alpha = (1 + Bi^{-1})\,\delta = 1.667\,\delta$. The tangent to the fluid temperature plot at $x = \delta$ intersects the line $\vartheta = \vartheta_F$ at $x_F = \delta + \lambda_F/\alpha = (1 + Nu^{-1})\,\delta = 1.100\,\delta$.

2.3: a) The plate heats up, because $\partial^2 \vartheta^+/\partial x^{+2} > 0$.
b) $x_T^+ = 1/2$; $(\partial \vartheta/\partial t)_{max} = 4.40$ K/s. c) $x_{min}^+ = 0.3778$.
d) $B(t) = 0.850 \exp[-0.03454\,\mathrm{s}^{-1}\,(t - t_0)]$. For $t \to \infty$ we have $\vartheta^+ = x^+$.

2.4: a) $\dot{q}(R) = \dot{W}_R R/(m + 2)$; $\dot{W}_R = \dot{W}_0(1 + m/2)$.
b) $\Theta(r) = \dfrac{R^2 \dot{W}_R}{\lambda\,(m+2)^2}\,[1 - (r/R)^{m+2}]$. c) $\Theta_{max}/\Theta_{max}^0 = 2/(m + 2)$.

2.5: a) $\vartheta_0 = 55.39\,°C$, $\vartheta_L = 37.39\,°C$. b) $\dot{Q}_0 = 3.204$ W, $\dot{Q}_L = 0.0622$ W.
c) The results do not differ within the numbers given. The simple calculation with the replacement bolts of length L_C is very exact.

2.6: $\eta_f = 0.603$

2.7: a) 87 fins/m. This means the heat flow increases by a factor 6.019. b) 1.40.

2.8: $\dot{Q}/L = 149$ W/m

2.9: a) $Q/A = 1027$ kJ/m^2. b) $\vartheta = 50.6\,°C$

2.10: Surface temperature: 73.1 °C; at 10 cm depth: 27.2 °C.

2.11: Surface temperature: 45.8 °C; at 10 cm depth: 24.3 °C.

2.12: a) The amplitude at a depth of 1 m is only $3.7 \cdot 10^{-4}$ K.
b) Highest temperature 18.6 °C on 2nd October, lowest temperature 1.4 °C on 2nd April.

2.13: a) $\alpha = 19.3$ W/m^{2}K. b) The temperatures of the insulated surface are $\vartheta(t_1) = 146.4\,°C$ and $\vartheta(t_2) = 112.3\,°C$.

2.14: a) $\vartheta = 37.5\,°C$. b) $\vartheta_m = 39.1\,°C$; $\vartheta_{surf} = 37.9\,°C$; $\vartheta_{centre} = 40.9\,°C$

2.15: $w = 0.206$ m/s

2.16: $t(s = 15\mathrm{mm}) = 3.26$ h; $t(s = 20\mathrm{mm}) = 3.57$ h.

2.18:

$i = x/\Delta x$	0	1	2	3	4	5	6	7	8	9
$(\vartheta^+)_i^6$	1.000	0.412	0.140	0.037	0.007	0.001	0.000			
$\vartheta^+(x, t^*)$	1.000	0.419	0.151	0.046	0.012	0.002	0.000			
$(\vartheta^+)_i^{12}$	1.000	0.681	0.414	0.216	0.098	0.038	0.012	0.003	0.001	0.000
$\vartheta^+(x, 2t^*)$	1.000	0.679	0.409	0.217	0.102	0.042	0.016	0.005	0.001	0.000

2.19: a) $M = 0.375$; $Bi^* = 0.2087$. The stability criterium

$$M \leq \frac{1}{2\,(1 + Bi^*)} = 0.4137$$

is satisfied. Temperature profile at time $t^* = 15$ min: $\vartheta_1^{15} = 80.0\,°C$; $\vartheta_2^{15} = 63.9\,°C$; $\vartheta_3^{15} = 50.7\,°C$; $\vartheta_4^{15} = 40.8\,°C$; $\vartheta_5^{15} = 34.9\,°C$; $\vartheta_6^{15} = 30.8\,°C$.
b) Steady-state temperature profile $(t \to \infty)$: $\vartheta_1^\infty = 80.0\,°C$; $\vartheta_2^\infty = 71.6\,°C$; $\vartheta_3^\infty = 64.1\,°C$; $\vartheta_4^\infty = 57.4\,°C$; $\vartheta_5^\infty = 51.3\,°C$; $\vartheta_6^\infty = 45.7\,°C$.

2.20: a) $-\left(2+m^2\Delta x^2\right)\vartheta_1^+ + \vartheta_2^+ + \vartheta_6^+ = 0,$

$\vartheta_1^+ - \left(3+m^2\Delta x^2\right)\vartheta_2^+ + \vartheta_3^+ + \vartheta_7^+ = 0,$

$\vartheta_2^+ - \left(3+m^2\Delta x^2\right)\vartheta_3^+ + \vartheta_4^+ + \vartheta_8^+ = 0,$

$\vartheta_3^+ - \left(3+m^2\Delta x^2\right)\vartheta_4^+ + \vartheta_5^+ + \vartheta_9^+ = 0,$

$\vartheta_4^+ - \left(1+0.5m^2\Delta x^2\right)\vartheta_5^+ = 0,$

$\vartheta_1^+ - \left(3+m^2\Delta x^2\right)\vartheta_6^+ + \vartheta_7^+ + \vartheta_{10}^+ = 0,$

$\vartheta_2^+ + \vartheta_6^+ - \left(4+m^2\Delta x^2\right)\vartheta_7^+ + \vartheta_8^+ + \vartheta_{11}^+ = 0,$

$\vartheta_3^+ + \vartheta_7^+ - \left(4+m^2\Delta x^2\right)\vartheta_8^+ + \vartheta_9^+ + \vartheta_{12}^+ = 0,$

$\vartheta_4^+ + \vartheta_8^+ - \left(2+0.5m^2\Delta x^2\right)\vartheta_9^+ = 0,$

$1.90384\,\vartheta_6^+ - \left(43.5992+0.56837m^2\Delta x^2\right)\vartheta_{10}^+ + (4/3)\vartheta_{11}^+ = -40.35875,$

$(4/3)\vartheta_7^+ + \vartheta_{10}^+ - \left(6+0.92069m^2\Delta x^2\right)\vartheta_{11}^+ + \vartheta_{12}^+ = -8/3$

$\vartheta_8^+ + \vartheta_{11}^+ - \left(2+0.5m^2\Delta x^2\right)\vartheta_{12}^+ = 0.$

b) $\vartheta_1^+ = 0.54791; \ \vartheta_2^+ = 0.50806; \ \vartheta_3^+ = 0.44968; \ \vartheta_4^+ = 0.39775;$

$\vartheta_5^+ = 0.36828; \ \vartheta_6^+ = 0.67543; \ \vartheta_7^+ = 0.60786; \ \vartheta_8^+ = 0.51519;$

$\vartheta_9^+ = 0.43891; \ \vartheta_{10}^+ = 0.97859; \ \vartheta_{11}^+ = 0.83003; \ \vartheta_{12}^+ = 0.64674.$

$\eta_R = 0.5756.$ Approximation from (2.81): $\eta_R = 0.5758,$ from (2.82): $\eta_R = 0.5721.$

2.21: We have

$$\frac{\xi_A(x=0,t)-\xi_{AS}}{\xi_{A\alpha}-\xi_{AS}} = \frac{4}{\pi}\exp\left(-\frac{\pi^2}{4}t_D^+\right) - \frac{4}{3\pi}\exp\left(-\frac{9\pi^2}{4}t_D^+\right) + \frac{4}{5\pi}\exp\left(-\frac{25\pi^2}{4}t_D^+\right) - \ldots + \ldots$$

Using the first term of the series we obtain $t_D^+ = 1.039$ and $t = 14.1$ days. The other series terms are negligibly small, which can be simply checked, so that it is actually sufficient to just use the first series term.

2.22: The diffusion now occurs in the direction of the x- and y-coordinate. In the treatment of the corresponding heat conduction problem, section 2.3.5, has shown that for a block with side lengths $2X, 2Y$, the temperature profile is given by (2.191)

$$\vartheta^+ = \frac{\vartheta-\vartheta_S}{\vartheta_0-\vartheta_S} = \vartheta_{Pl}^+\left(\frac{x}{X},\frac{at}{X^2},\frac{\alpha X}{\lambda}\right)\cdot\vartheta_{Pl}^+\left(\frac{y}{Y},\frac{at}{Y^2},\frac{\alpha'Y}{\lambda}\right) \ .$$

For the diffusion problem, in the centre of the rod $x=y=0$ and for $\beta X/D = \beta'Y/D \to \infty$ it correspondingly holds that

$$\xi_A^+ = \frac{\xi_A(x=y=0)-\xi_{AS}}{\xi_{A\alpha}-\xi_{AS}} = c_{Pl}^+\left(\frac{Dt}{X^2}\right)\cdot c_{Pl}^+\left(\frac{Dt}{Y^2}\right) \ .$$

Under the assumption still to be checked, i.e. that the first term of the series from the solution of the previous exercise is satisfactory, follows

$$\xi_A^+ = \left[\frac{4}{\pi}\exp\left(-\frac{\pi^2}{4}t_D^+\right)\right]^2 \ .$$

Which gives $t_D^+ = 0.569$ and $t = 7.7$ days. It is easy to prove that the remaining series terms are actually negligible in comparison with the first term.

2.23: a) We have $\dot{M}_{La} = 0.7$ kg/s, $\dot{M}_{Ga} = 6.0$ kg/s and $\dot{M}_W = 0.532$ kg/s.

b) The required time is obtained from (2.389). In which $c_{A\alpha} = 0$ and so

$$c_{Am}/c_{AS} = \xi_{Aa}^L/\xi_{A0} = 0.6 \ .$$

It then follows from (2.390) that

$$0.65797 = \sum_{i=1}^{\infty}\frac{1}{i^2}\exp\left(-i^2\pi^2 t_D^+\right) \ .$$

The solution is found by trial and error to be $t_D^+ = 0.0485$. The first three terms of the series suffice. They give $t = 72.7$ s.

c) $L = tw = 7.27$ m.

2.24: We can approximately say that $Bi_D \to \infty$. In addition to this the surface of the spheres are immediately completely immersed in the water, $\xi_{A0} = 1$. Therefore

$$c_{A0} = \varrho \, \xi_{A0} / \tilde{M}_A = 55.5 \, \text{kmol/m}^3 \ .$$

Furthermore

$$t_D^+ = 0.072.$$

From (2.390)

$$c_{Am}^+(t_D^+) = \frac{c_{Am} - c_{A0}}{c_{A\alpha} - c_{A0}} = \frac{6}{\pi^2} \sum_{i=1}^{\infty} \frac{1}{i^2} \exp\left(-i^2 \pi^2 t_D^+\right)$$

follows, with $c_{A\alpha} = 0$: $c_{Am}/c_{A0} = 0.6927$ and $c_{Am} = 38.42$ kmol/m^3. Each sphere takes in 0.241 g of water.

Chapter 3: Convective heat and mass transfer. Single phase flows

3.1: From $-\dot{q}_j = \lambda_{ji} \partial \vartheta / \partial x_i$ follows, under consideration from $\lambda_{12} = \lambda_{21}$,

$$
\begin{aligned}
-\dot{q}_1 &= \lambda_{11} \frac{\partial \vartheta}{\partial x_1} + \lambda_{12} \frac{\partial \vartheta}{\partial x_2} \\
-\dot{q}_2 &= \lambda_{12} \frac{\partial \vartheta}{\partial x_1} + \lambda_{22} \frac{\partial \vartheta}{\partial x_2} \\
-\dot{q}_3 &= \qquad\qquad \lambda_{33} \frac{\partial \vartheta}{\partial x_3} \ .
\end{aligned}
$$

In steady heat conduction, we generally have $\partial \dot{q}_j / \partial x_j = 0$ as the plate is thin in the x_2-direction, then $\partial \vartheta / \partial x_2 = 0$ and the differential equation for steady heat transfer looks like

$$\frac{\partial \dot{q}_1}{\partial x_1} + \frac{\partial \dot{q}_2}{\partial x_2} + \frac{\partial \dot{q}_3}{\partial x_3} = \lambda_{11} \frac{\partial^2 \vartheta}{\partial x_1^2} + \lambda_{33} \frac{\partial^2 \vartheta}{\partial x_3^2} = 0.$$

3.2: For the model (index M) and the original (index O) it has to hold that

$$Nu = f(Re, Pr) \ .$$

a) It has to be $Pr_M = Pr_O$. With $Pr_O = 4.5$, the associated temperature is $T = 311$ K.

b) We have $(w_M)_1 = 0.0355$ m/s and $(w_M)_2 = 0.355$ m/s.

c) Because $Nu_M = Nu_O$ or $(\alpha_M d_M)/\lambda_M = (\alpha_O d_O)/\lambda_O$, $\alpha_O = 484$ W/m^{2}K.

3.3: For the model (subscript M) and the original (subscript O), according to (3.330), it holds that

$$Nu_m \sim Gr^{1/4}$$

with $Nu_m = \alpha_m L / \lambda$ and $Gr = \beta_\infty \, (\Delta \vartheta) \, g \, L^3 / \nu^2$. This then gives

$$\frac{\alpha_{mO}}{\alpha_{mM}} = \left(\frac{L_M}{L}\right)^{1/4} \left(\frac{\Delta \vartheta_O}{\Delta \vartheta_M}\right)^{1/4} \ ,$$

and

$$\frac{\dot{Q}_O}{\dot{Q}_M} = \frac{\alpha_{mO} A_O \Delta \vartheta_O}{\alpha_{mM} A_M \Delta \vartheta_M} = \left(\frac{L_O}{L_M}\right)^{7/4} \left(\frac{\Delta \vartheta_O}{\Delta \vartheta_M}\right)^{5/4} \ ,$$

because of $A_O/A_M = L_O^2/L_M^2$. We obtain $\dot{Q}_O = 469.5$ W.

3.4: Firstly, a dimension matrix is set up. **L** indicates the dimension of a length:

$$\mathbf{L} = \dim L \; ;$$

correspondingly $\mathbf{t} = \dim t$, $\mathbf{T} = \dim \vartheta$, $\mathbf{M} = \dim M$. The dimension matrix looks like

	L	w_0	ϱ	λ	ν	c	α_m
L	1	1	−3	1	2	2	0
t	0	−1	0	−3	−1	−2	−3
M	0	0	1	1	0	0	1
T	0	0	0	−1	0	−1	−1

Only the rank r of the matrix is determined, using equivalent transformations; these are linear combinations of rows (or columns). For this we form linear combinations of rows until the diagonal of a sub-matrix only contains ones, with the neighbouring diagonals all zeros. The first row is indicated by Z_1, the second by Z_2 and so on, forming a new matrix, the rows of which will be indicated by dashes. So $Z'_4 = -Z_4$, $Z'_3 = Z_3 + Z_4$, $Z'_2 = -Z_2 + 3Z_4$. The new matrix looks like

	L	w_0	ϱ	λ	ν	c	α_m
	1	1	−3	1	2	2	0
	0	1	0	0	1	−1	0
	0	0	1	0	0	−1	0
	0	0	0	1	0	1	1

It already contains only ones in the front main diagonal. By a further transformation $Z''_1 = Z'_1 - Z'_2 + 3Z'_3 - Z'_4$ we get

	L	w_0	ϱ	λ	ν	c	α_m
	1	0	0	0	1	−1	−1
	0	1	0	0	1	−1	0
	0	0	1	0	0	−1	0
	0	0	0	1	0	1	1

The left hand sub-matrix now contains, as demanded, only ones in the main diagonal, in the neighbouring diagonals only zeros. The rank of the matrix is $r = 4$, as there are four rows that are linearly independent of each other: By equivalent transformation no more rows can be converted into another. According to Buckingham [1.20], the number m of π-quantities $m = n - r$, where n is the number of original variables, in our case is $n = 7$. There are

$$m = 7 - 4 = 3$$

π-quantities. These are yielded from the above matrix as

$$\pi_1 = \nu L^{-1} w_m^{-1} \varrho^{-0} \lambda^{-0} = \frac{\nu}{w_m L} = 1/Re$$

$$\pi_2 = c L^1 w_m^1 \varrho^1 \lambda^{-1} = \frac{c\varrho}{\lambda} w_m L = \frac{w_m L}{a} = Re\,Pr$$

$$\pi_3 = \alpha_m L^1 w_m^0 \varrho^0 \lambda^{-1} = \frac{\alpha_m L}{\lambda} = Nu$$

We have $f(\pi_1, \pi_2, \pi_3) = 0$ or $f(Nu, Re, Pr) = 0$.

3.5: a) The dimension matrix is

	d	w	ϱ_A	ϱ_W	g
L	1	1	−3	−3	1
t	0	−1	0	0	−2
M	0	0	1	1	0

By equivalent transformations, new rows Z_i' are obtained from the original rows $Z_i (i = 1, 2, 3)$. The following equivalent transformation of the rows $Z_1' = Z_1 + Z_2 + 3Z_3$, $Z_2' = -Z_2$, $Z_3' = Z_3$ is carried out yielding

d	w	ϱ_A	ϱ_W	g
1	0	0	0	-1
0	1	0	0	2
0	0	1	1	0

The rank of the matrix is $r = 3$. This gives $m = n - r = 5 - 3 = 2$. The dimensionless quantities are

$$\pi_1 = \varrho_W d^0 w^0 \varrho_L^{-1} = \frac{\varrho_W}{\varrho_L}$$

$$\pi_2 = g d^1 w^{-2} \varrho_L^0 = \frac{gd}{w^2} \ .$$

b) We have $\pi_2 = f(\pi_1)$ or $gd/w^2 = f(\varrho_W/\varrho_L)$.

3.6: Introducing the velocity profile in the integral condition that follows from (3.165) for the momentum

$$\frac{\mathrm{d}}{\mathrm{d}x}\left[w_\delta^2 \delta \int_0^1 \frac{w_x}{w_\delta}\left(1 - \frac{w_x}{w_\delta}\right) \mathrm{d}\left(\frac{y}{\delta}\right)\right] = \frac{\nu w_\delta}{\delta}\left(\frac{\partial w_x/\partial w_\delta}{\partial y/\delta}\right)_{y=0}$$

yields, with the abbreviation $z = \pi/2(y/\delta)$:

$$\frac{2}{\pi} w_\delta^2 \frac{\mathrm{d}}{\mathrm{d}x}\left[\delta \int_0^{\pi/2} \sin z(1 - \sin z)\,\mathrm{d}z\right] = \frac{\nu w_\delta}{\delta}\frac{\pi}{2} \ .$$

The integral has the value $1 - \pi/4$. With that, after integration

$$\delta = \pi \left(\frac{2}{4-\pi}\right)^{1/2}\left(\frac{\nu x}{w_\delta}\right)^{1/2} = 4.795\, x\, Re_x^{-1/2}.$$

This result differs from (3.170) by the fact that in place of the factor 4.64 in (3.170), the factor 4.795 appears.

3.7: In order to ascertain whether the flow is turbulent, the Reynolds number is calculated at the end of the plate $Re = w_\infty L/\nu$. As the viscosity is only found at a pressure 0.1 MPa, we make use of $\eta = \varrho\nu = \text{const}$ for $\vartheta = \text{const}$. It follows from this, that $p_1/(RT_1)\nu_1 = p_2/(RT_2)\nu_2$ or $\nu_2 = \nu_1 p_1/p_2$ at $\vartheta = \text{const}$. We have $\nu_1(p_1, \vartheta_m) = 30.84 \cdot 10^{-6}$ m^2/s with $p_1 = 0.1$ MPa, and the mean temperature $\vartheta_m = (\vartheta_\infty + \vartheta_0)/2 = (300 + 25)/2\,°\text{C} = 162.5\,°\text{C}$. This then gives $\nu_2 = 30.84 \cdot 10^{-5}$ m^2/s and the associated Reynolds number as $Re = 3.243 \cdot 10^4$. The flow is laminar to the end of the plate. The mean Nusselt number is, see also Example 3.8, $Nu_m = 105.5$. This yields $\alpha_m = 3.84$ W/m^{2}K and $\dot{q} = 1056$ W/m^2.

3.8: The Reynolds number at the end of the lake is $Re = 2.606 \cdot 10^6$. This means that the flow is initially laminar and becomes turbulent after a distance of $x_{cr} = 3.84$ m. The Sherwood number is calculated from (3.208), in which the Nusselt number is replaced by the Sherwood number. We have $Sh_{m,lam} = 904.1$ and $Sh_{m,turb} = 3592$ from (3.207). This then gives

$$Sh_m = \sqrt{Sh_{m,lam}^2 + Sh_{m,turb}^2} = 3704, \quad \beta_m = 4.741 \cdot 10^{-4}\text{m/s}$$

$$\dot{m}_A = \beta_m \varrho \left(\xi_{A0} - \xi_{A\infty} \right) = \beta_m \frac{1}{RT} \left(p_{A0} - p_{A\infty} \right) = \beta_m \frac{1}{RT} \left(p_{A0} - \varphi p_{A0} \right) \ .$$

We also have $p_{A0} = p_{Ws}$ and therefore

$$\dot{m}_A = \beta_m \frac{p_{Ws}}{RT} (1 - \varphi) = \beta_m \varrho'' (1 - \varphi) \ ,$$
$$\dot{m}_A = 4.099 \cdot 10^{-6} \, \mathrm{kg/m^2 s}$$
$$\dot{M}_A = 147.5 \, \mathrm{kg/h} \ .$$

3.9: $\dot{M} = \varrho \dot{V} = 0.25$ kg/s. Further, the energy balance holds

$$\dot{Q} = \dot{M} c_p \left(\vartheta_e - \vartheta_i \right) = \alpha_m d \pi L \Delta \vartheta_m$$

with

$$\Delta \vartheta_m = \frac{\left(\vartheta_0 - \vartheta_i \right) - \left(\vartheta_0 - \vartheta_e \right)}{\ln \left[\left(\vartheta_0 - \vartheta_i \right) / \left(\vartheta_0 - \vartheta_e \right) \right]} = 57.71 \ ^\circ \mathrm{C} \ .$$

This yields

$$\alpha_m = \frac{\dot{M} c_p \left(\vartheta_e - \vartheta_i \right)}{d \pi L \Delta \vartheta_m} = 768.2 \, \mathrm{W/m^2 K} \ .$$

3.10: The solar energy caught by the reflector is transferred to the absorber tube

$$\dot{q}_s s L = \dot{q} d_o \pi L / 2 \ .$$

The heat flux absorbed by the absorber tube is therefore $\dot{q} = 1.567 \cdot 10^4$ W/m². It serves to heat the water:

$$\dot{q} d_o \pi L = \dot{M} c_p \left(\vartheta_i - \vartheta_e \right)$$

with $\dot{M} = \varrho w_m d_i^2 \pi / 4 = 8.468 \cdot 10^{-2}$ kg/s.

$$L = \frac{\dot{M} c_p \left(\vartheta_i - \vartheta_e \right)}{\dot{q} d_o \pi} = 13.3 \, \mathrm{m} \ .$$

The wall temperature at the outlet follows from $\dot{q} = \alpha \left(\vartheta_0 - \vartheta_F \right) = \alpha \left(\vartheta_0 - \vartheta_e \right)$ as $\vartheta_0 = \dot{q}/\alpha + \vartheta_e = 139.5 \ ^\circ$C.

3.11: The heat losses are yielded from the energy balance as $\dot{Q} = \dot{M} c_p \left(\vartheta_i - \vartheta_e \right) = 11.97$ kW. The heat flux transferred at the end of the tube is

$$\dot{q} = k(\vartheta_e - \vartheta_0) \quad \text{with} \quad \frac{1}{k} = \frac{1}{\alpha} + \frac{1}{\alpha_e} \ .$$

Here, α is the heat transfer coefficient of the superheated steam on the inner tube wall. We set $d_i \approx d_o$ here. For the calculation of α, the Reynolds number has to be found first

$$Re = \frac{w_m d}{\nu} = \frac{4 \dot{M}}{\pi d \eta} = 7.48 \cdot 10^5 \ .$$

The flow is turbulent. In addition $L/d > 100$. We obtain from (3.259) $Nu_m = 1145$, $\alpha_m = 1267$ W/m²K and $\dot{q} = 1779$ W/m². Furthermore, it holds that $\dot{q} = \alpha_m \left(\vartheta_e - \vartheta_0 \right)$ and therefore $\vartheta_0 = \vartheta_e - \dot{q}/\alpha_m = 118.6 \ ^\circ$C.

3.12: The specific surface area of the particle from (3.266) is $a_P = 6(1 - \varepsilon)/d = 180$ m²/m³. The arrangement factor from (3.268) is $f_\varepsilon = 1.9$, the Reynolds number

$$Re = \frac{w_m d}{\varepsilon \nu} = 6.098 \cdot 10^3 \ .$$

The Nusselt number is calculated according to section 3.7.4, page 336, No. 5, from

$$Nu_\mathrm{m} = 2 + \sqrt{Nu_\mathrm{m,lam}^2 + Nu_\mathrm{m,turb}^2}$$

with $Nu_\mathrm{m,lam} = 46.47$ and $Nu_\mathrm{m,turb} = 35.57$ to be $Nu_\mathrm{m} = 60.52$. This yields $\alpha_\mathrm{m} = 79.28$ W/m²K. The total particle surface area is, according to (3.264), $nA_\mathrm{P} = a_\mathrm{P}V = a_\mathrm{P}A_0H = 117$ m².

a) With that $\dot{Q} = \alpha_\mathrm{m}nA_\mathrm{P}(\vartheta_\mathrm{A} - \vartheta_0) = 171601$ W ≈ 172 kW.

b) The amount of water evaporated due to the heat fed is

$$\dot{M}_\mathrm{W} = \dot{Q}/\Delta h_\mathrm{v} = 7 \cdot 10^{-2} \, \mathrm{kg/s}$$

The amount of water evaporated due to the partial pressure drop

$$\dot{M}_\mathrm{W} = \beta_\mathrm{m}\frac{p}{R_\mathrm{L}T}(X_\mathrm{WS} - X)$$

is around two orders of magnitude smaller, and can be neglected. This can be checked using the mass transfer coefficients and specific humidity $X = 0.622\,p_\mathrm{WS}/(p/\varphi - p_\mathrm{WS})$ from section 3.7.4, page 336, No. 5.

3.13: The height H_mf of a fluidised bed at the fluidisation point follows from the condition of constant sand mass

$$A_0H_\mathrm{mf}(1 - \varepsilon_\mathrm{mf})\varrho_\mathrm{S} = A_0H_\mathrm{S}(1 - \varepsilon_\mathrm{S})\varrho_\mathrm{S} \ .$$

It is

$$H_\mathrm{mf} = \frac{1 - \varepsilon_\mathrm{S}}{1 - \varepsilon_\mathrm{mf}}H_\mathrm{S} = 0.57\mathrm{m} \ .$$

The total pressure drop, according to (3.271), is $\Delta p = [\varrho_\mathrm{S}(1 - \varepsilon_\mathrm{mf}) + \varrho_\mathrm{G}\varepsilon_\mathrm{mf}]gH \cong \varrho_\mathrm{S}(1 - \varepsilon_\mathrm{mf})gH$. As $(1 - \varepsilon_\mathrm{mf})H = $ const holds, the pressure drop in the fluidised bed is practically constant, which can also be confirmed by experiment. We can also put, for $(1 - \varepsilon_\mathrm{mf})H$ at the fluidisation point, $(1 - \varepsilon_\mathrm{S})H_0$ of the quiescent sand layer. With that, we get $\Delta p \cong 7848$ Pa. This pressure drop has to be summoned up by the blower. The pressure p_2 at the blower outlet is equal to the pressure at the inlet of the fluidised bed $p_2 = p_1 + \Delta p = 107848$ Pa. The mean pressure of the air in the fluidised bed is $\varrho_\mathrm{mG} = p_\mathrm{m}/RT = 0.322$ kg/m³. The fluidisation velocity follows from (3.272). This includes the Archimedes number formed with the mean density ϱ_mG

$$Ar = \frac{\varrho_\mathrm{S} - \varrho_\mathrm{mG}}{\varrho_\mathrm{mG}}\frac{d_\mathrm{P}^3 g}{\nu^2} = 415.8 \ .$$

According to (3.272) it is $Re_\mathrm{mf} = 0.310$; and $w_\mathrm{mf} = 0.094$ m/s. The actual velocity is $w_\mathrm{m} = 10w_\mathrm{mf} = 0.94$ m/s. The mass flow rate of the air at the inlet is

$$\dot{M}_\mathrm{G} = \varrho_\mathrm{G}w_\mathrm{m}A_0 = \frac{p_2}{RT}w_\mathrm{m}A_0 = 2.22 \, \mathrm{kg/s}.$$

The required blower power is

$$P = \dot{M}_\mathrm{G}\frac{\kappa}{\kappa - 1}\frac{RT_1}{\eta_\mathrm{V}}\left[\left(\frac{p_2}{p_1}\right)^{\frac{\kappa-1}{\kappa}} - 1\right] = 20.4 \, \mathrm{kW}.$$

It is

$$\eta_\mathrm{V} = \frac{P_\mathrm{rev}}{P} = \frac{\dot{M}c_p(\vartheta_{1'} - \vartheta_1)}{P}$$

when $\vartheta_{1'}$ is the final temperature of the compression. It follows from this, that $\vartheta_{1'} = 299.5$K $= 26.4$ °C. The heat flow fed in is $\dot{Q} = \dot{M}_\mathrm{L}c_\mathrm{pmG}(\vartheta_2 - \vartheta_{1'}) = 2126$ kW.

3.14: The density of the air over the ground is

$$\varrho_{A1} = \frac{p_1}{R_A T_{A1}} = 1.2084 \,\text{kg/m}^3 \ ,$$

that of the waste gases is

$$\varrho_{G1} = \frac{p_1}{R_G T_{G1}} = 1.1946 \,\text{kg/m}^3 \ .$$

$\varrho_{G1} < \varrho_{A1}$; the exhaust gases can rise. They would no longer rise if $\varrho_{G1} \geq \varrho_{A1}$ or

$$T_{G1} \leq \frac{R_A}{R_G T_{A1}} = 438.1\text{K} = 164.9\,°\text{C} \ ,$$

if the exhaust gas temperature was to lie below 165 °C.

It holds for the air that $\mathrm{d}p = -\varrho g\,\mathrm{d}x$ and so $v\,\mathrm{d}p = -g\,\mathrm{d}x$. Mit $v = R_A T_A/p$ follows

$$\frac{\mathrm{d}p}{p} = -\frac{g}{R_A T_A}\,\mathrm{d}x \ .$$

Through integration the barometric height formula is obtained

$$p_2 = p_1 \exp\left(-\frac{g\Delta x}{R_A T_A}\right) \ .$$

The air pressure at 100m is $p_2 = 0.09882$ MPa. The density of the air at 100 m height is

$$\varrho_{A2} = \frac{p_2}{R_A T_A} = 1.194 \,\text{kg/m}^3 \ .$$

The density of the exhaust gases at 100m follows from

$$\varrho_{G2} = \left(\frac{p_2}{p_1}\right)^{1/\kappa} \varrho_{G1} = 1.184 \,\text{kg/m}^3 \ .$$

The exhaust gases are lighter than air at 100m height, they can rise further.

3.15: We have

$$Gr = \frac{\beta_\infty(\vartheta_0 - \vartheta_\infty)gL^3}{\nu^2} = 4.49 \cdot 10^9$$

and $Ra = GrPr = 3.20 \cdot 10^9$. Then, from (3.328) the mean Nusselt number for free flow is found to be $Nu_{mF} = 168.8$. Furthermore

$$Re = \frac{w_0 L}{\nu} = 5.794 \cdot 10^4 \ .$$

Giving $Nu_{m,lam} = 142.7$ and $Nu_{m,turb} = 203.99$. The mean Nusselt number for forced convection is obtained as

$$Nu_{mC} = \sqrt{Nu^2_{m,lam} + Nu^2_{m,turb}} = 248.9 \ .$$

As free and forced flow are directed against each other, the minus sign in (3.334) holds, $Nu_m = 219.7$ and $\alpha_m = 5.97$ W/m²K. The two sides of the plate release the heat flow $\dot{Q} = \alpha_m A(\vartheta_0 - \vartheta_\infty) = 477$ W.

3.16: In this section 3.9.3, page 381, No. 5 is used. We have

$$Ra = GrPr = \frac{\beta_\infty \dot{q}gx^4}{\nu^2\lambda}Pr = 7.176 \cdot 10^{10} \left(\frac{x}{\text{m}}\right)^4 \ .$$

At the end of the plate $x_1 = 0.4$ m, $Ra = 1.84 \cdot 10^9$. The flow at the end of the plate is just about still laminar. It is

$$Nu_x = 78.63 \left(\frac{x}{\text{m}}\right)^{4/5} \quad \text{and} \quad \alpha = Nu_x\frac{\lambda}{x} = 1.961\, x^{-1/5}\frac{\text{W}}{\text{m}^{9/5}\text{K}} \ .$$

It further follows from $\dot{q} = \alpha(\vartheta_0 - \vartheta_\infty)$:

$$\vartheta_0 = \frac{\dot{q}}{\alpha} + \vartheta_\infty = 7.649\frac{\text{W}}{\text{m}^{1/5}}x^{1/5} + 283.15\,\text{K} \ .$$

The wall temperature increases with $x^{1/5}$, and at the end of the plate $x_1 = 0.4$ m is

$$\vartheta_0 = 289.5\,\text{K} = 16.4 \ ^\circ\text{C} \ .$$

3.17: The cooling is determined by the heat transfer in free flow. (3.328) holds for the vertical cylinder, for the horizontal cylinder, equation No. 3 in section 3.9.3, page 380. The Rayleigh number Ra_{ver} for the vertical cylinder is formed with the cylinder height, that for the horizontal Ra_{hor} with the cylinder diameter. It is

$$Ra_{\text{ver}} = \frac{\beta_\infty(\vartheta_o - \vartheta_\infty)gL^3}{\nu^2}Pr = 8.81 \cdot 10^6 \quad \text{and} \quad Ra_{\text{hor}} = 5.64 \cdot 10^5 \ .$$

From (3.328) $Nu_{\text{mver}} = 30.14$ and $\alpha_{\text{mver}} = Nu_{\text{mver}}\lambda/L = 4.72$ W/m²K. From No. 3 in section 3.9.3, page 380, $Nu_{\text{mhor}} = 12.40$ and $\alpha_{\text{mhor}} = Nu_{\text{mhor}}\lambda/d = 4.86$ W/m²K. This gives $\alpha_{\text{mhor}} > \alpha_{\text{mver}}$. In addition to this heat is also released via the ends of the horizontal cylinder. The can therefore cools more rapidly if it is lying down.

3.18: The temperature of the outer skin T_0 is practically equal to the eigentemperature, as the heat flux released from the outer skin, according to (3.369), is $\dot{q} = \alpha(T_0 - T_e) = k'(T_i - T_0)$ mit $1/k' = 1/\alpha_i + \delta/\lambda$. It follows from this, that $T_0 - T_e = k'/\alpha(T_i - T_0)$. Here, $T_i - T_0$ is a few K, and the external heat transfer coefficient is $\alpha \gg k'$. Therefore, $T_0 \cong T_e$. The eigentemperature is calculated from (3.352) with $w_{S\delta} = \sqrt{\kappa R T_\delta} = 299.5$ m/s and $Ma_\delta = w_\delta/w_{S\delta} = 0.649$ to be

$$T_e = T_0 = T_\delta\left[1 + \frac{\kappa-1}{2}Ma_\delta^2\right] = 241.95\,\text{K} = -31.2\ ^\circ\text{C} \ .$$

Approximately the same results are obtained from (3.354) with $r = \sqrt[3]{Pr}\ Pr(-50\ ^\circ\text{C}) = 0.727$, $c_p = 1.007$ kJ/kgK. We obtain $T_0 = 240$ K $= -33,1\ ^\circ$C. The heat flux is $\dot{q} = k(T_i - T_0) = 170.7$ W/m² with

$$\frac{1}{k} = \frac{1}{\alpha_i} + \frac{\delta}{\lambda} \quad \text{and} \quad k = 3.33\,\text{W/m}^2\text{K} \ .$$

3.19: As soon as the temperature T_0 of the outer skin rises above an eigentemperature T_e of 300 °C cooling must occur. Therefore, $T_0 \leq T_e$ so that cooling is not necessary. According to (3.352)

$$\frac{T_e}{T_\delta} = 1.8 \ .$$

We should have $T_0 \leq T_e = 1.8T_\delta$, that is $T_\delta \geq T_0/1.8 = 45.3\ ^\circ$C. As soon as the air temperature rises above 45.3 °C cooling has to start.

Chapter 4: Convective heat and mass transfer. Flows with phase change

4.1: According to (4.15) $\alpha_{m,hor} = 5028.8$ W/m^{2}K. The temperature at the condensate surface, is yielded approximately from (4.35) as

$$\vartheta_I - 333.15 = \frac{30\,\text{W/m}^2\text{K} \cdot 2257.3\,\text{kJ/kgK}}{5028.8\,\text{W/m}^2\text{K} \cdot 1.93\,\text{kJ/kgK}} \ln \frac{0.101325 - p_I}{0.101325 - 0.0948}.$$

Here, $(p_{1G} - \tilde{y}_{1G})p = (1 - 0.0647)0.101325 = 0.0948$ MPa is set. For the determination of ϑ_I from the equation above, ϑ_I is guessed first, and then with the value $p_I(\vartheta_I)$ from the steam tables, the estimate is checked for its correctness. We find $\vartheta_I \cong 348.8$ K $= 75.7$ °C, for this we get $p_I = 0.0397$ MPa and the equation above is well satisfied. It follows from (4.37) that $\dot{q}_G/\dot{q} = 0.39$. The heat flux released falls to 39 % of the heat flux that would be released if pure, saturated steam was present. The area has to be increased by a factor of $1/0.39 = 2.56$, if the condensator power is to stay the same.

4.2: We get $\Delta\vartheta = \vartheta_0 - \vartheta_s = 200$ °C, according to Fig. 4.37 this is in the region of film boiling. The heat transfer coefficients are yielded from section 4.2.7, page 463, No. 6 as $\alpha_R = 25.1$ W/m^{2}K and $\alpha_{mG} = 688.8$ W/m^{2}K. With that we get $\alpha_{mG}/\alpha_R = 27.4$ and $\alpha_m/\alpha_{mG} = 1.029$, $\alpha_m = 708.8$ W/m^{2}K. The heating power is $\dot{Q} = \alpha_m A \Delta\vartheta = 1113.4$ W.

4.3: The heating power amounts to $\dot{Q} = \dot{M}_{oil}c_{poil}\Delta\vartheta = 3$ MW. The amount of steam generated $\dot{M}_W$ follows from $\dot{Q} = \dot{M}_W\Delta h_v$ as $\dot{M}_W = 1.39$ kg/s $= 5 \cdot 10^3$ kg/h. On the other hand, the heat flow transferred is

$$\dot{Q} = kA\Delta\vartheta_m \quad \text{with} \quad \Delta\vartheta_m = \frac{(\vartheta_{oili} - \vartheta_s) - (\vartheta_{oile} - \vartheta_s)}{\ln[(\vartheta_{oili} - \vartheta_s)/(\vartheta_{oile} - \vartheta_s)]} .$$

It is $\Delta\vartheta_m = 42.84$ K and from $1/(kA) = 1/(\alpha_i A_i) + 1/(\alpha_o A_o)$ mit $A = A_o \cong A_i$ follows $1/k = 1/\alpha_i + 1/\alpha_o$. According to (4.97), $\alpha_o = 2.43\dot{q}^{0.72}$ W/m^{2}K and so

$$\frac{1}{k} = \frac{1}{700} + \frac{1}{2.43\,\dot{q}^{0.72}}\text{m}^2\text{K/W} .$$

This gives

$$\frac{\dot{Q}}{A} = \dot{q} = k\Delta\vartheta_m = \left(\frac{1}{700} + \frac{1}{2.43\dot{q}^{0.72}}\right)^{-1} \text{W/m}^2\text{K} \cdot 42.84\,\text{K} .$$

This leads to a transcendental equation for $\dot{q}$:

$$3.3347 \cdot 10^{-5}\dot{q} + 9.606 \cdot 10^{-3}\dot{q}^{0.28} - 1 = 0 .$$

The solution is $\dot{q} = 25080$ W/m^2. The transfer area is follows from $\dot{Q} = \dot{q}A$ as $A \cong 120$ m^2.

4.4: a) The length follows from the energy balance

$$\dot{q}d\pi\Delta z = \dot{m}\frac{d^2\pi}{4}(h' - h_1) \quad \text{as} \quad \Delta z = \frac{\dot{m}d(h' - h_1)}{4\dot{q}} = 0.97\,\text{m} .$$

b) The quality at the outlet is yielded from (4.115) as $x^* = 0.399$.

4.5: According to (4.152),

$$X_{tt} = 0.2259 \left(\frac{1 - x^*}{x^*}\right)^{0.9} .$$

From (4.126) with (4.128) we get

$$\left(\frac{dp}{dz}\right)_r = -\Phi_L^2\left(\frac{dp}{dz}\right)_L = \Phi_L^2\zeta_L\frac{1}{d}\frac{\dot{m}^2}{2\varrho_L}(1 - x^*)^2$$

$$\left(\frac{dp}{dz}\right)_r = -2061.6\,\text{N/m}^3\,\Phi_L^2\,(1 - x^*)^2 .$$

With the values given above for X_{tt}, we get from (4.153)

$$\Phi_L^2 = 1 + \frac{20}{X_{tt}} + \frac{1}{X_{tt}^2} = 1 + 88.53 \left(\frac{x^*}{1-x^*}\right)^{0.9} + 19.596 \left(\frac{x^*}{1-x^*}\right)^{1.8}$$

and therefore

$$\left(\frac{dp}{dz}\right)_r dx^* = -2061 \, \text{N/m}^3 \left[(1-x^*)^2 + 88.53 x^{*0.9}(1-x^*)^{1.1}\right.$$
$$\left. + 19.596 x^{*1.8}(1-x^*)^{0.2}\right] dx^* \ .$$

As the heating is with constant heat flux, the flow quality x^* changes linearly with the distance z. This follows from the energy balance

$$x^* = \frac{1}{\Delta h_v} \frac{\dot{q} d\pi (z - \Delta z)}{\dot{m} d^2 \pi/4} = \frac{\dot{q} \cdot 4}{\Delta h_v \dot{m} d}(z - \Delta z) \ .$$

With $\Delta z = 0.97$ m, $x^* = 0.1574(z/\text{m} - 0.97)$. So $dx^*/dz = 0.1574 \, 1/\text{m}$ and in the previous equation for the pressure drop

$$\left(\frac{dp}{dz}\right)_r dx^* = (dp)_r \, 0.1574 \, 1/\text{m} \ .$$

It therefore holds that

$$(dp)_r = -13094 \, \text{N/m}^2 \left[(1-x^*)^2 + 88.53 \, x^{*0.9}(1-x^*)^{1.1}\right.$$
$$\left. + 19.596 \, x^{*1.8}(1-x^*)^{0.2}\right] dx^* \ .$$

Integration between $x^* = 0$ and $x^* = 0.399$ yields $(\Delta p)_r = -863 \, \text{hPa} = -0.863 \, \text{MPa}$.

4.6: a) The length is obtained from the energy balance

$$\dot{q} d\pi L + \dot{M} h_1 = \dot{M}_G h'' + \dot{M}_L h'$$
$$\frac{\dot{q} d\pi L}{\dot{m} d^2 \pi/4} = x^* h'' + (1 - x^*) h' - h_1$$

$$L = \frac{\dot{m} d}{4\dot{q}} \left[x^* h'' + (1 - x^*) h' - h_1\right] = 3.25 \, \text{m} \ .$$

b) The wall temperature is yielded from $\vartheta_0 = \vartheta_s + \dot{q}/\alpha_{2Ph}$. It is $\vartheta_s(5.95 \, \text{MPa}) = 275 \, °\text{C}$. The heat transfer coefficient α_{2Ph} follows from (4.159) with (4.160) and (4.161). It is

$$Re = \frac{\dot{m}(1 - x^*) d}{\eta_L} = 1.534 \cdot 10^5$$
$$Bo = \frac{\dot{q}}{\dot{m} \Delta h_v} = 5.09 \cdot 10^{-4} \ ,$$
$$X_{tt} = 0.64 \ .$$

Then, according to (4.161) $F = 6.64$ and according to (4.160) we get $S = 1.66 \cdot 10^{-2}$. It further follows from (4.100) and (4.103) that $F(p^+) = 2.523$ and $\alpha = \alpha_B = 106833 \, \text{W/m}^2\text{K}$. It is $\alpha_C = \lambda_L/d \cdot 0.023 \cdot Re^{0.7} Pr_L^{1/3} = 2726.6 \, \text{W/m}^2\text{K}$. Giving $\alpha_{2Ph} = 19878 \, \text{W/m}^2\text{K}$ and $\vartheta_0 = 315.2 \, °\text{C}$.

5.2: $\lambda_1 = 2.3\ \mu\text{m}$, $\lambda_2 = 4.7\ \mu\text{m}$; $L = 115.2\ \text{W}/(\text{m}^2\text{sr})$; $M = 361.9\ \text{W}/\text{m}^2$; $\Delta M/M - 1/32$.

5.3: $\Delta M/M = 1/4$

5.4: $L_{\text{ref}} = 38.2\ \text{W}/(\text{m}^2\text{sr})$; $\Phi_{\text{b,abs}}/A = 680\ \text{W}/\text{m}^2$

5.5: $T = 1139\ \text{K}$

5.6: $T = 1400\ \text{K}$; $M = 218\ \text{kW}/\text{m}^2$

5.7: $T = 4330\ \text{K}$ und $T = 11810\ \text{K}$. There are two temperatures!

5.8: a) $\varepsilon = 0.677$; b) $a = 0.375$

5.9: 0.332

5.10: $E = 6375\ \text{W}/\text{m}^2$; $\varepsilon = 0.864$

5.11: $T = 740\ \text{K}$; $\varepsilon = 0.0964$; $\varepsilon_{\text{n}} = 0.0789$

5.12: $F_{11} = 0$; $F_{12} = 1$; $F_{21} = 0.6366$; $F_{22} = 0.3634$

5.13: $F_{12} = 0.5$

5.14: $F_{11} = 0$; $F_{12} = 0.8500$; $F_{13} = 0.1500$; $F_{21} = 0.3400$; $F_{22} = 0$; $F_{23} = 0.6600$; $F_{31} = 0.0833$; $F_{32} = 0.9167$; $F_{33} = 0$.

5.15: $T(0) = 507\ \text{K}$; $T(h) = 439\ \text{K}$; $T(2h) = 375\ \text{K}$; $T(10h) = 305\ \text{K}$; $T(\infty) = 300\ \text{K}$.

5.16: $T = 953\ \text{K}$

5.17: $\dot{Q} = 436\ \text{W}$

5.18: a) The heat flow is independent of the orientation of the radiation protection shield.
b) If the side with the larger emissivity ($2.5\varepsilon_{\text{S}}$) is directed towards the plate with T_1, the higher temperature T_{S} is yielded.

5.19: a) $T_{\text{A}} = 679.6\ \text{K}$; $T_{\text{B}} = 576.1\ \text{K}$. b) $\dot{q} = 4063\ \text{W}/\text{m}^2$. c) $T_1 = 576.1\ \text{K} = T_{\text{B}}$ from part a).

5.20: a) $\lambda_{\text{eff}} = 0.2545\ \text{W}/\text{K m}$; $\dot{q} = 33.82\ \text{W}/\text{m}^2$. b) $\dot{q} = 13.1\ \text{W}/\text{m}^2$

5.21: $\alpha = 39.9\ \text{W}/(\text{m}^2\text{K})$; $\alpha_{\text{Str}} = 23.8\ \text{W}/(\text{m}^2\text{K})$. The gas radiation participates by around 37 % in the heat transfer.

Literature

Chapter 1: Introduction. Technical Applications

[1.1]   Baehr, H.D.: Thermodynamik, 9. Aufl. insbes. S. 56–57, Berlin: Springer-Verlag 1996

[1.2]   Stephan, K.; Mayinger, F.: Thermodynamik. Grundlagen und technische Anwendungen. Bd. 1, 14. Aufl. insbes. S. 68, Berlin: Springer-Verlag 1992

[1.3]   Churchill, S.W.; Chu, H.H.S.: Correlating equations for laminar and turbulent free convection from a horizontel cylinder. Int. J. Heat Mass Transfer 18 (1975) 1049–1053

[1.4]   Krischer, O.; Kast, W.: Wärmeübertragung und Wärmespannungen bei Rippenrohren. VDI-Forschungsheft 474. Düsseldorf: VDI-Verlag 1959

[1.5]   Sparrow, E.M.; Hennecke, D.H.: Temperature depression at the base of a fin. J. Heat Transfer 92 (1970) 204–206

[1.6]   Sparrow, E.M.; Lee, L.: Effects of fin base-temperature in a multifin array. J. Heat Transfer 96 (1975) 463–465

[1.7]   Hausen, H.: Wärmeübertragung im Gegenstrom, Gleichstrom und Kreuzstrom. 2. Aufl. Berlin: Springer-Verlag 1976

[1.8]   Martin, H.: Wärmeübertrager. Stuttgart: G. Thieme Verlag 1988

[1.9]   Roetzel, W.; Heggs, R.J.; Butterworth, D. (Eds.): Design and Operation of Heat Exchangers. Berlin: Springer-Verlag 1992

[1.10] Hausen, H.: Über die Theorie des Wärmeaustausches in Regeneratoren. Z. angew. Math. Mech. 9 (1929) 173–200; —: Vervollständigte Berechnung des Wärmeaustausches in Regeneratoren. VDI-Z. Beiheft „Verfahrenstechnik" (1942) Nr. 2, S. 31–43

[1.11] VDI-Heat Atlas. Section N1–N14. Düsseldorf: VDI-Verlag 1992

[1.12] Roetzel, W.; Spang, B.: Heat transfer. VDI-Heat Atlas. Section Cb1–Cb6, Düsseldorf: VDI-Verlag 1992

[1.13] Ahrendts, J.; Baehr, H.D.: Die thermodynamischen Eigenschaften von Ammoniak. VDI-Forschungsheft 596, Düsseldorf: VDI-Verlag 1979

[1.14] Smith, D.M.: Mean temperature difference in cross flow. Engng. 138 (1934) 479–481 u. 606–607

[1.15] Schedwill, H.: Thermische Auslegung von Kreuzstromwärmeaustauschern. Fortsch.-Ber. VDI. Reihe 6, Nr. 19. Düsseldorf: VDI-Verlag 1968

[1.16] Roetzel, W.; Spang, B.: Design of heat exchangers. VDI-Heat Atlas. Section Ca1–Ca 39. Düsseldorf: VDI-Verlag 1992

[1.17] Nußelt, W.: Eine neue Formel für den Wärmeübergang im Kreuzstrom. Techn. Mech. u. Thermodynamik 1 (1930) 417–422

[1.18] Roetzel, W.; Spang, B.: Verbessertes Diagramm zur Berechnung von Wärmeübertragern. Wärme- u. Stoffübertragung 25 (1990) 259–264

[1.19] Stichlmair, J.: Anwendung der Ähnlichkeitsgesetze bei vollständiger und partieller Ähnlichkeit. Chem.-Ing. Techn. 63 (1991) 38–51

[1.20] Pawlowski, J.: Die Ähnlichkeitstheorie in der physikalisch-technischen Forschung. Berlin: Springer 1971

[1.21] Haase, R.: Thermodynamik der irreversiblen Prozesse. Darmstadt: Dr. Dietrich Steinkopff Verlag 1963

[1.22] de Groot, S. R.: Thermodynamics of irreversible processes. Amsterdam: North-Holland 1951

[1.23] Sherwood, Th.K.; Pigford, R.L.; Wilke, Ch.: Mass Transfer. New York: McGraw Hill, Kogakusha Ltd., 1975

[1.24] Perry, R. H.; Green, D. W.: Chemical Engineers Handbook. 2. Aufl. New York: McGraw Hill 1984

[1.25] Brauer, H.; Mewes, D.: Stoffaustausch einschließlich chemischer Reaktionen. Hanau: Verlag Sauerländer 1971

[1.26] Mersmann, A.: Stoffübertragung. Berlin: Springer-Verlag 1986

[1.27] Lewis, W.K.; Whitman, W.G.: Principles of gas absorption. Ind. Eng. Chem. 116 (1924) 1215–1220

[1.28] Stefan, J.: Über das Gleichgewicht und die Bewegung, insbesondere die Diffusion von Gasgemengen. Sitzungsb. Akad. Wiss. Wien 63 (1871) 63–124

[1.29] VDI-Heat Atlas. Düsseldorf: VDI-Verlag 1992

[1.30] Bird, R.B.; Stewart, W.E.; Lightfoot, E.N.: Transport Phenomena. New York: John Wiley 1960

[1.31] Higbie, R.: The rate of absorption of a pure gas into a still liquid during short periods of exposure. Trans. Am. Inst. Chem. Eng. 31 (1935) 36–38

[1.32] Danckwerts, P.V.: Gas-Liquid-Reactions. New York: McGraw Hill 1970

[1.33] Stephan, K.; Mayinger, F.: Thermodynamik: Grundlagen und technische Anwendungen. Bd. 2. 13. Aufl., Berlin: Springer-Verlag 1992, S. 15

Chapter 2: Heat conduction and mass diffusion

[2.1] Carslaw, H.S.; Jaeger, J.C.: Conduction of heat in solids. 2nd ed. Oxford: Clarendon Press 1986

[2.2] Jakob, M.: Heat transfer. Vol. 1. New York: J. Wiley & Sons 1949

[2.3] Höchel, J.; Saur, G.; Borchers, H.: Strukturveränderungen in kurzzeitig behandeltem Urandioxyd mit und ohne Neutronenbestrahlung als Hilfe zur Ermittlung der Temperaturverteilung in Brennelementen für Kernreaktoren. J. Nucl. Mat. 33 (1969) 225–241

[2.4] Chen, S.-Y.; Zyskowski, G.L.: Steady-state heat conduction in a straight fin with variable film coefficient. ASME-Paper No. 63-HT-12 (1963)

[2.5] Han, L.S.; Lefkowitz, S.G.: Constant cross-section fin efficiencies for nonuniform surface heat-transfer coefficients. ASME-Paper No. 60-WA-41 (1960)

[2.6]   Ünal, H.C.: Dertermination of the temperature distribution in an extended surface with
        a non-uniform heat transfer coefficient. Int. J. Heat Mass Tranfer 28 (1985) 2279–2284

[2.7]   Harper, D.R.; Brown, W.B.: Mathematical equations for heat conduction in the fins of
        air-cooles engines. Natl. Advisory Comm. Aeronautics, Report no. 158 (1922)

[2.8]   Schmidt, E.: Die Wärmeübertragung durch Rippen. Z. VDI 70 (1926) 885–889 u. 947–
        1128 u. 1504

[2.9]   Gardner, K.A.: Efficiency of extended surfaces. Trans. ASME 67 (1945) 621–631

[2.10] Focke, R.: Die Nadel als Kühlelement. Forschung Ing.-Wes. 13 (1942) 34–42

[2.11] Kern, D.Q.; Kraus, A.D.: Extended surface heat transfer. New York: McGraw-Hill 1972

[2.12] Ullmann, A.; Kalman, H.: Efficiency and optimized dimensions of annular fins of different
        cross-section shapes. Int. J. Heat Mass Transfer 32 (1989) 1105–1110

[2.13] Brandt, F.: Wärmeübertragung in Dampferzeugern und Wärmeaustauschern. Essen:
        Vulkan-Verlag 1985

[2.14] Schmidt, Th.E.: Die Wärmeleistung von berippten Oberflächen. Abh. Deutsch. Kältetechn.
        Verein Nr. 4, Karlsruhe: C.F. Müller 1950

[2.15] Sparrow, E.M.; Lin, S.H.: Heat-transfer characteristics of polygonal and plate fins. Int.
        J. Heat Mass Transfer 7 (1964) 951–953

[2.16] Baehr, H.D.; Schubert, F.: Die Bestimmung des Wirkungsgrades quadratischer Scheiben-
        rippen mit Hilfe eines elektrischen Analogieverfahrens. Kältetechnik 11 (1959) 320–325

[2.17] Kakaç, S.; Yener, Y.: Heat conduction. 2nd ed. Washington: Hemisphere Publ. Comp.
        1985

[2.18] Grigull, U.; Sandner, H.: Heat conduction. Berlin: Springer 1992

[2.19] Elgeti, K.: Der Wärmeverlust eines in einer Wand verlegten Rohres. Heizung-Lüftung-
        Haustechn. 22 (1971) 109–113

[2.20] Keune, F.; Burg, K.: Singularitätenverfahren der Strömungslehre. Karlsruhe: Braun
        1975

[2.21] Elgeti, K.: Der Wärmeverlust einer erdverlegten Rohrleitung im stationären Zustand
        unter dem Einfluß der Wärmeübertragungszahl an der Erdoberfläche. Forsch. Ing.-Wes.
        33 (1967) 101–105

[2.22] Nußelt, W.: Die Temperaturverteilung in der Decke einer Strahlungsheizung. Gesundh.-
        Ing. 68 (1947) 97–98

[2.23] Hahne, E.; Grigull, U.: Formfaktor und Formwiderstand der stationären mehrdimen-
        sionalen Wärmeleitung. Int. J. Heat Mass Transfer 18 (1975) 751–767

[2.24] Elgeti, K. et al.: VDI-Heat Atlas, Section Ea 4–12. Düsseldorf: VDI-Verlag 1992

[2.25] Baehr, H.D.: Die Lösung nichtstationärer Wärmeleitungsprobleme mit Hilfe der Laplace-
        Transformation. Forsch. Ing.-Wes. 21 (1955) 33–40

[2.26] Tautz, H.: Wärmeleitung und Temperaturausgleich. Weinheim: Verlag Chemie 1971

[2.27] Doetsch, G.: Anleitung zum praktischen Gebrauch der Laplace-Transformation und der
        Z-Transformation. 3. Aufl. München: R. Oldenbourg 1967

[2.28] Spanier, J.; Oldham, K.B.: An atlas of functions. Washington: Hemisphere Publ. Comp.
        Berlin: Springer 1987

[2.29] Chu, H.S.; Chen, C.K.; Weng, C.: Applications of Fourier series technique to transient heat transfer problems. Chem. Eng. Commun. 16 (1982) 215–225

[2.30] Jahnke-Emde-Lösch: Tafeln höherer Funktionen. 6. Aufl. Stuttgart: Teubner 1960

[2.31] Berlyand, O.S.; Gavrilovo, R.I.; Prudnikov, A.P.: Tables of integral error functions and Hermite polynomials. Oxford: Pergamon Press 1962

[2.32] Abramowitz, M.; Stegun, I.A.: Handbook of mathematical functions with formulas, graphs and mathematical tables. Washington: U.S. Gouvernment Printing Off. 1964

[2.33] Jänich, K.: Analysis für Physiker und Ingenieure. 2. Aufl. Berlin: Springer 1990

[2.34] Grigull, U.: Temperaturausgleich in einfachen Körpern. Berlin: Springer 1964

[2.35] Schneider, P.J.: Temperature response charts. New York: J. Wiley & Sons 1963

[2.36] Grigull, U.; Bach, J.; Sandner, H.: Näherungslösungen der nichtstationären Wärmeleitung. Forsch. Ing.-Wes. 32 (1966) 11–18

[2.37] Goldstein, S.: The application of Heaviside's operational method to the solution of a problem in heat conduction. Z. angew. Math. Mech. 12 (1932) 234–243, sowie: On the calculation of the surface temperature of geometrically simple bodies. Z. angew. Math. Mech. 14 (1934) 158–162

[2.38] Baehr, H.D.: Die Berechnung der Kühldauer bei ein- und mehrdimensionalem Wärmefluß. Kältetechnik 5 (1953) 255–259

[2.39] Stefan, J.: Über die Theorie der Eisbildung, insbesondere über die Eisbildung im Polarmeer. Wiedemann Ann. Phys. u. Chem. 42 (1891) 269–286

[2.40] Ockendon, J.R.; Hodgkins, W.R. (Hrsg.): Moving boundary problems in heat flow and diffusion. Oxford: Clarendon Press 1975

[2.41] Wilson, D.G.; Solomon, A.D.; Boggs, P.T. (Hrsg.): Moving boundary problems. New York: Academic Press 1978

[2.42] Plank, R.: Über die Gefrierzeit von Eis und wasserhaltigen Lebensmitteln. Z. ges. Kälteindustrie 39 (1932) 56–58

[2.43] Nesselmann, K.: Systematik der Gleichungen für Gefrieren und Schmelzen von Eisschichten nebst Anwendung auf Trommelgefrierapparate und Süßwasserkühler. Kältetechnik 1 (1949) 169–172

[2.44] Nesselmann, K.: Die Trennung flüssiger Gemische durch kältetechnische Verfahren. Forsch. Ing.-Wes. 17 (1951) 33–39

[2.45] Goodman, T.R.: The heat balance integral and its application to problems involving a change of phase. J. Heat Transfer 80 (1958) 335–342

[2.46] Cho, S.H.; Sunderland, J.E.: Heat conduction problems with melting or freezing. J. Heat Transfer 91 (1969) 421–426

[2.47] Charach, Ch.; Zoglin, P.: Solidification in a finite, initially overheated slab. Int. J. Heat Mass Transfer 28 (1985) 2261–2268

[2.48] Megerlin, F.: Geometrisch eindimensionale Wärmeleitung beim Schmelzen und Erstarren. Forsch. Ing.-Wes. 34 (1968) 40–46

[2.49] Stephan, K.: Schmelzen und Erstarren geometrisch einfacher Körper. Kältetechnik-Klimatisierung 23 (1971) 42–46

[2.50] Stephan, K.; Holzknecht, B.: Die asymptotischen Lösungen für Vorgänge des Erstarrens. Int. J. Heat Mass Transfer 19 (1976) 597–602

[2.51] Stephan, K.; Holzknecht, B.: Wärmeleitung beim Erstarren geometrisch einfacher Körper. Wärme- u. Stoffübertragung 7 (1974) 200–207

[2.52] Myers, G.E.: Analytical methods in conduction heat transfer. New York: McGraw-Hill 1971

[2.53] Marsal, D.: Finite Differenzen und Elemente. Numerische Lösung von Variationsproblemen und partiellen Differentialgleichungen. Berlin: Springer 1989

[2.54] Bathe, K.-J.: Finite-Elemente-Methoden. Berlin: Springer 1990

[2.55] Knothe, K.; Wessels, H.: Finite Elemente. Eine Einführung für Ingenieure. 2. Aufl. Berlin: Springer 1992

[2.56] Nasitta, K.; Hagel, H.: Finite Elemente. Mechanik, Physik und nichtlineare Prozesse. Berlin: Springer 1992

[2.57] Smith, G.D.: Numerical solution of partial differential equations. Finite difference methods. 3. Aufl. Oxford: Clarendon Press 1985

[2.58] Incropera, F.P.; De Witt, D.P.: Fundamentals of heat and mass transfer. 3. Aufl. New York: J. Wiley & Sons 1990

[2.59] Lick, W.J.: Difference equations from differential equations. Lecture notes in engineering; No. 41. Berlin: Springer 1989

[2.60] Forsythe, G.E.; Wasow, W.R.: Finite-difference methods for partial differential equations. New York: J. Wiley & Sons 1960

[2.61] Binder, L.: Über äußere Wärmeleitung und Erwärmung elektrischer Maschinen. Dissertation TH München 1910

[2.62] Schmidt, E.: Über die Anwendung der Differenzenrechnung auf technische Aufheiz- und Abkühlprobleme. Beiträge zur techn. Mechanik u. techn. Physik (Föppl-Festschrift). Berlin 1924

[2.63] Schmidt, E.: Das Differenzenverfahren zur Lösung von Differentialgleichungen der nichtstationären Wärmeleitung, Diffusion und Impulsausbreitung. Forsch. Ing.-Wes. 13 (1942) 177–185

[2.64] Baehr, H.D.: Beiträge zur graphischen Bestimmung nichtstationärer Temperaturfelder mit Hilfe des Differenzenverfahrens. Forsch. Ing.-Wes. 20 (1954) 16–19

[2.65] Crank, J.; Nicolson, P.: A practical method for numerical evaluation of solutions of partial differential equations of the heat conduction type. Proc. Camb. Phil. Soc. 43 (1947) 50–67

[2.66] Törning, W.; Spelluci, P.: Numerische Mathematik für Ingenieure und Physiker. 2. Aufl. 2 Bde. Berlin: Springer 1988 u. 1990

[2.67] Stoer, J.: Numerische Mathematik 1. 5. Aufl. Berlin: Springer 1989

[2.68] Stoer, J.; Bulirsch, R.: Numerische Mathematik 2. 3. Aufl. Berlin: Springer 1990

[2.69] Peaceman, P.W.; Rachford, H.H.: The numerical solution of parabolic and elliptic differential equations. J. Soc. Industr. Appl. Math. 3 (1955) 28–41

[2.70] Douglas, J.: On the numerical integration of $\partial^2 u/\partial x^2 + \partial^2 u/\partial y^2 = \partial u/\partial t$ by implicit methods. J. Soc. Industr. Appl. Math. 3 (1955) 42–65

[2.71] Douglas, J.; Rachford, H.H.: On the numerical solution of heat conduction problems in two and three space variables. Trans. Amer. Math. Soc. 82 (1956) 421–439

[2.72] Douglas, J.: Alternating direction methods for three space variables. Numer. Math. 4 (1962) 41–63

[2.73] Young, D.M.: Iterative solution of large linear systems. New York: Academic Press 1971

[2.74] Stupperich, F.R.: Instationäre Wärmeleitung. Vereinfachte Berechnung der Aufheizung bzw. Abkühlung von Körpern in konstanter Umgebung. Brennst.-Wärme-Kraft 45 (1993) 247–258

[2.75] Haase, R.: Thermodynamik der irreversiblen Prozesse. Darmstadt: Dr.-Dietrich-Steinkopff-Verlag 1963, S. 300

[2.76] Wilke, C.R.; Chang, P.C.: Correlation of diffusion coefficients in dilute solutions. Amer. Inst. Chem. Eng. J. 1 (1955) 264–270

[2.77] Fick, A.: Über Diffusion. Poggendorffs Ann. Phys. Chemie 94 (1855) 59–86

[2.78] Crank, J.: The mathematics of diffusion. 2. Aufl. Oxford: Clarendon Press 1975, reprint 1990

[2.79] Landolt-Börnstein: Zahlenwerte und Funktionen aus Physik, Chemie, Astronomie, Geophysik und Technik. Bd. II, Teil 2 b. 6. Aufl. Berlin: Springer 1962

[2.80] Krischer, O.; Kast, W.: Die wissenschaftlichen Grundlagen der Trocknungstechnik. 3. Aufl., Bd. 1, Berlin: Springer 1978

[2.81] Aris, R.: On shape factors for irregular particles. Chem. Eng. Sci. 6 (1957) 262–268.

[2.82] Kohlrausch, F.: Praktische Physik. 23. Aufl., herausgegeben von D. Hahn u. S. Wagner, Band 1. Stuttgart: B.G. Teubner 1985

Chapter 3: Convective heat and mass transfer. Single phase flows

[3.1] Stephan, K.; Mayinger, F.: Thermodynamik, Grundlagen und technische Anwendungen. Bd. 2, 13. Aufl., Berlin: Springer-Verlag 1992, S. 35

[3.2] Stephan, K.; Mayinger, F.: Thermodynamik, Grundlagen und technische Anwendungen. Bd. 1, 14. Aufl., Berlin: Springer-Verlag 1992, S. 254

[3.3] Haase, R.: Thermodynamik der irreversiblen Prozesse. Darmstadt: Dr. Dietrich Steinkopff-Verlag 1963

[3.4] Prandtl, L.: Über Flüssigkeitsbewegung bei sehr kleiner Reibung. Verhandlungen der 3. Intern. Math. Kongr., Heidelberg 1904

[3.5] Tollmien, W.: Über die Entstehung der Turbulenz. 1. Mitteilung. Nachr. Ges. Wiss. Göttingen, Math. Phys. Klasse 21–24 (1929)

[3.6] Blasius, H.: Grenzschichten in Flüssigkeiten mit kleiner Reibung. Z. Math. u. Phys. 56 (1908) 1–37

[3.7] Töpfer, C.: Bemerkungen zu dem Aufsatz von H. Blasius, „Grenzschichten in Flüssigkeiten mit kleiner Reibung". Z. Math. u. Phys. 60 (1912) 197–398

[3.8] Howarth, L.: On the solution of the laminar boundary layer equations. Proc. Roy. Soc. London A 164 (1938) 547–579

[3.9] Schlichting, H.: Grenzschicht-Theorie. 5. Aufl., Karlsruhe: G. Braun 1965, S. 592

[3.10] Colburn, A.P.: A method of corrrelating forced convection, heat transfer data and a comparison with fluid friction. Trans. Am. Inst. Chem. Eng. 29 (1933) 174–210

[3.11] Chilton, T.H.; Colburn, A.P.: Mass transfer (absorption) coefficients. Ind. Eng. Chem. 26 (1934) 1138–1187

[3.12] Petukhov, B.J.; Popov, N.V.: Theoretical calculation of heat exchange and frictional resistance in turbulent flow in tubes of an incompressible fluid with variable physical properties. High Temperature 1 (1963) 69–83

[3.13] Gnielinski, V.: Berechung mittlerer Wärme- und Stoffübergangskoeffizienten an laminar und turbulent überströmten Einzelkörpern mit Hilfe einer einheitlichen Gleichung. Forsch. Ing. Wes. 41 (1975) 145–153

[3.14] Giedt, W. H.: Investigation of variation of point-unit heat transfer coefficient around a cylinder normal to an air stream. Trans. Am. Soc. Mech. Eng. 71 (1949) 375–381

[3.15] Žukauskas, A.A.; Žingžda J.: Heat transfer of a cylinder in cross flow. Washington: Hemisphere Publ. Comp. 1986, S. 162

[3.16] Žukauskas, A. A.; Makayawižus, V.I.; Žlantžauskas, A. K.: Wärmeübergang in Rohrbündeln bei Querausströmung von Fluiden (russ.). Vilnjus: Mintis 1968

[3.17] Hofmann, E.: Wärme- und Stoffübertragung. In: Planck, R. (Hrsg.): Handbuch der Kältetechnik. Bd. 3, Berlin: Springer-Verlag 1959

[3.18] Gnielinski, V.: Heat transfer in cross-flow around individual tube rows and through tube bundles. VDI-Heat Atlas, Section Gf, Düsseldorf: VDI-Verlag 1992

[3.19] Hirschberg, H.G.: Wärmeübergang und Druckverlust an quer angeströmte Rohrbündeln. Abh. Dtsch. Kältetech. Ver., Nr. 16. Karlsruhe: C.F. Müller 1961

[3.20] Hausen, H.: Gleichungen zur Berechnung des Wärmeübergangs im Kreuzstrom an Rohrbündeln. Kältetech. Klim. 23 (1971) 86–89

[3.21] Grimison, E.D.: Correlation and utilization of new data on flow resistance and heat transfer for crossflow of gases over tube banks. Trans. Amer. Soc. Mech. Eng. 59 (1937) 583–594

[3.22] Slipčević, B.: Wärmeübertragung durch Leitung und Konvektion. In: Steimle, F.; Stephan, K. (Hrsg.): Handbuch der Kältetechnik. Bd. VIB, Berlin: Springer-Verlag 1988, S. 61–62

[3.23] Stephan, K.: Wärmeübergang und Druckabfall laminarer Strömungen im Einlauf von Rohren und ebenen Spalten. Diss. T.H. Karlsruhe 1959

[3.24] Kays, W.M.; Crawford, M.E.: Convective heat and mass transfer. New York: McGraw Hill 1980

[3.25] Kakač, S.; Shah, R. K.; Aung, W.: Handbook of single-phase convective heat transfer. New York: John Wiley 1987, S. 3.122–3.125

[3.26] Graetz, L.: Über die Wärmeleitfähigkeit von Flüssigkeiten. Ann. Phys. Neue Folge 18 (1883) 79–94 und 25 (1885) 337–357

[3.27] Nußelt, W.: Die Abhängigkeit der Wärmeübergangszahl von der Rohrlänge. Z. Ver. Dtsch. Ing. 54 (1910) 1154–1158

[3.28] Brown, G. M.: Heat or mass transfer in a fluid in laminar flow in a circular duct or flat conduit. Amer. Inst. Chem. Eng. J. 6 (1960) 179–183

[3.29] Stephan, K.: Thermodynamik. In: Dubbel, Handbook of mechanical engineering. Berlin: Springer-Verlag 1994

[3.30] Lévêque M. A.: Les lois de transmission de la chaleur par convection. Ann. des Mines 12 (1928) 201–299, 305–362, 381–415

[3.31] Shah, R.K.; London, A.L.: Laminar flow forced convection in ducts. Suppl. 1 to Advances in Heat Transfer. New York: Academic Press 1978

[3.32] Schlichting, H.: Grenzschicht-Theorie. 5. Aufl., Karlsruhe: G. Braun 1965, S. 567

[3.33] Kraussold, H.: Die Wärmeübertragung bei zähen Flüssigkeiten in Rohren. VDI-Forschungsheft Nr. 351 (1931)

[3.34] McAdams, W.: Heat transmission. 2. Aufl., New York: McGraw Hill 1942, S. 168

[3.35] Hufschmidt, W.; Burck, E.: Der Einfluß temperaturabhängiger Stoffwerte auf den Wärmeübergang bei turbulenter Strömung von Flüssigkeiten in Rohren bei hohen Wärmestromdichten und Prandtlzahlen. Int. J. Heat Mass Transf. 41 (1968) 1041–1048

[3.36] Sieder, E.N.; Tate, G.E.: Heat transfer and pressure drop of liquids in tubes. Ind. Eng. Chem. 28 (1936) 1429–1436

[3.37] Petukhov, B.S.; Kirilov, V.V.: The problem of heat exchange in the turbulent flow of liquids in tubes. Teploenergetika 4 (1968) 63–68

[3.38] Gnielinski, V.: New equations for heat and mass transfer in turbulent pipe and channel flow. Int. J. Chem. Eng. 16 (1976) 359–368

[3.39] Gillespie, B.M.; Crandall, E.D.; Carberry, J.J.: Local and average interphase heat transfer coefficients in a randomly packed bed of spheres. Amer. Inst. Chem. Eng. J. 14 (1968) 483–490

[3.40] Schlünder, E.U.: Einführung in die Wärme- und Stoffübertragung. Braunschweig: Vieweg 1975, S. 75

[3.41] Wirth, T.: Flow patterns and pressure drops in fluidised beds. VDI-Heat Atlas, Section Lf, Düsseldorf: VDI-Verlag 1992

[3.42] Ihme, H.; Schmidt-Traub, H.; Brauer, H.: Theoretische Untersuchung über die Umströmung und den Stoffübergang an Kugeln. Chem. Ing. Techn. 44 (1972) 306–319

[3.43] Martin, H.: Wärme- und Stoffübertragung in der Wirbelschicht. Chem. Ing. Techn. 52 (1980) 199-209

[3.44] Reh, L.: Verbrennung in der Wirbelschicht. Chem. Ing. Techn. 40 (1968) 509–515

[3.45] Oberbeck, A.: Über die Wärmeleitung der Flüssigkeiten bei Berücksichtigung der Strömung infolge von Temperaturdifferenzen. Ann. Phys. Chem. 7 (1879) 271–292

[3.46] Boussinesq, J. M.: Théorie analytique de la chaleur. 2. Aufl., Paris: Gauthier-Villars 1903

[3.47] Ostrach, S.: An analysis of laminar free convection flow and heat transfer about a flat plate parallel to the direction of the generating body force. NACA, Techn. Report 1111 (1953)

[3.48] Pohlhausen, E.: Der Wärmeaustausch zwischen festen Körpern und Flüssigkeiten mit kleiner Reibung. Z. Angew. Math. Mech. 1 (1921) 115–121

[3.49] Le Fèvre, E.J.: Laminar free convection from a vertical plane surface. Proc. 9th Int. Congr. Appl. Mech., Brüssel, 4 (1956) paper I-168

[3.50] Churchill, S.W.; Chu, H.H.S.: Correlating equations for laminar and turbulent free convection from a vertical plate. Int. J. Heat Mass Transf. 18 (1975) 1323–1329

[3.51] Saville D.A.; Churchill, W.S.: Simultaneous heat and mass transfer in free convection boundary layers. Amer. Inst. Chem. Eng. J. 16 (1970) 268–273

[3.52] Eckert, E.R.G.; Drake, M.: Analysis of heat and mass transfer. New York: McGraw Hill 1972, S. 421

[3.53] Eckert, E.R.G.: Engineering relations for heat transfer and friction in high-velocity laminar and turbulent boundary-layer flow over surfaces with constant pressure and temperature. Trans. Amer. Soc. Mech. Eng., J. Heat Transf. 78 (1956) 1273–1283

Chapter 4: Convective heat and mass transfer. Flows with phase change

[4.1]   Butterworth, C.: Condensers: Basic heat transfer and fluid flow. In: Kakaç, S.; Bergles, A.E.; Mayinger, F.: Heat exchangers, thermal-hydraulic fundamentals and design. Washington: Hemisphere 1981, 289–313

[4.2]   Nußelt, W.: Die Oberflächenkondensation des Wasserdampfes, VDI-Z. 60 (1916) 541–546, 569–575

[4.3]   Chen, M.M.: An analytical study of laminar film condensation. Part 1. Flat plates. Part 2. Single and multiple horizontal tubes. Trans. Am. Soc. Mech. Eng., Ser. C. J. Heat Transfer 83 (1961) 48–60

[4.4]   Dukler, H.E.; Bergelin, O.P.: Characteristics of flow in falling liquid films. Chem. Eng. Progr. 48 (1952) 557–563

[4.5]   Kirschbaum, E.: Neues zum Wärmeübergang mit und ohne Änderung des Aggregatzustandes. Chem. Ing. Tech. 24 (1952) 393–402

[4.6]   Brauer, H.: Strömung und Wärmeübergang bei Rieselfilmen. VDI-Forschungsheft 457, Berlin: VDI-Verlag 1956

[4.7]   Grimley, L.S.: Liquid flow conditions in packed towers. Trans. Inst. Chem. Eng. (London) 23 (1945) 228–235

[4.8]   Van der Walt, J.; Kröger, D.G.: Heat transfer resistances during film condensation. Proc. Vth Int. Heat Transfer Conf., Tokyo 3 (1974) 284–288

[4.9]   Voskresenskij, K.D.: Heat transfer in film condensation with temperature dependent properties of the condensate (russ.). Izv. Akad. Nauk USSR (1948) 1023–1028

[4.10] Rohsenow, W.M.: Heat transfer and temperature distribution in laminar film condensation. Trans. Am. Soc. Mech. Eng., Ser. C. J. Heat Transfer 78 (1956) 1645–1648

[4.11] Ackermann, G.: Wärmeübergang und molekulare Stoffübertragung im gleichen Feld bei großen Temperatur- und Partialdruckdifferenzen. VDI-Forschungsheft 382, Berlin: VDI-Verlag 1937, 1–16

[4.12] Stephan, K.; Laesecke, A.: The influence of suction in condensation of mixed vapors. Wärme Stoffübertrag. 13 (1980) 115–123

[4.13] Sparrow, E.M.; Minkowycz, W.J.; Saddy, M.: Forced convection condensation in the presence of noncondensables and interfacial resistance. Int. J. Heat Mass Transfer 10 (1967) 1829–1845

[4.14] Grigull, U.: Wärmeübergang bei der Kondensation mit turbulenter Wasserhaut. Forsch. Ing.-Wes. 13 (1942) 49–57 und VDI Z. 86 (1942) 444–445

[4.15] Mostofizadeh, Ch.; Stephan, K.: Strömung und Wärmeübergang bei der Oberflächenverdampfung und Filmkondensation. Wärme Stoffübertrag. 15 (1981) 93–115

[4.16] Henstock, W.H.; Hanratty, Th.J.: The interfacial drag and the height of the wall layer in annular flows. Am. Inst. Chem. Eng. J. 6 (1976) 990–1000

[4.17] Isashenko, V.P.: Heat transfer in condensation (russ.). Moskau: Energia 1977

[4.18] Labunzov, D.A.: Heat transfer in film condensation of pure vapours on vertical plates and horizontal tubes (russ.). Teploenergetika 7 (1957) 72–79

[4.19] Rohsenow, W.M.; Webber, J.H.; Ling, A.T.: Effect of vapor velocity on laminar and turbulent film condensation. Trans. Am. Soc. Mech. Eng., Ser. C. J. Heat Transf. 78 (1956) 1637–1643

[4.20] Dukler, A.E.: Fluid mechanics and heat transfer in vertical falling film systems, Chem. Eng. Prog. Symp. Ser. 56 (1960) 1–10

[4.21] Shah, M.M.: A general correlation for heat transfer during film condensation inside pipes. Int. J. Heat Mass Transfer 22 (1979) 547–556

[4.22] Watson, R.G.H.; Birt, D.C.P.; Honour, G.W.: The promotion of dropwise condensation by Montan wax. Part I. Heat transfer measurements. J. Appl. Chem. 12 (1962) 539–552

[4.23] Woodruff, D.W.; Westwater, J.W.: Steam condensation on electroplated gold: Effect of plating thickness. Int. J. Heat Mass Transfer 22 (1979) 629–632

[4.24] Krischer, S.; Grigull, U.: Mikroskopische Untersuchung der Tropfenkondensation. Wärme Stoffübertrag. 4 (1971) 48–59

[4.25] Hampson, H.; Özisik, N.: An investigation into the condensation of steam. Proc. Inst. Mech. Eng. 1, Ser. B, (1952) 282–293

[4.26] Wenzel, H.: Versuche über Tropfenkondensation. Allg. Wärmetech. 8 (1957) 53–39

[4.27] Welch, J.F.; Westwater, J.W.: Microscopic study of dropwise condensation. Int. Dev. Heat Transfer 2 (1961) 302–309

[4.28] Kast, W.: Theoretische und experimentelle Untersuchung der Wärmeübertragung bei Tropfenkondensation. Fortschrittsber. VDI-Z., Reihe 3, Nr. 6, Düsseldorf 1965

[4.29] Le Fèvre, E.J.; Rose, J.W.: An experimental study of heat transfer by dropwise condensation. Int. J. Heat Mass Transfer 8 (1965) 1117–1133

[4.30] Tanner, D.W.; Pope, D.; Potter, C.J.; West, D.: Heat transfer in dropwise condensation at low pressures in the absence and presence of non-condensable gas. Int. J. Heat Mass Transfer 11 (1968) 181–190

[4.31] Griffith, P.; Lee, M.S.: The effect of surface thermal properties and finish on dropwise condensation. Int. J. Heat Mass Transfer 10 (1967) 697–707

[4.32] Brown, A.R.; Thomas, M.A.: Filmwise and dropwise condensation of steam at low pressures, Proc. 3rd Int. Heat Transfer Conf., Chicago 1966, Vol. II, 300–305

[4.33] Eucken, A.: Energie- und Stoffaustausch an Grenzflächen. Naturwissenschaften 25 (1937) 209–218

[4.34] Wicke, E.: Einige Probleme des Stoff- und Wärmeaustausches an Grenzflächen. Chem. Ing. Tech. 23 (1951) 5–12

[4.35] Kast, W.: Wärmeübergang bei Tropfenkondensation. Chem. Ing. Tech. 35 (1963) 163–168

[4.36] Jakob, M.: Heat transfer. Vol. I. New York: Wiley 1949, S. 694

[4.37] Umur, A.; Griffith, P.: Mechanism of dropwise condensation. Trans. Am. Soc. Mech. Eng., Ser. C. J. Heat Transfer 87 (1965) 275–282

[4.38] McCormick, J.L.; Baer, E.: On the mechanism of heat transfer in dropwise condensation. J. Colloid Sci. 18 (1963) 208–216

[4.39] Claude, G.: Air liquide, oxygène, azote, rare gases. 2. ed. Paris: Dunod 1926

[4.40] Lucas, K.: Die laminare Filmkondensation binärer Dampfgemische. Habil.-Schrift. Ruhr-Univ. Bochum 1974

[4.41] Tamir, A.; Taitel, Y.; Schlünder, E.U.: Direct contact condensation of binary mixtures. Int. J. Heat Mass Transfer 17 (1974) 1253–1260

[4.42] Prüger, W.: Die Verdampfungsgeschwindigkeit von Flüssigkeiten. Z. Phys. 115 (1949) 202–244

[4.43] Jakob, M.; Fritz, W.: Versuche über den Verdampfungsvorgang. Forsch. Ing-Wes. 2 (1931) 435–447

[4.44] Jakob, M.: Kondensation und Verdampfung. VDI Z. 76 (1931) 1161–1170

[4.45] Jakob, M.; Linke, W.: Der Wärmeübergang von einer waagerechten Platte an siedendes Wasser. Forsch. Ing.-Wes. 4 (1933) 75–81

[4.46] Jakob, M.; Linke, W.: Der Wärmeübergang beim Verdampfen von Flüssigkeiten an senkrechten und waagerechten Flächen. Phys. Z. 36 (1935) 267–280

[4.47] Jakob, M.: Heat transfer in evaporation and condensation. Mech. Eng. 58 (1936) 643–660, 729–739

[4.48] Fritz, W.: Wärmeübergang an siedende Flüssigkeiten. VDI Z. Beihefte Verfahrenstech. 5 (1937) 149–155

[4.49] Fritz, W.: Verdampfen und Kondensieren. VDI Z. Beihefte Verfahrenstech. 1 (1943) 1–14

[4.50] Stephan, K.: Wärmeübergang beim Kondensieren und beim Sieden. Berlin: Springer-Verlag 1988, S. 125–126

[4.51] Bashfort, F.; Adams, J.: An attempt to test the theory of capillary action. Cambridge: Cambridge University Press 1883

[4.52] Fritz, W.: Berechnung des Maximalvolumens von Dampfblasen. Phys. Z. 36 (1935) 379–384

[4.53] Fritz, W.; Ende, W.: Über den Verdampfungsvorgang nach kinematographischen Aufnahmen an Dampfblasen. Phys. Z. 37 (1936) 391–401

[4.54] Kabanow, W.; Frumkin, A.: Nachtrag zu der Arbeit „Über die Größe elektrisch entwickelter Gasblasen". Z. Phys. Chem. (A) 166 (1933) 316–317

[4.55] König, A.: Der Einfluß der thermischen Heizwandeigenschaften auf den Wärmeübergang bei der Blasenverdampfung. Wärme Stoffübertrag. 6 (1973) 38–44

[4.56] von Ceumern, W. C.: Abreißdurchmesser und Frequenzen von Dampfblasen in Wasser und wässrigen NaCl-Lösungen beim Sieden an einer horizontalen Heizfläche. Diss. TU Braunschweig 1975

[4.57] Mitrović, J.: Das Abreißen von Dampfblasen an festen Heizflächen, Int. J. Heat Mass Transfer 26 (1983) 955–963

[4.58] Stephan, K.: Bubble formation and heat transfer in natural convection boiling. In: Hahne, E.; Grigull, U.: Heat transfer in boiling. Washington: Hemisphere 1977, S. 3–20

[4.59] Tokuda, N.: Dynamics of vapor bubbles in binary liquid mixtures with translatory motion. 4th Int. Heat Transfer Conf., Paris 1970, B 7.5

[4.60] Jakob, M.; Linke, W.: Der Wärmeübergang von einer waagerechten Platte an siedendes Wasser. Forsch. Ing-Wes. 4 (1933) 75–81

[4.61] Zuber, N.: Nucleate boiling. The region of isolated bubbles and the similarity with natural convection. Int. J. Heat Mass Transfer 6 (1963) 53–78

[4.62] McFadden, P.; Grassmann, P.: The relation between bubble frequency and diameter. Int. J. Heat Mass Transfer 5 (1962) 169–173

[4.63] Cole, R.: Bubble frequencies and departure volumes at subatmospheric pressures. Am. Inst. Chem. Eng. J. 13 (1967) 779–783

[4.64] Ivey, H. J.: Relationships between bubble frequency, departure diameter and rise velocity in nucleate boiling. Int. J. Heat Mass Transfer 10 (1967) 1023–1040

[4.65] Malenkov, I. G.: The frequency of vapor-bubble separation as a function of bubble size. Fluid Mech. Sov. Res. 1 (1972) 36–42

[4.66] Jakob, M.; Linke, W.: Der Wärmeübergang beim Verdampfen von Flüssigkeiten an senkrechten und waagerechten Flächen. Phys. Z. 36 (1935) 267–280

[4.67] Nukijama, S.: Maximum and minimum values of heat transmitted from metal to boiling water under atmospheric pressure. J. Soc. Mech. Eng. Jpn. 37 (1934) 53–54, 367 - 374; vgl. auch Int. J. Heat Mass Transfer 9 (1966) 1419–1433

[4.68] Stephan, K.: Stabilität beim Sieden. Brennst.-Wärme-Kraft 17 (1965) 571–578

[4.69] Stephan, K.; Abdelsalam, M.: Heat transfer correlations for natural convection boiling. Int. J. Heat Mass Transfer 23 (1980) 73–87

[4.70] Stephan, K.; Preußer, P.: Wärmeübergang und maximale Wärmestoffdichte beim Behältersieden binärer und ternärer Flüssigkeitsgemische. Chem. Ing. Tech. 51 (1979) 37 (Synopse MS 649/79)

[4.71] Fritz, W.: Blasenverdampfung (nucleate boiling) im Sättigungszustand der Flüssigkeit an einfachen Heizflächen. In VDI-Wärmeatlas. Düsseldorf: VDI 1963, Abschn. Hb 2

[4.72] Danilowa, G. N.: Einfluß von Druck und Temperatur auf den Wärmeübergang an siedende Freone (russ.). Cholod. Techn. 4 (1965) 36–42

[4.73] Gorenflo, D.: Behältersieden. In VDI-Wärmeatlas, 6. Aufl., Düsseldorf: VDI-Verlag 1991, Abschn. Ha

[4.74] Gorenflo, D.: Zur Druckabhängigkeit des Wärmeübergangs an siedende Kältemittel bei freier Konvektion. Chem. Ing. Tech. 40 (1968) 757–762 und Diss. TH Karlsruhe 1966

[4.75] Slipčević, B.: Wärmeübergang bei der Blasenverdampfung von Kältemitteln an glatten und berippten Rohren. Klima + Kältetech. 3 (1975) 279–286

[4.76] Kutateladze, S.S.: Kritische Wärmestromdichte bei einer unterkühlten Flüssigkeitsströmung. Energetica 7 (1959) 229-239, und Izv. Akad. Nauk. Otd. Tekh. Nauk 4 (1951) 529

[4.77] Zuber, N.: On the stability of boiling heat transfer. Trans. Am. Soc. Mech. Eng., Ser. C. J. Heat Transfer 80 (1958) 711

[4.78] Roetzel, W.: Berechnung der Leitung und Strahlung bei der Filmverdampfung an der ebenen Platte. Wärme- und Stoffübertrag. 12 (1979) 1–4

[4.79] Collier, J.G.: Convective boiling and condensation. New York: McGraw-Hill 1972

[4.80] Baker, O.: Simultaneous flow of oil and gas. Oil Gas J. 53 (1954) 184–195

[4.81] Hewitt, H.F.; Roberts, D.N.: Studies of two phase flow patterns by simultaneous X-ray and flash photography. UK At. Energy Agency Rep. No. AERE-M 2159, H.M.S.O., 1969

[4.82] Taitel, Y.; Dukler, A.E.: A model for predicting flow regime transitions in horizontal and near horizontal gas-liquid flow. Am. Inst. Chem. Eng. J. 22 (1976) 47–55

[4.83] Schlünder, E.U. (Hrsg.): Heat Exchanger Design Handbook. Washington: Hemisphere Publishing 1986, Vol. 2, Kap. 2.3.

[4.84] Lockhart, R.W.; Martinelli, R.C.: Proposed correlation data for isothermal two-phase, two-component flow in pipes. Chem. Eng. Prog. 45 (1949) 38–48

[4.85] McAdams, W.H.; Wood, W.K.; Heroman, L.C.: Vaporization inside horizontal tubes-II-benzene-oil mixtures. Trans. Am. Soc. Mech. Eng., 64 (1942) 193–200

[4.86] Cicchitti, A.; Lombardi, C.; Silvestri, M.; Soldaini, G.; Zavatarlli, R.: Two-phase cooling experiments-pressure drop, heat transfer and burnot measurements. Energia Nucleare 7 (1960) 407–425

[4.87] Dukler, A.E.; Wicks, M.; Cleveland, R.G.: Pressure drop and hold-up in two-phase flow. Part A: A comparison of existing correlations, Part B: An approach through similarity analysis. Am. Inst. Chem. Eng. J. 10 (1964) 38–51

[4.88] Chawla, J.M.: Wärmeübergang und Druckabfall in waagerechten Rohren bei der Strömung von verdampfenden Kältemitteln. VDI-Forschungsh. 523 (1967)

[4.89] Stephan, K.; Auracher, H.: Correlations for nucleate boiling heat transfer in forced convection. Int. J. Heat Mass Transfer 24 (1981) 99–107

[4.90] Pujol L.: Boiling heat transfer in vertical upflow and downflow tubes. Ph. D. Thesis. Lehigh Univ. 1968

[4.91] Steiner, D.: Heat transfer in boiling of saturated liquids. Section Hbb, in: VDI-Heat Atlas, Düsseldorf: VDI-Verlag 1992

[4.92] Chen, J.C.: Correlation for boiling heat transfer to saturated liquids in convective flow. Ind. Eng. Chem. Proc. Des. Dev. 5 (1966) 322–329

[4.93] Forster, H.K.; Zuber, N.: Dynamics of vapour bubbles and boiling heat transfer. Am. Inst. Chem. Eng. J. 4 (1955) 531–535

[4.94] Jallouk, P.A.: Two-phase pressure drop and heat transfer characteristics of refrigerants in vertical tubes. Ph. D. Thesis, Univ. Tennessee, U. Microfilms 75-11-171, 1974

[4.95] Gungor, K.E.; Winterton, R.H.S.: A general correlation for flow boiling in tubes and in annuli. Int. J. Heat Mass Transfer 29 (1986) 351–358

[4.96] Davis, E.J.; Anderson, G.H.: The incipience of nucleate boiling in forced convection flow. Am. Inst. Chem. Eng. J. 12 (1966) 774–780

[4.97] Bergles, F.E.; Rohsenow, W.M.: The determination of forced- convection surface boiling heat transfer. Amer. Soc. Mech. Eng. Trans. Sec. V (1964) 365–372

[4.98] Shah, M.M.: A new correlation for heat transfer during boiling flow through pipes. Trans. Am. Soc. Heating Refrig. Air Cond. Eng. (ASHRAE) 82 (1976) 66–86

[4.99] Shah, M.M.: Chart Correlation for saturated boiling heat transfer: Equations and further study. Trans. Am. Soc. Heating Refrig. Air Cond. Eng. (ASHRAE). Preprint no. 2673, 1982

[4.100] Bonilla, C.F.; Perry, C.W.: Heat transmission to boiling binary mixtures. Am. Inst. Chem. Eng. J. 37 (1941) 685-705

[4.101] Preußer, P.: Wärmeübergang beim Verdampfen binärer und ternärer Flüssigkeitsgemische. Diss. Ruhr-Univ. Bochum 1978

[4.102] Stephan, K.; Körner, M.: Berechnung des Wärmeübergangs verdampfender binärer Flüssigkeitsgemische. Chem. Ing. Tech. 41 (1969) 409–417

Chapter 5: Thermal radiation

[5.1] DIN 1315: Winkel. Begriffe, Einheiten. Ausg. August 1982. Berlin: Beuth-Verlag

[5.2] Lambert, J.H.: Photometria, sive de mensura et gradibus luminis, colorum et umbrae. Augsburg 1760. Deutsch in Ostwalds Klassiker d. exakten Wissensch. Nr. 31–33, Leipzig: Engelmann 1892

[5.3] Wien, W.: Über die Energievertheilung im Emissionsspektrum eines schwarzen Körpers. Ann. Phys. Ser. 3, 58 (1896) 662–669

[5.4] Siegel, R.; Howell, J.R.; Lohrengel, J.: Wärmeübertragung durch Strahlung. Teil 1. Berlin: Springer 1988

[5.5] Kirchhoff, G.: Über das Verhältnis zwischen dem Emissionsvermögen und dem Absorptionsvermögen der Körper für Wärme und Licht. Ann. Phys. 19 (1860) 275–301

[5.6] Planck, M.: Ueber das Gesetz der Energieverteilung im Normalspektrum. Ann. Phys. 4 (1901) 553–563

[5.7] Reif, F.: Statistische Physik und Theorie der Wärme, S. 437–454. Berlin: W. de Gruyter 1987

[5.8] Cohen, E.R.; Taylor, B.N.: The 1986 adjustment of the fundamental physical constants. CODATA Bulletin No. 63, Nov. 1986

[5.9] Draper, J.W.: On the production of light by heat. Philos. Mag. Ser. 3, 30 (1847) 345–360

[5.10] Wien, W.: Temperatur und Entropie der Strahlung. Ann. Phys. Ser. 2, 52 (1894) 132–165

[5.11] Viskanta, R.; Anderson, E.E.: Heat transfer in semitransparent solids. In: Hartnett, J.P.; Irvine, T. (eds.): Advances in heat transfer, Vol. 11, New York: Academic Press 1975

[5.12] Quinn, T.J.; Martin, J.E.: A radiometric determination of the Stefan-Boltzmann constant and thermodynamic temperatures between $-40\,°C$ and $+100\,°C$. Philos. Trans. R. Soc. London, Ser. A 316 (1985) 85–189

[5.13] Stefan, J.: Über die Beziehung zwischen der Wärmestrahlung und der Temperatur. Sitzungsber. Akad. Wiss. Wien 79, Teil 2 (1879) 391–428

[5.14] Boltzmann, L.: Ableitung des Stefanschen Gesetzes, betreffend die Abhängigkeit der Wärmestrahlung von der Temperatur aus der elektromagnetischen Lichttheorie. Ann. Phys. Ser. 2, 22 (1884) 291–294

[5.15] Pivovonsky, M.; Nagel, M.R.: Tables of Blackbody Radiation Functions. New York: Macmillan Comp. 1961

[5.16] Wiebelt, J.A.: Engineering Radiation Heat Transfer. New York: Holt, Rinehart and Winston 1966

[5.17] Schmidt, E.: Die Wärmestrahlung von Wasser und Eis, von bereiften und benetzten Oberflächen. Forsch. Ing.-Wes. 5 (1934) 1–5

[5.18] Schmidt, E.; Eckert, E.: Über die Richtungsverteilung der Wärmestrahlung von Oberflächen. Forsch. Ing. Wes. 6 (1935) 175–183

[5.19] Sieber, W.: Zusammensetzung der von Werk- und Baustoffen zurückgeworfenen Wärmestrahlung. Z. techn. Physik 22 (1941) 130–135

[5.20] Sparrow, E.M.; Cess, R.D.: Radiation Heat Transfer. Belmont, Cal.: Brooks/Cole Publ. Comp. 1966, S. 68

[5.21] Dunkle, R.V.: Emissivity and inter-reflection relationships for infinite parallel specular surfaces. In: Symp. on Thermal Radiation of solids (S. Katzoff, Ed.) NASA SP-55 (1955) 39–44

[5.22] Drude, P.: Physik des Aethers auf elektromagnetischer Grundlage. 1. Aufl. 1894, 2. Aufl. bearb. v. W. König. Stuttgart: Enke 1912

[5.23] Hagen, E.; Rubens, H.: Das Reflexionsvermögen von Metallen und belegten Glasspiegeln. Ann. Phys. 1 (1900) 352–375 sowie: Emissionsvermögen und elektrische Leitfähigkeit der Metallegierungen. Verh. Dtsch. Phys. Ges. (1904) 128–136

[5.24] Eckert, E.: Messung der Reflexion von Wärmestrahlen an technischen Oberflächen. Forschung Ing.-Wes. 7 (1936) 265–270

[5.25] Kleemann, M.; Meliß, M.: Regenerative Energiequellen. 2. Aufl. Insbes. S. 50–69. Berlin: Springer 1993

[5.26] DIN 1349: Durchgang optischer Strahlung durch Medien. Ausg. Juni 1972. Berlin: Beuth Verlag

[5.27] Hale, G.M.; Querry, M.R.: Optical constants of water in the 200-nm to 200-μm wavelength region. Appl. Optics 12 (1973) 555–563

[5.28] Winter, C.-J.; Sizmann, R.L.; Vant-Hull, L.L. (Eds.): Solar Power Plants. Berlin: Springer 1991

[5.29] Duffie, J.A.; Beckmann, W.A.: Solar Engineering of Thermal Processes. New York: J. Wiley 1980

[5.30] Kreider, J.F.; Kreith, F. (Eds.): Solar Energy Handbook. New York: McGraw-Hill 1981

[5.31] Spencer, J.W.: Fourier series representation of the position of the sun. Search 2 (1971) 172

[5.32] German, S.; Drath, P.: Handbuch SI-Einheiten, insbes. S. 17, Braunschweig: Fr. Vieweg u. Sohn 1979

[5.33] Fröhlich, C.; Brusa, R.W.: Solar radiation and its variation in time. Sol. Phys. 74 (1981) 209–215

[5.34] Iqbal, M.: An Introduction to Solar Radiation. Toronto: Academic Press 1983

[5.35] Kasten, F.; Young, A.T.: Revised optical air mass tables and approximation formula. Applied Optics, 28 (1989) 4735–4738

[5.36] Strutt, J.W. (Lord Rayleigh): On the light from the sky, its polarisation and colour appendix. Phil. Mag. 41 (1871) 107–120, und: On the scattering of light by small particles. Phil. Mag. 41 (1871) 447–454

[5.37] Siegel, R.; Howell, J.R.: Thermal Radiation Heat Transfer, 3rd ed. Washington: Hemisphere Publishing Corp. 1992

[5.38] Bird, R.E.; Riordan, C.: Simple solar spectral model for direct and diffuse irradiance on horizontal and tilted planes at the earth's surface for cloudless atmospheres. J. Climate and Appl. Meteorology 25 (1986) 87–97

[5.39] Ångström, A.: On the atmospheric transmission of sun radiation and on dust in the air. Geografis. Annal. 2 (1929) 156–166 und 3 (1930) 130–159

[5.40] Lohrer, W. (Hrsg.): Verzicht aus Verantwortung. Maßnahmen zur Rettung der Ozon-schicht. Berlin: E. Schmidt 1989

[5.41] Robinson, N. (Eds.): Solar Radiation. New York: American Elsevier 1966

[5.42] Bird, R.E.; Hulstrom, R.L.: Direct insolation models. Trans. ASME, J. Sol. Energy Eng. 103 (1981) 182–192, —: A simplified clear sky model for direct and diffuse insola-tion on horizontal surfaces, SERI/TR-642-761, Solar Energy Research Institute, Golden, Colorado, (1981)

[5.43] Robinson, G.D.: Absorption of solar radiation by atmospheric aerosol as revealed by measurements from the ground. Arch. Meterol. Geophys. Bioclimatol. B 12 (1962) 19–40

[5.44] Kondratyev, K.Y.: Radiation in the Atmosphere. New York: Academic Press 1969. —: Radiation processes in the atmosphere. World Meteorological Organization, Nr. 309, 1972

[5.45] Siegel, R.; Howell, J.R.; Lohrengel, J.: Wärmeübertragung durch Strahlung. Teil 2: Strahlungsaustausch zwischen Oberflächen und Umhüllungen. Berlin: Springer-Verlag 1991

[5.46] Vortmeyer, D.: View factors. Section Kb in VDI-Heat Atlas, Düsseldorf: VDI-Verlag 1992

[5.47] Howell, J.R.: A Catalog of Radiation Configuration Factors. New York: McGraw-Hill 1982

[5.48] Hottel, H.C.; Sarofim, A.F.: Radiative Transfer. New York: McGraw-Hill 1967

[5.49] Poljak, G.: Analysis of heat interchance by radiation between diffuse surfaces (russ.). Techn. Phys. USSR 1 (1935) 555–590

[5.50] Oppenheim, K.A.: Radiation analysis by the network method. Trans. Amer. Soc. Mech. Engrs. 78 (1956) 725–735

[5.51] Eckert, E.R.G.: Einführung in den Wärme- und Stoffaustausch. 3. Aufl. Berlin: Springer-Verlag 1966

[5.52] Schack, A.: Über die Strahlung der Feuergase und ihre praktische Berechnung. Z. techn. Phys. 5 (1924) 267–278

[5.53] Schmidt, E.: Messung der Gesamtstrahlung des Wasserdampfes bei Temperaturen bis 1000 °C. Forsch. Ing.-Wes. 3 (1932) 57–70

[5.54] Schmidt, E.; Eckert, E.: Die Wärmestrahlung von Wasserdampf in Mischung mit nicht-strahlenden Gasen. Forsch. Ing.-Wes. 8 (1937) 87–90

[5.55] Eckert, E.: Messung der Gesamtstrahlung von Wasserdampf und Kohlensäure in Mis-chung mit nichtstrahlenden Gasen bei Temperaturen bis 1300 °C. VDI-Forschungsheft 387, Berlin: 1937

[5.56] Hottel, H.C.; Mangelsdorf, H.G.: Heat transmission by radiation from nonluminous gases. Trans. Amer. Inst. Chem. Eng. 31 (1935) 517–549

[5.57] Hottel, H.C.; Egbert, R.B.: The radiation of furnace gases. Trans. ASME 63 (1941) 297–307

[5.58] Hottel, H.C.; Egbert, R.B.: Radiant heat transmission from water vapor. Trans. Amer. Inst. Chem. Eng. 38 (1942) 531–568

[5.59] Vortmeyer, D.: Gas radiation-radiation from gas mixtures. Section Kc in VDI-Heat Atlas, Düsseldorf: VDI-Verlag 1992

[5.60] Schack, A.: Berechnung der Strahlung von Wasserdampf und Kohlendioxid. Chemie-Ing. Techn. 42 (1970) 53–58

[5.61] Schack, A.: Zur Berechnung der Wasserdampfstrahlung. Chemie-Ing. Techn. 43 (1971) 1151–1153

[5.62] Kohlgrüber, K.: Formeln zur Berechnung des Emissionsgrades von CO_2- und H_2O-Gasstrahlung bei Industrieöfen, Brennkammern und Wärmeaustauschern. Gaswärme International 35 (1986) 412–417

[5.63] Kostowski, E.: Analytische Bestimmung des Emissionsgrades von Abgasen. Gaswärme International 40 (1991) 529–534

[5.64] Elgeti, K.: Ein neues Verfahren zur Berechnung des Strahlungsaustausches zwischen einem Gas und einer grauen Wand. Brennst.-Wärme-Kraft 14 (1962) 1–6

[5.65] Edwards, D.K.: Absorption by infrared bands of carbon dioxide gas at elevated pressures and temperatures. J. Optical Soc. Amer. 50 (1960) 617–626

[5.66] Görner, K.; Dietz, U.: Strahlungsaustauschrechnungen mit der Monte-Carlo-Methode. Chem.-Ing. Techn. 62 (1990) 23–29

[5.67] Biermann, P.; Vortmeyer, D.: Wärmestrahlung staubhaltiger Gase. Wärme- und Stoffübertr. 2 (1969) 193–202

[5.68] Brummel, H.-G.; Kakaras, E.: Wärmestrahlungsverhalten von Gas-Feststoffgemischen bei niedrigen, mittleren und hohen Staubbeladungen. Wärme- und Stoffübertr. 25 (1990) 129–140

[5.69] Vortmeyer, D.; Brummel, H.-G.: Thermal radiation from gas-solid mixtures. Section Kd in VDI-Heat Atlas, Düsseldorf: VDI-Verlag 1992

[5.70] Richter, W.; Michelfelder, S.: Wärmestrahlung in Brennräumen. Abschnitt Ke in VDI-Wärmeatlas, 6. Aufl. Düsseldorf: VDI-Verlag 1991

Index

Printing: Mercedesdruck, Berlin
Binding: Buchbinderei Lüderitz & Bauer, Berlin